T0236122

Textbook of Clinical Trials in Oncology

A Statistical Perspective

Textbook of Clinical Trials in Oncology
A Statistical Perspective

Edited by
Susan Halabi
Department of Biostatistics and Bioinformatics,
Duke University, Durham, North Carolina

Stefan Michiels
Unit of Biostatistics and Epidemiology, Gustave Roussy,
CESP Inserm, University Paris-Sud,
University Paris-Saclay, Villejuif, France

CRC Press
Taylor & Francis Group
Boca Raton London New York

CRC Press is an imprint of the
Taylor & Francis Group, an **informa** business

A CHAPMAN & HALL BOOK

CRC Press
Taylor & Francis Group
6000 Broken Sound Parkway NW, Suite 300
Boca Raton, FL 33487-2742

First issued in paperback 2020

ISBN 13: 978-0-367-72957-8 (pbk)
ISBN 13: 978-1-138-08377-6 (hbk)

Library of Congress Cataloging-in-Publication Data

Names: Halabi, Susan, editor. | Michiels, Stefan, editor.
Title: Textbook of clinical trials in oncology : a statistical perspective /
edited by Susan Halabi, Stefan Michiels.
Description: Boca Raton, Florida : CRC Press, [2019] | Includes
bibliographical references and index.
Identifiers: LCCN 2019006740| ISBN 9781138083776 (hardback : alk. paper) |
ISBN 9781315112084 (e-book)
Subjects: LCSH: Cancer--Research--Statistical methods. | Clinical trials--Statistical methods.
Classification: LCC RC267 .T49 2019 | DDC 616.99/400727--dc23
LC record available at https://lccn.loc.gov/2019006740

Visit the Taylor & Francis Web site at
http://www.taylorandfrancis.com

and the CRC Press Web site at
http://www.crcpress.com

We dedicate this book to our mentors, collaborators, researchers and, most importantly, those patients who participate in clinical trials. The patients are not only scientific collaborators but, in many ways, are the ones who make the greatest contribution to the advancement of our collective quest to conquer cancer.

To Nicholas Robins and Saskia Ooms, for their unwavering support.

Contents

vii

Acknowledgment

This book is principally targeted to intermediate level statisticians who are in graduate programs in biostatistics. It evolved from the need for training the next generation of statisticians to become clinical trialists. In addition, we hope this book will serve as a guide for statisticians with minimal clinical trial experience, who are interested in pursuing a career in clinical trials.

We offer special thanks to our contributors for being a central part of this project. We would also like to express our gratitude to Aurélie Bardet, Sarah Flora Jonas, and Rachid Abbas for reading draft chapters.

Editors

Susan Halabi, PhD, is a Professor of Biostatistics and Bioinformatics, Duke University. She has extensive experience in the design and analysis of clinical trials in oncology. Dr. Halabi is a fellow of the American Statistical Association, the Society of Clinical Trials, and the American Society of Clinical Oncology. She serves on the Oncologic Drugs Advisory Committee for the Food and Drug Administration.

Stefan Michiels, PhD, is the Head of the Oncostat team of the Center for Research in Epidemiology and Population Health (INSERM U1018, University of Paris-Saclay, University Paris-Sud) at Gustave Roussy. His areas of expertise are clinical trials, meta-analyses, and prediction models in oncology. Dr. Michiels is the currently chair of the biostatisticians at Unicancer, a French collaborative cancer clinical trials group.

Contributors

William T. Barry, PhD
Rho Federal Systems Division
Chapel Hill, North Carolina

Robert A. Beckman, MD
Professor
Department of Oncology
and
Department of Biostatistics, Bioinformatics,
    and Biomathematics
Lombardi Comprehensive Cancer Center
    and Innovation Center for Biomedical
    Informatics
Georgetown University Medical Center
Georgetown, Washington, District of
    Columbia

Aurélie Bertrand, PhD
Institute of Statistics, Biostatistics and
    Actuarial Sciences
Louvain Institute of Data Analysis and
    Modeling in Economics and Statistics
Université catholique de Louvain
Louvain-la-Neuve, Belgium

Jan Beyersmann, PhD
Professor
Institute of Statistics
Ulm University
Ulm, Germany

Cong Chen, PhD
Executive Director
Early Oncology Statistics BARDS
Merck & Co., Inc
Kenilworth, New Jersey

Cody Chiuzan, PhD
Assistant Professor
Department of Biostatistics
Mailman School of Public Health
Columbia University
New York, New York

Catherine M. Crespi, PhD
Professor
Department of Biostatistics
University of California
Los Angeles, California

James J. Dignam, PhD
Professor
Department of Public Health Sciences
The University of Chicago
Chicago, Illinois

Keyue Ding, PhD
Canadian Cancer Trials Group
Department of Public Health Sciences
Queen's University
Kingston, Ontario, Canada

Alex Dmitrienko, PhD
Mediana, Inc.
Overland Park, Kansas

Paul Gallo, PhD
Novartis Pharmaceuticals
East Hanover, New Jersey

Ekkehard Glimm, PhD
Novartis Pharma AG
Basel, Switzerland

Susan Halabi, PhD
Professor
Department of Biostatistics and
  Bioinformatics
School of Medicine
Duke University
Durham, North Carolina

Jay Herson, PhD
Senior Associate
Department of Biostatistics
Johns Hopkins Bloomberg School of Public
  Health
Johns Hopkins University
Baltimore, Maryland

Chen Hu, PhD
Assistant Professor
Division of Biostatistics and
  Bioinformatics
Sidney Kimmel Comprehensive Cancer
  Center
Johns Hopkins University School of
  Medicine
Baltimore, Maryland

Masataka Igeta, PhD
Department of Biostatistics
Hyogo College of Medicine
Hyogo, Japan

Sin-Ho Jung, PhD
Professor
Department of Biostatistics and
  Bioinformatics
School of Medicine
Duke University
Durham, North Carolina

Kelley M. Kidwell, PhD
Associate Professor
Department of Biostatistics
University of Michigan School of Public
  Health
Ann Arbor, Michigan

Andrea Knezevic, MS
Research Biostatistician
Department of Epidemiology &
  Biostatistics
Memorial Sloan Kettering Cancer Center
New York, New York

Aya Kuchiba, PhD
Biostatistics Division
CRAS
National Cancer Center
Tokyo, Japan

Nicholas R. Latimer, PhD
Reader
School of Health and Related Research
University of Sheffield
Sheffield, United Kingdom

Aurelien Latouche, PhD
Professor
Conservatoire National des Arts
  et Métiers
Institut Curie
Paris, France

Catherine Legrand, PhD
Professor
Institute of Statistics, Biostatistics and
  Actuarial Sciences
Louvain Institute of Data Analysis and
  Modeling in Economics and Statistics
Université catholique de Louvain
Louvain-la-Neuve, Belgium

Gang Li, PhD
Professor
Department of Biostatistics and
  Biomathematics
University of California
Los Angeles, California

Chen-Yen Lin, PhD
Senior Research Scientist
Eli Lilly and Company
Indianapolis, Indiana

Ilya Lipkovich, PhD
Eli Lilly and Company
Indianapolis, Indiana

Shigeyuki Matsui, PhD
Professor
Department of Biostatistics
Nagoya University Graduate School of
    Medicine
Nagoya, Japan

Stefan Michiels, PhD
INSERM CESP—Oncostat Team
Biostatistics and Epidemiology
    Department
Paris Saclay University/UVSQ
Paris, France

Koji Oba, PhD
Associate Professor
Interfaculty Initiative in Information
    Studies
and
Department of Biostatistics
School of Public Health
Graduate School of Medicine
The University of Tokyo
Tokyo, Japan

Chris O'Callaghan, PhD
Canadian Cancer Trials Group
Department of Public Health Sciences
Queen's University
Kingston, Ontario, Canada

Nathaniel O'Connell, PhD
Department of Biostatistical Sciences
Wake Forest School of Medicine
Wake Forest University
Winston-Salem, North Carolina

Megan Othus, PhD
Associate Member
Fred Hutchinson Cancer
    Research Center
Seattle, Washington

Katherine S. Panageas, DrPH
Associate Attending Biostatistician
Department of Epidemiology &
    Biostatistics
Memorial Sloan Kettering Cancer Center
New York, New York

Xavier Paoletti, PhD
INSERM CESP—Oncostat Team
Biostatistics and Epidemiology
    Department
Paris Saclay University/UVSQ
Paris, France

Lira Pi, PhD
Duke University
Durham, North Carolina

Martin Posch, PhD
Center for Medical Statistics, Informatics,
    and Intelligent Systems
Medical University of Vienna
Vienna, Austria

Stephanie Pugh, PhD
Director of Statistics
American College of Radiology
Philadelphia, Pennsylvania

Bohdana Ratitch, PhD
Eli Lilly and Company
Montreal, Quebec, Canada

Federico Rotolo, PhD
Biostatistician
Innate Pharma
Marseille, France

Claudia Schmoor, PhD
Clinical Trials Unit
Faculty of Medicine and Medical
    Center
University of Freiburg
Freiburg, Germany

Juned Siddique, DrPH
Associate Professor
Division of Biostatistics
Department of Preventive Medicine
Northwestern University Feinberg School
    of Medicine
Chicago, Illinois

Richard Simon, DSc
R Simon Consulting
Potomac, Maryland

Kelly Speth, MS, PhD
Department of Biostatistics
University of Michigan School of Public
    Health
Ann Arbor, Michigan

Nils Ternès, PhD
Biostatistician
Sanofi-Aventis
Chilly-Mazarin, France

Kiichiro Toyoizumi, PhD
Biometrics
Shionogi Inc
Florham Park, New Jersey

Stephen Walters, PhD
Professor
Medical Statistics and Clinical Trials
School of Health and Related Research
University of Sheffield
Sheffield, United Kingdom

James Wason, PhD
Professor
Institute of Health and Society
Newcastle University
Newcastle, United Kingdom

and

Medical Research Council Biostatistics Unit
University of Cambridge
Cambridge, United Kingdom

Ian R. White, PhD
Professor
Statistical Methods for Medicine
Medical Research Council Clinical Trials
    Unit at University College London
London, United Kingdom

Yuan Wu, PhD
Assistant Professor
Department of Biostatistics and
    Bioinformatics
School of Medicine
Duke University
Durham, North California

Dong Xi, PhD
Novartis Pharmaceuticals
East Hanover, New Jersey

Qing Yang, PhD
Assistant Research Professor
School of Nursing
Duke University
Durham, North California

Sarah Zohar, PhD
INSERM, UMRS 1138, Team 22, CRC
University Paris 5
University Paris 6
Paris, France

1

Introduction to Clinical Trials

Susan Halabi and Stefan Michiels

CONTENTS

The first documented clinical trial dates back to May 20, 1747, to a study conducted on board the British naval vessel HMS *Salisbury* [1]. During the voyage, 12 sailors suffered from scurvy, prompting the vessel's surgeon, James Lind, to experiment by assigning groups of two patients to six different treatments [1]. One of those treatments included the consumption of oranges and lemons, which was found to be effective [1]. Despite this and other early achievements, it was not until 1926 that the modern concept of randomization was pioneered by Sir Ronald A. Fisher in his agricultural research [2,3]. Building on these methodological accomplishments, subsequent researchers introduced randomization in human experiments in 1936 [4]. The first trial to be conducted by the Medical Research Council was on the effectiveness of streptomycin in patients with tuberculosis, in 1948 [4].

Clinical trials have come a long way since then, and have become a keystone when evaluating the effectiveness of therapies [4,5]. Moreover, they are an engine of drug development and approval, which contributes to improved treatment and patient care. As such, they serve as the main conduit that regulatory agencies utilize to approve therapies in humans. We expect that clinical trials will play an ever increasing, and global, role in twenty-first-century medicine.

According to ClinicalTrials.gov, "a clinical trial is a research study to answer specific questions about vaccines or new therapies or new ways of using known treatments" [6]. We follow the clinical oncology paradigm in classifying clinical trials broadly into four phases: phase I (or early development), which examines an experimental treatment in a small group of patients in which the dose and safety are evaluated; phase II trials (or middle development), which refer to experimental therapies and their efficacy in tumor shrinkage in a larger group of patients; phase III trials (or late), which investigate and compare the experimental therapy to a placebo or control; and lastly, phase IV trials, which are conducted after an experimental treatment has been licensed and marketed [4,7,8].

Phase III clinical trials are crucial in determining the efficacy of innovative therapies and are considered the gold standard in assessing the efficacy of a new experimental arm or device. The basic principles of design in phase III trials are to minimize bias and increase precision in the estimation of the treatment effect [4,7]. Although the central objective of a phase III trial is to change medical practice, it is not always attained. Furthermore, results from a single phase III trial may not be sufficient for the intervention to be considered definitive or to change medical practice. Green and Byar confirmed that phase III trials establish the strongest evidence for approving an intervention, but other supporting trials are needed [9].

In recent years, clinical trials have become increasingly sophisticated as they incorporate genomic studies and embed quality of life objectives [4,7]. Historically, trials with cytotoxic agents were conducted sequentially, starting with a phase I trial and advancing to phase II and phase III trials. With the advancement in genetic and molecular technologies, this paradigm has increasingly been challenged, resulting in more efficient designs [10–13]. We have seen an upsurge of innovative trial designs ranging from seamless phase II/III trials to adaptive studies [14–16]. Basket and umbrella trials have also evolved so that trials are conducted efficiently across genetic variants or cancer histologies [17–20]. While well-conducted phase III studies have and will continue to form the foundation for drug approval, well-designed phase II trials may also play a role in regulatory approval.

1.1 Scope and Motivation

There is an evolving need for educational resources for statisticians and investigators. Reflecting this, the goal of this book is to provide readers with a sound foundation for the design, conduct, and analysis of clinical trials. Furthermore, this book is intended as a guide for statisticians and investigators with minimal clinical trial experience who interested in pursuing a career in clinical trials.

The following chapters provide both the theoretical justification and practical solutions to the problems encountered in the design, conduct, and analysis of a clinical trial. We also seek to encourage and advance the development of novel statistical designs or analytical methods. Additionally, this book may be of interest for public-health students and public-health workers involved with clinical trials, reflecting its focus on practical issues encountered in clinical trials exemplified by real-life examples.

The development, design, and conduct of a trial require a multidisciplinary approach. Altman describes the general sequence of steps in a research project as follows: planning, design, execution (data collection), data processing, data analysis, presentation, interpretation, and publication [21]. Statistical thinking is vital at each of these steps, and statisticians play a fundamental role in ensuring the objectivity of clinical trials and that the trials produce valid and interpretable results.

This book focuses on human studies in oncology, ranging from early, middle, and late phase trials to advanced topics relevant in the era of precision medicine and immunotherapy. Developing an understanding of these issues is imperative for statisticians to contribute effectively in a dynamic, twenty-first-century environment. This work is divided into four sections. In the first section, the chapters focus on the early to middle development designs that new investigators need to consider as they prepare for phase I through II/III trials. Throughout these chapters, the authors share their experiences in trial design and provide examples and software to illustrate their points. The URL of the companion website is **https://www.crcpress.com//9781138083776**. This section begins with Panageas highlighting the salient issues in choosing an endpoint for a clinical trial. Responding to the rich literature on the drawbacks of the 3 + 3 designs, Chiuzan and O'Connell promote innovative designs in phase I settings, while providing R functions and a tutorial for the design of phase I trials. Following this, Jung discusses phase II designs, focusing on two-stage designs and advocating for reporting p-values in phase II settings. Othus outlines the challenges in the design and conduct of trials with immunotherapies, while Barry reviews adaptive designs, which have played a vital role in recent years.

Section II explores late phase III trials. While Oba et al. describe the design of phase III trials, the gold-standard design in clinical trials, Ding and O'Callaghan focus on the principles of non-inferiority trial design. Wason examines multi-stage and multi-arm trials which have been successful in approving several drugs, and Glimm and colleagues consider multiple comparison and co-primary endpoints. Once the study is underway, the focus shifts to the conduct of the trial. Crespi provides an in-depth discussion on cluster randomization, which she illustrates with several examples. Monitoring of studies is taken up by Herson and Hu, who discuss the role of interim analysis and data monitoring during the conduct of the trial.

The third section deals with cutting-edge topics in personalized medicine. Matsui and colleagues detail the design of biomarker-driven trials, while Simon describes the statistical issues that arise in genomic studies. Beckmann et al. examine the challenges of designing a trial for rare diseases, and Lipkovich and colleagues review subgroup analysis and methods for biomarker analysis. The development and validation of prognostic models and genomic signatures has been embedded into many trials. While Halabi and colleagues share their expertise in developing and validating prognostic models in cancer studies, Rotolo et al. provide a detailed discussion on how signatures are developed and validated. This section ends with a thorough exposition by Kidwell and colleagues of the dynamic treatment regimen.

The final section is dedicated to advanced topics related to the analysis phase of clinical trials. Surrogate endpoints are examined by Paoletti and colleagues, reflecting the interest in decreasing the time for a new drug to be approved. Latouche and colleagues outline methods for competing risks analysis and when they should be implemented. We have seen an increased proportion of cancer patients surviving, and Legrand demonstrates the importance of the cure-rate models. Wu and Halabi discuss interval censoring, which is underutilized in clinical trials. Patients often are offered other therapies after they progress or relapse, and Latimer and White give an in-depth review of methods adjusting for treatment switches. Assessing and reporting the adverse events encountered during a trial are critical tasks, tackled by Beyersmann and Schmoor, who offer methods for the analysis of adverse events. Walters describes the design of a quality of life study, which increasingly are integrated into clinical trials, while Pugh and colleagues discuss how to deal with missing data in trials.

In summary, statistical input and thinking is critical at each stage of the process of a trial. The challenging issues that statisticians face from the design stage to the analysis stage are described in these chapters. Moreover, the contributors present theoretical and analytical solutions and highlight the practical approaches. We hope that this book will help advance the design of trials as they continue to evolve with the changes in therapeutic landscape in cancer and improve the delivery of treatment and care for oncology patients.

1.2 Resources

There are many resources dedicated to clinical trials (Table 1.1), and this list is not comprehensive. One great resource for statisticians and clinical trialists is the Society of

TABLE 1.1

Web-Based Resources

https://www.aacr.org/
https://www.asco.org/
http://www.cancer.gov/
https://www.canada.ca/en/services/health/drug-health-products.html
www.clinicaltrials.gov
www.cochrane.org
www.consort-statement.org
https://www.ctu.mrc.ac.uk/
http://www.ema.europa.eu/ema/
http://eng.sfda.gov.cn/WS03/CL0755/
http://www.esmo.org/
https://www.fda.gov/
https://latampharmara.com/mexico/cofepris-the-mexican-health-authority/
https://www.pmda.go.jp/english/
http://journals.sagepub.com/home/ctj
http://www.sctweb.org/public/home.cfm
https://seer.cancer.gov/
https://stattools.crab.org/
http://www.tga.gov.au/

Clinical Trials (www.sctweb.org), which is an organization dedicated to the study, design, and analysis of clinical trials, with a peer-reviewed journal (*Controlled Trials*).

1.3 Conclusion

Clinical trials are becoming increasingly sophisticated, are costly and time-consuming, and require expertise in their planning, execution, and reporting. It is our responsibility as statisticians and investigators to ensure that trials are rigorously planned, well conducted, and properly analyzed. This will enable them to continue to answer important clinical questions while improving the care of patients and their quality of life.

References

1. Bown SR. *Scurvy: How a Surgeon, a Mariner, and a Gentleman Solved the Greatest Medical Mystery of the Age of Sail.* Toronto: Thomas Allen Publishers, 2005.
2. Fisher RA and Bennett JH. *Statistical Methods, Experimental Design, and Scientific Inference: A Re-Issue of Statistical Methods for Research Workers, the Design of Experiments, and Statistical Methods and Scientific Inference.* Oxford: Oxford University Press, 2003.
3. Fisher RA. *The Design of Experiments.* New York: Hafner, 1974.
4. Crowley J and Hoering A. *Handbook of Statistics in Clinical Oncology*, 3rd Edition. Boca Raton, FL: CRC, 2012.

5. Friedman LM et al. *Fundamentals of Clinical Trials*. Cham: Springer International Publishing: Springer e-books, 2015.
6. http://www.clinicaltrials.gov.
7. Kelly WK and Halabi S. *Oncology Clinical Trials: Successful Design, Conduct, and Analysis*, 2nd Edition. New York: Demos Medical Publishing, 2018.
8. Piantadosi S. *Clinical Trials: A Methodologic Perspective*, 3rd Edition. Hoboken, NJ: Wiley and Sons, 2017.
9. Green SB and Byar DP. Using observational data from registries to compare treatments: The fallacy of omnimetrics. *Statist Med*. 1984;3(4):361–70.
10. Chapman PB. Improved survival with vemurafenib in melanoma with BRAF V600E mutation. *N Engl J Med*. 2011;364(26):2507–16.
11. Hainsworth JD et al. Targeted therapy for advanced solid tumors on the basis of molecular profiles: Results from MyPathway, an open-label, phase IIa multiple basket study. *JCO*. 2018;36(6):536–42.
12. https://www.lung-map.org/.
13. Herbst RS et al. Lung Master Protocol (Lung-MAP)–A biomarker-driven protocol for accelerating development of therapies for squamous cell lung cancer: SWOG S1400. *Clin Cancer Res*. 2015;21(7):1514–24.
14. Gillessen S et al. Repurposing metformin as therapy for prostate cancer within the STAMPEDE trial platform. *Eur Urol*. 2016;70(6):906–8.
15. https://www.ispytrials.org/.
16. https://www.cancer.gov/about-cancer/treatment/clinical-trials/nci-supported/nci-match.
17. Mangat PM et al. Rationale and design of the targeted agent and profiling utilization registry study. *JCO Precis Oncol*. 2018;2:1–14.
18. Massard C et al. High-throughput genomics and clinical outcome in hard-to-treat advanced cancers: Results of the MOSCATO 01 trial. *Cancer Discov*. 2017;7(6):586–95.
19. Le Tourneau C et al. Molecularly targeted therapy based on tumour molecular profiling versus conventional therapy for advanced cancer (SHIVA): A multicentre, open-label, proof-of-concept, randomised, controlled phase 2 trial. *Lancet Oncol*. 2015;16(13):1324–34.
20. Hyman DM et al. Vemurafenib in multiple nonmelanoma cancers with BRAF V600 mutations. *N Engl J Med*. 2015;373(8):726–36.
21. Altman DG. *Practical Statistics for Medical Research*, 1st Edition. London: Chapman & Hall, 2011.

Section I

Early to Middle Development

2

Selection of Endpoints

Katherine S. Panageas and Andrea Knezevic

CONTENTS

2.1 Introduction

The overall objective of the conduct of a research study is to make inferences about hypothesized relations within a population. Clinical research involves the evaluation of medications, devices, diagnostic products, and treatment regimens intended for human use [1]. Research studies are conducted in human subjects to answer questions pertaining to *interventions*, including new treatments such as novel drug combinations, immuno-therapies, or molecular targeted therapies; *disease prevention* such as smoking cessation studies or evaluation of cancer screening programs; *symptom relief* and pain management; *palliative treatments* to improve quality of life; or standard of care treatments that warrant further study or comparisons [1]. The goals of a study may include the evaluation of proposed medical treatments for safety and efficacy, assessment of the relative benefits of competing therapies, or determination of the optimal treatment schedule or drug combinations, in specific patients.

Clinical research encompasses *observational studies* (prospective or retrospective), *case-control studies*, and *clinical trials*. Some research questions, such as the effects of sun exposure on skin cancer, can best be answered through observational studies [2,3]. However, when evaluating treatment effects, control of treatment allocation is not possible in an observational study. Although observational studies can provide meaningful information, they often are criticized because of potential biased and erroneous conclusions [4,5]. Clinical trials are systematic experiments in which human subjects are assigned to interventions, and outcomes are evaluated prospectively, as defined in detail before the start of the study [6,7].

Clinical trials are the foundation of drug development in oncology. Clinical trials can definitively answer questions such as "Are recurrence rates lower in breast cancer patients who receive hormonal therapy plus chemotherapy versus hormonal therapy alone?" or "Is the incidence of serious adverse events greater in patients receiving combination immunotherapy or single agent immunotherapy?" The goal of conducting clinical trials in oncology is to determine the most effective and safe way to treat patients who have cancer [7].

Clinical trials have enabled new and promising treatments, resulting in medical breakthroughs in oncology. The development of both targeted or molecular therapy that interferes with specific molecular targets and immunotherapy that targets the immune system has contributed to the recent expansion of clinical trials in oncology. In 2013, *Science* declared cancer immunotherapy to be the breakthrough of the year based on the positive results of clinical trials of antibodies that block negative regulators of T-cell function, and the development of genetically modified T cells (chimeric antigen receptor T-cell therapy) [8]. Evaluation of novel agents typically progresses through a three-phase system of clinical trials (phase I, II, and III). A new treatment that is successful in one phase will continue to be tested in subsequent phases.

The goals of *phase I* trials in oncology are to determine whether the investigative drug is safe, characterize the toxicity profile, and determine the dose for further study. *Phase II* trials typically assess initial measures of drug efficacy. *Phase III* trials can result in changes in clinical practice and drug approval by the United States Food and Drug Administration (FDA), and study objectives focus on whether the experimental treatment is more effective than current treatments [6,7]. Phase III trials, also known as *randomized controlled trials* (RCTs), are the gold-standard study design for clinical research and the most likely to minimize any inherent biases [6,7].

2.2 Key Definitions and Endpoint Selection

Study *objectives* outline the research question related to the aims of a clinical trial: they should be concrete, be clearly specified, and include measurable outcomes. Some examples of study objective terms include "greater response," "improved survival," "less toxicity," and "reduced postoperative complications."

Typically, a study has one *primary objective*, and the statistical design, including designation of the sample size and study power, relates to the primary objective. Primary objectives may determine the maximum tolerated dose of a drug, or establish the acceptable minimum activity level of a drug. *Secondary objectives* also are important but do not affect trial design. Analyses of secondary objectives have the specified power determined based on the sample size that had been selected to meet the primary objective. Secondary objectives may include describing the pharmacokinetic properties of a drug, frequency of drug toxicities, or biomarker levels.

The trial *endpoint* is the quantitative measurement or data point needed to meet the trial objectives. Endpoints are data captured for each study participant, and may be continuous, binary, ordinal, event times, biomarker levels, or response to items from a questionnaire. Endpoints may be measured at one time point or multiple follow-up times (repeated measurements) per patient. Measurements of endpoints should be objective, reproducible, and free of bias (Table 2.1).

The primary endpoint is used to address the primary objective of the study by means of a *statistic* to be used for descriptive or inferential analyses. Note that an endpoint is not an objective. Continuous endpoints are typically summarized using standard descriptive statistics including means, standard deviations, medians, and ranges, whereas categorical variables are summarized with frequencies and percentages. Statistics used for inferential analyses may include odds ratios or hazard ratios derived from regression modeling techniques. Study objectives often are accompanied by hypotheses, for example, "the complete response rate of dacarbazine will be $\geq 35\%$" or "the combination treatment will result in a 20% improvement in survival as compared to monotherapy." Examples are provided in Table 2.2 [9–11].

Event times are common endpoints in oncologic clinical trials and measure the length of time before the occurrence of an event (Figure 2.1). Examples include "survival time" (time from the start of a treatment to death) and "time to progression of disease" (time from diagnosis to tumor progression). For these endpoints, time is measured from entry into the clinical trial until observation of the event.

Event times include two important pieces of information for each patient: length of time during which no event was observed and whether or not the patient experienced the event by the end of the observation period [12,13]. If the event is not observed during follow-up, such as when a patient is lost to follow-up or begins another therapy, event time is defined as a *censored observation* [12,13].

A clinical trial might use a *clinical endpoint* or a *surrogate endpoint*. A clinical endpoint is an outcome that represents direct clinical benefit, such as survival, decreased pain, or absence of disease. A surrogate endpoint is a substitute for a clinical endpoint when use of the clinical endpoint might not be possible or practical, such as a when measurement of

TABLE 2.1

Endpoints in Cancer Clinical Trials

Type of Endpoint Data	Example	Example Value
Continuous	Quality-of-life score	Scores from 0 to 100 points
	Laboratory measurements	Levels from 0.1 to 500 mg/dL
Binary	Postoperative complication	Yes or no
Ordinal	Pathologic staining	None, mild, moderate, severe
	Tumor response	Complete response, partial response, stable disease, disease progression
Categorical (unordered)	Toxicities	Common toxicity categories
Event times	Time to death	Days from treatment start to death
	Time to tumor progression	Days from treatment start to tumor progression

TABLE 2.2

Cancer Clinical Trial Definitions and Examples

Definition	Example 1	Example 2	Example 3
Objective	To compare efficacy (objective response rate) of sorafenib versus bevacizumab and erlotinib in hepatocellular carcinoma patients	To determine overall safety and toxicity of paclitaxel and trastuzumab in women who have HER2/neu-positive breast cancer	To compare efficacy (overall survival) of dacarbazine versus temozolomide in unresectable metastatic melanoma
Endpoint	Best overall response defined by complete response, partial response, stable disease, or disease progression by RECIST criteria	Grade 3 or higher toxicity assessed by the NCI CTCAE	Survival time for each patient, defined as the difference in time from the start of treatment to death or last contact
Statistic	Response frequency defined as the proportion of all treated patients who have complete or partial response	Frequency of grade 3 or higher toxicity, defined as the proportion of patients who have grade 3 or higher toxicity	Overall survival distributions estimated using Kaplan–Meier method, and median survival time for each treatment group; hazard ratio estimated as a summary of treatment effect on risk of death

Source: National Cancer Institute Common Terminology Criteria for Adverse Events v5.0 NCI, NIH, DHHS. November 27, 2017. Retrieved February 15, 2018, from https://ctep.cancer.gov/protocoldevelopment/electronic_applications/ctc.htm#ctc_50; Therasse P et al. J Natl Cancer Inst. 2000;92:205–16; Eisenhauer E et al. Eur J Cancer. 2009;45:228–47.

Abbreviations: CTCAE, Common Terminology Criteria for Adverse Events; NCI, National Cancer Institute; RECIST, Response Evaluation Criteria in Solid Tumors.

the direct benefit of a drug may require many years. Surrogate endpoints do not represent direct clinical benefit but correlate with clinical benefit. A true surrogate endpoint must be valid, highly correlated with the clinical outcome measured, and yield the same statistical inference as the definitive endpoint [6,7]. Tumor shrinkage may be a surrogate endpoint for longer survival in clinical trials for anticancer drugs. The use of surrogate endpoints may shorten clinical trial duration, but imprecise association between the surrogate and clinical endpoints may cause misleading results [6,7].

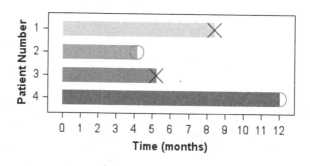

FIGURE 2.1

Time-to-event data showing the length of time from study entry to final follow-up for 4 patients. Patients 1 and 3: the event occurred at the designated times (X). Patients 2 and 4: still alive at last follow-up (O). Patient 4 remained in the study until study termination at month 12.

ILLUSTRATIVE TRIALS

In this chapter, we will describe seminal studies from the oncology literature to illustrate the use of various endpoints.

The study endpoint is chosen based on the clinical trial design, drug or drug combination under study, drug mechanism of action, and patient population. Chemotherapy agents are cytotoxic and kill rapidly proliferating cells. The dose of a cytotoxic agent selected for further study assumes that there is a proportional increase between dose, efficacy, and toxicity. In contrast, targeted therapies are more likely to be cytostatic and inhibit cell growth. Cytostatic drugs may not shrink the tumor, but instead, delay disease progression, and selection of the appropriate endpoint to measure disease progression is important for trial success. Immunotherapeutic agents have a different mechanism of action than cytotoxic drugs and do not target the cancer cells directly, but activate the immune system to fight the cancer. Therefore, immune cells, such as T cells, indirectly mediate the cytotoxic effects of immunotherapy agents. The traditional assumption of a proportional relation between efficacy and dose may not apply with immunotherapy [14].

Knowledge about the mechanism of action of a drug may guide the objectives and endpoints of the research study. Furthermore, objectives and endpoints may vary with the patients under investigation, such as tumor shrinkage in early-stage cancer patients versus symptom relief in advance-stage cancer patients. It is necessary to have clinical and statistical guidance at all stages of a study, including the development and conduct of the study, to ensure clinical trial success. Although clinical trials are essential in advancing patient care, they are expensive and time- and labor-intensive for investigators and may require major dedication by patients. It is essential to carefully consider endpoint selection; patient study sample; and data management, analysis, and interpretation to achieve high-quality results.

2.3 Patient-Centered Endpoints

2.3.1 Overall Survival

Survival time is the most specific and objective endpoint in cancer studies. A patient's survival time is defined as the time from randomization or start of treatment to death due to any cause, including death from the trial disease or unrelated conditions. Therefore, in contrast with other endpoints, overall survival is affected by treatment effect after relapse and the effect of adverse events on survival [7].

Survival distributions may be estimated with censored survival times, even when not all patients have died. Censored survival time is defined as the time from the start of treatment to the date of last contact with patients who are alive. Evaluation of overall survival requires a clinical trial with a large sample size that spans over a long period of time. Overall survival also includes non-cancer deaths that may skew the study results, especially with larger and older populations in which death may be due to other causes [15]. With diseases

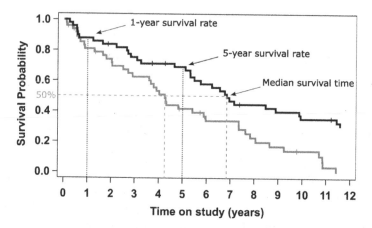

FIGURE 2.2
Kaplan–Meier curves for 2 hypothetical study arms showing *median survival* and 1-year and 5-year *survival rates*. Censored observations are denoted on the curve by tick marks.

such as metastatic breast cancer, in which numerous chemotherapy combination regimens often are utilized, it may be difficult to determine the regimen that contributed to overall survival.

Overall survival is determined by plotting the proportion of patients who remain alive over time using a survival curve known as a *Kaplan–Meier curve* (Figure 2.2) [16].

2.3.2 Adverse Events and Toxicity

An *adverse event* (AE) is any unfavorable and unintended sign (such as laboratory finding), symptom, or disease that occurs while a patient is enrolled in a clinical trial. An AE may or may not be related to the study intervention, but they are considered relevant and reportable regardless of relatedness to treatment [17,18].

The National Cancer Institute issues the Common Terminology Criteria for Adverse Events (CTCAE), a descriptive terminology utilized for AE reporting [9]. The CTCAE is widely accepted throughout the oncology community as the standard classification and severity grading scale for adverse events in cancer therapy clinical trials and other oncology settings. In the CTCAE, AEs are grouped by System Organ Class, and a *grade* (severity) scale is provided for each AE term. The CTCAE displays grades 1 through 5 with unique clinical descriptions of severity for each AE based on general guidelines (Table 2.3) [9].

A *serious adverse event* (SAE) is defined by the FDA as any AE that results in death, illness requiring hospitalization, events deemed life-threatening, persistent or significant disability/incapacity, a congenital anomaly/birth defect, or a medically important condition [17,18].

2.3.2.1 Dose-Limiting Toxicity

The main objectives of phase I trials are to determine the drug dose that is appropriate for use in phase II and III trials and to gather information about drug pharmacokinetics. Patients who have advanced disease that is resistant to current therapy but who have normal organ function usually are included in phase I trials.

TABLE 2.3

National Cancer Institute Common Terminology Criteria for Adverse Events (version 5.0): Grading Severity Scale

Grade	Definition
1: Mild	Asymptomatic or mild symptoms Clinical or diagnostic observations only Intervention not indicated
2: Moderate	Minimal, local, or noninvasive intervention indicated Limiting age-appropriate instrumental ADL[a]
3: Severe	Severe or medically important Not immediately life-threatening Hospitalization: admission or prolongation indicated Disabling Limiting self-care ADL[b]
4: Life-threatening	Life-threatening consequences Urgent intervention indicated
5: Death	Death related to adverse event

Source: National Cancer Institute Common Terminology Criteria for Adverse Events v5.0 NCI, NIH, DHHS. November 27, 2017. Retrieved February 15, 2018, from https://ctep.cancer.gov/protocoldevelopment/electronic_applications/ctc.htm#ctc_50.

[a] Instrumental activities of daily living (ADL): preparing meals, shopping for groceries or clothes, using the telephone, managing money.

[b] Self-care ADL: bathing, dressing and undressing, feeding self, using the toilet, and taking medications. Patient is not bedridden.

A drug in a phase I trials is started at a low dose that is not expected to result in toxicity. The dose is increased for subsequent patients according to a series of previously planned steps, and only after sufficient time has passed to observe acute toxic effects for patients treated at the lower dose. The endpoint *dose-limiting toxicity* (DLT) is pre-specified and observed for a patient when the side effects of the treatment are serious enough to prevent an increase in dose of the treatment [17–20]. Typically, cohorts of 3–6 patients are treated at each dose and dose increase is stopped if DLT occurs in an unacceptable number of patients at a given dose. Phase I trial designs will be discussed in detail in Chapter 3.

2.3.3 Health-Related Quality of Life

Quality of life refers to overall enjoyment of life as affected by social, environmental, mental, and physical factors [19]. *Health-related quality of life* (HRQOL) focuses on an individual's well-being related to the state of health. In health studies, terms *quality of life* and *HRQOL* are used interchangeably [20]. In cancer research, HRQOL is the effect of cancer and treatment on the well-being of an individual patient [19,20]. Clinical trials with HRQOL endpoints measure aspects of an individual's sense of well-being and ability to perform various activities [21]. These endpoints may include disease symptoms, adverse events from treatment, functional status, and general perceptions of well-being and life satisfaction [22]. Drugs that may improve HRQOL may provide substantial benefits because many current cancer treatments have negative adverse events.

The HRQOL is a type of *patient-reported outcome*. Patient-reported outcomes are outcomes questionnaires or scales that may include any information reported directly by a patient [23]. Patient-reported outcomes may include information beyond the scope of HRQOL, such as treatment adherence or health behaviors. Patient-reported outcomes such as HRQOL are assessed using *measurement instruments* that typically include self-reported questionnaires. Measurement instruments must be rigorously developed in a process that includes multiple studies including testing and validation to ensure that the instrument is reliable and reproducible [24]. Several measurement instruments that are used commonly in oncology research are reviewed in this chapter. HRQOL is measured directly from the patient using HRQOL measurement instruments. The *instrument* includes the questionnaire and instructions for administration, scoring, and interpretation [24].

2.3.3.1 European Organization for Research and Treatment of Cancer Quality-of-Life Questionnaire Core 30 Items

The Quality-of-Life Questionnaire Core 30 Items (QLQ-C30, version 3.0) was developed by the European Organization for Research and Treatment of Cancer (EORTC) to assess the quality of life of cancer patients [23]. The EORTC QLQ-C30 has been translated and validated in over 100 languages and used in many published studies. It can be used in patients with any type of cancer.

The EORTC QLQ-C30 contains scales for functional status (physical, role, cognitive, emotional, social), symptoms (fatigue, pain, nausea, and vomiting), global health status, and quality of life, and several single-item symptom measures. The items are scaled from 1 (not at all) to 4 (very much) or 1 (very poor) to 7 (excellent). The total score is from 0 to 100 points. For functional and global quality-of-life scales, higher scores mean a better level of functioning. For symptom-oriented scales, a higher score means more severe symptoms [25].

2.3.3.2 Functional Assessment of Cancer Therapy – General Version

The Functional Assessment of Chronic Illness Therapy (FACIT) Measurement System provides questionnaires that measure HRQOL for people who have chronic illnesses [26]. The Functional Assessment of Cancer Therapy – General version (FACT-G) (current version 4) is a 27-item compilation of general questions that may be used in patients who have any tumor type [27].

The FACT-G is divided into four HRQOL domains: physical, social/family, emotional, and functional well-being. The FACIT scales are constructed to complement the FACT-G by addressing relevant issues related to disease, treatment, or condition that are not included in the general questionnaire. Each scale is intended to be as specific as necessary to cover the clinically relevant problems associated with a given health condition or symptom, but sufficiently general to enable comparison between diseases and extension to other chronic medical conditions (Figure 2.3) [28].

2.3.3.3 Short-Form 36 Survey

Developed by the RAND Corporation, the Short-Form 36 survey (SF-36) is a set of 36 generic questions related to HRQOL measures [29,30]. It is not specific to cancer patients but is commonly used in oncology (Figure 2.4) [29,30].

FACT-G (Version 4)

Below is a list of statements that other people with your illness have said are important. **Please circle or mark one number per line to indicate your response as it applies to the past 7 days.**

	PHYSICAL WELL-BEING	Not at all	A little bit	Some-what	Quite a bit	Very much
GP1	I have a lack of energy	0	1	2	3	4
GP2	I have nausea	0	1	2	3	4
GP3	Because of my physical condition, I have trouble meeting the needs of my family	0	1	2	3	4
GP4	I have pain	0	1	2	3	4
GP5	I am bothered by side effects of treatment	0	1	2	3	4
GP6	I feel ill	0	1	2	3	4
GP7	I am forced to spend time in bed	0	1	2	3	4

FIGURE 2.3
Functional Assessment of Cancer Therapy – General version (FACT-G) Questionnaire. Physical well-being domain, questions 1–7 (total, 27 questions). (FACIT Measurement System. Functional Assessment of Cancer Therapy – General (FACT-G). Retrieved February 15, 2018, from http://www.facit.org/FACITOrg/Questionnaires.)

RAND > RAND Health > Surveys > RAND Medical Outcomes Study > 36-Item Short Form Survey (SF-36) >

36-Item Short Form Survey Instrument (SF-36)

RAND 36-Item Health Survey 1.0 Questionnaire Items

Choose one option for each questionnaire item.

1. In general, would you say your health is:

  ○ 1 - Excellent

  ○ 2 - Very good

  ○ 3 - Good

  ○ 4 - Fair

  ○ 5 - Poor

2. **Compared to one year ago**, how would you rate your health in general **now**?

  ○ 1 - Much better now than one year ago

  ○ 2 - Somewhat better now than one year ago

  ○ 3 - About the same

  ○ 4 - Somewhat worse now than one year ago

  ○ 5 - Much worse now than one year ago

FIGURE 2.4
Short-Form 36 survey (SF-36). Questions 1–2 (total, 36 questions). (Ware JE Jr, Sherbourne CD. *Med Care.* 1992;30(6):473–83; RAND Corporation. 36-Item Short Form Survey (SF-36). Retrieved February 15, 2018, from https://www.rand.org/health/surveys_tools/mos/36-item-short-form.html.)

Illustrative Trial: Symptom Monitoring with Patient-Reported Outcomes during Routine Cancer Treatment: a Randomized Controlled Trial

Basch E et al. (2016) [31]

Study Description

In this randomized trial, patients with advanced solid tumors (genitourinary, gynecologic, breast, and lung) who received outpatient chemotherapy were randomly assigned to report their symptoms via tablet computers (intervention arm) or have their symptoms monitored and documented by clinicians (usual care, control arm). Patients in both arms were assessed for health-related quality of life during study participation, and arms were compared to determine whether or not clinical benefits were associated with symptom self-reporting during routine cancer care.

Study Endpoint

The primary endpoint of the study was change in health-related quality of life from baseline to 6 months as measured by the EuroQoL EQ-5D Index.

The EQ-5D was designed to measure general health status. It was a 5-item questionnaire measuring mobility, self-care, usual activities, pain/discomfort, and anxiety/depression,

By placing a tick in one box in each group, please indicate which statements best describe your health today.

Mobility
I have no problems in walking about — ✓
I have some problems in walking about — ☐
I am confined to bed — ☐

Self-Care
I have no problems with self-care — ✓
I have some problems washing or dressing myself — ☐
I am unable to wash or dress myself — ☐

Usual Activities *(e.g. work, study, housework, family or leisure activities)*
I have no problems with performing my usual activities — ☐
I have some problems with performing my usual activities — ✓
I am unable to perform my usual activities — ☐

Pain/Discomfort
I have no pain or discomfort — ☐
I have moderate pain or discomfort — ☐
I have extreme pain or discomfort — ✓

Anxiety/Depression
I am not anxious or depressed — ☐
I am moderately anxious or depressed — ✓
I am extremely anxious or depressed — ☐

FIGURE 2.5
EuroQoL EQ-5D Questionnaire.

each rated on 3 levels (EQ-5D-3L) or 5 levels (EQ-5D-5L). The questionnaire produces a 5-digit code that represents the health state, which is then converted to a score (range 0–100 points) using a scoring algorithm available from EuroQoL.

Lower scores represented worse health-related quality of life. A score change of 6 points on the 0–100 scale was considered clinically meaningful in United States cancer populations (Figure 2.5).

Results and Conclusion

Health-related quality of life scores improved in more participants in the intervention (34%) than usual care arm (18%), and worsened health-related quality of life scores were less frequent in the intervention (38%) than usual care arm (53%). More participants in the intervention (21%) than usual care arm (11%) had an improvement in health-related quality of life by ≥ 6 points compared with usual care, and fewer patients in the intervention arm experienced a ≥ 6-point worsening (intervention, 28%; usual care, 37%). The authors concluded that clinical benefits were associated with symptom self-reporting during routine cancer care [31,32].

2.4 Tumor-Centered Endpoints

2.4.1 Assessment of Response in Tumor-Centered Endpoints

Unlike patient-centered endpoints such as survival, tumor-centered endpoints require more complex and standardized definitions. Assessment of tumor response to therapy is based on tumor measurement from a radiographic scan. The definitions of tumor-related responses, such as tumor shrinkage or progression of disease, should be defined precisely for any study that uses tumor-centered endpoints.

Tumor endpoints are based on an assessment of anatomic *tumor burden*. Tumor burden refers to the amount of cancer in the body, such as the size of a solid tumor. Tumor burden, and *tumor response*, in which tumor burden is compared to a baseline measurement after treatment, are assessed in a standardized way to enable comparison between research studies.

In response to problems of non-standardized tumor response criteria in cancer research, an international working group was formed that led to the publication of Response Evaluation Criteria in *Solid Tumors (RECIST)* [10,11]. RECIST is the most commonly used set of definitions and criteria for assessment of solid-tumor response, and it describes a standardized approach to solid-tumor measurement and definitions for objective assessment of change in tumor size.

The RECIST provides definitions for tumor response categories including *complete response, partial response, progressive disease*, and *stable disease*. *Progressive disease* is defined as a 20% increase in the sum of the longest diameters of target lesions, unequivocal progression of nontarget lesions, and/or the development of new lesions. A *complete response* is the disappearance of all target lesions. A *partial response* is at least a 30% decrease in the sum of the diameters of target lesions. *Stable disease* is disease other than progressive disease, complete response, or partial response [10,11,14]. These definitions of tumor response are relevant to all tumor-centered endpoints discussed in this chapter. The RECIST guidelines are not applicable to all types of cancer. Separate criteria are available for other cancers, such as lymphomas, and for assessment of response in brain tumors [33,34].

2.4.2 Progression-Free Survival and Time to Progression

Progression-free survival (PFS) is defined as the time from start of treatment or randomization until disease progression or death. It is a composite endpoint based on patient survival and tumor assessment. The PFS may shorten drug-development time because it includes an event that occurs earlier than overall survival. Therefore, PFS may enable more rapid availability of efficacious therapies. In addition, unlike overall survival, PFS is not affected by second-line treatment choices [35].

A key consideration in the design of studies with PFS as an endpoint is the timing of study assessments because timing has an important effect on the determination of the outcome. The date of clinical or radiographic evaluation when progression is first evident is a surrogate for true progression. The precise date of progression would be unknown unless assessments were made daily; therefore, the true progression date is during the interval between two assessments. In statistical analysis, this is known as *interval-censored data* [35].

When progression is evident from clinical symptoms, dates of progression coincide with appointment dates; therefore, progression is a function of the frequency of scheduled clinical assessments. Similarly, when progression is determined from radiographic scans, the timing of the scans can affect the documented dates of progression. In response to the relation between time of assessment and determination of progression date, FDA guidelines include specific recommendations for determination of dates of progression and censoring (Table 2.4) [36].

TABLE 2.4

United States Food and Drug Administration Recommendations for Determination of Dates of Cancer Progression and Censoring

Situation	Date of Progression or Censoring	Outcome
No baseline tumor assessments	Randomization	Censored
Progression documented between scheduled visits	Earliest of: • Date of radiographic assessment showing new lesion (if progression is based on new lesion) or • Date of last radiographic assessment of measured lesions (if progression is based on increase in sum of measured lesions)	Progressed
No progression	Date of last radiographic assessment of measured lesions	Censored
Treatment discontinuation for undocumented progression	Date of last radiographic assessment of measured lesions	Censored
Treatment discontinuation for toxicity or other reason	Date of last radiographic assessment of measured lesions	Censored
New anticancer treatment started	Date of last radiographic assessment of measured lesions	Censored
Death before first progressive disease assessment	Date of death	Progressed
Death between adequate assessment visits	Date of death	Progressed
Death or progression after >1 missed visit	Date of last radiographic assessment of measured lesions	Censored

Source: Adapted from U.S. Department of Health and Human Services Food and Drug Administration Center for Drug Evaluation and Research (CDER), Center for Biologics Evaluation and Research (CBER). Guidance for industry: clinical trial endpoints for the approval of cancer drugs and biologics. http://www.fda.gov/downloads/Drugs/.../Guidances/ucm071590.pdf. Published May 2007. Accessed February 22, 2016.

Time to progression is defined as the duration between start of treatment or randomization and disease progression. In contrast with PFS, time to progression does not include deaths. In time-to-progression analysis, deaths are censored at the time of death.

Illustrative Trial: Overall Survival with Combined Nivolumab and Ipilimumab in Advanced Melanoma

Wolchok et al. (2017) [37]

Study Description

In this double-blind phase III trial, patients who had advanced melanoma were randomly assigned to receive ipilimumab and nivolumab, ipilimumab and placebo, or nivolumab and placebo. The trial was designed with 2 primary endpoints—overall survival and progression-free survival—and multiple secondary endpoints including objective response, toxicity, and tumor PD-L1 expression.

Study Endpoints

Overall survival was defined as time from randomization to death. Median survival time and survival rate at 3 years were reported.

Progression-free survival was defined as time from randomization to first documented disease progression or death (whichever occurred first). Tumor response was assessed according to RECIST 1.1 every 12 weeks in each group until progression was detected or treatment was discontinued.

Objective response rate was defined as the proportion of patients who had a best overall response of partial or complete response.

Results and Conclusion

A total of 945 patients were randomized to the 3 groups. The median overall survival was 37.6 months in the nivolumab group and 19.9 months in the ipilimumab group. Median overall survival was not reached in the nivolumab and ipilimumab group by the minimum follow-up (36 months). The overall 3-year survival was 58% in the nivolumab and ipilimumab groups, 52% in the nivolumab group, and 34% in the ipilimumab group.

The median progression-free survival was 11.5 months in the nivolumab and ipilimumab group, 6.9 months in the nivolumab group, and 2.9 months in the ipilimumab group.

The objective response rate was 58% in the nivolumab and ipilimumab group, 44% in the nivolumab group, and 19% in the ipilimumab group.

The authors concluded that in patients with previously untreated advanced melanoma, significantly longer overall survival occurred with nivolumab and ipilimumab combination therapy or nivolumab alone than ipilimumab alone. Survival outcomes were better with nivolumab and ipilimumab or nivolumab alone than ipilimumab alone.

2.4.3 Disease-Free Survival

Disease-free survival (DFS) is defined as the duration between treatment start and relapse of disease or death from any cause. DFS is an endpoint used in settings with treatments such as surgery that leave patients without any detectable signs of disease. DFS often is used as a surrogate for overall survival in which disease recurrence represents a major reason for death in the treated population. In such cases, therapies used to treat cancer recurrence may prolong survival but are unlikely to result in a cure. The DFS is particularly appropriate when the interval between recurrence and death is lengthy and would otherwise require longer follow-up for evaluation of overall survival [38].

2.4.4 Time to Treatment Failure

Time to treatment failure is defined as time from start of treatment or randomization to treatment discontinuation for any reason, including treatment toxicity, patient or physician withdrawal from study, tumor progression, or death. Time to treatment failure is not considered a valid endpoint from a regulatory perspective because it fails to distinguish treatment efficacy from other factors that are not related to the effectiveness of a treatment [39].

2.4.5 Objective Response Rate and Duration of Response

Objective response rate is defined as the proportion of patients who have tumor size reduction of a predefined amount for a minimum time period [36]. The objective response rate typically is defined as the total number of complete responses and partial responses; with this definition, objective response rate includes treatment responses with major antitumor activity and is a measure of direct therapeutic effect. The objective response rate commonly is used in single-arm studies.

Duration of response is defined as the time from initial tumor response (complete or partial response) to tumor progression and often is reported with objective response rate.

Illustrative Trial: Open-Label, Multicenter, Phase II Study of Ceritinib in Patients with Non-Small-Cell Lung Cancer Harboring ROS1 Rearrangement
Lim SM et al. [40]

Study Description

Patients with a distinct molecular subset of non-small-cell lung cancer were enrolled in this study to test the safety and efficacy of ceritinib, a promising and more potent alternative to existing chemotherapy.

Study Endpoints

The primary endpoint was objective response rate (ORR), defined as complete response (CR) or partial response (PR) by RECIST 1.1.

Drug toxicity was a secondary endpoint. AEs that occurred at grades 1–2 in 10% or more patients, or at grades 3–5 in any patients, were reported in the manuscript.

Other secondary endpoints included duration of response (DoR) and disease control rate (DCR), defined as CR + PR + Stable disease (SD) by RECIST 1.1.

Several exploratory endpoints were reported, including results from next-generation sequencing (NGS), immunohistochemistry (IHS), and fluorescence *in situ* hybridization (FISH).

Results and Conclusion

A total of 32 patients were enrolled in the study and received at least one dose of ceritinib. The ORR was 62% (95% CI 45–77), with 1 CR and 19 PRs in 32 patients. Stable disease was seen in 6 patients, for a DCR of 81% (95% CI 65–91). The median duration of response was 21.0 months (95% CI 17–25).

All 32 patients experienced at least one AE during the course of the study, and 16 (50%) experienced an SAE, with 7 (22%) experiencing SAEs suspected to be related to the study drug. The most common grade 1–2 AEs were diarrhea (78%), nausea (59%), and anorexia (56%), and the most common grade 3–4 AE was fatigue (16%). Three patients died while on study treatment and all deaths were determined not to be related to study drug.

The authors concluded that ceritinib showed clinically meaningful efficacy with respect to ORR and DoR endpoints. The safety profile of ceritinib was considered manageable and consistent with its established safety profile. This phase II study supports the use of ceritinib in this patient population.

2.5 Endpoints under Evaluation

2.5.1 Pathologic Complete Response (pCR)

Pathologic complete response (pCR, also known as pathologic complete remission) is defined as the lack of evidence of cancer in tissue samples removed during surgery or biopsy, after treatment with radiation or chemotherapy [19]. It is a surrogate endpoint that is increasingly used as an early marker of potential treatment efficacy.

Pathologic complete response is an important endpoint for assessing anticancer treatment in the neoadjuvant (preoperative) setting, particularly in breast cancer, where neoadjuvant treatment is becoming increasingly common. There is evidence that pCR following neoadjuvant treatment may predict long-term outcomes in certain types of breast cancer [41,42]. As with other emerging outcomes, pCR does not have a uniform definition across clinical trials and the published literature [41,42]. The FDA has recently published guidance on the definition of pCR in neoadjuvant breast cancer treatment and its use as an endpoint in clinical trials for neoadjuvant drugs targeting high-risk, early-stage breast cancer [41].

2.5.2 Immune-Related Response Criteria (irRC)

Although RECIST criteria provide a framework when measuring the response to traditional therapies such as cytotoxic therapy, it may not be applicable for response to immunotherapy. It is not uncommon that a decrease in lesion size occurs after an initial increase that was caused by inflammatory cell infiltrates or necrosis, confirmed by biopsy [14]. These atypical patterns of response have been labeled *pseudo-progression* and may be caused by T-cell infiltration [14]. These atypical patterns of response can result in RECIST categorization of progression of disease and will usually result in discontinuation of experimental treatment, but patients who have pseudo-progression derive clinical benefit from therapy. Alternative immune-related response criteria have been developed for the evaluation of immune therapy activity in solid tumors (Table 2.5) [43].

TABLE 2.5

Commonly used efficacy endpoints in oncology clinical trials: advantages and limitations

Endpoints	Definition	Advantages	Limitations
Overall survival (OS)	Time from randomization[a] until death from any cause	• Universally accepted measure of direct benefit • Easily and precisely measured	• May require a larger trial population and longer follow-up to show statistical difference between groups • May be affected by crossover or subsequent therapies • Includes deaths unrelated to cancer
Progression-free survival (PFS)	Time from randomization[a] until disease progression or death	• Requires small sample size and shorter follow-up time compared with OS • Includes measurement of stable disease (SD)	• Validation as a surrogate for survival can be difficult in some treatment settings • Not precisely measured (i.e., measurement may be subject to bias) • Definition may vary among trials • Requires frequent radiologic or other assessments • Requires balanced timing of assessment among treatment arms
Time to progression (TTP)	Time from randomization[a] until objective tumor progression; does not include deaths	• Not affected by crossover or subsequent therapies • Generally based on objective and quantitative assessment	
Time to treatment failure (TTF)	Time from randomization[a] to discontinuation of treatment for any reason, including disease progression, treatment toxicity, and death	• Useful in settings in which toxicity is potentially as serious as disease progression (e.g., allogeneic stem cell transplant)	• Does not adequately distinguish efficacy from other variables, such as toxicity
Event-free survival (EFS)	Time from randomization[a] to disease progression, death, or discontinuation of treatment for any reason (e.g., toxicity, patient preference, or initiation of a new treatment without documented progression)	• Similar to PFS; may be useful in evaluation of highly toxic therapies	• Initiation of next therapy is subjective. Generally not encouraged by regulatory agencies because it combines efficacy, toxicity, and patient withdrawal
Time to next treatment (TTNT)	Time from end of primary treatment to institution of next therapy	• For incurable diseases, may provide an endpoint meaningful to patients	• Not commonly used as a primary endpoint • Subject to variability in practice patterns
Objective response rate (ORR)	Proportion of patients with reduction in tumor burden of a predefined amount	• Can be assessed in single-arm trials • Requires a smaller population and can be assessed earlier, compared with survival trials	• Not a comprehensive measure of drug activity
Duration of response (DoR)	Time from documentation of tumor response to disease progression	• Effect is attributable directly to the drug, not the natural history of the disease	

Source: Adapted from U.S. Department of Health and Human Services Food and Drug Administration Center for Drug Evaluation and Research (CDER), Center for Biologics Evaluation and Research (CBER). Guidance for industry: clinical trial endpoints for the approval of cancer drugs and biologics. http://www.fda.gov/downloads/Drugs/.../Guidances/ucm071590.pdf. Published May 2007. Accessed February 22, 2016.

[a] Not all trials are randomized. In non-randomized trials, time from study enrollment is commonly used.

References

1. U.S. Department of Health and Human Services National Institutes of Health. October 20, 2017. Retrieved February 15, 2018, from https://www.nih.gov/health-information/nih-clinical-research-trials-you/basics.

2. Hönigsmann H, Diepgen TL. UV-induced skin cancers. *J Dtsch Dermatol Ges*. 2005 Sep;3(Suppl 2):S26–31.

3. Chang C, Murzaku EC, Penn L, Abbasi NR, Davis PD, Berwick M, Polsky D. More skin, more sun, more tan, more melanoma. *Am J Public Health*. 2014 Nov;104(11):e92–9.

4. Hammer GP, du Prel J-B, Blettner M. Avoiding Bias in Observational Studies: Part 8 in a Series of Articles on Evaluation of Scientific Publications. *Dtsch Ärztebl Int*. 2009;106(41), 664–8.

5. Jepsen P, Johnsen, SP, Gillman MW, Sørensen HT. Interpretation of observational studies. *Heart*. 2004;90(8), 956–960.

6. Piantadosi S. *Clinical Trials: A Methodologic Perspective*, Third Edition. New York, NY: John Wiley & Sons, Inc, 2017.

7. Green S, Benedetti J, Smith A, Crowley J. *Clin Trials Oncol*, Third Edition. Boca Raton, FL: CRC Press, 2016.

8. Couzin-Frankel J. Breakthrough of the year 2013. Cancer Immunotherapy. *Science*. 2013 Dec 20;342(6165):1432–3.

9. National Cancer Institute Common Terminology Criteria for Adverse Events v5.0 NCI, NIH, DHHS. November 27, 2017. Retrieved February 15, 2018, from https://ctep.cancer.gov/protocoldevelopment/electronic_applications/ctc.htm#ctc_50.

10. Therasse P, Arbuck S, Eisenhauer E et al. New guidelines to evaluate the response to treatment in solid tumors. *J Natl Cancer Inst*. 2000;92:205–16.

11. Eisenhauer E, Therasse P, Bogaerts J et al. New response evaluation criteria in solid tumours: Revised RECIST guideline (version 1.1). *Eur J Cancer*. 2009;45:228–47.

12. Lee HP. On clinical trials and survival analysis. *Singapore Med J*. 1982;23:164–7.

13. Cox DR, Oakes D. *Analysis of Survival Data*. London, England: Chapman and Hall, 2001.

14. Friedman CF, Panageas KS, Wolchok JD. *Special Considerations in Immunotherapy Trials. In Oncology Clinical Trials*, Second Edition. Kelly WK, Halabi S, eds. New York, NY: Demos Medical Publishing, 2018.

15. FDA. Guidance for Industry. Clinical Trial Endpoints for the Approval of Cancer Drugs and Biologics. May, 2007. Retrieved February 15, 2018 from, https://www.fda.gov/downloads/Drugs/Guidances/ucm071590.pdf.

16. Kaplan EL, Meier P. Nonparametric estimation from incomplete observations. *J Amer Statist Assn*. 1958;53(282):457–81.

17. FDA. What is a serious adverse event? Retrieved February 15, 2018, from https://www.fda.gov/Safety/MedWatch/HowToReport/ucm053087.htm.

18. The International Conference on Harmonisation of Technical Requirements for Registration of Pharmaceuticals for Human Use (ICH). *Guideline for Good Clinical Practice*. Jun 10, 1996.

19. NCI Dictionary of Cancer Terms. Bethesda, MD: National Cancer Institute. https://www.cancer.gov/publications/dictionaries/cancer-terms

20. Armitage P, Colton T. 2005. *Encyclopedia of Biostatistics*. Hoboken, NJ: Wiley Interscience.

21. Cella D, Stone AA. Health-related quality of life measurement in oncology: Advances and opportunities. *Am Psychol*. 2015 Feb–Mar;70(2):175–85.

22. Osaba D. Health-related quality of life and cancer clinical trials. *Ther Adv Med Oncol*. 2011;3(2):57–71.

23. Aaronson NK, Ahmedzai S, Bergman B et al. The European Organization for Research and Treatment of Cancer QLQ-C30: A quality-of-life instrument for use in international clinical trials in oncology. *J Natl Cancer Inst*. 1993 Mar 3;85(5):365–76.

24. Kimberlin CL, Winterstein AG. Validity and reliability of measurement instruments used in research. *Am J Health Syst Pharm*. 2008;65(23):2276–84.

25. EORTC Quality of Life Study Group. EORTC QLQ-C30. Retrieved February 15, 2018, from http://groups.eortc.be/qol/eortc-qlq-c30.
26. Cella D. Manual of the Functional Assessment of Chronic Illness Therapy (FACIT) Measurement System. Center on Outcomes, Research and Education (CORE), Evanston Northwestern Healthcare and Northwestern University, Evanston IL, Version 4, 1997.
27. Cella DF, Tulsky DS, Gray G. The Functional Assessment of Cancer Therapy scale: Development and validation of the general measure. *J Clin Oncol.* 1993;11(3):570–9.
28. FACIT Measurement System. Functional Assessment of Cancer Therapy – General (FACT-G). Retrieved February 15, 2018, from http://www.facit.org/FACITOrg/Questionnaires.
29. Ware JE Jr, Sherbourne CD. The MOS 36-item short-form health survey (SF-36). I. Conceptual framework and item selection. *Med Care.* 1992;30(6):473–83.
30. RAND Corporation. 36-Item Short Form Survey (SF-36). Retrieved February 15, 2018, from https://www.rand.org/health/surveys_tools/mos/36-item-short-form.html.
31. Basch E. Deal AM, Kris MG et al. Symptom monitoring with patient-reported outcomes during routine cancer treatment: A randomized controlled trial. *J Clin Oncol.* 2016 Feb 20;34(6):557–65.
32. EuroQol Group. EuroQol-a new facility for the measurement of health-related quality of life. *Health Policy.* 1990;16:199–208.
33. Cheson BD, Pfistner B, Juweid ME et al. Revised Response Criteria for Malignant Lymphoma. *J Clin Oncol.* 2007;25(5):579–86.
34. Macdonald DR, Cascino TL, Schold SC Jr, Cairncross JG. Response criteria for phase II studies of supratentorial malignant glioma. *J Clin Oncol.* 1990;8(7):1277–80.
35. Panageas KS, Ben-Porat L, Dickler MN, Chapman PB, Schrag D. When you look matters: The effect of assessment schedule on progression-free survival. *J Natl Cancer Inst.* 2007 Mar 21;99(6):428–32.
36. U.S. Department of Health and Human Services Food and Drug Administration Center for Drug Evaluation and Research (CDER), Center for Biologics Evaluation and Research (CBER). Guidance for industry: clinical trial endpoints for the approval of cancer drugs and biologics. http://www.fda.gov/downloads/Drugs/.../Guidances/ucm071590.pdf. Published May 2007. Accessed February 22, 2016.
37. Wolchok JD, Chiarion-Sileni V, Gonzalez R et al. Overall Survival with Combined Nivolumab and Ipilimumab in Advanced Melanoma. *N Engl J Med.* 2017 Oct 5;377(14):1345–56.
38. Gill S, Sargent D. End Points for adjuvant therapy trials: Has the time come to accept disease-free survival as a surrogate endpoint for overall survival? *Oncologist.* 2006;11:624–9.
39. Pazdur R. Endpoints for assessing drug activity in clinical trials. *Oncologist.* 2008;13(suppl 2):19–21.
40. Lim SM, Kim HR, Lee JS et al. Open-label, multicenter, phase II study of ceritinib in patients with non-small cell lung cancer harboring ROS1 rearrangement. *J Clin Oncol.* 2017;35(23):2613–8.
41. FDA. Guidance for Industry. Pathological Complete Response in Neoadjuvant Treatment of High-Risk Early-Stage Breast Cancer: Use as an Endpoint to Support Accelerated Approval. October 2014. Retrieved February 15, 2018 from https://www.fda.gov/downloads/drugs/guidances/ucm305501.pdf.
42. von Minckwitz G, Untch M, Blohmer JU et al. Definition and Impact of Pathologic Complete Response on Prognosis After Neoadjuvant Chemotherapy in Various Intrinsic Breast Cancer Subtypes. *J Clin Oncol.* 2012;30:1796–804.
43. Wolchok JD, Hoos A, O'Day S et al. Guidelines for the evaluation of immune therapy activity in solid tumors: Immune-related response criteria. *Clin Cancer Res.* 2009;15(23):7412–20.

3

Innovative Phase I Trials

Cody Chiuzan and Nathaniel O'Connell

CONTENTS

3.1 Early-Phase Designs for Cytotoxic Agents

3.1.1 Designs Based on Safety Endpoints

Clinical cancer research has made remarkable advances in methods of treatment over the past decade. The cytotoxic drugs (e.g., chemotherapy) continue to be the predominant part of most therapies, but in the forthcoming years, it is expected that innovative treatments such as molecularly targeted agents (MTAs) or immunotherapies will be incorporated more into the standard of care. Starting in the 2000s, the number of Food and Drug Administration (FDA) approved targeted drugs (65) has increased approximately five times compared to the number of cytotoxic drugs (13) [1]. These agents have demonstrated impressive clinical activity across many tumor types (e.g., melanoma, breast cancer, bladder cancer, ovarian cancer, lymphoma), but also revealed different toxicity profiles and mechanisms of action. This section focuses mainly on cytotoxic agents: underlying assumptions and classical statistical designs. Section 3.2 describes the new challenges brought by MTAs and immunotherapies and provides alternative endpoints and modeling approaches with illustrative examples.

Initially developed for testing cytotoxic agents, most of the phase I designs have focused on establishing the maximum tolerated dose (MTD) of a therapeutic regimen by assessing safety as the primary outcome. The main endpoint of interest is usually binary: presence or absence of a dose-limiting toxicity (DLT). The definition of DLT is defined in the preparatory stages of the trial, consisting of some grade (e.g., grade 3) of adverse event according to the National Cancer Institute (NCI) Common Terminology Criteria for Adverse Events (CTCAE) [2].

These dose-finding designs are based on the assumption that both dose-toxicity and dose-efficacy relationships are monotonically increasing. Hence, the highest (maximum) dose with an acceptable safety profile is also considered to be the most likely to be efficacious in subsequent phase II trials. Under these premises, the "classical" dose-escalation methods for phase I cancer trials fall into two main categories: parametric and non-parametric. The parametric designs assume one or more parameter models for dose-toxicity relationship. The non-parametric or model-free designs assume no parametric representation of the dose-response relationship and the treatment allocation follows a rule-based (algorithmic) scheme.

3.1.1.1 Rule-Based Algorithms: "A + B" Designs

Rule-based or algorithmic designs assign patients sequentially using pre-specified criteria based on observations of target events (e.g., the DLT) from clinical data. Typically, the largest dose that produces a DLT in a limited proportion (less than 20%, 33%, or 50% of the patients) is considered the MTD. No cumbersome calculations are involved and no underlying dose-toxicity model is needed.

At first, patients are enrolled in a cohort of size A receiving the same dose. This first dose is determined from preclinical studies, and it is expected to be relatively nontoxic. In general, dose escalation follows a modified Fibonacci sequence in which dose increments become smaller as the dose increases (by 100%, 67%, 50%, 40%, and 30%–35% of the preceding dose). The scheme of escalation is as follows.

Consider n_A the number of DLTs in cohort A. If $n_A \leq a_E$ patients (E, escalate), the next cohort will be assigned to the next highest dose. If $n_A \geq a_T$ patients (T, terminate), the trial is stopped with MTD declared as the dose below the current assignment. If $a_E < n_A < a_T$ with $a_T - a_E \geq 2$, an additional cohort B is enrolled at the same dose. Consider now n_B, the number of DLTs in cohort B. If $n_B \leq b_E$ patients, the next cohort will be assigned to the next highest dose. If $n_B \geq b_T$ patients, the trial is stopped with MTD declared as the dose below the current assignment. A famous "A + B" design is known as "3 + 3" where $A = B = 3$, $a_E = b_E = 0$, $a_T = 2$, and $b_T = 1$.

The traditional 3 + 3 algorithm wishes to sample around the 33rd percentile [3]. In spite of its practical simplicity, the standard design has several important limitations including short memory (i.e., the decision rules are based on the most recent cohort), slow dose escalation (leading to patients treated at dose levels less likely to be efficacious), and high error rates (recommend the incorrect MTD) [3–5]. Moreover, it has been theoretically proven that on average, the 3 + 3 actually targets doses DLT rates between 16% and 27%, and not the intended 33% [6].

3.1.1.2 Dose-Expansion Cohorts (DECs)

The rapid development of non-cytotoxic agents motivated the inclusion of additional endpoints in phase I trials such as efficacy, pharmacokinetics (PK), or pharmacodynamics (PD). Many studies have used a modification of the traditional 3 + 3 by including one or more dose-expansion cohort (DEC) after completion of dose escalation. Dose-finding trials may consider DECs with the goal of further exploring toxicity, assessment of additional

PK or PD, alternative dosage regimens, evaluation of predictive biomarkers, or even preliminary evaluation of antitumor activity. The sample size in DECs is based on a pre-specified number of patients, usually 6–15, but larger cohorts have been recently used, especially for trials testing immunotherapy agents. A well-known example of a trial that utilized multiple expansion cohorts is the Merck pembrolizumab development program that had an initial sample size of 18 patients with additional two disease-specific expansion cohorts of 14 patients [7]. By the time pembrolizumab was given accelerated FDA approval, the trial enrolled more than 1,000 patients with nine distinct cohorts [8].

Even though DECs have been repeatedly criticized because of the lack of statistical justification, they continue to be frequently used in dose-finding studies. Iasonos and O'Quigley emphasize that most protocols do not contain any rationale to justify the sample size used in DECs [9]. While for a small cohort (e.g., 10–15 patients) the standard power calculation might not be feasible, for studies with several and larger DECs, investigators should provide probabilistic justifications based on observed activity in comparison with historical rates.

3.1.1.3 Model-Based Designs

In the last two decades, a number of model-based designs have been proposed and implemented in phase I clinical trials. These methods assume a dose-response model and use toxicity data from all enrolled patients to generate a more precise estimation of the target probability of DLT at the recommended phase II dose (RP2D). The superiority of model-based methods compared to algorithmic approaches has already been shown in a multitude of studies. Despite their major gains such as a higher proportion of patients treated at levels closer to the MTD, a reduced total sample size, and less biased estimates of the target DLT, the adoption of model-based designs continues to be slow [10]. Rogatko et al. reported that 98.4% of dose-finding cancer trials published from 1991 to 2006 utilized a rule-based design [11]. An updated review of phase I studies published between 2008 and 2014 showed that only 5.4% of the trials adopted a model-based design, of which the most frequently used was the Continual Reassessment Method (CRM) or its variations. However, the number of trials testing MTAs of immunotherapies has drastically increased across the years and agents tend to favor the use of model-based designs [12].

a. Continual Reassessment Method (CRM) and Variations

The CRM was the first Bayesian model-based method introduced by O'Quigley et al. in [4]. This adaptive model was originally designed for dose-finding based on binary toxicity outcomes and aimed to identify the MTD from among a set of preselected dose levels. One of the most important issues of the design is choosing the functional form of the dose-toxicity curve. The most-used models are the "power" model (commonly referred to as the "empiric" model) and the one- or two-parameter logistic models [13,14]. Initially, the CRM was described with a one-parameter model, which requires less information in estimating the curve fluctuation, but the choice of the mathematical model is usually a joint decision of the statistician and clinician. Several models have been tested via simulations, and results show that the MTD estimation is robust to model misspecification, with efficiency being the most affected by increasing the number of parameters in the model [15]. A comprehensive, practical CRM tutorial with simulated examples and guidance for choosing design parameters was provided by Garret-Mayer [16].

A complexity of the CRM is its Bayesian paradigm. Some of the barriers to implement a Bayesian design involve expertise in the relevant statistical methods and software and consensus regarding prior and posterior distributions. Thus, some investigators are reluctant in using this statistical approach. Piantadosi et al. proposed a "practical" approach of the CRM by allowing *pseudodata* to be incorporated in dose estimation [13]. Before the trial, preclinical information (pseudodata) about toxicities is used to select the doses expected to produce low (e.g., 10%) and high (e.g., 90%) DLT rates. These two "anchor" points are used to estimate the dose-toxicity curve and then the target dose for the DLT. A dose-response model is selected *a priori* (e.g., one- or two-parameter logistic model) and updated based on the observed outcomes and the pseudodata. As the trial progresses and more information is being collected, pseudodata can be down weighted or even dropped.

The original CRM usually treats the first patient enrolled into a trial on a prior guessed MTD which is not necessarily the lowest dose level (i.e., might skip lower levels without testing them). These safety concerns have been addressed by the two-stage CRM design that starts at the lowest dose, follows a pre-specified, rule-based assignment and after the first DLT occurrence switches to a conventional CRM [17]. Goodman et al. predefined the dose levels for escalation as if for a 3 + 3 design, assigned more than one subject at a time to each dose, and limited each dose increase to one level [14].

Late-onset or cumulative toxicities have been frequently encountered due to prolonged treatment periods or in trials testing immune-stimulatory monoclonal antibodies (imAbs). An early-phase trial in metastatic melanoma patients identified four late-onset DLTs that were not subsequently taken into account in the dose-recommendation process [18]. This limitation has been addressed by the TITE-CRM design [19]. The authors extend the CRM by incorporating time-to-event (toxicity) information for each patient. The TITE-CRM utilizes data in real time including data from patients still in follow-up, allows for enrollment of patients before previously enrolled patients complete the DLT evaluation window, and thus can significantly reduce the length of the trial.

b. Beyond a Binary Safety Endpoint

As noted in Section 3.1.1, conventional methods often simplify the primary toxicity endpoint into a binary outcome. However, CTCAE guidelines define hundreds of toxicities across 26 system organ classes, each of which is individual graded on an ordinal scale from 0 to 5. The simplification of the toxicity endpoint ignores a great deal of potentially useful information that could help better estimate the MTD.

During the course of a trial, patients often experience several different types of toxicities of varying grades. Simplifying the toxicity outcome to a binary endpoint treats all non-DLT toxicities as irrelevant, which inherently ignores the possibility of using low- and moderate-grade toxicities (that are not individually dose limiting) to predict higher-grade toxicities. Furthermore, the binary endpoint ignores potentially harmful combinations of co-occurring low- and moderate-grade toxicities that together may present a toxicity burden that is too severe for a patient. For example, suppose a patient experiences five distinct grade 2 and grade 3 toxicities simultaneously, but none individually constitute a DLT. Utilizing the binary DLT endpoint, the toxicity burden experienced by this patient would be treated equivalently to that of a patient experiencing no toxicities of any kind.

These issues have served as motivation for the development of dose-finding methodology to make better use of patient toxicity data. Several methods have been proposed that make use of ordinal toxicity outcomes. Among the first designs to account for the natural ordering of toxicities is the accelerated titration design proposed by [20]. In an algorithmic framework, decision rules for dose escalation are extended to account for the presence of multiple grade 2 toxicities, such that two grade 2 toxicities observed in a cohort result in the same decision rule as a single grade 3 toxicity. In a model based design framework, the CRM was extended to model ordinal toxicity outcomes using a proportional odds model [21] and a continuation ratio model [22], where a patient's toxicity response is equal to the highest grade of toxicity experienced, and the dose-finding model is used to estimate the probability of a toxic response for each grade of toxicity. These methods demonstrate that using the range of ordinal grades can lead to improved performance in MTD estimation compared to the CRM using the binary DLT when respective model assumptions are upheld (e.g., proportional odds), and low grade toxicities are predictive of higher grade toxicities at succeeding dose levels.

Extending from this, to accommodate the occurrence of multiple toxicities across varying grades, additional methods have been proposed using composite patient toxicity scores in place of binary or ordinal outcomes. Bekele and Thall first proposed the concept of the total toxicity burden (TTB) score. [23] To calculate a patient's toxicity score, first a set of toxicity types deemed likely to occur in a trial need to be established. Next, toxicity "severity weights" are defined for each grade of each toxicity type to represent the relative burden each contributes to a patient's overall toxicity burden if it occurs. A patient's toxicity score is calculated as a function of toxicity severity weights and the toxicities experienced by the patient. Elicitation of toxicity severity weights are generally subjectively determined by expert clinicians in close collaboration with trial statisticians. This process is generally an unformal one (as detailed by Hobbs et al. [24]), although Lee et al. propose a statistical approach that may be used to help elicit them [25].

To illustrate toxicity severity weights and their use in calculating a toxicity score, suppose for a trial there are four defined toxicity categories: neutropenia, myelosuppression, hematological, and other, where the 'other' category is defined to give weight to all other toxicities not specifically defined. Suppose the toxicity severity weights elicited for this hypothetical trial are as listed in Table 3.1.

Next, suppose a patient presents the given toxicity profile, where an 'X' represents the occurrence of the toxicity from the corresponding row and grade of the corresponding column in Table 3.2.

TABLE 3.1

Examples of Severity Grades

	Severity Weight by Grade			
	1	2	3	4
Neutropenia	0.2	0.5	1.1	2
Myelosuppression	0.3	0.75	1.2	2
Hematological	0.1	0.4	0.8	1.5
Other	0.2	0.5	1.1	2

TABLE 3.2

Observed Toxicities by Grade

	Observed Toxicities			
	1	2	3	4
Neutropenia		X		
Myelosuppression	X			
Hematological			X	
Other				

Usually, grade 5 toxicity, which indicates a patient death due to toxicity, requires special consideration from the data and safety monitoring board (DSMB), possibly resulting in stoppage of the trial. Therefore grade 5 toxicities are not assigned severity weights for the purpose of calculating toxicity scores. The toxicity score for this patient is calculated as a function of the toxicity weights that correspond to the toxicities observed in the patient. Assuming an additive function, the toxicity score for this patient would be calculated as the sum of the weights corresponding to the toxicities that are observed, i.e., $0.5 + 0.3 + 0.8 = 1.6$.

Differing from the methods that use a binary toxicity outcome, which seek to estimate the probability of toxicity, dose-finding methods for toxicity scores attempt to model the continuous toxicity score directly as a function of dose. In the method established by Bekele and Thall [23], a Bayesian model is used to estimate the posterior mean TTB score at each dose based on all patient data available at the time. Rather than targeting a DLT rate based on the probability of toxicity, a target TTB score is established. Using CRM-like criteria, successive patients/cohorts are assigned the dose with the poster mean TTB score closest to a predefined target TTB score.

Since the proposal of the TTB score, various other dose-finding methods have been proposed using a toxicity score endpoint and, in recent years, have received greater attention. Chen et al. propose the "equivalent toxicity score" (ETS) [26], and model the ETS using a modified isotonic regression dose-finding model first proposed by Leung and Wang [27]. Rather than assigning subjectively defined severity weights as in the method of Van Meter et al. [22], their ETS uses a predefined function that calculates an ETS based on CTCAE defined severity grades, the number of unique toxicities, and user specified parameters. Lee et al. propose a dose-finding model that can handle both toxicity scores and ordinal toxicity grades and define the MTD in terms similar to that of the traditional DLT rate, i.e., the probability of the toxicity outcome at each dose level exceeding a predefined level of tolerance [28]. Additionally, Hobbs et al. extend toxicity scores to a group sequential trial [24], and Yin et al. propose a Bayesian dose-finding method that uses toxicity scores to account for toxicities occurring over multiple cycles of treatment [29]. Tseng et al. [30] published the first completed phase I trial utilizing toxicity scores and statistical methodology based on the method of Bekele and Tall [23].

In the context of the CRM, Yuan et al. extend CRM methodology to accommodate toxicity scores in the context of the Bayesian CRM framework, called the quasi-CRM [31]. Ezzalfani et al. further extended the quasi-CRM in a likelihood framework [32], and Yin and Yuan, 2009 merged the quasi-CRM framework with the Bayesian model averaging CRM [33]. In general, as long as the method quantifying toxicity scores is clinically logical, their use demonstrates improved performance over binary endpoint methods in terms of MTD estimation and the number of patients treated at optimal doses during a trial. Section 3.2.1 will focus on the quasi-CRM defined in a statistical framework to illustrate the use of toxicity scores.

3.1.2 Designs Based on Safety and Efficacy Endpoints

Traditionally, phase I trials are followed by a phase II trial to confirm the safety profile of the MTD/RP2D and to assess signals of efficacy. Usually, these two phases have been conducted sequentially. Although clinical efficacy data might be collected in phase I studies, they may not be sufficient to inform decisions in the subsequent phase II trials, mostly due to the patient heterogeneity and small sample sizes. Also, implementing two distinct trials increases the duration and cost for the entire drug development process. In recent years, more attention has focused on developing seamless phase I/II trials.

A phase I/II design assesses both toxicity and efficacy. In some, the goal is to select a dose with high efficacy among dose levels with acceptable toxicity, or the dose with the optimal efficacy-toxicity profile. Most of these designs are Bayesian-based methods and take one of two approaches: (1) approach where the decision criterion is based on trade-offs between probabilities of treatment efficacy and toxicity or (2) approach using elicited joint utilities of ordinal (efficacy, toxicity) outcomes.

The efficacy-toxicity trade-off method is based on a family of contours partitioning the two-dimensional set of possible outcome probability pairs $\pi = (\pi_E, \pi_T)$ [34–36]. For each contour, the trade-offs between π_E (efficacy probability) and π_T (toxicity probability) are characterized by a single number, the desirability of the pair. The contours are first constructed by selecting equally desirable target probability pairs with input from a physician or determining trade-offs. A target efficacy-toxicity trade-off contour is obtained by fitting a curve to the elicited target pairs, such that as π_T increases, π_E also must increase. The design is illustrated through several clinical applications, including a trial of graft-versus-host disease treatment in allogeneic bone marrow transplantation [35]. The operating characteristics are promising, but the design relies on crucial information on the predefined equally desirable pairs.

Extensions of the standard CRM also allow for modeling of both toxicity and efficacy outcomes. The first approach, the bivariate CRM (bCRM), maintains the bivariate structure of the outcomes through a joint modeling of toxicity and efficacy [37]. It extends the CRM to a marginal logit dose-toxicity curve and a marginal logit dose-disease progression curve with a flexible bivariate distribution of toxicity and progression. This procedure was described with application in a bone marrow transplant study. Bekele and Shen [38] and Dragalin and Fedorov [39] utilize bivariate probit models for toxicity and efficacy, with the former measuring efficacy based on the expression of a continuous biomarker.

Another approach deals with an observed clinical outcome considered as "trinary". Each patient's response falls into one of the following categories: no toxicity and no efficacy, no toxicity but with efficacy, or toxicity (with or without efficacy). In this case, the joint distribution of the binary toxicity and efficacy outcomes can be collapsed into an ordinal trivariate variable and is modeled by the TriCRM approach using a continuation ratio model [40].

3.2 Early-Phase Designs: Moving Beyond Cytotoxic Agents

Innovative cancer research in the last 10 years has led to the discovery of many novel and promising agents. As of October 2017, the FDA has approved over 60 molecular targeted agents (MTAs) and immune-oncology (I-O) strategies such as monoclonal antibodies (e.g., checkpoint inhibitors: pembrolizumab, nivolumab), cancer vaccines (sipuleucel-T, T-VEC), or adoptive T-cell therapies (CAR T-cells) [41]. These new forms of treatment harness the power of the immune system to control cancer growth using a patient's own body. Although these regimens have shown unprecedented results (e.g., remission rates), they have also raised new biological and statistical challenges.

Compared to cytotoxic treatments, these novel regimens have different mechanisms of action and less severe toxicity. Traditional definitions including, DLT and MTD, were based on the empirical monotonic relationships dose-toxicity and dose-efficacy seen in chemotherapies and some MTAs. With cytotoxic treatments, hematological toxicities are far more common than with MTAs, while MTAs induce more organ-specific toxicities [42,43].

Furthermore, toxic side effects of MTAs are often delayed, occurring in later cycles of treatment [44]. To date, the majority of the phase I trials in immune-stimulatory monoclonal antibodies have not reached a DLT dose [45]. For most trials, the RP2D was based on the maximum administered dose (MAD) or pharmacokinetic (PK) data. This is one of the major changes in early drug development in I-O. Significant early efficacy signals are often seen in I-O agents, but not considered in the dose selection process. Thus, there is an acute need for developing new designs that (1) capture early, late, and/or multiple grade toxicities; (2) incorporate evidence of antitumor activity or other immunologic parameters; and (3) impose no monotonicity on the dose-response relationship. Moreover, since the MTD might not be the most effective dose to induce antitumor activity, the primary goal of a dose-finding design should be finding the optimal biological dose (OBD), i.e., the lowest safe dose that has the highest efficacy rate at the beginning of the plateau of the dose-efficacy curve.

3.2.1 The Bayesian Quasi-CRM for Continuous Toxicity Endpoints

As discussed in Section 3.1.1.3, part b., there exist several methods for incorporating toxicity scores into dose-finding trials. This section illustrates one of those methods, the quasi-CRM [31] following the original CRM notation [4]. The quasi-CRM is chosen as the illustrative example due to its similarities to CRM, which is among the most widely accepted model-based designs in the statistical community.

For patients $i = 1, 2, \ldots, n$, let θ_i denote the toxicity score observed from patient i, where θ_i is normalized to a continuous domain from [0,1]. Note that it is not necessary that the original scale of toxicity scores falls on this domain; a normalized value for a patient's toxicity scores may be simply obtained by dividing their unadjusted toxicity score by the maximum toxicity score possible. Let dose $d_k, k = 1, 2, \ldots, K$ administered to patient i, be denoted by X_i, which takes on the value $x_i \in \{d_1, d_2, \ldots, d_K\}$. Just like the original CRM, the quasi-CRM assumes a dose-toxicity response function to model toxicity scores as a function of dose. Let this dose-toxicity response function be denoted by $F(x_i, \alpha)$, where α is a scaling parameter to be estimated. $F(x_i, \alpha)$ may take on one of the several functional forms used by the traditional binary CRM, such as the power or one-parameter logistic models. Note that doses do not take on clinical values such as mg/L, but are represented in the dose-response function by the initial guesses for the toxicity score at each dose level, thus $d_k \in [0,1]$. This is called the model skeleton; review CRM literature for more details. Using the power model as an example, the expected toxicity score for patient i is modeled by

$$E(\theta_i \mid x_i) = F(x_i, \alpha) = x_i^{\exp(\alpha)}, \quad \text{for } 0 \leq x_i \leq 1. \tag{3.1}$$

Just as in the binary CRM, a model skeleton must be established *a priori* to represent dose levels, except instead of using initial guesses for the probability of a DLT at each dose level, initial guesses for toxicity scores at each dose level are defined. Let the toxicity scores and doses up to the ith patient be denoted by $\Theta_i = \{x_1, \theta_1, \ldots, x_i, \theta_i\}$. Toxicity scores are modeled via the quasi-Bernoulli likelihood [46] given by

$$L(\alpha \mid \Theta_i) = \prod_{j=1}^{i} \left\{ F(x_j, \alpha) \right\}^{\theta_j} \left\{ 1 - F(x_j, \alpha) \right\}^{1-\theta_j}, i = 1, 2, \ldots, n. \tag{3.2}$$

The quasi-Bernoulli likelihood is a specific "quasi-likelihood" distribution based on "Pseudo-Maximum Likelihood Theory" [47]. The outcome, θ_i, does not need to be binary;

it can take on values in the range of [0,1] without affecting the consistency of the quasi maximum likelihood estimator (quasi-MLE). In context of the CRM, this means that the functional form of the likelihood used in the quasi-CRM is identical to the standard Bernoulli likelihood used in the binary CRM, allowing the dose-finding model to operate similarly.

If the quasi-CRM is being defined in a Bayesian framework, a prior distribution must be specified on the scaling parameter α, which will be denoted as $h(\alpha)$. Guidelines for prior distributions established for the binary CRM may simply be used since the domain for probability of toxicity estimated using the conventional binary method is the same as the domain of toxicity scores. Thus, the quasi-posterior density is given by

$$g(\alpha \mid \Theta_i) = \frac{L(\alpha \mid \Theta_i)h(\alpha)}{\int\limits_{-\infty}^{\infty} \alpha L(\alpha \mid \Theta_i)d\alpha}, i = 1, 2, \ldots, n, \tag{3.3}$$

and the posterior expectation for α after i patients is given by

$$\hat{\alpha}_i = \int\limits_{-\infty}^{\infty} \alpha g(\alpha \mid \Theta_i)d\alpha. \tag{3.4}$$

The posterior estimated toxicity score at each dose level is found by plugging in $\hat{\alpha}_i$ in the dose-response function and solving for each dose level:

$$F(d_k, \hat{\alpha}_i), k = 1, 2, \ldots, K. \tag{3.5}$$

Let τ denote a predefined target toxicity score (akin to a desired DLT rate in the binary CRM). The dose administered to the next patient, $d(i + 1)$, is the dose with an estimated toxicity score closest to the target toxicity score. That is,

$$d(i+1) = \mathrm{argmin}_{d_k} \left[F(d_k, \hat{\alpha}_i) - \tau \right], \forall k = 1, 2, \ldots, K. \tag{3.6}$$

Once the total sample size has been reached or stopping criteria are met, the dose estimated to be the MTD is the final dose recommended using all observed patient data.

3.2.1.1 *Illustrative Example of Modeling Toxicity Scores: Quasi-CRM versus Conventional CRM*

To underline the relevance of using severity toxicity weights, the quasi-CRM and the CRM will be compared in a simulation. For the purposes of this simulation, toxicity scores will be calculated using severity weights as discussed in Section 3.1.1.3, part b., where toxicity scores are the sum of severity weights corresponding to observed toxicities. In this hypothetical example, assume there are three unique toxicity categories identified by expert clinicians for which toxicity severity weights are established: myelosuppression, hematological, and other. The hypothetical toxicity weights in this example are given in Table 3.3, normalized to a scale of 0–1.

Six dose levels are being tested in this example. The target toxicity score is defined to be $\tau = 0.18$, so that the estimated MTD based on toxicity scores is the dose with an estimated

TABLE 3.3

Toxicity Weights Used in the Quasi-CRM Tutorial

	Severity Weight by Grade			
	1	2	3	4
Myelosuppression	0.04	0.09	0.22	0.36
Hematological	0.04	0.07	0.20	0.36
Other	0.02	0.05	0.15	0.27

mean toxicity score closest to 0.18. To simulate data, "true" probabilities for each toxicity of each grade must be defined. For this example, probabilities are defined so that when the severity weight matrix shown in Table 3.3 is applied and toxicity scores are calculated, the true mean toxicity scores for dose levels 1–6 are 0.02, 0.06, 0.13, 0.20, 0.25, and 0.28, respectively. Thus, dose 4 would be the true MTD since its true mean toxicity score of 0.20 is closest to the target rate of 0.18. Cohort sizes of 1 patient are enrolled sequentially, with a total trial sample size set to 36 patients. No skipping dose and no stopping rules are employed. The estimated MTD is the final recommended dose after all 36 patients have been followed. A power model, as defined in equation 3.1, is assumed as the dose-response function for both the toxicity score model with a commonly used normal prior distribution $N(0.1.34)$ for the scaling parameter [48]. The model skeleton (prior guesses) for the toxicity score model at each dose level is {0.06, 0.11, 0.18, 0.27, 0.36, 0.46}. Therefore, the initial guess of the MTD is dose 3, which is misspecified by 1 dose level (i.e., the true MTD is dose 4).

To start the trial, patient 1 is enrolled at the first dose and no toxicities are observed, thus patient 1 has a toxicity score of 0. Fitting this in the model yields the updated estimate for the scaling parameter: $\hat{\alpha}_1 = 0.046$. Plugging this into the dose response function, $d_k^{exp(0.046)}$, gives the estimated toxicity scores for doses 1–6 to be 0.05, 0.10, 0.17, 0.25, 0.35, and 0.44, respectively. The estimated MTD is dose 3 since it is closest to the target score. However, since dose 2 has not yet been tried, the second patient is administered dose 2 and experiences a grade 1 "other" toxicity, yielding a toxicity score of 0.04. Now fitting the data from the first two patients, the updated scaling parameter is $\hat{\alpha}_1 = 0.122$, and the updated estimated toxicity scores at each dose level are 0.04, 0.08, 0.15, 0.23, 0.32, and 0.42, respectively. Thus the updated MTD remains 3, which is now administered to patient 3. This process continues until all 36 patients have been evaluated, after which the final recommended dose is the estimated MTD. Figure 3.1 demonstrates how the mean toxicity score estimates converge to the correct dose (dose 4) as the number of patients increases. In this exact scenario evaluated over 2,000 simulations, the quasi-CRM correctly identifies dose 4 as the MTD 76% of the time.

To compare these results to the conventional binary CRM, consider a DLT to be defined as a grade 4 hematological toxicity or any grade 3 toxicity of another type. Defined in this way, any toxicity that is considered a DLT is also considered too toxic based on its severity weight used in the earlier example. However, the binary model will not account for combinations of non-DLT toxicities like the toxicity score model. The target DLT rate for the binary model will be defined as 30% and using the same "true" probabilities for each grade of each toxicity used for toxicity scores in the previous example, the probability of a DLT for each dose level is computed. For the purposes of this tutorial, these "true" probabilities of toxicity were designed so that the true MTD is the same for both the toxicity score model and binary DLT model, which are given in Table 3.4.

The skeleton for the binary DLT model representing the prior guesses for the probability of DLT at each dose level is {0.12, 0.20, 0.30, 0.40, 0.50, 0.59}. These differ from the toxicity-score

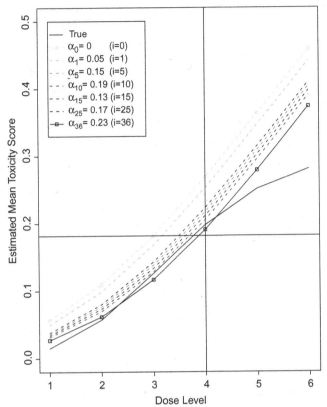

FIGURE 3.1
Quasi-CRM simulations: Estimated toxicity scores as function of sample size. In this example, dose level 4 represents the true MTD.

model because the target toxicity score is not numerically equal to the target DLT rate, and the range of plausible toxicity scores is less than that for the probability of a DLT. Nevertheless, the initial guess for the MTD is dose 3, as in the previous example.

By design, simulations for both models are based on the exact same "true" probability distributions for toxicity, dose 4 is the true MTD in both scenarios, and the true toxicity score and probability of DLT are both just slightly higher than their respective target score and DLT rate. With this in consideration, when evaluating the binary CRM model over 2,000 simulations, dose 4 is correctly identified as the MTD 52% of the time (compared to 76% by the quasi-CRM). Furthermore, out of 36 patients in the trial, the quasi-CRM treated 19.3 patients at the correct dose on average, whereas the binary-CRM treated 13.0.

TABLE 3.4

Mean Toxicity Score and DLT Probability

	Dose Level					
	1	2	3	4	5	6
Mean Toxicity Score	0.02	0.06	0.13	0.2	0.25	0.28
Probability of DLT	0.01	0.1	0.21	0.33	0.45	0.51

This tutorial illustrates the benefit of utilizing innovative trial designs that incorporate toxicity scores. The more comprehensive use of patient toxicity data through toxicity scores allows for potentially harmful combinations of moderate toxicities to be considered excessively toxic in the dose-finding model and, in addition, has the potential to improve the MTD estimation and treat more patients at optimal doses.

R-code to simulate the quasi-CRM in a single agent setting. Please see the companion website for functions https://www.crcpress.com//9781138083776.

NOTE: A more comprehensive collection of functions is being developed and will soon be made available in the R-package "**qcrm**".

```
###########################################################################
#                          Define Scenario                                #
###########################################################################

# Number of toxicities to be assigned toxicity weights
ntox <- 3

# Number of pre-specified dose levels
ndose <- 6

# Array of true toxicity probabilities for each toxicity grade across dose levels

TOX <- array(NA, c(ndose, 5, ntox))

# Probability of toxicity for grades 0-4 (columns) across doses (rows) for
toxicity 1
TOX[,,1] <- matrix(c(0.791, 0.172, 0.032, 0.004, 0.001,
                     0.738, 0.195, 0.043, 0.015, 0.009,
                     0.635, 0.230, 0.078, 0.044, 0.013,
                     0.592, 0.260, 0.088, 0.046, 0.014,
                     0.505, 0.273, 0.131, 0.071, 0.020,
                     0.390, 0.307, 0.201, 0.074, 0.028),
                   nrow=ndose, byrow=T)

# Probability of toxicity for grades 0-4 (columns) across doses (rows) for toxicity 2
TOX[,,2] <- matrix(c(0.968, 0.029, 0.002, 0.001, 0.000,
                     0.763, 0.192, 0.006, 0.039, 0.000,
                     0.652, 0.223, 0.091, 0.030, 0.004,
                     0.552, 0.265, 0.168, 0.010, 0.005,
                     0.397, 0.258, 0.276, 0.061, 0.008,
                     0.260, 0.377, 0.281, 0.073, 0.009),
                   nrow=ndose, byrow=T)

# Probability of toxicity for grades 0-4 (columns) across doses (rows) for toxicity 3
TOX[,,3] <- matrix(c(0.907, 0.070, 0.010, 0.008, 0.005,
                     0.602, 0.281, 0.035, 0.040, 0.042,
                     0.306, 0.258, 0.121, 0.181, 0.134,
                     0.015, 0.134, 0.240, 0.335, 0.276,
                     0.005, 0.052, 0.224, 0.372, 0.347,
                     0.004, 0.022, 0.223, 0.345, 0.406),
                   nrow=ndose, byrow=T)

# Define Toxicity Weight Matrix (defined on arbitrary scale)
W <- matrix(c(0.2, 0.5, 1.2, 2,
              0.2, 0.4, 1.1, 2,
              0.1, 0.3,  0.8, 1.5), nrow=ntox, byrow=T)
```

```
# Scale weights to [0,1]; required for the quasi-CRM
W <- W/sum(W[,4])

# True mean toxicity score at each dose level (function of W and TOX)
ms <- score.means(ntox, W, TOX)

# Target Toxicity Score
target.score <- 1/5.5

# Model Skeleton (prior guesses for mean toxicity score at each dose level)
skel <- c(0.06, 0.11, 0.18, 0.27, 0.36, 0.46)

###################################################################
# Estimate current MTD/Next dose given vector of patient toxicity scores   #
# and administered doses                                                    #
###################################################################

# Vector of patient toxicity scores (for 10 observed patients)
y <- c(0, 0, 0.15, 0, 0.27, 0.38, 0.02, 0.02, 0.15, 0.02)

# Vector of doses administered (for 10 observed patients)
n <- c(1,2,3,3,4,4,3,3,4,4)

# Get the current estimate of the MTD (dose level for next patient)
run.qcrm(y, n, skel, target.score, weights)

> run.qcrm(y, n, skel, target.score, weights)

$`current.MTD.est`# Gives the current dose estimated to be the MTD
[1] 4

###################################################################
#                     Single Trial Simulation                       #
###################################################################

# cohort size
cohortsize <- 1

# Number of Cohorts (if cohortsize=1, this is sample size)
ncohort <- 30

# Run Simulation
qcrm(TOX, skel, target.score, W, cohortsize, ncohort)

> qcrm(TOX, skel, target.score, W, cohortsize, ncohort)
$`dose.selected` # Estimated MTD indicated by "1"s position in vector
[1] 0 0 0 1 0 0

$mean.toxscore # Mean observed toxicity score in patients treated at each dose
[1] 0.036 0.000 0.118 0.196 0.000 0.000

$pt.per.dose # Number of patients treated at each dose over the trial
[1]  1  2  6 21  0  0

$est.scores # Final estimated toxicity scores of each dose level
[1] 0.04 0.07 0.13 0.20 0.28 0.38

$pt.toxscores # Observed patient toxicity scores over the trial
 [1] 0.04 0.00 0.31 0.00 0.05 0.18 0.04 0.07 0.18 0.09 0.18 0.09 0.35 0.38
0.31 0.05 0.05 0.15 0.18 0.15 0.05 0.31 0.09 0.31 0.18 0.35 0.15 0.15 0.05
0.36

$doses.assigned # Doses assigned to patients over the trial
 [1] 1 2 3 2 3 3 3 3 4 4 4 4 4 4 4 3 4 4 4 4 4 4 4 4 4 4 4 4 4 4
```

```
###############################################################################
#                      Simulation over iterative trials                       #
###############################################################################

# Load Required Packages
library(foreach)
require(doSNOW)

# Number of trials to run (default is 1000)
ntrial <- 1000

# Number of CPU Cores to run simulation (default is 1)
ncores <- 4

# Run Simulation
qcrmsim(TOX, skel, target.score, ms, W, cohortsize, ncohort, ntrial, ncores)

$`true.scores` #True mean toxicity score at each dose level
[1] 0.016 0.057 0.126 0.198 0.251 0.281

$est.MTD.perc # Percent of times each dose is estimated as the MTD over simulation
[1] 0.000 0.000 0.284 0.714 0.002 0.000

$mean.toxscore # Mean observed toxicity score of patients treated at each dose level
[1] 0.013957 0.039258 0.098420 0.187124 0.018118 0.000000

$mean.pt.per.dose # Mean number of patients treated at each dose over trials
[1]   1.090  1.894 12.248 14.550  0.218  0.000

###############################################################################
#                   Comparison to conventional binary CRM                     #
###############################################################################

#load library 'dfcrm'
library('dfcrm')

# Must define the grades of each toxicity that qualify as a DLT
# Example: Define a DLT to be a grade 3 of toxicities 1 and 2 (rows 1
#   and 2 in the weight matrix) and a grade 4 of toxicity 3 (row 3 in
#   the weight matrix).
tox.dlt <- c(3, 3, 4)

# Elicit the probability of a DLT by dose level based on the previous rule
dlt.probs <- dlt.prob(TOX, ntox, tox.dlt)

# specify a model skeleton
skeleton.dlt <- c(0.12, 0.2, 0.3, 0.4, 0.5, 0.59)

# Target DLT Rate
theta <- 0.33

#Run Simulation
crmsim(dlt.probs, skeleton.dlt, theta, n=ncohort, x0=1 , nsim=ntrial, method="bayes",
model="empiric")
```

3.2.2 Novel Endpoints in Early-Phase Trials

3.2.2.1 Dose-Finding Designs Incorporating Pharmacokinetics (PK) Measures

Under the new assumptions of milder, non-monotone toxicity profiles, determining the OBD is an attractive goal for the early-phase studies. Practically speaking, dose-finding studies incorporating multiple (biological) endpoints have become frequently used approaches for evaluation of targeted, non-cytotoxic drugs. However, several barriers limit their potential to just exploratory endpoints. The inclusion of biological endpoints and determination of an optimal dose based on some biomarker occurrence should rely on pre-specified thresholds such as targeted plasma, blood drug concentration, or other immunologic parameter. Assessing pharmacodynamic (PD) markers as primary endpoints can be challenging, as they not only require a strong scientific rationale, but also a noninvasive reproducible assay that can track PD markers with minimal harm to the patient [49].

Integration of clinical PK and preclinical PD has provided an additional modality of augmenting early clinical data with animal data, but nothing is relevant in the absence of definitive correlations between the target inhibition in PK or PD biomarkers and clinical efficacy (e.g., tumor response). As long as repeated measurements can be recorded in real time, incorporating PK information in the dose-finding process can provide a better estimation of the dose-toxicity curve while maintaining the performance in terms of MTD selection. However, in most phase I trials, the dose-finding and pharmacokinetics (PK) analyses are considered separately, which for small populations might impact the estimation of both the toxicity and PK parameters.

Ursino et al. developed methods that take into account PK measurements in sequential Bayesian adaptive early-phase designs [50]. The authors propose six different models that include PK measures either as covariates, as dependent variables, or in a hierarchical model. The motivating study is based on a PK model for TGF-β signaling to block tumor growth in patients with glioma. In simulations, toxicity (binary) is related to the area under the concentration curve (AUC), which is considered a measure of PK exposure. PK data is generated first using a simplified one-compartment model with first-order absorption rate, and then toxicity values are generated based on AUCs. The PK-toxicity relationship is modeled by a linear function with the assumption that toxicity occurs when the value of a function of AUC is above a certain given threshold.

Let $D = \{d_1, \ldots, d_k\}, d_1 < d_2 < \cdots < d_k$ be the set of K possible doses, y_i be a binary variable indicating presence (1) or absence (0) of a DLT, and z_i be the logarithm of the AUC of the drug concentration in blood plasma for the ith patient, $i = 1, 2, \ldots, n$, where n denotes the total sample size. The same dose-escalation rule applies to all methods, i.e., the next dose assignment is the one with probability of toxicity nearest to the target probability selected *a priori* by the trial investigators. The recommended MTD is declared by the dose that would have been administered to the $(n + 1)$ subject enrolled in the trial. If the posterior probability of toxicity of the first dose is greater than a specified threshold, then no dose is suggested and the trial is stopped.

Below we provide a brief description for each of the methods, with emphasis on the model specifications. More details regarding choice of priors and dose allocation rules can be found in the original article [50].

1. **PKCOV** is a modification of the method proposed by Piantadosi and Liu [51].

   The AUC is modeled as a covariate for probability of toxicity (p_T) through the logit link, with the following dose-toxicity model:

   $$\text{logit}(p_T(d_k, \Delta z_{d_k}, \beta)) = -\beta_0 + \beta_1 \log(d_k) + \beta_2 \Delta z_{d_k}, \forall d_k \in D, \tag{3.7}$$

   where β_0 is a constant selected from prior information, Δz_{d_k} represents the difference between the logarithm of the population AUC at dose d_k and z, the logarithm of AUC of the subject at the same dose.

2. **PKLIM** is a method proposed by Patterson et al. [52] and Whitehead et al. [53]. The PK-toxicity relationship is given by

$$z_i \,|\, \beta, \nu \sim N(\beta_0 + \beta_1 \log(d_i), \nu^2),$$ (3.8)

where regression parameters $\beta = (\beta_0, \beta_1)$ are modeled via bivariate normal distributions and ν is the standard deviation following a beta distribution. The probability of toxicity at each dose is calculating using

$$P(z > L \,|\, d_k, \beta = \hat{\beta}, \nu = \hat{\nu}), \forall d_k \in D,$$ (3.9)

where L is a threshold used for the assumption that DLTs occur when AUCs exceed a certain value. For cases in which L is unknown, one can use the PKCRM method.

3. **PKCRM** is a combination of PKLIM and CRM using a power working model with a normal prior on the scale parameter.

4. **PKLOGIT**, inspired by Whitehead et al. [54], models the probability of toxicity (p_T) as a function of AUC alone using

$$\text{logit}(p_T(z, \beta)) = -\beta_2 + \beta_3 z,$$ (3.10)

where β_2 and β_3 are specified with independent uniform prior distributions. The probability of toxicity is obtained by using the estimated parameters of each regression model in the following expected value formula:

$$P(y = 1 \,|\, d_k, \beta = \hat{\beta}, \nu = \hat{\nu}) = E\left[\frac{1}{1 + e^{\hat{\beta}_2 - \hat{\beta}_3 z}}\right] = \int \frac{1}{1 + e^{\hat{\beta}_2 - \hat{\beta}_3 z}} g(z) dz,$$ (3.11)

where $g(z)$ is the distribution of the logarithm of AUC given dose d_k obtained from equation 3.8.

5. **PKPOP** is a variation of PKLOGIT by replacing the observed value AUC for the patient with the population mean value, i.e., substitute z with $z_{k,pop}$.

6. **PKTOX** method is essentially PKLOGIT with a probit link, that is

$$p_T(z, \beta) = \Phi(-\beta_2 + \beta_3 z),$$ (3.12)

where Φ represents the cumulative normal distribution.

All methods are evaluated and compared to a model that considered toxicity only (DTOX) in terms of MTD percentage of correct selection (PCS) and the ability to estimate the dose-response curve. Operating characteristics are presented for a fixed sample size of 30 subjects, six predefined dose levels, seven toxicity scenarios, and target toxicity probability of 0.20. The main finding is that good prior knowledge about PK can help reduce the percentage of overdosing without altering the MTD selection. The PKCRM is recommended when non-monitorable toxicity has been observed in preclinical studies or when the L threshold is known from the literature. The PKLOGIT and PKTOX methods are useful when a more precise dose-response curve estimation is required. The PKCOV achieves the best PCS for the scenario in which MTD is at the highest dose level, for a larger number of DLTs (maximum 11 instead of 7–9 compared to other methods).

All methods are implemented in a Bayesian framework, with parameters estimation carried out using the R package (rstan). To ease the implementation, authors also developed a user-friendly R package (dfpk) that can be used to (1) determine the next recommended dose or the MTD (at the end of the trial) and (2) run simulations for implementation in a new trial [55].

3.2.2.2 Dose-Finding Designs for Immunotherapies

Adoptive T-cell therapy is a rapidly emerging immunotherapeutic approach that consists of an infusion of genetically engineered T cells expressing a specific antigen on their cell membrane. In 2017, the FDA approved the first chimeric antigen receptor (CAR-T) cell therapy (tisagenlecleucel) for children and young adults with B-cell acute lymphocytic leukemia (ALL) [56]. With an 83% remission rate, this therapy has demonstrated an early and durable response, but much remains to be learned regarding cell proliferation, persistence, and mechanisms of relapse. An important predictor of the efficacy of CAR-T cells is their ability to expand *in vivo* in response to recognition of CD19+ target cells, and therefore, patients that failed to respond in prior studies typically had poor accumulation of CAR-T cells. Interestingly, recent studies investigating CD19 CAR-T cells demonstrated a correlation between cell dose levels (magnitudes of 10^5 cells/kg), earlier/higher peak expansion and clinical response [57], or correlation between clinical response and persistence of administered cells at one month [58]. Although CD19 CAR-T cells showed a therapeutic effect in B-cell ALL patients, significant toxicities have occurred, especially after infusion of higher CAR-T-cell doses. Data imply that an optimal dosing strategy to minimize toxicity would be to initially give low CAR-T-cell doses to patients with higher tumor burden, whereas those with low tumor burden may require higher or repeated doses. Early-phase trials should start incorporating more immunological information and randomize patients toward doses with higher predicted efficacy.

3.2.2.2.1 A Novel Early-Phase Adaptive Design Incorporating Immunological Outcomes

Chiuzan et al. propose an adaptive dose-finding design for assessing toxicity and continuous efficacy outcomes in cancer immunotherapy trials [59]. The two-stage design considers both a binary toxicity measure and a continuous efficacy outcome, modeled independently. The model imposes monotonicity on the dose-toxicity relationship, but it is flexible in allowing a non-monotone dose-efficacy curve. Fit in a frequentist framework, the model is easy to implement and it can be tailored to different toxicity and activity profiles, especially in testing immuno-oncology agents for which identifying the MTD might be less relevant in selecting the RP2D.

The goal is to select the dose that optimizes some efficacy criterion while maintaining safety. The first stage establishes the safety profile of each dose, with escalation decisions based on likelihood principles. Continuous immunologic outcomes are used to evaluate the relative efficacy of the doses. The second stage employs an adaptive randomization to assign patients to doses showing higher efficacy. Safety is being continuously monitored throughout stage 2, where some doses may be "closed" due to unacceptable toxicity. The proposed design is compared to the modified toxicity probability interval (mTPI) design [60] in terms of percentage dose allocation and estimation of outcomes under different dose-response scenarios. The operating characteristics tested in extensive simulations show that by using an efficacy-driven adaptive randomization with safety constraints, the allocation distribution is skewed towards more efficacious doses, and thus limit the number of patients exposed to toxic or non-therapeutic doses.

3.2.2.2.2 *Illustrative Example of the Early-Phase Design for Immunological Outcomes*

Let $d_j, j = 1, 2, \ldots, 5$, be the set of the five ordered dose levels, and p_j be the true, unknown DLT rate at each dose. For each subject assigned to a dose d_j, a single binary variable T indicates whether the patient has experienced (1) or not (0) a DLT.

Let $p_1 = 0.40$ and $p_2 = 0.15$ be the unacceptably high and acceptable toxicity rates, respectively, decided *a priori* by the investigators. Suppose that cohorts of $m_j = 3$ patients are initially assigned sequentially to increasing doses, with no escalation occurring until acceptability has been determined in the previous cohort. The two-stage design is implemented in the following order:

Stage 1: Establish safety profile and collect immunological outcomes.
The following hypotheses are to be tested at each dose level:

$$H_1 : p_j = 0.40 \; vs \; H_2 : p_j = 0.15, \tag{3.13}$$

with the likelihood ratio (LR_j) being computed given observed toxicity data t_j:

$$LR_j = \left(\frac{p_2}{p_1}\right)^{t_j} \left(\frac{1-p_2}{1-p_1}\right)^{m_j - t_j}, j = 1, 2, \ldots, 5. \tag{3.14}$$

Then, the likelihood ratio is compared to benchmark $k = 2$ (see [61]), and safety for each dose is determined using the following rule: if $LR_j > 1/k$, dose j is regarded as acceptably safe and escalation is allowed; if $LR_j \leq 1/k$ dose j is thought to be unacceptably toxic and "closed" for allocation.

Stage 2: Employ an efficacy-driven adaptive randomization based on continuous efficacy endpoints. After stage 1 has been completed and all acceptable doses have been selected, the remaining patients are adaptively randomized (cohorts of size 1) based on efficacy outcomes. The efficacy outcome chosen for demonstration is inspired by an adoptive T-cell therapy (ACT) trial in metastatic melanoma patients (NIH P01 CA154778-01). It represents percent persistence of transduced T cells at 30 days relative to baseline.

Let Y_{ij} define percent persistence (immunological/efficacy endpoint) collected from patient i assigned to dose j. A linear model is fit to the efficacy data collected in stage 1 as a function of "acceptable" doses:

$$Y_{ij} \mid d_j = \beta_0 + \sum_{j=1}^{D-1} \beta_j d_j + e_{ij}. \tag{3.15}$$

The estimated efficacy outcomes are then used to calculate randomization probabilities:

$$\pi_j = \frac{\hat{Y}_j}{\sum_r \hat{Y}_r}. \tag{3.16}$$

   The next patient will be randomized to dose d_j with probability π_j, where doses with higher percent persistence (efficacy) are more likely to be assigned. If a DLT is observed, the dose safety is re-evaluated using the likelihood approach from stage 1. As data become available, the model is refitted and new randomization probabilities are being generated for accruing patients. The process is repeated until the maximum sample size $N = 25$ is achieved.

   R code example to generate toxicity, efficacy data, and dose allocations for the two-stage adaptive design with immunological outcomes. Please see the companion website for functions https://www.crcpress.com//9781138083776.

NOTE: The full collection of functions will be available soon in the R package "**iAdapt**".

```
###################################################################
#                       Input parameters                         #
###################################################################

# Number of pre-specified dose levels
  dose <- 5

# Vector of true toxicities associated with each dose
  dose.tox <- c(0.05, 0.10, 0.15, 0.20, 0.30)

# Acceptable (p2) and unacceptable (p1) DLT rates used for establishing
safety
  p1 <- 0.40
  p2 <- 0.15

# Likelihood-ratio (LR) threshold
  K <- 2

# cohort size used in stage 1
  coh.size <- 3

# Vector of true mean efficacies per dose (here mean percent persistence
per dose)
  m <- c(5, 15, 40, 65, 80)

# Efficacy variance per dose
  v <- rep(0.01,5)

# Total sample size (stages 1&2)
  N <- 25

# Stopping rule: if dose 1 is the only safe dose, allocate up to 9 pts.
  stop.rule <- 9

###################################################################
#                          Stage 1                               #
###################################################################

# Function to generate and tabulate toxicities per dose level
> tox.profile(dose, dose.tox, p1, p2, K, coh.size)
# Output: dose assignment(col1), toxicities per dose(col2), coh.number(col3),
LR value(col4)
```

```
     [,1] [,2] [,3] [,4]
[1,]    1    0    1 2.84
[2,]    2    0    2 2.84
[3,]    3    0    3 2.84
[4,]    4    0    4 2.84
[5,]    5    2    5 0.20
```

```
# Dose 5 is unacceptably toxic (LR ≤ 1/2), thus it won't be used in stage 2

# Function to select only the acceptable toxic doses
> safe.dose(dose, dose.tox, p1, p2, K, coh.size)
# Output: safe dose assignment(col1), toxicities per dose(col2)

$alloc.safe
     [,1] [,2]
[1,]    1    0
[2,]    2    0
[3,]    3    0
[4,]    4    0

# Other output
$alloc.total
 [1] 1 1 1 2 2 2 3 3 3 4 4 4 5 5 5        # complete allocation vector
used in stage 1

$n1
[1] 15                                    # no. subjects used in stage 1

# Function to generate efficacy outcomes (here percent persistence) for
each dose
> gen.eff.stg1(dose, dose.tox, p1, p2, K, coh.size, m, v, nbb=100)

# Output
# Efficacy outcomes only for safe doses to be used in stage 2
$Y.safe
 [1]   9   1   0 34 10 27 38 42 60 75 48 62

# Safe doses to be used in stage 2
$d.safe
 [1] 1 1 1 2 2 2 3 3 3 4 4 4

# Number of toxicities for safe doses
$tox.safe
[1] 0 0 0 0


################################################################################
#                                  Stage 2                                     #
################################################################################


# Function to fit a linear regression for the continuous efficacy outcomes,
# compute the randomization probabilities per dose and allocate the next
# subject to an acceptably safe dose that has the highest randomization probability
efficacy
# (cohort of size 1)
> rand.stg2(dose, dose.tox, p1, p2, K, coh.size, m, v, N, stop.rule=9,
cohort=1, samedose=T, nbb=100)
```

```
# Output: vector of all efficacy outcomes (N=25)
$ Y.final
 [1]   9   1   0 34 10 27 38 42 60 75 48 62 90 89 73 50 47 55 68 64 48 80 45 28 57

# 9 1 0 34 10 27 38 42 60 75 48 62 90 89 73 - efficacy outcomes from stage 1
# Note: only efficacy from acceptable doses are used to compute
randomization probabilites
# Here: 9   1   0 34 10 27 38 42 60 75 48 62 (see output from safe.dose above)
# 50 47 55 68 64 48 80 45 28 57 - efficacy outcomes from stage 2

# Output: vector of all dose allocations (N=25)
$ d.final
 [1] 1 1 1 2 2 2 3 3 3 4 4 4 5 5 5 3 4 4 4 4 3 4 4 2 3

# 1 1 1 2 2 2 3 3 3 4 4 4 5 5 5 - dose allocation from stage 1 (cohorts
of 3 subj. per dose)
# 3 4 4 4 4 3 4 4 2 3 - dose allocation from stage 2 (cohorts of 1) - no
dose 5, b/c it was considered unacceptably toxic in stage 1
```

The complete vectors of dose allocations and efficacy outcomes can be used to compute the operating characteristics of the design in repeated simulations.

Simulation results are shown in the paper for different combinations of toxicity and efficacy scenarios. Toxicity profiles vary from low (0.02) to high (0.50) DLT rates, constant or monotonically increasing trends. Efficacy scenarios include linear, curvilinear, and plateau trends (see Figure 3.2). In each scenario, the optimal dose is the dose producing the highest percent persistence of transduced T cells at follow-up among all acceptable doses.

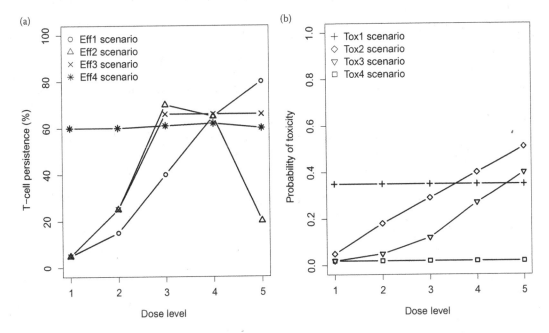

FIGURE 3.2
Simulated efficacy (a) and toxicity (b) scenarios as a function of dose level. The efficacy outcome is defined as persistence (%) of transduced T cells at follow-up compared to baseline.

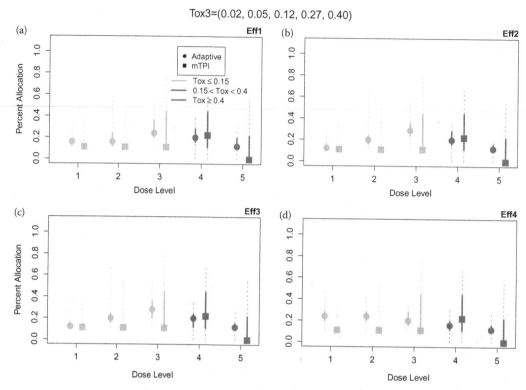

FIGURE 3.3

Percent allocation per dose for the two-stage adaptive design versus mTPI design ($N = 25$). Results of 5,000 simulated trials are summarized as median (filled circle/square), 25th–75th percentiles (solid line), and 2.5th–97.5th percentiles (dashed line). (a) Eff1 = (0.05, 0.15, 0.40, 0.65, 0.80), (b) Eff2 = (0.05, 0.25, 0.70, 0.65, 0.20), (c) Eff3 = (0.05, 0.25, 0.65, 0.65, 0.65), (d) Eff4 = (0.60, 0.60, 0.61, 0.62, 0.60).

Overall, the two-stage adaptive design demonstrates good operating characteristics, even for relatively small sample sizes ($N = 25$). The main conclusion is that by using an efficacy-driven randomization with safety constraints, the allocation distribution is skewed toward doses that are more promising. For example, when the first three doses have low toxicity (≤ 0.12 as in Tox3), the adaptive design assigns most of the patients to the dose with the highest T-cell percent persistence, i.e., dose 3 for efficacy scenarios Eff1-Eff3, and dose 1 and 3 for the plateau trend Eff4 (see Figure 3.3). Also, at doses with higher efficacy, there is an improvement in estimation of outcomes (less bias, greater precision).

While there is still much to be discovered and learned about immunotherapies, it is acknowledged that this is the future of cancer treatment and study designs should be tailored to their new characteristics and challenges.

3.3 Conclusion

The field of cancer research has changed dramatically in the last decades, with new biologic therapies starting to gradually replace cytotoxic treatments. The main goal of this chapter is to discuss statistical challenges imposed by the new agents (MTAs and immunotherapies) and to present illustrative model alternatives that can be used as reference in future trial implementation. While all the described methods involve single-agent treatments, we need to acknowledge the increasingly popular trend of testing drug combinations. Some of the recommended methods can be generally adapted and applied to combination studies; however, at this point, more scientific research is needed to better understand the design challenges associated with novel combinations, especially those involving immunotherapies.

References

1. Sun J, Wei Q, Zhou Y, Wang J, Liu Q, Xu H. A systematic analysis of FDA-approved anticancer drugs. *BMC Syst Biol.* 2017;11(Suppl 5):87.
2. National Cancer Institute. Common terminology criteria for adverse events version 4.0. cancer therapy evaluation program, 2009. NIH.
3. Storer BE. Design and analysis of phase I clinical trials. *Biometrics.* 1989;45(3):925–37.
4. O'Quigley J, Pepe M, Fisher L. Continual reassessment method: A practical design for phase I clinical trials in cancer. *Biometrics.* 1990;46(1):33–48.
5. Reiner E, Paoletti X, O'Quigley J. Operating characteristics of the standard phase I clinical trial design. *Comput Stat Data Anal.* 1999;30(3):303–15.
6. Ivanova A. Escalation, group and A+B designs for dose-finding trials. *Stat Med.* 2006;25(21):3668–78.
7. Robert C, Ribas A, Wolchok JD et al. Anti-programmed-death-receptor-1 treatment with pembrolizumab in ipilimumab-refractory advanced melanoma: A randomised dose-comparison cohort of a phase 1 trial. *The Lancet.* 2014;384:1109–17.
8. Theoret MR, Pai-Scherf LH, Chuk MK et al. Expansion cohorts in first-in-human solid tumor oncology trials. *Clin Cancer Research.* 2015;21:4545–51.
9. Iasonos A, O'Quigley J. Early phase clinical trials—are dose expansion cohorts needed? *Nature Reviews. J Clin Oncol.* 2015;12(11):626–8.
10. Le Tourneau C, Lee J, Siu LL. Dose escalation methods in phase I cancer clinical trials. *J Natl Cancer Inst.* 2009;101(10):708–20.
11. Rogatko A, Schoeneck D, Jonas W et al. Translation of innovative designs into phase I trials. *J Clin Oncol.* 2007;25(31):4982–6.
12. Chiuzan C, Shtaynberger J, Manji GA et al. Dose-finding designs for trials of molecularly targeted agents and immunotherapies, *J Biopharm Stat.* 2017;27:3, 477–494.
13. Piantadosi S, Fisher JD, Grossman S. Practical implementation of a modified continual reassessment method for dose-finding trials. *Cancer Chemother Pharmacol.* 1998;41(6):429–36.
14. Goodman SN, Zahurak ML, Piantadosi S. Some practical improvements in the continual reassessment method for phase I studies. *Stat Med.* 1995;14(11):1149–61.
15. Shen L, O'Quigley J. Consistency of CRM under model misspecification. *Biometrika.* 1996; 83(2):395–405.
16. Garrett-Mayer E. The continual reassessment method for dose-finding studies: A tutorial. *Clin Trials.* 2006;3:57–71.
17. Cheung YK. *Dose-Finding by the Continual Reassessment Method*, First Edition Boca Raton, FL: Chapman & Hall, 2011.
18. Camacho LH, Antonia S, Sosman J et al. Phase I/II trial of tremelimumab in patients with metastatic melanoma. *J Clin Oncol.* 2009 Mar 1;27(7):1075–81.

19. Cheung YK, Chappell R. Sequential designs for phase I clinical trials with late-onset toxicities. *Biometrics.* 2000;56:1177–82.
20. Simon R, Rubinstein L, Arbuck SG, Christian MC, Freidlin B, Collins J. Accelerated titration designs for phase I clinical trials in oncology, *J Natl Cancer Inst.* 1997;89:1138–47.
21. Van Meter EM, Garrett-Mayer E, Bandyopadhyay D. Proportional odds model for dose-finding clinical trial designs with ordinal toxicity grading. *Stat Med.* 2011;30:2070–80.
22. Van Meter EM, Garrett-Mayer E, Bandyopadhyay D. Dose-finding clinical trial design for ordinal toxicity grades using the continuation ratio model: An extension of the continual reassessment method. *Clin Trials.* 2012;9(3):303–13.
23. Bekele BN, Thall PF. Dose-finding based on multiple toxicities in a soft tissue sarcoma trial. *J American Stat Assoc.* 2004;99:26–35.
24. Hobbs BP, Thall PF, Lin SH. Bayesian group sequential clinical trial design using total toxicity burden and progression-free survival. *J Royal Stat Society: Series C (Applied Statistics).* 2016;65:273–97.
25. Lee SM, Hershman DL, Martin P, Leonard JP, Cheung YK. Toxicity burden score: A novel approach to summarize multiple toxic effects. *Ann Oncol.* 2012;23:537–41.
26. Chen Z, Krailo MD, Azen SP, Tighiouart M. A novel toxicity scoring system treating toxicity response as a quasi-continuous variable in Phase I clinical trials. *Contemporary Clin Trials.* 2010;31:473–82.
27. Leung DH, Wang Y. Isotonic designs for phase I trials. *Control Clin Trials.* 2001;22(2):126–38.
28. Lee SM, Cheng B, Cheung YK. Continual reassessment method with multiple toxicity constraints. *Biostatistics.* 2011;12:386–98.
29. Yin J, Qin R, Ezzalfani M, Sargent DJ, Mandrekar SJ. A Bayesian dose-finding design incorporating toxicity data from multiple treatment cycles. *Stat Med.* 2017;36:67–80.
30. Tseng WW, Zhou S, To CA et al. Phase 1 adaptive dose-finding study of neoadjuvant gemcitabine combined with radiation therapy for patients with high-risk extremity and trunk soft tissue sarcoma. *Cancer.* 2015;121:3659–67.
31. Yuan Z, Chappell R, Bailey H. The continual reassessment method for multiple toxicity grades: A Bayesian quasi-likelihood approach. *Biometrics.* 2007;63:173–9.
32. Ezzalfani M, Zohar S, Qin R, Mandrekar SJ, Deley M-CL. Dose-Finding designs using a novel quasi-continuous endpoint for multiple toxicities. *Stat Med.* 2013;32:2728–46.
33. Yin G, Yuan Y. Bayesian model averaging continual reassessment method in phase I clinical trials. *J Am Stat Assoc.* 2009;104:954–68.
34. Thall PF, Russell KE. A strategy for dose-finding and safety monitoring based on efficacy and adverse outcomes in phase I/II clinical trials. *Biometrics.* 1998;54(1):251–64.
35. Thall PF, Cook JD. Dose-finding based on efficacy-toxicity trade-offs. *Biometrics.* 2004;60(3):684–93.
36. Thall PF, Cook JD, Estey EH. Adaptive dose selection using efficacy-toxicity tradeoffs: Illustrations and practical considerations. *J Biopharm Stat.* 2006;16(5):623–38.
37. Braun TM. The bivariate continual reassessment method. Extending the CRM to phase I trials of two competing outcomes. *Control Clin Trials.* 2002;23(3):240–56.
38. Bekele BN, Shen Y. A Bayesian approach to jointly modeling toxicity and biomarker expression in a phase I/II dose-finding trial. *Biometrics.* 2005;61(2):343–54.
39. Dragalin V, Fedorov V. Adaptive designs for dose-finding based on efficacy and toxicity response. *J Stat Planning Inf.* 2006;136:1800–23.
40. Zhang W, Sargent DJ, Mandrekar S. An adaptive dose-finding design incorporating both toxicity and efficacy. *Stat Med.* 2006;25(14):2365–83.
41. National Cancer Institute NCI. 2017. Fact Sheet - Targeted Cancer Therapies. Web: https://www.cancer.gov/about-cancer/treatment/types/targeted-therapies/targeted-therapies-fact-sheet.
42. Penel N, Adenis A, Clisant S, Bonneterre J. Nature and subjectivity of dose-limiting toxicities in contemporary phase 1 trials: Comparison of cytotoxic versus non-cytotoxic drugs. *Invest New Drugs.* 2011;29:1414–9.
43. Gupta S, Hunsberger S, Boerner SA et al. Meta-analysis of the relationship between dose and benefit in phase I targeted agent trials. *J Natl Cancer Inst.* 2012;104:1860–6.

44. Le Tourneau C, Diéras V, Tresca P, Cacheux W, Paoletti X. Current challenges for the early clinical development of anticancer drugs in the era of molecularly targeted agents. *Target Oncol.* 2010;5:65–72.
45. Golan T, Milella M, Ackerstein A, Berger R. The changing face of clinical trials in the personalized medicine and immuno-oncology era: Report from the international congress on clinical trials in Oncology & Hemato-Oncology (ICTO 2017). *J Exper & Clinl Cancer Res.* 2017;CR, 36:192.
46. Papke LE, Wooldridge JM. Econometric methods for fractional response variables with an application to 401k plan participation rates. *J Applied Econ.* 1996;11:619–32.
47. Gourieroux C, Monfort A, Trognon A. Pseudo maximum likelihood methods: Theory. Econometrica. *J Econometric Society.* 1984;52(3):681–700.
48. O'Quigley J, Shen LZ. Continual reassessment method: A likelihood approach, *Biometrics.* 1996;52(2):673–84.
49. Cook N, Hansen AR, Siu LL et al. Early phase clinical trials to identify optimal dosing and safety. In *Molecular Oncology*, Mendelsohn J, Ringborg U, Schilsky R, eds. 2015;9:997–1007.
50. Ursino M, Zohar S, Lentz F et al. Dose-finding methods for phase I clinical trials using pharmacokinetics in small populations. *Biom J.* 2017;59:804–25.
51. Piantadosi S, Liu G. Improved designs for dose escalation studies using pharmacokinetic measurements, *Stat Med.* 1996 15;15:1605–18.
52. Patterson S, Francis S, Ireson M, Webber D, Whitehead J, A novel Bayesian decision procedure for early-phase dose-finding studies. *J Biopharm Stat.* 1999;9(4):583–97.
53. Whitehead J, Zhou Y, Hampson L, Ledent E, Pereira A. A Bayesian approach for dose-escalation in a phase I clinical trial incorporating pharmacodynamic endpoints. *J Biopharm Stat.* 2007;17(6):1117–29.
54. Whitehead J, Patterson S, Webber D, Francis S, Zhou Y. Easy-to-implement Bayesian methods for dose-escalation studies in healthy volunteers. *Biostatistics.* 2001;2(1):47–61.
55. Toumazi A, Comets E, Alberti C et al. dfpk: An R-package for Bayesian dose-finding designs using pharmacokinetics (PK) for phase I clinical trials. *Comput Methods Programs Biomed.* 2018;157:163–77.
56. Maude SL, Laetsch TW, Buechner J et al. Tisagenlecleucel in children and young adults with B-cell lymphoblastic leukemia. *N Engl J Med.* 2018;378:439–48.
57. Turtle CJ, Hanafi LA, Berger C et al. CD19 CAR-T cells of defined $CD4^+$: $CD8^+$ composition in adult B cell ALL patients. *J Clin Invest.* 2016;126:2123–38.
58. Johnson LA, Morgan RA, Dudley ME et al. Gene therapy with human and mouse T-cell receptors mediates cancer regression and targets normal tissues expressing cognate antigen. *Blood.* 2009;114:535–46.
59. Chiuzan C, Garrett-Mayer E, Nishimura M. An adaptive dose-finding design based on both safety and immunologic responses in cancer clinical trials. *Stat in Biopharm Res.* 2018;10(3):185–95.
60. Ji Y, Wang SJ. Modified toxicity probability interval design: A safer and more reliable method than the 3+3 design for practical phase I trials. *J Clin Oncol.* 2013;31:1785–91.
61. Chiuzan C, Garrett-Mayer E, Yeatts S. A likelihood-based approach for computing the operating characteristics of the 3+3 phase I clinical trial design with extensions to other A+B designs. *Clinical Trials.* 2015;12(1):24–33.

4

Current Issues in Phase II Cancer Clinical Trials

Sin-Ho Jung

CONTENTS

4.1 Introduction

Cancer clinical trials investigate the efficacy and toxicity of experimental cancer therapies. If an appropriate dose level of an experimental drug is determined from a phase I trial, the drug's anticancer activity is assessed through phase II clinical trials. Phase II clinical trials screen out inefficacious experimental therapies before they proceed to further investigation through large-scale phase III trials. In order to expedite this process, phase II trials traditionally use a single-arm design to treat patients with experimental therapies only. The efficacy of an experimental therapy is compared with that of a standard therapy using historical controls. The most popular primary endpoint of phase II cancer clinical trials is tumor response, which is measured by the change in tumor size before and during treatment. If the size of a target tumor, defined as the largest diameter of the tumor, decreases by at least 30% compared to that at the baseline, a partial response is declared.

A complete response is defined as complete disappearance of the tumor. Overall response is defined as partial or complete response.

Phase II trials generally require shorter study periods than phase III trials. Consequently, phase II trials have small sample sizes, so that exact statistical methods are preferable to asymptotic methods for their design and analysis. Various exact methods have been published for phase II trials with binary outcomes such as tumor response. For ethical reasons, two-stage designs are commonly used for phase II cancer clinical trials. A typical single-arm two-stage trial with futility (a_1) and superiority (b_1) stopping values and a stage 2 rejection value (a) are conducted as follows.

- Stage 1: Treat n_1 patients and count the number of responders X_1.
  - If $X_1 \leq a_1$, then reject the experimental therapy and stop the trial.
  - If $X_1 \geq b_1$, then accept the experimental therapy and stop the trial.
  - If $a_1 < X_1 < b_1$, then proceed to the second stage.
- Stage 2: Treat an additional n_2 patients and count the number of responders X_2.
  - If $X_1 + X_2 \leq a$, then reject the experimental therapy.
  - Otherwise, accept the experimental therapy for further investigation.

We usually do not stop the trial early for superiority since there is no ethical issue to continue treating patients with an efficacious therapy and we want to collect as much data as possible to use when designing a subsequent phase III trial. In this case, we choose $b_1 = n_1 + 1$.

The design and analysis of phase II trials require exact statistical methods accounting for the two-stage design and small sample sizes. The most popular design for phase II cancer clinical trials has been a two-stage design with futility stopping only with $b_1 = n_1 + 1$ based on [1] the minimax or the optimality criterion. But most publications reporting the results of phase II trials fail to appropriately address these issues. When the above two-stage phase II trial is completed, we should be able to accept or reject the experimental therapy based on the critical values and sample sizes, (a_1, b_1, a, n_1, n_2). In a usual clinical trial, we would recruit slightly more patients than required to make up for possible attrition due to ineligibility and dropout. As a result, the observed sample size tends to be different from that specified by the design. In this case, the critical values chosen at the stopping stage become meaningless. This is one reason why investigators are not able to draw clear conclusions from their phase II trials. To address this issue, we extend the two-stage testing rule by calculating a p-value or a confidence interval of the true response rate (RR). These methods exactly coincide with the two-stage testing rule if the observed sample sizes are identical to the prespecified (n_1, n_2). In this chapter, we discuss further design and analysis methods for single-arm phase II trials with tumor response as the primary endpoint.

These design and analysis methods assume that the RR is identical for all patients in the study population. Oftentimes, however, a study population may have subpopulations, usually defined by some baseline characteristics, with different expected RRs. We investigate single-arm designs using a stratified one-sample binomial test.

In spite of these adjustments, single-arm phase II trials have intrinsic shortcomings. Single-arm phase II trials are appropriate only when reliable and valid data for an existing standard therapy are available for the same patient population. Furthermore, the response assessment method in the historical control data should be identical to the one that will be used for a new study. If no historical control data satisfying these conditions exist or the existing data are too small to represent the whole patient population, we have to consider a randomized phase II clinical trial with a prospective control to be compared with the experimental therapy under investigation. Cannistra [2] recommends a randomized phase II trial if a single-arm design is subject to any of the above issues. Readers may refer to Gan et al. [3] about more issues associated with which design to choose between a single-arm phase II trial and a randomized phase II trial. In this chapter, we discuss randomized phase II trials based on Fisher's [4] exact test.

4.2 Single-Arm Phase II Trials

4.2.1 Optimal Two-Stage Designs

In this section, we focus on the popular two-stage, single-arm phase II trial designs with a futility interim test only, i.e., $b_1 = n_1 + 1$. These designs are defined by the number of patients to be treated during stages 1 and 2, n_1 and n_2, and rejection values a_1 and a $(a_1 < a)$, so that we specify them by $(a_1/n_1, a/n)$, where $n = n_1 + n_2$ is the maximal sample size. The values of $(a_1/n_1, a/n)$ are determined based on prespecified design parameters, as described below. Let p_0 denote the maximum unacceptable probability of response, which is usually chosen by the RR of a historical control, and p_1 be the minimum acceptable probability of response with $p_0 < p_1$. For the true RR p of the experimental therapy, we want to test $H_0 : p \leq p_0$ against $H_1 : p > p_0$. Given (p_0, p_1), we can calculate the type I error probability α and power $1 - \beta$ of a two-stage design $(a_1/n_1, a/n)$ based on the fact that the number of responders from the two stages, X_1 and X_2, are independent binomial random variables.

Suppose that we want to find a two-stage design with a type I error probability no larger than α^* and a power no smaller than $1 - \beta^*$ for given (p_0, p_1) values. We call $(p_0, p_1, \alpha^*, \beta^*)$ design parameters. Given (p_0, p_1), there are infinitely many 2-stage designs satisfying the $(\alpha^*, 1 - \beta^*)$-constraint. Noting that the sample size of a two-stage trial is a random variable taking n_1 or n, we can calculate the expected sample size of the trial given a true RR of the experimental therapy. Simon [1] proposes two criteria to select a good two-stage design among these candidate designs. The minimax design minimizes the maximal sample size, n, among the designs, satisfying the $(\alpha^*, 1 - \beta^*)$-constraint. On the other hand, the so-called "optimal" design minimizes the expected sample size under $H_0 : p = p_0$, denoted as EN.

While the minimax and the optimal designs have both been widely used, other two-stage designs have been largely ignored. Oftentimes, the two criteria result in very different two-stage designs, i.e., the minimax design may have an excessively large EN as compared to the optimal design and the optimal design may have an excessively large maximal sample size n as compared to the minimax design. This results from the discrete nature of the exact binomial method.

EXAMPLE 4.1

For the design parameters $(p_0, p_1, \alpha^*, 1 - \beta^*) = (0.1, 0.3, 0.05, 0.85)$, the minimax design is given by $(a_1/n_1, a/n) = (2/18, 5/27)$ and the optimal design by $(1/11, 6/35)$. The maximal sample size n for the minimax design is less than that for the optimal design by 8. However, the expected sample size EN under H_0 for the optimal design is 18.3, which is only slightly smaller than EN $= 20.4$ for the minimax design. Thus, we may consider choosing the minimax design since, compared to the optimal design, it largely saves the maximal sample size while sacrificing the expected sample size by only about 2. In this case, there exists a practical compromise without changing the statistical operating characteristics appreciably. For the same design parameters $(p_0, p_1, \alpha^*, 1 - \beta^*) = (0.1, 0.3, 0.05, 0.85)$, the design given by $(a_1/n_1, a/n) = (1/13, 5/28)$ requires only one more patient in the maximal sample size n than the minimax design, but its expected sample size EN under H_0 is very comparable to that of the optimal design (18.7 vs. 18.3).

This design is a good compromise between the minimax design and the optimal design [5]. There can be multiple compromising designs, and an algorithm to identify these designs, called admissible two-stage designs, has been proposed [6]. Simon's minimax and optimal designs belong to the family of admissible designs. An interactive computer program to find admissible designs is available at impact.unc.edu/impact7/CTDSystems.

Figure 4.1 is a snapshot of this program for Example 4.1. It displays the plot of EN against n for various designs with $n \leq 37$. Simon's minimax design given by $(a_1/n_1, a/n) = (2/18, 5/27)$ is the leftmost point in the figure and the optimal design by $(1/11, 6/35)$ is at the very bottom. The program provides specification of a design $(a_1/n_1, a/n)$ along with $(\alpha, 1-\beta)$, EN, and the probabilities of early termination for $p = p_0$ and p_1 when the circle representing a candidate design, actually (n, EN), is clicked with a pointer.

In this section, we have considered two-stage designs with a futility stopping value only. Also available are optimal multi-stage designs with both futility and superiority stopping boundaries by minimizing the average of expected sample sizes under H_0 and H_1 [7]. The computer program mentioned above also provides minimax, optimal, and admissible designs with both futility and superiority stopping boundaries.

4.2.2 Estimation of Response Rate

In a two-stage phase II trial, let M ($=1$ or 2) denote the stopping stage and S the cumulative number of responders by the stopping stage, i.e., $S = X_1$ if $M = 1$ and $S = X_1 + X_2$ if $M = 2$. The most popular estimator of RR p for $(M, S) = (m, s)$ is the sample proportion, i.e.,

$$\hat{p} = \begin{cases} s/n_1 & \text{if } m = 1 \\ s/(n_1 + n_2) & \text{if } m = 2 \end{cases}$$

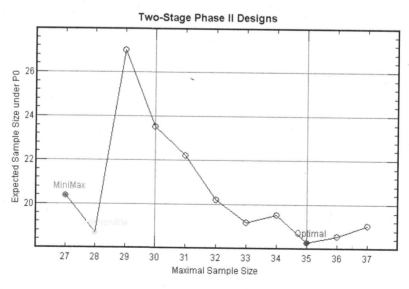

FIGURE 4.1
Two-stage designs for $(p_0, p_1, \alpha, \beta) = (0.1, 0.3, 0.05, 0.15)$ with $N = 37$.

By not reflecting the two-stage design aspect of the trial, this estimator, called the maximum likelihood estimator (MLE), is always negatively biased for standard two-stage trials with futility stopping only [8]. As a result, if p_0 of a future trial is chosen by the sample proportion of a historical control taken from a previous two-stage phase II trial, then it will underestimate the true RR of the historical control and result in a higher chance of false positivity for the future trial.

For $(M, S) = (m, s)$, the uniformly minimum variance unbiased estimator (UMVUE) of p for two-stage phase II trials is given by

$$\tilde{p} = \begin{cases} s/n_1 & \text{if } m = 1 \\[2ex] \dfrac{\displaystyle\sum_{x_1=(a_1+1)\vee(s-n_2)}^{s\wedge(b_1-1)} \binom{n_1-1}{x_1-1}\binom{n_2}{s-x_1}}{\displaystyle\sum_{x_1=(a_1+1)\vee(s-n_2)}^{s\wedge(b_1-1)} \binom{n_1}{x_1}\binom{n_2}{s-x_1}} & \text{if } m = 2 \end{cases} \tag{4.1}$$

where $a \wedge b = \min(a,b)$ and $a \vee b = \max(a, b)$ [8]. Note that the UMVUE and the MLE are identical if the trial stops after stage 1, i.e., $m = 1$.

For a true RR of p, the probability mass function (PMF) of (M, S), $f(m, s|p) = Pr(M = m, S = s)$, is given as

$$f(m,s|p) = \begin{cases} p^s(1-p)^{n_1-s}\binom{n_1}{s} & \text{if } m = 1, 0 \leq s \leq a_1 \text{ or } b_1 \leq s \leq n_1 \\[2ex] p^s(1-p)^{n_1+n_2-s}\displaystyle\sum_{x_1=a_1+1}^{(b_1-1)\wedge s}\binom{n_1}{x_1}\binom{n_2}{s-x_1} & \text{if } m = 2, a_1+1 \leq s \leq b_1-1+n_2 \end{cases} \tag{4.2}$$

Since the UMVUE of p is a function of (M, S), its PMF is derived from $f(m, s|p)$.

EXAMPLE 4.2

Under $(p_0, p_1, \alpha^*, 1-\beta^*) = (0.2, 0.4, 0.05, 0.8)$, Simon's optimal two-stage design with a futility stopping value only is given as $(a_1/n_1, a/n) = (3/13, 12/43)$. Table 4.1 gives the UMVUE and the MLE for observations (m, s) for the two-stage design and their PMF for various RR values. When $m = 1$, two estimates are exactly the same as noted earlier. When $m = 2$, the MLE is much smaller than UMVUE for small s values. Jung and Kim [8] show that the UMVUE has a comparable variance as compared to the MLE overall.

In order to account for ineligible or unevaluable patients, we often accrue slightly more (or possibly less) patients than the planned sample size at the stopping stage, especially in multicenter trials (e.g., [9,10]). But, this does not become any issue in the calculation of the UMVUE, since it does not require specification of the critical values (a_m, b_m) at the stopping stage m, where $a_2 = a$ and $b_2 = a+1$. If the study is terminated after stage 1 (i.e., $m = 1$), then the UMVUE is calculated by regarding the observed number of stage 1 patients as n_1. If the study is proceeded to the second stage (i.e., $m = 2$), then the UMVUE in (Equation 4.1) depends only on the rejection values for stage 1 (a_1, b_1) and the number of patients for the two stages (n_1, n_2). When $m = 2$, the interim test is always conducted using (a_1, b_1) when exactly n_1 patients have response data as planned. However, the number of stage 2 patients may be slightly different from n_2. In this case, we calculate the UMVUE by regarding the observed number of patients for stage 2 as n_2, with (n_1, a_1, b_1) fixed by the design.

TABLE 4.1

UMVUE, MLE, and Probability Mass Function for True p for Each Observation in a Two-Stage Design with $(a_1/n_1, a/n) = (3/13, 12/43)$

m	s	UMVUE	MLE	\multicolumn{5}{c}{$f(m, s	p)$ for p}
				0.1	0.2	0.3	0.4	0.5
1	0	0.000	0.000	0.254	0.055	0.010	0.001	0.000
1	1	0.077	0.077	0.367	0.179	0.054	0.011	0.002
1	2	0.154	0.154	0.245	0.268	0.139	0.045	0.010
1	3	0.231	0.231	0.100	0.246	0.218	0.111	0.035
2	4	0.308	0.093	0.001	0.000	0.000	0.000	0.000
2	5	0.312	0.116	0.004	0.002	0.000	0.000	0.000
2	6	0.317	0.140	0.007	0.006	0.001	0.000	0.000
	7	0.322	0.163	0.008	0.015	0.002	0.000	0.000
2	8	0.328	0.186	0.006	0.027	0.006	0.000	0.000
2	9	0.335	0.209	0.004	0.038	0.015	0.001	0.000
2	10	0.343	0.233	0.002	0.043	0.030	0.003	0.000
2	11	0.351	0.256	0.001	0.041	0.049	0.008	0.000
2	12	0.360	0.279	0.000	0.033	0.068	0.018	0.001
2	13	0.371	0.302	0.000	0.023	0.081	0.033	0.003
2	14	0.382	0.326	0.000	0.014	0.084	0.054	0.006
2	15	0.395	0.349	0.000	0.007	0.076	0.076	0.013
2	16	0.409	0.372	0.000	0.003	0.062	0.096	0.025
2	17	0.424	0.395	0.000	0.001	0.044	0.107	0.042
2	18	0.440	0.419	0.000	0.001	0.029	0.108	0.063
2	19	0.458	0.442	0.000	0.000	0.017	0.098	0.085
2	20	0.477	0.465	0.000	0.000	0.009	0.080	0.105
2	21	0.496	0.488	0.000	0.000	0.004	0.059	0.116
2	22	0.517	0.512	0.000	0.000	0.002	0.040	0.118
2	23	0.538	0.535	0.000	0.000	0.001	0.025	0.108
2	24	0.560	0.558	0.000	0.000	0.000	0.014	0.091
2	25	0.582	0.581	0.000	0.000	0.000	0.007	0.069
2	26	0.605	0.605	0.000	0.000	0.000	0.003	0.048
2	27	0.628	0.628	0.000	0.000	0.000	0.001	0.030
2	28	0.651	0.651	0.000	0.000	0.000	0.001	0.017
2	29	0.674	0.674	0.000	0.000	0.000	0.000	0.009
2	30	0.698	0.698	0.000	0.000	0.000	0.000	0.004
2	31	0.721	0.721	0.000	0.000	0.000	0.000	0.002
2	32	0.744	0.744	0.000	0.000	0.000	0.000	0.001
2	33	0.767	0.767	0.000	0.000	0.000	0.000	0.000
2	34	0.791	0.791	0.000	0.000	0.000	0.000	0.000
2	35	0.814	0.814	0.000	0.000	0.000	0.000	0.000
2	36	0.837	0.837	0.000	0.000	0.000	0.000	0.000
2	37	0.861	0.861	0.000	0.000	0.000	0.000	0.000
2	38	0.884	0.884	0.000	0.000	0.000	0.000	0.000
2	39	0.907	0.907	0.000	0.000	0.000	0.000	0.000
2	40	0.930	0.930	0.000	0.000	0.000	0.000	0.000
2	41	0.954	0.954	0.000	0.000	0.000	0.000	0.000
2	42	0.977	0.977	0.000	0.000	0.000	0.000	0.000
2	43	1.000	1.000	0.000	0.000	0.000	0.000	0.000

EXAMPLE 4.2 (Revisited)

Suppose that a phase II trial was designed for Simon's optimal two-stage design with a futility stopping value only as $(a_1/n_1, a/n) = (3/13, 12/43)$ under $(p_0, p_1, \alpha^*, 1-\beta^*) = (0.2, 0.4, 0.05, 0.8)$, but it was completed with $s = 7$ responders from a total of 45 patients after stage 2. Then, by using $(a_1, n_1) = (3, 13)$ and conditioning n_2 at $32 (= 45 - 13)$, we have $(a_1 + 1) \vee (s - n_2) = (3 + 1) \vee (7 - 32) = 4$ and $s \wedge (b_1 - 1) = 7 \wedge (14 - 1) = 13$. Hence, from (Equation 4.1), the UMVUE is calculated by

$$\tilde{p} = \frac{\sum_{x_1=4}^{7} \binom{12}{x_1-1}\binom{32}{7-x_1}}{\sum_{x_1=4}^{7} \binom{13}{x_1}\binom{32}{7-x_1}} = \frac{1,362,988}{4,241,380} = 0.321$$

while the MLE is only $\hat{p} = 7/45 = 0.156$. Jennison and Turnbull [11] proposed a confidence interval method for RR for multi-stage clinical trials. Jung and Kim [8] prove that the ordering of the sample space [11] used to calculate their confidence intervals is identical to that of (M, S) in terms of the magnitude of the UMVUE.

4.2.3 Confidence Interval

Jung and Kim [8] prove that the ordering of the sample space for (M, S) in terms of the magnitude of the UMVUE is identical to that by Jennison and Turnbull [11]; see also Armitage [12] and Tsiatis et al. [13]. In other words, we have

$$\tilde{p}1,0 < \tilde{p}(1,1) < \cdots < \tilde{p}(1,a_1)$$
$$< \tilde{p}(2,a_1+1) < \cdots < \tilde{p}(2,b_1-1+n_2) \tag{4.3}$$
$$\tilde{p}(1,b_1) < \cdots < \tilde{p}(1,n_1)$$

where $\tilde{p}(m,s)$ denotes the UMVUE for $(M, S) = (m, s)$. Hence, the confidence intervals calculated according to Clopper and Pearson [14] and the stochastic ordering based on the magnitude of the UMVUE are identical to those by Jennison and Turnbull [11].

With the UMVUE $\tilde{p}(m,s)$, an exact $100(1-\alpha)\%$ equal tail confidence interval (p_L, p_U) for p is given by

$$\Pr(\tilde{p}(M,S) \geq \tilde{p}(m,s) \mid p = p_L) = \alpha/2$$

and

$$\Pr(\tilde{p}(M,S) \leq \tilde{p}(m,s) \mid p = p_U) = \alpha/2,$$

where the probabilities are calculated using the PMF of (M, S) as defined in (Equation 4.2). Confidence limits p_L and p_U can be obtained by the bisection method to solve the equations.

EXAMPLE 4.3

Suppose that we observed $(m, s) = (2, 7)$ from a two-stage study with lower boundaries only, $(n_1, n_2, a_1, a) = (13, 30, 3, 12)$. In this case, we have $b_1 = n_1 + 1 = 14$,

$(a_1 + 1) \vee (s - n_2) = (3 + 1) \vee (7 - 30) = 4$, and $s \wedge (b_1 - 1) = 7 \wedge (14 - 1) = 7$. From (Equation 4.1), the UMVUE is

$$\tilde{p}(2,7) = \frac{\sum_{x_1=4}^{7} \binom{13-1}{x_1-1}\binom{30}{7-x_1}}{\sum_{x_1=4}^{7} \binom{13}{x_1}\binom{30}{7-x_1}} = 0.322.$$

Using (Equation 4.2), we have

$$\Pr(\tilde{p}(M,S) \geq 0.322 \mid p = 0.103) = 0.025,$$

$$\Pr(\tilde{p}(M,S) \leq 0.322 \mid p = 0.538) = 0.025$$

so that a 95% confidence interval on p is given as (0.103, 0.538), which is the same as the one according to Jennison and Turnbull [11]. In contrast, a naive exact 95% confidence interval by Clopper and Pearson [14], ignoring the two-stage aspect of the study design, is given as (0.068, 0.307). Note that the latter is narrower than the former by ignoring the group sequential feature of the study. Furthermore, the former is slightly shifted to the right from the latter to reflect the fact that the study has been continued to stage 2 after observing more responders than $a_1 = 3$ in stage 1. The popularly used asymptotic confidence interval $\hat{p} \pm 1.96\sqrt{\hat{p}(1-\hat{p})/n} = (0.052, 0.273)$ with 95% significance level is even narrower and further shifted to the left than the naive exact confidence interval.

The Jennison-Turnbull confidence interval based on the stochastic ordering (Equation 4.3) has a desirable property: given (p_0, α^*), the testing result based on the two-stage design $(a_1/n_1, a/n)$ exactly coincides with that based on the one-sided confidence interval with significance level $100(1 - \alpha^*)\%$, i.e., we reject $H_0 : p = p_0$ by the two-stage design if and only if the lower confidence limit is larger than p_0. This property is very valuable, especially when the observed sample size for the stopping stage is different from the sample size specified by the study design. In this case, the two-stage design does not provide us a proper testing rule any more since the rejection values, (a_1, b_1) or a, are meaningful only when the number of patients are identical to the prespecified sample size. But we can calculate a Jennison-Turnbull confidence interval of RR by treating the observed sample size at the stopping stage like the planned sample size. This is possible since Equations 4.2 and 4.3 do not require specification of the rejection values at the stopping stage. Various confidence interval methods have been proposed for multi-stage designs (e.g., [15,16]), but no other confidence intervals have this property.

EXAMPLE 4.4

Kim et al. [17] published a phase II trial of concurrent chemoradiotherapy for newly diagnosed patients with limited-stage extranodal natural killer/T-cell lymphoma, nasal type. The primary endpoint is complete response (CR) rate. The study had the [1] optimal two-stage design $(a_1/n_1, a/n) = (4/6, 22/27)$ under $(p_0, p_1, \alpha^*, 1 - \beta^*) = (0.7, 0.9, 0.05, 0.8)$. The trial proceeded to the second stage to treat a total of 30 eligible and evaluable patients, of which 27 achieved CR. Since the observed maximal sample size is different from the planned one, the rejection value $a = 22$ was of no use to determine the positivity of the study. The authors concluded the study to be positive without statistical ground. Noting this, Shimada and Suzuki [18] claimed that this trial was a negative study by calculating an asymptotic confidence interval with 2-sided 95% significance level and showing that it covers $p_0 = 0.7$. This claim is misleading since their interval (1) has a two-sided $100(1 − \alpha) = 95\%$ confidence level while the two-stage design has 1-sided $\alpha = 5\%$, (2) ignores the two-stage feature of the trial, and (3) is using a large-sample approximation. Kim et al. [19] defended their conclusion by showing that the 1-sided 95% Jennison-Turnbull confidence interval (0.762, 1] is above $p_0 = 0.7$.

4.2.4 P-Value Calculation

In a phase II trial, we conduct a statistical test to reject or accept the experimental therapy. If we reject or fail to reject the null hypothesis, we should be able to provide a p-value as a measure of how strong the evidence of decision is against the null hypothesis. However, publications from phase II trials rarely report p-values to support their conclusions. In most of the publications, it is not even clear if the investigators decide to accept their therapies or not.

Using the stochastic ordering of UMVUE (Equation 4.3), Jung et al. [6] propose a p-value method for 2-stage phase II clinical trials. Noting that a p-value is defined as the probability of observing an extreme test statistic value toward the direction of H_1 when H_0 is true, they propose to calculate the probability of observing an UMVUE value larger than that obtained from the study under H_0. Let $\tilde{p}$ denote the UMVUE of the RR observed from a two-stage phase II trial specified by (a_1, b_1, a, n_1, n_2). Given $(M, S) = (m, s)$, the p-value $= Pr\{\tilde{p}(M, S) \geq \tilde{p}(m, s) \mid p_0\}$ based on UMVUE can be calculated as

$$\text{p-value} = \begin{cases} \sum_{j=s}^{n_1} f(1, j \mid p_0) & \text{if } m = 1, s \geq b_1 \\ 1 - \sum_{j=0}^{s-1} (1, j \mid p_0) & \text{if } m = 1, s \leq a_1 \\ \sum_{j=b_1}^{n_1} f(1, j \mid p_0) + \sum_{j=s}^{b_1-1+n_2} f(2, j \mid p_0) & \text{if } m = 2 \end{cases} \tag{4.4}$$

Jung et al. [20] show that if the observed sample size at each stage is identical to that specified by the design, the decision by the two-stage design exactly matches with that by the p-value compared with the α value. As discussed in the previous section, the confidence interval by Jennison and Turnbull [11] also has this property.

EXAMPLE 4.5

For $H_0 : p_0 = 0.4$ vs. $H_1 : p_1 = 0.6$, the two-stage design $(n_1, n_2, a_1, b_1, a) = (34, 5, 17, 35, 20)$ is the Simon's minimax design among those with $(\alpha^*, 1 − \beta^*) = (0.05, 0.8)$ and a futility stopping boundary only. Table 4.2 displays the p-values and the PMF at $p_0 = 0.4$ for each sample point in the descending order for UMVUE. For example, for $(m, s) = (2, 19)$, we calculate the UMVUE by

$$\tilde{p} = \frac{\sum_{x_1=(17+1)\vee(19-5)}^{19\wedge(35-1)} \binom{34-1}{x_1-1}\binom{5}{19-x_1}}{\sum_{x_1=(17+1)\vee(19-5)}^{19\wedge(35-1)} \binom{34}{x_1}\binom{5}{19-x_1}} = \frac{\binom{33}{17}\binom{5}{1} + \binom{33}{18}\binom{5}{0}}{\binom{34}{18}\binom{5}{1} + \binom{34}{19}\binom{5}{0}} = 0.5337.$$

TABLE 4.2

UMVUE and MLE, and p-Value Using These Estimators for a Two-Stage Design of $(n_1, n_2, a_1, b_1, a) = (34, 5, 17, 35, 20)$ with $p_0 = 0.4$

			Estimate		p-value		
m	s	$f(m, s\|p_0)$	UMVUE	MLE	UMVUE	MLE	Naive
1	0	0.0000	0.0000	0.0000	1.0000	1.0000	1.0000
1	1	0.0000	0.0294	0.0294	1.0000	1.0000	1.0000
1	2	0.0000	0.0588	0.0588	1.0000	1.0000	1.0000
1	3	0.0001	0.0882	0.0882	1.0000	1.0000	1.0000
1	4	0.0003	0.1176	0.1176	0.9999	0.9999	0.9999
1	5	0.0010	0.1471	0.1471	0.9997	0.9997	0.9997
1	6	0.0034	0.1765	0.1765	0.9986	0.9986	0.9986
1	7	0.0090	0.2059	0.2059	0.9952	0.9952	0.9952
1	8	0.0203	0.2353	0.2353	0.9862	0.9862	0.9862
1	9	0.0391	0.2647	0.2647	0.9659	0.9659	0.9659
1	10	0.0652	0.2941	0.2941	0.9268	0.9268	0.9268
1	11	0.0948	0.3235	0.3235	0.8617	0.8617	0.8617
1	12	0.1211	0.3529	0.3529	0.7669	0.7669	0.7669
1	13	0.1366	0.3824	0.3824	0.6458	0.6458	0.6458
1	14	0.1366	0.4118	0.4118	0.5092	0.5092	0.5092
1	15	0.1214	0.4412	0.4412	0.3726	0.3726	0.3726
1	16	0.0961	0.4706	0.4706	0.2512	0.2478	0.2512
1	17	0.0679	0.5000	0.5000	0.1550	0.1388	0.1550
2	18	0.0033	0.5294	0.4615	0.0872	0.2512	0.2653
2	19	0.0129	0.5337	0.4872	0.0838	0.1517	0.1713
2	20	0.0219	0.5403	0.5128	0.0709	0.0709	0.1021
2	21	0.0217	0.5508	0.5385	0.0490	0.0490	0.0559
2	22	0.0145	0.5672	0.5641	0.0273	0.0273	0.0280
2	23	0.0075	0.5897	0.5897	0.0128	0.0128	0.0128
2	24	0.0033	0.6154	0.6154	0.0053	0.0053	0.0053
2	25	0.0013	0.6410	0.6410	0.0020	0.0020	0.0020
2	26	0.0005	0.6667	0.6667	0.0007	0.0007	0.0007
2	27	0.0002	0.6923	0.6923	0.0002	0.0002	0.0002
2	28	0.0000	0.7179	0.7179	0.0001	0.0001	0.0001
2	29	0.0000	0.7436	0.7436	0.0000	0.0000	0.0000
2	30	0.0000	0.7692	0.7692	0.0000	0.0000	0.0000
2	31	0.0000	0.7949	0.7949	0.0000	0.0000	0.0000
2	32	0.0000	0.8205	0.8205	0.0000	0.0000	0.0000
2	33	0.0000	0.8462	0.8462	0.0000	0.0000	0.0000
2	34	0.0000	0.8718	0.8718	0.0000	0.0000	0.0000
2	35	0.0000	0.8974	0.8974	0.0000	0.0000	0.0000
2	36	0.0000	0.9231	0.9231	0.0000	0.0000	0.0000
2	37	0.0000	0.9487	0.9487	0.0000	0.0000	0.0000
2	38	0.0000	0.9744	0.9744	0.0000	0.0000	0.0000
2	39	0.0000	1.0000	1.0000	0.0000	0.0000	0.0000

Hence, when $(m, s) = (2, 19)$ is observed, we calculate the p-value based on UMVUE as

$$Pr(\tilde{p}(M,S) \geq 0.5337 \mid p_0) = \sum_{j=19}^{39} f(2, j \mid p_0)$$

$$= 0.0129 + 0.0219 + 0.0217 + 0.0145 + 0.0075$$
$$+ 0.0033 + 0.0013 + 0.0005 + 0.0002 + 0.0000 + \cdots + 0.0000$$
$$= 0.0838$$

Since p-value $> \alpha^* (= 0.05)$, we fail to reject H_0. Since $s = 19$ after stage $2(= m)$ is smaller than $a = 20$, we fail to reject H_0 by the two-stage testing rule also. Table 4.2 also reports the p-values based on the ordering by MLE values and the p-values calculated by ignoring the fact that the two-stage design aspect, denoted as "naive." Note that the p-values based on the UMVUE and MLE orderings are different for large s values with $m = 1$ and for small s values with $m = 2$. This occurs because of the difference between the UMVUE-ordering and the MLE-ordering for the small s values with $m = 2$. The naive p-values are quite different from the p-values based on the UMVUE-ordering too.

The strength of the above p-value method is that it can be extended to the cases where the observed sample size at the stopping stage is different from that specified by the design. This becomes possible because the PMF $f(m, s \mid p)$ and the UMVUE depend on the stopping boundaries only up to stage $m - 1$, i.e., $\{(a_k, b_k), 1 \leq k \leq m - 1\}$, where $a_2 = b_2 - 1$ denotes a, the rejection value at the second stage. Suppose that a study stopped after stage $m = 1$ with $S = s$ responders from n_1 patients that may be different from the stage 1 sample size specified by the design. Then, Equation 4.4 can be modified to calculate a p-value

$$\text{p-value} = \begin{cases} \left| \sum_{j=s}^{n_1} p^s (1-p)^{n_1 - s} \binom{n_1}{s} \right. & m = 1 \text{ and the study is stopped for superiority} \\ \left| 1 - \sum_{j=0}^{s-1} p^s (1-p)^{n_1 - s} \binom{n_1}{s} \right. & m = 1 \text{ and the study is stopped for futility} \end{cases}$$

Note that the p-value with $m = 1$ is calculated as if the study had a single-stage design with sample size n_1, so that it does not require specification of stopping values (a_1, b_1).

On the other hand, suppose that the study proceeded to the second stage, i.e., $m = 2$. In this case, the stage 1 decision will be made based on (a_1, b_1, n_1) as specified by the design, but the sample size for the second stage may be slightly different from n_2 that is specified by the design. In this case, based on observed values of (n_2, s), we calculate

$$\text{p-value} = \sum_{j=b_1}^{n_1} p^s (1-p)^{n_1 - s} \binom{n_1}{s} + \sum_{j=s}^{b_1 - 1 + n_2} p^s (1-p)^{n_1 + n_2 - s} \sum_{x_1 = a_1 + 1}^{(b_1 - 1) \wedge s} \binom{n_1}{x_1} \binom{n_2}{s - x_1}$$

when $m = 2$. This p-value does not require specification of a either. We can reject H_0 and accept the study therapy when p-value $< \alpha^*$ regardless of the observed sample size at the stopping stage.

It is easy to show that the ordering of MLE depends on the critical values at the stopping stage. Hence, we can not calculate an MLE-based p-value if the sample size at the stopping stage is different from that specified by the design. Furthermore, the decision rule based on the MLE-based p-value may not match with that based on the two-stage testing rule whether the observed sample is identical to the planned one or not.

EXAMPLE 4.4 (Revisited)

In Example 4.4, with 27 GRs out of $n = 30$ patients after the second stage, the p-value is 0.0093 by Jung et al. [6] for $p_0 = 0.7$ and $(a_1, n_1) = (4, 6)$. Since the p-value $< \alpha^*$, this is another justification that Kim et al. [25] is a positive study.

In summary, it is obvious from Equation 4.3 that UMVUE gives a p-value satisfying the properties: (1) the p-values in the acceptance region of H_0 are larger than those in the rejection region and (2) the p-value for the critical value matches with the type I error probability α of the two-stage testing. As mentioned above, the critical values of a two-stage design can not be used for testing if the realized sample size at the stopping stage is different from that specified at the design. In this case, we can conduct a statistical test by calculating the p-value with the sample size at the stopping stage conditioned on the observed value and checking if it is smaller than the prespecified α level or not. In Section 4.2.3, we show that the Jennison-Turnbull confidence interval also can be used for this purpose. This property makes the p-value, together with the Jennison-Turnbull confidence interval, very useful for the analysis of two-stage phase II trials. Other p-value methods for multi-stage design [20,27] do not have these desirable properties. Green and Dahlberg [9] and Herndon [10] have also considered under- or over-accrual of patients in each stage and proposed some ad hoc approaches to the selection of rejection values depending on the realized sample sizes. Chen and Ng [21] considered a set of combinations for possible sample sizes for stages 1 and 2 (n_1, n_2) and provided the rejection values (a_1, a) to minimize the maximal sample size or the average expected sample, assuming that each combination in the set has the same probability to be observed. The latter approach has some undesirable properties: (1) if the combination of realized sample sizes does not belong to the prespecified set, it does not provide valid rejection values; and (2) for each combination of (n_1, n_2), rejection values (a_1, a) are given. For each n_1 value, there are multiple n_2 values with their own a_1 values. So, it is not clear which rejection value a_1 should be used for a realized n_1 after stage 1. Furthermore, suppose that we observed a combination (n_1, n_2) through two stages. Then, we will use a that is assigned to this combination, but a_1 that have been used after stage 1 may be different from the stage 1 rejection value assigned to this combination of sample sizes.

4.3 Phase II Trials with Heterogeneous Patient Populations

So far, we have assumed that the patients of a phase II trial population have identical response rates. Oftentimes, however, a study population consists of multiple subpopulations (also called cohorts) which have different response rates. For example, suppose that we want to evaluate the tumor response of CD30 antibody, SGN-30, combined with GVD (gemcitabine, vinorelbine, pegylated liposomal doxorubicin) chemotherapy in patients with relapsed or refractory classical Hodgkin's lymphoma (HL) through a phase II trial. In a previous study, GVD only has led to responses in 65% of patients with relapsed or refractory HL for patients who never had a transplant and 75% in the transplant group. About 50% of patients in the previous study never had a transplant. Combining the data from the two cohorts, the response rate (RR) for the whole patient population is estimated as 70%($= 0.5 \times 0.65 + 0.5 \times 0.75$).

Using this outcome as historical control data, the new study is designed as a single-arm trial for testing

$$H_0 : p \leq 70\% \quad \text{against} \quad H_a : p > 70\%, \tag{4.5}$$

where p denotes the true RR of the combination therapy in the patient population combining the two subgroups, one for those with prior transplants and the other for those without one.

A standard design to account for the heterogeneity of the patient population is a single-arm trial based on a specified prevalence for each cohort for testing hypotheses (Equation 4.5). For the example study, we consider an increase in RR by 15% or larger clinically significant for each cohort. So, we will not be interested in the combination therapy if the true RR, p, is lower than $p_0 = 70\%$ and will be strongly interested if the true RR is higher than $p_a = 85\%$. Then, the Simon's[1] two-stage optimal design for testing

$$H_0 : p_0 = 70\% \quad \text{against} \quad H_a : p_a = 85\%$$

with type I error no larger than $\alpha^* = 0.1$ and power no smaller than $1 - \beta^* = 0.9$, described as follows.

Stage 1 Accrue $n_1 = 20$ patients. If $\bar{a}_1 = 14$ or fewer patients respond, then we stop the trial, concluding that the combination therapy is inefficacious. Otherwise, the trial proceeds to stage 2.

Stage 2 Accrue an additional $n_2 = 39$ patients. If more than $\bar{a} = 45$ patients out of the total $n = 59(= n_1 + n_2)$ respond, then the combination therapy will be accepted for further investigation.

Assuming that the number of responders from the two stages are independent binomial random variables with probability of "success" p_l under H_l $(l = 0, a)$, we obtain the exact type I error and power of the two-stage design as 0.0980 and 0.9029, respectively.

In developing such a standard design, an accurate specification of the prevalence of each cohort is critical. If the prevalence is erroneously specified, the type I error of the statistical testing can not be accurately controlled. Even when the prevalence is accurately specified, the observed prevalence from the new study may be quite different from the true one. This can easily happen in phase II trials with mostly small sample sizes. If a new study accrues a larger number of high-risk (low-risk) patients than expected, then the trial will have a higher false negativity (positivity).

London and Chang [22] resolve this issue by choosing rejection values based on a stratified analysis method. They adopt early stopping boundaries for both low and high efficacy cases based on a type I error rate and power spending function approach. Sposto and Gaynon [23] propose a two-stage design with a lower stopping value only based on large sample approximations that may not hold well for phase II trials with small sample sizes. Wathen et al. [24] propose a Bayesian method to test the efficacy for each subgroup. Jung et al. [25] consider a similar design situation to those by London and Chang [22] and Sposto and Gaynon [23]. The sample sizes are determined by a standard design, such as Simon's minimax or optimal, based on a specified prevalence of each cohort, but the rejection value is adjusted depending on the observed prevalence from the trial. Conditioning on the observed prevalence, the rejection values are chosen by calculating the type I error rate and the power using the accurate probability distributions, accounting for the small sample sizes of typical phase II trials. Using some real examples, we show that our design accurately controls the conditional type I error rate and power in a wide range of prevalences. In contrast, the conditional type I error of a standard design with fixed rejection values wildly fluctuates around the pre-specified level, depending on the observed prevalence. Furthermore, the marginal type I error and power of the standard design can be heavily biased if the specified prevalence is different from the true prevalence.

A p-value method is presented for phase II trial with heterogeneous patient populations. This will be useful for statistical testing when the realized sample size of the trial is different from the sample sizes determined by the design.

4.3.1 Single-Stage Designs

Suppose that we want to design a phase II trial on a new therapy with respect to a patient population with two cohorts of patients, called the high-risk cohort and the low-risk cohort. Cases with more than two cohorts will be discussed later. For cohort $j(= 1, 2)$, let p_j denote the RR of the therapy and γ_j denote the prevalence ($\gamma_1 + \gamma_2 = 1$). The RR for the combined population is given as $p = \gamma_1 p_1 + \gamma_2 p_2$. Based on some historical control data, we will not be interested in the new therapy if its RR for cohort j is p_{0j} or lower, and we will be highly interested in it if its RR is $p_{aj}(= p_{0j} + \Delta_j$ for $\Delta_j > 0)$ or higher. Let $p_0 = \gamma_1 p_{01} + \gamma_2 p_{02}$ and $p_a = \gamma_1 p_{a1} + \gamma_2 p_{a2}$.

For a stratified single-stage design, an integer a is chosen to satisfy the α^*-condition given the observed m_1 value while fixing $n(= m_1 + m_2)$ at the sample size of a standard design. Given $M_1 = m_1$ $(m_2 = n - m_1)$, the conditional type I error for a rejection value a is calculated as

$$\alpha(m_1) = P(X_1 + X_2 > a \,|\, p_{01}, p_{02}, m_1) = \sum_{x_1=0}^{m_1} \sum_{x_2=0}^{m_2} I(x_1 + x_2 > a) b(x_1 \,|\, m_1, p_{01}) b(x_2 \,|\, m_2, p_{02}).$$

Given m_1, we want to choose the maximal $a = a(m_1)$ such that $\alpha(m_1) \leq \alpha^*$. For the chosen rejection value $a = a(m_1)$, the conditional power is calculated as

$$1 - \beta(m_1) = P(X_1 + X_2 > a \,|\, p_{a1}, p_{a2}, m_1) = \sum_{x_1=0}^{m_1} \sum_{x_2=0}^{m_2} I(x_1 + x_2 > a) b(x_1 \,|\, m_1, p_{a1}) b(x_2 \,|\, m_2, p_{a2}). \quad (4.6)$$

In summary, a stratified single-stage design for a population with two cohorts is chosen as follows:

Step 1. Specify γ_1, $(p_{01}, p_{02}, p_{a1}, p_{a2})$, and $(\alpha^*, 1 - \beta^*)$.

Step 2. Choose a reasonable n as follows.

   a. Calculate $p_0 = \gamma_1 p_{01} + \gamma_2 p_{02}$ and $p_a = \gamma_1 p_{a1} + \gamma_2 p_{a2}$.

   b. Choose a standard single-stage design $(n, \bar{a})$ for testing

$$H_0 : p = p_0 \text{ vs. } H_a : p = p_a$$

      under the $(\alpha^*, 1 - \beta^*)$-condition. We choose this n (or a little larger number) as the sample size of the stratified design.

Step 3. For $m_1 \in [0, n]$, choose the maximum $a = a(m_1)$ satisfying $\alpha(m_1) \leq \alpha^*$.

Step 4. Given (n, m_1, a), calculate the conditional power $1 - \beta(m_1)$ by (4.6).

The study protocol using a stratified design may provide a table of $\{a(m_1), \alpha(m_1), 1 - \beta(m_1)\}$ for $0 \leq m_1 \leq n$. When the study is over, we observe m_1 and $x(= x_1 + x_2)$ and reject the study therapy if $x \leq a(m_1)$.

Noting that $M_1 \sim b(n, \gamma_1)$, we can calculate the marginal type I error and power of the stratified design by

$$\alpha = \mathrm{E}\{\alpha(M_1)\} = \sum_{m_1=0}^{n} \alpha(m_1)b(m_1 \mid n, \gamma_1)$$

and

$$1 - \beta = \mathrm{E}\{1 - \beta(M_1)\} = \sum_{m_1=0}^{n} \{1 - \beta(m_1)\}b(m_1 \mid n, \gamma_1),$$

respectively. Since, for each $m_1 \in [0, ..., n]$, we choose $a = a(m_1)$ so that its conditional type I error does not exceed α^*, the marginal type I error will not exceed α^*.

4.3.2 Example 4.6

Let's consider the study discussed in Section 4.1 using $\Delta_1 = \Delta_2 = 0.15$. Under $\gamma_1 = \gamma_2 = 0.5$ and response rates $(p_{01}, p_{02}) = (0.65, 0.75)$, the hypotheses in terms of the population RR are expressed as $H_0 : p_0 = 0.7$ and $H_a : p_1 = 0.85$. For $(\alpha^*, 1 - \beta^*) = (0.1, 0.9)$, the standard (unstratified) design with the minimal sample size is $(n, \bar{a}) = (53, 41)$, which has $\alpha = 0.0906$ and $1 - \beta = 0.9093$. The type I error and power are valid only when the true prevalence is $\gamma_1 = \gamma_2 = 0.5$.

Suppose that the study observed $(x_1, x_2) = (28, 13)$ and $m_1 = 36$. Note that the observed prevalence for the high-risk cohort, $\hat{\gamma}_1 = 36 / 53 = 0.68$, is much larger than the expected $\gamma_1 = 0.5$. By the unstratified design, $x = 41$ equals the rejection value $\bar{a} = 41$, so that the therapy will be rejected. However, noting that $m_1 = 36$ is much larger than expected, the stratified design lowers the rejection value to $a = 40$, so that, with observation $x = 41$, the therapy will be accepted for further investigation. Similarly, the unstratified Simon's design may falsely accept the therapy if $\hat{\gamma}_1$ is much lower than the specified prevalence $\gamma_1 = 0.5$.

Table 4.3 lists the conditional type I error and power of the standard unstratified design for each $m_1 \in [0, n]$. Note that if m_1 is much larger than $n\gamma_1$, i.e., too many cohort 1 (high-risk) patients are accrued, then the standard rejection value $\bar{a} = 41$ is so anti-conservative that the conditional type I error and power become smaller than the specified $\alpha^* = 0.1$ and $1 - \beta^* = 0.9$, respectively. On the other hand, if m_1 is too small compared to $n\gamma_1$, i.e., too many cohort 2 (low-risk) patients are accrued, then the standard rejection value $\bar{a} = 41$ is so conservative that the conditional type I error becomes larger than the specified $\alpha^* = 0.1$ level. Figure 4.2a displays the conditional type I error rate and power of the standard (unstratified) design. We observe that the conditional type I error of the standard design widely varies between 0.0182 for $m_1 = 53$ and 0.2961 for $m_1 = 0$. Its conditional power also widely varies around $1 - \beta^* = 0.9$.

The second part of Table 4.3 reports the conditional rejection value $a(m_1)$ and its $\{a(m_1), 1 - \beta(m_1)\}$ for each $m_1 \in [0, n]$. The conditional rejection value $a(m_1)$ decreases from 44 to 39 as m_1 increases. Note that $\bar{a} = a(m_1) = 41$ for m_1 values around $n\gamma_1 = 26.5$. Figure 4.2a also displays the conditional type I error rate and power of the stratified design. While the conditional type I error of the stratified design $\alpha(m_1)$ is closely controlled below α^*, the conditional power is also well controlled around $1 - \beta^* = 0.9$. If we want $1 - \beta$ to be larger than $1 - \beta^*$ for all $m_1 \in [0, n]$, we have to choose a slightly larger n than 53.

TABLE 4.3

Conditional Type I Error and Power of Single-Stage Standard (Unstratified) and Stratified Designs with $n = 53$ for $(p_{01}, p_{02}, \Delta) = (0.65, 0.75, 0.15)$ and $(\alpha^*, 1 - \beta^*) = (0.1, 0.9)$

	Unstratified		Stratified				Unstratified		Stratified		
m_1	α	$1-\beta$	a	α	$1-\beta$	m_1	α	$1-\beta$	a	α	$1-\beta$
0	0.2961	0.9947	44	0.0606	0.9215	27	0.0869	0.9081	41	0.0869	0.9081
1	0.2852	0.9939	44	0.0569	0.9142	28	0.0823	0.9011	41	0.0823	0.9011
2	0.2746	0.9930	44	0.0535	0.9065	29	0.0780	0.8938	41	0.0780	0.8938
3	0.2641	0.9919	43	0.0972	0.9517	30	0.0738	0.8862	41	0.0738	0.8862
4	0.2540	0.9908	43	0.0920	0.9467	31	0.0699	0.8782	41	0.0699	0.8782
5	0.2440	0.9896	43	0.0870	0.9414	32	0.0661	0.8699	41	0.0661	0.8699
6	0.2343	0.9882	43	0.0822	0.9357	33	0.0625	0.8612	41	0.0625	0.8612
7	0.2249	0.9866	43	0.0776	0.9296	34	0.0590	0.8523	41	0.0590	0.8523
8	0.2157	0.9849	43	0.0733	0.9232	35	0.0557	0.8430	41	0.0557	0.8430
9	0.2067	0.9830	43	0.0691	0.9163	36	0.0526	0.8334	40	0.0961	0.9049
10	0.1980	0.9810	43	0.0651	0.9091	37	0.0496	0.8236	40	0.0913	0.8981
11	0.1895	0.9787	43	0.0614	0.9015	38	0.0468	0.8134	40	0.0867	0.8909
12	0.1813	0.9763	43	0.0578	0.8935	39	0.0441	0.8029	40	0.0822	0.8835
13	0.1734	0.9736	43	0.0544	0.8851	40	0.0415	0.7922	40	0.0780	0.8757
14	0.1656	0.9707	42	0.0969	0.9368	41	0.0391	0.7812	40	0.0739	0.8677
15	0.1582	0.9676	42	0.0919	0.9311	42	0.0367	0.7699	40	0.0701	0.8593
16	0.1510	0.9642	42	0.0870	0.9250	43	0.0346	0.7584	40	0.0664	0.8506
17	0.1440	0.9606	42	0.0823	0.9186	44	0.0325	0.7467	40	0.0628	0.8417
18	0.1373	0.9566	42	0.0779	0.9119	45	0.0305	0.7347	40	0.0594	0.8325
19	0.1308	0.9525	42	0.0736	0.9048	46	0.0287	0.7225	40	0.0562	0.8230
20	0.1245	0.9480	42	0.0696	0.8973	47	0.0269	0.7102	39	0.0955	0.8886
21	0.1185	0.9432	42	0.0657	0.8895	48	0.0252	0.6977	39	0.0908	0.8813
22	0.1127	0.9382	42	0.0620	0.8813	49	0.0237	0.6850	39	0.0864	0.8738
23	0.1071	0.9328	42	0.0585	0.8727	50	0.0222	0.6722	39	0.0820	0.8659
24	0.1017	0.9271	42	0.0551	0.8638	51	0.0208	0.6592	39	0.0779	0.8578
25	0.0966	0.9211	41	0.0966	0.9211	52	0.0195	0.6462	39	0.0739	0.8495
26	0.0916	0.9148	41	0.0916	0.9148	53	0.0182	0.6330	39	0.0701	0.8408

Note: The standard design has a fixed critical value $\bar{a} = 41$.

If the difference of the response probabilities between two cohorts $|p_{01} - p_{02}|$ is larger, then the range of the rejection values for the stratified design will be wider, and the conditional type I error and power of the standard design will vary more widely. Let's consider $(p_{01}, p_{02}) = (0.6, 0.8)$ and $\Delta_1 = \Delta_2 = 0.15$. Under $(\gamma_1, \alpha^*, 1 - \beta^*) = (0.5, 0.1, 0.9)$, the standard design will be the same as above, $(n, \bar{a}) = (53, 41)$, but the stratified rejection value $a(m_1)$ decreases from 47 to 37 as m_1 increases from 0 to 53. Figure 4.2b displays the conditional type I error rate and power of the standard and stratified designs. Comparing Figure 4.1a and b, we observe that the stratified design controls its conditional type I error rate and power closely to their nominal levels regardless of $|p_{01} - p_{02}|$ value, but those of the standard design change further away from their specified levels with a larger difference. We also observe that, with a larger $|p_{01} - p_{02}|$ value, the conditional type I error and power of the stratified design fluctuate more often because the conditional critical value changes more frequently; see Figure 4.2b.

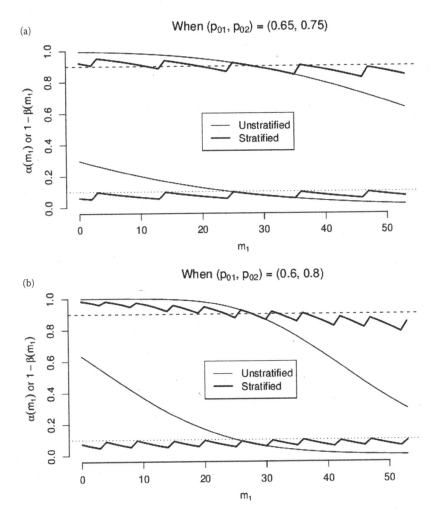

FIGURE 4.2
Conditional type I error and power of unstratified and stratified single-stage designs with $n = 53$ for $(\alpha^*, 1 - \beta^*, \Delta) = (0.1, 0.9, 0.15)$. The standard design has a fixed critical value $\bar{a} = 41$. The upper lines are conditional powers and the lower lines are conditional type I error.

Let's investigate the impact of an erroneously specified prevalence on the study design. Suppose that the true prevalence is $\gamma_1 = 0.3$, but the study is designed under a wrong specification of $\gamma_1 = 0.5$. Let's assume $(p_{01}, p_{02}) = (0.65, 0.75)$, $\Delta_1 = \Delta_2 = 0.15$, and $(\alpha^*, 1 - \beta^*) = (0.1, 0.9)$, as above. Under the erroneously specified prevalence, the standard and stratified designs will be the same as above, as shown in Table 4.3. The standard design has marginal type I error and power $(\alpha, 1 - \beta) = (0.1530, 0.9631)$ and the stratified design has $(\alpha, 1 - \beta) = (0.0767, 0.9116)$. Under the true $\gamma_1 = 0.3$, $p_0 = \gamma_1 p_{01} + \gamma_2 p_{02} = 0.72$ and $p_1 = \gamma_1 p_{a1} + \gamma_2 p_{a2} = 0.87$ are farther away from 1/2 than those under the specified $\gamma_1 = 0.5$, so that the marginal power for the stratified design is still larger than $1 - \beta^*$ even though the marginal type I error is much below $\alpha^* = 0.1$. The marginal type I error for the standard design is much larger than the specified $\alpha^* = 0.1$. Under a wrong projection of the prevalence, the type I error of a standard design can be heavily biased, but that of the stratified design will be always controlled below α^*.

Now, suppose that the true prevalence is $\gamma_1 = 0.7$, but the study is designed under an erroneously specified $\gamma_1 = 0.5$. In this case, the standard design has marginal type I error and power $(\alpha, 1 - \beta) = (0.0501, 0.8209)$, and the stratified design has $(\alpha, 1 - \beta) = (0.0768, 0.8762)$. The power for the stratified design is slightly smaller than $1 - \beta^*$ because of the conservative adjustment of conditional type I error. However, the power for the standard design is much smaller than $1 - \beta^*$. The impact of erroneously specified prevalence on the bias of marginal type I error and power will be larger with a larger difference between p_{01} and p_{02}.

4.3.3 Two-Stage Designs

Under a two-stage design, we accrue n_k patients during stage $k(= 1, 2)$. Let $n = n_1 + n_2$. For stage $k(= 1, 2)$ and cohort $j(= 1, 2)$, let M_{kj} and X_{kj} be random variables denoting the number of patients and the number of responders, respectively. Note that $n_k = m_{k1} + m_{k2}$.

Given $(M_{11}, M_{21}) = (m_{11}, m_{21})$, a design (n_1, n_2, a_1, a) has conditional type I error

$$\alpha(m_{11}, m_{21}) = P(X_{11} + X_{12} > a_1, X_{11} + X_{12} + X_{21} + X_{22} > a \mid p_{01}, p_{02})$$

$$= \sum_{x_{11}=0}^{m_{11}} \sum_{x_{12}=0}^{m_{12}} \sum_{x_{21}=0}^{m_{21}} \sum_{x_{22}=0}^{m_{22}} I(x_{11} + x_{12} > a_1, x_{11} + x_{12} + x_{21} + x_{22} > a)$$

$$\times b(x_{11} \mid m_{11}, p_{01}) b(x_{12} \mid m_{12}, p_{02}) b(x_{21} \mid m_{21}, p_{01}) b(x_{22} \mid m_{22}, p_{02})$$

and power

$$1 - \beta(m_{11}, m_{21}) = P(X_{11} + X_{12} > a_1, X_{11} + X_{12} + X_{21} + X_{22} > a \mid p_{a1}, p_{a2})$$

$$= \sum_{x_{11}=0}^{m_{11}} \sum_{x_{12}=0}^{m_{12}} \sum_{x_{21}=0}^{m_{21}} \sum_{x_{22}=0}^{m_{22}} I(x_{11} + x_{12} > a_1, x_{11} + x_{12} + x_{21} + x_{22} > a)$$

$$\times b(x_{11} \mid m_{11}, p_{a1}) b(x_{12} \mid m_{12}, p_{a2}) b(x_{21} \mid m_{21}, p_{a1}) b(x_{22} \mid m_{22}, p_{a2}). \quad (4.7)$$

We want to find a two-stage stratified design $\{n_1, n_2, a_1(m_{11}), a(m_{11}, m_{21})\}$ whose conditional type I error is smaller than or equal to α^* for each combination of (m_{11}, m_{21}) in $m_{k1} \in [0, n_k]$. In order to simplify the computation associated with the search procedure, we fix (n_1, n_2) at the first and second stage sample sizes for a standard two-stage design based on a specified prevalence γ_1, such as Simon's [1] minimax or optimal design, or admissible design by Jung et al. [6] as discussed in Section 4.2.1. Given $M_{11} = m_{11}$, we also propose to fix $a_1 = a_1(m_{11})$ at $[m_{11}p_{01} + m_{12}p_{02}]$, where $[c]$ denotes the largest integer not exceeding c. In other words, we reject the experimental therapy early if the observed number of responders from stage 1 is no larger than the expected number of responders under H_0. Now, the only design parameter we need to choose is a, the rejection value for stage 2. Given $\{\alpha^*, n_1, m_{11}, n_2, m_{21}, a_1(m_{11})\}$, we choose the largest $a = a(m_{11}, m_{21})$ satisfying $\alpha(m_{11}, m_{21}) \le \alpha^*$. Its conditional power, $1 - \beta(m_{11}, m_{21})$, is calculated by Equation 4.7.

If the observed prevalence is close to the specified one (i.e., $m_{11}/n_1 \approx \gamma_1$ and $m_{21}/n_2 \approx \gamma_1$), then the conditional rejection values $\{a_1(m_{11}), a(m_{11}, m_{21})\}$ will be the same as the unstratified rejection values $(\bar{a}_1, \bar{a})$. As in single-stage designs, the conditional power may be smaller than $1 - \beta^*$ for some (m_{11}, m_{21}). If we want to satisfy $1 - \beta \ge 1 - \beta^*$ for all combinations of $\{(m_{11}, m_{21}), 0 \le m_{11} \le n_1, 0 \le m_{21} \le n_2\}$, then we have to choose a slightly larger n than that of a standard design.

When the true prevalence of cohort 1 is γ_1, M_{k1} for $k = 1, 2$ are independent random variables following $b(n_k, \gamma_1)$. Given $(M_{11}, M_{21}) = (m_{11}, m_{21})$, let $\alpha(m_{11}, m_{21})$ and $1 - \beta(m_{11}, m_{21})$ denote the conditional type I rate and power for conditional rejection values $\{a_1(m_{11}), a(m_{11}, m_{21})\}$, respectively. Then, the marginal (unconditional) type I error and power are obtained by

$$\alpha = \sum_{m_{11}=0}^{n_1} \sum_{m_{21}=0}^{n_2} \alpha(m_{11}, m_{21}) b(m_{11} \mid n_1, \gamma_1) b(m_{21} \mid n_2, \gamma_1)$$

$$1 - \beta = \sum_{m_{11}=0}^{n_1} \sum_{m_{21}=0}^{n_2\,\cdot} \{1 - \beta(m_{11}, m_{21})\} b(m_{11} \mid n_1, \gamma_1) b(m_{21} \mid n_2, \gamma_1),$$

respectively. In summary, a phase II trial with a stratified two-stage design is conducted as follows.

Step 1. Specify $(p_{01}, p_{02}, p_{a1}, p_{a2})$ and $(\alpha^*, 1 - \beta^*)$.

Step 2. Choose sample sizes for two stages (n_1, n_2) by:
a. Specify γ_1, the prevalence for cohort 1.
b. For $p_0 = \gamma_1 p_{01} + \gamma_2 p_{02}$ and $p_a = \gamma_1 p_{a1} + \gamma_2 p_{a2}$, choose a standard (unstratified) two-stage design for testing

$$H_0 : p = p_0 \text{ vs. } H_a : p = p_a$$

that satisfies the $(\alpha^*, 1 - \beta^*)$-condition. We use (n_1, n_2) for the chosen standard design as the stage 1 and 2 sample sizes of the stratified design.

Step 3. After stage 1, calculate $a_1 = a_1(m_{11}) = [m_{11}p_{01} + m_{12}p_{02}]$ based on the observed m_{11}. We reject the therapy if $x_1 = x_{11} + x_{12}$ is smaller than or equal to $a_1(m_{11})$. Otherwise, we proceed to stage 2.

Step 4. After stage 2, choose the maximum $a = a(m_{11}, m_{21})$ satisfying $\alpha(m_{11}, m_{21}) \leq \alpha^*$ based on (m_{11}, m_{21}). Accept the therapy if $x = x_{11} + x_{12} + x_{21} + x_{22}$ is larger than $a(m_{11}, m_{21})$.

Step 5. The conditional power $1 - \beta(m_{11}, m_{21})$ for a two-stage design $(n_1, m_{11}, n_2, m_{21}, a_1, a)$ is calculated by (4.7).

Note that the description of the whole procedure and the design parameters listed in Steps 1 and 2 should be included in the study protocol.

4.3.4 Example 4.7

Let's consider the design setting of Example 4.6 with $(p_{01}, p_{02}, \Delta_1, \Delta_2) = (0.65, 0.75, 0.15, 0.15)$, $\gamma_1 = 0.5$ and $(\alpha^*, 1 - \beta^*) = (0.1, 0.9)$. Under the setting, the Simon's optimal two-stage design is given as $(n_1, n, \bar{a}_1, \bar{a}) = (20, 59, 14, 45)$. We choose $(n_1, n) = (20, 59)$ for our stratified two-stage design.

Suppose that the study observed $(x_1, x) = (15, 45)$ and $(m_{11}, m_{21}) = (14, 28)$. Note that a much larger number of patients than expected are accrued from the high-risk group, cohort 1. By the Simon's design, $x = 45$ equals $\bar{a} = 45$, so that the therapy will be rejected. However, the stratified critical values for $(m_{11}, m_{21}) = (14, 28)$ are given as $(a_1, a) = (13, 44)$, so that, with observations $(x_1, x) = (15, 45)$, the therapy will be accepted for further investigation.

Figure 4.3a displays the conditional type I error and power of the Simon's optimal design (marked as 'Unstratified') and the stratified design under the design settings. While the conditional type I error of the stratified design is closely controlled below α^*, that of the unstratified design wildly fluctuates between 0.0185 and 0.3110 depending on (m_{11}, m_{21}). Also, the conditional power of the stratified design is closely maintained around $1 - \beta^*$, but that of the Simon's design widely changes between 0.6447 and 0.9876. In the x-axis of Figure 4.3a (4.3b also), only m_{11} values are marked, but actually m_{21} values run from 0 to $n_2 = 39$ between consecutive m_{11} values. Consequently, the conditional type I error rate and power, especially for the standard unstratified design, regularly fluctuate between consecutive m_{11} values.

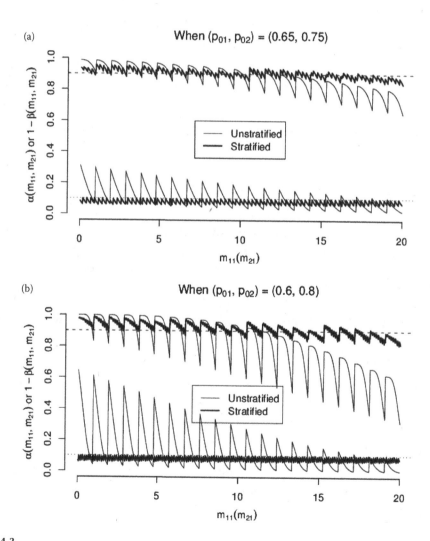

FIGURE 4.3

Conditional type I error and power of two-stage unstratified and stratified designs under $(\alpha^*, 1 - \beta^*, \Delta) = (0.1, 0.9, 0.15)$. The unstratified design has $(n_1, n, \bar{a}_1, \bar{a}) = (20, 59, 14, 45)$. The upper lines are conditional powers and the lower lines are conditional type I error.

Figure 4.3b displays the conditional type I error and power of the two designs when the two cohorts have a larger difference in RR, $(p_{01}, p_{02}) = (0.6, 0.8)$, with other parameters fixed at the same values as above. Note that, with $\gamma_1 = 0.5$, Simon's optimal design will be identical to that for $(p_{01}, p_{02}) = (0.65, 0.75)$. As in the single-stage design case (Figure 4.3b), we observe that the conditional type I error and power of the the unstratified design fluctuate more wildly than those with $(p_{01}, p_{02}) = (0.65, 0.75)$, whereas the performance of the stratified design is almost the same.

If the true prevalence is accurately specified, then the Simon's optimal design has marginal type I error and power of $(\alpha, 1 - \beta) = (0.0954, 0.9010)$, and the stratified design has $(\alpha, 1 - \beta) = (0.0792, 0.9044)$ if $(p_{01}, p_{02}) = (0.65, 0.75)$ and $(\alpha, 1 - \beta) = (0.0788, 0.9159)$ if $(p_{01}, p_{02}) = (0.6, 0.8)$. Both designs satisfy $(\alpha^*, 1 - \beta^*) = (0.1, 0.9)$. However, if the true prevalence of cohort 1 is $\gamma_1 = 0.3$, but $\gamma_1 = 0.5$ is specified in designing the study, then the Simon's design has $(\alpha, 1 - \beta) = (0.1618, 0.9521)$ if $(p_{01}, p_{02}) = (0.65, 0.75)$ and $(\alpha, 1 - \beta) = (0.2548, 0.9798)$ if $(p_{01}, p_{02}) = (0.6, 0.8)$. Note that the Simon's design has a more biased marginal type I error when two cohorts are more different in RR. On the other hand, the stratified design always controls the marginal type I error below α^* and power close to $1 - \beta^*$ even under an erroneously specified prevalence, e.g., $(\alpha, 1 - \beta) = (0.0776, 0.9203)$ if $(p_{01}, p_{02}) = (0.65, 0.75)$ and $(\alpha, 1 - \beta) = (0.0782, 0.9481)$ if $(p_{01}, p_{02}) = (0.6, 0.8)$.

4.3.5 Conditional P-Value

So far, a stratified two-stage design is determined by the sample sizes (n_1, n_2) and the rejection value (a_1, a) conditioning on the number of patients from each cohort at each stage at the design stage. When the trial is completed, however, the number of patients accrued to the study may be slightly different from the predetermined sample size. This happens since often some patients drop out or turn out to be ineligible after registration. Because of this, we usually accrue a slightly larger number of patients than the planned sample size, say 5% more. So, the total number of eligible patients at the end of a trial may be different from the planned n. In this case, sample size is a random variable, and the rejection value chosen for the planned sample size may not be valid anymore. As a flexible testing method for two-stage phase II trials, we propose to calculate the p-value conditioning on the observed sample size as well as the observed prevalence from each cohort, and to reject H_0 when the conditional p-value is smaller than the pre-specified α^* level.

If a trial is stopped due to lack of efficacy after stage 1, then usually we are not interested in p-value calculation. Suppose that the trial has proceeded to stage 2 to observe (x_1, x) together with $(n_1, m_{11}, n_2, m_{21})$. Then, the interim testing after stage 1 will be conducted using the rejection value $a_1 = [m_{11}p_{01} + m_{12}p_{02}]$. Given m_{kj} $(m_{k1} + m_{k2} = n_k)$, $X_{kj} \sim b(m_{kj}, p_{0j})$ under H_0. Hence, the p-value for an observation $(x_{11}, x_{12}, x_{21}, x_{22})$ conditioning on $(n_1, m_{11}, n_2, m_{21})$ is obtained by

$$\text{p-value} = \sum_{i_{11}=0}^{m_{11}} \sum_{i_{12}=0}^{m_{12}} \sum_{i_{21}=0}^{m_{21}} \sum_{i_{22}=0}^{m_{22}} I(i_{11} + i_{12} > a_1, i_{11} + i_{12} + i_{21} + i_{22} \geq x) \prod_{j=1}^{2} \prod_{k=1}^{2} b(i_{kj} \mid m_{kj}, p_{0j}).$$

We reject H_0 if p-value $< \alpha^*$. Note that the calculation of a conditional p-value does not require specification of the true prevalence. In order to avoid an informative sampling issue, the final sample size should be determined without knowing the number of responders from the study.

Let's revisit Example 4.7 with $(p_{01}, p_{02}) = (0.65, 0.75)$. Suppose that, at the design stage, we chose $(n_1, n) = (20, 59)$ based the Simon's optimal design, but the study accrued a slightly larger number of patients $(n_1, n_2) = (20, 40)$, among whom $(m_{11}, m_{21}) = (12, 25)$ were from cohort 1 and $(x_1, x) = (15, 46)$ responded. For the original sample size $(n_1, n) = (20, 59)$, the stratified rejection values are $(a_1, a) = (13, 45)$ with respect to $(m_{11}, m_{21}) = (12, 24)$ or $(12, 25)$. Hence, we could accept the therapy if the number of responders $(x_1, x) = (15, 46)$ was observed from the design as originally planned, $(n_1, n) = (20, 59)$. However, by having one more eligible patient from stage 2, it became unclear whether we should accept the therapy or not. To resolve this issue, we calculate the p-value for $(x_1, x) = (15, 46)$ conditioning on $(n_1, n_2) = (20, 40)$ and $(m_{11}, m_{21}) = (12, 25)$, p-value $= 0.1089$. The conditional p-value is marginally larger than $\alpha^* = 0.1$, so that we may consider accepting the therapy for further investigation.

4.4 Randomized Phase II Trials

Let p_x and p_y denote the RRs of the experimental and control arms, respectively. In a randomized phase II trial, we want to test $H_0 : p_x \leq p_y$ against $H_1 : p_x > p_y$. The null distribution of the binomial test statistic depends on the common RR $p_x = p_y$, see [26]. Consequently, if the true RRs are different from the specified ones, the testing based on binomial distributions may not maintain the type I error close to the specified design value. In order to avoid this issue, [26] proposes to control the type I error rate at $p_x = p_y = 1/2$. This results in a strong conservativeness when the true RR is different from 50%. Asymptotic tests avoid specification of $p_x = p_y$ by replacing them with their consistent estimators, but the sample sizes of phase II trials usually are not large enough for a good large sample approximation.

The Fisher [4] exact test has been a popular testing method for comparing two sample binomial proportions with small sample sizes. In a randomized phase II trial setting, Fisher's exact test is based on the distribution of the number of responders on one arm conditioning on the total number of responders, which is a sufficient statistic of $p_x = p_y$ under H_0. Hence, the rejection value of Fisher's exact test does not require specification of the common RRs $p_x = p_y$ under H_0. In this section, we propose two-stage randomized phase II trial designs based on Fisher's exact test.

4.4.1 Single-Stage Design

If patient accrual is fast or it takes a lengthy time (say, longer than 6 months) for response assessment, we may consider using a single-stage design. Suppose that n patients are randomized to each arm, and let X and Y denote the number of responders in arms x (experimental) and y (control), respectively. Let $q_k = 1 - p_k$ for arm $k (= x, y)$. Then the frequencies (and RRs in the parentheses) can be summarized as in Table 4.4.

At the design stage, n is pre-specified. Fisher's exact test is based on the conditional distribution of X given the total number of responders $Z = X + Y$ with a probability mass function

$$f(x \mid z, \theta) = \frac{\binom{n}{x}\binom{n}{z-x}\theta^x}{\sum_{i=m_-}^{m_+}\binom{n}{i}\binom{n}{z-i}\theta^i}$$

for $m_- \leq x \leq m_+$, where $m_- = \max(0, z-n)$, $m_+ = \min(z, n)$, and $\theta = p_x q_y/(p_y q_x)$ denotes the odds ratio.

TABLE 4.4

Frequencies (and RRs in the Parentheses) of a Single-Stage Randomized Phase II Trial

		Arm 1	Arm 2	Total
Response	Yes	x (p_x)	y (p_y)	z
	No	$n-x$ (q_x)	$n-y$ (q_y)	$2n-z$
Total		n	n	

Suppose that we want to limit the type I error rate to be no larger than α^*. Given $X+Y=z$, we reject $H_0 : p_x = p_y$ (i.e. $\theta = 1$) in favor of $H_1 : p_x > p_y$ (i.e. $\theta > 1$) if $X - Y \geq a$, where a is the smallest integer satisfying

$$\alpha(z) \equiv P(X - Y \geq a \mid z, H_0) = \sum_{x=\langle (z+a)/2 \rangle}^{m_+} f(x \mid z, \theta = 1) \leq \alpha^*,$$

where $\langle c \rangle$ is the round-up integer of c. Hence, the critical value a depends on the total number of responders z. Under $H_1 : \theta = \theta_1 (>1)$, the power conditional on $X+Y=z$ is given by

$$1 - \beta(z) \equiv P(X - Y \geq a \mid z, H_1) = \sum_{x=\langle (z+a)/2 \rangle}^{m_+} f(x \mid z, \theta_1).$$

We propose to choose n so that the marginal power is no smaller than a specified power level $1 - \beta^*$, i.e.,

$$E\{1 - \beta(Z)\} = \sum_{z=0}^{2n} \{1 - \beta(z)\} g(z) \geq 1 - \beta^*$$

where $g(z)$ is the probability mass function of $Z = X+Y$ under $H_1 : p_x > p_y$ that is given as

$$g(z) = \sum_{x=m_-}^{m_+} \binom{n}{x} p_x^x q_x^{n-x} \binom{n}{z-x} p_y^{z-x} q_y^{n-z+x}$$

for $z = 0, 1, \ldots, 2n$. Note that the marginal type I error rate is controlled below α^* since the conditional type I error rate is controlled below α^* for any z value.

Given a type I error rate and a power $(\alpha^*, 1 - \beta^*)$ and a specific alternative hypothesis $H_1 : (p_x, p_y)$, we find a sample size n as follows.

Algorithm for Single-Stage Design

1. For $n = 1, 2, \ldots$,
   a. For $z = 0, 1, \ldots, 2n$, find the smallest $a = a(z)$ such that

   $$\alpha(z) = P(X - Y \geq a \mid z, \theta = 1) \leq \alpha^*$$

and calculate the power conditional on $X + Y = z$ for the chosen $a = a(z)$

$$1 - \beta(z) = P(X - Y \geq a \mid z, \theta_1).$$

   b. Calculate the marginal power $1 - \beta = E\{1 - \beta(Z)\}$.

2. Find the smallest n such that $1 - \beta \geq 1 - \beta^*$.

  Given a fixed n, Fisher's test, which is based on the conditional distribution, is valid under $\theta = 1$ (i.e. controls the type I error rate exactly), and its power conditional on the total number of responders depends only on the odds ratio θ_1 under H_1. However, the marginal power, and hence the sample size n, depends on (p_x, p_y), so that we need to specify (p_x, p_y) at the design stage. If (p_x, p_y) are mis-specified, the trial may be over- or under-powered, but the type I error in data analysis will always be appropriately controlled.

4.4.2 Two-Stage Design

For ethical and economical reasons, clinical trials are often conducted using multiple stages. Phase II trials usually enter small numbers of patients, so that the practical number of stages is two at the most. We consider designs with the same features as popular two-stage phase II trial designs, with an early stopping rule when the experimental therapy has a low probability achieving additional benefits to the patients.

  Suppose that n_l ($l = 1, 2$) patients are randomized to each arm during stage $l(= 1, 2)$. Let $n_1 + n_2 = n$ denote the maximal sample size for each arm, X_l and Y_l denote the number of responders during stage l in arms x and y, respectively, $X = X_1 + X_2$ and $Y = Y_1 + Y_2$.

  At the design stage, n_l are appropriately prespecified. Note that X_1 and X_2 are independent, and, given $X_l + Y_l = z_l$, X_l has the conditional probability mass function

$$f_l(x_l \mid z_l, \theta) = \frac{\binom{n_l}{x_l}\binom{n_l}{z_l - x_l}\theta^{x_l}}{\sum_{i=m_{l-}}^{m_{l+}} \binom{n_l}{i}\binom{n_l}{z_l - i}\theta^{i}}$$

for $m_{l-} \leq x_l \leq m_{l+}$, where $m_{l-} = \max(0, z_l - n_l)$ and $m_{l+} = \min(z_l, n_l)$.

  We consider a two-stage randomized phase II trial whose rejection values are chosen conditional on z_1 and z_2 as follows.

Stage 1: Randomize n_1 patients to each arm; observe x_1 and y_1.
  a. Given $z_1 (= x_1 + y_1)$, find a stopping value $a_1 = a_1(z_1)$.
  b. If $x_1 - y_1 \geq a_1$, proceed to stage 2.
  c. Otherwise, stop the trial.

Stage 2: Randomize n_2 patients to each arm; observe x_2 and y_2 ($z_2 = x_2 + y_2$).
  a. Given (z_1, z_2), find a rejection value $a = a(z_1, z_2)$.
  b. Accept the experimental arm if $x - y \geq a$.

Now, the question is how to choose rejection values (a_1, a) conditioning on (z_1, z_2).

4.4.2.1 Choice of a_1 and a_2

In this section, we assume that n_1 and n_2 are given. As the first stage stopping value, we propose to use $a_1 = 0$. Most standard optimal two-stage phase II trials also stop early when the observed RR from stage 1 is no larger than the specified RR under H_0, refer to Simon [1] and Jung et al. [6] for single-arm trial cases and Jung [26] for randomized trial cases.

With a_1 fixed at 0, we choose the second-stage rejection value a conditioning on (z_1, z_2). Given type I error rate α^*, a is chosen as the smallest integer satisfying

$$\alpha(z_1, z_2) \equiv P(X_1 - Y_1 \geq a_1, X - Y \geq a \mid z_1, z_2, \theta = 1) \leq \alpha^*.$$

We calculate $\alpha(z_1, z_2)$ by

$$P(X_1 \geq (a_1 + z_1)/2, X_1 + X_2 \geq (a + z_1 + z_2)/2 \mid z_1, z_2, \theta = 1)$$

$$= \sum_{x_1 = m_{1-}}^{m_{1+}} \sum_{x_2 = m_{2-}}^{m_{2+}} I\{x_1 \geq (a_1 + z_1)/2, x_1 + x_2 \geq (a + z_1 + z_2)/2\} f_1(x_1 \mid z_1, 1) f_2(x_2 \mid z_2, 1),$$

where $I(\cdot)$ is the indicator function.

Under $H_1 : \theta = \theta_1$, the power conditional on (z_1, z_2) is obtained by

$$1 - \beta(z_1, z_2) = P(X_1 - Y_1 \geq a_1, X - Y \geq a \mid z_1, z_2, \theta_1)$$

$$= \sum_{x_1 = m_{1-}}^{m_{1+}} \sum_{x_2 = m_{2-}}^{m_{2+}} I\{x_1 \geq (a_1 + z_1)/2, x_1 + x_2 \geq (a + z_1 + z_2)/2\} f_1(x_1 \mid z_1, \theta_1) f_2(x_2 \mid z_2, \theta_1).$$

Note that, as in the single-stage case, the calculation of type I error rate $\alpha(z_1, z_2)$ and rejection values (a_1, a) does not require specification of the common RR $p_x = p_y$ under H_0, and that the conditional power $1 - \beta(z_1, z_2)$ requires specification of the odds ratio θ_1 under H_1, but not the RRs for the two arms, p_x and p_y.

4.4.2.2 Choice of n_1 and n_2

We now investigate how to choose sample sizes n_1 and n_2 at the design stage based on some optimality criteria.

Given (α^*, β^*), we propose to choose n_1 and n_2 so that the marginal power is maintained above $1 - \beta^*$ while controlling the conditional type I error rate for any (z_1, z_2) below α^* as described in Section 4.2.1. For stage $l(= 1, 2)$, the marginal distribution of $Z_l = X_l + Y_l$ has a probability mass function

$$g_l(z_l) = \sum_{x_l = m_{l-}}^{m_{l+}} \binom{n_l}{x_l} p_x^{x_l} q_x^{n_l - x_l} \binom{n_l}{z_l - x_l} p_y^{z_l - x_l} q_y^{n_l - z_l + x_l}$$

for $z_l = 0, \ldots, 2n_l$. Under $H_0 : p_x = p_y = p_0$, this is expressed as

$$g_{0l}(z_l) = p_0^{z_l} q_0^{2n_l - z_l} \sum_{x_l = m_{l-}}^{m_{l+}} \binom{n_l}{x_l} \binom{n_l}{z_l - x_l}.$$

Noting that Z_1 and Z_2 are independent, we choose n_1 and n_2 so that the marginal power is no smaller than a specified level $1 - \beta^*$, i.e.

$$1 - \beta \equiv \sum_{z_1=0}^{2n_1} \sum_{z_2=0}^{2n_2} \{1 - \beta(z_1, z_2)\} g_1(z_1) g_2(z_2) \geq 1 - \beta^*.$$

The marginal type I error is calculated by

$$\alpha \equiv \sum_{z_1=0}^{2n_1} \sum_{z_2=0}^{2n_2} \alpha(z_1, z_2) g_{01}(z_1) g_{02}(z_2).$$

Since the conditional type I error rate is controlled below α^* for any (z_1, z_2), the marginal type I error rate is no larger than α^*.

Although we do not have to specify the RRs for testing, we need to do so when choosing (n_1, n_2) at the design stage. If the specified RRs are different from the true ones, then the marginal power may be different from that expected. But in this case, our proposed test is still valid in the sense that it always controls both the conditional and marginal type I error rates below the specified level. Given (n_1, n_2) and $a_1 = 0$ and under H_0, the conditional probability of early termination given z_1 and the marginal probability of early termination

$$\mathrm{PET}_0(z_1) = \mathrm{P}\left(X_1 - Y_1 < a_1 \mid z_1, H_0\right) = \sum_{x_1 = m_{1-}}^{[(a_1 + z_1)/2] - 1} f_1(x_1 \mid z_1, \theta = 1)$$

and

$$\mathrm{PET}_0 = \mathrm{E}\{\mathrm{PET}_0(Z_1) \mid H_0\} = \sum_{z_1=0}^{2n_1} \mathrm{PET}_0(z_1) g_{01}(z_1),$$

respectively. Then, among those (n_1, n_2) satisfying the $(\alpha^*, 1 - \beta^*)$-condition, the Simon-type [1] minimax and the optimal designs can be chosen as follows.

- *Minimax design* chooses (n_1, n_2) with the smallest maximal sample size $n(= n_1 + n_2)$.
- *Optimal design* chooses (n_1, n_2) with the smallest marginal expected sample size EN under H_0, where

$$\mathrm{EN} = n_1 \times \mathrm{PET}_0 + n \times (1 - \mathrm{PET}_0).$$

Tables 4.5 through 4.8 report the sample sizes (n, n_1) of the minimax and optimal two-stage designs for $\alpha^* = 0.15$ or 0.2, $1 - \beta^* = 0.8$ or 0.85, and various combinations of (p_x, p_y) under H_1. For comparison, we also list the sample size n of the single-stage design under each setting. Note that the maximal sample size of the minimax is slightly smaller than or equal to the sample size of the single-stage design. If the experimental therapy is inefficacious, however, the expected sample sizes of minimax and optimal designs are much smaller than the sample size of the single-stage design. We also observe from Tables 4.5 and 4.8 that the sample sizes under $(\alpha^*, 1 - \beta^*) = (0.15, 0.8)$ are similar to those under $(\alpha^*, 1 - \beta^*) = (0.2, 0.85)$. Jung [26] proposed a randomized phase II design method based on the binomial test, called

TABLE 4.5

Single-Stage Designs, and Minimax and Optimal Two-Stage Fisher Designs for $(\alpha^*, 1 - \beta^*) = (0.15, 0.8)$

			Single-Stage Design			Minimax Two-Stage Design				Optimal Two-Stage Design			
p_y	p_x	θ	n	α	$1-\beta$	(n, n_1)	α	$1-\beta$	EN	(n, n_1)	α	$1-\beta$	EN
0.05	0.15	3.353	79	0.0827	0.8005	(78, 40)	0.0823	0.8001	63.00	(81, 26)	0.0836	0.8008	60.83
	0.2	4.750	45	0.0631	0.8075	(44, 17)	0.0620	0.8033	35.11	(44, 17)	0.0620	0.8033	35.11
	0.25	6.333	29	0.0450	0.8109	(29, 11)	0.0448	0.8014	23.96	(29, 11)	0.0448	0.8014	23.96
0.1	0.25	3.000	56	0.0884	0.8033	(56, 25)	0.0896	0.8003	43.46	(58, 19)	0.0925	0.8016	42.80
	0.3	3.857	36	0.0747	0.8016	(36, 16)	0.0783	0.8009	28.41	(37, 12)	0.0800	0.8024	28.03
0.15	0.3	2.429	65	0.1036	0.8023	(65, 36)	0.1036	0.8004	52.42	(69.22)	0.1060	0.8009	49.48
	0.35	3.051	41	0.0879	0.8025	(41, 19)	0.0919	0.8010	32.01	(42, 14)	0.0925	0.8016	30.99
0.2	0.35	2.154	74	0.1076	0.8054	(74, 42)	0.1097	0.8001	59.74	(79, 26)	0.1133	0.8003	56.17
	0.4	2.667	46	0.0953	0.8074	(46, 23)	0.1016	0.8007	36.19	(49, 14)	0.1061	0.8019	34.81
0.25	0.4	2.000	83	0.1149	0.8022	(81, 37)	0.1147	0.8005	61.35	(84, 30)	0.1155	0.8006	60.21
	0.45	2.455	47	0.0986	0.8056	(47, 27)	0.1007	0.8005	38.25	(50, 17)	0.1053	0.8003	36.10
0.3	0.45	1.909	85	0.1112	0.8005	(85, 65)	0.1115	0.8000	75.76	(95, 27)	0.1210	0.8004	65.02
	0.5	2.333	53	0.1121	0.8015	(49, 26)	0.1100	0.8007	38.88	(52, 19)	0.1113	0.8014	37.82
0.35	0.5	1.857	86	0.1155	0.8006	(86, 66)	0.1153	0.8000	76.73	(95, 32)	0.1186	0.8002	66.78
	0.55	2.270	54	0.1009	0.8026	(54, 21)	0.1169	0.8016	39.62	(55, 19)	0.1168	0.8012	39.43
0.4	0.55	1.833	87	0.1228	0.8032	(87, 59)	0.1216	0.8004	74.05	(94, 35)	0.1205	0.8000	67.36
	0.6	2.250	54	0.1015	0.8013	(54, 22)	0.1150	0.8017	39.95	(56, 18)	0.1147	0.8006	39.56
0.45	0.6	1.833	87	0.1265	0.8032	(87, 59)	0.1252	0.8004	74.03	(94, 35)	0.1234	0.8000	67.32
	0.65	2.270	54	0.1043	0.8026	(54, 21)	0.1151	0.8016	39.53	(55, 19)	0.1148	0.8012	39.33
0.5	0.65	1.857	86	0.1263	0.8006	(86, 66)	0.1260	0.8000	76.69	(95, 32)	0.1235	0.8002	66.63
	0.7	2.333	53	0.1033	0.8015	(49, 26)	0.1147	0.8007	38.77	(52, 19)	0.1126	0.8014	37.62
0.55	0.7	1.909	85	0.1237	0.8005	(85, 65)	0.1234	0.8000	75.70	(96, 26)	0.1203	0.8010	64.87
	0.75	2.455	47	0.1114	0.8056	(47, 27)	0.1240	0.8005	38.09	(50, 17)	0.1221	0.8003	35.75
0.6	0.75	2.000	83	0.1173	0.8022	(81, 37)	0.1134	0.8005	61.08	(84, 30)	0.1193	0.8006	59.83
	0.8	2.667	46	0.1208	0.8074	(46, 23)	0.1163	0.8007	35.87	(50, 12)	0.1145	0.8011	34.13
0.65	0.8	2.154	74	0.1150	0.8054	(74, 42)	0.1214	0.8001	59.46	(81, 23)	0.1213	0.8006	55.56
	0.85	3.051	41	0.1020	0.8025	(41, 19)	0.1004	0.8010	31.48	(43, 12)	0.1060	0.8024	30.12
0.7	0.85	2.429	65	0.1088	0.8023	(65, 36)	0.1083	0.8004	51.98	(69, 22)	0.1139	0.8009	48.57
	0.9	3.857	36	0.1023	0.8016	(36, 16)	0.1068	0.8009	27.53	(38, 9)	0.1075	0.8004	26.45
0.75	0.9	3.000	56	0.1098	0.8033	(56, 25)	0.1099	0.8003	42.52	(59, 17)	0.1110	0.8016	41.30
	0.95	6.333	29	0.0929	0.8109	(29, 11)	0.0919	0.8014	21.76	(30, 7)	0.0949	0.8022	21.32
0.8	0.95	4.750	45	0.0948	0.8075	(44, 17)	0.1036	0.8033	32.81	(46, 11)	0.1041	0.8007	32.23
0.85	0.95	3.353	79	0.1065	0.8005	(78, 40)	0.1071	0.8001	61.38	(83, 22)	0.1098	0.8011	57.67

MaxTest in this chapter, by controlling the type I error rate at $p_x = p_y = 50\%$. Since the type I error rate of the two-sample binomial test is maximized at $p_x = p_y = 50\%$, this test will be conservative if the true RR under H_0 is different from 50%.

4.4.3 Numerical Studies

Jung and Sargent [27] compare the performance of our Fisher's exact test with that of MaxTest. All the calculations in this section are based on exact distributions, not on simulations. Figure 4.4 displays the type I error rate and power in the range of $0 < p_y < 1 - \Delta$

TABLE 4.6

Single-Stage Designs, and Minimax and Optimal Two-Stage Fisher Designs for $(\alpha^*, 1-\beta^*) = (0.15, 0.85)$

p_y	p_x	θ	Single-Stage Design			Minimax Two-Stage Design				Optimal Two-Stage Design			
			n	α	$1-\beta$	(n, n_1)	α	$1-\beta$	EN	(n, n_1)	α	$1-\beta$	EN
0.05	0.15	3.353	92	0.0868	0.8502	(92, 48)	0.0870	0.8500	74.20	(94, 35)	0.0881	0.8502	71.17
	0.2	4.750	51	0.0677	0.8526	(51, 24)	0.0676	0.8506	41.27	(52, 18)	0.0678	0.8505	40.61
	0.25	6.333	35	0.0531	0.8581	(34, 11)	0.0516	0.8506	27.56	(34, 11)	0.0516	0.8506	27.56
0.1	0.25	3.000	65	0.0952	0.8529	(65, 37)	0.0950	0.8502	53.18	(68, 24)	0.0963	0.8501	50.29
	0.3	3.857	41	0.0785	0.8515	(41, 21)	0.0823	0.8500	33.09	(42, 16)	0.0847	0.8502	32.14
0.15	0.3	2.429	78	0.1058	0.8521	(78, 48)	0.1067	0.8502	64.71	(82, 29)	0.1091	0.8506	59.40
	0.35	3.051	49	0.0948	0.8541	(49, 23)	0.0976	0.8506	38.15	(51, 17)	0.1004	0.8527	37.28
0.2	0.35	2.154	88	0.1113	0.8518	(88, 43)	0.1120	0.8502	67.92	(93, 32)	0.1147	0.8500	66.49
	0.4	2.667	52	0.0990	0.8506	(52, 29)	0.0992	0.8506	42.01	(56, 18)	0.1062	0.8506	40.16
0.25	0.4	2.000	94	0.1132	0.8511	(94, 68)	0.1133	0.8502	82.03	(102, 37)	0.1203	0.8503	72.98
	0.45	2.455	59	0.1067	0.8535	(59, 30)	0.1122	0.8503	46.22	(62, 19)	0.1129	0.8506	43.71
0.3	0.45	1.909	104	0.1210	0.8504	(100, 55)	0.1200	0.8501	79.37	(107, 39)	0.1209	0.8503	76.34
	0.5	2.333	60	0.1034	0.8524	(60, 39)	0.1062	0.8501	50.53	(65, 22)	0.1138	0.8510	46.31
0.35	0.5	1.857	106	0.1148	0.8508	(106, 79)	0.1148	0.8501	93.40	(114, 39)	0.1252	0.8503	80.04
	0.55	2.270	61	0.1089	0.8551	(61, 37)	0.1082	0.8507	50.16	(65, 24)	0.1112	0.8504	46.96
0.4	0.55	1.833	107	0.1178	0.8520	(107, 74)	0.1170	0.8500	91.60	(115, 45)	0.1205	0.8500	83.00
	0.6	2.250	61	0.1147	0.8538	(61, 38)	0.1132	0.8500	50.57	(66, 23)	0.1138	0.8502	47.07
0.45	0.6	1.833	107	0.1214	0.8520	(107, 74)	0.1203	0.8500	91.59	(115, 45)	0.1196	0.8500	82.95
	0.65	2.270	61	0.1184	0.8551	(61, 37)	0.1163	0.8507	50.11	(65, 24)	0.1156	0.8504	46.86
0.5	0.65	1.857	106	0.1215	0.8508	(106, 79)	0.1210	0.8501	93.36	(114, 39)	0.1240	0.8503	79.88
	0.7	2.333	60	0.1176	0.8524	(60, 39)	0.1163	0.8501	50.45	(65, 22)	0.1151	0.8510	45.07
0.55	0.7	1.909	104	0.1180	0.8504	(100, 55)	0.1159	0.8501	79.22	(107, 39)	0.1234	0.8503	76.08
	0.75	2.455	59	0.1145	0.8535	(59, 30)	0.1104	0.8503	46.00	(62, 19)	0.1114	0.8506	43.28
0.6	0.75	2.000	94	0.1205	0.8511	(94, 68)	0.1256	0.8502	81.91	(102, 37)	0.1277	0.8503	72.57
	0.8	2.667	52	0.0999	0.8506	(52, 29)	0.1071	0.8506	41.72	(56, 18)	0.1147	0.8506	39.56
0.65	0.8	2.154	88	0.1180	0.8518	(88, 43)	0.1146	0.8502	67.52	(93, 32)	0.1178	0.8500	65.68
	0.85	3.051	49	0.1158	0.8541	(49, 23)	0.1160	0.8506	37.60	(52, 15)	0.1155	0.8519	36.31
0.7	0.85	2.429	78	0.1141	0.8521	(78, 48)	0.1190	0.8502	64.33	(84, 26)	0.1201	0.8503	58.49
	0.9	3.857	41	0.0980	0.8515	(41, 21)	0.1000	0.8500	32.34	(43, 14)	0.1050	0.8514	30.87
0.75	0.9	3.000	65	0.1082	0.8529	(65, 37)	0.1102	0.8502	52.50	(69, 22)	0.1150	0.8500	48.76
	0.95	6.333	35	0.1019	0.8581	(34, 11)	0.0997	0.8506	24.75	(35, 7)	0.1016	0.8505	24.43
0.8	0.95	4.750	51	0.0983	0.8526	(51, 24)	0.1021	0.8506	39.45	(53, 15)	0.1073	0.8518	37.47
0.85	0.95	3.353	92	0.1065	0.8502	(92, 48)	0.1101	0.8500	72.52	(98, 27)	0.1140	0.8503	67.92

for single-stage designs with $n = 60$ per arm, $\Delta = p_x - p_y = 0.15$ or 0.2 under H_1 and $\alpha = 0.1$, 0.15 or 0.2 under $H_0 : p_x = p_y$. The solid lines are for Fisher's test and the dotted lines are for MaxTest; the lower two lines represent type I error rate and the upper two lines represent power. As is well known, Fisher's test controls the type I error conservatively over the range of p_y. The conservativeness gets slightly stronger with small p_y values close to 0. MaxTest controls the type I error accurately around $p_y = 0.5$, but becomes more conservative for p_y values far from 0.5, especially with small p_y values. For $\alpha = 0.1$, Fisher's test and MaxTest have similar power around $0.2 \leq p_y \leq 0.4$, except that MaxTest is slightly more powerful for $p_y \approx 0.4$. Otherwise, Fisher's test is more powerful. The difference in power between

TABLE 4.7

Single-Stage Designs, and Minimax and Optimal Two-Stage Fisher Designs for $(\alpha^*, 1-\beta^*) = (0.2, 0.8)$

			Single-Stage Design			Minimax Two-Stage Design				Optimal Two-Stage Design			
p_y	p_x	θ	n	α	$1-\beta$	(n, n_1)	α	$1-\beta$	EN	(n, n_1)	α	$1-\beta$	EN
0.05	0.15	3.353	65	0.1078	0.8031	(65, 36)	0.1121	0.8005	53.73	(68, 24)	0.1152	0.8004	52.14
	0.2	4.750	38	0.0799	0.8040	(38, 15)	0.0813	0.8007	30.73	(38, 15)	0.0813	0.8007	30.73
	0.25	6.333	26	0.0509	0.8050	(25, 10)	0.0481	0.8005	20.98	(25, 10)	0.0481	0.8005	20.98
0.1	0.25	3.000	48	0.1200	0.8043	(47, 19)	0.1222	0.8002	36.08	(47, 19)	0.1222	0.8002	36.08
	0.3	3.857	30	0.0991	0.8012	(30, 19)	0.1022	0.8005	25.71	(31, 13)	0.1066	0.8024	24.43
0.15	0.3	2.429	54	0.1343	0.8005	(54, 40)	0.1349	0.8003	47.88	(60, 18)	0.1445	0.8015	42.94
	0.35	3.051	35	0.1220	0.8057	(34, 15)	0.1209	0.8022	26.46	(35, 13)	0.1221	0.8020	26.44
0.2	0.35	2.154	63	0.1512	0.8003	(62, 29)	0.1491	0.8011	47.66	(65, 23)	0.1507	0.8002	47.09
	0.4	2.667	39	0.1264	0.8005	(39, 26)	0.1342	0.8005	33.40	(40, 14)	0.1408	0.8001	29.46
0.25	0.4	2.000	67	0.1424	0.8028	(67, 47)	0.1467	0.8001	54.95	(73, 24)	0.1582	0.8003	51.75
	0.45	2.455	41	0.1295	0.8101	(40, 35)	0.1292	0.8000	37.78	(45, 12)	0.1477	0.8001	31.59
0.3	0.45	1.909	68	0.1518	0.8006	(68, 55)	0.1517	0.8001	62.04	(75, 30)	0.1570	0.8003	55.02
	0.5	2.333	41	0.1392	0.8034	(41, 28)	0.1389	0.8002	35.25	(45, 16)	0.1445	0.8003	32.72
0.35	0.5	1.857	69	0.1631	0.8012	(69, 54)	0.1625	0.8001	62.10	(76, 32)	0.1602	0.8010	56.29
	0.55	2.270	42	0.1515	0.8092	(42, 25)	0.1479	0.8004	34.50	(46, 16)	0.1491	0.8003	33.20
0.4	0.55	1.833	70	0.1713	0.8043	(70, 50)	0.1692	0.8005	60.81	(77, 32)	0.1656	0.8010	56.78
	0.6	2.250	42	0.1581	0.8079	(42, 26)	0.1545	0.8007	34.90	(45, 18)	0.1528	0.8007	33.32
0.45	0.6	1.833	70	0.1751	0.8043	(70, 50)	0.1727	0.8005	60.80	(77, 32)	0.1685	0.8010	56.75
	0.65	2.270	42	0.1618	0.8092	(42, 25)	0.1572	0.8004	34.46	(46, 16)	0.1549	0.8003	33.11
0.5	0.65	1.857	69	0.1746	0.8012	(69, 54)	0.1737	0.8001	62.07	(76, 32)	0.1686	0.8010	56.19
	0.7	2.333	41	0.1601	0.8034	(41, 28)	0.1581	0.8002	35.19	(45, 16)	0.1542	0.8003	32.53
0.55	0.7	1.909	68	0.1716	0.8006	(68, 55)	0.1711	0.8001	62.00	(77, 28)	0.1651	0.8022	55.11
	0.75	2.455	41	0.1589	0.8101	(41, 23)	0.1559	0.8000	37.74	(45, 12)	0.1505	0.8001	31.17
0.6	0.75	2.000	67	0.1660	0.8028	(67, 47)	0.1638	0.8001	57.84	(73, 24)	0.1601	0.8003	51.37
	0.8	2.667	39	0.1491	0.8005	(39, 26)	0.1472	0.8005	33.23	(40, 14)	0.1430	0.8001	28.98
0.65	0.8	2.154	63	0.1522	0.8003	(62, 29)	0.1531	0.8011	47.31	(65, 23)	0.1599	0.8002	46.58
	0.85	3.051	35	0.1311	0.8057	(34, 15)	0.1425	0.8022	25.94	(36, 11)	0.1458	0.8003	25.71
0.7	0.85	2.429	54	0.1466	0.8005	(54, 40)	0.1518	0.8003	47.68	(60, 18)	0.1590	0.8015	42.03
	0.9	3.857	30	0.1386	0.8012	(30, 19)	0.1427	0.8005	25.27	(33, 7)	0.1428	0.8008	22.99
0.75	0.9	3.000	48	0.1446	0.8043	(47, 19)	0.1427	0.8002	35.09	(48, 17)	0.1458	0.8009	35.09
	0.95	6.333	26	0.1305	0.8050	(25, 10)	0.1249	0.8005	19.04	(26, 6)	0.1267	0.8007	18.65
0.8	0.95	4.750	38	0.1392	0.8040	(38, 15)	0.1406	0.8007	28.60	(40, 9)	0.1419	0.8017	28.16
0.85	0.95	3.353	65	0.1411	0.8031	(65, 36)	0.1410	0.8005	52.42	(70, 20)	0.1483	0.8010	49.45

the two methods becomes larger with $\Delta = 0.15$. We observe similar trends overall, but the difference in power becomes smaller with $\Delta = 0.2$, especially when combined with a large $\alpha(= 0.2)$.

Figure 4.5 displays the type I error rate and power of two-stage designs with $n_1 = n_2 = 30$ per arm. We observe that, compared to MaxTest, Fisher's test controls the type I error more accurately in most range of p_y values. If $\alpha = 0.1$, Fisher's test is more powerful than MaxTest over the range of $p_y < 0.2$ or $p_y > 0.6$. But with a larger α, such as 0.15 or 0.2, MaxTest is more powerful in the range of p_y between 0.2 and 0.6. As in the single-stage design case, the difference in power diminishes as Δ and α increase. While Fisher's exact test is more

TABLE 4.8

Single-Stage Designs, and Minimax and Optimal Two-Stage Fisher Designs for $(\alpha^*, 1-\beta^*) = (0.2, 0.85)$

p_y	p_x	θ	Single-Stage Design			Minimax Two-Stage Design				Optimal Two-Stage Design			
			n	α	$1-\beta$	(n, n_1)	α	$1-\beta$	EN	(n, n_1)	α	$1-\beta$	EN
0.05	0.15	3.353	81	0.1148	0.8535	(78, 37)	0.1197	0.8503	62.00	(81, 29)	0.1214	0.8501	61.52
	0.2	4.750	44	0.0897	0.8459	(44, 19)	0.0924	0.8504	35.50	(44, 19)	0.0924	0.8504	35.50
	0.25	6.333	30	0.0618	0.8542	(30, 11)	0.0621	0.8519	24.68	(30, 11)	0.0621	0.8519	24.68
0.1	0.25	3.000	56	0.1292	0.8511	(56, 36)	0.1301	0.8501	47.58	(59, 21)	0.1321	0.8503	43.97
	0.3	3.857	35	0.1046	0.8545	(35, 19)	0.1097	0.8521	28.76	(37, 12)	0.1132	0.8502	28.03
0.15	0.3	2.429	65	0.1411	0.8523	(65, 35)	0.1411	0.8506	52.01	(69, 26)	0.1453	0.8508	50.85
	0.35	3.051	43	0.1294	0.8577	(42, 19)	0.1344	0.8508	32.60	(43, 16)	0.1345	0.8511	32.19
0.2	0.35	2.154	74	0.1469	0.8528	(74, 51)	0.1497	0.8502	63.64	(80, 30)	0.1575	0.8507	58.22
	0.4	2.667	45	0.1298	0.8559	(45, 27)	0.1349	0.8501	37.22	(48, 18)	0.1443	0.8505	35.50
0.25	0.4	2.000	80	0.1575	0.8515	(78, 50)	0.1530	0.8500	65.29	(84, 36)	0.1567	0.8515	62.60
	0.45	2.455	46	0.1402	0.8514	(46, 32)	0.1398	0.8501	39.81	(50, 20)	0.1453	0.8508	37.18
0.3	0.45	1.909	87	0.1531	0.8506	(87, 68)	0.1561	0.8501	78.21	(92, 36)	0.1670	0.8503	66.87
	0.5	2.333	53	0.1468	0.8501	(50, 29)	0.1568	0.8510	40.70	(52, 21)	0.1547	0.8500	38.57
0.35	0.5	1.857	89	0.1541	0.8533	(89, 63)	0.1537	0.8502	76.97	(93, 40)	0.1660	0.8501	68.97
	0.55	2.270	54	0.1392	0.8523	(53, 27)	0.1589	0.8503	41.47	(55, 23)	0.1599	0.8506	40.96
0.4	0.55	1.833	89	0.1600	0.8508	(89, 71)	0.1595	0.8501	80.61	(96, 39)	0.1686	0.8516	70.12
	0.6	2.250	54	0.1405	0.8510	(53, 28)	0.1569	0.8501	41.85	(57, 21)	0.1563	0.8502	41.25
0.45	0.6	1.833	89	0.1637	0.8508	(89, 71)	0.1632	0.8501	80.60	(96, 39)	0.1713	0.8516	70.08
	0.65	2.270	54	0.1437	0.8523	(53, 27)	0.1564	0.8503	41.41	(55, 23)	0.1559	0.8506	40.88
0.5	0.65	1.857	89	0.1649	0.8533	(89, 63)	0.1627	0.8502	76.92	(92, 50)	0.1604	0.8502	72.67
	0.7	2.333	53	0.1426	0.8501	(50, 29)	0.1548	0.8510	40.60	(52, 21)	0.1559	0.8500	38.40
0.55	0.7	1.909	87	0.1610	0.8506	(87, 68)	0.1603	0.8501	78.15	(92, 36)	0.1664	0.8503	66.64
	0.75	2.455	46	0.1437	0.8514	(46, 32)	0.1616	0.8501	39.70	(50, 20)	0.1659	0.8508	36.89
0.6	0.75	2.000	80	0.1485	0.8515	(78, 50)	0.1621	0.8500	65.14	(85, 34)	0.1692	0.8502	62.01
	0.8	2.667	45	0.1664	0.8559	(45, 27)	0.1615	0.8501	36.99	(48, 18)	0.1576	0.8505	35.02
0.65	0.8	2.154	74	0.1672	0.8528	(74, 51)	0.1679	0.8502	63.45	(82, 27)	0.1627	0.8501	57.62
	0.85	3.051	43	0.1544	0.8577	(42, 19)	0.1439	0.8508	32.05	(44, 14)	0.1473	0.8505	31.35
0.7	0.85	2.429	65	0.1480	0.8523	(65, 35)	0.1503	0.8506	51.56	(69, 26)	0.1591	0.8508	50.09
	0.9	3.857	35	0.1286	0.8545	(35, 19)	0.1362	0.8521	28.13	(37, 12)	0.1461	0.8502	26.71
0.75	0.9	3.000	56	0.1536	0.8511	(56, 36)	0.1545	0.8501	47.08	(59, 21)	0.1538	0.8503	42.70
	0.95	6.333	30	0.1408	0.8542	(30, 11)	0.1373	0.8519	22.36	(31, 7)	0.1386	0.8519	21.94
0.8	0.95	4.750	44	0.1290	0.8549	(44, 19)	0.1429	0.8504	33.53	(46, 14)	0.1451	0.8520	33.02
0.85	0.95	3.353	81	0.1506	0.8535	(78, 37)	0.1496	0.8503	60.17	(83, 26)	0.1516	0.8506	58.94

powerful than the binomial test in single-stage designs, their performance is comparable in two-stage designs. With a_1 value fixed at 0, the two-stage designs based on Fisher's exact test are not fully optimal. A fully optimal choice of a_1 value will depend on z_1, and the two-stage design using the optimal a_1 value is believed to outperform the two-stage designs based on binomial test in a single-stage design case. We will develop an efficient algorithm to identify fully optimal two-stage designs using Fisher's exact test in our future study.

Readers may refer to Jung and Sargent [27] for the cases where two arms have an unbalanced allocation.

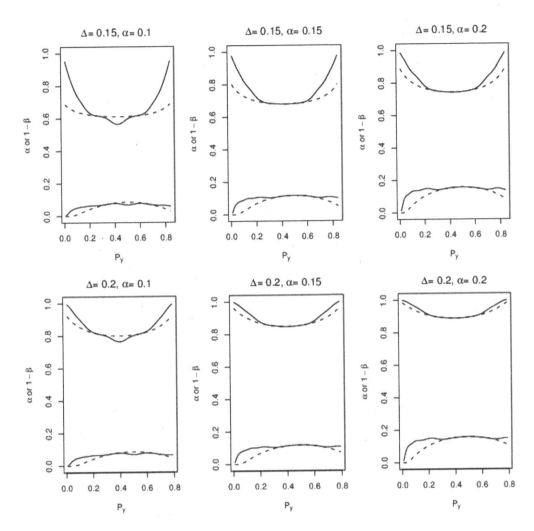

FIGURE 4.4
Single-stage designs with $n = 60$ per arm: Type I error rate and power for Fisher's test (solid lines) and MaxTest (dotted lines).

4.5 Conclusion

Traditionally, phase II cancer clinical trials have been conducted using single-arm, two-stage design. Most publications reporting the results of these trials do not use appropriate statistical methods for data analysis. These trials are specified by sample sizes and critical values for two stages. Often the number of patients treated at each stage is different from that specified by the design. In this case, the critical values become meaningless. Possibly due to this reason, publications reporting phase II trials often are not clear if they accept or reject the treatments under study, or they accept or reject them without a solid statistical ground.

When analyzing a phase III clinical trial, this issue is avoided by conducting statistical analysis conditioning on the observed sample size. Phase III trials usually have large

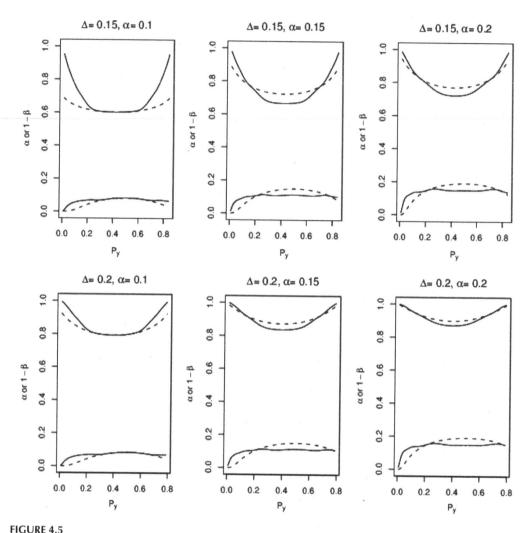

FIGURE 4.5
Two-stage designs with $n_1 = n_2 = 30$ per arm: Type I error rate and power for Fisher's test (solid lines) and MaxTest (dotted lines).

sample sizes, and the statistical methods using large sample approximation do not require pre-specification of a sample size. In fact, we can use the same approach for the analysis of phase II trials with small sample sizes, but most investigators are not familiar with them. To this end, we can use p-value and confidence interval methods for two-stage phase II clinical trials. The theory behind these methods is identical to that of two-stage design of phase II clinical trials, so that any of these methods lead to the same conclusions if the sample size at the stopping stage is the same as that specified by the design. This is possible because all of these methods are based on the UMVUE of RR. If the observed sample size for the stopping stage is different from that specified by the design, then the testing based on two-stage critical values is of no use while one can calculate unbiased p-value and confidence interval by extending the UMVUE calculation to this case.

We have extended the search algorithm of two-stage phase II trial designs beyond Simon's minimax and optimal designs. The new designs, called admissible designs, closely satisfy both the minimax and optimality criteria of Simon. These methods are appropriate when the study population is so homogeneous that the expected RR is identical for each patient in the population.

If the expected RR is different for some subpopulations, then the traditional approach is to specify the RR of the whole population based on the estimated prevalence and use the standard [1] designs with respect to the RR of the whole population. Due to the small sample sizes of single-arm phase II trials, however, the realized prevalence can be different from the estimated one at the design stage, and the type I error rate of the resulting test can be seriously biased. In order to avoid this issue, we have studied a stratified analysis method that will reflect the number of patients from each subpopulation.

We also have investigated randomized phase II trials for evaluating the efficacy of an experimental therapy compared to a prospective control. Although we demonstrate the proposed method using response as the endpoint, the method can be applied to any binomial endpoint, e.g., the proportion of patients who are progression-free at a fixed time point (say 6 months). We have limited our discussions to single-stage and two-stage designs that are most commonly used for phase II cancer clinical trials, but all of the methods discussed in this chapter can be extended to phase II trials with any number of stages.

References

1. Simon R. Optimal two-stage designs for phase II clinical trials. *Control Clin Trials*. 1989;10:1–10.
2. Cannistra SA. Phase II trials in *Journal of Clinical Oncology*. *J Clin Oncol*. 2009;27(19):3073–6.
3. Gan HK, Grothey A, Pond GP, Moore MJ, Siu LL, Sargent DJ. Randomized phase II trials: Inevitable or inadvisable? *J Clin Oncol*. 2010;28(15):2641–7.
4. Fisher RA. The logic of inductive inference (with discussion). *J R Stat Soc*. 1935;98:39–82.
5. Jung SH, Carey M, Kim KM. Graphical search for two-stage phase II clinical trials. *Control Clin Trials*. 2001;22:367–72.
6. Jung SH, Lee TY, Kim KM, George SL. Admissible two-stage designs for phase II cancer clinical trials. *Stat Med*. 2004;23:561–9.
7. Chang MN, Therneau TM, Wieand HS et al. Designs for group sequential phase II clinical trials. *Biometrics*. 1987;43:865–74.
8. Jung SH, Kim KM. On the estimation of the binomial probability in multistage clinical trials. *Stat Med*. 2004;23:881–96.
9. Green SJ, Dahlberg S. Planned vs attained design in phase II clinical trials. *Stat Med*. 1992;11:853–62.
10. Herndon J. A design alternative for two-stage, phase II, multicenter cancer clinical trials. *Control Clin Trials*. 1998;19:440–50.
11. Jennison C, Turnbull BW. Confidence intervals for a binomial parameter following a multistage test with application to MIL-STD 105D and medical trials. *Technometrics*. 1983;25:49–58.
12. Armitage P. Numerical studies in the sequential estimation of a binomial parameter. *Biometrika*. 1958;45:1–15.
13. Tsiatis AA, Rosner GL, Mehta CR. Exact confidence intervals following a group sequential test. *Biometrics*. 1984;40:797–803.
14. Clopper CJ, Pearson ES. The use of confidence or fiducial limits illustrated in the case of the binomial. *Biometrika*. 1934;26:404–13.
15. Duffy DE, Santner TJ. Confidence intervals for a binomial parameter based on multistage tests. *Biometrics*. 1987;43:81–93.
16. Tsai WY, Chi Y, Chen CM. Interval estimation of binomial proportion in clinical trials with a two-stage design. *Stat Med*. 2008;27:15–35.

17. Kim SJ, Kim K, Kim BS et al. Phase II trial of concurrent radiation and weekly cisplatin followed by VIPD chemotherapy in newly diagnosed, stage IE to IIE, nasal, extranodal NK/T-cell lymphoma: Consortium for Improving Survival of Lymphoma study. *J Clin Oncol.* 2009;27:6027–32.
18. Shimada K, Suzuki R. Concurrent chemoradiotherapy for limited-stage extranodal natural Killer/T-cell lymphoma, nasal type. *J Clin Oncol.* 2010;28(14):e229.
19. Kim SJ, Jung SH, Kim WS. *J Clin Oncol.* 2010;28(14):e230.
20. Jung SH, Owzar K, George SL et al. P-value calculation for multistage phase II cancer clinical trials (with discussion). *J Biopharm Stat.* 2006;16:765–83.
21. Chen TT, Ng TH. Optimal flexible designs in phase II clinical trials. *Stat Med.* 1998;17:2301–12.
22. London WB, Chang MN. One- and two-stage designs for stratified phase II clinical trials. *Stat Med.* 2005;24:2597–611.
23. Sposto R, Gaynon PS. An adjustment for for patient heterogeneity in the design of two-stage phase II trials. *Stat Med.* 2009;28:2566–79.
24. Wathen JK, Thall PF, Cook JD, Estey EH. Accounting for patient heterogeneity in phase II clinical trials. *Stat Med.* 2008;27:2802–15.
25. Jung SH, Chang M, Kang S. Phase II cancer clinical trials with heterogeneous patient populations. *J Biopharm Stat.* 2012;22:312–28.
26. Jung SH. Randomized phase II trials with a prospective control. *Stat Med.* 2008;27:568–83.
27. Jung SH, Sargent DA. Randomized phase II cancer clinical trials. *J Biopharm Stat.* 2014; 24(4):802–816.

5

Design and Analysis of Immunotherapy Clinical Trials

Megan Othus

CONTENTS

5.1 Introduction

The idea of using a patient's immune system to target their cancer is not a recent development. Allogeneic stem cell transplant was first used in the 1950s [1], the BCG cancer vaccine in the 1960s [2], and cytokine therapy, including interferon-α and interleukin-2 (IL2), in the 1970s and 1980s [3,4]. Newer immunotherapies include oncolytic viruses [5], adoptive T-cell therapy [6], monoclonal antibodies [7], and the particular class of monoclonal antibodies known as check-point inhibitors [6,8].

Some of the older therapies have shown limited efficacy in a small number of cancers, while newer therapies are showing improved efficacy across a broader range of cancers. There are a number of characteristics of immunotherapies that should be accounted for in clinical trial designs, and we can use our decades of prior experience of treating cancers with immunotherapies to understand some ways to incorporate these characteristics into clinical trials of new agents.

In this chapter, we will review and discuss four specific features of clinical trials using immunotherapies: (1) immune-related toxicity, (2) delayed treatment benefit (both for tumor response and survival endpoints), (3) marker stratification, and (4) treatment benefit in a subset of patients.

5.2 Immune-Related Toxicity

Immunotherapies work by turning on the immune system to fight cancer, though when the immune system is turned on it can attack organ systems uninfected by cancer and generate serious toxicities for patients. The specific toxicities, and associated supportive care, vary depending on the mechanism of the immunotherapies. Some monoclonal antibodies and adoptive T-cell therapies are associated with cytokine-release syndrome (CRS), which presents a constellation of inflammatory symptoms. CRS symptoms can range from mild (flu-like) to severe (leading to multi-organ failure and potentially death). Management of CRS is an active area of research, from understanding optimal supportive care, to re-engineering the therapies to be mediate CRS [9,10]. Some monoclonal antibodies, including checkpoint inhibitors, have associated autoimmune toxicities because the targeted antigen is expressed on non-cancer cells and tissue, leading the immune system to attack some non-cancer organ systems. With CTLA-4 and PDL1 checkpoint inhibitor therapies, skin toxicity and fatigue are commonly reported immune-related toxicities [11]. There have been deaths due to immunue-related adverse events with immune checkpoint inhibitors [12].

With many immunotherapies, there is a positive correlation between immune-related adverse events and efficacy. Finding optimal (tolerable) balancing of toxicity and efficacy is an ongoing research area [13,14]. For clinical trials using these therapies, it is important to collect detailed adverse event data, including an attribution of whether the adverse event was immune-related or not. For many trials, additional data is needed for immune-related adverse events, specifically if steroids were used and, if so, the specific doses and durations. In addition, there may need to be ongoing assessment of the toxicity burden on the trial and pre-specified rules to stop accrual early if the adverse event profile is too toxic for the patient population.

5.3 Delayed Treatment Benefit

Some immunotherapies, including checkpoint inhibitors, exhibit a delayed treatment effect, which can impact assessments of both tumor response and time-to-event endpoints.

For solid tumors, RECIST is one of most commonly used methodologies for measuring tumor burden [15]. RECIST was developed and optimized for the ways in which tumors respond to chemotherapy agents, but immunotherapies have different mechanisms of action and tumors change in ways that RECIST cannot characterize well. Several specific complicating issues are that tumors can have an initial flare and become larger before shrinking, and new tumors can appear and then later disappear [16,17]. Using standard RECIST methodology these features would be deemed a progression and considered a therapy failure, but allowing patients to remain on the the checkpoint therapy has shown that a non-negligible proportion of patients who progress by RECIST have a favorable outcome when followed for more time. These progressions were deemed "psuedo-progressions" and alternative response criteria were proposed that allowed for a potential "psuedo-" or "unconfirmed progression" [17,18]. A confirmed progression was deemed to be necessary to declare treatment failure. The field is still evolving and new or revised criteria are published regularly [19]. When designing a trial with immunotherapies with a response endpoint, it is important to consider what response criteria are appropriate for the trial (for example iRECIST and RECIST) and to ensure that adequate data are collected to be able to characterize potentially complicated response patterns.

Delayed treatment effects are a feature of many types of clinical data, not just immunotherapies. But they are a feature that is seen in many immunotherapies, including checkpoint inhibitors [20, for example, Figure 1]. While a lag in treatment benefit may be new to some areas of oncology research, there is a deep statistical literature on methods to account for or accommodate delayed treatment benefits. Many methods account for the treatment lag by weighting a test statistic in some way [21,22]. Other methods model the time lag through piece-wise or similar models [23]. When choosing what class of methods to use, the trial designer needs to think through whether the time-lag is known and can be well characterized in advance of the trial [24,25], or if the time lag needs to be assumed to be random across patients, or minimax type of statistic that reflects the patterns observed in the trial would be appropriate [26].

5.4 Marker Stratification

Many immunotherapies, and in particular monoclonal antibodies, are targeted therapies. When designing a trial with such an agent, one needs to consider how to incorporate that information in the design of the trial. There is a rich literature on clinical trial designs that assess and incorporate marker information [27–30, for example] (see Chapters 13, 14, and 15 of this book). There are several practical considerations when deciding how to incorporate a marker assessment into a trial. First, the study team needs to decide whether to enroll patients regardless of biomarker status, or if the trial will only allow biomarker-positive patients to be eligible. If all patients will be allowed on the trial regardless of biomarker status, the study team needs to decide whether the biomarker status should be required before randomization (if the trial is randomized) or before treatment starts, or if the biomarker status can be assessed at the end of the trial. Assessing a biomarker before randomization allows for randomization to be stratified by biomarker status, though any assessment takes time and, depending on how long the biomarker assay takes and the specific emergent need of treatment in the patient population, assessment before randomization may not be feasible. Many biomarkers are assessed quantitatively, but the quantitative values would need to be categorized to be used a randomization stratification factor (for example, high versus low). Such categorization can be subject to bias. Assessing biomarker status at the end of the trial is usually less expensive and allows patients to get access to the trial immediately. Regardless of whether the design calls for assessing biomarker status prospectively or retrospectively, the study team needs to decide how to analyze patients whose biomarker status is unknown, either due to sample or assay failure or another reason. Such patients may be excluded from the primary analysis, ineligible for the trial, or treated in a third biomarker category (positive, negative, unknown).

5.5 Treatment Benefit in a Subset of Patients

One of the most promising benefits of immunotherapies is that immunotherapies are one of the few cancer therapies that have shown success in terms of long-term durable remissions and possible cures [31–33]. Currently there does not exist an immunotherapy that cures all patients, rather the benefit is mixed with some patients being cured and some patients potentially deriving no benefit to the drug. Complicating this issue is that currently there do not exist biomarkers or other patient characteristics that can predict with high accuracy which patients will have long-term benefit from a specific immunotherapy. Without a biomarker to identify a population most likely to benefit from the therapy, trials need to enroll broadly, and designs need to account for the potentially heterogenous mixture of patients outcomes. This can make design of trials with time-to-event endpoints, such as progression-free survival and overall survival, more complicated than in a trial that does not need to account for long-term survivors.

The most common parametric assumption made when designing oncology trials with a time-to-event endpoint is an assumption of the endpoint following an exponential distribution. When a trial population may include patients who are long-term survivors or cured, the exponential distribution can over estimate the number of events expected on the study. This leads to trials that take much longer than anticipated to reach full information or trials that are underpowered. In order to avoid this scenario, the design needs to account for the fact that some patients will not be observed to have an event during the finite follow-up of the trial. There exists a rich literature on cure modeling and with methods to analyze data exhibiting this characteristic and methods for sample size calculations [23,34–37, among many others].

5.6 Conclusion

This chapter has covered four complicating features shared by many clinical trial study immunotherapies: immune-related toxicities, delayed treatment benefit, marker stratification, and treatment benefit in an unidentifiable subset of patients. These issues are new to many areas in which trials are being conducted. New immunotherapies have the promise of efficacy in cancer types that do not have a history with immunotherapies. Fortunately, many of these issues have been studied and characterized in other trial areas. Immunotherapy trials provide a unique opportunity to design trials that need to pull design elements from a wide swath of biostatistical research.

Acknowledgment

This work was supported in part by NIH/NCI/NCTN grant CA180819.

References

1. Thomas ED, Blume KG. Historical markers in the development of allogeneic hematopoietic cell transplantation. *Biol Blood Marrow Transplant*. 1999;5(6):341–6.
2. Herr HW, Morales A. History of bacillus Calmette-Guerin and bladder cancer: An immunotherapy success story. *J Urol*. 2008;179(1):53–6.

3. Dinarello CA. Historical insights into cytokines. *Eur J Immunol.* 2007;37(S1).
4. Goldstein D, Laszlo J. Interferon therapy in cancer: From imaginon to interferon. *Cancer Res.* 1986;46(9):4315–29.
5. Kelly E, Russell SJ. History of oncolytic viruses: Genesis to genetic engineering. *Mol Ther.* 2007;15(4):651–9.
6. Hinrichs CS, Rosenberg SA. Exploiting the curative potential of adoptive T-cell therapy for cancer. *Immunol Rev.* 2014;257(1):56–71.
7. Liu JKH. The history of monoclonal antibody development–progress, remaining challenges and future innovations. *Ann Med Surg.* 2014;3(4):113–6.
8. Pardoll DM. The blockade of immune checkpoints in cancer immunotherapy. *Nat Rev Cancer.* 2012;12(4):252–64.
9. Lee DW, Gardner R, Porter DL et al. Current concepts in the diagnosis and management of cytokine release syndrome. *Blood.* 2014;124(2):188–95.
10. Maude SL, Barrett D, Teachey DT et al. Managing cytokine release syndrome associated with novel T cell-engaging therapies. *Cancer J (Sudbury, Mass.).* 2014;20(2):119.
11. Haanen JBAG, Carbonnel F, Robert C et al. Management of toxicities from immunotherapy: Esmo clinical practice guidelines for diagnosis, treatment and follow-up. *Ann Oncol.* 2017;28(suppl 4):iv119–42.
12. Tarhini AA, Lee SJ, Hodi F et al. A phase iii randomized study of adjuvant ipilimumab (3 or 10 mg/kg) versus high-dose interferon alfa-2b for resected high-risk melanoma (us intergroup e1609): Preliminary safety and efficacy of the ipilimumab arms. 2017.
13. Weber JS, Hodi FS, Wolchok JD et al. Safety profile of nivolumab monotherapy: A pooled analysis of patients with advanced melanoma. *Int J Clin Oncol.* 2016;35(7):785–92.
14. Xu X-J, Tang Y-M. Cytokine release syndrome in cancer immunotherapy with chimeric antigen receptor engineered t cells. *Cancer Lett.* 2014;343(2):172–8.
15. Eisenhauer EA, Therasse P, Bogaerts J et al. New response evaluation criteria in solid tumours: Revised recist guideline (version 1.1). *Eur J Cancer.* 2009;45(2):228–47.
16. Hodi FS, Hwu W-J, Kefford R et al. Evaluation of immune-related response criteria and RECIST v1.1 in patients with advanced melanoma treated with pembrolizumab. *J Clin Oncol.* 2016;34(13):1510–7.
17. Wolchok JD, Hoos A, O'Day S et al. Guidelines for the evaluation of immune therapy activity in solid tumors: Immune-related response criteria. *Clin Cancer Res.* 2009;15(23):7412–20.
18. Nishino M, Giobbie-Harder A, Gargano M et al. Developing a common language for tumor response to immunotherapy: Immune-related response criteria using unidimensional measurements. *Clin Cancer Res.* 2013;19(14):3936–43.
19. Seymour L, Bogaerts J, Perrone A et al. iRECIST: Guidelines for response criteria for use in trials testing immunotherapeutics. *Lancet Oncol.* 2017; 18(3):e143–52.
20. Hodi FS, O'day SJ, McDermott DF et al. Improved survival with ipilimumab in patients with metastatic melanoma. *N Engl J Med.* 2010;2010(363):711–23.
21. Hasegawa T. Sample size determination for the weighted log-rank test with the fleming–harrington class of weights in cancer vaccine studies. *Pharm Stat.* 2014;13(2):128–35.
22. Zucker DM, Lakatos E. Weighted log rank type statistics for comparing survival curves when there is a time lag in the effectiveness of treatment. *Biometrika.* 1990;77(4):853–64.
23. Kim HT, Gray R. Three-component cure rate model for nonproportional hazards alternative in the design of randomized clinical trials. *Clin Trials.* 2012;9(2):155–63.
24. Chen T-T. Statistical issues and challenges in immuno-oncology. *J Immunother Cancer.* 2013;1(1):18.
25. Mick R, Chen T-T. Statistical challenges in the design of late-stage cancer immunotherapy studies. *Cancer Immunol Res.* 2015;3(12):1292–8.
26. Zhang D, Quan H. Power and sample size calculation for log-rank test with a time lag in treatment effect. *Stat Med.* 2009;28(5):864–79.
27. Freidlin B, McShane LM, Korn EL. Randomized clinical trials with biomarkers: Design issues. *J Natl Cancer Inst.* 2010;102(3):152–60.

28. Hoering A, LeBlanc M, Crowley JJ. Randomized phase iii clinical trial designs for targeted agents. *Clin Cancer Res.* 2008;14(14):4358–67.
29. Mandrekar SJ, Sargent DJ. Clinical trial designs for predictive biomarker validation: Theoretical considerations and practical challenges. *J Clin Oncol.* 2009;27(24):4027–34.
30. McShane LM, Hunsberger S, Adjei AA. Effective incorporation of biomarkers into phase ii trials. *Clin Cancer Res.* 2009;15(6):1898–905.
31. Mac Cheever MA. Twelve immunotherapy drugs that could cure cancers. *Immunol Rev.* 2008;222(1):357–68.[Copyeditor: Plase check this author group.]
32. Othus M, Bansal A, Koepl L et al. Accounting for cured patients in cost-effectiveness analysis. *Value in Health.* 2017;20(4):705–9.
33. Rosenberg SA. Raising the bar: The curative potential of human cancer immunotherapy. *Sci Transl Med.* 2012;4(127):127ps8.
34. Gray RJ, Tsiatis AA. A linear rank test for use when the main interest is in differences in cure rates. *Biometrics.* 1989;45(3):899–904.
35. Peng Y, Dear KBG. A nonparametric mixture model for cure rate estimation. *Biometrics.* 2000;56(1):237–43.
36. Wang S, Zhang J, Lu W. Sample size calculation for the proportional hazards cure model. *Stat Med.* 2012;31(29):3959–71.
37. Xiong X, Wu J. A novel sample size formula for the weighted log-rank test under the proportional hazards cure model. *Pharm Stat.* 2017;16(1):87–94.

6

Adaptive Designs

William T. Barry

CONTENTS

6.1 Introduction

Clinical trials are designed to rigorously monitor and assess health interventions in a given population, whether as an observational study or as a prospective experimental study that allocates patients to interventions. Prospective studies are further classified by whether patients are assigned to one intervention (single-arm) or to several interventions in a random or non-random manner (multi-arm). In the context of a prospective study, adaptation is broadly defined as making planned, well-defined changes in clinical trial design parameters during trial execution, where changes are based on data from that trial and are to achieve goals of validity, scientific efficiency, or patient safety.

The type of clinical trial and general study design should depend first and foremost on the underlying research question of interest. The general approach to developing a trial is to enumerate primary and secondary objectives which reflect the research aim and the overall goals of investigating the intervention. To make proper inferences under each objective, outcome measures are selected to be study endpoints which need to be both clinically relevant and ascertainable in the study population. Finally, the trial should have a target sample size defined *a priori*, and is dependent on the type of analysis planned for the primary objective(s). Conventional sample size determinations are made under a framework of hypothesis testing or a desired level of precision in estimation. With hypothesis testing, sample size requirements are derived to control the probability of making an error, whether as a false-positive result when in truth there is no effect (termed a type I error, and the alpha level of the testing procedure) or as a false-negative result when the intervention is effective (termed a type II error, and the complement to power). These statistical principles have been ratified by the International Conference on Harmonisation of Technical Requirements for Registration of Pharmaceuticals for Human Use (ICH) as a means of adhering to good clinical practice in the development and conduct of clinical trials. ICH has released a set of guidelines for good clinical practice which provides a critical framework for considering the advantages and challenges in choosing an adaptive design [1].

The general scheme of an adaptive design is to activate the study under an initial set of rules for enrollment, allocation to interventions, and the schedule of assessments. Once pre-specified study milestones are met, the available data are analyzed and decision criteria regarding adaption are followed to either (1) stop the study; (2) continue the study under current rules for enrollment, allocation, and follow-up; or (3) continue the study under revised rules, accordingly. Under this broad definition, group sequential testing procedures that are used widely in the interim analysis of phase III studies could be considered as adaptive designs restricted to options (1) and (2). However, general consensus is that this is an overly broad definition of adaptive designs when discussing motivations and classes of methods. As such, studies with interim plans that only make a "go versus no-go" decision will be considered as fixed designs here. The principles behind conducting interim analyses and the statistical methods for group sequential testing in fixed designs are well characterized elsewhere [2,3]. Likewise, it is important to note that unplanned modifications to a study, even if enacted through a protocol amendment, are sometimes mischaracterized as adaptive designs. This does not meet the criteria of pre-specification, and ad-hoc changes to the design violate the principle of a well-controlled study; in particular, they disrupt integrity of a study and there is not a proper framework to control study-wide error (both type I and II). This concern was highlighted by the Pharmaceutical Research and Manufacturers of America (PhRMA) Working Group on Adaptive Designs in Clinical Drug Development [4], and is the basis for developing rigorous statistical methods specific to the adaptive procedures implemented in one's trial.

A common motivation for an adaptive design is that there may be insufficient information about the disease, study population, or intervention to inform the assumptions made about the primary endpoint and distribution of the linked variable. Adaptive designs are promoted as one way to overcome this limitation, if the use of early information from the trial can refine the study aims and analysis plans without inflating study-wide error rates. Another positive quality attributed to adaptive designs is that they are more flexible. Flexibility allows a study to potentially address more-complex research questions, and ultimately could provide more information about what is important to the investigation. Adaptive trials are often promoted as more efficient, resulting in either smaller trials or more accurate conclusions. Finally, some adaptations may be beneficial to study participants, such as adaptive randomization schemes where allocation is directed more toward the superior treatment under investigation.

Challenges with adaptive designs fall into several domains regarding the conduct of a clinical trial. First, with the stipulation that the adaptive schemes are pre-specified, the statistical considerations may need to reflect the operating characteristics of the design under a wider set of scenarios than single null and alternative hypotheses. Second, adaptations will often disrupt exchangeability, a mathematical property of a random sample. As a consequence, analysis plans may require switching to alternative statistical methods to make valid inferences. Bayesian approaches to statistical inference, while not required for adaptive designs, are a popular choice for modeling data with the loss of exchangeability. Under the Bayesian paradigm, a model is updated over the course of a study to reflect both new information and any change in the structure of data induced by an adaptive procedure. Conversely, if a frequentist hypothesis testing procedure is planned, martingales and other stochastic processes can be used to partition data around the discrete time point an adaptation is made. A third challenge to adaptive trials is that study conduct will have an increased operational burden. This includes procedures for rapid data collection and data cleaning that can maintain quality, requirements of specialized software to integrate adaptive procedures into registration and randomization systems, and complex governance structures to prevent leakage of confidential information when adaptations are implemented. Finally, adaptive designs may make it more difficult to develop the protocol and informed consent documents, to meet regulatory requirements, and for a general audience to interpret study results.

There has been a long history to developing methods for adaptive designs, despite limitations to deriving the statistical properties and implementing iterative procedures until more recent computing environments. A famous early incarnation of a response-adaptive allocation scheme is the "play-the-winner" rule proposed by Zelen, whereby treatment assignment is defined by the clinical outcome achieved by the previous subject [5]. If a response was observed, the same treatment is assigned; if no response was observed, the alternative treatment is assigned. In the decades that followed, adaptive procedures were extended to randomized allocation schemes and to many other design elements of a clinical trial. While the majority of clinical trials continue to utilize fixed designs, modern registries, including ClinicalTrials.gov and the World Health Organization platform, have shown a sharp increase in the use of adaptive designs during the twenty-first century [6]. With increased adoption in all phases of drug development, the Food and Drug Administration (FDA) has issued guidelines on the use of adaptive designs in both academic and industry-sponsored trials. In their review of the FDA guidelines, Cook and DeMets stressed that "proper implementation of adaptive designs requires an adequate understanding of the inherent trade-offs that accompany their use" [7].

The remainder of this chapter is devoted to the major types of adaptive designs that have been proposed for oncology clinical trials. Table 6.1 lists the general setting of each

TABLE 6.1

Adaptive Designs Proposed for Oncology Drug Development Trials

Adaptive Design	Setting	Parameter	Clinical Endpoint	Methods (Selected)
Dose level	Phase I	Dose toxicity relationship	Dose-limiting toxicity (DLT)	• 3 + 3 and other rule-based methods • Continual reassessment method • Escalation with overdose control
Target population	Phase II	Subgroup(s)	Baseline biomarker status and response rates	• Adaptive Simon two-stage test • Adaptive enrichment design and weighted test statistics • Bayesian hierarchical models
Randomization	Phase II	Allocation ratio	Contrast of response rates	• Polya urn models • Bayesian hierarchical models
Sample size	Phase III	Nuisance parameters: • Distribution of control arm • Variances	Response rate (control arm or pooled)	• Internal pilot designs
		Treatment effect	Contrast of response rates	• Conditional power and weighted test statistics

type of adaptive design, the phase(s) of drug development which overlap with the research question, the parameters of interest in the model, corresponding endpoints which inform the likelihood, and a selected and abbreviated list of methods. Each section will provide motivation for making the adaptation and an overview of statistical principles which characterize the methods for analysis. Examples of use of adaptive designs in oncology will also be given, as well as careful consideration about how the properties of each design reflect back to good clinical practice.

6.2 Adaptive Designs for Dose-Finding Studies

In oncology clinical trials and early drug development, the primary aim of a phase I trial is predominantly to assess the safety of a new agent and to determine the maximum tolerated dose (MTD). In addition to finding the MTD, a phase I study may have a pharmacokinetic or pharmacodynamics focus. Likewise, they can document preliminary evidence of efficacy, but historically were not designed to make statistical inferences on this endpoint. This is in part driven by the target population for a phase I trial, which is often times a heterogeneous group of terminally ill cancer patients who have failed all conventional treatments, where the ethical framework of the study is with the intention of therapeutic benefit regardless of whether it will be evaluable in the study population. For many anticancer therapies, the hypothesis is that there is a dose-response relationship to both efficacy and toxicity, but that the area between the two provides a therapeutic window for benefit with the agent. Based on this assumption, dose-finding designs are driven by the requirement that they start at a low dose of the agent, particularly with first-in-human studies of a novel compound, and establish safety before escalating to higher doses. The desirable operating characteristics in finding the MTD are to escalate rapidly yet safely through sub-therapeutic doses while minimizing overdoses. Because the assigned dose level will depend on the observed outcomes of prior patients on study, all phase I dose-finding studies meet the definition of an adaptive design. Design choices are generally classified as "rule based" or "model based" depending on how the dose assignment is determined [8].

Among rule-based designs, the conventional "3 + 3" design has been most commonly used in oncology phase I trials. Here, the justification of cohort size is driven by the probability of observing one or fewer dose-limiting toxicities (DLT) under a hypothesized true rate of DLT. For example, under the assumptions that the true DLT rate is 30% and independence among subjects, there is near equal probability of escalation under the 3 + 3 design (0.494, to be precise). Alternative rule-based designs have been proposed for when greater levels of toxicity would be accepted (e.g., 3 + 3 + 3 design) or for when lower levels of toxicity would be accepted (e.g., 5 + 5 design) [9]. Although the probability of escalation is constant in each design, it is important to remember that the probability of determining the true MTD will also depend on (1) the choice of dose levels and (2) a model of the dose-toxicity relationship. Parametric models and simulation studies can be used to illustrate the properties given the dose levels to be explored.

Model-based designs have been proposed as an alternative approach to assigning dose levels of patients over the course of a trial. The continual reassessment method (CRM) was first proposed by O'Quigley et al., where a Bayesian model defines the dose-toxicity relationship, and the prior distribution is updated by the data collected during the trial [10].

For instance, the probability of a DLT for a given dose, d, can be defined under a logistic or a power function as follows:

$$\text{Logistic: } \Pr(\text{DLT}) = p(d) = \frac{e^{(3 + \alpha \cdot d)}}{1 + e^{(3 + \alpha \cdot d)}}$$

$$\text{Power: } \Pr(\text{DLT}) = p(d) = d^{e^{\alpha}}$$

Convention in the CRM is to treat the occurrence of a DLT in the ith patient as a dichotomous event: $y_i \in \{0,1\}$, such that the likelihood after treating n patients corresponds to the binomial distribution as follows:

$$L_n(\text{DLT}) = \prod_{i=1}^{n} (p(d_i))^{y_i} (1 - p(d_i))^{1 - y_i}$$

The likelihood and prior model of the dose-toxicity relationship are then combined to determine the posterior distribution for the MTD, and treatment of the next patient on trial is allocated to the closest dose level. The rationale for model-based designs is that they are a more efficient method for determining the MTD, particularly when a large number of dose levels are to be explored. This can result in fewer total patients and a greater percentage of patients treated near the true MTD [9]. However, one criticism of the original CRM method is that there was too high a probability that a patient would receive drug at an unsafe and toxic level. This led to development of a different Bayesian model of the dose-toxicity relationship which controls the probability and degree of escalation, termed escalation with overdose control (EWOC) [11]. Model-based methods for determining the MTD have been developed that consider toxicity as a time-to-event endpoint (TITE-CRM) [12], and practical extensions to the model have been made to monitor both early- and late-onset adverse events [13]. Finally, methods have applied Bayesian decision theory, which incorporates all information with the priors on the dose-toxicity relationship in order to make the treatment assignment [14].

When exploring new treatment options in oncology, it's important to point out that many phase I trials will evaluate combinations of two or more investigational agents and seek to define maximum tolerated doses of each agent. It is relatively straightforward to extend the rule- and model-based designs to explore multiple dose levels of each agent. Yin et al. proposed using a bivariate model of the dose-toxicity relationship and a Bayesian approach to selecting the dosage of the next cohort of patients based on the prior distribution and observed clinical outcomes [15]. Likewise, the probability theory for rule-based designs remains constant, but more careful considerations should be given to the strategy of escalation and de-escalation agents to optimize both efficiency and patient safety. For instance, the safety of a novel combination of a PARP inhibitor (olaparaib) and HSP90 inhibitor (onalespib) has been investigated in gynecologic and breast cancer by the National Cancer Institute (NCI) Experimental Therapeutics Clinical Trials Network (NCT02898207), exploring four doses of the first agent and six doses of the second. The study uses the conventional $3 + 3$ scheme with rules for escalation and de-escalation that can result in one of 21 possible combinations as the MTD. In order to justify the design minimizes overdosing while having a reasonable probability of determining the true MTD, a simulation study was performed under the bivariate one-parameter logistic model that is utilized in Bayesian approaches.

Although the methods used in phase I dose-finding studies are broadly classified as rule or model based, in practice studies will adopt elements of each approach. For instance, a popular choice for first-in-human studies of anticancer therapeutics is the accelerated titration design, which was characterized using a simulation study based on data from previous phase I trials [16]. The proposed method incorporates several rule-based algorithms to move from a rapid escalation phase into the conventional 3 + 3 scheme, and then a model-based procedure for estimating the true MTD at the end of the trial. Likewise, with the concerns of overdosing noted in the original CRM design, a number of rules were added to the procedure that set a minimum number of patients treated at a dose level and that restrict escalation to one dose level at a time [17]. These two examples illustrate the pragmatic approach to defining adaptive procedures under either framework that achieve the desired properties of efficiency and patient safety.

6.3 Population Finding

Once a recommended phase II dose is established from preclinical data and the safety information collected from phase I studies, the next step in drug development has traditionally been to look for preliminary evidence of clinical activity. This was done in single-arm studies that enrolled a more narrowly defined study population where the treatment is hypothesized to be effective. When a binary endpoint of response is used, such as radiographic evidence of tumor shrinkage defined according to RECIST criteria, Simon proposed a two-stage design to efficiently stop early for futility if inadequate clinical activity is observed [18]. These two-stage designs have been widely adopted in phase II studies over the past several decades. Recently, many new drugs are in development which target specific molecular alterations or dysregulated cellular pathways which have been identified in one or more tumor types. These scientific discoveries have led to the concept of biomarker-driven trial designs, including enrichment designs that focus on select subpopulations and basket trials that evaluate activity across multiple subpopulations [19]. However, when the subpopulation that would receive clinical benefit is not fully established, adaptive designs can be a valuable mechanism for moving from a broader eligible study population to the specific subgroups where activity is observed.

A natural extension to the Simon two-stage design in a biomarker-driven trial is to conduct parallel tests of clinical activity within the two subgroups defined by marker status [20]. Here, the null hypotheses for the true event rates is specified $H_{0,i}$: $\pi_i = \pi_{0,i}$ $i \in \{1,2\}$ and sample sizes are determined under a set of target rates under the alternative hypothesis: $\pi_{1,i} > \pi_{0,i}$ $i \in \{1,2\}$. If one assumes that activity is likely to be in a target subgroup (say, $i = 1$) or adequate in the entire population, Jones and Holmgren suggested an adaptive enrichment design to improve efficiency. This restricts the possible outcomes to either rejecting both null hypotheses and continuing investigation in all-comers, or only rejecting $H_{0,1}$ and continuing investigation in the target subgroup, but does not allow continuation in the untargeted subgroup alone [21]. The two-stage design has 10 parameters related to sample sizes and the number of responses at each decision point, and optimal designs can be identified which control the family-wise error rate of the joint hypotheses [22]. There have been other approaches for adaptively enriching a study population during a trial, such as deriving the predictive value of the biomarker that defines the target subgroup in an interim analysis, say by using the odds ratio $\dfrac{\pi_1 \cdot (1 - \pi_2)}{(1 - \pi_1) \cdot \pi_2}$ and setting rules to restrict eligibility in the later phase of recruitment [23].

In scenarios where a predictive biomarker is less established and assay outputs are quantitative, the adaptive signature design was proposed to both develop and validate a classifier under a single study population. Despite the moniker of an "adaptive" method, analyses were originally fully retrospective and refer to options in the split-sample approach and use cross-validation procedures to avoid bias from overfitting [24]. Later, the principal of searching for an optimal threshold for a continuous biomarker was applied prospectively to update the entry criteria to a prospective trial. The methods can be applied iteratively in a group sequential test to transitioning from an all-comers design to enrolling only a marker-enrichment population for final testing of efficacy. The principal of adaptive enrichment also applies to multi-arm randomized controlled trials where the primary objective is to compare outcomes between arms (say, A and B) [25]. Adaptive procedures can model the probability of a positive treatment effect given a set of known baseline covariates, z, and after interim analyses restrict enrollment to the entry criteria where there is predicted benefit based on interim data. Formally, the null hypothesis is that there is no treatment effect in any subpopulation

$$H_0: \pi_A(z) - \pi_B(z) = 0 \quad \forall z$$

and the study is designed to have sufficient power to detect a treatment effect in the target subpopulation, $f(z) = I(\pi_A(z) - \pi_B(z) > 0)$. Since enrichment is dependent on outcome data, methods for statistical inference will have to adjust for multiplicity and adaptations to control the overall type I error. While the framework of controlling the family-wide error from multiple comparisons with interim analyses applies, the error spending functions of Lan and DeMets for early stopping for efficacy do not control the alpha level [3]. This is because there is no longer independence of enriched cohorts from earlier interim data. One alternative method is to partition data and use a weighted sum for the test statistic at final analysis. The principles and notation are presented in more detail in Section 6.5.

Finally, recent basket and platform trials have used Bayesian hierarchical models to define criteria for early stopping in discrete subgroups of the study population. The same goal of population finding can be applicable to both comparative and non-comparative multi-arm trials. Hierarchical models were also proposed to adapt the allocation ratios using randomization and are discussed in more detail in Section 6.4. Ultimately, with the vast heterogeneity of the disease and the multiple classes of therapies that target specific molecular alternations in the tumor, adaptive enrichment methods have great potential for a more efficiently designed trial. Adaption under the staged designs broadly adopted in phase II studies also minimally increase operational burdens. However, one needs to have the same mindfulness of an increased risk of false discovery as with evaluating subgroups, and specialized analysis plans are needed to make valid and unbiased inferences on an enriched population, regardless of whether a trial employs a frequentist or Bayesian approach [25,26].

6.4 Response-Adaptive Randomization

One of the first adaptive methods proposed for conducting a multi-arm clinical trial was the stochastic play-the-winner process by Zelen [5], as noted in the Introduction. Since that time, there have been several approaches to extending the play-the-winner rule to random allocation, by allowing for an imbalance in the allocation ratios toward the treatment arm where improved outcomes are seen. Wei and Durham extended Zelen's concept to Polya urn models for generating a random treatment assignment [27], and the strategy has been applied by many others, such as the drop-the-loser rule by Ivanova [28]. Other rule-based approaches to generating response-adaptive random assignments in multi-arm trials include optimal mapping of the treatment effect to allocation ratios in a way that minimizes the number of non-responders [29] and the doubly adaptive biased-coin design [30].

The Bayesian paradigm has also been applied to response-adaptive randomization. For instance, allocation ratios for the nth patient, r_n, to one of J treatments, can be proportional to the superiority of each agent $\{\theta_j\}$ $j = 1 \ldots J$, conditional on the data and prior distributions:

$$r_{j,n} \propto Pr\left(\prod_{j \neq j'} \theta_j \geq \theta_{j'} \middle| \text{data}_{n-1}\right)^{c_n}$$

Early criticisms of adaptive randomization schemes were that the methods were unstable with small n, and an important element of the model is the monotonic function, c_n, for diverging from equal randomization at $n = 1$ to the estimated superiority of a given treatment. Thall and Wathen suggested $c_n = n/2m$, where m is the total sample size of the study [31], and a number of other monotonic functions have been proposed [32]. More recently, higher order Bayesian models have been proposed for phase II oncology trials with integral biomarkers. Kass and Steffey [63] established as a class "conditionally independent hierarchical models" with an original intent of recruitment from distinct units, whether hospital sites or geographic regions. Using their formulation, the variable related to the primary endpoint is a random vector, y_n, observed in n subjects, and is independent given parameters, θ. Conditioning on hyperparameters, φ, as the next hierarchical level, the individual $\{\theta_{ij}\}$, are identical and independently distributed, and the elements of y_n have a common probability model, p:

$$p(y_n | \theta) = \prod_{i=1}^{n} p(y_i | \theta_i)$$

$$p(\theta | \varphi) = \prod_{i=1}^{n} p(\theta_i | \varphi)$$

In a single-arm phase II study of imatinib in a heterogeneous population of patients with sarcoma, Thall et al. applied a binary Bayesian hierarchical model so that information could be borrowed across subgroups when estimating the activity of therapy and to derive early stopping rules [33,34]. For multi-arm studies, a second hierarchical level can be added to a binary model to evaluate j treatments in k subpopulations, such that

$$y_i = \begin{cases} 1 & \text{if the ith patient had a treatment response} \\ 0 & \text{otherwise} \end{cases}$$

$$\theta_i \sim N(\mu_{jk}, 1)$$

$$\mu_{jk} \sim N(\phi_j, \sigma^2)$$

$$\phi_j \sim N(\alpha, \tau^2)$$

with a link function, $\theta_i = f(\pi_i)$ mapping to the probability of achieving a response in the ith patient. The hyperparameters σ^2 and τ^2 control the extent of borrowing within a treatment arm and across treatment arms, respectively. Lastly, for master protocols where multiple targeted therapies are investigated both across traditional disease types and with molecularly defined subpopulations, Ventz et al. extended the hierarchical structure to a third level and adopted an iterative procedure for tuning continuous stopping boundaries such that nominal type I error levels are adequately controlled [32].

Bayesian approaches to response-adaptive randomization have been utilized in two high-profile biomarker-driven trials that have enrolled heterogeneous patient populations and investigated multiple experimental treatments. First, in advanced non-small cell lung cancer, the Biomarker-integrated Approaches of Targeted Therapy for Lung Cancer Elimination Trial (BATTLE, NCT00409968) evaluated four experimental treatments within five subgroups of patients defined according to molecular phenotypes that were the most clinically relevant to lung cancer at the time [35]. Second, the I-SPY 2 trial, Neoadjuvant and Personalized Adaptive Novel Agents to Treat Breast Cancer (NCT01042379), is a platform trial evaluating the addition of multiple investigational agents to standard upfront treatment of early stage breast cancer [36]. Each trial uses a Bayesian method both for stratified randomization and for making inferences on treatment effects in multiple subpopulations. A side-by-side display of the major design characteristics is given in Table 6.2.

In BATTLE, four experimental regimens (erlotinib, vandetanib, erlotinib plus bexarotene, and sorafenib) were evaluated in five marker-defined subgroups of non-small cell lung cancer (sequentially positive for EGFR, KRAS/BRAF, VEGF, RXR/CycD1, or otherwise marker negative). A probit link function and non-informative priors $\sigma^2 = \tau^2 = $ 1e6 were used in the hierarchical model denoted above. The primary endpoint was disease control at 8 weeks, and within arm/subgroup assessments used benchmarks of $\pi_{jk} = 0.5$ and 0.3 for inferring whether sufficient or insufficient clinical activity is seen. The study defined early stopping rules for futility as a lower posterior probability of sufficient clinical activity for a given arm/subgroup. Randomization ratios were proportional to the posterior mean rate of disease control, and the randomization scheme used $c_n = \prod_{jk} I(n_{jk} \geq 1)$ so that it began once one patient in each combination was evaluable for response. A simulation study demonstrated that with 200 evaluable subjects, desired operating characteristics of the Bayesian adaptive design were achieved under a variety of scenarios [37].

As an early adopter of the Bayesian approach for a master protocol, the BATTLE trial serves as an important case study to the flexibility and the limitations of response-adaptive randomized studies. The trial opened in 2006 and closed in 2009 after registering 341 patients; 255 patients were molecularly profiled, of which 241 were treated and evaluable for response. As a first critique, the percentage of subjects in each marker-defined subgroup {36%, 11%, 34%, 2%, 17%} differed substantially from the anticipated prevalence {10%, 20%, 30%, 25%, 10%}, with the size of the RXR/CycD1 subgroup ending up being less than 1/10th of what was assumed. Because of this, adaptive randomization did not begin until the 97th patient was allocated to treatment. Moreover, inferences on disease control in the RXR/CycD1 subgroup were based on as little as a single patient treated with a given regimen. No minimum sample size was defined *a priori* in the analysis plan, and with the choice of non-informative priors, the posterior distribution from one responder exceeded the

TABLE 6.2

Design Elements of two Multi-arm Response-Adaptive Randomized Trials

Design Element	BATTLE (NCT00409968)	I-SPY 2 (NCT01042379)
Disease setting	Advanced stage non-small cell lung cancer	Early-stage breast cancer
Primary endpoint	Disease control (DC) at 8 weeks	Pathologic complete response (pCR)
Parameter for efficacy	DC rate of each experimental arm	Difference in pCR rate of each experimental arm versus a common control arm
Subpopulations	5 mutually exclusive subgroups: EGFR+, else KRAS/BRAF+, else VEGF+ else RXR/CycD1+, else marker–	9 overlapping subgroups: HR+, HR–, HER2+, HER2–, HR+/HER2–, HR–/HER2+, HR+/HER2+, HR–/HER2–, and high risk by MammaPrint
Study arms	4 experimental arms: Erlotinib, vandetanib, erlotinib plus bexarotene, and sorafenib	16 experimental arms to date: Figitumumab[a], neratinib[b], veliparib–carboplatin[b], AMG 386[b], conatumumab[a], AMG 479 plus metformin[b], MK-2206[b], T-DM1 and pertuzumab[b], pertuzumab, ganetespib[b], PLX3397[b], pembrolizumab (2 treatment plans), talazoparib plus irinotecan[b], patritumab, SGN-LIV1A One control arm: paclitaxel, doxorubicin, and cyclophosphamide with or without trastuzumab
Bayesian model	Probit model with non-informative priors	Logistic model with informative priors
Allocation ratio	Proportional to the relative rates within each subgroup	Proportional to the superiority of each experimental agent
Allowance for early stopping	Futility within each subgroup	Efficacy in one or more subgroups, futility across all subgroups

[a] Removed from listing of Arms and Interventions on ClinicalTrials.gov by 2012.
[b] Listed as closed to recruitment as of January 1, 2018.

threshold for declaring sufficient activity. Despite these challenges to making reliable and valid conclusions, the flexibility provided by the adaptive design is noted by the studies completion despite the observed biomarker prevalence. Likewise, the study team was able to perform molecular assays in real time at study entry and to capture data on disease control in order to operationalize the adaptive randomization, and to detect early signals of clinical activity with varying precision in a heterogeneous and heavily pretreated population.

The I-SPY 2 study similarly uses a hierarchical model to evaluate the pathologic complete response rate at time of surgery for early stage breast cancer. Here, the study population is stratified by a factorial model of three dichotomous molecular phenotypes (estrogen receptor status, HER2 status, and risk stratum defined by the MammaPrint assay). Comparisons of efficacy are made against a concurrent control arm of standard chemotherapy alone, and the hierarchical model uses a logistic link function. Unlike in the BATTLE trial, informative priors were elicited from a pilot cohort and allow for borrowing of information between subgroups. A unique decision rule for declaring success/failure of an experimental treatment was derived from the predictive probability that a future two-arm phase III study that enrolled 300 patients of a given subtype of breast cancer would reach statistical significance. While this is distinct from traditional frequentist hypothesis testing procedures, each is derived from the parameters π_{jk} and π_{0k}, where 0 denotes the control arm, and one could map the relationship between predicted probabilities and a frequentist p-value with the Bayesian model fully specified for a given sample size. Under their novel decision theory, an experimental regimen "graduates" for further investigation in a phase III study after 60 patients are evaluated if the predicted probability of success is >85% in any subgroup. Also, an arm can be declared as futile after 20 patients are treated if there is less than a 10% predicted probability of success in every subgroup. Finally, a unique element of the I-SPY 2 trial is that it is designed as a platform trial, where new experimental arms can be added to the study at any given point as other arms are removed from the study due to graduation or futility.

The study activated in 2010, and as of January 1, 2018, 16 experimental arms have been listed on clinicaltrials.gov, going back to the first version on record (see Table 6.2). Over the previous eight years, the status of arms and interventions was changed 14 times, and all but four experimental arms were listed as closed in the most recent version prior to 2018. Over that same time period, only a single arm was reported in conference proceedings to have closed for failure to reach the criteria for graduation [38]. Conversely, an important milestone was reached in 2015 when two investigational agents graduated from the study, veliparib–carboplatin [39] and neratinib [40], and additional experimental agents were reported as having graduated up to 2018. Up to that point, specifics of the study design were not publically available, but consistent with International Committee of Medical Journal Editors (ICMJE) policy [41], the protocol was provided as supplemental material to the two publications and included the Bayesian hierarchical model and formulation of the prior. However, the prior is dependent on patient characteristics and clinical outcomes from a pilot study (I SPY 1 [42]) that are not publicly available and prevents external evaluation of the trial design. Another concern is that Rugo et al. do not report the actual number of responses among treated subjects, stating "we do not report the raw data within biomarker subtypes or signatures; our analysis carries greater precision than would a raw-data estimate" [39]. Lastly, software for the adaptive procedure and Bayesian decision theory is not made available to investigators as intellectual property. These choices by the I-SPY 2 study team have the unfortunate effect of limiting the ability to give critical review of this innovative method of an adaptive platform study and drawing inferences of success in a phase II setting.

Ultimately, response-adaptive randomization will likely remain one of the most controversial adaptive procedures [43,44]. However, with the increasing popularity of multi-arm and platform trials and with the large number of oncology drugs in development in the era of precision medicine, the value of dynamic allocation to more beneficial therapies is likely to increase. For this reason, the continued development and critical evaluation of innovative statistical methods is important to oncology clinical trials. In order for this to have maximum impact to both current and future patients of oncology trials, the fundamental principles of transparency, sharing of data which can reproduce the parameters used in the study adaptations, and open source statistical software must be embraced by the community of investigators.

6.5 Sample Size Re-Estimation

One of the fundamental principles of good clinical practice in a trial, is that the number of subjects be pre-specified and justified as sufficient for reliably answering the research question of interest [45]. The usual method for determining the sample size requires identifying the variable for the primary endpoint, a testing procedure for a treatment effect which includes a null and alternative hypothesis, and acceptable error rates. Sometimes, a lack of prior knowledge about a clinically meaningful effect size or outcomes in the study population can impact the reliability of a sample size determination. This motivates the use of an adaptive procedure to update the target sample size based on information accumulated during the course of the trial. Historically, the choice of an adaptive procedure and the level of controversy around the method depended on whether the parameter of interest and the corresponding information relates to (1) the target treatment effect versus, (2) the expected outcome of patients on the control arm, or (3) other nuisance parameters in the distribution of the variable.

If one wants to make a more reliable assumption of a nuisance parameter, the adaptive procedure is less controversial, and an internal pilot study can be conducted whereby one first enrolls a small cohort in the first stage and modifies the total sample size under updated assumptions. Often times an unadjusted final analysis is performed on all subjects enrolled on the study. An early simulation study by Wittes and Brittain demonstrated that for the variance of a continuous outcome and for the event rate on the control arm of a dichotomous outcome, implementing a small pilot can substantial correct a study's overall power with negligible effects on the alpha level [46]. In the case where total sample sizes are restricted, a bounded test that modifies the critical value can control study-wide type I error [47]. This work demonstrates error rates can be properly controlled, but the FDA guidelines still suggest internal pilots use pooled estimates of nuisance parameters in order to keep allocation masked as much as possible [7].

Adaptive procedures have also been developed to adjust the total sample size based on an interim analysis of the primary endpoint and the observed treatment effect. The motivation for adapting due to uncertainty of nuisance parameters does not directly apply here, in that the target effect size under the alternative hypothesis should be implicit to the underlying research question. That said, protocol development can be a delicate balance between identifying a minimum clinically meaningful treatment effect versus practical constraints in the study population and trial costs. Thus while controversial, there has been some willingness to allow for sample size re-estimation when using methods that control the overall type I error. When the primary analysis is a frequentist hypothesis test of the treatment effect, this can be achieved by utilizing a test statistic that partitions the data into the information before and after each modification of the sample size. Assuming partitioned test statistics $\{Z_k\}$ are asymptotically independent and normally distributed, Chi, Hung, and Wang applied preplanned weights to combine the information for final analysis

$$Z = \frac{\sqrt{n_1} \cdot Z_1 + \sqrt{n_2} \cdot Z_2}{\sqrt{n_1 + n_2}}$$

to allow for stopping boundaries from the original design to be applied [48]. However, when increases in sample sizes are made (i.e., $N_2 > n_2$), this effectively decreases the contribution to the likelihood by subjects enrolled in the second stage. An alternative approach would be to apply weights according to the actual sample sizes in each partition, but this requires recalculating the critical value in order to maintain the overall alpha level [49]. Next, an adaptive design needs to define the procedure for modifying the total sample size based on the treatment effect observed at the interim analysis. A popular procedure is to use conditional power computed at the interim analysis as the summary statistic [50], and to divide the scale into "favorable," "promising," and "unfavourable" zones. Figure 6.1 shows

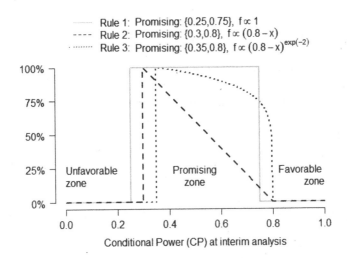

Rule 1: Promising: {0.25,0.75}, $f \propto 1$
Rule 2: Promising: {0.3,0.8}, $f \propto (0.8-x)$
Rule 3: Promising: {0.35,0.8}, $f \propto (0.8-x)^{\exp(-2)}$

FIGURE 6.1
Rules for sample size re-estimation using conditional power.

mock examples of rules for increasing the sample size by up to 100% within various bounds for a promising zone, $\{cp_l, cp_u\}$, and with various functions, $f(\cdot)$, defining the magnitude of the increase in sample size.

In order to select the appropriate design, one uses a simulation study to compute the conditional power, overall power, expected sample size, and other design characteristics (e.g., duration of the study) across a range of possible true effect sizes. While gains in power will be achieved relative to the unadjusted sample size, it is important to note that the adaptive design will be less efficient than a group sequential design powered for the maximum adjusted sample size [51].

In addition to using conditional power for re-estimating the sample size, many other approaches have been taken to allow for flexibility in the total sample size of the study. For example, to evaluate the non-inferiority of capecitabine versus standard chemotherapies in the treatment of early stage breast cancer in an elderly population, Muss et al. implemented a Bayesian adaptive design which was similar to a group sequential test [52]. Predictive probabilities were derived for the margin of inferiority, and applied to the information available at study milestones when sample sizes reached 600–1800 patients. Rather than use the critical values as early stopping bounds in a conventional manner, a decision was made on discontinuation versus further recruitment, but all patients continued to be followed for disease recurrence. Because of the unique use of Bayesian predicted probabilities with sequential testing, simulation was used to demonstrate the error rates under the null (hazard ratio of 1.24) and alternative hypotheses (hazard ratio of 1.0) under a proportional hazard model. Ultimately, the study stopped recruitment at 633 patients and then followed patients for an additional 16 months before reporting a statistically significant level of inferiority treating with capecitabine as compared to standard adjuvant chemotherapy (hazard ratio of 2.1). This case study and others where conditional power is used to increase *or* decrease the total sample size [53] illustrate the flexibility provided by the adaptive procedure.

When implementing an adaptive procedure which utilizes an interim analysis of the treatment effect, the integrity of a trial can be threatened if results can be inferred as from the resulting change to the study design. For instance, if the revised sample size falls on a unique point of a conditional power function, as illustrated for two of the sets of rules displayed in Figure 6.1, then anyone with access to the design can back-calculate the observed treatment effect. For this reason, it is important to have proper governance structure to the trial and make use of data safety and monitoring boards (DSMBs) or other independent committees to review the interim data and recommend any re-estimated sample sizes. In multicenter studies, the exact adaptive procedure can be redacted from the full protocol and restricted to a statistical analysis plan to avoid broad dissemination. Even with details kept confidential, an announced change to a sample size can result in speculation of a positive treatment effect. This is thought to have occurred in the VALOR trial, which investigated the use of vosaroxin for acute myeloid leukemia (NCT01191801), when a press release was issued by the sponsor following an interim analysis and DSMB recommendation that the total sample size be increased by 225 subjects. This could have disrupted clinical equipoise regarding the investigational agent, although any bias would be difficult to ascertain without surveying sites and potential study participants. Ultimately, a more modest improvement in overall survival with vosaroxin was observed than what was targeted in the study, and the test failed to reach statistical significance [54]. Although preventing the release of information on the treatment effect is critical with any trial that uses sample size re-estimation, it is important to recognize that a similar risk is present with traditional group sequential testing procedures. There, continuation of a study after an announced interim analysis will likewise leak partial information about the observed treatment effect, although this will generally be muted unless stopping boundaries are aggressively narrow. For this reason, strict adherence to procedures that maintain blind to study data and selection of functions for increasing sample size which are not monotonic to the treatment effect can help maintain the integrity of an adaptive design.

6.6 Adaptive Seamless Designs

Adaptive designs have also been proposed as a means of spanning the traditional phases of drug development. One framework for combining the objectives of a phase I and phase II study is to make joint inferences on the toxicity and efficacy of an intervention. Thall, Simon, and Estey introduced a fixed design that uses a Bayesian factorial model for the two binary events and a Dirchlet prior for marginal probabilities [55]. The design includes decision criteria and stopping rules that allow for sequential assessments with one of four possible outcomes: (1) continuing enrollment, (2) stopping for excess toxicity, (2) stopping for futility, or (4) stopping for efficacy. Similar methods with bivariate endpoints have been applied to dose-finding schemes which will identify a recommended dose and derive posterior probabilities for the safety and efficacy of an intervention [56,57].

Seamless phase II/III trials have been proposed for sequentially evaluating the efficacy of an intervention. One scenario is to use a randomized controlled study with fixed design and two clinical endpoints: an early signal of clinical activity (say, RECIST response) to make a go versus no-go decision during the phase II part and then a long-term outcome (say, overall survival) for demonstrating superiority in the phase III part. If the only decision in the first part is to stop the study early for futility, then a nominal alpha level for the phase III test of superiority can be used to control type I error [58]. However, to solve the overall study power, one needs to make assumptions about both marginal treatment effects and correlation between the endpoints. Another strategy in seamless phase II/III studies, is to adaptively use the first portion of the trial to evaluate several experimental agents or dosing strategies, and then select the most active therapy for continuation to the second portion as a head-to-head comparison against a concurrent control arm. In this case, control of study-wide type I error must account for "winner's curse" from the selection made in the phase II portion of the trial. A traditional approach to control of type I error is to correct for multiple testing under an assumption any experimental arm can continue to the phase III portion [59] but can be very conservative with increasing numbers of arms and assuming similar treatment effects. Inoue et al. proposed a Bayesian bivariate model for applications where there is a binary endpoint in the phase II and time-to-event endpoint in the phase III portion of the trial [60]. One uses simulation of various scenarios to demonstrate the operating characteristics of the seamless phase II/III design, and to identify stopping bounds that approximately control study-wide type I error under the frequentist paradigm.

The primary goal of adaptive seamless designs is to improve the efficiency of trials and the drug development process. This has motivated also other adaptive procedures for necessary elements for an adequate and well-controlled trial, such as selection of the endpoint for the primary analysis [61]. As noted with other adaptive procedures, one must be concerned about leakage of information about the treatment effect with dissemination of the results of the phase II portion, and the risk of disruption of clinical equipoise [7,62]. Moreover, selection based upon interim data will require good quality data at that study milestone and that there is uniformity before and after the adaptive procedure and design modifications. Some of the challenges in conducting adaptive seamless designs were highlighted in the review and case studies presented by Bauer et al. [53].

6.7 Conclusion

Ultimately, the concepts and theory behind an adaptive versus a fixed design for clinical trials were established in the last century and have spanned all phases of drug development and the variety of study objectives therein. More recently, modern methods of electronic data capture and access to computational resources for quickly solving iterative algorithms have allowed for application of the theory of adaptive designs. As such, they have begun to have a substantial impact on cancer clinical research. Challenges remain, including broader integration of adaptive procedures into clinical trial operations, the systems for enrollment and allocation of subjects, and data management. Moreover, adaptive platform trials will need special consideration in how to comply with international laws and policies on trial registration, reporting, and data sharing to promote transparency and accountability in clinical research. Adaptation can offer improved operating characteristics for established and relevant scenarios in the conduct of cancer clinical trials, but this benefit should be carefully weighed against costs and feasibility of implementation in the disease setting and by a study sponsor.

References

1. Guideline IHT. Statistical principles for clinical trials. *International Conference on Harmonisation E9 Expert Working Group.* Stat. Med., 1999;18: 1905–1942.

2. Evans SR, and Barry, WT. Interim Analysis and Data Monitoring. In *Oncology Clinical Trials—Successful Design, Conduct, and Analysis*, Kelly WK, Halabi S, eds. Springer Publishing Company: New York City, NY, 2018: 290–302.

3. Jennison C, and Turnbull, BW. *Group Sequential Methods with Applications to Clinical Trials*. Boca Raton: Chapman & Hall/CRC, 2000, xviii, 390 pp.

4. Gallo P, Chuang-Stein C, Dragalin V et al. Adaptive designs in clinical drug development—an executive summary of the PhRMA working group. *J Biopharm Stat*.2006;16(3):275–83.

5. Zelen M. Play the winner rule and the controlled clinical trial. *J A Stat Assoc*. 1969;64(325):131–46.

6. Hatfield I, Allison A, Flight L et al. Adaptive designs undertaken in clinical research: A review of registered clinical trials. *Trials*. 2016;17(1):150.

7. Cook T, and DeMets DL. Review of draft FDA adaptive design guidance. *J Biopharm Stat*. 2010;20(6):1132–42.

8. Le Tourneau C, Lee JJ, Siu LL. Dose escalation methods in phase I cancer clinical trials. *J Natl Cancer Inst*. 2009;101(10):708–20.

9. Storer BE. An evaluation of phase I clinical trial designs in the continuous dose–response setting. *Stat Med*. 2001;20(16):2399–408.

10. O'Quigley J, Pepe M, and Fisher L, Continual reassessment method: A practical design for phase 1 clinical trials in cancer. *Biometrics*. 1990;46(1):33–48.

11. Babb J, Rogatko A, and Zacks S. Cancer phase I clinical trials: Efficient dose escalation with overdose control. *Stat Med*. 1998;17(10):1103–20.

12. Cheung YK, Chappell R. Sequential designs for phase I clinical trials with late-onset toxicities. *Biometrics*. 2000;56(4):1177–82.

13. Braun TM. Generalizing the TITE-CRM to adapt for early-and late-onset toxicities. *Stat Med*. 2006;25(12):2071–83.

14. Whitehead J, Thygesen H, Whitehead A. A Bayesian dose-finding procedure for phase I clinical trials based only on the assumption of monotonicity. *Stat Med*. 2010;29(17):1808–24.

15. Yin G, Li Y, Ji Y. Bayesian dose-finding in phase I/II clinical trials using toxicity and efficacy odds ratios. *Biometrics*. 2006;62(3):777–87.

16. Simon R, Rubinstein L, Arbuck SG et al. Accelerated titration designs for phase I clinical trials in oncology. *J Natl Cancer Inst*. 1997;89(15):1138–47.

17. Goodman SN, Zahurak ML, Piantadosi S. Some practical improvements in the continual reassessment method for phase I studies. *Stat Med*. 1995;14(11):1149–61.

18. Simon R. Optimal 2-stage designs for phase-II clinical-trials. *Control Clin Trials*. 1989;10(1):1–0.

19. Barry WT. Trial designs and biostatistics for molecular-targeted agents. In *Breast Cancer—Innovations in Research and Management*, Veronesi U, Goldhirsch A, Veronesi P, Gentilini OD, Leonardi MC, eds. Cham, SW: Springer International Publishing, 2017: pp. 915–24.

20. Mandrekar SJ, Sargent DJ. Clinical trial designs for predictive biomarker validation: Theoretical considerations and practical challenges. *J Clin Oncol*. 2009;27(24):4027–34.

21. Jones CL, Holmgren E. An adaptive Simon Two-Stage Design for Phase 2 studies of targeted therapies. *Contemp Clin Trials*. 2007;28(5):654–61.

22. Parashar D, Bowden J, Starr C et al. An optimal stratified Simon two-stage design. *Pharm Stat*. 2016;15(4):333–40.

23. Karuri SW, Simon R. A two-stage Bayesian design for co-development of new drugs and companion diagnostics. *Stat Med*. 2012;31(10):901–14.

24. Freidlin B, Jiang W, Simon R. The cross-validated adaptive signature design. *Clin Cancer Res*. 2010;16(2):691–8.

25. Simon N, Simon R. Adaptive enrichment designs for clinical trials. *Biostatistics*. 2013;14(4):613–25.

26. Muller P, Parmigiani G, Rice K. FDR and Bayesian multiplecomparisons rules. Johns Hopkins University, Dept. of Biostatistics Working Papers. Working Paper 115, 2006.

27. Wei LJ, Durham S. Randomized play-winner rule in medical trials. *J Am Stat Assoc*. 1978;73(364):840–3.

28. Ivanova A. A play-the-winner-type urn design with reduced variability. *Metrika*. 2003;58(1):1–3.

29. Rosenberger WF, Stallard N, Ivanova A et al. Optimal adaptive designs for binary response trials. *Biometrics.* 2001;57(3):909–13.
30. Hu FF, Zhang LX. Asymptotic properties of doubly adaptive biased coin designs for multitreatment clinical trials. *Ann Stat.* 2004;32(1):268–301.
31. Thall PF, Wathen JK. Practical Bayesian adaptive randomisation in clinical trials. *Eur J Cancer.* 2007;43(5):859–66.
32. Ventz S, Barry WT, Parmigiani G et al. Bayesian response-adaptive designs for basket trials. *Biometrics.* 2017;73(3):905–15.
33. Thall PF, Wathen JK, Bekele BN et al. Hierarchical Bayesian approaches to phase II trials in diseases with multiple subtypes. *Stat Med.* 2003;22(5):763–80.
34. Chugh R, Wathen JK, Maki RG et al. Phase II multicenter trial of imatinib in 10 histologic subtypes of sarcoma using a Bayesian hierarchical statistical model. *J Clin Oncol.* 2009;27(19):3148–53.
35. Kim ES, Herbst RS, Wistuba II et al. The BATTLE Trial: Personalizing therapy for lung cancer. *Cancer Discov.* 2011;1(1):44–53.
36. Barker AD, Sigman CC, Kelloff GJ et al. I-SPY 2: An adaptive breast cancer trial design in the setting of neoadjuvant chemotherapy. *Clin Pharmacol Ther.* 2009;86(1):97–100.
37. Zhou X, Liu S, Kim ES et al. Bayesian adaptive design for targeted therapy development in lung cancer—a step toward personalized medicine. *Clin Trials.* 2008;5(3):181–93.
38. Albain KS, Leyland-Jones B, Symmans F et al. Abstract P1-14-03: The evaluation of trebananib plus standard neoadjuvant therapy in high-risk breast cancer: Results from the I-SPY 2 TRIAL. *American Association for Cancer Research.* 2016; 76(4):P1-14-03–P1-14-03.
39. Rugo HS, Olopade OI, DeMichele A et al. Adaptive randomization of veliparib–carboplatin treatment in breast cancer. *N Engl J Med.* 2016;375(1):23–34.
40. Park JW, Liu MC, Yee D et al. Adaptive randomization of neratinib in early breast cancer. *N Engl J Med.* 2016;375(1):11–22.
41. De Angelis C, Drazen JM, Frizelle FA et al. *Clinical Trial Registration: A Statement from the International Committee of Medical Journal Editors.* 2004, Mass Medical Soc.
42. Esserman LJ, Berry DA, Cheang MCU et al. Chemotherapy response and recurrence-free survival in neoadjuvant breast cancer depends on biomarker profiles: Results from the I-SPY 1 TRIAL (CALGB 150007/150012; ACRIN 6657). *Breast Cancer Res Treat.* 2012;132(3):1049–62.
43. Berry DA. Adaptive clinical trials: The promise and the caution. *J Clin Oncol.* 2011;29(6):606–9.
44. Korn EL, Freidlin B. Outcome--adaptive randomization: Is it useful? *J Clin Oncol.* 2011;29(6):771–6.
45. Lewis JA. Statistical principles for clinical trials (ICH E9): An introductory note on an international guideline. *Stat Med.* 1999;18(15):1903–42.
46. Wittes J, Brittain E. The role of internal pilot studies in increasing the efficiency of clinical trials. *Stat Med.* 1990;9(1–2):65–72.
47. Coffey CS, Muller KE. Controlling test size while gaining the benefits of an internal pilot design. *Biometrics.* 2001;57(2):625–31.
48. Chi L, Hung H, Wang SJ. Modification of sample size in group sequential clinical trials. *Biometrics.* 1999;55(3):853–7.
49. Chen Y, DeMets DL, Gordon Lan K. Increasing the sample size when the unblinded interim result is promising. *Stat Med.* 2004;23(7):1023–38.
50. Denne JS. Sample size recalculation using conditional power. *Stat Med.* 2001;20(17–18):2645–60.
51. Mehta CR, Pocock SJ. Adaptive increase in sample size when interim results are promising: A practical guide with examples. *Stat Med.* 2011;30(28):3267–84.
52. Muss HB, Berry DA, Cirrincione CT et al. Adjuvant chemotherapy in older women with early-stage breast cancer. *N Engl J Med.* 2009;360(20):2055–65.
53. Bauer P, Bretz F, Dragalin V et al. Twenty-five years of confirmatory adaptive designs: Opportunities and pitfalls. *Stat Med.* 2016;35(3):325–47.
54. Ravandi F, Ritchie EK, Sayar H et al. Vosaroxin plus cytarabine versus placebo plus cytarabine in patients with first relapsed or refractory acute myeloid leukaemia (VALOR): A randomised, controlled, double-blind, multinational, phase 3 study. *The Lancet Oncol.* 2015;16(9):1025–36.

55. Thall PF, Simon RM, Estey EH. Bayesian sequential monitoring designs for single-arm clinical trials with multiple outcomes. *Stat Med.* 1995;14(4):357–9.
56. Thall PF, Cook JD. Dose-finding based on efficacy–toxicity trade-offs. *Biometrics.* 2004;60(3):684–93.
57. Ivanova A. A new dose-finding design for bivariate outcomes. *Biometrics.* 2003;59(4):1001–7.
58. Korn EL, Freidlin B, Abrams JS et al. Design issues in randomized phase II/III trials. *J Clin Oncol.* 2012;30(6):667.
59. Dunnett CW. A multiple comparison procedure for comparing several treatments with a control. *J Am Stat Assoc.* 1955;50(272):1096–121.
60. Inoue LY, Thall PF, Berry DA. Seamlessly expanding a randomized phase II trial to phase III. *Biometrics.* 2002;58(4):823–31.
61. Sugitani T, Hamasaki T, Hamada C. Partition testing in confirmatory adaptive designs with structured objectives. *Biom J.* 2013;55(3):341–59.
62. Wang SJ, James Hung H, O'Neill RT. Impacts on type I error rate with inappropriate use of learn and confirm in confirmatory adaptive design trials. *Biom J.* 2010;52(6):798–810.
63. Kass RE, Steffey D. Approximate Bayesian inference in conditionally independent hierarchical models (parametric empirical Bayes models). *J Am Stat Assoc.* 1989;84(407):717–26.

Section II

Late Phase Clinical Trials

7

Sample Size Calculations for Phase III Trials in Oncology

Koji Oba and Aya Kuchiba

CONTENTS

7.1 Introduction

One day, I received the following e-mail from a collaborator:

"Dear Dr Oba,

Thank you for your support of the research. Could you do me a favor?

Last week, new results comparing AA vs BB in adult patients were published in *New England Journal of Medicine*. In the overall population, the study did not detect a significant difference in survival between patients treated with AA and BB but a subgroup analysis suggested an increased treatment benefit in the younger population. It is very interesting, and I would like to conduct a new randomized trial targeting the young population.

So, how many patients do we need for the trial?

Best regards,

XXX"

Although I was very happy to receive such an exciting e-mail, my answer was "I am sorry, but I do not know the answer without receiving extra information."

Sample size calculation is one of several crucial points when researchers plan a new trial because the number of patients directly relates to the feasibility of the study (e.g., how many investigators can actually enroll patients, or can the investigators acquire the necessary funding). In addition, designing trials that are too large or too small (relative to the objective) may be considered unethical [1]. Especially in phase III clinical trials (sometimes defined as a therapeutic confirmatory trial [2]), it is important to assure high probability (statistical power) to confirm an assumed difference in each study hypothesis as a statistically significant difference, if such a difference exists.

To calculate the sample size is not a simple matter of obtaining a number using a mathematical formula. It should be based on several discussions between investigators and biostatisticians. During the preliminary stage, the investigators establish the study null hypothesis that they want to test in the clinical trial and the clinically relevant treatment effect to detect. Biostatisticians can determine a rough estimate of the sample size needed, depending on the study hypothesis. If the study hypothesis has been defined, it will be important to refine the supportive evidence for the sample size calculation until a consensus is reached regarding the final number of patients selected for the study, which should then be included and justified in the final protocol.

The global reporting guidelines for randomized controlled trials, i.e., the Consolidated Standards of Reporting Trials (CONSORT), state that the description of sample size calculation is a mandatory component when publishing a randomized controlled trial [3,4]. However, several authors have pointed out inadequate reporting in several therapeutic areas [5,6]. In phase III oncology trials, the quality of reporting regarding the sample size calculation is also discussed frequently. Bariani et al. [7] reviewed the frequency with which reports of phase III oncology trials specified the parameters required for sample size calculation. Firstly, approximately 64% of primary endpoints in phase III oncology trials were time to event in nature (e.g., overall survival, progression-free survival, etc.). Secondly, 93% of trials were superiority trials, although non-inferiority trials have become increasing prevalent in recent years. Lastly, only 28% of trials provided all the required parameters for proper sample size calculations.

To design a well-crafted trial and improve the quality of reporting, it is important to share and understand among investigators, including biostatisticians, what kinds of parameters are needed to calculate the sample size for phase III trials in oncology. This chapter focuses on randomized parallel controlled (1:1) phase III trials whose primary endpoint is time-to-event data. In addition, this chapter addresses two types of phase III trials whose objectives are superiority or non-inferiority, with examples from actual trials. Programming codes in SAS (SAS Institute, Cary, NC, USA), which are available on the companion website, and software for sample size calculation and online sample size calculators are also introduced. (see https://www.crcpress.com//9781138083776).

7.2 Basics of Sample Size Calculation in Phase III Oncology Trials

7.2.1 Required Parameters and Settings

Several parameters are needed to calculate a sample size for clinical trials. As shown in Table 7.1, such parameters correspond to the variable type of the primary endpoint of the trial.

Inspection of Table 7.1 shows that an α error, a power ($1-\beta$ error), and specification of a one-sided or two-sided test are common for all types of primary endpoints. If the primary endpoint is set to a time-to-event data, as is typically done in phase III oncology trials, then the assumed hazard ratio (HR) in the planned trial, absolute risk of the reference group, duration of accrual, and duration of follow-up (or total trial duration) are additionally required.

Such additional parameters (target or clinically relevant HR to detect in the planned trial, absolute risk of the reference group, duration of accrual, and duration of follow-up) are required because the amount of "information" obtained from the trial depends on the number of events in the survival analysis. Therefore, the number of events is calculated first using the target HR, and the total sample size needs to be back-calculated using the absolute risk of the control group, accrual, and duration of the follow-up.

7.2.2 Relationships among Survival Parameters

Under the assumption of an exponential curve, absolute risks are variously expressed, and they relate to the *HR*. The simplest relationship is between the target *HR* and hazard

TABLE 7.1

Required Parameters for Sample Size Calculations under a 1:1 Randomized
Phase III Trial

Type of the Primary Endpoint	Required	Optionally Required
Continuous	α error	Drop-out rate
	Power (= 1–β error)	Planned statistical test
	Difference of means between groups (or mean of each group)	
	Pooled standard deviation (SD)	
	One-sided or two-sided test	
Binary	α error	Drop-out rate
	Power (= 1–β error)	Planned statistical test
	Proportion in each group	
	One-sided or two-sided test	
Time-to-event	α error	Drop-out rate
	Power (= 1–β error)	Accrual rate
	Hazard ratio between groups	Planned statistical test
	Absolute risk of the control group	
	Duration of accrual	
	Duration of follow-up	
	One-sided or two-sided test	

rates of the control (λ_c) and experimental (λ_e) groups. If the investigator wants to calculate the HR of the experimental group compared to the control group, the target HR equals to λ_e/λ_c. Secondly, the survival probability at a specific time point is also converted to the HR. Setting the assumed survival probability at time t as $S_c(t)$ and $S_e(t)$ for each group, the relationship between each survival probability and each hazard rate is defined as

$$S_c(t) = e^{-\lambda_c \times t} \leftrightarrow \lambda_c = -\log\{S_c(t)\}/t$$

$$S_e(t) = e^{-\lambda_e \times t} \leftrightarrow \lambda_e = -\log\{S_e(t)\}/t$$

From these two formulas, the relationship between the assumed *HR* and survival probability is obtained as

$$HR = \lambda_e/\lambda_c = \log\{S_e(t)\}/\log\{S_c(t)\}$$

The cumulative incidence function $I_e(t)$ and $I_c(t)$ can be obtained instead of the survival probability. In the absence of competing risks (see Chapter 21), it is easily deduced by replacing $S_e(t)$ and $S_c(t)$ with $1 - I_e(t)$ and $1 - I_c(t)$, respectively, in the above formulas. In some medical publications only the median survival times (*MSTs*) are reported. If the survival curve can be assumed to follow an exponential distribution, the *MST* and target HR also share a simple relationship:

$$HR = MST_c/MST_e$$

These conversions are often required, depending on the reported statistics in a previous publication, when calculating the sample size for a new trial.

7.2.3 Basic Parameters: α Error, Power, and HR

Clinical trials utilize probabilistic decisions, based on the clinical data obtained, for statistical hypothesis testing under the pre-specified "null hypothesis." For example, the null hypothesis H_0, which the researcher would like to reject, will be set before calculating the sample size:

H_0: there is no difference in survival functions between treatment A and treatment B (i.e., $HR = 1.00$).

After obtaining the data, the probability of obtaining the study result or a more extreme result if the null hypothesis is true is calculated. This is the well-known p-value. If the calculated p-value is under a pre-specified threshold (usually 5% for a two-sided test), then the investigator considers that the observed difference is unlikely if there genuinely was no difference between treatment A and treatment B, and thereby rejects the null hypothesis and concludes that a statistically significant difference occurred. Although this decision may seem to be explicit, it should be noted that it is possible that 5% of the time (if 5% was chosen as the threshold), the investigator wrongly concludes that a survival difference occurs between treatment A and B, even if both treatments are the same (false-positive decision). This is an inevitable error, based on the probabilistic decision. In this case, this error is referred to as an α error (or type I error).

Alternatively, investigators may expect or assume a certain effect before planning the trial based on the accumulated evidence to date. In general, an alternative (or working) hypothesis H_1 will be set corresponding to the null hypothesis, as follows:

H_1: there is a difference in survival functions between treatment A and treatment B (i.e., $HR \neq 1.00$).

This hypothesis is sometimes called a "composite" alternative hypothesis because all hypothesized values (except for the null value) are included in this hypothesis. Actually, investigators need to determine a target HR under the alternative hypothesis to power a study based on previous evidence. This target value should also be considered clinically relevant to patient management or be based on a judgement concerning the anticipated effect of the new treatment [8]. In addition to the type I, or α, error, a false-negative β error (or type II error) occurs when, at the end of the study, the null hypothesis is not rejected when a genuine difference in survival functions exist. This value is typically controlled at 10%–20% in the phase III trial by controlling the sample size. The errors based on the probabilistic decision are summarized in Figure 7.1.

The sample size calculation can be considered to obtain adequate statistical power to detect a statistically significant difference under the alternative hypothesis for a particular target HR. It is also important to recognize that if a larger target HR is used, the calculated sample size becomes smaller. If a smaller true HR is likely to be detected with adequate power, then a larger sample size will be required in such a phase III trial.

If an exponential curve can be assumed for the survival curve, the following simple sample size formula can be derived using the above parameters only (sometimes referred

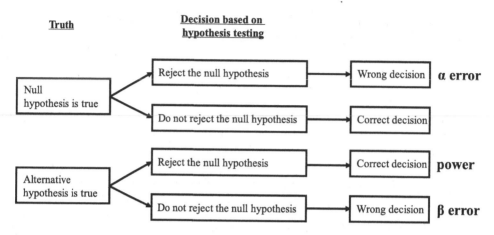

FIGURE 7.1
Summary of error types, based on statistical testing.

to as Schoenfeld's formula) [9]. Letting $z_{1-\alpha/2}$ and $z_{1-\beta}$ denote, respectively, $100(1-\alpha/2)\%$ and $100(1-\beta)\%$ quantiles of the standard normal distribution, and $\log(HR)$ be set as an assumed natural log-transformed value of the target HR, the required total number of events d (not the sample size) for two groups is calculated as follows:

$$d = 2 \times \frac{2(z_{1-\alpha/2} + z_{1-\beta})^2}{\log(HR)^2}$$

In general, the α error is fixed at the 5% level for two-sided tests in major journals or regulatory agencies. This means that $z_{1-0.05/2} = 1.96$ could be substituted in the above formula. Since the β error is often controlled at 0.2 or 0.1 (as explained above), this results in values of $z_{0.8} = 0.84$ and $z_{0.9} = 1.26$. Thus, if we choose the parameters of $\alpha = 0.05$ and $\beta = 0.20$, then $(z_{1-0.05/2} + z_{1-0.20})^2 = 7.84 \approx 8$. If we choose the parameters of $\alpha = 0.05$ and $\beta = 0.10$, then $(z_{1-0.05/2} + z_{1-0.10})^2 = 10.5$. These approximations make Schoenfeld's formula much simpler (it is sometimes called a "quick" formula):

$$d = \frac{42}{\log(HR)^2} \quad \text{if power was set 90\%;}$$

$$d = \frac{32}{\log(HR)^2} \quad \text{if power was set 80\%}$$

If readers have a smart phone and can calculate natural log of the HR, then the required total number of events can be calculated by hand (Table 7.2).

Please note again that the required events, and not the sample size, are calculated for the time-to-event endpoints based on the above formula. Total sample size will be back-calculated using the incidence of the events during the trial, as explained in the Section 7.2.4. Of course, since the "quick" formula uses an approximation, it is mainly recommended for use in quick discussions among investigators. Other parameters related to the sample size calculation also need to be considered in practice.

TABLE 7.2

Required Number of Events Based on the "Quick" Formula with $\alpha = 0.05$

True Hazard Ratio (HR)	{log(HR)}²	Power (%)	Total Events
0.5	$(-0.693)^2 = 0.48$	80	67
0.6	$(-0.511)^2 = 0.26$	80	123
0.7	$(-0.357)^2 = 0.13$	80	252
0.8	$(-0.223)^2 = 0.05$	80	643
0.9	$(-0.105)^2 = 0.01$	80	2,883
0.5	$(-0.693)^2 = 0.48$	90	88
0.6	$(-0.511)^2 = 0.26$	80	161
0.7	$(-0.357)^2 = 0.13$	80	330
0.8	$(-0.223)^2 = 0.05$	80	844
0.9	$(-0.105)^2 = 0.01$	80	3,784

7.2.4 Sample Size Calculations Using Additional Parameters

In a trial whose primary endpoint is a time-to-event data, all participants do not experience the event by the end of the trial. Such participants are censored during or at the end of follow-up. Accordingly, investigators need to consider the incidence probability (or survival probability) under the total study period. For example, assuming the target HR is set at 0.7 with $\alpha = 0.05$ and $\beta = 0.10$, 330 total events will be required for two groups, based on the "quick" formula. If readers want to calculate the total sample size, then the incidence probability of each group should be determined by back-calculation. Based on previous evidence, for example, the 5-year overall survival may be assumed as 60% in the control group (5-year probability of event of 40%). In this case, a hazard rate λ_c of the control group is calculated as $0.4 = 1 - e^{-\lambda_c \times 5} \leftrightarrow \lambda_c = -\log(0.60)/5 = 0.1022$ if an exponential distribution is assumed. Thus, the hazard λ_e of the experimental group is back-calculated as $0.7 \times 0.1022 = 0.0715$, and the 5-year probability of an event in the experimental group becomes $(1 - e^{-0.0715 \times 5}) \times 100 = 30\%$. In this example, the sample size is calculated as

$$330 \times \frac{2}{0.4 + 0.3} = 942.9 (\approx 944) \text{ for two groups (472 participants per group).}$$

However, the above calculation assumes that all participants were followed for 5 years and censored at the end of follow-up. In a real clinical trial, participants enter the trial in order, which is sometimes referred to as a staggered entry [10]. The total study period represents a total of the accrual and follow-up periods. Therefore, the above calculation tends to underestimate the true requirement for the sample size. Collett derived the expected probability of event for a given accrual and follow-up period [11]. If the duration of accrual, noted as a, the duration of follow-up after enrollment was completed, noted as f, the estimated probability of no event until time t, noted as $S_c(t)$ and $S_e(t)$ in the control and experimental groups, respectively, and the accrual rate is constant during the accrual period, then the cumulative incidence probability, Pr(event), is approximately described using the following formula (see Collett's textbook for a detailed derivation of the formula):

$$\text{Pr(event)} = 1 - \frac{1}{6} \left\{ \frac{S_c(f) + S_e(f)}{2} + 4 \times \frac{S_c(0.5a + f) + S_e(0.5a + f)}{2} + \frac{S_c(a + f) + S_e(a + f)}{2} \right\}$$

Returning to the above hypothesized example, if the accrual period was 2 years, the follow-up period after enrollment was completed was 3 years (total study period becomes 5 years), and the other conditions were the same, then how many participants are required? If Collett's approximation formula is applied, then it is necessary to prepare estimated 3-year, 4-year, and 6-year survival probability values for each group. In our example, the survival probability of each time point is estimated as

$$\frac{0.73 + 0.81}{2} = 0.77 \text{ for 3-year overall survival}$$

$$\frac{0.66 + 0.75}{2} = 0.71 \text{ for 4-year overall survival}$$

$$\frac{0.60 + 0.70}{2} = 0.65 \text{ for 5-year overall survival}$$

Thus, the Pr(event) during the study course (accrual plus follow-up) can be calculated as 0.29, and the required sample size becomes $330/0.29 = 1133$ (≈ 1134) for two groups (567 participants per group). The required sample size became larger than the previous determination because the observed person years are shorter than in the previous example, which increased the total sample size needed to observe the required number of events of 330.

7.2.5 Sample Size Calculations Based on the Log-Rank Test

The log-rank test is a non-parametric test that does not require any functional assumptions for the survival curve or proportional hazards assumption; thus, this test is often considered as a gold standard for survival analysis in phase III oncology trials. Therefore, some readers may consider that Schoenfeld's formula is applicable for trials in which the primary analysis is a log-rank test. Interestingly, the required number of events based on log-rank test statistics is identical to the results obtained with Schoenfeld's formula, which assumes an exponential distribution for the survival curve and proportional hazards between groups [11]. Therefore, Schoenfeld's formula corresponds to the planned statistical test (i.e., the log-rank test). However, this does not mean that the log-rank test requires the same assumptions set in Schoenfeld's formula. It should be noted that the log-rank test has an adequate power to detect differences between groups if the assumptions of Schoenfeld are fulfilled but cannot achieve the nominal level of statistical power if such assumptions are violated. If some assumptions of Schoenfeld's formula are not met, it is necessary to find another extended solution that relaxes them.

If the assumption of an exponential distribution for the survival curve is questionable, it is recommended to employ another method that does not assume such a distribution. Lakatos proposed a sample size formula for the log-rank test under more complex conditions (e.g., non-exponential type survival curve, loss to follow-up, lag time, etc.) [12]. For example, Lakatos's method allows for a piecewise linear model for the survival-curve data and relaxing the exponential assumption. Recently, this method was implemented for sample size calculation for the log-rank test in many commercial sample size-calculation software programs, which we introduce in the next Section 7.3 (e.g., SAS Power procedure, PASS, nQuery, etc.).

7.3 Software for Sample Size Calculations

Although it is possible to calculate a rough estimation of the sample size by hand or with an electronic calculator, it is reasonable to use commercial or validated software developed for sample size calculations. Several analytical software programs (SAS, JMP, SPSS, Stata, R, etc.) have their own sample size/power calculation functions. PASS (https://www.ncss.com/software/pass/) and nQuery (https://www.statsols.com/nquery) are also well-known software programs for sample size calculations. They involve a moderate cost, and they provide several sample size tools for many types of statistical tests. Last, some free online calculators are available for researchers. The South West Oncology Group (SWOG) statistics and data management center periodically releases a statistical tool for sample size calculations (https://stattools.crab.org/). This website is well known in the oncology field, and some groups have used the tool for designing phase III clinical trials. However, it is sometimes unclear whether these online tools have been validated. Thus, researchers should be careful to select a valid system when such tools are applied. Lee and Chen reviewed software for sample size calculations in oncology [13], and their result is helpful for understanding the software and tools in detail.

Regardless of the software or website that is used for sample size calculations, it is important for users to understand which method of sample size calculations has been applied. In the next section, the sample size calculation methods used for survival analysis will be examined for the (1) SAS Power procedure with the TWOSAMPLESURVIVAL statement, (2) PASS survival procedures, and (3) SWOG statistical tool.

7.4 Superiority Trials

7.4.1 Purpose of Superiority Trials

Determining the primary objective of a clinical phase III trial is the first step in designing the trial. Superiority trials are defined as trials with the primary objective of showing that the response to the investigational product is superior to a comparative agent (active or placebo control) [8]. The judgement of superiority usually relies upon hypothesis testing. Thus, the methods of sample size calculation explained so far are applicable to superiority trials.

7.4.2 The Sample Size Calculation Methods Used in Various Software Programs

7.4.2.1 SAS Power Procedure: TWOSAMPLESURVIVAL Statement

SAS has a function for calculating the sample size for a superiority trial involving time-to-event data. This function is referred to as the TWOSAMPLESURVIVAL statement in the Power procedure. This procedure uses Lakatos's method. Not only a log-rank test, but also Gehan (generalized Wilcoxon) [14] or Tarone–Ware test [15] is applicable as a statistical test. If a single parameter for the curve is specified, an exponential curve is assumed. However, survival curves can be customized using the piecewise linear model if a multipoint curve is specified in the program. Various options presented in Table 7.3 are available. The latest version is SAS version 9.4 and SAS/STAT version 14.3 (as of May 31, 2018).

TABLE 7.3

Results of the Required Sample Size Calculated by Each Software for the EAGLE Trial

Software	Formula	Power (%)	Total Sample Size	Total Events
SAS power procedure	Lakatos's method	90	366	333
PASS log-rank test	Lakatos's method	90	367	334
PASS Cox model	Schoenfeld's formula	90	363	331
SWOG	Bernstein's method	90	361	N.D.
By hand	"Quick" formula	90	N.D.	330

Abbreviation: N.D., not determined.

7.4.2.2 PASS: Log-Rank Tests and Tests for Two Survival Curves Using Cox's Proportional Hazards Model

PASS software has 32 sample size calculation procedures for a superiority trial, based on time-to-event data. Among these procedures, two different types are discussed in this section. One is the Log-rank Tests procedure, which is based on the Lakatos's method (like the SAS Power procedure). PASS provides a user-friendly interface to enter the required parameters, along with a detailed manual and examples. PASS has four Log-rank Test procedures, depending on how HR and the related survival parameters are inputted. The second procedure is the Two Survival Curves Using Cox's Proportional Hazards Model. This procedure is based on Schoenfeld's formula and the output returns the sample size if the assumed survival probabilities were observed at the end of the trial (it does not consider the accrual and follow-up durations). The latest version is PASS 13 (as of May 31, 2018).

7.4.2.3 SWOG Statistical Tool: Two-Arm Survival

The SWOG statistical tool has one module for sample size calculations for superiority trials based on time-to-event data. The method utilized in the website was proposed by Bernstein [16]. This method is similar to Schoenfeld's formula, except that this approach considers some strata of the prognostic factors. In addition, the total sample size is calculated using a uniform accrual time and a fixed follow-up time. The website was accessed on May 31, 2018.

7.4.3 Example of a Superiority Trial (the EAGLE Trial)

The EAGLE trial was a randomized, multicenter, open-label, two-arm phase III study designed to evaluate the superiority of bevacizumab (10 mg/kg) plus FOLFIRI compared with bevacizumab (5 mg/kg) plus FOLFIRI as second-line therapy in patients with metastatic colorectal cancer who were previously treated with first-line bevacizumab plus an oxaliplatin-based regimen [17]. The sample size calculation was described in the original article as follows:

> Median progression free survival (PFS) of arm A in this trial is assumed to be 5.0 months based on previous studies and it is considered as a clinically relevant prolongation if the median PFS of arm B is 7.0 months (risk reduction 30%). At the start of this trial, the planned sample size was 280 patients to detect 30% risk reduction with 80% power for a log-rank test comparing two survival curves with a two-sided significance level of 0.05, assuming an accrual time of 2 years and a follow-up time of 1 year. This calculation was carried out by employing nQuery Advisor 7.0 software (Statistical Solutions,

Saugus, MA, USA). On 8 April 2011, an independent data monitoring committee of the EAGLE trial recommended that the statistical power be amended from 80 to 90% with the consideration of the promising enrollment of patients. As a result, 358 patients (330 events) will be needed to detect 90% power under the same assumption. Taking some dropouts into account, the sample size to be accrued was set at 370 patients in total.

Patients with advanced/metastatic colorectal cancer were enrolled in the EAGLE trial; thus, the incidence of PFS events was relatively high. The accrual rate was assumed constant and no lost to follow-up patients were assumed. Based on these assumptions, the clinically relevant target *HR* was set to 0.70 for the EAGLE trial.

7.4.4 Comparison of the Sample Size Calculated with Each Software Program

7.4.4.1 SAS Power Procedure

The program used for the Power procedures, based on the parameters assumed in the EAGLE trial (power set to 90%) was as follows:

```
proc power;
    TWOSAMPLESURVIVAL
    ALPHA=0.05 SIDES=2 POWER=0.9
    ACCRUALTIME= 24
    FOLLOWUPTIME= 12
    GROUPMEDSURVTIMES= (5 7.14)
    NTOTAL= .
    TEST= LOGRANK;
run;
```

The median PFS in the experimental group was calculated based on a target $HR = 0.70$ and a median PFS = 5 months in the control group, assuming an exponential curve. This code returns the required total sample size as 366 participants (183 per group). If NTOTAL is replaced by EVENTSTOTAL, the output returns the number of total events required. In this case, the total required number of events was 333. The sample size and event numbers are slightly larger than the results described in the explanation of the EAGLE trial, based on nQuery (which used Schoenfeld's formula with the accrual period and the follow-up period).

7.4.4.2 PASS: Log-Rank Tests (Input Median Survival Times)

In this module of PASS, it is necessary to input the power, α error, median survival of the control group, HR, accrual time with the accrual pattern, and total time (accrual time plus follow-up time). In addition, the proportion of patients lost to follow-up or that underwent a treatment switch can be inputted, although these parameters are optional. If power = 0.9, α error = 0.05, median survival of control group = 5 months, target HR = 0.70, accrual time with accrual pattern = 24 months constantly, and total time = 36 months are inputted, then the output returns 367 participants as the total sample size with 334 required events. Since the formula used in PASS: Logrank Tests is the same as with SAS, the same results are obtained, after consideration of a rounding error.

7.4.4.3 PASS: Tests for Two Survival Curves Using Cox's Proportional Hazards Model

This module is based on the proportional hazards model and only has five input fields (power, α error, probability of control event, probability of treatment event, and target *HR*). The overall probability of events during the course of the trial is given by 91.1% using Collett's approximation. In this setting, the output returns 363 participants as the total sample size, with 331 required events. This is almost the same result compared to the results from PASS: Logrank tests, although the assumed observed period is different.

7.4.4.4 SWOG Statistical Tool Website

Finally, the SWOG website requires the input of several strata, including the proportion in the standard group (proportion of allocation), α error, years of accrual, years of follow-up, target HR, power, and hazard rate of the control group if type calculation (power or sample size), input type (hazard rate or survival proportion), and the sides were set as "sample size," "hazard rate," and "2-sided," respectively. The website asks to provide the accrual and follow-up times in years, but no problems are encountered if the time scale is changed to the month. A total of 361 participants are required if the following parameters are inputted: number of strata = 1, proportion in standard group = 0.5, α error = 0.05, year accrual = 24 months, months follow-up = 12, HR = 0.7, power = 0.9, and hazard rate of standard group = 0.138.

7.4.4.5 Interpretation of the Results

The results calculated using each software program, plus results calculated based on the "quick" formula are summarized in Table 7.3.

In the case of the EAGLE trial, the assumed incidence was relatively high (most participants experienced an event) and, thus, the calculated numbers of events among software programs were very similar. The total sample size is only slightly influenced by how the accrual period is considered.

7.5 Non-Inferiority Trials

7.5.1 Purpose of Non-Inferiority Trials and Formulas to Calculate the Sample Size

Non-inferiority phase III oncology trials are becoming increasingly prevalent [18] are have a dedicated chapter in this book (Chapter 8). A non-inferiority trial is defined as a trial with the objective of demonstrating that an investigational treatment is not clinically worse than the standard of treatment by more than a pre-specified boundary (i.e., the non-inferiority margin) [19]. Such trials are useful for comparing new treatments that are thought to bring advantages over the standard treatment in terms of costs, safety, or more convenient schedules of administration, but with a clinically acceptable loss of efficacy. According to ICH E9, a one-sided confidence interval should be used in non-inferiority trials (we discuss this matter in Section 7.6.1) [8].

Usually, the null and alternative hypotheses of non-inferiority trials can be written as follows:

H_0: *HR* (of experimental/control) $\geq M$

H_1: *HR* (of experimental/control) $< M$

The M value is referred to as the non-inferiority margin. For sample size calculations, it is necessary to consider the non-inferiority margin in addition to the parameters used in superiority trials. Based on Schoenfeld's formula, the simplest form of the sample size calculation (sometimes referred to as Chow's formula [20]) is

$$d = 2 \times \frac{2(Z_{1-\alpha} + Z_{1-\beta})^2}{\{\log(HR) - \log(M)\}^2}$$

In this formula, the α value is one-sided. If we use the approximations $(z_{1-0.025} + z_{1-0.20})^2 = 7.84 \approx 8$ and $(z_{1-0.025} + z_{1-0.10})^2 = 10.5$ for the one-sided $\alpha = 2.5\%$, then the above formula becomes simpler (we refer to the simplified version as the "quick" non-inferiority formula), as with superiority trials, as follows:

$$d = \frac{42}{\{\log(HR) - \log(M)\}^2} \quad \text{if the power was set to 90\%}$$

$$d = \frac{32}{\{\log(HR) - \log(M)\}^2} \quad \text{if the power was set to 80\%}$$

In non-inferiority trials, the target HR is often assumed to be 1.0. In that case, the above formulas can be simplified to $d = 42 / \{\log(M)\}^2$ and $d = 32 / \{\log(M)\}^2$. If we set the non-inferiority margin to 1.33, then the required number of events becomes 516 for a power of 90% and 393 for a power of 80%, considering that log (1.33) = 0.0813. The total sample size can be back-calculated using an assumed event rate like in a superiority trial. Please note again that this formula is for generating a very rough estimate of the required number of events.

7.5.2 Specification of the Non-Inferiority Margin, M

Specification of the non-inferiority margin is not the main topic discussed here, but we will briefly introduce the basic idea. Two components should be considered: (1) the effect of the active control treatment in the new non-inferiority trial and (2) the fraction of the effect of the active control treatment that needs to be preserved during the investigational treatment.

Figure 7.2 shows the relationships between the effects of a placebo, an active control treatment, and an investigational treatment. Here, the historical evidence suggests that the active control treatment is better than the placebo by $M1$ as shown in Figure 7.2. The clinical non-inferiority margin should be smaller than $M1$, e.g., $M1/2$; otherwise a non-inferiority trial can accept a treatment that is less effective than a placebo. A large non-inferiority margin reflects a situation where a large loss of effect would be clinically acceptable.

Assay sensitivity is an essential property of a non-inferiority trial, as discussed in detail in ICH E10 [21]. Assay sensitivity reflects the ability of the trial to detect a pre-specified difference between the treatments. A trial with low assay sensitivity cannot reliably distinguish an effective treatment from a less effective treatment group, which can lead to a wrong conclusion regarding non-inferiority. In addition, the effect of the active control treatment versus the placebo should be at least the presumed size of the design (i.e., $M1$), if the trial had a placebo group. Trial data cannot be validated empirically unless a

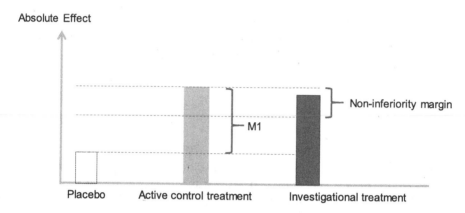

FIGURE 7.2
The relationships between the effects of a placebo, active control treatment, and investigational treatment, presuming a non-inferiority trial.

placebo group is included. A well-designed trial is always critical, but poor quality in a non-inferiority trial generally causes bias toward the alternative hypothesis, which means it incorrectly provides evidence of non-inferiority. For more details we refer to the non-inferiority chapter (Chapter 8).

7.5.3 The Sample Size-Calculation Methods Used in Each Software Program

7.5.3.1 SAS

As of May 31, 2018, SAS does not support sample size calculations for non-inferiority trials for survival analysis in the Power procedure. Thus, investigators need to write a code for the non-inferiority trial. If Schoenfeld's formula [9,22] or Freedman's formula [23] is selected, then it is possible to input simple code for the data step in SAS software.

7.5.3.2 PASS: Non-Inferiority Log-Rank Tests and Tests for Two Survival Curves Using Cox's Proportional Hazards Model

PASS has three sample size calculation procedures for non-inferiority trials based on time-to event data. One is the Non-inferiority Log-rank Tests procedure. This procedure uses the sample size calculation proposed by Jung et al., which is not explained here, but this method is considered the standard calculation method for simple non-inferiority trials [24]. The second procedure is the Non-inferiority Tests for Two Survival Curves Using Cox's Proportional Hazards Model. This procedure is based on Chow et al.'s textbook [20], which summarizes the work of Schoenfeld [9,22]. The output returns the sample size if the assumed survival probabilities were observed at the end of the trial (it does not consider the accrual and follow-up durations). The latest version of PASS is version 13 (as of May 31, 2018).

7.5.3.3 SWOG Statistical Tool: Two-Arm Survival

The SWOG statistical tool has one module for sample size calculations for non-inferiority trials based on time-to-event data. The formula used in the SWOG calculator is the same as the one explained above (Section 7.5.1). The website was accessed on May 31, 2018.

7.5.4 Example Trial (JCOG0404 Trial)

Although the benefits of laparoscopic surgery (compared with open surgery) have been suggested, the long-term survival of patients undergoing laparoscopic surgery for colon cancer requiring Japanese D3 dissection remains unclear. The JCOG0404 trial was a prospective randomized, open label, phase III trial intended to establish the non-inferiority of laparoscopic surgery compared to open surgery in patients with stage II or III colon cancer [25]. The following description of the sample size calculation was provided in the report by Kitano et al. [25]:

> The initial sample size needed to observe 254 deaths at the primary analysis was 818 patients, which was determined with a one-sided alpha of 0.05, power of 0.80, and non-inferiority margin for a *HR* of 1.366. This power calculation was based on an expected 5-year overall survival of 75%. However, during the patient accrual period, it became apparent that this was an underestimate of 5-year overall survival for this patient population; therefore, the sample size was increased to 1050 patients based on a projected 5-year overall survival of 82% to maintain the required statistical power. We planned to recruit patients for 4.5 years and follow-up patients for 5 years.

Compared to the EAGLE trial, the event probability in the JCOG0404 trial was relatively low because it was conducted in an adjuvant setting after curative resection. In addition, this trial used a one-sided value of $\alpha = 5\%$. Thus, $z_{1-\alpha} = z_{1-0.05} = 1.68$ should be used instead of $z_{1-0.025} = 1.96$. The non-inferiority "quick" formulas are $d = 35/\{\log(M)\}^2$ and $d = 25/\{\log(M)\}^2$ for a power of 90% or 80%, respectively.

7.5.5 Comparison of Sample Sizes Calculated with Each Software Program

7.5.5.1 SAS Power Procedure

The following SAS code has been adapted from a sample size textbook [26].

```
data sample;
    alpha=0.05; beta=0.20; *alpha is set as one-sided;
    Spc=0.82; Spe=0.82; time=5; *Spc and Spe: survival probability at
    time t for both group;
    accrual=4.5; followup=5;
    margin=1.366; *margin: noninferiority margin;

    lambdac=-log(Spc)/time;lambdae=-log(Spe)/time;
    hr=lambdae/lambdac;
    za=probit(1-alpha); zb=probit(1-beta);

    a=((exp(-lambdac*followup))+(exp(-lambdae*followup)))/2;
    b=((exp(-lambdac*(accrual/2+followup)))+(exp(-lambdae*(accrual/2+foll
    owup))))/2;
    c=((exp(-lambdac*(accrual+followup)))+(exp(-lambdae*(accrual+follo
    wup))))/2;
    Pr_event=1-(a+4*b+c)/6;

    Dtotal_S=4*(za+zb)**2/((log(hr)-log(margin))**2);
    Ntotal_S=Dtotal_S/Pr_event;

    Dtotal_F=((za+zb)**2)*((hr+margin)**2)/((hr-margin)**2);
    Ntotal_F=Dtotal_F/Pr_event;
run;
proc print;run;
```

Dtotal_S and Ntotal_S are, respectively, the required number of events and the sample size based on Schoenfeld's formula. This code returns the required total sample size and number of events as 1,020.8 ($\approx$ 1,022) participants and 254.2 ($\approx$ 255) events. In addition, Dtotal_F and Ntotal_F are, respectively, the required number of events and the sample size based on Freedman's formula. The required total sample size and number of events become 1,037.4 ($\approx$ 1,038) participants and 258.4 ($\approx$ 259) events

7.5.5.2 PASS: Non-Inferiority Log-Rank Tests

In this module of PASS, it is necessary to input the power, α error, non-inferiority margin or "*HR* of equivalence" as denoted by the program, hazard rate of control group, accrual time with accrual pattern, and total time (accrual time plus follow-up time). In addition, the proportion of patients lost to follow-up or undergoing a treatment switch can be inputted, although these parameters are optional. If power = 0.8, one-sided α error = 0.05, *HR* of equivalence = 1.366, hazard rate of control group = 0.00331, accrual time with accrual pattern = 54 months constantly, and total time = 114 months are inputted, then the output returns 1,022 participants as the total sample size with 255 required events. These results are identical to the results based on the Schoenfeld's formula in the SAS Data step.

7.5.5.3 PASS: Non-Inferiority Tests for Two Survival Curves Using Cox's Proportional Hazards Model

Similar to the comparator in the superiority trial, this module is based on the proportional hazards model and only involves five input fields (power, α error, probability of control event, probability of treatment event, and *HR*) plus the non-inferiority *margin*. If power = 0.8, one-sided α error = 0.05, probability of control event = 0.25, probability of treatment event = 0.25, and HR = 1.00, and "HR of equivalence" = 1.366, then the output returns 1,018 participants as the total sample size with 255 required events.

7.5.5.4 SWOG Statistical Tool Website

Finally, the SWOG website requires the input of the proportion of patients in the experimental group (proportion of allocation), α error, years of accrual, years of follow-up, target HR, hazard rate of the experimental group, the non-inferiority margin called "ratio defining equivalence," and power (if the sample size was selected as a target). Similar to the other example, if the proportion in experimental group = 0.5, α error = 0.05, year accrual = 54, year follow-up = 60, target HR = 1.0, hazard rate of experimental group = 0.00331, ratio defining equivalence = 1.366, and power = 80%, then 1,022 participants are required in total.

7.5.6 Interpretation of the Results

All results give very similar sample sizes (or required numbers of events) for this setting (see Table 7.4). It is known that Schoenfeld's formula can be slightly biased in terms of power with an unbalanced randomization and that the bias increases as two arms become more unbalanced or the projected non-inferiority margin becomes farther from 1 [24].

TABLE 7.4

Results of the Required Sample Size Calculated by Each Software for the JCOG0404 Trial

Software	Formula	Power (%)	Total Sample Size	Total Events
SAS Data Step	Schoenfeld's formula	80	1,022	255
PASS log-rank test	Jung's method	80	1,022	255
PASS Cox model	Schoenfeld's formula	80	1,018	255
SWOG	Schoenfeld's formula	80	1,022	N.S.
By hand	Non-inferiority "Quick" formula	80	N.D.	257

Abbreviation: N.D., not determined.

7.6 Other

7.6.1 Consideration for One-Sided or Two-Sided Tests

The choice of a one-sided or two-sided test when planning a clinical trial is closely related to the research question that the trial addresses. Here, we will discuss the choice of one-sided versus two-sided tests, while considering typical situations in clinical trials. Suppose that two treatments are in fact equivalent, e.g., both can be considered as current standards. We are interested in which treatment is statistically better, because it is better if one treatment can serve as the only standard. We should also confirm whether either one is significantly worse, because if either one is significantly worse, then such treatment can no longer be used as the standard. If no statistical significance is observed in either case (i.e., better or worse), then both two treatments can be considered as standard. In this situation, a two-sided test is appropriate (Figure 7.3a).

Another situation can be considered, in which the purpose of the trial is to compare the new treatment to the current standard treatment. It is important to decide whether the new

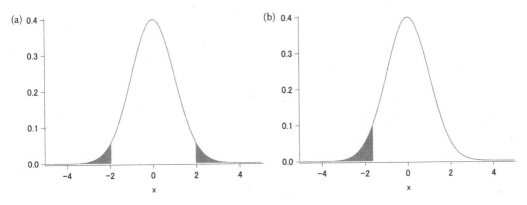

FIGURE 7.3

The distribution of test statistics under the null hypothesis and the rejection regions of two-sided and one-sided tests. Both curves are based on a standard normal distribution. The gray area is the rejection region corresponding to the significance level of 5% for a two-sided test (a) or a one-sided test (b). In Figure (a), sum of the gray areas is 5%. In Figure (b), the rejection region is defined in lower (left) tail only.

treatment can replace the standard treatment. In this situation, the question of whether the new treatment is better requires statistical justification. It may not be necessary to confirm whether the new treatment performs worse than the standard treatment because if the new treatment offers no significant improvement over the standard treatment, regardless of how much worse the new treatment is, then the current standard treatment can remain as the standard. In other words, the statistical justification for worse performance will not affect clinical decisions. In this situation, a one-sided test may be appropriate (Figure 7.3b).

Note that one-sided tests have a seemingly lenient threshold for judging statistical significance under the same significance level, but this does not provide a rationale for choosing which test should be used. As we have seen above, choosing a one-sided or two-sided test depends on the objective of the trial. If the statistical design of clinical trials is linked to the research questions, then the results can contribute to decision-making for further research or clinical practice.

7.6.2 Violation of the Proportional-Hazards and Exponential-Curve Assumptions

For the EAGLE trial (superiority trial) and the JCOG0404 trial (non-inferiority trial), the planned sample sizes were very close for all calculation methods. However, this is only valid under the assumption of proportional hazards and exponential distribution with a constant hazard rate, as noted above. Recently, such assumptions have been considered questionable. For example, Robert et al. evaluated immunotherapy with ipilimumab plus dacarbazine compared to dacarbazine alone in patients with previously untreated metastatic melanoma [27]. In this trial, the survival curves for overall survival (OS) and PFS was almost the same for the first 3 or 4 months but began to diverge after 4 months. Such a phenomenon is referred to as a delayed treatment effect in the literature. Such situations clearly challenge the proportional-hazards and exponential-curve assumptions. In this case, the traditional sample size calculation fails to achieve a planned nominal power and it is dangerous to start the phase III trial. Biostatisticians need to consider another solution for not only the analytical method, but also the sample size calculation. When planning such trials, it is advisable to use formula-based approaches only for a first orientation and to always examine various alternative scenarios using computer simulations.

7.7 Conclusion

In this chapter, we sought to share knowledge regarding how sample sizes are calculated in phase III oncology trials using a time-to-event data as a primary endpoint. Finally, we want to stress that the sample size calculation is associated with the feasibility. Even if the study hypothesis is mature, the evidence needed for the sample size calculation is somewhat vague. Thus, sample size calculations are frequently based on inaccurate assumptions for the control group, and the resulting calculations are often erroneous. Therefore, it is important to confirm statistical powers under several scenarios while referring to the feasibility of the trial. If most scenarios suggest feasibility (e.g., above 70% power for most scenarios), it is reasonable to start the trial. However, if most scenarios correspond to low power, then it still needs to be discussed whether testing another study hypothesis is feasible and statistically acceptable.

References

1. Altman DG. Statistics and ethics in medical research: III How large a sample? *Br Med J.* 1980;281:1336–8.
2. ICH Topic E8. General consideration, 17 July 1997. Available from: http://www.ich.org/fileadmin/Public_Web_Site/ICH_Products/Guidelines/Efficacy/E8/Step4/E8_Guideline.pdf.
3. Moher D, Hopewell S, Schulz KF et al. CONSORT 2010 explanation and elaboration: Updated guidelines for reporting parallel group randomised trials. *Br Med J.* 2010;340:c869.
4. Schulz KF, Altman DG, Moher D et al. CONSORT 2010 Statement: updated guidelines for reporting parallel group randomised trials. *Br Med J.* 2010;340:c332.
5. Charles P, Giraudeau B, Dechartres A, Baron G, Ravaud P. Reporting of sample size calculation in randomised controlled trials: review. *Br Med J.* 2009;338:b1732.
6. Halpern SD, Karlawish JH, Berlin JA. The continuing unethical conduct of underpowered clinical trials. *JAm Med Assoc.* 2002;288:358–62.
7. Bariani GM, de Celis Ferrari AC, Precivale M, Arai R, Saad ED, Riechelmann RP. Sample size calculation in oncology trials: quality of reporting and implications for clinical cancer research. *Am J Clin Oncol.* 2015;38:570–4.
8. ICH Topic E9. Statistical principles for clinical trials, 5 February 1998. Available from: http://www.ich.org/fileadmin/Public_Web_Site/ICH_Products/Guidelines/Efficacy/E9/Step4/E9_Guideline.pdf.
9. Schoenfeld D. The asymptotic properties of nonparametric tests for comparing survival distributions. *Biometrika.* 1981;68:316–9.
10. Johnson LL, Shih JH. An introduction of survival analysis. In *Principles and Practice of Clinical Research*, Third Edition, Gallin JI. Ognibene FP, eds. Cambridge: Academic Press, 2012: 285–93.
11. Collett, D. Sample size requirements for a survival study. In *Modelling Survival Data in Medical Research*, Third Edition, Boca Raton: Chapman & Hall/CRC Press, 2014: 471–85.
12. Lakatos E. 1988. Sample sizes based on the log-rank statistic in complex clinical trials. *Biometrics.* 44:229–41.
13. Lee JJ, Chen N. Software for design and analysis of clinical trials. In *Handbook of Statistics in Clinical Oncology*, Third Edition, Crowley J, Hoeling A, eds. Boca Raton: Chapman & Hall/CRC Press, 2012: 305–53.
14. Gehan EA. A generalized Wilcoxon test for comparing arbitrarily singly censored samples. *Biometrika.* 1967;52:203–23.
15. Tarone RE, Ware J. On distribution-free tests for equality of survival distributions. *Biometrika.* 1977;64:156–60.
16. Bernstein D, Lagakos SW. Sample size and power determination for stratified clinical trials. *J Stat Comput Sim.* 1978;8:65–73.
17. Iwamoto S, Takahashi T, Tamagawa H et al. FOLFIRI plus bevacizumab as second-line therapy in patients with metastatic colorectal cancer after first-line bevacizumab plus oxaliplatin-based therapy: the randomized phase III EAGLE study. *Ann Oncol* 2015;26:1427–33.
18. Tanaka S, Kinjo Y, Kataoka Y, Yoshimura K, Teramukai S. Statistical issues and recommendations for noninferiority trials in oncology: A systematic review. *Clin Cancer Res.* 2012;18:1837–47.
19. Fleming TR, Odem-Davis K, Rothmann MD, Li Shen Y. Some essential considerations in the design and conduct of non-inferiority trials. *Clin Trials.* 2011;8:432–9.
20. Chow SC, Shao J, Wang H. Comparing time-to-event data. In *Sample Size Calculation in Clinical Research*, Second Edition, Boca Raton: Chapman & Hall/CRC Press, 2008: 163–85.
21. ICH Topic E10. Choice of control group in clinical trials, 20 July 2000. Available from: http://www.ich.org/fileadmin/Public_Web_Site/ICH_Products/Guidelines/Efficacy/E10/Step4/E10_Guideline.pdf.
22. Schoenfeld DA. Sample-size formula for the proportional-hazards regression model. *Biometrics.* 1983;39:499–503.

23. Freedman LS. Tables of the number of patients required in clinical trials using the logrank test. *Stat Med*. 1982;1:121–9.
24. Jung SH, Kang SJ, McCall LM, Blumenstein B. Sample size computation for two-sample noninferiority log-rank test. *J Biopharm Stat*. 2005;15:969–79.
25. Kitano S, Inomata M, Mizusawa J et al. Survival outcomes following laparoscopic versus open D3 dissection for stage II or III colon cancer (JCOG0404): A phase 3, randomised controlled trial. *Lancet Gastroenterol Hepatol*. 2017;2:261–8.
26. Ohashi Y, Hamada C, Uozumi R. (eds.) Sample size calculation in survival analysis. *Advanced Survival Analysis: Biostatistics Using SAS [Japanese]*, Tokyo: University of Tokyo Press, 2016: 153–96.
27. Robert C, Thomas L, Bondarenko I et al. Ipilimumab plus dacarbazine for previously untreated metastatic melanoma. *N Engl J Med*. 2011;364:2517–26.

8

Non-Inferiority Trial

Keyue Ding and Chris O'Callaghan

CONTENTS

8.1 Introduction

Randomized, double-blind, placebo-controlled trials are the gold standard in establishing the efficacy of new therapeutics, as they are the most efficient in determination of effect and risk-benefit profile of an investigational drug. However, for many diseases, effective treatments already exist. In the face of many emerging new therapies that may affect survival or serious morbidity, the recent revision of the Declaration of Helsinki [1] concerning the ethical principles of human experimentation, increased interest has been generated in active-control trials, particularly those intended to show equivalence or non-inferiority of a potential new therapy to an active control.

To establish the efficacy of a new treatment, phase III studies are usually conducted to compare the treatment to a control. Efficacy can be established by showing either

1. The new treatment has superior efficacy to the control, or
2. The new treatment has efficacy equivalent to that of the active control, or
3. The new treatment is not substantively worse than the active control.

The first is commonly called a superiority trial; the second, an equivalence trial; and the third, a non-inferiority (NI) trial.

Let T, C, and P denote the experimental treatment, the control treatment, and the placebo, respectively, in the patient population investigated in a clinical trial study, and let Δ be the difference in treatment effect between T and C with larger values the better; then we can illustrate the objectives of each as follows:

Superiority trials: The superiority study design aims to show that a new drug is better than the control with respect to the primary efficacy variable of interest. The statistical hypotheses of the superior study are

$$H_0: \Delta \leq 0 \text{ vs. } H_a: \Delta > 0. \tag{8.1}$$

The hypothesis test designed to reject no difference in effect between T and C is usually used to confirm the efficacy of T.

NI trials: The NI study design aims to show that a new drug is not substantively worse than an active control with respect to the efficacy variable of interest.

The statistical hypotheses of the NI study are

$$H_0: \Delta \leq -M \text{ vs. } H_a: \Delta > -M, \tag{8.2}$$

where $M > 0$ is the pre-specified NI margin, the largest clinically acceptable difference between the T and C that T is still considered as acceptable. A confidence interval that estimates the difference in effect between the treatment and control is usually used to confirm this hypothesis, whereby if the upper $1 - \alpha$ confidence bound for $-\Delta$ is less than M, NI of T to C is confirmed with $1 - \alpha$ confidence.

Interpretation of non-inferiority trials can be much more difficult than that of superiority trials. Some specific and important issues to consider are (1) the quality of the trial and compliance of the subjects to the trial and (2) the selection and justification of M (the acceptable difference).

What is the difference between a non-inferiority trial and an equivalence trial? Although the terminology is often used synonymously, a subtle difference is that equivalence implies that the interest is in showing that the two therapies are similar; i.e., the differences between T and C are not large in either direction, a two-sided problem of showing two treatments are not very different. Whereas non-inferiority allows T to be better than C, a one-sided problem. A trial can be designed to show both non-inferiority and superiority. If the definition of NI (particularly M) is pre-specified and accepted, then it is generally appropriate for a trial to lead to either conclusion. Strict definitional equivalence is used more often for bio-equivalence trials that compare different formulations of the same drug.

Figure 8.1 illustrates the potential outcome of a trial with a point estimate treatment difference between experimental and control treatment based on confidence intervals of the estimate. From (a) to (d), NI of T to C are shown. For case (a), T is in fact superior to C,

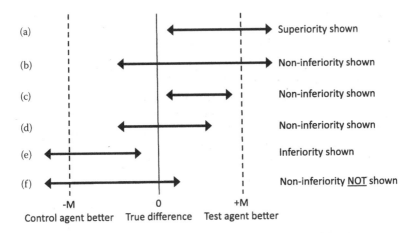

FIGURE 8.1
Potential outcomes with specified non-inferiority margin.

and the same for case (c); however, for (c), the difference may not be considered clinically significant. Case (d) demonstrates non-inferiority between T and C, while case (e) shows C is better than T, and (f) does not support NI of T to C.

NI clinical trials are conducted more often in medical research in recent years to establish the effectiveness of an experimental therapy. The number of randomized trials assessing NI increased by a factor of six in a decade—in 2005, just under 100 trials were listed in MEDLINE under the general rubric of "non-inferiority," whereas in 2015, there were almost 600 such trials [2]. A lot of research has been published on design and analysis of non-inferiority clinical trials [3–10], and the United States Food and Drug Administration (FDA) has issued, and recently updated, a draft guideline on conducting non-inferiority trials [11].

What is the rationale for performing an NI trial? When active treatments exist, it would be considered unethical to withhold, and an NI trial may be used to indirectly confirm the efficacy of the new treatment. As pointed out by Siegel [12] and Fleming [5], a new treatment similar to or somewhat less effective than the control may not be desirable unless it has other noteworthy benefits that outweigh the seeming loss of efficacy. In the situation where a new treatment may provide only a small advantage over the standard, showing the superiority over the active control would require a very large study, while NI could be shown with a much smaller sample size. Perhaps more commonly, NI trials are used to show that an alternative treatment with advantages regarding safety, convenience, or cost has an efficacy similar to, or at least not much worse than, the active-control treatment. To decide if an NI design is reasonable, we must question whether the purported advantages of the new treatment over the standard would justify adopting the new treatment should NI be shown. Typically, the advantages are price, ease of use, or a reduced number of side effects. If the answer is "no," then the NI design should be abandoned.

In this chapter, we will describe the key issues underlying the design and analysis of NI trials. Our explicit goals are to (1) review the key issues related to the design of non-inferiority trials, (2) review the statistical approaches and illustrate the degree to which the various assumptions in the statistical methodology influence the non-inferiority conclusions, and (3) suggest practical standards for conducting and reporting of these trials that improve the accuracy of their interpretation by clinicians and regulators alike. At the end, we present two cancer clinical trials (Canadian Cancer Trials Group trials LY12 and PR7) as exemplars with respect to these issues.

8.2 Assumptions for NI Trials

The term of NI test is well established for NI trial. It places a limit on the amount of treatment effect of the experimental therapy (T) is allowed to be inferior to the active control (C) and still be considered "effective" or clinically acceptable. The NI implies that T is effective compared to placebo/no therapy, and it is not inferior to C by a "clinically important" amount, so that clinical application of T in place of C would be ethical and acceptable.

If we intend to embark on an NI trial, then we must make the following two assumptions:

i. The constancy of the control effect and

ii. NI trial has assay sensitivity.

8.2.1 The Constancy of the Control Effect

An NI trial provides an indirect approach to demonstrating efficacy of a test treatment, i.e., there is no direct comparison to placebo/no treatment. One of the key assumptions in the NI test is the constancy assumption, that is, the effect of the reference treatment C is the same in the current NI trial as that in historical superiority trials, which demonstrated the efficacy of C. It has been shown that violations of the constancy assumption can result in a dramatic increase in the rate of incorrectly concluding non-inferiority in the presence of ineffective or even harmful treatment [8]. The constancy assumption is the key assumption in determination of the appropriate NI margin.

Historical evidence of sensitivity to drug effects (HESDE) [13]: The foundation of the NI trial is that one or more prior randomized trials had established the superiority of C over placebo/ no treatment. An NI trial which fails to show a difference in treatment effects between T and C may means that either both drugs were effective or neither drug was effective. A historically based conclusion that appropriately designed, sized, and conducted trials in a particular disease, with C (or group of related drugs) reliably showing an effect of at least some defined size on a particular endpoint, is thus imperative. Usually this is established by showing that appropriately powered and well-conducted trials in a specified population regularly distinguish C from placebo/no treatment for the particular endpoint (i.e., efficacy outcome used in establishing the treatment effect of C). Using data from those historical trials, the expected effect of C is derived. The constancy assumption is that the effect of C relative to placebo/no treatment is unchanged over time; otherwise, it may not be possible to interpret the results of an NI trial as indicative of true efficacy of T. HESDE applies only to trials of a particular design (patient population, selection criteria, endpoints, dose, and background therapy). Thus, the similarity of the current NI trial to past studies is also a consideration. The constancy condition cannot be verified internally (without a concurrent placebo/no treatment arm), but one can compare the primary endpoint in C to that from historical studies of C. In addition, the conditions under which the active control's effect was evaluated should be relatively recent to ensure that clinical practice has not changed substantially since the placebo-controlled trial was conducted. If clinical practice has not changed substantially, it is expected that the effect of the active control compared with a placebo will be consistent with historical data. The constancy assumption may be problematic in the case of subjective and highly variable responses, such as animal pain or behavioral endpoints that result in subjective clinical scores. If the effect of the active control is not relatively constant and consistent from study to study, it cannot be used for an NI study because of the inability to establish a reliable effect of the active control and thus ascertain the assay sensitivity of any trial conducted. In the case where the effect size can be inconsistent, perhaps attributable to a subjective and variable clinical endpoint, the disease state itself may not lend itself to an NI study. There is a risk that a new drug could be demonstrated as non-inferior to an active drug that itself is ineffective in the clinical trial and lead to the incorrect conclusion that the new drug is effective. Wide clinical use of an ineffective treatment introduces a much broader ethical dilemma than conducting a single placebo-controlled study. For this reason, regulatory bodies have required placebo-controlled superiority trials to increase the degree of confidence in the decision about the effectiveness of treatment.

8.2.2 Assay Sensitivity

The other basic assumptions for NI trials conducted without a placebo/no treatment arm is the so-called assay sensitivity [13–15], i.e., the ability of the trial to distinguish an effective treatment from a less effective or ineffective treatment if such a difference truly exists. In the case of a superiority trial, if it successfully demonstrates superiority, it has *de facto* simultaneously demonstrated assay sensitivity. However, an NI trial that successfully finds the effects of the treatments to be similar has not necessarily demonstrated assay sensitivity. Explicitly, a well-executed NI clinical trial that correctly demonstrates the treatments to be similar cannot be distinguished, on the basis of the data alone, from a poorly executed trial that fails to find a true difference. It could mean that (1) both treatments are similarly effective, (2) both treatments are similarly ineffective (failure of constancy assumption), or (3) the trial lacks assay sensitivity. Therefore, in addition to the constancy assumption, an NI trial must also rely on an assumption of assay sensitivity, the latter evaluable on the basis of information external to the trial, such as the quality-control procedures.

The International Conference on Harmonization (ICH) guidelines [13] list a number of factors that can reduce assay sensitivity. These include poor compliance with the study medication, poor diagnostic criteria, excessive variability of measurements, and biased endpoint assessment. In order to be credible, therefore, NI trials must attempt to avoid these factors to every possible extent and, even then, might not be able to escape suspicion without evidence of independent quality assurance. For example, a successful superiority trial can be very credible despite a moderately large rate of discontinuation from study drug, but a successful NI trial would be less so, because discontinuations can obscure a true treatment effect and thus reduce assay sensitivity. In conclusion, to consider an NI trial valid requires a combination of (1) historical information of the active control effect, (2) assurance of similarity of the new trial to historical trials, and (3) information about the quality of the new trial, thereby confirming the assumptions of consistency and assay sensitivity.

8.3 Design

In this section, we focus on issues in designing an NI trial and the relevant statistical issues. The most popularly used methods are the fixed-margin and the synthesis approaches.

To illustrate the idea, as before, we use T, C, and P to denote the experimental treatment, the control treatment, and the placebo/no treatment, respectively, in the patient population targeted by the active-control study. Let C_0 and P_0 be the control treatment and placebo, respectively, in the historical trial populations. (Note that C and C_0 may differ and so may

P and P_0 because of potential heterogeneity in the trial populations, possible changes in event rates over time, or other sources of heterogeneity.)

Endpoints: An NI trial intends to show the effect of T is either better than or not too much worse than the effect of C (active control) on the same endpoint; then an additional design element is that the endpoint employed in the NI trial should be the same one that efficacy of the active control was demonstrated previously. In cancer clinical trials, the typical endpoint is either the objective response rate (ORR, an assessment of reduction in tumor size/presence) or, alternatively, a time-to-event outcome, such as overall survival (OS = time to death), disease-free survival (DFS = time to disease recurrence/death), or progression-free survival (PFS = time to disease progression/death).

8.3.1 Selecting the Active Control

An NI trial employs a selected active-control C to serves as a comparator for evaluating the effectiveness of an experimental treatment T. The choice of C is very critical in designing a clinical trial. It affects trial feasibility, inferences, credibility, and many other features of trial design, conduct, and interpretation. In general, the effect of C must have been established in historical placebo-controlled trials.

One of the theoretical issues in selecting C arises in the context of the so-called "bio-creep" [4,8,16], i.e., after a series of non-inferiority trials in which each new drug meets the pre-specified NI criteria, but is in fact slightly worse than the previous control, eventually, an ineffective or even potentially harmful therapy may falsely be deemed efficacious, but be selected as the control in new trials. For example, suppose A, B, and D are three different drugs. Let us consider the following scenario:

1. Drug A is a proven efficacious drug for disease X versus a placebo control.
2. Drug B is shown to be non-inferior to drug A.
3. Drug D is shown to be non-inferior to drug B.

According to the conclusions (1) and (2), we can say that drug B is better than placebo. However, caution must be exercised when extending the interpretation to drug D. Although, drug D is shown to be non-inferior to drug B, it does not absolutely follow that we can conclude with certainty that drug D is better than placebo. In fact, it is not prudent to conclude even that drug D is non-inferior to drug A. This phenomenon is known as "bio-creep" [4,8]. Everson-Stewart and Emerson [16] investigated factors that may contribute to the occurrence of bio-creep. Those factors include the distribution of the effects of the new agents being tested, the choice of active comparator, the method used to account for the variability of the estimate of the effect of the active comparator, and changes in the effect of the active comparator from one trial to the next (i.e., violations of the constancy assumption).

To avoid using an ineffective control, the effectiveness of the selected control must be well established in previous studies [13]. There must be convincing prior evidence of the effectiveness of the active control compared with placebo; its effectiveness should be consistently demonstrated [17]. It must be clear that the active control is effective in the specific application, ideally within the specific population, used in the current NI study. The conditions of the trial (e.g., setting, dose, duration) should not unfairly favor one treatment over another [18]. For the purpose of the NI trial, we need to choose an active control with a well-established evidence base. Therefore, the best active therapy with reliable estimate of its effect in the population is generally recommended.

8.3.2 Determining the NI Margin

A key consideration in the design of NI trials is to select *a priori* the NI margin (M), defined as a prespecified maximum amount of difference in treatment effectiveness that can be allowed. The determination of the NI margin is the most critical issue in designing an NI study. It is a matter of a combination of statistical reasoning and clinical judgment [11,13,19].
ICH E10 [13] provides the two guidelines:

1. The determination of the margin in a non-inferiority trial is based on both statistical reasoning and clinical judgment, and should reflect uncertainties in the evidence on which the choice is based, and should be suitably conservative.

2. This non-inferiority margin cannot be greater than the smallest effect size that the active drug would be reliably expected to have compared with placebo in the setting of a placebo-controlled trial.

The process of selecting an appropriate M may be broken down into two steps. First, use statistical reasoning to determine the smallest reliable effect size of the active control compared with placebo. Second, uses clinical judgment to determine the largest loss of effectiveness of the new treatment relative to the active-control that would still be considered as clinically insignificant while maintaining the smallest effect size. The resolution from these two steps results in a single clinically and statistically relevant NI margin.

Determine the Smallest Reliable Effect Size of C: In designing an NI trial, we have to quantify the size of the effect of the active control based on historical evidence. As noted above, without certainty of active-control effectiveness, we risk concluding from an NI trial that an ineffective new treatment works. Determining the smallest reliable effect size of the active control is an objective step of determining the NI margin, although some subjectivity may still be incorporated into the process. If we assume that M_1 is the entire known effect of the active control relative to placebo [11], and this has been determined based on historical data derived from comparison of active control versus placebo or no treatment, then M_1 should be the smallest reliable effect size demonstrable. The determination of the smallest effect size for C is critical, because if the new treatment is less effective than the active control and the magnitude of this difference is greater than the smallest reliable effect size of the active control, the new treatment may have no effect at all.

What Should be M_1? Different approaches have been proposed for establishing it based on statistical reasoning; however, all of them require using the results of one or more historical placebo-controlled studies of the active treatment to determine the effect of the active control versus placebo. The most straightforward and probably most commonly used procedure is to calculate the lower bound of the two-sided 90% or 95% confidence interval (CI) of the estimated effect of active control versus placebo. The CI can be constructed from a single well-designed study, or from multiple placebo-controlled studies of the active treatment that have been conducted. It can be constructed from meta-analysis using a random/fixed effects model approach. The use of the lower bound of the 95% CI conservatively incorporates the uncertainty of effect of the active treatment, thus ensuring with a high degree of confidence that if the magnitude of the difference in treatments is less than this value, the new treatment has an effect relative to the placebo. FDA Guidance suggests M_1, the effect of active control, as the lower bound of 95% CI of the estimated active control's effect versus placebo, while acknowledging that it is conservative, and M, largest clinically acceptable difference, as a fraction of M_1 (say, $M = 0.5 * M_1$).

Clinical Judgment: In determining the NI margin M, clinical judgment as well as statistical reasoning needs to be incorporated into its selection. This margin incorporating clinical judgment also is sometimes referred to as margin 2 (M_2) [11]. Incorporating clinical judgment into M is subjective and little has been published on a recommended procedure. The clinically relevant margin of indifference often is described as determining the largest loss of effectiveness of the new treatment that would be considered clinically insignificant. Obviously, the selected value may vary substantially among individuals, and may likely vary within an individual on a case-by-case basis. Furthermore, the severity of the disease state being treated strongly determines the magnitude of M. To ensure that the new treatment has some effect, the selected M cannot be greater than the smallest reliable effects size (i.e., $M \leq M_1$). Therefore, M_1 must be determined before selecting M. Ensuring that M is $\leq M_1$ also allows for straightforward incorporation of statistical reasoning into the selection of the NI margin. The most common procedure for selecting M is for it to be half of M_1 to maintain at least half of the clinical effect of the active control. An alternative, less-conservative approach is to be half of the point estimate of the effect of active control over placebo, and then select the smaller of this value and M_1 to ensure that M is $\leq M_1$. However, in reality, there is no "one-size-fits-all" approach to selecting M because of inherent subjectivity in using clinical judgment.

What should be M? The smaller the margin, the lower the upper bound of the 95% two-sided confidence interval for $C - T$ must be, and the larger the sample size needed to establish non-inferiority. Showing that the upper bound of the 95% CI of $C - T$ is less than M_1 demonstrates that the test drug has some effect (i.e., an effect > 0). The margin of interest, however, as noted above, is usually smaller than M_1 (to show that an adequate portion of the clinical benefit of the control is preserved).

FDA Guidance suggests [11] M_1: effect of active control (recommend bound of 95% CI versus placebo, acknowledging conservative) and M: largest clinically acceptable difference:

$$M = M_1^* (1 - \lambda) \tag{8.3}$$

where λ is a given proportion of retention (50%, 75% of M_1).

One strategy is to use a fixed-margin approach (preserve a fraction of the effect), e.g., set the margin to be half of the estimated effect that the active control had over placebo. (Note that this approach does not consider the fact that the estimate from historical data is measured with uncertainty.) If the non-inferiority margin is too small, it leads to an extremely large study and will be difficult to conduct. On the other hand, if the non-inferiority margin is too large, it may lead to the approval of some treatments which have only slight or no benefit to patients.

The choice of M involves several considerations:

i. The effect of the control in the new non-inferiority study. This is not measured in the trial but must be known or estimated from previous studies of C, generally versus placebo (P). The effect represented by M should be no larger than the entire effect of the drug. If $C - T$ were larger than M, then a non-inferiority finding could represent a loss of all of the effect versus the control.

ii. The fraction of the effect of the control drug that needs to be preserved. This is a clinical judgment. If the desired retention is very high, for example, 90%, as noted above, studies become very large or, as a practical matter, only drugs that are actually superior will be successful. On the other hand, as the usual reason for using a two-arm non-inferiority design is the need to avoid exposure of patients to inferior treatment (that is, a placebo) when effective treatment exists, loss of much of the control agent's effect is unacceptable.

The margin for NI trials. In general, a more stringent margin is required for testing a new (different) drug, while a relative wider margin can be used if the objective of the NI trial is to test two same drugs with different doses, routes of administration, etc.

8.3.3 Statistical Algorithm for Assessing Non-Inferiority

The NI study intends to demonstrate the effectiveness of a study drug by ruling out that the experimental therapy is worse than the active control by a smallest clinically important difference M (the NI margin), statistically shown by the study. By demonstrating that the new treatment is no less effective than the control within the margin, the efficacy of the new treatment would have been established over placebo had the placebo arm existed in the trial. There are two major types of statistical methods for NI testing, i.e., the fixed-margin and synthesis approaches [9,11,20–26].

8.3.3.1 The Fixed-Margin Approach

The method entails a pre-specified NI margin against which comparison of the new treatment with the active control is then performed. It is often called the indirect confidence interval comparison (ICIC) method. The statistical algorithms for assessing NI are described by Blackwelder [3]. The hypotheses can be expressed as.

$$\text{H}_0: \Delta \leq -M \text{ versus } \text{H}_a: \Delta > -M, \tag{8.4}$$

where Δ is the difference in treatment effect between the test treatment (T) and the control treatment (C). Depending on the primary endpoint used, Δ could be the difference in mean for continuous outcome, risk difference (RD) or relative risk (RR, with log transformation), odd ratio (OR, with log transformation) for binary outcome, or hazard ratio (HR, with log transformation) for the time-to-event outcome. M is a pre-specified small, positive quantity defining the NI margin, that is, how much C can exceed T with T still being considered non-inferior to C, which is usually a fraction of the estimated effect of the control relative to placebo (for example, 50%). In order to assess if NI is met, we can either using confidence intervals or apply variations to the analytic strategy of null hypothesis testing. For the confidence intervals approach, we compute a $100(1-2\alpha)\%$ two-sided confidence interval for the difference ($T - C$); if the confidence interval's lower bound is greater than $-M$, then with $100(1-\alpha)\%$ confidence, we say the active control is more efficacious than the investigational product by no more than M, hence allowing us to claim non-inferiority of the experimental product as compared to the active control at an α significance level. The nominal significance level α for this one-sided test is generally set at half of the conventional significance level for a two-sided test for the difference in treatment effect between comparison groups (usually one-sided $\alpha = 0.025$ or 0.05). This approach has been adopted in regulatory environments, as suggested in the International Conference on Harmonization E9 (ICH E9) guidelines [27].

For the testing approach, we test the null hypothesis that $\Delta \leq -M$ at a 1-sided α level; if the null hypothesis is rejected, the NI is confirmed. We present the most common cases in oncology trials in the following subsections.

8.3.3.1.1 A Binary Endpoint

In this subsection, we illustrate the statistical algorithm for assessing non-inferiority for a binary outcome (success or failure). In cancer clinical trials, this outcome could be the objective response status. P_T and P_C are probabilities of a good outcome. The most commonly used ones are the risk difference (RD) in proportions $\Delta = P_T - P_C$ or the relative risk (RR) $\Delta = P_T/P_C$.

Measure using RD: We shall first consider the RD of proportions. The hypotheses can be expressed as follows with selected NI margin M:

$$H_0: \Delta = P_T - P_C \leq -M \text{ versus } H_a: \Delta = P_T - P_C > -M \tag{8.5}$$

Accept non-inferiority of T to C if the lower α confidence bound of $P_T - P_C$ is $> -M$; i.e.,

$$P_C - P_T + Z_{1-\alpha} * \left(\sqrt{\frac{(P_C * (1 - P_C))}{n_C} + \frac{P_T * (1 - P_T)}{n_T}} \right) < M \tag{8.6}$$

where $Z_{1-\alpha}$ is the $1 - \alpha$ quantile of standard normal distribution.

Or perform a test with the test statistic:

$$Z = \frac{\widehat{P_T} - \widehat{P_C} + M}{\sqrt{(\widehat{P_C} * (1 - \widehat{P_C}))/n_C + (\widehat{P_T} * (1 - \widehat{P_T}))/n_T}}. \tag{8.7}$$

If $Z > Z_{1-\alpha}$, we reject the null hypothesis of $\Delta = P_T - P_C \leq -M$ and accept NI of T to C.

Measure using Relative Risk: When probability for outcome is relatively small or varies very much, the RR is a more stable estimate of difference between groups. The hypotheses can be expressed as follows with the selected NI margin M:

$$H_0: \Delta = P_T/P_C \leq M \text{ versus } H_a: \Delta = P_T/P_C > M, \tag{8.8}$$

where M is the NI margin, a positive number that is less than 1. Accept non-inferiority of T to C if the lower α confidence bound of P_T/P_C is $> M$, i.e.,

$$\frac{P_T}{P_C} * \exp \left(-Z_{1-\alpha} * \left(\sqrt{\frac{(1 - P_C)}{n_C P_C} + \frac{(1 - P_T)}{n_T PT}} \right) \right) > M. \tag{8.9}$$

Or perform a test with the test statistic:

$$Z = \frac{\log\left(\frac{\widehat{P_T}}{P_C}\right) - \log(M)}{\sqrt{\frac{1 - \widehat{P_C}}{n_C P_C} + \frac{1 - \widehat{P_T}}{n_T P_T}}}, \tag{8.10}$$

If $Z > Z_{1-\alpha}$, we reject the null hypothesis of $\Delta = P_T/P_C \leq M$ and accept NI of T to C.

8.3.3.1.2 A Time-to-Event Endpoint

Many clinically meaningful outcomes are time-to-event endpoints in oncology, such as overall survival, progression-free survival, etc. Hazard ratio between groups is usually used to measure the difference and are compared using the log-rank test. The hypotheses can be expressed as follows with selected NI margin M (>1, high value represents worse effect):

$$H_0: \Delta = HR(T/C) \geq M \text{ versus } H_a: \Delta = HR(T/C) < M. \tag{8.11}$$

Accept non-inferiority of T to C if the upper $1 - \alpha$ confidence bound of $HR(T/C)$ is less than M; i.e.,

$$\widehat{HR\left(\frac{T}{C}\right)} * \exp\left(Z_{1-\alpha} \middle/ \sqrt{\sum_{k=1}^{k=K} \frac{Y_{Tk} Y_{Ck}}{(Y_{Tk} + Y_{Ck})^2}}\right) < M, \tag{8.12}$$

where $Z_{1-\alpha}$ is the $1 - \alpha$ quantile, $\widehat{HR(T/C)}$ is the estimate of hazard ratio of study treatment T versus the active control C, K is the total number of events, Y_{Tk} is the number of subjects at risk just prior to the kth event in the study treatment arm, Y_{Ck} is the number of subjects at risk just prior to the kth event in the control treatment arm, and I_{Tk} is the indicator that the kth event is from study treatment arm.

Or perform a test with a test statistic:

$$L = \sum_{k=1}^{k=K} \left(I_{Tk} - \frac{Y_{Tk}M}{Y_{Tk} + Y_{Ck}}\right) \middle/ \sqrt{\sum_{k=1}^{k=K} \frac{Y_{Tk} Y_{Ck} M}{(Y_{Tk} + Y_{Ck})^2}} \tag{8.13}$$

where K, Y_{Tk}, Y_{Ck}, and I_{Tk} are as in above.

If $L < -Z_{1-\alpha}$, we reject the null hypothesis of $\Delta = HR(T/C) \geq M$ and accept NI of T to C.

8.3.3.2 Synthesis Approach

The synthesis method, also being called the virtual comparison (VC) method, synthesizes the estimated effect of the new treatment versus the control from the active-controlled trial and the estimated effect of the control relative to placebo from the historical trials. The synthetic estimate or test is used to assess the efficacy of the new treatment versus a putative placebo and to estimate the fraction of the control effect that may have been preserved by the new treatment had the placebo been present in the active-controlled trial. The effect of the active control is obtained from historical placebo-controlled trials that are independent of the active-controlled trial to be undertaken. Several authors, including Holmgren [23], Hasselblad and Kong [24], and Fisher [25], had proposed the method. Simon [26] gave a Bayesian analog of this method. The test procedures are generally formulated to control the type I error rate under the assumption that the estimates of the effect of the active control and its corresponding variance in the NI trial and historical trials are unbiased.

Let $\hat{\beta}_{TC}$ and $\widehat{Var}_{TC}$ be the estimates of the effect and its corresponding estimate of variance of study treatment to active control in the active control trial, while $\hat{\beta}_{CP}$ and $\widehat{Var}_{CP}$ are the estimate effect and its corresponding estimate of variance of active control to placebo in the historical placebo-controlled trial (assume the higher value the better). The estimates of the effect could the risk difference; logarithm of relative risk for a binary outcome, or logarithm of hazard ratio for a time-to-event outcome. Thus, $\hat{\beta}_{TC} + \hat{\beta}_{CP}$ is an estimate of the relative effect of the study treatment T versus placebo P under the constancy assumption.

Let $\lambda = (\hat{\beta}_{TC} + \hat{\beta}_{CP}) / \hat{\beta}_{CP}$ denote the proportion of active treatment effect of retention. The hypothesis of the experimental therapy maintains more than λ_0 proportion $(0 \leq \lambda_0 \leq 1)$ active control's effect can be expressed as

$$H_0 : \beta_{TC} + (1-\lambda_0)\beta_{CP} < 0 \text{ vs. } H_a : \beta_{TC} + (1-\lambda_0)\beta_{CP} > 0, \tag{8.14}$$

with the test statistic

$$Z = \frac{(\hat{\beta}_{TC} + (1-\lambda_0)\hat{\beta}_{CP})}{\sqrt{\widehat{Var}_{TC} + (1-\lambda_0)^2\,\widehat{Var}_{CP}}}. \tag{8.15}$$

Reject the null in favor of NI of T to C if $Z > Z_{1-\alpha}$.
For a binary endpoint with RD, the test statistic:

$$Z = \frac{(\widehat{RD}_{TC} + (1-\lambda_0)\widehat{RD}_{CP})}{\sqrt{\widehat{Var}_{TC} + (1-\lambda_0)^2\,\widehat{Var}_{CP}}}, \tag{8.16}$$

where $\widehat{RD}_{TC}$ is $\widehat{P_T} - \widehat{P_C}$ and its variance $\widehat{Var}_{TC}$ estimated from a current active-control trial, while $\widehat{RD}_{CP}$ is $\widehat{P_C} - \widehat{P_P}$ and its variance $\widehat{Var}_{CP}$ estimated from a historical trial of active control versus placebo.
For a binary endpoint with RR, the test statistic

$$Z = \frac{(\log(\widehat{RR}_{TC}) + (1-\lambda_0)\log(\widehat{RR}_{CP}))}{\sqrt{\widehat{Var}_{TC} + (1-\lambda_0)^2\,\widehat{Var}_{CP}}}, \tag{8.17}$$

where $\widehat{RR}_{TC}$ is $\widehat{P_T}/\widehat{P_C}$ and the variance of $\log((\widehat{RR}_{TC}))$, $\widehat{Var}_{TC}$ estimated from a current active-control trial, while $\widehat{RR}_{CP}$ is $\widehat{P_C}/\widehat{P_P}$ and the variance of $\log((\widehat{RR}_{CP}))$, $\widehat{Var}_{CP}$ estimated from a historical trial of active control versus placebo.
For time-to-event endpoint with hazard ratio, the test statistic is

$$Z = \frac{(\log(\widehat{HR}_{CT}) + (1-\lambda_0)\log(\widehat{HR}_{PC}))}{\sqrt{\widehat{Var}_{CT} + (1-\lambda_0)^2\,\widehat{Var}_{PC}}}, \tag{8.18}$$

where $\log(\widehat{HR}_{CT})$ and its variance $\widehat{Var}_{CT}$ are estimates of log hazard ratio of C versus T and its variance from a current active-control trial, while $\log(\widehat{HR}_{PC})$ and its variance $\widehat{Var}_{PC}$ are estimates of log hazard ratio of P versus C and its variance from a historical trial of active control versus placebo.

8.3.4 Sample Size

In the planning of a clinical trial, estimation of sample size is one of the most fundamental steps. A proper sample size provides required power to detect a clinically meaningful difference among groups for superiority trials, or to rule out the test treatment is inferior to the control for the given NI margin. Insufficiently sized trials would lead to statistically inclusive results, failing to give a conclusive decision. Sample size derivation should be based on the statistical procedure used to draw the conclusion of the study, reflecting the primary study objective. Different types of study and primary endpoints lead to different sample size estimation methods. As usual, the information on parameters of study design are required. Those include hypothesis on primary endpoint, metric used to measure the primary endpoint (mean and variance of the summary statistic under null and alternative hypotheses), significance level of test, and the power to rule out the difference with the specified NI margin (M) or proportion of preservation (λ_0) of the active effect.

We first present the sample size formula for NI trials with a fixed margin. Let Δ denote the difference in the treatment effect of the experimental therapy and the active control ($T - C$) and Δ_a be the assumed true difference in the effects of the experimental and the active-control therapy under the alternative. α is the significance level of the confidence interval, and $1 - \beta$ is the power at alternative Δ_a that we can rule out the difference is within the non-inferiority margin M. Let σ^2 be the common variance population variance of the treatment effect summary statistic for each study treatment arm, $Z_{1-\gamma}$ be the $100(1 - \gamma)$ percentile of a standard normal distribution. Let r be the ratio of patients randomized between the experimental and the control arm.

Power and sample size consideration for fixed margin: The sample size for an NI trial depends on the confidence level $(1 - \alpha)$ to rule out an NI margin of M, the power $(1 - \beta)$, σ^2 is the variance of the test statistic, and Δ_a is the target difference in effect of T and C designed with specified power to detect. For an NI trial, usually Δ_a is chosen to be 0, assuming T and C are equally effective. For a 1-to-1 ratio of randomization, the sample size is estimated as

$$n = 4 * (Z_{1-\alpha} + Z_{1-\beta})^2 \sigma^2 / (\Delta_a - M)^2. \tag{8.19}$$

For an r:1 ratio of randomization, the total sample size will be the $n * (r+1)^2 / 4 * r$. It should be pointed out that the sample size should be sufficient for both ITT and PP analyses.

Sample size for binary outcome with RD, $\Delta = P_T - P_C$:
For r:1 ratio of randomization:

$$n_T = r \, n_C;$$

$$n_C = (Z_{1-\alpha} + Z_{1-\beta})^2 (P_T * (1 - P_T)/r + P_C * (1 - P_C)) / (\Delta_a - M)^2. \tag{8.20}$$

Sample size for binary outcome with RR, $\Delta = P_T / P_C$:

$$n_T = r \, n_C;$$

$$n_C = (Z_{1-\alpha} + Z_{1-\beta})^2 ((1 - P_T)/rP_T + (1 - P_C)/P_C) / (\log(\Delta_a) - \log(M))^2. \tag{8.21}$$

Sample size for time to event outcome, the number of events required is:
For r:1 ratio of randomization, the minimum required number of events from both groups are

$$n = \frac{(r+1)^2}{r} * (Z_{1-\alpha} + Z_{1-\beta})^2 / (\log(\mathrm{HR}_a) - \log(M))^2. \tag{8.22}$$

To get the sample size on number of patients needed for the study, it would be determined by the length of study; i.e., patient's accrual period/accrual speed and the follow-up time after completion of patients' accrual and event rates. Many authors have investigated the method for determining the sample size, the methods and formulas devised by Freedman [28], Schoenfeld [29], and Schoenfeld and Richter [30] which have particularly simple assumptions and are arguably the tools most widely used. It should be pointed out that, for the time-to-event endpoint, the number of events is the key to ensure the power of the study. Due to the changes in patients care, the projected event rate at the design stage might not be very accurate; the duration of the trial might vary to get the required number of events.

The sample size formula for NI trials with the synthesis approach: The sample size for the synthesis testing method depends on the estimated C's effect and its corresponding estimated variance. The sample size for the NI trial is to find a sample size which makes the estimated variance of $\widehat{Var_{TC}}$ for the estimated effect of T relative to C of $\hat{\beta}_{TC}$ satisfy the equation:

$$Z_{1-\beta} * \widehat{Var_{TC}} = (\hat{\beta}_{TC} + (1-\lambda_0)\hat{\beta}_{CP}) - Z_{1-\alpha} * \sqrt{\widehat{Var_{TC}} + (1-\lambda_0)^2 \widehat{Var_{CP}}}. \tag{8.23}$$

where the sample size or the number of events (for time-to-event outcomes) is in the estimated variance of $\widehat{Var_{TC}}$. There is an explicit formula for the sample size. It should be pointed out that the sample size depends on the historical data. There may be no solution for equation 8.23 [31].

8.3.5 Other Design Alternatives and Issues

8.3.5.1 Three-Arm Studies

Two-arm NI studies are preferred when the placebo arm is unethical or problematic. However, when the efficacy of the active control in the NI trial is in doubt, or great heterogeneity of active treatment effect is found in historical trial(s), or the active treatment effect could be small in the NI study, it may be ethical to add a placebo arm. Pigenot et al. [32] presented the cases that the placebo group could be added if

1. The use of a placebo control is ethical;
2. The use of a placebo control will not preclude potential study subjects from giving informed consent to participate the study;
3. An active control is available that will provide additional information about the benefit and/or risk of the experimental medication compared with other available therapy.

In situations in which inclusion of both a placebo and an active control are ethical and scientifically defensible, the three-arm design for the NI trial is a better choice [33]. The three-arm study can directly check the trial's assay sensitivity and constancy assumption by comparing both T and C to P and compare T to C to check if T is substantially worse than C. Potential issues for the three-arm trial could be the multiple comparisons in the trial's analysis. Analysis strategy must be pre-specified according the study's primary objectives. The Hierarchical Test Procedures [34] are first to establish the efficacy of T over P, then demonstrate T is NI to C. Meanwhile, show that C's effect compared to P is larger than the NI margin used in the study design.

8.3.5.2 *Switching between NI and Superiority*

For the NI trials, when the NI is established, it is natural to further check if T is superior to C with the same primary endpoint and other pre-specified endpoints. It is generally acceptable for fixed-sequence hierarchical tests without adjusting for the multiple tests [35]. However, the test will stop once a hypothesis is not rejected. On the other hand, it is generally not acceptable to test NI after failing the superiority test unless it is pre-planned at the design stage, i.e., with a pre-specified NI margin, as it is unlikely to choose an NI margin objectively after seeing the results. However, multiple tests in the further secondary analyses should be adjusted among themselves to control the overall type I error in those secondary analyses.

8.3.5.3 *Interim Analyses*

Interim analyses (IA) are routinely planned for large clinical trials to assess the efficacy, futility, and safety of the experimental therapy at the design stage, and group sequential designs [36] are the statistical tools for it. The group sequential designs allow interim analyses based on the cumulative data to specified time points with a decision rule to either continue or stop the study [37]. Ethical considerations have been used for the rationale for interim analyses for efficacy, futility, or harm. If the experimental therapy is highly effective, it is imperative to allow patients access to such treatment; on the other hand, if the experimental therapy is lacking efficacy or is harmful, early evidence could stop the study early, restricting patients from continuing the futile/harmful treatment. However, for NI studies, when early determination of NI is not a conformed superiority, there may be no ethical reason to stop the study early unless the control therapy has significance other risks. However, if experimental therapy is worse than the control, stopping the study early is justified. As interim analyses are planned based on the realistic objectives, and NI determination needs a relatively large amount of information, usually one or two interim analyses are planned. When a trial continues after demonstrating NI at interim analysis to test for superiority, there is no need to adjust for multiple tests.

8.4 Trial Conduction

"Sloppiness Obscures Differences." The conclusion of NI of the experimental therapy to active control implies it is superior to the placebo, which requires the inference has external validity as well. For the historical comparison to have operational validity in the current trial, the critical assumption of constancy must be met. A proper trial design and high quality of conduction (e.g., maximal compliance, minimization of protocol deviations and outcome misclassifications, adherence to identical experimental protocol, etc.) is crucial for this, as well as for assay sensitivity assumptions to hold. The NI trial should be adequately executed to ascertain outcomes. Incomplete or inaccurate ascertainment of outcomes because of loss to follow-up, treatment crossover or nonadherence, or outcomes that are difficult to measure or are subjective may cause the treatments being compared to falsely appear similar. The conclusion of NI of the experimental therapy to active control should not be due to the trial's lack of assay sensitivity.

The NI trials assume that the active control is effective in the study population. Several particularly important design features must be followed to gain confidence that the conducted NI trial maintains assay sensitivity and consistent effect. The first is that the NI trial must utilize a similar treatment regimen as the prior trial(s) of the active control against the placebo. For example, data from a single-dose study of the active control would not be supportive of a planned NI trial examining multiple doses over time. Second, the planned NI trial also must study a similar patient population as the placebo-controlled trial of the active control. If the active control was studied in a population of patients with mild disease, this would not be supportive in the planned NI study in a population of patients with severe disease. Third, the selected primary outcome measure and the time of evaluation in the NI trial must be the endpoint measured and reported in a placebo-controlled trial of the selected active control because this is the outcome that has been shown to demonstrate sensitivity to treatment effects.

The quality of the trial(s) conducted with the selected active control compared with the placebo is also important. A high-quality, well-described trial demonstrating the effectiveness of the active control ensures a high degree of confidence that the effect is reproducible (i.e., assay sensitivity) and that trial design can be recreated. If a single low-quality study of the active control is all that exists, the strength of evidence from the NI trial will be undermined, no matter how well the NI trial is conducted.

8.5 Analyses

8.5.1 Analysis Populations

The "gold standard" approach to analysis in a superiority trial is based on the intention-to-treat (ITT) principle, where participants are analyzed in the groups to which they were originally randomized [38]. As the randomization is essential for achieving the comparability and the basis for statistical inference, this approach is favored as it preserves randomization and, in the case of departures from randomized treatment, makes treatment groups appear more similar; therefore, producing a conservative estimate of treatment effect. ICH E10 [13] and CONSORT [39] guidelines recommend that NI trials conduct an intent-to-treat (ITT) analysis. It should be noted that although the importance of ITT analysis in a traditional superiority trial is well established, the role of the ITT population in a non-inferiority trial is not equivalent to that of a superiority trial [40]. An ITT analysis in a superiority trial tends to reduce the treatment effect, minimizing the difference between groups—in essence favoring the null hypothesis of no difference. This is a conservative way to view the results. However, in the non-inferiority setting, because the null and alternative hypotheses are reversed, a dilution of the treatment effect actually favors the alternative hypothesis, making it more likely that true inferiority is masked. An alternative approach is to use a per-protocol (PP) population, defined as only participants who comply with the protocol. A PP analysis excludes participants with departures from randomized treatment, but assumes that the group of participants who are excluded are similar to those who are included on both observed and unobserved variables, an assumption that is usually deemed implausible. The per-protocol analysis can potentially have bias as well. Because non-completers are not included, it can distort the reported effectiveness of a treatment. If, for example, a significant percentage of participants dropped out of the experimental treatment because they felt the treatment received was ineffective, the per-protocol analysis would not adequately capture that. Another approach is to use a modified ITT, which excludes participants who never actually received the treatment, but includes non-compliant participants who started the treatment but did not fully complete it. While selection bias is thought to be minimized in trials with blinding, and modified definitions of these populations that adjust for observed confounders can be used, selection bias can never be completely discounted from any analyses that makes post-randomization exclusions or manipulations. Thus, recommendations are that a PP analysis should be conducted alongside an ITT analysis for NI trials [41]. Regardless of the approach used, the reported findings should specify and fully describe the type analysis population. Similar results from both analyses are expected, and contraction findings may revere issues in trial conduction.

The statistical analysis of NI can be conducted either by using confidence intervals or by applying variations to the analytic strategy of null hypothesis testing. Under conventional hypothesis testing in a comparative study of two interventions, the goal is to reject the null in favor of a difference between the two interventions. If this approach is extended to an NI study, then can NI be claimed when the test fails to reject the null hypothesis of no difference? Some argue that it is acceptable assuming that a strict level of type II error (failing to reject the null when it is false) is provided [36,50]. However, others argue that it is logically impossible to conclude NI on the basis of failing to reject the null hypothesis [3,41]. Rather, failure to reject the null means that there is not sufficient evidence to accept the alternative hypothesis. Alternately, if the null hypothesis states that the true difference is greater than or equal to the pre-specified NI margin, then failing to reject it can be interpreted as insufficient evidence to conclude that the difference of the two procedures is less than the pre-specified NI margin.

To avoid misinterpretation of null hypothesis testing, some investigators, including those who contributed to the CONSORT guidelines, favor the confidence interval approach to show NI of two treatments [39–42]. The width of the interval signifies the extent of NI, which is a favorable characteristic of this approach. If the confidence interval for the difference between two interventions lies below the NI margin, then NI can be concluded. If the interval crosses the boundary (contains the value of the margin) then NI cannot be claimed. Some investigators prefer the confidence interval approach to examine the precision of NI. Others prefer hypothesis testing. We recommend that both confidence interval and p-values be provided to allow the audience to interpret the extent of the findings.

8.5.2 Missing Data

Missing data are common in clinical studies and introduce additional uncertainty in results, may result in biased estimates, and could seriously compromise inferences from clinical trials. The potential effect of missing data must be assessed.

An important consideration in choosing a missing data approach is the missing data mechanism, and different approaches have different assumptions about the mechanism.

The data missing mechanism describes the possible relationship between the propensity of data to be missing and values of the data, both missing and observed. Little and Rubin [43] established the foundations of missing data theory. Central to missing data theory is their classification of missing data problems into three categories: (1) missing completely at random (MCAR), (2) missing at random (MAR), and (3) missing not at random (MNAR).

Missing completely at random (MCAR) means there is no relationship between the missingness of the data and any values, observed or missing. Those missing data points are a random subset of the data. There is no systematic trend that makes some data more likely to be missing than others.

Missing at random (MAR) means there is a systematic relationship between the propensity of missing values and the observed data, but not the missing data.

Whether an observation is missing has nothing to do with the missing values, but it does have to do with the values of an individual's observed variables. So, for example, if men are more likely to tell you their weight than women, weight is MAR.

Missing not at random (MNAR) means there is a relationship between the propensity of a value to be missing and its values. This is a case where the people with the lower education are missing on education or the sickest people are most likely to drop out of the study.

MNAR is called "non-ignorable" missing, because the missing data mechanism itself has to be modeled as you deal with the missing data. You have to include some model for why the data are missing and what the likely values are.

MCAR and MAR are both considered "ignorable" because we don't have to include any information about the missing data itself when we deal with the missing data.

Multiple imputation and maximum likelihood [43,44] assume the data are at least missing at random. So, the important distinction here is whether the data are MAR as opposed to MNAR.

It should be pointed out that in most datasets, more than one variable can have missing data, and they may not all have the same mechanism. It's worthwhile diagnosing the mechanism for each variable with missing data before choosing an approach.

The best approach of dealing with missing data is to prevent missing data, so high quality of trial conduction is key for reliable results.

Analysis in the presence of missing data are typically with two objectives: unbiased estimation of treatment effect and making valid inference under the null hypothesis. For handling ignorable missing, multiple imputation under the null hypothesis is recommended [44]. For example, continuous endpoint: impute reasonable expected value (m) for the active control and ($m-M$) for the new intervention. For binary data: impute expected proportion (p) for active control and ($p-M$) for new intervention. However, to assess the robustness of the outcome, certain biased imputation can be used. For example, for binary endpoint: impute success for the active control and failure for new intervention. If you still show NI, then it is not because of missing data.

8.5.3 NI and Superiority

Switching between NI and Superiority: It is generally okay to test for superiority after showing NI, and there is no multiplicity adjustment necessary (closed testing procedure), as the NI test includes the superiority. However, the primary analysis population for the superiority test should be the ITT population. However, if possible, the superiority test after NI would be better pre-planned in the study protocol.

Switching from Superiority to NI: It is generally not acceptable to go from failing to demonstrate superiority to then evaluating NI as it causes multiple issues. First, is the control group an appropriate control for an NI trial? It is obviously not appropriate if the control is a placebo. Was the efficacy displayed by the control group similar to that shown in trials versus placebo (constancy)? Most important, the post-hoc definition of NI margin is difficult to justify, as the choice of NI margin needs to be independent of trial data. Meanwhile, trial quality must be high, as poor adherence, drop-out, etc. could bias toward NI.

On the other hand, it may be feasible if the NI margin was pre-specified, or it can be justified based on external information and not chosen to fit the data, which is difficult to do. Meanwhile, the trial was of conducted with high quality with few drop-outs and good adherence, the control group displayed similar efficacy to trials versus placebo, and results from ITT and PP analyses are similar. The trial was sensitive enough to detect effects, i.e., has the assay sensitivity.

Changing the NI margin: It is generally okay to decrease the NI margin; however, an increase should be justified from external data (independent of trial). It is usually difficult to justify.

8.6 Reporting

Piaggio et al. [39] for the CONSORT (Consolidated Standards of Reporting Trials) Group provided a checklist and a flow diagram for the Reporting of NI and Equivalence Randomized Trials, Extension of the CONSORT 2010 Statement [45]. The intent is to improve reporting of NI and equivalence trials, enabling readers to assess the reliability of the reported results and conclusions. More specifically, in addition to the CONSORT 2010 Statement [45], updated guidelines for reporting parallel group randomized trials, a specific feature for the NI trial, should be reported as well. First, the title of the report should reflect the NI randomized trial design. In the background, the rationale for using an NI design, the hypotheses concerning NI, and the method for specifying the NI margin should be provided. To ensure the assay sensitivity and the consistency assumptions, the study design should also provide information on whether participants in the NI trial are similar to those in any trial(s) that established efficacy of the active-control treatment, and whether the treatment in the NI trial is identical (or very similar) to that in any trial(s) that established efficacy. Meanwhile, specify whether the primary and secondary outcomes in the NI trial are identical (or very similar) to those in any trial(s) that established efficacy of the active control, and specify whether hypotheses for main and secondary outcome(s) are NI or superiority. For the sample size justification, it should specify whether the sample size was calculated based on an NI criterion and the NI margin used. Finally, it should specify whether a one- or two-sided confidence interval approach was used for the NI criterion.

8.7 Examples

In this section, we present two examples of NI trials conducted by the Canadian Cancer Trials Group (CCTG, formerly, NCIC CTG).

NCIC CTG LY.12: LY.12 was a multicenter, international, randomized controlled trial in patients with relapsed/refractory aggressive lymphoma. The trial was designed to compare the regimen of combination of gemcitabine, cisplatin, and dexamethasone (GDP) with the combination regimen of dexamethasone, cytarabine, and cisplatin (DHAP) as second-line chemotherapy prior to autologous transplantation (first randomization) and to evaluate the efficacy of post-transplantation treatment with the anti-CD20 antibody rituximab (second randomization). We only focus on the first randomization, which is an NI design. Detailed information on the study can be found in the publication of the trial by Crump et al. [46]. The trial was designed to show that the well-tolerated regimen GDP is NI to DHAP with the primary endpoint of response rate after two cycles of treatment. If GDP was shown to be non-inferior, further testing would be performed to check whether or not GDP had a superior transplantation rate (hierarchical test).

Historical trials provided an estimated response rate of 50% with DHAP. A fixed NI margin of 10% was chosen. To use a one-sided 5% level test, with 80% power under the alternative hypothesis that the response rates are equal, we needed to accrue a total of 630 patients. An IA for futility around the NI endpoint was planned after 320 patients were enrolled; the protocol was amended to add a second IA after 480 patients were entered with O'Brien-Fleming type of stopping boundaries [47], resulting in a final sample size of 637 patients.

Due to slow accrual rate, the Data and Safety Monitoring Committee (DSMC) approved the trial committee's request to stop the trial early with a total 619 patients for the final analysis. The primary per-protocol analyses, the response rates after two cycles of treatment were 46.2% for GDP(R) versus 44.7% for DHAP(R), with estimated upper 95.6% confidence bound of 6.4% ($p = 0.004$ for the NI test). For the intention-to-treat analysis, the response rate after two cycles of treatment was 45.2% in the GDP(R) group and 44.0% in the DHAP(R) group. The upper boundary of the one-sided 95.6% (adjusted for the two IAs) confidence bound for the difference in response rates was 5.7% ($p = 0.005$ for NI test). Both ITT and PP analyses confirm the NI of GDP(R) to DHAP(R).

Transplantation rate is defined as the number of patients who respond sufficiently to protocol salvage chemotherapy to be planned for transplantation minus those who do not meet the endpoint of successful transplantation, divided by the number of all randomized patients (ITT population). The transplantation rate was not significantly different with 52.1% in the GDP(R) group and 49.3% in the DHAP(R) group (Risk Difference: 2.5%, 95% CI −5.5%, 10.5%, $p = 0.44$).

NCIC CTG PR7: PR7 was a randomized trial, which assess the equivalence in overall survival between the intermittent androgen suppression (IAS) and continuous androgen deprivation (CAD) in patients with PSA-evidence of progression in the absence of distant metastases following previous radical radiotherapy treatment of prostatic cancer [48]. The trial was designed using an NI margin of hazard ratio of 1.25. In order to rule out the difference with 80% power using a one-sided 5% level test assuming that we are using a point of indifference [49] at a hazard ratio of 1.12, the final analysis required 800 deaths. A formal IA was planned using the method by Freedman, Lowe, and Macaskill [7] with the O'Brien and Fleming boundary [47]. An early-stopping decision in favor of IAS would be considered if we were 99.5% sure that the true hazards ratio was less than 1.25 (i.e., the upper bound of the 99% confidence interval for the hazards ratio was less than 1.25 and the lower bound of the 99% CI was less than one). The IA would be performed when around 400 deaths happened in the trial. The results of the analysis were presented to the DSMC of the NCIC CTG. The IA was performed with 446 events (deaths) for the interim analysis. The results demonstrated that IAS is NI to CAD, with an estimated *HR* of 0.971 and a 99% CI from 0.756 to 1.248. The *p*-value for test non-inferiority (*HR* [IAS vs. CAD]: ≥1.25) was 0.0047. The interim analysis report was submitted to the NCIC DSMC for discussion, and the committee suggested reporting the outcome. Based on the DSMC suggestion, the final analysis of the trial was performed with 524 deaths. For the ITT analysis, the estimated hazard ratio (IAS vs. CAD) was 1.02 with a 90% CI from 0.89 to 1.19 (95% CI 0.86–1.21). The *p*-value for testing non-inferiority (HR [IAS vs. CAD] ≥1.25) is 0.009. Per-protocol analysis provided similar results with the estimated hazard ratio (IAS vs. CAD) of 1.03 and a 90% CI from 0.90 to 1.19 (95% CI 0.87–1.22), which supports that IAS is non-inferior to CAD in the study population.

References

1. World Medical Association. Declaration of Helsinki: Ethical principles for medical research involving human subjects. *JAMA*. 2013;310(20):2191–2194.
2. Mauri L, D'Agostino RB Sr. Challenges in the design and interpretation of noninferiority trials. *N Engl J Med*. 2017;377:1357–67.
3. Blackwelder WC. Proving the null hypothesis in clinical trials. *Control Clin Trials*. 1982;3:345–353.
4. D'Agostino RB, Massaro JM, Sullivan L. Non-inferiority trials: Design concepts and issues— The encounters of academic consultants in statistics. *Stat Med*. 2003;22:169–186.
5. Fleming TR. Design and interpretation of equivalence trials. *Am Heart J*. 2000;139:S171–S176.
6. Fleming TR. Treatment evaluation in active control studies. *Cancer Treat Rept*. 1987;71:1061–1064.
7. Freedman LS, Lowe D, Macaskill P. Stopping rules for clinical trials incorporating clinical opinion. *Biometrics*. 1984;40:575–586.
8. Fleming TR. Current issues in non-inferiority trials. *Stat Med*. 2008;27:317–332.
9. Rothmann M, Li N, Chen G, Temple R, Tsou R. Design and analysis of non-inferiority mortality trials in oncology. *Stat Med*. 2003;22:239–264.
10. Hung HMJ, Wang SJ. Statistical considerations for noninferiority trial designs without placebo. *Stat Biopharm Res*. 2013;5:239–247.
11. Office of Communications, Division of Drug Information Center for Drug Evaluation and Research Food and Drug Administration: Non-Inferiority Clinical Trials to Establish Effectiveness, Guidance for Industry. http://www.fda.gov/Drugs/GuidanceComplianceRegulatoryInformation/Guidances/default.htm. November 2016.
12. Siegel JP. Equivalence and noninferiority trials. *Am Heart J*. 2000;139:S166–S170.
13. International Conference on Harmonisation. July 2000. Guidance E10: Choice of Control Group and Related Issues in Clinical Trials.
14. Pater C. Equivalence and noninferiority trials: Are they viable alternatives for registration of new drugs? (III) *Curr Control Trials Cardiovasc Med*. 2004;5(1):8.
15. Snapinn SM. Noninferiority trials. *Curr Control Trials Cardiovasc Med*. 2000;1:19–21.
16. Everson-Stewart S, Emerson SS. Bio-creep in non-inferiority clinical trials. *Stat Med*. 2010;29:2769–2780.
17. Blackwelder WC. Showing a treatment is good because it is not bad: When does non-inferiority imply effectiveness? *Control Clin Trials*. 2002;23:52–54.
18. Hwang IK, Morikawa T. Design issues in noninferiority/equivalence trials. *Drug Inf J*. 1999;33:1205–1218.
19. Committee for Medicinal Products for Human Use, the European Medicines Agency. 2005. Guideline on the choice of the non-inferiority margin [online]. Available from: http://www.ema.europa.eu/docs/en_GB/document_library/Scientific_guideline/2009/09/WC500003636.pdf (last accessed 01 Feb 2016).
20. Chow SC, Shao J. On non-inferiority margin and statistical tests in active control trial. *Stat Med*. 2006;25:1101–1113.
21. Committee for Medicinal Products for Human Use (CHMP). Guideline on the choice of the non-inferiority margin. *Stat Med*. 2006;25:1628–1638.
22. Hung HMJ, Wang SJ, O'Neill RT. A regulatory perspective on choice of margin and statistical inference issue in non-inferiority trials. *Biom J*. 2005;47:28–36.
23. Holmgren EB. Establishing equivalence by showing that a specified percentage of the effect of the active control over placebo is maintained. *J Biopharm Stat*. 1999;9:651–659.
24. Hasselblad V, Kong D. Statistical methods for comparison to placebo in active-control trials. *Drug Inf J*. 2001;35:435–449.
25. Fisher LD, Gent M, Bullet HR. Active controlled trials: what about a placebo? A method illustrated with clopidogrel, aspirin, and placebo. *Am Heart J* 2001;141:26–32.
26. Simon R. Bayesian design and analysis of active control clinical trials. *Biometrics*. 1999;55:484–487.

27. ICH Expert Working Group. 1998. ICH harmonised tripartite guideline: statistical principles for clinical trials (E9) [online]. Available from: http://www.ich.org/fileadmin/Public_Web_Site/ICH_Products/Guidelines/Efficacy/E9/Step4/E9_Guideline.pdf (last accessed 01 Feb 2016).

28. Freedman LS. Tables of the number of patients required in clinical trials using the logrank test. *Stat Med.* 1982;1:121–129.

29. Schoenfeld DA. Sample size formula for the proportional hazards regression model. *Biometrics.* 1983;39:499–503.

30. Schoenfeld DA, Richter JR. Nomograms for calculating the number of patients needed for a clinical trial with survival as an endpoint. *Biometrics.* 1982;38:163–170.

31. Rothmann MD, Wiens BL, Chan ISF. *Design and analysis of Non-inferiority Trials.* Chapman & Hall/CRC, 2011.

32. Pigeot I, Schafer J, Rohmel J, Hauschke D. Assessing the therapeutic equivalence of two treatments in comparison with a placebo group. *Stat Med.* 2003;22:883–899.

33. Hauschke D, Pigeot I. Establishing efficacy of a new experimental treatment in the 'Gold Standard' design. *Biom J.* 2005;47(6):782–786.

34. Liu J-T, Tzeng C-S, Tsou H-H. Establishing non-inferiority of a new treatment in a three-arm trial: Apply a step-down hierarchical model in a papulopustular acne study and an oral prophylactic antibiotics study. *Int J Stats Med & Bio.* 2014;3:11–20.

35. Committee for Proprietary Medicinal Products (CPMP) 1999. Switching between superiority and non-inferiority.

36. Jennison C, Turnbull BW. *Group Sequential Methods with Applications to Clinical Trials.* London: Chapman & Hall/CRC, 2000.

37. DeMets DL, Lan KK. Interim analysis: The alpha spending function approach. *Stat Med.* 1994;13:1341–1352.

38. Fisher LD, Dixon DO, Herson J, Frankowski RK, Hearron MS, Peace KE. Intention to treat in clinical trials. In: *Statistical Issues in Drug Research and Development*, Peace K.E., ed. New York: Marcel Dekker, 331–50, 1990.

39. Piaggio G, Elbourne DR, Pocock SJ et al. Reporting of noninferiority and equivalence randomized trials: Extension of the CONSORT 2010 statement. *JAMA.* 2012;308(24):2594–2604.

40. Brittain E, Lin D. A comparison of intent-to-treat and per-protocol results in antibiotic non-inferiority trials. *Stat Med.* 2005;24:1–10.

41. Jones B, Jarvis P, Lewis JA, Ebbutt AF. Trials to assess equivalence: The importance of rigorous methods. *Br Med J.* 1996;313:36–39.

42. Durrleman R, Simon R. Planning and monitoring of equivalence studies. *Biometrics.* 1990;46(2):329–36.

43. Little RJ, Rubin D. *Statistical Analysis with Missing Data.* Hoboken: John Wiley & Sons, Inc, 2002.

44. Schafer JL. *Analysis of Incomplete Multivariate Data.* New York: Chapman and Hall, 1997.

45. Schulz KF, Altman DG, Moher D, the CONSORT Group. CONSORT 2010. Statement: Updated guidelines for reporting parallel group randomised trials. *Trials.* 11:32.

46. Crump M, Kuruvilla J, Couban S et al. Randomized comparison of gemcitabine, dexamethasone, and Cisplatin versus Dexamethasone, Cytarabine, and Cisplatin Chemotherapy before autologous stem-cell transplantation for relapsed and refractory aggressive lymphomas: NCIC-CTG LY.12. *J Clin Oncol.* 2014;32:3490–3496.

47. O'Brien PC, Fleming TR. A multiple testing procedure for clinical trials. *Biometrics.* 1979;35:549–556.

48. Crook JM, O'Callaghan CJ, Duncan G, et al. Intermittent androgen suppression for rising PSA level after radiotherapy. *N Engl J Med.* 2012;367:895–903.

49. Willan AR. Power function arguments in support of an alternative approach for analyzing management trials. *Control Clin Trials.* 1994;15:211–219.

50. Ng TH. Issues of Simultaneous Tests for Noninferiority and Superiority. *Journal of Biopharmaceutical Statistics.* 2003; 13:629–639.

9

Design of Multi-Arm, Multi-Stage Trials in Oncology

James Wason

CONTENTS

9.1 Introduction

In many tumor types, there are multiple new treatments available for testing. As of a few years ago there were more than 1,500 new pharmaceutical treatments for oncology in the pipeline [1]. If one also considers potentially testing of multiple doses of a particular treatment, multiple combinations of distinct treatments, and multiple schedules of treatments, it is easy to see that there are a huge number of potential treatment arms to test in clinical trials. Conducting randomized controlled trials of each available treatment arm may lead to too many trials competing for the limited number of patients that can be enrolled. If traditional two-arm randomized controlled trials are to be used, then either the number of treatments to be tested must be reduced, or trials will risk being underpowered.

An alternative approach is to test several experimental treatments within a single trial. This is known as a *multi-arm trial*. There are several main advantages to a multi-arm trial, compared to several separate trials, as described in Parmar et al. [2]. First, in the case that a control group is included in the trial, fewer patients are required for a multi-arm trial compared to testing the same number of treatment arms in separate randomized trials: this is due to fewer patients being required for the control arm. Second, a multi-arm trial will generally permit head-to-head comparison between experimental treatments without the heterogeneity arising from multiple trials. Thirdly, it is generally less effort to run one multi-arm trial compared to running several separate trials. Fourth, a multi-arm trial increases the chance of a patient getting a new treatment instead of control, which might be appealing to patients and thus improve recruitment [3].

There are also some drawbacks of multi-arm trials. Generally, a multi-arm trial is larger and more expensive than a typical trial, meaning it may be more difficult to secure funding. It also may be difficult in practice to cope with the demands of running a large multi-arm trial. Other disadvantages include having to specify common inclusion/exclusion criteria and a primary endpoint for each arm and requiring patients to consent to being randomized to all available treatments. However, these disadvantages may be possible to address with future methodological research.

Some additional advantages arise from considering an adaptive design within a multi-arm trial. Adaptive designs allow modifications to be made during a trial using information collected, in a statistically robust way [4]. Within a multi-arm trial, this may include stopping recruitment to arms that have not performed well on patients enrolled to them, or potentially changing the allocation to prioritize arms that have performed well. Although referred to by different names in the literature, I will refer to an adaptive multi-arm trial as a multi-arm, multi-stage (MAMS) trial throughout this chapter.

There has been much methodological work for multi-arm and MAMS trials over the past decade, as well as work that reviews different methodology available. The aim of this chapter is not to recreate this previous work, but instead to provide guidance on how one can use this existing methodology to actually design multi-arm and MAMS trials. Thus the focus will be on how to implement methods, with R code referenced available in the companion website to the book https://www.crcpress.com//9781138083776.

Although this chapter restricts attention to frequentist approaches, most methods described can be implemented in a Bayesian way, with more relevant information available from Berry et al. [5].

9.2 Notation

9.2.1 Multi-Arm Trial

We will assume that the trial is testing K experimental treatments against a single shared control treatment. The effect of each treatment is measured through an endpoint, which can be assumed to follow a parametric family distribution (in this chapter we will restrict attention to normally distributed, binary, and time-to-event outcomes). We use $\delta_k, k = 1, \ldots,$ K to represent the difference in effect (as measured by the endpoint) between experimental treatment k and the control. For example:

1. For normally distributed endpoints, δ_k would represent the difference between the mean outcome on the kth experimental arm and the mean outcome on the control arm.

2. For binary endpoints, δ_k could represent the difference between the endpoint probabilities of the kth experimental arm and control, or it could represent the log odds ratio between experimental arm k and control.

3. For time to event endpoints, δ_k could represent the log hazard ratio between arm k and control.

We will consider null hypotheses indexed by k, of the form: $H_0^{(k)} : \delta_k \leq 0$ for $k = 1, \ldots, K$, with δ_k values parameterized so that positive values represent experimental treatment k being superior to control and negative values represent inferiority. That is, we are testing null hypotheses representing the experimental treatments being equal or inferior to the control treatment, with rejecting a null hypothesis meaning the respective experimental treatment is recommended as superior over control.

To test these hypotheses, data is gathered on how the patients respond to treatment. The outcome data for the ith patient allocated to experimental arm k is Y_{ki}, which will be assumed to be univariate and coming from one of the three distributions mentioned above. The number of patients allocated to arm k is denoted as n_k ($k = 0$ representing the control arm), with $N = \Sigma_{k=0}^{K} n_k$ representing the total number of patients.

Once the trial is complete and the data $\{Y_{ki}; k = 0, \ldots, K, i \in 1, \ldots, n_k\}$ are observed, test statistics are calculated for testing each of the null hypotheses. We denote the test statistic for testing hypothesis $H_0^{(k)}$ as Z_k. Typically, Z_k will be chosen so that its distribution is asymptotically standard normal when $\delta_k = 0$ and has a positive mean when $\delta_k > 0$. Thus, the null hypothesis will be rejected when Z_k is above some critical value, c_k which is pre-specified and chosen such that the probability of incorrectly rejecting $H_0^{(k)}$ when it is actually true, is $\leq \alpha_k$. Here, α_k represents the maximum type I error rate for $H_0^{(k)}$. Unless otherwise said, the same critical value, c, and type I error rate, α, will be used for each arm $k = 1, \ldots, K$.

The power of a multi-arm trial is more complex than in an RCT. As there are multiple hypotheses of interest, we have to consider what is the objective of the trial. Is it to find all effective treatments that exist or just to find one sufficiently promising treatment? Do we wish to demonstrate that a particular treatment is truly best, or just whether it is better than control? These issues may be even more complicated if there are subgroups present for which different treatment effects may be expected. This last issue is not considered in this chapter except briefly in the Conclusion.

There are different ways of considering the power when multiple hypotheses are present. Two of these are the *disjunctive* and *conjunctive* powers. The disjunctive power is the probability of rejecting one or more false null hypotheses, whereas the conjunctive power is the probability of rejecting all false null hypotheses. The conjuctive power will typically require higher sample sizes to control at a particular level than the disjunctive power.

9.2.2 Multi-Arm, Multi-Stage

For MAMS, we will use the notation and concepts described in Section 9.2.1, but require some additional concepts. These generally involve denoting the information gathered at the time of each interim analysis for each treatment.

We will assume that the maximum number of stages of the trial is J, which means there are $J - 1$ interim analyses and one final analysis. The maximum likelihood estimator of the parameter δ_k using all information gathered up to analysis j is denoted as $\hat{\delta}_{jk}$. Thus, in the MAMS case, we can form a vector of MLE estimates for treatment k at the different analyses, $\hat{\delta}_k = (\hat{\delta}_{1k}, \hat{\delta}_{2k}, \ldots, \hat{\delta}_{Jk})^T$. As in group-sequential designs (see Jennison and Turnbull [6]), it is helpful to consider the *information* for parameter δ_k at analysis j, I_{jk}, which is $1/\text{SE}(\hat{\delta}_{jk})^2$.

The Wald test represents a convenient way to test the hypotheses $H_0^{(1)}, H_0^{(2)}, \ldots, H_0^{(K)}$. We denote the Wald test for hypothesis $H_0^{(k)}$ using all outcome data gathered up to stage j as

$$Z_{jk} = \frac{\hat{\delta}_{jk}}{\text{SE}(\hat{\delta}_{jk})} = \sqrt{I_{jk}}\,\hat{\delta}_{jk} \tag{9.1}$$

As described further in Jennison and Turnbull [6], the asymptotic distribution (under assumptions which are met for cases considered in this chapter) of the Wald statistics for testing hypothesis k is:

$$(Z_{1k}, Z_{2k}, \ldots, Z_{Jk})^T \sim MVN\left(\delta_k\left(\sqrt{I_{1k}}, \sqrt{I_{2k}}, \ldots, \sqrt{I_{Jk}}\right), \Sigma\right)$$

$$\Sigma_{lm} = \frac{\sqrt{I_{lk}}}{\sqrt{I_{mk}}} \quad \text{for } l = 1, \ldots, m-1; m = 2, \ldots, JK \tag{9.2}$$

$$\Sigma_{lm} = \Sigma_{ml} \quad \text{for } l > m$$

$$\Sigma_{ll} = 1$$

As in Reference [7], it is useful, for representing the events that can occur in MAMS designs, to define random variables $\phi = (\phi_1, \ldots, \phi_K)$ and $\psi = (\psi_1, \ldots, \psi_K)$. First, ϕ represents the last stage of the trial that a treatment is present ($\phi_k = J$ if experimental treatment k still remains in the trial until the end). Second, ψ represents whether or not each experimental treatment was recommended ($\psi_k = 1$ if $H_0^{(k)}$ is rejected and 0 otherwise). All possible outcomes of the trial can be summarized with different values taken by ϕ and ψ.

As an example of the notation, imagine a MAMS trial with three experimental arms and two stages. If $\phi = (1, 2, 1)$ and $\psi = (0, 1, 1)$, then this means that experimental treatment 1 was dropped for futility after stage 1, treatment 2 continued to the second stage and ended up being recommended, and treatment 3 was stopped for efficacy after stage 1.

9.3 Determining Statistical Quantities for Multi-Arm Trials

To choose a suitable sample size for a multi-arm trial, we need to first be able to work out the statistical operating characteristics for a design with a given sample size and critical value determining whether null hypotheses should be rejected or not. In this section, we will first examine the distribution of the test statistics Z for different endpoint types. We will then consider how to calculate different error rates and power. Lastly, we will consider how to choose a design that controls the error rate and power at a desirable level.

9.3.1 Distribution of Test Statistics from a Multi-Arm Trial

Recall from Section 9.2.1 that null hypotheses for each experimental treatment $H_0^{(1)}, \ldots, H_0^{(K)}$ are tested with test statistics $Z = (Z_1, \ldots, Z_K)$. Knowing the theoretical distribution of this vector Z will be useful for determining a suitable design. We consider normal outcomes, binary outcomes, and time-to-event outcomes.

9.3.1.1 Normal Outcomes

The most straightforward case is when the outcome is normally distributed with a known variance. That is, Y_{ki}, the outcome of the ith patient on the kth arm, is $N\left(\mu_k, \sigma_k^2\right)$, where σ_k^2 is known. The test statistic for arm k will be

$$Z_k = \frac{1}{\sqrt{\left(\sigma_k^2/n_k\right) + \left(\sigma_0^2/n_0\right)}} \left(\frac{\sum_{i=1}^{n_k} Y_{ki}}{n_k} - \frac{\sum_{i=1}^{n_0} Y_{0i}}{n_0} \right). \tag{9.3}$$

Since the the vector of test statistics, $Z = (Z_1, \ldots, Z_K)$ can be expressed as an affine transformation (i.e., of the form $MY + b$) of a vector of independent univariate normal distributions, it is exactly distributed as a multivariate normal with mean μ and covariance Σ. The value of μ and Σ will be

$$\mu = \left(\frac{1}{\sqrt{\left(\sigma_1^2/n_1\right) + \left(\sigma_0^2/n_0\right)}} \delta_1, \ldots, \frac{1}{\sqrt{\left(\sigma_K^2/n_K\right) + \left(\sigma_0^2 n_0\right)}} \delta_K \right)$$

$$\Sigma_{ii} = 1 \tag{9.4}$$

$$\Sigma_{ij} = \frac{1}{\sqrt{\left(\sigma_i^2/n_i\right) + \left(\sigma_0^2/n_0\right)}} \frac{1}{\sqrt{\left(\sigma_j^2/n_j\right) + \left(\sigma_0^2/n_0\right)}} \frac{\sigma_0^2}{n_0}$$

Note in Equation 9.4 that if the variance and sample size of each arm is identical, the covariance matrix simplifies to a matrix with 1s along the diagonal and 0.5 in each other entry.

In the case where the variance parameters are unknown and to be estimated during the trial, the joint distribution is more complicated. The test statistics in this case would individually be two-sample t tests:

$$Z_k = \frac{\left(\left(\sum_{i=1}^{n_k} Y_{ki}/n_k \right) - \left(\sum_{i=1}^{n_0} Y_{0i}/n_0 \right) \right)}{\bar{s}_k}, \tag{9.5}$$

where $\bar{s}_k^2 = \sqrt{\left(s_k^2/n_k\right) + \left(s_0^2/n_0\right)}$ is an estimator of the variance of the numerator. Here, $s_k^2 = (1/(n-1))\sum_{i=1}^{n_k}(y_{ki} - \bar{y}_k)^2, k = 0, \ldots K.$

Although the multivariate t distribution exists, the vector of test statistics will not generally form a multivariate t distribution unless the variance is assumed to be the same in each arm and a pooled estimator is used [8]. Asymptotically, the distribution of Z is multivariate normal; however it is recommended that the critical value for testing the hypothesis is adjusted when the variance is estimated, as described in Wason et al. [9].

9.3.1.2 Binary Outcome

When the outcome is binary, with $Y_{ki} \sim Bi(1, p_k)$, the difference in effect between arm k and control, δ_k, could be expressed in a number of different ways. For example, δ_k could be defined as $p_k - p_0$, the difference in the binary proportions. It could also be defined as the log odds ratio $\delta_k = \log(p_k/(1 - p_k)) - \log(p_0/(1 - p_0))$.

In both cases, Z, the vector of standardized test statistics will be asymptotically normally distributed. We consider each case separately.

First, if $\delta_k = p_k - p_0$, the MLE for δ_k will be $\hat{p}_k - \hat{p}_0$. The statistic for testing hypothesis $H_0^{(k)}$ will then be

$$Z_k = \frac{\hat{p}_k - \hat{p}_0}{\sqrt{Var(\hat{p}_k) + Var(\hat{p}_0)}}$$

$$= \frac{\hat{p}_k - \hat{p}_0}{\sqrt{(p_k(1-p_k)/n_k) + (p_0(1-p_0)/n_0)}} \tag{9.6}$$

The covariance between Z_j and $Z_k, j \neq k$ is:

$$Cov(Z_j, Z_k) = \frac{(p_0(1-p_0)/n_0)}{\sqrt{(p_j(1-p_j)/n_j) + (p_0(1-p_0)/n_0)}\sqrt{(p_k(1-p_k)/n_k) + (p_0(1-p_0)/n_0)}} \tag{9.7}$$

Unlike the normally distributed case, the covariance does depend on the mean effect of the treatments. Note that if the sample size per arm is the same ($n_1 = \cdots = n_K$) and all treatments have the same effect ($p_1 = \cdots = p_K$), then the covariance between test statistics will be 0.5.

If the log odds ratio is used as the parameter of interest instead of the difference between proportions, then it is helpful to use the approximation for the variance of the estimated log odds:

$$Var\left(\log\left(\frac{\hat{p}_k}{1-\hat{p}_k}\right)\right) \approx \frac{1}{n_k p_k} + \frac{1}{n_k(1-p_k)}, \tag{9.8}$$

which is a result of applying the delta method to a normal approximation of $\hat{p}_k$.

With this result, the test statistic for testing $H_0^{(k)}$ is:

$$
\begin{aligned}
Z_k &= \frac{\log(\hat{p}_k/(1-\hat{p}_k)) - \log(\hat{p}_0/(1-\hat{p}_0))}{\sqrt{\mathrm{Var}(\log(\hat{p}_k/(1-\hat{p}_k))) + \mathrm{Var}(\log(\hat{p}_0/(1-\hat{p}_0)))}} \\
&\approx \frac{\log(\hat{p}_k/(1-\hat{p}_k)) - \log(\hat{p}_0/(1-\hat{p}_0))}{\sqrt{\dfrac{1}{n_k\hat{p}_k} + \dfrac{1}{n_k(1-\hat{p}_k)} + \dfrac{1}{n_0\hat{p}_0} + \dfrac{1}{n_0(1-\hat{p}_0)}}}
\end{aligned}
\tag{9.9}
$$

The covariance between Z_j and Z_k is approximately:

$$
\mathrm{Cov}(Z_j, Z_k) \approx \frac{(1/n_0 p_0) + (1/n_0(1-p_0))}{\left[\begin{array}{l} \sqrt{(1/n_j p_j) + (1/n_j(1-p_j)) + (1/n_0 p_0) + (1/n_0(1-p_0))} \\ \times \sqrt{(1/n_k p_k) + (1/n_k(1-p_k)) + (1/n_0 p_0) + (1/n_0(1-p_0))} \end{array}\right]}
\tag{9.10}
$$

As before, if all binary proportions and sample sizes are the same, then this asymptotic covariance is 0.5.

9.3.1.3 *Time-to-Event Outcome*

In the time-to-event setting, the null hypotheses being tested will generally use the hazard ratio between experimental arms and control. Since the hazard ratio may vary over the course of the trial, here we will only assume settings where all hazards are proportional. In that case, we define δ_k as the log of the hazard ratio between control and experimental k. Given that events are undesirable outcomes, this setup keeps our null hypotheses of the same form as previously, i.e., $H_0^{(k)} : \delta_k = \log(\lambda_0) - \log(\lambda_k) \leq 0$.

If events represent desirable outcomes, such as going into remission, then one might consider the hazard ratio between the experimental and control, so that the null hypothesis is of the same form.

At the design stage, it is common to use the distribution of the log-rank test for choosing a sample size. The log-rank test for comparing arm k to control can be written as

$$
Z_k = \frac{\sum_{i=1}^{I}(O_{ki} - E_{ki})}{\sqrt{\sum_{i=1}^{I} V_{ki}}},
\tag{9.11}
$$

where $i = 1, \ldots, I$ represents the distinct times of observed events in arm k or control; O_{ki} is the observed number of events in arm k at time i; and E_{ki} is the expected number under the null hypothesis.

There has been some literature on the joint distribution of log-rank tests from multi-arm trial settings such as Barthel et al. [10]. The marginal asymptotic distribution of Z_k is:

$$
Z_k \sim N\left(\delta_k \frac{\sqrt{(n_0 d_0 + n_k d_k)}}{4}, 1\right)
$$

where n_k is the number of participants recruited to arm k and d_k is the proportion of participants in arm k that have events by the analysis. The correlation is a more complicated formula, but we will use a formula that provides a good approximation under proportional hazards:

$$\text{Cor}(Z_j, Z_k) \approx \frac{(1/n_0 d_0)}{\sqrt{(1/n_0 d_0)+(1/n_j d_j)}\sqrt{(1/n_0 d_0)+(1/n_k d_k)}} \tag{9.12}$$

9.3.2 Evaluating the Operating Characteristics of a Multi-Arm Design

In the previous subsection, we found the joint distribution of the test statistics given a specified treatment effect and sample size. We now will move on to showing how this joint distribution can be used to find the statistical properties of a design, including different type I error rates and power definitions that may be of interest.

9.3.2.1 Type I Error Rate

As mentioned in Section 9.2.1, the type I error rate of a multi-arm trial is more complex than the RCT case. Here we will show how to find the type I error rate per arm and discuss how to control for multiple testing.

First we consider the type I error rate for arm k, i.e., the probability of falsely recommending that arm k is better than control when $H_0^{(k)}$ is true. To find this, we simply need to find the probability $\mathbb{P}\left(Z_k > c | H_0^{(k)}\right)$. This is straightforward in all of the multi-arm cases we consider here as the marginal distribution of Z_k is not influenced by the results of experimental arms other than k.

Before continuing, it is worth commenting on situations when this might not be true, as this demonstrates the need to be careful about how the data is collected. As an example, imagine one is conducting a multi-arm trial with a time-to-event endpoint, and the patients are followed until a certain number of events are observed. This may lead to situations where the distribution of Z_k is affected by how good a different experimental arm is. For example, if other experimental arms are very poor, so events are collected quickly, then less information (events) will be collected on arm k than if other arms are very good. Either the design can be changed so a fixed follow-up time is used for all arms, or one would have to consider the maximum probability of $\mathbb{P}(Z_k > c)$ when $H_0^{(k)}$ is true, across all other possible effects of other experimental arms, which may be non-trival.

We will continue to assume we are in the situation where the distribution of Z_k is not affected by other arms (i.e., $\mathbb{P}(Z_k > c | \delta, n) = \mathbb{P}(Z_k > c | \delta_k, n_k, n_0)$); it suffices to consider the maximum probability $\mathbb{P}(Z_k > c | H_0)$, which will generally be $\mathbb{P}(Z_k > c | \delta = 0)$. This is straightforward to evaluate by considering the tail probability of a normal distribution.

We might then choose c so that the maximum type I error rate across each arm is α. This would then mean that the *per-arm* type I error rate is controlled at α.

In the above, we have made the assumption that the highest type I error rate is when $\delta = 0$. Asymptotically for the cases we consider in this chapter, this will be the case. However, for smaller sample sizes, it may not be the true. If controlling the type I error rate exactly is vital, then it may be prudent to do simulations to confirm that the choice of c from the above procedure gives the correct type I error rate.

If it is desirable to consider the total chance of making a type I error across all arms tested, then we need to consider the joint (multivariate) distribution of Z. It is common in trials to consider the *family-wise error rate* (FWER). This is the total probability of making a type I error across a set (or family) of null hypotheses. Here the family of hypotheses would be $H_0^{(1)}, \ldots, H_0^{(K)}$. The FWER depends on which null hypotheses are actually true (and, more specifically, the actual values of δ). There are two commonly used definitions for control of the family-wise error rate. *Weak control* of the FWER at level α means that the FWER is below α for every value of δ such that all null hypotheses are true. That is, all values of δ such that $\delta_k \leq 0 \, \forall k = 1, \ldots, K$. *Strong control* of the FWER at level α is that the FWER is below α for all possible values of δ (i.e., regardless of which null hypotheses are actually true). Strong control of FWER implies weak control, although the converse is not the case. Generally, regulatory agencies encourage confirmatory trials testing multiple hypotheses to strongly control the FWER, although there is controversy over whether or not it is always necessary [11].

In many cases, controlling the FWER under the *global null hypothesis* at level α provides strong control at level α. The global null hypothesis, H_G, is $\delta_1 = \delta_2 = \cdots = \delta_K = 0$. A sufficient condition for this result to be true is that increasing δ_k increases $\mathbb{P}\left(\text{Reject } H_0^{(k)}\right)$ but does not change the probabilities of rejecting any of the other null hypotheses. Intuitively, this is because if one begins at H_G and increases one or more of the δ components, then the respective null hypotheses become false and the probability of rejecting a true null hypothesis falls. If, starting from H_G, one or more of the δ components decreases then the probability of rejecting these (true) null hypotheses falls, but the probability of rejecting other null hypotheses does not change, then this causes the total chance of making a type I error to fall.

To actually calculate the FWER under H_G, we use the fact that the probability of rejecting at least one null hypothesis is equal to 1 minus the probability of not rejecting any, which has a convenient expression:

$$\mathbb{P}(\text{Reject at least one null} \mid H_G) = 1 - \mathbb{P}(\text{Reject no null hypotheses} \mid H_G)$$
$$= 1 - \mathbb{P}(Z_1 \leq c, Z_2 \leq c, \ldots, Z_K \leq c \mid H_0)$$
$$= 1 - \int_{-\infty}^{c} \int_{-\infty}^{c} \cdots \int_{-\infty}^{c} f_Z((x_1, \ldots, x_k)) dx_1 dx_2 \ldots dx_K, \tag{9.13}$$

where f_Z is the probability density function of the joint distribution of Z under H_G.

In the cases we consider, where Z is asymptotically multivariate normal, there are efficient ways to evaluate the multidimensional integral in Equation 9.13, such as the method of Genz and Bretz [12], which is implemented in standard software packages such as R and Stata.

In multiarm.R, a function called fwer.hg evaluates the FWER of a design with critical value c and asymptotic covariance matrix Σ. The following code finds the FWER of a design with three experimental arms for different critical values:

```
fwer=fwer.hg(seq(1.5,3,length=100),K=3)
```

Figure 9.1 uses these results to show how the FWER changes as the critical value increases. Compared to the critical value that gives a one-sided type I error rate of 5% (1.65), the critical value needed for 5% FWER in this case is 2.06.

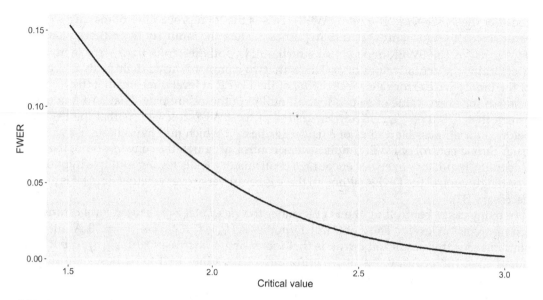

FIGURE 9.1
Family-wise error rate of a multi-arm trial with three experimental arms as the critical value changes.

9.3.3 Power

As discussed in Section 9.2, there are several definitions of power for a multi-arm trial. Here we provide expressions for them and demonstrate code that can be used in R. We first introduce notation for *interesting* and *uninteresting* treatment effects, which are used in different definitions. The *interesting* treatment effect, $\delta^{(1)}$, is a clinically relevant treatment effect that we would like the trial to have high power to detect. The *uninteresting* treatment effect, $\delta^{(0)}$ is a non-interesting treatment effect representing a treatment that we would not wish to consider further. In some cases, $\delta^{(0)}$ might be set to 0, so that any improvement is seen as important; in other cases it might be set to a low positive value.

9.3.3.1 Conjunctive Power

The conjuctive power is the probability of rejecting all false null hypotheses. We will assume that a false null hypothesis is represented by the treatment effect being $\delta^{(1)}$. As an example, if all nulls are false, then the conjuctive power is the probability of all test statistics being above c:

$$\mathbb{P}(Z_1 > c, \ldots, Z_K > c \mid \delta_1 = \delta^{(1)}, \ldots, \delta_K = \delta^{(1)}) = \int_c^\infty \cdots \int_c^\infty f_Z(x_1, \ldots, x_K) dx_1 \ldots dx_K,$$

where here, f_Z represents the pdf of the joint distribution of Z when $\delta_1 = \cdots = \delta_K = \delta^{(1)}$.

9.3.3.2 Disjunctive Power

The disjunctive power is the probability of rejecting at least one false null hypothesis. If all nulls are false, then this probability is similar in appearance to the FWER in Equation 9.13, except with the distribution of Z under the case where all nulls are false:

$$\mathbb{P}(\text{Reject at least one null} \mid \delta_1 = \delta^{(1)}, \ldots, \delta_K = \delta^{(1)})$$
$$= 1 - \mathbb{P}(\text{Reject no null hypotheses} \mid \delta_1 = \delta^{(1)}, \ldots, \delta_K = \delta^{(1)})$$
$$= 1 - \mathbb{P}(Z_1 \le c, Z_2 \le c, \ldots, Z_K \le c \mid \delta_1 = \delta^{(1)}, \ldots, \delta_K = \delta^{(1)}) \qquad (9.14)$$
$$= 1 - \int_{-\infty}^{c} \int_{-\infty}^{c} \cdots \int_{-\infty}^{c} f_Z((x_1, \ldots, x_k)) dx_1 dx_2 \ldots dx_K,$$

where f_Z represents the pdf of the joint distribution of Z when $\delta_1 = \cdots = \delta_K = \delta^{(1)}$.

9.3.3.3 Least Favorable Configuration

The least favorable configuration (LFC) was proposed by Dunnett [13] as an appropriate way to power multi-arm trials. It uses the definitions $\delta^{(1)}$ and $\delta^{(0)}$ above. The LFC is defined as the power of recommending a particular experimental treatment, without loss of generality of experimental treatment 1, when $\delta_1 = \delta^{(1)}$ and $\delta_2 = \ldots = \delta_K = \delta^{(0)}$. For the settings we consider in this chapter, this will simply be the marginal probability that $Z_1 > c$ when $\delta_1 = \delta^{(1)}$, as the other treatment effects do not affect this probability. We will see later on that the LFC can be more complex when a MAMS study is done.

9.3.3.4 Comparison of Power

To demonstrate how the power varies, Table 9.1 shows how the three different powers described above vary as the number of experimental arms and number of false nulls/effective experimental treatments vary. Effective treatments are defined by a standardized effect size $\delta_1 = 0.5$, with ineffective treatments having a mean of 0. It is assumed that 60 participants are recruited per arm (so that the mean of the test statistic for an effective treatment is $\left(\sqrt{60}/\sqrt{2}\right) \times 0.5$ and that the critical value is chosen to control the FWER at 0.05.

TABLE 9.1

Different Powers as the Number of Experimental Arms (K) and Number of Effective Treatments Vary for a Continuous Outcome

K	Number Effective Treatments	Disjunctive Power	Conjuctive Power	Power at LFC
3	1	0.75	0.75	0.75
3	2	0.88	0.62	0.75
3	3	0.93	0.54	0.75
6	1	0.67	0.67	0.67
6	3	0.88	0.43	0.67
6	6	0.95	0.30	0.67

As one would expect, the conjuctive power decreases as the number of effective treatments increases, as all respective null hypotheses have to be rejected. On the other hand, the disjunctive power increases as just one false null needs to be rejected. The power at the LFC remains the same as it is the power to recommend a specific effective treatment, which does not depend on the effect of the other treatments in this case.

9.3.4 Case Study

We consider how one might choose the sample size and critical value for an oncology case study. We will use the example of the Neosphere trial [14] which was a multi-arm trial in 407 women with HER2 positive breast cancer. The trial evaluated four arms: trastuzumab plus docetaxel (arm 0), pertuzumab plus docetaxel (arm 1), pertuzumab with trastuzumab (arm 2), and pertuzumab plus docetaxel plus trastuzumab (arm 3). The primary endpoint is pathological complete response (pCR) at surgery, which is a binary outcome. The actual trial was powered to compare arm 0 with arms 1 and 2, and arm 1 against arm 3. To simplify our illustration, we imagine how we might choose the sample size if the trial was to compare arms 1–3 versus arm 0.

The trial was powered to detect a difference of 15% in pCR rate (from 25% in arm 0 to 40% in arms 1 and 2). Here we will assume that the number of patients per arm is the same and denote this n.

We consider the sample size and critical value needed when the difference in effect is represented by (1) the difference in response probabilities and (2) the log odds ratio. The test statistics for each experimental arm control are defined in (respectively) Equations 9.6 and 9.9. To avoid confusion, the test statistics using the difference in probabilities is denoted $Z^{(d)}$ and the test statistics using the log odds ratio is denoted $Z^{(\text{LOR})}$

The joint distribution of $\left(Z_1^{(d)}, Z_2^{(d)}, Z_3^{(d)}\right)$ is

$$Z^{(d)} \sim N(\mu^{(d)}, \Sigma^{(d)}) \tag{9.15}$$

$$\mu^{(d)} = \left(\frac{p_1 - p_0}{\sqrt{\frac{p_1(1-p_1)+p_0(1-p_0)}{n}}}, \frac{p_2 - p_0}{\sqrt{\frac{p_2(1-p_2)+p_0(1-p_0)}{n}}}, \frac{p_3 - p_0}{\sqrt{\frac{p_3(1-p_3)+p_0(1-p_0)}{n}}} \right) \tag{9.16}$$

$$\Sigma_{11}^{(d)} = \Sigma_{22}^{(d)} = \Sigma_{33}^{(d)} = 1 \tag{9.17}$$

$$\Sigma_{12}^{(d)} = \Sigma_{21}^{(d)} = \frac{p_0(1-p_0)}{\sqrt{p_1(1-p_1)+p_0(1-p_0)}\sqrt{p_2(1-p_2)+p_0(1-p_0)}} \tag{9.18}$$

$$\Sigma_{13}^{(d)} = \Sigma_{31}^{(d)} = \frac{p_0(1-p_0)}{\sqrt{p_1(1-p_1)+p_0(1-p_0)}\sqrt{p_3(1-p_3)+p_0(1-p_0)}} \tag{9.19}$$

$$\Sigma_{23}^{(d)} = \Sigma_{32}^{(d)} = \frac{p_0(1-p_0)}{\sqrt{p_2(1-p_2)+p_0(1-p_0)}\sqrt{p_3(1-p_3)+p_0(1-p_0)}}, \tag{9.20}$$

and the joint distribution of $\left(Z_1^{(LOR)}, Z_2^{(LOR)}, Z_3^{(LOR)}\right)$ is

$$Z^{(LOR)} \sim N\left(\mu^{(LOR)}, \Sigma^{(LOR)}\right)$$

$$\mu_1^{(LOR)} = \frac{\log(p_1/(1-p_1)) - \log(p_0/(1-p_0))}{\sqrt{\dfrac{1}{np_1} + \dfrac{1}{n(1-p_1)} + \dfrac{1}{np_0} + \dfrac{1}{n(1-p_0)}}}$$

$$\mu_2^{(LOR)} = \frac{\log(p_2/(1-p_2)) - \log(p_0/(1-p_0))}{\sqrt{\dfrac{1}{np_2} + \dfrac{1}{n(1-p_2)} + \dfrac{1}{np_0} + \dfrac{1}{n(1-p_0)}}}$$

$$\mu_3^{(LOR)} = \frac{\log(p_3/(1-p_3)) - \log(p_0/(1-p_0))}{\sqrt{\dfrac{1}{np_3} + \dfrac{1}{n(1-p_3)} + \dfrac{1}{np_0} + \dfrac{1}{n(1-p_0)}}}$$

$$\Sigma_{11}^{(LOR)} = \Sigma_{22}^{(LOR)} = \Sigma_{33}^{(LOR)} = 1$$

$$\Sigma_{12}^{(LOR)} = \Sigma_{21}^{(LOR)} = \frac{\dfrac{1}{np_0} + \dfrac{1}{n(1-p_0)}}{\sqrt{\dfrac{1}{np_1} + \dfrac{1}{n(1-p_1)} + \dfrac{1}{np_0} + \dfrac{1}{n(1-p_0)}}\sqrt{\dfrac{1}{np_2} + \dfrac{1}{n(1-p_2)} + \dfrac{1}{np_0} + \dfrac{1}{n(1-p_0)}}}$$

$$\Sigma_{13}^{(LOR)} = \Sigma_{31}^{(LOR)} = \frac{\dfrac{1}{np_0} + \dfrac{1}{n(1-p_0)}}{\sqrt{\dfrac{1}{np_1} + \dfrac{1}{n(1-p_1)} + \dfrac{1}{np_0} + \dfrac{1}{n(1-p_0)}}\sqrt{\dfrac{1}{np_3} + \dfrac{1}{n(1-p_3)} + \dfrac{1}{np_0} + \dfrac{1}{n(1-p_0)}}}$$

$$\Sigma_{23}^{(LOR)} = \Sigma_{32}^{(LOR)} = \frac{\dfrac{1}{np_0} + \dfrac{1}{n(1-p_0)}}{\sqrt{\dfrac{1}{np_2} + \dfrac{1}{n(1-p_2)} + \dfrac{1}{np_0} + \dfrac{1}{n(1-p_0)}}\sqrt{\dfrac{1}{np_3} + \dfrac{1}{n(1-p_3)} + \dfrac{1}{np_0} + \dfrac{1}{n(1-p_0)}}}.$$

As an example, if $p_0 = 0.25$, $p_1 = 0.4$, $p_2 = p_3 = 0.25$, $n = 100$, then

$$\mu^{(d)} = (2.294, 0, 0)$$

$$\Sigma^{(d)} = \begin{pmatrix} 1 & 0.468 & 0.468 \\ 0.468 & 1 & 0.500 \\ 0.468 & 0.500 & 1 \end{pmatrix}$$

$$\mu^{(LOR)} = (2.249, 0, 0)$$

$$\Sigma^{(LOR)} = \begin{pmatrix} 1 & 0.530 & 0.530 \\ 0.530 & 1 & 0.500 \\ 0.530 & 0.500 & 1 \end{pmatrix}.$$

To find the critical value for rejecting null hypotheses, we first have to decide whether to adjust for multiple testing or not. If we do not, then for a per-hypothesis (one-sided) type I error of α, we would pick a critical value of $\Phi^{-1}(1 - \alpha)$. If we would like to control the

TABLE 9.2

Different Powers for Case Study as Number of Effective Treatments Varies and for Different Test Statistic Forms

Test Statistic	Number Effective Treatments	Disjunctive Power	Conjuctive Power	Power at LFC
Difference	1	0.592	0.592	0.592
Difference	2	0.764	0.419	0.592
Difference	3	0.841	0.324	0.592
LOR	1	0.574	0.574	0.574
LOR	2	0.726	0.422	0.574
LOR	3	0.797	0.340	0.574

family-wise error rate at α, then we would first consider the joint distribution of the test statistics under the global null, $p_0 = p_1 = p_2 = p_3$. In this case, the joint distributions of $Z^{(d)}$ and $Z^{(LOR)}$ will both be asymptotically multivariate normal with mean 0 and covariance matrix $\begin{pmatrix} 1 & 0.5 & 0.5 \\ 0.5 & 1 & 0.5 \\ 0.5 & 0.5 & 1 \end{pmatrix}$. The methods in Section 9.3.2.1 can be used to find the critical value for a required FWER of α. For $\alpha = 0.05$, the critical value is 2.062.

We can then determine what the power is for $n = 100$. Defining an effective treatment as one that increases the pCR rate from 0.25 to 0.4, and an ineffective one as one that has pCR rate 0.25, Table 9.2 shows the different types of power for $n = 100$ and critical value 2.062.

As Table 9.2 shows, there is a large variety in the power depending on what definition is used. If a power of 80% is required, then $n = 100$ is slightly overpowered for the case where the difference in proportions is used to form the test statistic, the disjunctive power is used and all three arms are assumed to be effective. However in all other cases it is not. As an indication, if the conjuctive power was to be used instead in the previous example, a sample size of around $n = 225$ per arm would be needed for 80% power.

It is interesting to note the differences between using the difference in proportions to form the test statistic instead of the log odds ratio. In most cases, the former is more powerful. However, generally, the estimated log odds ratio is closer to normally distributed, so it should have correct properties (i.e., type I error rate) when the sample size is small. It is advisable to use the above procedures as a guide and then to conduct simulations using the actual analysis to be proposed to ensure the type I error rate and power are correct when asymptotic assumptions are not made.

9.4 Designing Multi-Arm Multi-Stage Trials

As for multi-arm trials, for MAMS trials if we can determine what the distribution of the test statistics is under different scenarios, then we can find the probability of the trial recommending a particular treatment, or set of treatments. In most cases, it is sufficient to consider the joint distribution of test statistics in the case that all treatment arms continue to the end of the trial, and then use multivariate integration to determine the chance of the test statistics taking values such that particular treatments are recommended.

However this approach may not always be sufficient; later we will briefly consider such situations. For example, one example would be if the sample size enrolled per treatment depended on how many treatments there are remaining in the trial.

We will first consider the distribution of the test statistics for each treatment at each stage, as introduced in Section 9.2.2 for different endpoint types. Then it will be shown how to use this distribution to find a design's operating characteristics. We will consider both group-sequential MAMS designs and drop-the-loser designs, which are defined later. Finally, an example of how to choose the design in practice is provided.

9.4.1 Distribution of Test Statistics

We follow the notation in Section 9.2.2, with J stages in the trial, K experimental treatments, and Z_{jk} representing the test statistic for testing $H_0^{(k)}$, using data from all patients assessed up to stage j. We will use the Wald test defined in Equation 9.1.

From the joint canonical distribution, we know that the test statistics from each experimental arm form a multivariate normal distribution with distribution given in Equation 9.2. However, for the joint distribution of $(Z_{11}, Z_{12}, \ldots , Z_{JK})$, that still leaves the covariance between test statistics from different arms. Here we consider an assumption that simplifies finding this covariance.

We consider the situation that the maximum likelihood estimator at stage j of δ_k, $\hat{\delta}_{jk}$ can be expressed as the difference between the estimates of parameters at stage j, reflecting the endpoint in experimental arm k and the other in the control arm, i.e,

$$\hat{\delta}_{jk} = \hat{\mu}_{jk} - \hat{\mu}_{j0}. \tag{9.21}$$

In that case, the covariance between the MLEs of distinct parameters δ_k and δ_l at stages j_1 and j_2, $\mathrm{Cov}(\hat{\delta}_{j_1k}, \hat{\delta}_{j_2l})$, can be written as:

$$\begin{aligned}
\mathrm{Cov}\left(\hat{\delta}_{j_1k}, \hat{\delta}_{j_2l}\right) &= \mathrm{Cov}\left(\hat{\mu}_{j_1k} - \hat{\mu}_{j_10}, \hat{\mu}_{j_2l} - \hat{\mu}_{j_20}\right) \\
&= \mathrm{Cov}\left(\hat{\mu}_{j_10}, \hat{\mu}_{j_20}\right),
\end{aligned} \tag{9.22}$$

which allows us a way to specify the entire joint distribution.

9.4.2 Group-Sequential MAMS

Given the joint distribution of Z, the vector of test statistics at each stage for testing each hypothesis $H_0^{(k)}$, we can now find the statistical properties of different trial designs. We first start with the group-sequential MAMS design, where experimental arms can be dropped according to whether test statistics are above or below *stopping boundaries*. There are two types of stopping boundaries: *futility* and *efficacy*.

The futility boundary, which at stage j we will denote f_j, determines when a treatment is showing insufficient promise to continue. More specifically, if treatment k is still in the trial at stage j, and Z_{jk} is below f_j, then this means that treatment k is not showing sufficient promise and should be dropped from the trial.

The efficacy boundary, denoted by e_j at stage j, determines when a treatment is showing sufficient promise to reject the associated null hypothesis. That is, if treatment k is still in the trial at stage j and Z_{jk} is above e_j, then we would reject $H_0^{(k)}$. Either futility or efficacy stopping can be disallowed by setting boundaries to $-\infty$ and ∞ respectively.

Stopping for efficacy is more complex in the multi-arm, multi-stage setting than in the group-sequential RCT setting. Stopping an individual arm for efficacy can either lead to stopping the whole trial or continuing with arms which have not stopped. Comparisons between these two settings in terms of the statistical properties are presented in Wason [7]. In this chapter, we concentrate on the latter choice as it simplifies the evaluation of the statistical properties, although a brief comment on the former case is made later in Section 9.4.2.2.

To find the statistical properties of the trial design using the joint distribution, we first consider in which situations a null hypothesis is rejected. For $H_0^{(k)}$ to be rejected at the third stage of a five stage trial, the following events need to occur:

1. $f_1 \leq Z_{1k} \leq e_1$
2. $f_2 \leq Z_{2k} \leq e_2$
3. $Z_{3k} > e_3$

Note that this event occuring does not depend on test statistics for other experimental arms, nor does it depend on Z_{4k} or Z_{5k}.

For more general cases, the ϕ and ψ notation introduced in Section 9.2.2 is useful. Recall that (1) $\psi_k = 1$ if the trial results lead to $H_0^{(k)}$ being rejected, and 0 otherwise and (2) $\phi_k \in (1, \ldots, J)$ represents the stage that treatment k was last present in the trial.

Then the probability of rejecting $H_0^{(k)}$ is simply $\mathbb{P}(\psi_k = 1)$, but calculating this probability is easier if the additional ϕ_k variable is introduced, so that $\mathbb{P}(\psi_k = 1) = \sum_{j=1}^{J} \mathbb{P}(\psi_k = 1, \phi_k = j)$. The terms in this sum can each be expressed as multivariate integrations. For example,

$$\mathbb{P}(\psi_k = 1, \phi_k = j) = \mathbb{P}(f_1 \leq Z_{1k} \leq e_1, \ldots, f_{j-1} \leq Z_{j-1k} \leq e_{j-1}, Z_{jk} > e_j)$$

$$= \int_{f_1}^{e_1} \int_{f_2}^{e_2} \cdots \int_{e_j}^{\infty} f_{Z_{[j]k}}(x) dx \qquad (9.23)$$

where $f_{Z_{[j]k}}$ represents the pdf of the multivariate normal distribution formed by test statistics for $H_0^{(k)}$ up to stage j, i.e., $(Z_{1k}, Z_{2k}, \ldots, Z_{jk})$. Equivalently, one can consider the full distribution of Z but integrate other components from $-\infty$ to ∞.

This process allows us to determine the probability of making a type I error. First, for working out the type I error rate, we determine the distribution under $\delta_k = 0$ and find

$$\mathbb{P}\left(\text{Reject } H_0^{(k)}\right) = \sum_{j=1}^{J} \mathbb{P}(\psi_k = 1, \phi_k = j \mid \delta_k = 0). \qquad (9.24)$$

If we are interested in the family-wise error rate, then we consider the full distribution of Z under the global null H_G (which strongly controls the FWER) and find:

$$\mathbb{P}(\text{Reject at least one null hypothesis} \mid H_G)$$
$$= 1 - \mathbb{P}(\text{Reject no null hypotheses} \mid H_G)$$
$$= 1 - \sum_{\phi \in \Phi} \mathbb{P}(\psi_1 = \cdots = \psi_K = 0, \phi \mid H_G), \qquad (9.25)$$

where Φ is the set of all valid realizations of ϕ. The individual terms within the sum in Equation 9.25 can be expressed as a multivariate integration. For a particular value of $\phi = (\phi_1, \dots, \phi_K)$, the event of rejecting no null hypotheses will occur if, for each k, treatment k reaches stage ϕ_k and then stops for futility. That is, each test statistic up to that stage is between the futility and efficacy boundaries, and then the test statistic at stage ϕ_k is below the futility boundary for that stage. This can be written as

$$\int_{f_1}^{e_1} \cdots \int_{-\infty}^{f_{\phi_1}} \int_{f_1}^{e_1} \cdots \int_{-\infty}^{f_{\phi_2}} \int_{f_1}^{e_1} \cdots \int_{-\infty}^{f_{\phi_K}} f_Z(x)dx, \tag{9.26}$$

where Z is the distribution of the test statistics under H_G.

Code provided in mams.R finds the type I error rates per arm, and the FWER for a multi-arm, multi-stage situation with specified asymptotic distribution of Z.

The same choices of different powers, as discussed in Section 9.3.3, are available in group-sequential MAMS trials. The disjunctive power, under the situation of all treatments being effective ($\delta_k = \delta^{(1)}\ \forall k$), is

$$1 - \mathbb{P}(\text{Reject no null hypotheses} \mid \delta_1 = \delta^{(1)}, \dots, \delta_K = \delta^{(1)})$$
$$= 1 - \sum_{\phi \in \Phi} \mathbb{P}(\psi_1 = \cdots = \psi_K = 0, \phi \mid \delta_1 = \delta^{(1)}, \dots, \delta_K = \delta^{(1)}). \tag{9.27}$$

The conjuctive power can be expressed as

$$\mathbb{P}(\text{Reject all null hypotheses} \mid \delta_1 = \delta^{(1)}, \dots, \delta_K = \delta^{(1)})$$
$$= \sum_{\phi \in \Phi} \mathbb{P}(\psi_1 = \psi_2 = \cdots = \psi_K = 1, \phi \mid \delta_1 = \delta^{(1)}, \dots, \delta_K = \delta^{(1)}). \tag{9.28}$$

The power under the least favorable configuration will be

$$\mathbb{P}(\text{Reject } H_0^{(1)} \mid \delta_1 = \delta^{(1)}, \delta_2 = \cdots = \delta_K = \delta^{(0)})$$
$$= \sum_{j=1}^{J} \mathbb{P}(\text{Reject } H_0^{(1)}, \phi_1 = j \mid \delta_1 = \delta^{(1)}, \delta_2 = \cdots = \delta_K = \delta^{(0)})$$
$$\sum_{j=1}^{J} \int_{f_1}^{e_1} \cdots \int_{e_j}^{\infty} f_{Z_{[j]1}}(x)dx. \tag{9.29}$$

Compared with the multi-arm case, finding the design is more complicated due to the higher number of parameters. Even assuming the number of patients recruited per arm per stage is the same, there are still a total of $2J$ parameters to choose: J efficacy boundary parameters, $J - 1$ futility boundary parameters (assuming the final efficacy and futility boundary is set to the same), and a group size parameter representing the number of patients to be recruited per arm per stage.

The least computationally intensive option is to use known group-sequential stopping boundaries such as those of O'Brien and Fleming [15], Pocock [16], or Whitehead [17]. One could start with the boundaries for the group-sequential RCT case and then multiply the boundaries by a constant that gives correct type I error rate. This is illustrated in the example in the Section 9.4.2.1.

Once the boundaries are chosen such that the type I error rate is as required, the group-size, n, can be chosen so that the required power is correct. As for the multi-arm case, this might require an iterative process in case the covariance matrix Σ (and therefore the error rate) depends on the group size.

A more computationally intensive way is to search for an optimal design, as described more fully in Wason and Jaki [18]. An optimal design is a group-sequential MAMS design that meets error rate constraints and has the lowest average sample for some configuration of δ. This has some advantages, such as being able to tailor the design's properties; however it is not considered further here due to its complexity and reliance on simulation and stochastic search techniques.

9.4.2.1 Example

We now consider a toy example. Suppose we are considering a normally distributed outcome and wish to find a MAMS design that gives FWER $= 0.05$ and conjuctive power 0.9 when $\delta^{(1)} = 0.5$ with a known value of $\sigma = 1$. We consider five stages and three experimental treatments ($J = 5$, $K = 3$). In Wason et al. [19], the Triangular test boundaries (which are triangular shaped on the score statistic scale as described in Reference [17]) for $J = 5$ and a two-arm RCT are $f = (-0.85, 0.30, 0.98, 1.49, 1.90)$, $e = (2.55, 2.10, 1.96, 1.91, 1.90)$, so we begin with these (with code from mams.R):

```
> fwer(J=5,K=3,f=c(-0.85,0.30,0.98,1.49,1.90),e=c(2.55,2.10,1.96,1.91,1.90))
[1] 0.121265
```

So, clearly these boundaries are not sufficiently stringent. We consider what happens when the two boundaries are multiplied by c, which we vary until we find boundaries that give a FWER of 5%. This is shown in Figure 9.2. This indicates that a value of c around 1.217 gives a FWER of 5%.

Given the resulting boundaries $f = (-1.03, 0.37, 1.19, 1.81, 2.31)$, $e = (3.10, 2.56, 2.39, 2.32, 2.31)$, Figure 9.3 shows the conjunctive power as the sample size per arm per stage varies. A value of $n = 29$ is needed for the power to be above 90% in this case. If all arms were to continue until the end without stopping, then this would mean a total sample size of 580 would be needed.

9.4.2.2 Extensions

The expressions for type I error rates and power are simplified by the assumption about efficacy stopping only resulting in the arm stopping, as opposed to it leading to the trial stopping. If the latter is the case, then for a null hypothesis to be rejected, the respective arm would have to stop for efficacy before (or at the same time that) any other arm stops for efficacy. This leads to much more complicated analytical formulas. For example, in Magirr et al. [20], the formulas (using the Wald test) involve multivariate normal integrations of an integrand, which is itself a function of a multivariate normal integral. It is not possible to use the efficient integration methods to find this, and as a result it is much slower to evaluate, either through numerical integration or simulation [18]. Some more recent work [21] has proposed methods based on using the distribution of the score test to increase the speed.

We have also assumed that the group-size, n, is constant. It is possible to allow different numbers of patients to be recruited to each arm as long as this is specified in advance. In fact

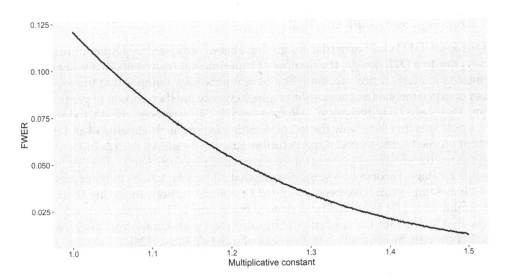

FIGURE 9.2
Family-wise error rate of a MAMS trial with $J = 5$, $K = 3$ with boundaries $f = (-0.85c, 0.30c, 0.98c, 1.49c, 1.90c)$, $e = (2.55c, 2.10c, 1.96c, 1.91c, 1.90c)$ as c varies between 1 and 1.5.

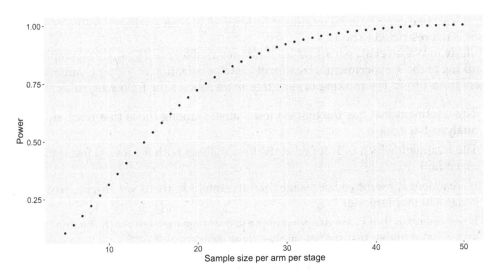

FIGURE 9.3
Conjunctive power of a MAMS trial with $J = 5$, $K = 3$ with boundaries $f = (-1.03, 0.37, 1.19, 1.81, 2.31)$, $e = (3.10, 2.56, 2.39, 2.32, 2.31)$ as n, the group size varies.

there is an efficiency advantage in allocating slightly more patients to the control than to each individual experimental arm [9]. More broadly, one might want to do a trial where the number of patients per stage is constant, with the number of patients per arm depending on the number of treatments that remain in the trial. Although this is a potentially useful design, it does mean that the existing methodology is not applicable. The methods would be complicated because the power for a particular arm would depend on the effects of other arms: if all arms were effective, then the number of patients allocated to a particular arm would on average be lower than in the situation where all other treatments are ineffective. Further work investigating this design, and whether controlling the FWER at H_G strongly controls the FWER, would be useful.

9.4.3 Drop-the-Loser Multi-Arm Trials

Drop-the-loser (DTL) multi-arm trial designs have been considered in the literature for several decades now. In a DTL design, the number of experimental treatments that will progress at each interim analysis is pre-specified. The design generally assumes that the pre-specified number of experimental treatments with highest efficacy are then chosen to progress at each interim. Two-stage DTL designs (e.g., References [22,23]) have been considered most; these have a single interim where only the top performing experimental treatment and the control treatment proceed to the second stage. A futility rule can be added, such as in Thall et al. [22], which means the chosen experimental treatment must be sufficiently effective to continue to the second stage. Flexible two-stage designs that allow adaptations in other aspects of the trial such as sample size have been proposed by several authors, including Bretz et al. [24] and Schmidli et al. [25]. Work has been done to extend the DTL to more than two stages and to show that (at least for the normally distributed case with known variance) the FWER is controlled strongly by controlling it at the global null hypothesis [26].

9.4.3.1 Notation and Operating Characteristics

At each stage, a fixed and pre-determined number of experimental treatments are dropped. Let $n^{(j)}$ denote the number of experimental treatments continuing into stage j. For J stages, the design is denoted as a $n^{(1)} : n^{(2)} : \cdots : n^{(J-1)} : n^{(J)}$ design, where $K > n^{(2)} > \cdots > n^{(J-1)} > n^{(J)}$. Thus, at least one experimental treatment is dropped at each analysis. Unlike Wason et al. [26], $n^{(J)}$ is not restricted to be 1.

Similarly to Wason et al., we introduce random variables $\gamma = (\gamma_1, \gamma_2, \dots, \gamma_K)$ representing the ranking of the K experimental treatments. Each realization of γ is a permutation of the integers from 1 to K. The ranking of each treatment follows the following rules:

1. The treatment that has the highest test statistic among those that reach the final analysis has rank 1.

2. The treatment which is dropped at the first analysis with the lowest test statistic is given rank, K.

3. If treatment k_1 reaches a later stage than treatment k_2, then $\gamma_{k_1} < \gamma_{k_2}$, i.e., treatment k_1 has a higher ranking.

4. If treatments k_1 and k_2 are dropped at the same stage (or both reach the final stage), and k_1 has a higher test statistic at that stage, then $\gamma_{k_1} < \gamma_{k_2}$.

Thus γ can be thought of as an injective function mapping the vector of test statistics Z to a permutation of $(1, 2, \dots, K)$.

This notation is useful for allowing us to succinctly write the probability of rejecting a particular null hypothesis. To reject $H_0^{(k)}$, treatment k must reach the final stage of the trial ($\gamma_k \leq n^{(J)}$) and the final test statistic must be above c ($Z_{Jk} > c$). In this way we can write the probability of rejecting a hypothesis as the sum of tail probabilities of multivariate normal distributions, which can be efficiently evaluated.

Without loss of generality, we consider the probability of recommending treatment 1 when the test statistics Z are multivariate normal with mean μ and covariance Σ. This will be the probability of $\gamma_1 \leq n^{(J)}$ and $Z_{J1} > c$, which is the sum

$$\sum_{\gamma \in \Gamma} \mathbb{P}(Z_{J1} > c, \gamma) I\{\gamma_1 \leq n^{(J)}\}.$$

We now write the probability of a particular ranking in terms of the tail probability of the distribution of a linear transformation of Z.

Without loss of generality, by relabeling the treatments (suitably shuffling the elements of μ and the rows/columns of Σ), we can just consider the probability of $\mathbb{P}(\gamma_k = k, k = 1, \dots, K)$. The event corresponding to this can be written as

1. (For $n^{(J)} > 1$) Treatment k has higher Jth stage test statistic than $k + 1$, $k = 1, \dots, n^{(J)} - 1$.

2. In stage j, treatments 1 to $n^{(j+1)}$ are selected, so have higher test statistics than treatment $n^{(j+1)} + 1$. For $k \in \{n^{(j+1)} + 1, \dots, n^{(j)} - 1\}$, the stage j test statistic for treatment k is higher than the one for treatment $k + 1$.

Thus, for stage $j \in \{1, \dots J\}$, there are $n^{(j)} - 1$ conditions. This gives a total of $(\Sigma_{j=1}^{J} n^{(j)}) - J$. Each of these conditions can be represented as a linear transform of Z being greater than 0. For example, the probability of treatment 1 having a higher Jth stage test statistic than treatment 2 is equivalent to $Z_{J1} - Z_{J2} > 0$. For a treatment to be recommended, it has to have final test statistic above c, which adds another element to the vector of linear transforms.

Thus, since a vector of linear transforms of a multivariate normal distribution is itself multivariate normal, we can express $\mathbb{P}(\gamma_k = k, k = 1, \dots, K)$ as the tail probability of a multivariate normal. To illustrate this, we consider an example of how to evaluate the probability of recommending each treatment when $J = 3$, $K = 4$. Given that we know the distribution of Z is asymptotically normal with mean μ and covariance matrix Σ, we can develop a matrix A, so that AZ is the vector of linear transformations where the tail probability of the resulting distribution is $\mathbb{P}(\gamma_k = k, k = 1, \dots, K)$. If $Z = (Z_{11}, Z_{21}, Z_{31}, Z_{12}, Z_{22}, Z_{32}, Z_{13}, Z_{23}, Z_{33}, Z_{14}, Z_{24}, Z_{34})^T$, then A is:

$$A = \begin{pmatrix} 1 & 0 & 0 & 0 & 0 & 0 & 0 & 0 & 0 & -1 & 0 & 0 \\ 0 & 0 & 0 & 1 & 0 & 0 & 0 & 0 & 0 & -1 & 0 & 0 \\ 0 & 0 & 0 & 0 & 0 & 0 & 1 & 0 & 0 & -1 & 0 & 0 \\ 0 & 1 & 0 & 0 & 0 & 0 & 0 & -1 & 0 & 0 & 0 & 0 \\ 0 & 0 & 0 & 0 & 1 & 0 & 0 & -1 & 0 & 0 & 0 & 0 \\ 0 & 0 & 1 & 0 & 0 & -1 & 0 & 0 & 0 & 0 & 0 & 0 \end{pmatrix}$$

The first three rows represent the first-stage requirement: that treatments 1, 2 and 3 are selected (i.e., that their first-stage test statistics are higher than treatment 4's first-stage test statistic). The next two rows represent the second-stage requirement: that treatments 1 and 2 are selected (i.e., their second-stage test statistics are higher than treatment 3's second-stage test statistic). The final row represents the third-stage requirement: that treatment 1 has a higher third-stage test statistic than treatment 2.

This gives the required matrix to represent the ranking of treatments. However, if we wish to also assess the probability of treatment 1 or 2 being recommended, we need to add a final row with a 1 in the column corresponding to the respective treatment's third-stage test statistic. If we wanted to evaluate the probability of both being recommended (conjunctive power), then we could add both of these rows. We assume it is of interest to evaluate the probability $\mathbb{P}(Z_{31} > c, \gamma_k = k, k = 1, \dots, K)$, which leads to the requirements of this event occurring being

$$(AZ)_i > 0 \quad \text{for } i = 1, \dots, 6 \quad \text{and} \quad (AZ)_7 > c.$$

Now, AZ is an affine transformation of a multivariate normal random variable, and so is normal with mean $A\mu$ and covariance matrix $A\Sigma A^T$. Thus, the event $(Z_{31} > c, \gamma_k = k, k = 1, \dots, K)$ can be expressed as the tail probability of a multivariate normal distribution with mean $A\mu$ and covariance matrix $A\Sigma A^T$. This can be evaluated efficiently using the method of Genz and Bretz [12].

To get the family-wise error rate under H_G, one can calculate the probability

$$\sum_{\gamma \in \Gamma} \mathbb{P}(\text{Reject any } H_0^{(k)}, \gamma \mid H_G). \tag{9.30}$$

We do not consider the conjuctive power as it is impossible to reject all $H_0^{(k)}$ with a drop-the-loser design where at least one treatment is forced to be dropped. The disjunctive power could be found using a similar expression, but when all δ values are equal to $\delta^{(1)}$.

9.4.3.2 Extensions

The methodology described above assumes that the design is followed exactly, in particular, that the treatments that get selected at the interim analysis have the highest test statistics and always continue. It may be of interest to deviate from this. For example, a futility rule could be introduced where a treatment has to meet a minimum bar to continue, with the trial stopping early for futility if no treatments continue at a particular stage. This futility rule could be built into the design by adding conditions to the list above and expanding the matrix A to include these additional conditions. A simpler approach would be to design the trial without a futility rule and then add it in a non-formal way. This would reduce the family-wise error rate as it would make it less likely for a null hypothesis to be rejected. However it would also reduce the power, although not by much if the futility rule is set so that an ineffective treatment is unlikely to be classed as futile.

It could also be of interest to allow criteria other than efficacy to be used in the decision as to which treatments should continue. For example, two treatments may have very similar efficacy and the design may only allow one to continue. In this case, one treatment may be preferred over the other due to showing more promise on an important secondary endpoint, having a better safety profile, or being cheaper. In this case, it would be possible to take forward a slightly less efficacious treatment. This should result in the family-wise error rate being lower, as well as the power. However this result is not known, and there may be situations where it could result in an inflation in the FWER. Nevertheless, it may still be considered worth doing, despite this small potential drawback.

9.4.4 Case Study

To illustrate and compare the group-sequential and DTL-MAMS designs, we will consider the FOCUS trial [27], which compared three treatment strategies for advanced colorectal cancer: strategy A (control), single-agent fluorouracil until failure followed by single-agent irinotecan; strategy B, flourouracil until failure followed by combination chemotherapy; and strategy C, combination chemotherapy from the beginning. In addition, strategies B and C were divided into two sub-strategies, with combination chemotherapy being either fluorouracil plus irinotecan (labeled as strategies B-ir and C-ir) or fluourouracil plus oxaliplatin (B-ox and C-ox). The planned sample size was 2,100 in total, with 700 in strategy A and 350 in each of B-ir, C-ir, B-ox and C-ox. This was powered to detect a difference when the control group's two-year survival was 15% and an effective arm had a two-year survival of 22.5%.

For ease of presentation, we consider the hypotheses of B versus A $(H_0^{(1)})$ and C versus A $(H_0^{(2)})$, with a two-stage MAMS trial used. We will compare a group-sequential approach with a drop-the-loser approach, both with the maximum number of patients set to 2,100. For the group-sequential MAMS approach, this means the group size is 350; for a DTL design that drops one treatment at the interim, the group size is 420. The test statistic will be assumed to be formed from the difference in estimated two-year survival between arms.

Assuming that all first-stage patients are followed up for two years before the second stage starts (clearly an unrealistic assumption for most trials), the test statistics of $H_0^{(1)}$ and $H_0^{(2)}$ after interim analyses 1 and 2 are

$$Z_{11} = \frac{\hat{p}_{11} - \hat{p}_{10}}{\sqrt{\frac{p_1(1-p_1)}{n} + \frac{p_0(1-p_0)}{n}}}$$

$$Z_{12} = \frac{\hat{p}_{12} - \hat{p}_{10}}{\sqrt{\frac{p_2(1-p_2)}{n} + \frac{p_0(1-p_0)}{n}}}$$

$$Z_{21} = \frac{\hat{p}_{21} - \hat{p}_{20}}{\sqrt{\frac{p_1(1-p_1)}{2n} + \frac{p_0(1-p_0)}{2n}}}$$

$$Z_{22} = \frac{\hat{p}_{22} - \hat{p}_{20}}{\sqrt{\frac{p_2(1-p_2)}{2n} + \frac{p_0(1-p_0)}{2n}}},$$

where $\hat{p}_{jk}$ represents the MLE for p_k, using all patients up to stage j.

We will consider three situations: (1) all treatments are equally (in)effective, with 15% two-year survival; (2) treatment B is effective, treatment C is not; and (3) treatment B and C are effective.

For the group-sequential and DTL designs, we first find the joint distribution of $Z = (Z_{11}, Z_{12}, Z_{21}, Z_{22})$ for the three different settings. To do this, we use the formulas for the covariances throughout the chapter (Equations 9.2, 9.7 and 9.22).

For scenario 1,

$$Z_{GS} \sim N\left((0,0,0,0)^T, \begin{pmatrix} 1 & 0.707 & 0.500 & 0.353 \\ 0.707 & 1 & 0.353 & 0.500 \\ 0.500 & 0.353 & 1 & 0.707 \\ 0.353 & 0.500 & 0.707 & 1 \end{pmatrix}\right)$$

$$Z_{DTL} \sim N\left((0,0,0,0)^T, \begin{pmatrix} 1 & 0.707 & 0.500 & 0.353 \\ 0.707 & 1 & 0.353 & 0.500 \\ 0.500 & 0.353 & 1 & 0.707 \\ 0.353 & 0.500 & 0.707 & 1 \end{pmatrix}\right)$$

For scenario 2,

$$Z_{GS} \sim N\left((2.55, 3.61, 0, 0)^T, \begin{pmatrix} 1 & 0.707 & 0.460 & 0.325 \\ 0.707 & 1 & 0.325 & 0.460 \\ 0.460 & 0.325 & 1 & 0.707 \\ 0.325 & 0.460 & 0.707 & 1 \end{pmatrix}\right)$$

$$Z_{DTL} \sim N\left((2.80, 3.96, 0, 0)^T, \begin{pmatrix} 1 & 0.707 & 0.460 & 0.325 \\ 0.707 & 1 & 0.325 & 0.460 \\ 0.460 & 0.325 & 1 & 0.707 \\ 0.325 & 0.460 & 0.707 & 1 \end{pmatrix}\right)$$

For scenario 3,

$$Z_{GS} \sim N\left((2.55, 3.61, 2.55, 3.61)^T, \begin{pmatrix} 1 & 0.707 & 0.422 & 0.299 \\ 0.707 & 1 & 0.299 & 0.422 \\ 0.422 & 0.299 & 1 & 0.707 \\ 0.299 & 0.422 & 0.707 & 1 \end{pmatrix}\right)$$

$$Z_{DTL} \sim N\left((2.80, 3.96, 2.80, 3.96)^T, \begin{pmatrix} 1 & 0.707 & 0.422 & 0.299 \\ 0.707 & 1 & 0.299 & 0.422 \\ 0.422 & 0.299 & 1 & 0.707 \\ 0.299 & 0.422 & 0.707 & 1 \end{pmatrix}\right)$$

We consider a GS-MAMS design that starts with Pocock efficacy boundaries (e.g., $e = (1.96, 1.96)$) and a first-stage futility boundary of 0 ($f = (0, 1.96)$). Under H_G, this gives a FWER of 0.0736. We thus will multiply the boundaries by a constant until the FWER is 0.05. This gives boundaries of $f = (0, 2.14)$, $e = (2.14, 2.14)$.

The DTL-MAMS design will simply select the experimental treatment with the highest test statistic at the first interim. The critical value at the end of the trial, c, is chosen to control the FWER at 5%; in this case, it is 1.83.

Given these designs, Table 9.3 shows the chance of rejecting each null hypothesis in the different scenarios.

TABLE 9.3

Power and Expected Sample Size for GS-MAMS and DTL-MAMS for the FOCUS Sample Size; Scenario 1, $p_0 = p_1 = p_2 = 0.15$; Scenario 2, $p_0 = 0.15$, $p_1 = 0.225$, $p_2 = 0.15$; Scenario 3, $p_0 = 0.15$, $p_1 = p_2 = 0.225$

Scenario	Design	Power Reject $H_0^{(1)}$	Power Reject $H_0^{(2)}$	Expected Sample Size
1	GS	0.027	0.027	1,622
1	DTL	0.025	0.025	2,100
2	GS	0.936	0.027	1,770
2	DTL	0.979	0.002	2,100
3	GS	0.936	0.936	1,461
3	DTL	0.497	0.497	2,100

Table 9.3 shows advantages and disadvantages of the two designs. When there is one effective treatment and one ineffective treatment, then the DTL design is considerably more powerful (note that going from 93.6% to 97.9% power generally takes a large sample size increase). However in the case that there is no effective treatments or two effective treatments, the GS-MAMS approach appears to be preferable. This is because it has a good chance of stopping the trial early when both treatments are ineffective or effective, which reduces the average sample size used. In addition, when both treatments are effective, it has good power for rejecting each null, whereas the DTL-MAMS design only allows one hypothesis to be rejected.

Generally in clinical trials, it is difficult or undesirable to pause recruitment. Assuming recruitment continues as interim analyses are being done, then both GS- and DTL-MAMS designs will provide less of an advantage over a multi-arm trial without interim analyses. In the FOCUS trial, a MAMS approach would have likely not been useful unless the patients were recruited over a very long time period. For example, if patients were recruited over a four- year period, then all patients would have been in the trial by the time the first-stage patients had been followed-up for two years. Thus, no dropping of ineffective treatments would have been possible.

9.5 Conclusion

In this chapter, we have examined the basics of how to design multi-arm trials that can evaluate multiple experimental treatments in comparison to a common control group. With the large number of treatments available for testing in oncology, multi-arm trials have great potential to help increase the number of treatments that can be robustly tested. Multi-arm trials provide notable advantages over separate randomized studies and can potentially be improved further through the use of an adaptive design that can drop ineffective treatments using accruing information. This is known as a multi-arm, multi-stage (MAMS) design. In the event that the endpoint is observed relatively quickly in comparison to the recruitment period of the trial, MAMS designs can provide further advantage over multi-arm designs, either through reducing the expected sample size used or through increasing the power given a fixed sample size.

In recent years, there has been much work on multi-arm and multi-arm multi-stage designs that has extended methodology discussed in this chapter. We will briefly discuss some of this work here.

First, single-arm trials are common in early phase II oncology settings, albeit their utility is controversial [28]. A multi-arm trial could potentially just have experimental arms and no shared control arm. Although this would not increase the statistical efficiency compared to separate single-arm trials, it could have logistical and administrative advantages. It would also allow unbiased comparisons between arms, although it would likely lack power for this. Methodology for how to power such trials is discussed in Magaret et al. [29].

Second, we have considered MAMS trials where the definitive endpoint is used to make decisions at interim analyses. Although this is ideal, as we saw in the FOCUS case study, it would not always work due to the length of time it takes observing the definitive endpoint. There have been several articles that have proposed a MAMS design that uses an intermediate (more quickly observed) time-to-event endpoint at interim analyses, with the definitive endpoint used at the end [30]. The method is somewhat similar to, but in fact pre-dates, the group-sequential MAMS approach described in this chapter, with treatment arms having to pass futility boundaries. This methodology is used in the STAMPEDE trial [31], which is a flagship example of a MAMS trial in oncology. There is software available for implementing this design in a Stata module called NSTAGE [32]; this software has been extended to allow the endpoints to be binary instead of time-to-event endpoints [33]. One has to be careful with using intermediate endpoints if strict control of the FWER is required; since the endpoint is different to the one the null hypothesis is testing, it might be that the FWER is not maximized when the experimental treatments have no benefit over control with the intermediate endpoint. Thus, considering the global null hypothesis for the definitive and intermediate endpoint might not be sufficient.

A third extension to multi-arm designs in oncology has been the use of biomarkers. In oncology there is often heterogeneity in the treatment effect that can be explained by tumor mutations. More modern treatments are often developed to target certain biological pathways and their effect is often restricted to patient subgroups. Umbrella studies are ones which test multiple experimental treatments in different biomarker subgroups. Their use has been highest in oncology trials, with an increasing number ongoing. Adaptive designs can also be applied to improve efficiency, such as in the UK National Lung Matrix trial [34], FOCUS 4 [35], and trials using Bayesian adaptive randomization [36–38] (see also Chapter 6 and 14).

Clearly there is much work in this area and increasing application of these approaches in oncology. Future methodology work will likely focus not only on statistical methods, but also on more practical methodology and guidance on how to apply methods in practice in the most efficient way. With the increasing number of case studies, there are now a lot of examples to choose from, and more guidance to follow. It is recommended, when embarking on planning a multi-arm trial, that much effort is put in at the design stage in order to ensure the design is fit for purpose and does not cause issues in the running and reporting of the trial [4].

These methods are being successfully used in practice and can provide much benefit; so to finish the chapter, although I would like to caution the reader on the amount of effort required, I would also like to encourage them to be ambitious and to consider novel designs for trials they are involved with.

References

1. Reed JC. Toward a new era in cancer treatment: Message from the new editor-in-chief. *Mol Cancer Ther.* 2012;11:1621–2.
2. Parmar MKB, Carpenter J, Sydes MR. More multiarm randomised trials of superiority are needed. *Lancet.* 2014;384(9940):283–4.
3. Dumville JC, Hahn S, Miles JNV, Togerson DJ. The use of unequal randomisation ratios in clinical trials: A review. *Contemp Clin Trials.* 2006;27:1–12.
4. Pallmann P, Bedding AW, Choodari-Oskooei B et al. Adaptive designs in clinical trials: Why use them, and how to run and report them. *BMC Med.* 2018;16(1):29.
5. Berry SM, Carlin BP, Lee JJ, Muller P. *Bayesian Adaptive Methods for Clinical Trials.* Boca Raton FL: CRC Press, 2010.
6. Jennison C, Turnbull BW. *Group Sequential Methods with Applications to Clinical Trials.* Boca Raton FL: Chapman and Hall, 2000.
7. Wason J. Multi-arm multi-stage designs for clinical trials with treatment selection. In *Modern Adaptive Randomized Clinical Trials: Statistical and Practical Aspects,* Sverdlov O, ed., 2015: 389–410.

8. Grayling MJ, Wason JMS, Mander AP. An optimised multi-arm multi-stage clinical trial design for unknown variance. *Contemp Clin Trials.* 2018;67:116–20.

9. Wason J, Magirr D, Law M, Jaki T. Some recommendations for multi-arm multi-stage trials. *Stat Methods Med Res.* 2016;25(2):716–27.

10. Barthel FM-S, Babiker A, Royston P, Parmar MKB. Evaluation of sample size and power for multi-arm survival trials allowing for non-uniform accrual, non-proportional hazards, loss to follow-up and cross-over. *Stat Med.* 2006;25(15):2521–42.

11. Wason JMS, Stecher L, Mander AP. Correcting for multiple-testing in multi-arm trials: Is it necessary and is it done? *Trials.* 2014;15(1):364.

12. Genz A, Bretz F. Methods for the computation of multivariate t-probabilities. *J Comput Graph Stat.* 2002;11:950–71.

13. Dunnett CW. A multiple comparison procedure for comparing several treatments with a control. *J Am Stat Assoc.* 1955;50:1096–121.

14. Gianni L, Pienkowski T, Im YH et al. Efficacy and safety of neoadjuvant pertuzumab and trastuzumab in women with locally advanced, inflammatory, or early HER2-positive breast cancer (neosphere): A randomised multicentre, open-label, phase 2 trial. *Lancet Oncol.* 2011;13:25–32.

15. O'Brien PC, Fleming TR. A multiple-testing procedure for clinical trials. *Biometrics.* 1979;35:549–56.

16. Pocock SJ. Group sequential methods in the design and analysis of clinical trials. *Biometrika.* 1977;64:191–9.

17. Whitehead J, Stratton I. Group sequential clinical trials with triangular continuation regions. *Biometrics.* 1983;39:227–36.

18. Wason JMS, Jaki T. Optimal design of multi-arm multi-stage trials. *Stat Med.* 2012;31:4269–79.

19. Wason JMS, Mander AP, Thompson SG. Optimal multi-stage designs for randomised clinical trials with continuous outcomes. *Stat Med.* 2012;31:301–12.

20. Magirr D, Jaki T, Whitehead J. A generalized Dunnett test for multiarm-multistage clinical studies with treatment selection. *Biometrika.* 2012;99:494–501.

21. Ghosh P, Liu L, Senchaudhuri P, Gao P, Mehta C. Design and monitoring of multi-arm multi-stage clinical trials. *Biometrics.* 2017.

22. Thall PF, Simon SS, Ellenberg R. A two-stage design for choosing among several experimental treatments and a control in clinical trials. *Biometrics.* 1989;45:537–47.

23. Sampson A, Sill M. Drop-the-losers design: Normal case. *Biom J.* 2005;47:257–68.

24. Bretz F, Schmidli H, Konig F, Racine A, Maurer W: Confirmatory seamless phase II/III clinical trials with hypotheses selection at interim: General concepts. *Biom J.* 2006;48:623–34.

25. Schmidli H, Bretz F, Racine A, Maurer W. Confirmatory seamless phase II/III clinical trials with hypotheses selection at interim: Applications and practical considerations. *Biom J.* 2006;48:635–43.

26. Wason J, Stallard N, Bowden J, Jennison C. A multi-stage drop-the-losers design for multi-arm clinical trials. *Stat Methods Med Res.* 2017;26(1):508–24.

27. Seymour MT, Maughan TS, Ledermann JA et al. Different strategies of sequential and combination chemotherapy for patients with poor prognosis advanced colorectal cancer (mrc focus): A randomised controlled trial. *Lancet.* 2007;370(9582):143–52.

28. Grayling MJ, Mander AP. Do single-arm trials have a role in drug development plans incorporating randomised trials? *Pharm Stat.* 2016;15(2):143–51.

29. Magaret A, Angus DC, Adhikari NKJ, Banura P, Kissoon N, Lawler JV, Jacob ST. Design of a multi-arm randomized clinical trial with no control arm. *Contemp Clin Trials.* 2016;46:12–7.

30. Royston P, Parmar MKB, Qian W. Novel designs for multi-arm clinical trials with survival outcomes with an application in ovarian cancer. *Stat Med.* 2003;22:2239–56.

31. James ND, Sydes MR, Clarke NW et al. Addition of docetaxel, zoledronic acid, or both to first-line long-term hormone therapy in prostate cancer (stampede): Survival results from an adaptive, multiarm, multistage, platform randomised controlled trial. *Lancet.* 2016;387(10024):1163–77.

32. Bratton DJ, Choodari-Oskooei B, Royston P et al. A menu-driven facility for sample size calculation in multi-arm multi-stage randomised controlled trials with time-to-event outcomes: Update. *Stata J*. 2015;15(2):350–68.

33. Bratton DJ, Phillips PPJ, Parmar MKB. A multi-arm multi-stage clinical trial design for binary outcomes with application to tuberculosis. *BMC Med Res Methodol*. 2013;13(1):139.

34. Middleton G, Crack LR, Popat S, Swanton C, Hollingsworth SJ, Buller R, Walker I, Carr TH, Wherton D, Billingham LJ. The national lung matrix trial: Translating the biology of stratification in advanced non-small-cell lung cancer. *Ann Oncol*. 2015;26(12):2464–9.

35. Kaplan R, Maughan T, Crook A, Fisher D, Wilson R, Brown L, Parmar M. Evaluating many treatments and biomarkers in oncology: A new design. *Am J Clin Oncol*. 2013;31:4562–70.

36. Kim ES, Herbst RS, Wistuba II et al. The BATTLE Trial: Personalizing Therapy for Lung Cancer. *Cancer Discov*. 2011;1(1):44–53.

37. Rugo HS, Olopade OI, DeMichele A et al. Adaptive randomization of veliparib–carboplatin treatment in breast cancer. *N Engl J Med*. 2016;375(1):23–34.

38. Wason JMS, Abraham JE, Baird RD, Gournaris I, Vallier A-L, Brenton JD, Earl HM, Mander AP. A bayesian adaptive design for biomarker trials with linked treatments. *Br J Cancer*. 2015;113(5):699.

10

Multiple Comparisons, Multiple Primary Endpoints and Subpopulation Analysis

Ekkehard Glimm, Dong Xi, and Paul Gallo

CONTENTS

10.1 Sources of Multiplicity in Oncology Trials

In confirmatory clinical trials in oncology, multiple primary endpoints as well as confirmatory subpopulation analyses are becoming increasingly common. A multitude of endpoints are typically assessed in oncology trials, e.g., overall survival (OS), tumor response rate, progression-free survival (PFS), disease-free progression (DFS), time to progression (TTP), tumor marker activity, etc. (see Chapter 1 in the book by George et al. [1]). Traditionally, in many confirmatory clinical trials, a single endpoint is chosen as the primary endpoint, and the trial is deemed successful if a beneficial treatment effect can be established, typically by a statistical significance test. In oncology trials, however, it has become common practice to declare more than one of the endpoints to be primary in the sense that a statistically significant positive effect of the experimental drug in just one of them suffices for claiming that the treatment is beneficial. In many oncology trials, PFS and OS are the two primary endpoints. (This is different from the situation where two *co-primary* endpoints both have to be statistically significant in order to claim that the treatment is beneficial.) In other cases, (key) secondary endpoints are included to the list of multiple endpoints which require adjustment for multiplicity. This is often mandated by health authorities if formal claims of treatment benefits are sought on such secondary endpoints in the product label.

Advances in targeted therapies and in molecular biology have also changed the practice of confirmatory clinical trials in oncology. Today, it has become very common to investigate treatment effects for biomarker-defined subpopulations within a single clinical trial in addition to the entire trial population ("full population"). The biomarkers used for subpopulation definition are typically genotypes (e.g., TP53, BRCA, or HERC) or markers of genetic activity or other molecular biomarkers related to tumor growth. Often, the full population is stratified into a biomarker-positive (B+) population (the patients having a certain genetic mutation or tumor growth due to the activity of a certain gene or pathway, for example) and a biomarker-negative (B−) population (all other patients). Sometimes, the dichotomization into B+ and B− can be subject to misclassification error, e.g., if genetic activity varies in time in a patient and thus its measurement is subject to random variation.

In both cases (investigation of more than one endpoint or investigation of more than one subpopulation), appropriate statistical methods are needed to deal with the fact that simultaneous multiple inference is done on a research question. In this chapter, we use the word *multiplicity* to describe such situations. If inference on a statistical question is not adjusted for multiplicity (that is, if every analysis, e.g., on a single endpoint, is performed and interpreted as if it were the only one that had been done), bias arises. For example, if two treatments are compared on two different endpoints with statistical tests both using a significance level of α, then the probability of at least one significant result is larger than α, if in reality there is no difference between the two treatments. As another reflection of the bias, assume that a point estimate and a confidence interval for the treatment effect are calculated separately in each of several subpopulations, based only on data from that subpopulation. Subsequently, only the largest point estimate and its confidence interval are reported. This is misleading, because the treatment effect estimate will be overestimated and the confidence interval "shifted upward" due to the selection of the reported result.

We have mentioned the two most common sources of multiplicity in oncology trials, but other sources are also relevant at times, for example,

- Multiple treatment arms, usually either different doses or different dosing regimens for a treatment,
- Simultaneous comparisons of an experimental drug with an active comparator and a placebo group,
- Combination drug therapies, e.g., targeted therapy plus chemotherapy versus targeted therapy plus immune therapy versus chemotherapy alone,
- Multiple comparisons of different types of interventions, e.g., chemotherapy versus surgery versus radiation therapy, possibly including combinations of these treatments,

- Assessing non-inferiority and superiority of a new treatment versus a comparator on multiple endpoints, and
- Combinations of these types of multiplicity.

Another common source of multiplicity is the repeated inspection of accruing data in group-sequential and adaptive trials. This typically involves repeated statistical testing of a hypothesis. From the point of view of statistical methodology, the approaches used for this share many similarities with multiple testing procedures used for the multiplicity situations just described. Specifically, in both fields, the joint distribution of test statistics are often assumed to be multivariate normal with some correlation structure [2]. The difference lies in the fact that in group-sequential approaches, one or more individual hypotheses are tested repeatedly with different amounts of information available. Group-sequential methods are the topic of Chapters 9 and 12 and thus not discussed here in detail. In contrast to Chapter 16 which deals with subpopulation search, we are dealing with confirmatory subpopulation analysis here, i.e., the situation where pre-specified subpopulations are investigated to confirm the effectiveness of the treatment.

In the next Section 10.1.1, we introduce an example of an actual clinical trial which will be used in subsequent sections to illustrate the discussed methodology. Testing for significant differences between treatment options is a key technique in confirmatory clinical trials and has been the focus of much research into multiplicity-adjustment methodology (e.g., [3,4,5]). We therefore continue with a review of multiple testing procedures including their graphical representation [6]. We conclude with some strategic and operational considerations for multiplicity adjustments in oncology. R code is available on the book's companion website https://www.crcpress.com//9781138083776.

10.1.1 Introductory Example

In the following, we will use the BELLE-4 study [7] to illustrate various aspects of multiplicity adjustment arising in oncology trials. BELLE-4 was a 1:1 randomized, double-blind, placebo-controlled, adaptive phase II/III study. The study compared the therapy of advanced HER2-positive breast cancer by the experimental treatment paclitaxel + burparlisib versus the standard of care by paclitaxel + placebo. Buparlisib is an oral PI3 K inhibitor and thus it was expected that its effect would be stronger in patients with strong PI3 K activation. Consequently, the study used stratified randomization according to PI3 K activation (active or non-active) and also stratified patients by hormone-receptor status (HR+ or HR−).

The study used progression-free survival (PFS) as the primary endpoint and overall survival (OS) as the key secondary endpoint. Furthermore, the study design included an interim analysis allowing the possibility of stopping for futility for the full population, or potentially the discontinuation of recruitment of patients in the non-activated PI3 K pathway stratum. The final analysis, if reached in the full population, would include statistical tests for superiority of the experimental treatment in the full patient population as well as in the subpopulation of patients with PI3 K activation. Hence, there are three sources of multiplicity in this example:

1. Two endpoints, PFS and OS,
2. An interim and a final analysis, and
3. The full population and the subpopulation of PI3 K activated patients.

The study was designed to control the family-wise error rate (FWER) across all these sources of multiplicity.

10.2 Multiple Testing Procedures

10.2.1 Basic Concepts

10.2.1.1 Error Rate in Confirmatory Clinical Trials

When testing multiple hypotheses, the desired overall one-sided type I error rate, often $\alpha = 0.025$, will be inflated if every hypothesis is tested at the targeted nominal significance level. For example, consider a new compound being investigated for efficacy claims in three disjoint and independent populations, with each hypothesis tested at level 0.025. The chance of falsely claiming that the compound has a positive treatment effect in at least one of the three populations is $1 - (1 - 0.025)^3 = 0.073$. In this case, the overall type I error rate of interest is the family-wise error rate (FWER), which is the probability to reject at least one true null hypothesis, or to make at least one type I error. Before conducting the trial, we do not know whether the compound is efficacious in all or some populations, or in another setting, for all or some endpoints. Thus in confirmatory clinical trials, it is desired by regulatory agencies [8,9] to control the FWER at level α under any configuration of true and false hypotheses, which is called FWER control in the strong sense. On the other hand, the weak sense only considers the configuration in which all null hypotheses are true, e.g., the compound is ineffective in all populations. Weak control of the FWER is usually considered non-sufficient in confirmatory clinical trials.

Another type of error rate is becoming more frequently considered, especially when exploring many hypotheses. The false discovery rate is the expected proportion of falsely rejected hypotheses among the rejected hypotheses. As the focus of this chapter involves confirmatory trials, we only consider the FWER in this chapter and multiple testing procedures that control it in the strong sense.

10.2.1.2 Single-Step and Stepwise Procedures

Multiple testing procedures can be categorized as either single-step or stepwise procedures. Single-step procedures are simple methods in which the rejection of a hypothesis does not depend on the decision for other hypotheses. An example of a single-step procedure is the Bonferroni test. For stepwise procedures, the rejection of a given hypothesis may affect the decision on other hypotheses and thus they are performed in a stepwise fashion. An example of a stepwise procedure is the Holm [10] procedure, which is also a stepwise extension of the Bonferroni test.

10.2.1.3 Closed Testing Procedures

The closed testing procedure (CTP) is a general approach for adjustment of multiple statistical tests [11]. As a method, the CTP is applied to a set of elementary hypotheses $\{H_1, \dots, H_m\}$ with $I = \{1, \dots, m\}$. It consists of performing statistical tests at level α for all $2^m - 1$ intersection hypotheses $H_J = \cap_{i \in J} H_i$ for all subsets of $J \subseteq I$. Hypothesis H_i is rejected if and only if all H_J with $i \in J$ are rejected. The relevance of the CTP is due to the fact that it controls the FWER strongly at level α.

To illustrate the CTP, consider three elementary hypotheses of no treatment effect H_1: $\mu_1 \leq 0$, H_2: $\mu_2 \geq 0$, and H_3: $\mu_3 \geq 0$ for the three endpoints tumor response rate, PFS, and OS, respectively. For tumor response rate, μ_1 is the mean difference between the responder rates of the treatment and the control group. For PFS and OS, μ_2 and μ_3 are the logarithms

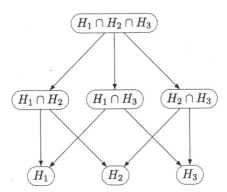

FIGURE 10.1
Closed testing procedure for three hypotheses.

of the respective hazard ratios. The family of all intersection hypotheses is displayed in Figure 10.1. All seven intersection hypotheses can be tested using an α-level test such as the Bonferroni test discussed in Section 10.2.2. For example, the elementary hypothesis H_1 is rejected if and only if $H_1 \cap H_2 \cap H_3$, $H_1 \cap H_2$, $H_1 \cap H_3$ and H_1 are all rejected.

If m is large, setting up as well as calculating all $2^m - 1$ tests in the CTP can be very burdensome. Therefore, shortcut CTPs are very useful, since they reach the decisions on elementary hypotheses in at most m or m^2 tests. In these approaches, the rejection of an intersection hypothesis H_J is "due to" one of the elementary p-values p_i, $i \in J$. Thus, following the rejection of H_J due to p_i, all H_{J^*} with $i \in J^* \subseteq J$ can be rejected without actually performing the corresponding test [12]. Apart from greatly reducing the computational burden, this also renders the CTP *consonant*, which means that for any intersection hypothesis that is rejected, at least one elementary hypothesis represented in this intersection will also be rejected [13]. Many of the procedures discussed below are shortcut procedures.

10.2.1.4 Adjusted Critical Values and Adjusted p-Values

Multiple testing procedures often apply adjustments to critical values which can be compared directly with p-values for all hypotheses. Alternatively, one can also adjust p-values so that they can be compared with the overall significance level α. Formally, the adjusted p-value is defined as the smallest significance level at which a hypothesis can be rejected by the chosen multiple testing procedure [14]. As an example, the Bonferroni test for three hypotheses has the adjusted critical value of $\alpha/3$. Hence, if the p-values for the three hypotheses are p_1, p_2, and p_3, then the Bonferroni test rejects H_i if $p_i \leq \alpha/3$, $i = 1, 2, 3$. Alternatively, the decision procedure can be described in terms of the adjusted p-values $3p_1$, $3p_2$, $3p_3$, which means H_i is rejected if $3p_i \leq \alpha$, $i = 1, 2, 3$.

10.2.1.5 Simultaneous Confidence Intervals

Simultaneous confidence intervals are an extension of univariate confidence intervals so that the coverage probability of the multiple true parameters is at least $1 - \alpha$. For single-step procedures, these intervals can be readily obtained because the decision for a hypothesis does not depend on another. Loosely speaking, for single-step procedures, the acceptance region of the multiple test defines a multiple confidence region for the treatment-effect estimates. This analogy breaks down for stepwise procedures, hence even at a conceptual level, it is not easy to derive a "matching" confidence region for a stepwise procedure. Further discussions are provided in Brannath and Schmidt [15] and Strassburger and Bretz [16].

10.2.2 Common Multiple Testing Procedures

10.2.2.1 Bonferroni Test

The Bonferroni test is a single-step procedure. In its simplest version, it splits the total significance level α equally among m null hypotheses and tests every null hypothesis at level α/m. A hypothesis is rejected if its p-value is less than or equal to α/m. From another angle, we can view the m hypotheses as being weighted equally with weights $1/m$. In practice, however, some hypotheses may be more relevant than others because of clinical importance, likelihood of rejecting, etc. To address this, weighted versions of the Bonferroni test assign different weights w_i to hypotheses H_i, where $\sum_{i=1}^{m} w_i = 1$. A hypothesis is rejected by the weighted Bonferroni test if its p-value is less than or equal to $w_i \alpha$. Bonferroni tests control the FWER strongly because of the Bonferroni inequality:

$$P\left[\bigcup_{i=1}^{m}(p_i \leq w_i \alpha)\right] \leq \sum_{i=1}^{m} P(p_i \leq w_i \alpha) = \sum_{i=1}^{m} w_i \alpha = \alpha.$$

Bonferroni tests are simple to implement and communicate. They also control the FWER under any joint distribution of test statistics. However, they tend to be quite conservative when the test statistics are highly correlated.

10.2.2.2 Holm Procedure

The Holm procedure is a stepwise extension of the Bonferroni test. It applies the Bonferroni test in each step [10]. Assume that the raw p-values are denoted by $p_1, \ldots, p_m$, respectively. The procedure first orders the p-values from the smallest to the largest: $p_{(1)} \leq p_{(2)} \leq \cdots \leq p_{(m)}$. Then it tests hypothesis $H_{(1)}$ and rejects it if $p_{(1)} \leq \alpha/m$. If $H_{(1)}$ is rejected, the procedure tests $H_{(2)}$ and rejects it if $p_{(2)} \leq \alpha/(m-1)$; otherwise, testing stops. If $H_{(2)}$ is also rejected, the procedure tests $H_{(3)}$ and rejects it if $p_{(3)} \leq \alpha/(m-2)$; otherwise, testing stops. In general, given that $H_{(1)}, \ldots, H_{(i)}$ are rejected, the procedure tests $H_{(i+1)}$ and rejects it if $p_{(i+1)} \leq \alpha/(m-i)$; otherwise, testing stops and no further hypothesis can be rejected. If the procedure proceeds to $H_{(m)}$, it will be tested at level α. It can be shown that the Holm procedure is a closed testing procedure which applies the Bonferroni test to each intersection hypothesis. As a result, the Holm procedure controls the FWER strongly [10].

10.2.2.3 Hochberg Procedure

The Hochberg procedure is a stepwise procedure which uses the same critical values as the Holm procedure but in an opposite order [17]. Given the ordered p-values $p_{(1)} \leq p_{(2)} \leq \cdots \leq p_{(m)}$ for hypotheses $H_{(1)}, \ldots, H_{(m)}$, respectively, the Hochberg procedure rejects all hypotheses if $p_{(m)} \leq \alpha$. If all are rejected, no further testing is needed. Otherwise, if the procedure fails to reject $H_{(m)}$, it tests $H_{(m-1)}, \ldots, H_{(1)}$ and rejects them if $p_{(m-1)} \leq \alpha/2$. In general, given that $H_{(m)}, \ldots, H_{(i+1)}$ are not rejected, the procedure tests $H_{(i)}, \ldots, H_{(1)}$ and rejects them if $p_{(i)} \leq \alpha/(m-i+1)$; otherwise, the procedure does not reject $H_{(i)}$ and proceeds to $H_{(i-1)}$. Because the Hochberg procedure starts with the "least significant" p-value, it is called a *step-up* procedure whereas the Holm procedure is called a *step-down* procedure since it starts with the "most significant" p-value. It can be shown that the Hochberg procedure rejects all hypotheses rejected by the Holm procedure, and possibly more [17]. As a trade-off, the Hochberg procedure controls the FWER strongly only if the joint distribution of test statistics satisfies a certain condition [18,19]. For example, if the joint distribution is a multivariate normal distribution, the condition requires that the correlations should be non-negative.

10.2.2.4 Numerical Illustration

Consider a clinical trial with three hypotheses for response rate, PFS, and OS. Assume that the p-values are $p_1 = 0.005$, $p_2 = 0.015$, $p_3 = 0.02$. To control the FWER at one-sided level $\alpha = 0.025$, the Bonferroni test rejects H_1 only because $p_1 \le \alpha/3 = 0.008$ but both p_2 and p_3 are greater than 0.008. Applying the Holm procedure, H_1 is first rejected because $p_1 \le \alpha/3 = 0.008$. Then the procedure tests H_2 at level $\alpha/2 = 0.0125$ but it fails to reject H_2 because $p_2 > 0.0125$. The procedure stops with only rejecting H_1. The Hochberg procedure starts with H_3 with the largest p-value. Since $p_3 \le \alpha = 0.025$, the Hochberg procedure rejects all three hypotheses.

10.2.3 Gatekeeping and Graphical Procedures Based on the CTP

In clinical trials, hypotheses are often prioritized differently based on clinical importance, likelihood of success, etc. In addition, often certain sets of claims are more valuable than others. For example, it may be more useful to demonstrate a treatment benefit on both OS and PFS for one treatment dose than to find a treatment benefit only on PFS in the high and the low dose.

These considerations lead to multiple testing strategies which take such considerations into account. Many of them are so-called gatekeeping structures. One simple but common gatekeeping structure is that key secondary hypotheses are tested only if the primary hypothesis has been rejected. In this case, the primary hypothesis is the "gatekeeper" for key secondary hypotheses.

Many gatekeeping procedures as well as other complex multiple testing strategies can be visualized in a graphical procedure [6]. This graphical approach defines a shortcut to the closed testing procedure. As already mentioned in Section 10.2.1.3, shortcut procedures are attractive because they reach the decision about rejection of hypotheses in only m or m^2 steps rather than $2^m - 1$ steps.

10.2.3.1 Bonferroni-Based Graphical Procedures

Weighted Bonferroni procedures are particularly intuitive in the graphical approach. In the case of two hypotheses tested at level $w_1\alpha$ and $w_2\alpha$, respectively, their graphical representation is illustrated in Figure 10.2. Two hypotheses are shown as nodes with weights attached.

FIGURE 10.2
Weighted Bonferroni test as a graph.

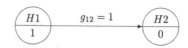

FIGURE 10.3
Fixed-sequence test as a graph.

Another simple example is the *fixed-sequence* (also called *strictly hierarchical*) test [20]. A fixed order of hypotheses is determined before data are collected, then testing proceeds sequentially in this order. For example, assume that the order is H_1: No treatment benefit on PFS, then H_2: No treatment benefit on OS. PFS would be tested first and be declared beneficial if its p-value $p_1 \leq \alpha$. Only if this condition is met, OS would be tested also, and H_2 would be rejected as well if $p_2 \leq \alpha$. Such methods are often called *gatekeeping* procedures with PFS being the "gatekeeper" for OS in this example. Notice also that this approach follows the CTP: $H_1 \cap H_2$ is rejected if and only if $p_1 \leq \alpha$. Hence, p_1 is used as the test statistic for both $H_1 \cap H_2$ and H_1.

A graphical representation of this procedure is shown in Figure 10.3. Here, $w_1 = 1$ in the node H_1 indicates that H_1 is initially tested at level $1 \cdot \alpha$, whereas H_2 is tested initially at level $0 \cdot \alpha$ (i.e., it is not tested initially). If H_1 can be rejected, $1 \cdot \alpha$ is shifted to H_2, as indicated by the edge weight $g_{12} = 1$ on top of the edge from H_1 to H_2.

Graphical displays such as the one given in Figure 10.3 turn out to be very valuable tools for investigating Bonferroni-based CTPs. As another simple example, the Holm procedure for two hypotheses is shown in Figure 10.4. Here, the CTP rejects the global intersection hypothesis $H_1 \cap H_2$ if $p_1 \leq (1/2)\alpha$ or $p_2 \leq (1/2)\alpha$. If any inequality is satisfied, the corresponding hypothesis (H_1 or H_2) is also rejected and $1 \cdot 1/2\alpha$ is propagated to the other hypothesis which is then tested at level $(1/2)\alpha + (1/2)\alpha = \alpha$. If both inequalities are satisfied, both H_1 and H_2 are rejected.

Bretz et al. [6] showed how to generalize this principle of visualizing Bonferroni-based CTPs. They gave an algorithm that allows one to work through the CTP in m steps and proved that graphical procedures are consonant by design. The initial weights fulfill $\sum_i w_i \leq 1$ and the edge weights also fulfill $\sum_j g_{ij} \leq 1$ for all nodes i, where g_{ij} is the edge weight associated with the edge from H_i to H_j. They also proved that the ultimate decision reached by the stepwise procedure does not depend on the order of the rejection steps. For example, in the Holm procedure, if both p_1 and p_2 are smaller than $\alpha/2$, it does not matter whether H_1 or H_2 is rejected first.

The graphical approach is implemented in the freely available R package gMCP [21].

10.2.3.1.1 *BELLE-4 Example*

In the BELLE-4 study, PFS and OS were the primary endpoints to be investigated in both the PI3 K-activated patients and the full population of patients. The clinical team had many discussions regarding the relative importance of a benefit claim for all of

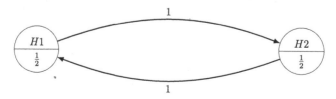

FIGURE 10.4
Holm test as a graph.

the possible combinations of endpoints and populations, and correspondingly many possible multiplicity-adjustment strategies were discussed. For illustrative purposes, we now discuss one of the considered strategies, though ultimately not implemented in the actual study. Figure 10.5 illustrates this strategy. Initially, $w_i \cdot \alpha = (\alpha/4)$ is assigned to each of the four hypotheses tests of PFS or OS in the full or the PI3K-activated subpopulation. Hence, a successful study claiming a statistically significant treatment benefit with strong FWER control at level α would require that $\min_{\{i=1, \ldots, 4\}} p_i \leq (\alpha/4)$ where p_i is the unadjusted p-value of the comparison of the experimental treatment versus the standard of care on the respective endpoint in the respective population (in this case, from a log-rank test).

If this requirement is met, additional hypotheses may be tested at possibly higher significance levels. For example, assume that $p_1 \leq (\alpha/4)$ so that a treatment benefit can be claimed for PFS in the full population. Then in the next step, H_2 (OS, full population) and H_3 (PFS, PI3 K population) would each be tested at level $(1/4) + (1/2) \cdot (1/4)\alpha = (3/8)\alpha$. H_4 (OS, PI3 K population) would still only be rejected if $p_4 \leq \alpha_4$ (because there is no edge from H_1 to H_4). The levels $w_i \cdot \alpha$, and the weights g_{ij} on the edges are updated according to the algorithm given by Bretz et al. [6]. If the strategy is specified in the R package gMCP and the p-values from the single tests are entered, these steps are by default performed internally, and only the final decision about rejection and non-rejection of hypotheses is reported. Optionally, adjusted p-values are reported. The user is also able to go through the updating algorithm step by step.

The strategy presented is of course only one of very many multiplicity-adjustment approaches that could be used within the framework of graphical procedures. For example, if significance in the full population is considered more important than in the subpopulation, an initial weight w_i higher than 1/4 could be given to H_1 and H_2. The weights for H_3 and H_4 would have to be lowered correspondingly such that $\sum_i w_i \leq 1$ is maintained. Likewise, if a claim of treatment benefit in both PFS and OS in a population is deemed a more convincing result than a significant result in both populations for only one of the endpoints, then the edge weights g_{12} and g_{21} could be set to 1 and the weights g_{13} and g_{24} could be set to 0. This would increase the probability of being able to claim a treatment benefit in PFS and OS for the full population, but decrease the probability of a claim on PFS in both populations.

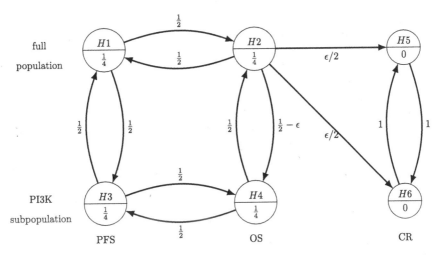

FIGURE 10.5
Graph for the BELLE-4 study.

Figure 10.5 contains a third endpoint, complete response (CR), which is a key secondary endpoint. In this example, it will be tested only if all four primary analyses have led to rejections. This is indicated by the fact that $w_5 = w_6 = 0$ and by the ϵs in the graph. In the graphical approach, ϵ represents a "quasi-zero." If a node i has an outgoing edge with an ϵ, then this is converted into a real weight only if other hypotheses connected from node i by real weights are rejected before.

If H_5 can be rejected in this framework, then a claim of statistical significance with strong FWER for this endpoint can be made. This may be advantageous if label claims for the corresponding endpoint are desired. However, the hurdle for achieving this result is very high since four gatekeepers potentially block testing of CR. In the BELLE-4 study, CR was not part of the multiple testing strategy.

10.2.3.2 *Procedures Based on Asymptotic Normality*

The graphical procedures discussed in Section 10.2.3 are very flexible. In addition to the ϵ-device discussed in Section 10.2.3.1, Maurer and Bretz [22] discuss convex combinations of graphs to allow memory in graphs. This, however, does not mean that all multiple testing procedures can be represented by these types of graphs. For example, the method of Hommel and Kropf [23] can not be displayed in this graphical form. Furthermore, the methods we discussed so far are completely agnostic to any correlation between the test statistics. If such correlations or the joint distribution are known, it is often possible to make use of them to devise more powerful multiple and multivariate tests.

Historically, the very first multiple tests were developed in the context of analysis of variance (ANOVA) models, predating the discovery of the CTP. These tests apply to (asymptotically) normally distributed endpoints. They deal with the simultaneous comparisons of more than two treatment arms in ANOVA models. Among the many suggestions from that era (see, for e.g., Dean et al. [24]), Tukey's all pairs comparison [25] and, in particular, Dunnett's many-to-one comparison method [26] are still in widespread use today since they control the FWER. It was later discovered that these methods can be combined with the closed testing procedure.

For the sake of illustration, let us assume that the elementary hypotheses are H_i: $\theta_i \leq 0$ with corresponding test statistics t_i where large values of t_i indicate deviations from H_i. Dunnett's many-to-one comparison uses normally or t-distributed test statistics t_i for the elementary hypotheses H_i. In contrast to the Bonferroni-based methods, the entire joint distribution of $t = (t_1, \dots, t_k)'$ is used to derive critical values. This distribution is a multivariate t-distribution or a multivariate normal distribution (see e.g., Dmitrienko [3]). In practical applications, these distributions are usually approximate (since test statistics are rarely really normally or t-distributed, and even if they are, the degrees of freedom in t tests are often approximate). In its original unweighted, single-step version, the critical value c for the Dunnett test is obtained by numerically solving $P\left[\cup_{i=1}^{k}(t_i \geq c)\right] = \alpha$. All hypotheses H_i which fulfill $t_i \geq c$ are rejected by the single-step Dunnett test. Since the known correlations between test statistics are used for the derivation of critical values, the Dunnett test is more powerful than the Bonferroni test. The mvtnorm package in R can be used to obtain critical values for Dunnett tests, but it is also built into many packages for linear and generalized linear models, for example, SAS PROC GLIMMIX.

The (unweighted) step-down Dunnett test arises if the ordinary single-step Dunnett test is applied sequentially to the ordered test statistics $t_{(1)} \geq t_{(2)} \geq \cdots \geq t_{(k)}$ with a recalculation of the critical value after every rejection. For example, if $t_{(1)} \geq c$, $H_{(1)}$ is rejected and a new c_2 is calculated for the remaining test statistics $t_{(2)}, \dots, t_{(k)}$ in such a way that $P\left[\cup_{i=2}^{k}(t_{(i)} \geq c_2)\right] = \alpha$.

Since $P\left[\cup_{i=2}^{k}(t_{(i)} \geq c)\right] \leq P\left[\cup_{i=1}^{k}(t_{(i)} \geq c)\right]$, we have $c_2 \leq c$ and hence the unweighted step-down Dunnett is a shortcut to a CTP. Obviously, the step-down Dunnett is more powerful than the single-step Dunnett test. It can also be viewed as an application of the Holm procedure to the Dunnett test, and hence it is also more powerful than the corresponding Holm test.

The (unweighted) step-up Dunnett test [27] proceeds in the opposite direction. The smallest value $t_{(k)}$ is compared to a critical value c_{uk} first. If $t_{(k)} \geq c_{uk}$, all hypotheses $H_1, \dots,$ H_k are rejected. Otherwise, $H_{(k)}$ is not rejected and $H_{(k-1)}$ is tested by comparing $t_{(k-1)}$ to a recalculated critical value $c_{u(k-1)} \geq c_{uk}$. If $t_{(k-1)} \geq c_{u(k-1)}$, all hypotheses except $H_{(k)}$ are rejected. Otherwise, the procedure continues in this fashion until the remaining hypotheses are rejected or until $H_{(1)}$ could not be rejected. The critical values $c_{uk} \leq \cdots \leq c_{u1}$ are calculated in such a way that the entire procedure maintains the FWER at α. This procedure can also be viewed as a CTP, but finding the critical values requires a complicated numerical search, and to our knowledge, the procedure is not implemented in readily available standard software. Figure 10.6 shows how Hochberg, step-down and step-up Dunnett relate to each other in the case of two hypotheses. The x- and y-axis correspond to the values of the correlated normally distributed test statistics for the first and the second hypothesis, respectively. The ellipse depicts a contour of constant density from the joint distribution of the two test statistics (which are positively correlated here). The light gray square is a region of possible values of the two test statistics where both the Hochberg and the step-up Dunnett test reject $H_1 \cap H_2$, but the step-down Dunnett test does not. In contrast, the two black stripes correspond to outcomes where both the step up- and the step-down Dunnett test reject $H_1 \cap H_2$, but the Hochberg test does not. Finally, the two dark gray stripes contain outcomes where only the step-down Dunnett test, but not the step-up Dunnett test or the Hochberg test reject $H_1 \cap H_2$. The graph illustrates that the step-up Dunnett test is uniformly more powerful than the Hochberg test, but not uniformly more powerful than the step-down Dunnett test. The Hochberg and step-down Dunnett tests do not dominate each other.

Weighted versions of the step-down Dunnett procedure have also been developed [28]. The joint distribution of the test statistics $t_1, \dots, t_k$ is still used to derive critical values, but rather than using the same critical value c for all hypotheses, a separate critical value is calculated for every H_i according to a pre-assigned weight for that hypothesis. This is repeated at each step in the step-down process. Utilizing the joint distribution to devise a

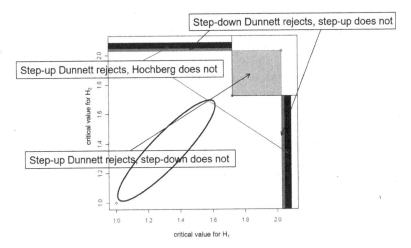

FIGURE 10.6
Comparison of rejection regions of the Hochberg and step-down and step-up Dunnett tests for $k = 2$ hypotheses.

more powerful graphical procedure may in general lead to the loss of consonance and thus the shortcut. If this occurs, one has to go through the CTP [29]. The R package gMCP also performs this task, but execution times can become rather long in case many hypotheses are investigated. For certain ways of updating the weights, the shortcut can be retained. For example, if the weights $w_j(J)$ for an intersection hypothesis $H_J, J \subseteq I$, are determined by "proportional upscaling" (that is, in such a way that they only depend on the initial weights w_i via $w_i(J) = \left(w_i / \sum_{i \in J} w_i \right)$ [30], then consonance and the shortcut still apply [28]. The Dunnett test and its step-down version are special cases of proportional upscaling.

The Dunnett procedures discussed in this chapter assume knowledge of the correlations between test statistics. (They do not necessarily assume knowledge of the error variance which may be estimated from the observed data, leading to a multivariate t-distribution.) In the case of linear models with identically normally distributed random errors, these correlations will often be known. An illustrative case is given by a Dunnett test of two regimens of an experimental treatment tested against a common control. If the sample sizes of the three groups are n_0 for the common control, n_1 and n_2, respectively, and normally distributed observations with equal error variance are assumed in all three groups, then the correlation between the two tests of experimental treatment regimens versus control is $\rho = \sqrt{(n_1/n_0 + n_1)(n_2/n_0 + n_2)}$. If $n_0 = n_1 = n_2$, we obtain $\rho = 1/2$. If in the BELLE-4 study, a continuous, normally-distributed endpoint would be tested, a similar argument could be made regarding the correlation between the full population test and the test in the biomarker positive subpopulation: the correlation between the two test statistics would be $\sqrt{n_S/n_F}$ where $n_S \leq n_F$ are the sample sizes in the sub- and the full population, respectively (we would condition on observed n_S in this formula if only the sample size n_F is fixed for all-comers in a trial).

Additional complications can arise if tests use asymptotically normal test statistics with correlations that depend on the available information, and this information is not merely equivalent to sample size. This is important in oncology because many studies use time-to-event endpoints such as OS and PFS, which are based on asymptotically normal test statistics with variances and correlations that depend on numbers of events rather than on numbers of patients.

Finally, one may wonder whether it is admissible to use an estimate of correlation instead of the (approximately known) correlation in cases where correlation is not governed by sample size or information. The typical example would be the case of two different endpoints measured on the same individual. If the two measurements are continuous, a bivariate normal distribution is often assumed. In this case, an estimate of the covariance matrix can often be determined and could be treated as fixed and known, similar as how in a z-test the estimate of error variance is assumed to be fixed and known. Although this approach performs quite well in simulation studies (α-inflation seems to be minor in almost all cases, e.g., in Su et al. [31]), it is rarely used in practice. Reasons are that power gains are only moderate unless correlations become very high and that health authorities may be reluctant to accept yet another element of approximation, especially in situations where the normal distribution assumption for the test statistics is already only asymptotic (such as in time-to-event analysis or the analysis of binary responses).

10.2.3.2.1 *BELLE-4 Example*

In the BELLE-4 study, we consider the effect of treatment on PFS in the full population F and in the PI3K-activated subpopulation S. Let us assume that a log-rank test is performed in both populations. To keep notation simple, we assume that event times are all distinct. The corresponding test statistics are

$$l_j = \frac{\sum_{k=1}^{d_j} (\delta_{jk} - p_{jk})}{\sqrt{\sum_{k=1}^{d_j} p_{jk}(1-p_{jk})}}$$

where d_j is the total number of events in population $j = S, F$, $\delta_{jk} \in \{0, 1\}$ is the indicator of event occurrence in the treatment group at the time t_k of the kth event in population j, $p_{jk} = (r_{jkT})/(r_{jkT} + r_{jkC})$, and r_{jkp} is the number of patients at risk in group $p = T, C$ at event time t_k. For the full population, this test could be replaced by a stratified log-rank test stratified by PI3K-activation/non-activation. If PI3K-pathway activation is prognostic for disease progression, this is actually a better choice because it provides more power.

It is easy to see by standard methods of time-to-event analysis that, under the global null hypothesis of no treatment benefit in either population, the joint distribution of $(l_S, l_F)'$ is asymptotically bivariate normal with mean 0 and variance 1 for both l_S and l_F and correlation $\sqrt{d_S/d_F}$. Hence, we can test the global null hypotheses by investigating whether $\max(l_S, l_F) \geq c_\rho(1 - \alpha)$ where $c_\rho(1 - \alpha)$ is the $(1 - \alpha)$-quantile of the bivariate standard normal distribution with correlation $\rho = \sqrt{d_S/d_F}$. In the BELLE-4 study, there were 131 progression events, of which 42 occurred in the PI3K-activated subpopulation. For one-sided $\alpha = 0.025$, the critical value for the Dunnett test is thus $c_{0.566}(0.975) = 2.204$ instead of 2.241 as for the Bonferroni test. Equivalently, the critical value for the p-value is $1 - \Phi^{-1}(2.204) = 0.0138$ instead of 0.0125. Had this threshold been achieved by the global test (which it did not in the BELLE-4 study), we could have proceeded to test the two populations S and F separately at level $\alpha = 0.025$ according to the step-down Dunnett approach.

Notice that $\rho = \sqrt{d_S/d_F}$ is an approximation to the correlation between the two test statistics, conditional on the total number of observed events in the two populations across the treatment groups. Under the global null hypothesis of no treatment effect, this converges to the true correlation. If the global null hypothesis is not true, it does not. However, the correlation under an alternative (e.g., a beneficial effect in the PI3K-population, but none in its complement) is not needed for the derivation of critical values of the global test. Notice also that in the absence of a parametric model for the hazard rates, the correlation between PFS and OS is much more difficult to approximate.

10.2.4 Multiplicity Adjustment for Other Types of Endpoints

So far, we have focused on normally distributed test statistics, since most statistical methodology for multiplicity adjustment has been derived for this case, or is generic (such as the CTP). This is justified by the fact that for many non-normally distributed endpoints (e.g., PFS and OS), the commonly used test statistics as well as point estimates (e.g., log hazard ratio estimates from a Cox-regression) are asymptotically normal. For some endpoints and types of trials, however, asymptotic methods may not be appropriate. In oncology, this usually arises in trials with binomial endpoints such as complete response (CR) or partial response (PR). If such trials are small or the events of interest are rare, it is preferable to work with exact methods for the binomial distribution rather than with normal approximations. There is an extensive literature on this topic (e.g., [32,33,34,35]) which is useful in this context. There are also close connections with permutation testing ([36,37]). More specifically, Tarone [38], Hommel and Krummenauer [39], and Ristl et al. [40] have investigated exact methods for multiple binary endpoints (such as partial and complete response measured after 6 weeks of treatment in all participating patients). Ristl et al. [40] also discuss exact methods for binary responses and subpopulation testing which is conceptually very similar to the case of multiple binomial endpoints.

10.3 Multiple Comparison Procedures in Oncology

After presenting a number of techniques for multiplicity adjustment in clinical trials in the previous sections, this section considers some strategic and operational aspects in oncology clinical trials.

10.3.1 The Scope of Multiplicity Adjustment

In complex clinical trials, it is sometimes unclear for how many of the investigated endpoints FWER control is really necessary. According to the US FDA [9], "primary endpoints are necessary to establish the effectiveness of a treatment," whereas secondary endpoints "may be selected to demonstrate additional effects after success on the primary endpoint." These are the two types of endpoints which would be placed under FWER control in a trial. All other endpoints are called "exploratory" and would not have to be treated with the same inferential statistical rigor. The European Medicines Agency [8] is not so strict on terminology, but essentially also requires that the rules for what to place under FWER control must be spelled out before the trial starts.

The decision about categorizing endpoints in a trial depends on many considerations. In oncology, PFS is often chosen as the primary endpoint with OS being only a secondary endpoint, in spite of the fact that OS is clinically more important in most instances. The reason is often that PFS allows an earlier readout or that an OS treatment benefit without a PFS benefit is almost impossible. The latter may be due to the fact that patients experiencing a progression are allowed to switch to another treatment. This may be ethically required, but if many patients on the standard-of-care treatment switch to the experimental treatment after progression, it of course makes it considerably more difficult to achieve a significant test result for OS.

Long "laundry lists" of primary and secondary endpoints should also be avoided. There have been instances of clinical trials where many secondary endpoints were generated by categorizing an underlying continuous endpoint into a series of binary responses, for example, by splitting patients into two groups based on their prostate-specific antigen (PSA) serum levels using cutpoints of 3, 5, 10, and 20 ng/mL. The resulting four response variables "PSA serum level ≥ 3 ng/mL, ... , PSA serum level ≥ 20 ng/mL" would then be tested for a treatment effect in some multiplicity-adjustment scheme, for example, hierarchically. While nothing is fundamentally technically wrong with such an approach, it nevertheless seems an overly complicated and formalistic assessment of the available information about the treatment effect.

An interesting current debate surrounds the handling of multiplicity arising from simultaneous investigation of several subpopulations. In the BELLE-4 study, it was decided to test the full and a biomarker-positive (PI3 K activated) population simultaneously with FWER control. This is certainly not the only possible option. As an alternative, one might have considered to do inference on the PI3 K activated and the PI3 K non-activated populations separately, seeking separate health authority approvals for the two populations. In this case, no multiplicity adjustment for the simultaneous testing of the two populations would be done. Approval for the full population would be a "by-product" of demonstrating a statistically significant benefit of the experimental treatment in both populations. This is an example of a "2-in-1" trial: two clinical trials that share the screening for patients, participating centers, inclusion criteria other than PI3 K status, visit schedules etc. An example of such a trial is the NeoPHOEBE trial [41].

In the case that more than two subpopulations are relevant, it is less clear how to deal with the multiplicity issue. In phase III trials, the number of such subpopulations will often be restricted to a very small number (often two) and FWER control will often be mandated by health authorities. In phase II studies, on one hand, requirements regarding multiplicity control will be less strict and formal FWER control may not be required. On the other hand, the danger of overinterpreting a randomly high treatment effect in one of many subpopulations is still a concern. Weaker forms of multiplicity control (e.g., control of the false discovery rate [42], or shrinkage methods [43] may be considered here; see also Chapter 16).

10.3.2 Multiple Endpoints Complications in Group Sequential Designs

Clinical trials in oncology, with endpoints such as PFS and OS, commonly utilize group-sequential trial designs. As alluded to in Section 10.1 and described in more depth in Chapters 6 and 9, these designs include interim analyses of accruing data as the trial proceeds, with repeated analyses addressing one or more of the study's hypotheses. They allow the possibility of stopping with a valid claim of success based on rejection of a primary null hypothesis (they may also result in stopping a trial if the results are sufficiently weak, commonly referred to as stopping for *futility*). Oncology trials with endpoints involving disease progression or mortality are obvious candidates for these designs for a number of reasons: the endpoints are very serious and there may be ethical imperatives that a trial not be continued if a standard of proof of efficacy can be achieved before the originally planned trial end (potentially resulting in earlier market access for a beneficial treatment), or if it is clear that a treatment is not effective. Such trials are commonly of quite long duration, and if a trial can be justifiably stopped early, this can result in cost savings to the trial sponsor or reallocation of resources to more promising research investigations.

As mentioned in Section 10.1, group-sequential designs induce a source of multiplicity arising from the multiple investigations of individual hypotheses. If we desire that the significance level for a null hypothesis be maintained at α, in a group-sequential design the chance that it would be rejected at least once among the multiple times that it might be examined must be controlled at that level. Necessarily, each test of that hypothesis must be performed using a level less than α. The repeated tests are correlated, and this correlation arises directly and precisely from the structure of the repeated tests: the data included in an interim analysis of a hypothesis are a subset of the data included in later analyses of the same hypothesis. For example, consider an interim analysis of a normally-distributed outcome based on n_1 patients where the design allows that the study could be stopped for success if the null hypothesis is validly rejected, and there is a later analysis based on n_2 patients (i.e., an additional $n_1 - n_2$ patients) with the same treatment allocation as the first analysis. Then the correlation between test statistics is $\sqrt{n_1/n_2}$. This is one source of multiplicity where it is commonly accepted to control exactly for this structural correlation, i.e., not to rely on Bonferroni-type alpha-splitting bounds. There are various approaches for doing this and determining analysis critical values, discussed elsewhere in this book, so these are not in the scope of this chapter. A very common approach involves the use of α-spending functions. However we will briefly mention two additional issues related to multiple testing that may arise in group sequential designs. These arise from the extra dimension of time, and the fact that there is not a single timepoint and dataset with which hypotheses are investigated.

One of these issues involves situations where a group-sequential plan is implemented to allow stopping at an interim analysis if valid criteria for its primary endpoints (e.g., PFS and/or OS) are achieved; however the trial analysis plan includes key secondary endpoints that would not be considered as a basis for stopping the trial on their own, but for which perhaps a label claim might be desirable if their null hypotheses can be rejected. Thus, they would be analyzed at a single point of the trial, the one at which the trial stops, whether at an interim analysis or the final one. One traditional possibility is to test such endpoints hierarchically behind the primary endpoint, at full α at the point where the primary analysis is performed. However, Hung et al. [44] have shown that this does not protect the FWER across the primary and key secondary endpoints. This can be addressed by imposing group-sequential schemes for those secondary endpoints as well. They need not use the same scheme as the primary endpoint (e.g., a different spending function may be used for key secondary endpoints). Glimm et al. [45] and Tamhane et al. [46] make suggestions for efficient schemes in this regard.

Another issue is that it may be planned for different endpoints to be addressed at different points of the trial. A frequent design in oncology can allow that there is a "final analysis" point for PFS, and the study is planned to continue for a later evaluation of OS (more generally, there may be group-sequential schemes for both endpoints, and at the final analysis of PFS, an interim analysis for OS is performed). The question of how to deal with this complication from a technical perspective is investigated in more detail by Chapter 3 of the book by George et al. [1] and the references therein.

From the perspective of result interpretation, significance in PFS alone would lead to so-called *conditional* or *accelerated* approval (terminology depends on the health authority), but if OS is not yet significant, the trial would all the same continue. This means that the treatment will be made available on the market, but this decision is subject to revision after the end of the study, and the market access authorization may be revoked. Even after conditional approval, treatment assignment will not be unblinded to the sponsor clinical trial team. There is of course some "unblinding" of information about the putative treatment effect in the mere fact that conditional approval was granted, but with a difficult-to-manipulate endpoint such as overall survival, this is deemed acceptable.

10.3.3 Outlook on Future Developments

We have focused on the handling of multiplicity in confirmatory clinical trials. In the past, relatively little attention had been given to biases arising from simultaneous inferences in earlier phases of drug development. However, the rise of personalized medicine, with its increased use of biomarkers for the identification of targeted therapies and the advances in genotyping, has brought about a proliferation of objectives in phase I and II trials. As further detailed in Chapters 13, 14, and 15, basket trials investigate different types of cancer (which are all affected by a common molecular pathway) with the same targeted therapy. Umbrella and platform trials try to identify biomarkers for different pathways of tumor growth and then identify appropriate targeted therapies for them. In the case of platform trials, multiple treatments enter or leave the platform in an adaptive manner [47]. The multiple objectives of these types of trials are often specified in a *master protocol*. It is obvious that if these tasks are performed simultaneously, there is a risk of overinterpreting "random highs" in the treatment benefit of a specific subpopulation or one of the many treatments investigated. At the same time, it seems overly pedantic to insist on FWER control in such an exploratory setting. There is currently an ongoing debate in the statistical community regarding the extent to which multiplicity must be contained in this situation. If a trial is strictly exploratory in the sense that it is not intended to obtain market authorization, but rather to just identify treatments to be investigated further, the trial sponsor may decide to protect against overinterpretation of results by picking from a wide range of methods, e.g., Bayesian shrinkage estimation, false discovery rate control, or more integrative benefit-risk calculations, regarding the selection of therapies for further investigation. A more problematic case arises if a sponsor intends to seek market authorization for a treatment with the results from a phase II trial if these results are very positive. Health authorities seem generally open to an "n-in-1" trial paradigm (no multiplicity adjustment for n), if "n" stands for distinct subpopulations in a basket trial. Even then, however, dealing with overlap in subpopulations characterized by different biomarkers, or how to deal with the case of a large n, are topics for more research. In platform trials with adaptive randomization of patients to changing sets of therapies, the challenge is even greater.

10.4 Conclusion

In this chapter, we present various methods for strong error-rate control in clinical trials where simultaneous inference is done on multiple research hypotheses. Typically, all these hypotheses are related to some kind of treatment benefit (longer survival, fewer disease progressions, slower tumor growth, more remissions, etc. under a new treatment than under the standard of care). For market authorization decisions, the availability of decision tools which strictly limit the probability of false claims regarding the benefit of a new treatment are of key importance. Multiple comparison procedures provide these tools. The graphical approach discussed in detail in this chapter facilitates a tailoring of these procedures to the requirements of the trial and guarantees FWER control even in complex situations with multiple endpoints, subpopulations, and treatment regimens.

Multiplicity issues also arise in earlier phases of clinical research, as well as regarding point and interval estimation. For a number of reasons, we only broached these topics here. In earlier phases of clinical research, no market authorization is sought yet, any new treatment emerging from early phases will still be subject to further investigations before being cleared for use in a wider patient population. Therefore, the risks associated with failure to adjust for multiplicity are to some extent contained. Regarding interval and point estimation, the role of and the remedies for multiplicity are less clear than in a decision-making context. While it is true that selective reporting of extreme results from univariate analyses (e.g., the largest treatment effect estimate from one of many investigated subpopulations) leads to biased point estimates and confidence intervals, the magnitude of this bias cannot be quantified without strong and often unverifiable additional assumptions. Regarding confidence intervals, this problem can to some extent be addressed by the use of simultaneous confidence regions. However, these regions tend to become very wide when many comparisons are involved, and their relation with stepwise testing procedures is awkward. In many cases, it will thus be acceptable to restrict the use of multiplicity adjustments to the decision making part of a clinical trial, while reporting treatment effect estimates and confidence intervals without adjustment—but remaining aware of the biases which may affect them.

References

1. George SL, Wang X, Pang H (eds.). *Cancer Clinical Trials: Current and Controversial Issues in Design and Analysis*. CRC Press, 2016.
2. Maurer W, Glimm E, Bretz F. Multiple and repeated testing of primary, coprimary, and secondary hypotheses. *Stat Biopharm Res*. 2011; 3(2):336–52.
3. Dmitrienko A, Tamhane AC, Bretz F. *Multiple Testing Problems in Pharmaceutical Statistics*. CRC Press, 2010.
4. Hochberg Y, Tamhane AC. *Multiple Comparison Procedures*. John Wiley & Sons, Inc., 1987.
5. Hsu J. *Multiple Comparisons: Theory and Methods*. CRC Press, 1996.
6. Bretz F, Maurer W, Brannath W, Posch M. A graphical approach to sequentially rejective multiple test procedures. *Stat Med*. 2009;28(4):586–604.
7. Martin M, Chan A, Dirix L et al. A randomized adaptive phase ii/iii study of buparlisib, a pan-class i-pi3k kinase inhibitor, combined with paclitaxel for the treatment of her2-advanced breast cancer(belle-4). *Ann Oncol*. 2017;28:313–20.
8. European Medicines Agency. *Guideline on Multiplicity Issues in Clinical Trials*. EMA, 2017.
9. Food and Drug Administration. *Multiple Endpoints in Clinical Trials - Draft Guidance*. U.S. Department of Health and Human Services, 2017.
10. Holm S. A simple sequentially rejective multiple test procedure. *Scand J Stat*. 1979;65–70.
11. Marcus R, Eric P, Gabriel KR. On closed testing procedures with special reference to ordered analysis of variance. *Biometrika*. 1976;63(3):655–60.
12. Hommel G, Bretz F, Maurer W. Powerful short-cuts for multiple testing procedures with special reference to gatekeeping strategies. *Stat Med*. 2007;26(22):4063–73.
13. Gabriel KR. Simultaneous test procedures–Some theory of multiple comparisons. *Ann Math Stat*. 1969;224–50.
14. Wright SP. Adjusted p-values for simultaneous inference. *Biometrics*. 1992;1005–13.
15. Brannath W, Schmidt S. A new class of powerful and informative simultaneous confidence intervals. *Stat Med*. 2014;33(19):3365–86.
16. Strassburger K, Bretz F. Compatible simultaneous lower confidence bounds for the holm procedure and other bonferroni-based closed tests. *Stat Med*. 2008;27(24):4914–27.
17. Hochberg Y. A sharper bonferroni procedure for multiple tests of significance. *Biometrika*. 1988;75(4):800–2.
18. Sarkar SK. Some probability inequalities for ordered mtp2 random variables: A proof of the simes conjecture. *Ann Stat*. 1998;494–504.
19. Sarkar SK, Chang C-K. The simes method for multiple hypothesis testing with positively dependent test statistics. *J Am Stat Assoc*. 1997;92(440):1601–8.
20. Maurer W, Hothorn LA, Lehmacher W. Multiple comparisons in drug clinical trials and preclinical assays: A-priori ordered hypotheses. *Biometrie in der chemisch-pharmazeutischen Industrie*. 1995;6:3–18.
21. Rohmeyer K, Klinglmüller F. gMCP: Graph Based Multiple Test Procedures, 2017. R package version 0.8-13. http://CRAN.R-project.org/package=gMCP.
22. Maurer W, Bretz F. Memory and other properties of multiple test procedures generated by entangled graphs. *Stat Med*. 2013;32(10):1739–53.
23. Hommel G, Kropf S. Tests for differentiation in gene expression using a data-driven order or weights for hypotheses. *Biom J*. 2005;47:554–62.
24. Dean A, Voss D, Draguljic D. *Design and Analysis of Experiments, Second Edition*. Springer, 2017.
25. Tukey JW. *The Problem of Multiple Comparisons*. Technical Report, Princeton, NJ: Department of Statistics, Princeton University, 1953.
26. Dunnett CW. A multiple comparison procedure for comparing several treatments with a control. *J Am Stat Assoc*. 1955;50(272):1096–121.
27. Dunnett XW, Tamhane AC. A step-up multiple test procedure. *J Am Stat Assoc*. 1992;87(417):162–70.

28. Xi D, Glimm E, Maurer W, Bretz F. A unified framework for weighted parametric multiple test procedures. *Biom J.* 2017;59(5):918–31.
29. Bretz F, Posch M, Glimm E, Klinglmueller F, Maurer W, Rohmeyer K. Graphical approaches for multiple comparison procedures using weighted bonferroni, simes, or parametric tests. *Biom J.* 2011;53(6):894–913.
30. Xie C. Weighted multiple testing correction for correlated tests. *Stat Med.* 2012;31(3):341–52.
31. Su T-L, Glimm E, Whitehead J, Branson M. An evaluation of methods for testing hypotheses relating to two endpoints in a single clinical trial. *Pharm Stat.* 2012;11:107–17.
32. Westfall PH, Troendle JF. Multiple testing with minimal assumptions. *Biom J.* 2008;50(5):745–55.
33. Gutman R, Hochberg Y. Improved multiple test procedures for discrete distributions: New ideas and analytical review. *J Stat Plan Inference.* 2007;137(7):2380–93.
34. Hirji KF. *Exact inference.* John Wiley & Sons, Inc., 2006.
35. Agresti A. *Categorical Data Analysis,* Third Edition. John Wiley and Sons, 2013.
36. Westfall PH, Young SS. *Resampling-Based Multiple Testing: Examples and Methods for p-Value Adjustment.* John Wiley & Sons, 1993.
37. Pesarin F, Salmaso L. *Permutation Tests for Complex Data: Theory, Applications and Software.* John Wiley & Sons, 2010.
38. Tarone RE. A modified bonferroni method for discrete data. *Biometrics.* 1990;515–22.
39. Hommel G, Krummenauer F. Improvements and modifications of tarones multiple test procedure for discrete data. *Biometrics.* 1998;54(2):673–81.
40. Ristl R, Xi D, Glimm E, Posch M. Optimal exact tests for multiple binary endpoints. *Comput Stat Data Anal.* 2018.
41. Loibl S, de la Pena L, Nekljudova V et al. Neoadjuvant buparlisib plus trastuzumab and paclitaxel for women with her 2+ primary breast cancer: A randomised, double-blind, placebo-controlled phase ii trial (neophoebe). *European Journal of Cancer.* 2017;85:133–45.
42. Benjamini Y, Hochberg Y. Controlling the false discovery rate: A practical and powerful approach to multiple testing. *J R Stat Soc Series B.* 1995;57:289–300.
43. Freidlin B, Korn EL. Borrowing information across subgroups in phase ii trials: Is it useful? *Clin Cancer Res.* 2013;19(6):1326–34.
44. Hung HMJ, Wang S-J, O'Neill R. Statistical considerations for testing multiple endpoints in group sequential or adaptive clinical trials. *J Biopharm Stat.* 2007;17(6):1201–10.
45. Glimm E, Maurer W, Bretz F. Hierarchical testing of multiple endpoints in group-sequential trials. *Stat Med.* 2010;29(2):219–28.
46. Tamhane AC, Mehta CR, Liu L. Testing a primary and a secondary endpoint in a group sequential design. *Biometrics.* 2010;66(4):1174–84.
47. Woodcock J, LaVange LM. Master protocols to study multiple therapies, multiple diseases, or both. *N Engl J Med.* 2017;377(1):62–70.

11

Cluster Randomized Trials

Catherine M. Crespi

CONTENTS

11.1 Introduction

In most randomized trials, individuals are randomized to study conditions and then followed to compare their outcomes. In a cluster randomized trial (CRT), also called a group randomized trial, preexisting groups of individuals such as clinics, schools, or communities are randomized to conditions, with all individuals in the same group receiving the same treatment. After a follow-up period, outcomes are typically measured at the individual level. Examples of interventions that have been studied using a cluster randomized design include an intervention to improve the patient-care experience for cancer patients that randomized nurses [1], an educational intervention to promote serologic testing for hepatitis B that randomized churches [2], and a cervical cancer screening program in rural India that randomized villages [3].

When individuals are randomized to conditions and do not interact with each other, their outcomes are generally regarded as independent. When preexisting groups are randomized, the outcomes of individuals in the same group cannot be considered independent. Rather, members of the same group share some commonalities—they may be patients with the same healthcare provider, children attending the same school, or residents of the same village—and also may interact during the treatment period, which will make the outcomes of individuals in the same group more similar than the outcomes of individuals from different groups. As explained in Section 11.3.1.1, this correlation of outcomes within groups makes cluster randomized trials less statistically efficient (that is, the intervention effect estimates have larger standard errors) than individually randomized trials in which clustering does not occur. As a result, cluster randomized trials require larger overall numbers of individuals to achieve the same level of statistical power. They also require the use of data analysis methods that account for clustering.

If cluster randomized trials are less efficient, why use them? Various considerations may motivate the selection of a cluster randomized design. The intervention may naturally be implemented at the group level, for example, group therapy or education sessions or a clinic-wide change in procedures. It may be less costly or logistically easier to implement the intervention at the cluster level. Cluster randomization can also prevent "contamination," that is, exposure of the control group to the intervention. Contamination tends to reduce differences in outcomes between conditions, making it more difficult to detect a treatment effect.

A key characteristic of cluster randomized trials is that they have multilevel data structure, with individuals at the lower level and clusters at the higher level. Thus design and analysis of cluster randomized trials are naturally handled using multilevel modeling. Multilevel models are extendable to accommodate other modeling features such as covariates and additional hierarchical structure such as repeated measures on individuals or additional levels. In this chapter we emphasize the multilevel modeling approach to cluster randomized trial design and analysis. For general treatments of multilevel model analysis, see Hox [4] and Snijders and Bosker [5].

This chapter is organized into three sections: randomization, analysis, and sample size and power. R code to implement methods is provided (see also companion website https://www.crcpress.com//9781138083776). Due to space constraints, we confine our attention to CRTs with two-level designs and a continuous or dichotomous outcome. Resources for CRTs with time-to-event or count outcomes or with more complex design elements such as additional levels, stepped-wedge, or crossover designs are provided in Section 11.5.

11.2 Randomization

Randomization helps to ensure balance across conditions on known and unknown prognostic factors. Due to randomization, we expect that the only systematic difference between two study arms will be that one received the intervention and the other did not; hence a comparison of the two conditions produces an unbiased estimate of the treatment effect.

Compared to an individually randomized trial, the number of randomized units in a cluster randomized trial is often relatively small, and simple randomization without restrictions (i.e., a coin flip) can result in chance imbalance between arms on important baseline covariates. For example, if 10 clinics are randomized to two conditions purely by chance, we may end up with most of the larger clinics in one arm and the smaller clinics in the other. Additionally, the number of clusters allocated to each arm may end up unequal. We briefly discuss several strategies for avoiding these problems, including matching and stratification, constrained randomization and minimization. For a more thorough discussion, see Hayes and Moulton [6] and Ivers et al. [7].

11.2.1 Matching and Stratification

In a matched or stratified design, clusters are sorted into groups or "strata" based on one or more prognostic factors, then clusters within strata are randomized to conditions [6,8,9]. A matched-pair design is the special case of strata of size two; clusters are paired and one cluster in each pair is assigned to each condition. Randomization within strata defined by one or more characteristics ensures balance between arms on these characteristics.

The Korean Health Study [2] provides an example of the use of stratified randomization in a cluster randomized trial. This study evaluated a church-based intervention to improve hepatitis B virus serological testing among Korean Americans in Los Angeles. Fifty-two Korean churches were stratified by size (small, medium, large) and location (Koreatown versus other) and randomized to intervention or control conditions within the six strata. This ensured balance between the intervention and control arms on size and location. Church location was considered potentially prognostic because of acculturation differences among participants attending churches inside versus outside Koreatown. Church size was considered potentially prognostic because of the potential for competing activities and resource differences at larger churches.

11.2.2 Constrained Randomization

Stratification or matching become difficult when there are many matching or stratification factors and a limited number of clusters. Constrained randomization, also called restricted randomization, is an alternative [6,10]. Constrained randomization involves generating all possible allocations of clusters to conditions, identifying the allocations that satisfy some predetermined balance criteria, then randomly selecting one allocation from the constrained set. This ensures acceptable balance on the predetermined criteria.

Constrained randomization was used in the implementation study reported by Maxwell et al. [11]. This study evaluated two strategies for implementing an evidence-based intervention to promote colorectal cancer screening in Filipino American community organizations. Twenty-two community organizations were randomized to either a basic or enhanced implementation strategy. Constrained randomization was used to ensure balance as well as to avoid contamination across arms. The investigators enumerated all two-group equal allocations of the 22 organizations that balanced the arms as to faith-based versus non-faith-based organizations, organizations with prior experience with the screening program versus organizations with no prior exposure, and zip code-level mean income and education, and also kept three organizations that were in close geographic proximity in the same arm (to prevent contamination), and randomly selected one of these allocations. The two groups were then randomly assigned to the basic or enhanced implementation strategy using a coin flip.

11.2.3 Minimization

Constrained randomization requires that all participating clusters be recruited and have relevant covariate information available at the beginning of the study. When clusters are recruited and randomized sequentially and/or there are many factors to balance, an alternative is minimization [12]. In minimization, the first few units (individuals in individually randomized trials, clusters in cluster randomized trials) are randomly assigned to conditions, and subsequent units are randomized to the arm that will minimize an imbalance measure that considers multiple covariates. Although minimization has not been widely used in cluster randomized trials [13], its ability to balance many covariates makes it an attractive option.

RANDOMIZATION IN CLUSTER RANDOMIZED TRIALS

Cluster randomized trials typically involve a relatively small number of clusters, and as a result, simple unrestricted randomization can result in chance imbalance between arms on important prognostic factors. Techniques such as matching, stratification, constrained randomization, and minimization can be useful for promoting balance across study arms.

11.3 Analysis

In this section, we discuss conducting the outcome analysis for cluster randomized trials. As discussed in the introduction, we focus on a multilevel modeling approach. We begin with a multilevel model for two-level data with a continuous outcome variable. This model introduces important concepts, including the intraclass correlation coefficient and the design effect. We then discuss estimation and inference for the intervention effect for continuous outcomes and for dichotomous outcomes.

Throughout this section, we assume that we have balanced data, meaning a two-arm trial with equal numbers of clusters in each condition and each cluster having an equal number of members. This assumption simplifies the derivation of key results. In practice, CRTs often have unequal numbers of clusters in each condition and clusters with varying numbers of members. In general, this does not alter the basic approach to estimation and inference using a general or generalized linear mixed-effects model. However, these factors can have an impact on statistical power and sample size requirements. For this reason, we defer discussion of these issues to Section 11.4.

11.3.1 Continuous Outcomes

Assume that we have continuous, normally distributed outcomes on individuals who are nested within clusters. For example, we may have pain scores on patients nested within hospital wards or depressive symptom scores on individuals nested within therapists. We set up a model for a single population of clusters and study some properties of the model. Then we add a covariate to encode cluster condition and discuss estimation and inference for the intervention effect.

11.3.1.1 Model

The basic model for a single population of clusters assumes that each cluster has its own mean and individuals within the cluster have outcomes that vary around that mean. The model for the outcome of individual i in cluster j, denoted Y_{ij}, is

$$Y_{ij} = \mu_j + \epsilon_{ij}, \tag{11.1}$$

where μ_j is the mean for cluster j and ϵ_{ij} is the error term indicating the discrepancy between the individual's observed outcome Y_{ij} and the cluster mean outcome μ_j.

We further assume that our clusters are sampled from a population of clusters that has an overall mean, with the cluster means varying around it. The model for the mean of cluster j, μ_j, is

$$\mu_j = \gamma_0 + u_j \tag{11.2}$$

where γ_0 denotes the population mean, assumed to be fixed, and u_j is a random effect representing the discrepancy between cluster j's mean and the population mean. Substituting Equation 11.2 into Equation 11.1 gives the single equation model

$$Y_{ij} = \gamma_0 + u_j + \epsilon_{ij}. \tag{11.3}$$

The random effects and error terms are assumed to be normal, with $u_j \sim N(0, \sigma_u^2)$ and $\epsilon_{ij} \sim N(0, \sigma_\epsilon^2)$, and to be independent of each other.

Model 11.3 is a simple model for two-level normally distributed data. Inspection of the model reveals that the total variance of an observation Y_{ij}, not conditional on cluster, can be decomposed as the sum of two independent variance components, one at the cluster level and the other at the individual level, namely,

$$Var\left(Y_{ij}\right) = \sigma_y^2 = \sigma_u^2 + \sigma_\epsilon^2. \tag{11.4}$$

Note that σ_ϵ^2 is the conditional variance of observations given that they are from the same cluster, and we expect that this variance will be lower than the total variance, that is, $\sigma_\epsilon^2 \leq \sigma_y^2$.

A useful quantity for characterizing the apportionment of the total variance of the outcome between the two levels of variation is the intraclass correlation coefficient (ICC). The ICC, commonly denoted ρ, is defined as

$$\rho = \frac{\sigma_u^2}{\sigma_y^2} = \frac{\sigma_u^2}{\sigma_u^2 + \sigma_\epsilon^2}. \tag{11.5}$$

The ICC quantifies the proportion of the total variance of the outcome that is attributable to clustering, or more precisely, to variance of the cluster-level means. Because σ_u^2 and σ_ϵ^2 are non-negative, we must have $0 \leq \rho \leq 1$.

It can be shown that, for Model 11.3, ρ also equals the correlation between two different observations from the same cluster, $Corr(Y_{ij}, Y_{i'j}), i \neq i'$. Furthermore, the covariance of two different observations from the same cluster, $Cov(Y_{ij}, Y_{i'j})$, is equal to $\rho \sigma_y^2$. By rearranging Equation 11.5, we also have that $\sigma_u^2 = \rho \sigma_y^2$ and $\sigma_\epsilon^2 = (1-\rho)\sigma_y^2$.

THE INTRACLASS CORRELATION COEFFICIENT

For two-level data following Model 11.3, the *intraclass correlation coefficient* (ICC), often denoted ρ, quantifies the proportion of the total variance of the outcome that is due to variance between clusters (i.e., variance in cluster-level means). The ICC is also equal to the correlation between two observations within the same cluster.

For most cluster randomized trials, ρ is small, typically in the range of 0.001–0.05. The value of the ICC in any specific trial will depend on the outcome variable, the type of cluster and other context-specific factors. Reporting guidelines recommend that cluster randomized trials report the observed ICC [14]. Reviews have compiled ICC values from various studies [15–17].

Now we consider sample means of data from cluster randomized trials and their variance. This will lead to some fundamental quantities and principles.

Suppose that we have a set of n_2 clusters with n_1 individuals in each cluster. This is our balanced data assumption. We discuss unequal allocation to conditions and varying cluster sizes in Section 11.4. The sample mean for cluster j can be calculated as

$$\bar{Y}_{\cdot j} = \frac{1}{n_1} \sum_{i=1}^{n_1} Y_{ij}. \tag{11.6}$$

What is the variance of the sample cluster mean? Using rules for the variance of the sum of correlated random variables, the variance can be found to be

$$Var(\bar{Y}_{\cdot j}) = \frac{1}{n_1}\left[Var(Y_{ij}) + (n_1 - 1)Cov(Y_{ij}, Y_{i'j})\right] = \frac{\sigma_y^2}{n_1}\left[1 + (n_1 - 1)\rho\right]. \tag{11.7}$$

Let $\bar{Y}_{\cdot\cdot} = (1/n_2)\sum_{j=1}^{n_2}\bar{Y}_{\cdot j} = (1/n_1 n_2)\sum_{j=1}^{n_2}\sum_{i=1}^{n_1} Y_{ij}$ be the overall sample mean across all observations. Because observations in different clusters are assumed to be independent, the variance of $\bar{Y}_{\cdot\cdot}$ is

$$Var(\bar{Y}_{\cdot\cdot}) = \frac{\sigma_y^2}{n_1 n_2}\left[1 + (n_1 - 1)\rho\right]. \tag{11.8}$$

If the $n_1 n_2$ observations had been independent, the variance of the sample mean would have been $\sigma_y^2/(n_1 n_2)$. The ratio of the variances is $1 + (n_1 - 1)\rho$, which is called the design effect for cluster randomized trials. We expect that the design effect will be greater than one when $\rho > 0$; hence another common term for the design effect is the variance inflation factor. The design effect will reduce to 1 when we have clusters of size 1 or when $\rho = 0$, that is, independent observations. The loss of statistical efficiency in cluster randomized trials is due to the design effect, which leads to larger standard errors.

THE DESIGN EFFECT

The term *design effect* comes from the field of survey sampling; see, for example, Kish [18]. When we conduct a cluster randomized trial, we can be regarded as collecting data from a *cluster sample* of individuals in each condition, rather than a simple random sample of individuals. The design effect or Deff quantifies the increase in the variance of the sample mean resulting from using a cluster sampling design:

$$\text{Deff} = \frac{\text{Variance for cluster sampling}}{\text{Variance for simple random sampling}}. \tag{11.9}$$

The design effect for a cluster randomized trial following Model 11.3 equals $1 + (n_1 - 1)\rho$ and represents the multiplicative factor by which the variance of the sample mean is increased due to cluster sampling.

Although ρ is typically small, the design effect also depends on cluster size and can be quite large. For example, an ICC of 0.02 and a cluster size of 100 leads to a variance inflation factor of 2.98, that is, almost a tripling of the variance of the sample mean compared to independent observations. This represents a substantial loss of statistical efficiency.

11.3.1.2 Estimation and Inference

Suppose now that our n_2 clusters are randomized to two conditions, with $n_2/2$ clusters in each condition. To accommodate different population means in each condition, we modify the model for the mean of cluster j to be

$$\mu_j = \gamma_0 + \gamma_1 w_j + u_j \tag{11.10}$$

where w_j is coded as -0.5 for the control condition and 0.5 for the intervention condition. Thus γ_0 is the grand mean (mean of the two means) and γ_1 is the difference in means between the two conditions. Note that the treatment indicator w_j is subscripted only by j and not by i, since treatment is assigned at the cluster level. The single equation model for the outcome Y_{ij} is

$$Y_{ij} = \gamma_0 + \gamma_1 w_j + u_j + \epsilon_{ij}. \tag{11.11}$$

We have allowed clusters in different conditions to have different cluster means, but this was done using a fixed effect; we have not altered the random effect terms in the model. The total variance of an observation is still $Var(Y_{ij}) = \sigma_u^2 + \sigma_\epsilon^2$ and we still express the ICC as $\rho = \sigma_u^2 / (\sigma_u^2 + \sigma_\epsilon^2)$, although it is possible that the magnitude of the variance components differs between conditions; we discuss this in Section 11.4.1.4.

Interest usually focuses on estimating the treatment effect γ_1. An unbiased estimate of γ_1 can be obtained as the difference of treatment group means. Indexing condition by $k = 1,2$ and the outcomes as Y_{ijk}, our estimate of the treatment effect is

$$\hat{\gamma}_1 = \bar{Y}_{.1} - \bar{Y}_{.2}$$

with variance

$$Var\left(\hat{\gamma}_1\right) = \frac{4\sigma_y^2}{n_1 n_2}\left[1 + (n_1 - 1)\rho\right] \tag{11.12}$$

(the factor of 4 arises because there are $n_2/2$ clusters per condition). Using Equations 11.4 and 11.5, we can also write this variance as

$$Var\left(\hat{\gamma}_1\right) = \frac{4}{n_1 n_2}\left(\sigma_\epsilon^2 + n_1 \sigma_u^2\right). \tag{11.13}$$

The test for a treatment effect is the test of the null hypothesis $H_0 : \gamma_1 = 0$ in Model 11.11. This test can be conducted using the test statistic

$$\frac{\hat{\gamma}_1}{SE\left(\hat{\gamma}_1\right)} \tag{11.14}$$

where $SE\left(\hat{\gamma}_1\right) = \sqrt{Var\left(\hat{\gamma}_1\right)}$. Under the null hypothesis, the test statistic in Expression 11.14 has approximately a t distribution. In general, for a two-level model, the number of degrees of freedom (df) associated with the regression parameter for a cluster-level covariate is $n_2 - q - 1$, where q equals the total number of cluster-level covariates. For testing γ_1 in Model 11.11, the df for the t statistic are $n_2 - 2$. When the df are large, the standard normal distribution can be used.

The parameter γ_1 and its standard error, as well as γ_0 and its standard error, can be estimated by maximum likelihood or restricted maximum likelihood (REML). When the number of clusters is small ($n_2 \leq 50$), REML estimation is recommended for estimating fixed-effect parameters; for datasets with a larger number of clusters, either method may be used and should give similar results [5,19]. REML estimates are preferred for estimating the variance components σ_ϵ^2 and σ_u^2, regardless of dataset size, because maximum likelihood estimators of the variance components have a downward bias [5].

If stratification is used, the model should include indicators for strata, which ensures that the treatment effect estimates are conditional on stratum. Stratification can increase power; see Section 11.4.1.8.

11.3.1.3 Example

To illustrate inference for a cluster randomized trial with a continuous outcome, we simulate and analyze data based loosely on a CRT of a pain self-management intervention for cancer patients reported in Jahn et al. [20]. The intervention was delivered in the hospital setting and involved a nurse-led counseling program focused principally on reducing patient-related cognitive barriers. To avoid contamination across conditions, the study was designed as a cluster randomized trial and the intervention was applied at the ward level. Nurses in wards assigned to the intervention condition received special training, while nurses in control wards did not. Outcomes were measured on patients in the wards. The primary outcome was patient score on the Barriers Questionnaire II.

While the actual trial had nine oncology wards in each condition and a variable number of patients per ward, for pedagogical reasons, we simulated data with 10 wards per condition and 10 patients per ward. Our simulated data were based on Model 11.11 with parameter values $\gamma_0 = 60$, $\gamma_1 = 10$, $\sigma_u^2 = 25$, and $\sigma_\epsilon^2 = 600$. These values imply that $\sigma_y^2 = 625$ and $\rho = 0.04$. The R commands to simulate the data and fit the model are

```
# set parameter values
n2 <- 20
n1 <- 10
gamma_0 <- 60
gamma_1 <- 10
sigma_u <- sqrt(25)
sigma_e <- sqrt(600)
sigma_y <- sqrt(625)

# simulate data
set.seed(96135)
u <- rep(rnorm(n2, sd=sigma_u), each=n1)
e <- rnorm(n1*n2, sd=sigma_e)
w <- c(rep(-0.5, n1*n2/2), rep(0.5, n1*n2/2))
y <- gamma_0 + gamma_1*w + u + e
j <- rep(seq(1:n2), each=n1)
pain.data <- data.frame(y, j, w)

# fit model
library(lme4)
paincrt <- lmer(y ~ w + (1|j), data=pain.data)
summary(paincrt)
...
Random effects:
 Groups Name          Variance Std.Dev.
 j        (Intercept)  30.3      5.51
 Residual              618.7     24.87
Number of obs: 200, groups: j, 20

Fixed effects:
             Estimate Std. Error    df t value Pr(>|t|)
(Intercept)     63.95       2.15 18.00   29.79  <2e-16 ***
w                9.90       4.29 18.00    2.31   0.033 *
```

The grand mean is estimated as $\hat{\gamma}_0 = 63.95$, while the treatment effect is estimated as $\hat{\gamma}_1 = 9.90$ and is significant at the 0.05 level. The df for both fixed effects parameters is $20 - 2 = 18$. The estimated variance components are $\hat{\sigma}_u^2 = 30.3$ and $\hat{\sigma}_\epsilon^2 = 618.7$, from which we can calculate $\hat{\sigma}_y^2 = 649.0$ and $\hat{\rho} = 0.047$. The estimated parameter values do not coincide with the true values due to sampling variability.

What would we have inferred if we had neglected to account for the clustering of observations within wards? The following R code fits the linear regression model $Y_{ij} = \gamma_0 + \gamma_1 w_j + e_{ij}$, which assumes independent observations, to the data:

```
# Wrong model!
summary(lm(y ~ w , data=pain.data))
...
```

```
Coefficients:
            Estimate Std. Error t value Pr(>|t|)
(Intercept)   64.0        1.8    35.58  <2e-16  ***
w              9.9        3.6     2.75  0.0064  **
```

We get the same point estimates for γ_0 and γ_1, but the standard errors are substantially smaller and hence the p-value for the treatment effect is also lower. Although in this case we would have reached the same conclusion about rejecting the null hypothesis of no treatment effect at level 0.05, in many data analyses, the deflated p-value would have led us to incorrectly reject the null. The regression parameter estimates are identical in the models fit with and without clustering because our data are balanced; in the case of unbalanced data, this will not always occur.

We can get confidence intervals for the variance components using the command

```
confint(paincrt)
```

which yields a 95% confidence interval for $\hat{\sigma}_u$ of $(0.00, 10.2)$. The interval appears to include zero. Should we drop this variance component? No, as explained in the accompanying box.

SHOULD NONSIGNIFICANT VARIANCE COMPONENTS BE DROPPED?

When analyzing data from a cluster randomized trial, if a variance component is not significantly different from zero, should it be dropped from the model?

Even a small ICC, if ignored, can inflate the type I error rate, that is, the probability that we erroneously reject the null hypothesis and declare the intervention to be effective. Furthermore, the standard errors for variance components are not well estimated when their true values are close to zero, and the degrees of freedom for such tests, which are based on the number of clusters, are usually limited, which limits the power of such tests. Therefore it is recommended that all random effects associated with the study design and sampling plan be retained in the model.

In the example, the true value of the variance of the cluster means is known to be non-zero because the data were simulated. Dropping this term would lead to a mis-specified model.

11.3.2 Dichotomous Outcomes

Now we consider cluster randomized trials with dichotomous outcomes. Examples of dichotomous outcomes include achieving a tumor response, receiving a cancer screening procedure, or acquiring an infection.

There are two common approaches to modeling dichotomous outcomes from cluster randomized trials [21]. One approach models the cluster-level proportions, and the second models the cluster-level log odds. We discuss both approaches. We spend some time discussing the intraclass correlation, which is more complicated for dichotomous data than it is for continuous data. We first discuss simple models for two-level clustered data, without covariates or different intervention conditions, in order to study important principles, and then discuss estimation and inference for an intervention effect.

11.3.2.1 Cluster-Level Proportions Model

Let Y_{ij} denote the dichotomous outcome of the ith individual in the jth cluster, where $Y_{ij} = 1$ for success and 0 for failure. Under the cluster-level proportions model, the individuals in cluster j have a probability of success that is specific to their cluster, denoted π_j. Thus the Y_{ij} are Bernoulli random variables with success probability π_j. The cluster-level success probabilities π_j are assumed to be random variables that follow a distribution with $E(\pi_j) = \pi$ and $Var(\pi_j) = \sigma_d^2$. The specific distribution does not affect the key results and we leave it unspecified. Under this model, the mean and variance of Y_{ij}, unconditional on cluster, are $E(Y_{ij}) = E(\pi_j) = \pi$ and $Var(Y_{ij}) = \pi(1-\pi)$. This leads to an expression for the ICC in the cluster-level proportions model as

$$\rho_d = \frac{Var(\pi_j)}{Var(Y_{ij})} = \frac{\sigma_d^2}{\pi(1-\pi)}. \tag{11.15}$$

Now we consider sample proportions and their properties. The sample proportion for cluster j can be calculated as

$$\hat{\pi}_j = \bar{Y}_j = \frac{1}{n_1}\sum_{i=1}^{n_1} Y_{ij}, \tag{11.16}$$

where n_1 is cluster size. The sample cluster proportion is an unbiased estimate of the true cluster proportion, $E(\hat{\pi}_j) = \pi_j$, and the variance of $\hat{\pi}_j$ can be found to be

$$Var(\hat{\pi}_j) = \frac{\pi(1-\pi)}{n_1}\left[1 + (n_1 - 1)\rho_d\right]. \tag{11.17}$$

This variance is the analogue of the variance of the sample cluster mean for continuous outcomes given in Equation 11.7.

Let $\hat{\pi} = \bar{Y}_{..} = \frac{1}{n_1 n_2}\sum_{j=1}^{n_2}\sum_{i=1}^{n_1} Y_{ij}$ be the overall sample proportion across all observations, assuming n_2 clusters of size n_1. The overall sample proportion provides an unbiased estimate of population proportion; $E(\hat{\pi}) = \pi$. Because observations in different clusters are assumed to be independent, the variance of $\hat{\pi}$ is

$$Var(\hat{\pi}) = \frac{\pi(1-\pi)}{n_1 n_2}\left[1 + (n_1 - 1)\rho_d\right]. \tag{11.18}$$

Had the $n_1 n_2$ observations been independent, the variance of the sample proportion would have been $(\pi(1-\pi))/(n_1 n_2)$. As for continuous outcomes, the ratio of the variances is the design effect, $1 + (n_1 - 1)\rho_d$, which reflects the increase in the variance of the sample proportion attributable to correlation of observations within clusters, or from another perspective, due to cluster sampling of observations.

Now suppose that our n_2 clusters are randomized to two conditions with $n_2/2$ clusters in each condition. Denote the success proportions in the two conditions as π_1 and π_2. Adding a subscript k to denote condition, we could estimate these proportions as $\hat{\pi}_k = \bar{Y}_{\cdot k} = \frac{1}{n_1 n_2/2} \sum_{j=1}^{n_2/2} \sum_{i=1}^{m_1} Y_{ijk}$ for $k = 1, 2$. The intervention effect can be estimated as the difference in sample proportions, $\hat{\pi}_1 - \hat{\pi}_2$, and its variance is

$$Var\left(\hat{\pi}_1 - \hat{\pi}_2\right) = \left[\frac{\pi_1\left(1-\pi_1\right)}{n_1 n_2/2} + \frac{\pi_2\left(1-\pi_2\right)}{n_1 n_2/2}\right]\left[1 + \left(n_1 - 1\right)\rho_d\right]. \tag{11.19}$$

This result is the basis for a commonly used sample size calculation approach for cluster randomized trials with dichotomous outcomes, which we discuss in Section 11.4.2.1. However, the cluster-level log-odds model, which we discuss next, is more commonly used for analysis.

11.3.2.2 Cluster-Level Log-Odds Model

The other approach for modeling dichotomous outcomes for CRTs is to use a random-effects logistic regression model, such as the logistic-normal model. This model assumes that Y_{ij} is Bernoulli with cluster-specific success probability π_j, and that the logits of cluster proportions π_j follow a normal distribution. The basic model without covariates can be expressed as

$$log\left(\frac{\pi_j}{1-\pi_j}\right) = \gamma_0 + u_j \tag{11.20}$$

where $u_j \sim N(0, \sigma_u^2)$.

Under this model, the between-cluster variance σ_u^2 is expressed on the log-odds scale. The model implicitly assumes an overall population proportion $\pi = 1/(1+e^{-\gamma_0})$, making the total population outcome variance equal to $\pi(1-\pi)$, which is on the proportions scale. The quantities σ_u^2 and $\pi(1-\pi)$ are not comparable because they are on different scales and it is not sensible to form an ICC as their ratio. One way of finding an ICC for Model 11.20 that is on the proportions scale is to use a Taylor expansion of logit(π_j), which yields an approximation of ρ_d as

$$\rho_d \approx \sigma_u^2[\pi\left(1-\pi\right)]^2; \tag{11.21}$$

see Turner et al. [22]. An alternative approach is to define the ICC on the log-odds scale. This leads to the expression

$$\rho_{d(l)} = \frac{\sigma_u^2}{\sigma_u^2 + \Pi^2/3} \tag{11.22}$$

where Π is the mathematical constant 3.14156...; see Eldridge et al. [21] and Snijders and Bosker [5]. The term $\Pi^2/3$ is the variance of the standard logistic distribution and plays the role of the within-cluster variance.

ICCS FOR DICHOTOMOUS DATA

The ICC for clustered dichotomous data can be defined on the proportions scale (ρ_d) or the log-odds scale ($\rho_{d(l)}$). These ICCs can take very different values for the same data. The proportional discrepancy between ρ_d and $\rho_{d(l)}$ is greater for larger values of ρ_d and when the prevalence π is farther from 0.5. For further discussion, see Eldridge et al. [21].

Researchers should be aware of the two different scales for the ICC for clustered dichotomous data and be careful to use the ICC on the correct scale for sample size and power calculations.

11.3.2.3 Estimation and Inference

To model data from a cluster randomized trial, we use the cluster-level log-odds model and expand the model to include a covariate encoding cluster condition. The model is

$$log\left(\frac{\pi_j}{1-\pi_j}\right) = \gamma_0 + \gamma_1 w_j + u_j \tag{11.23}$$

where w_j is coded as -0.5 for the control condition and 0.5 for the intervention condition. Thus γ_0 is the average log odds of success across all clusters and γ_1 is the difference in log odds for success between the two conditions. The intervention effect is typically reported as an odds ratio, obtained as e^{γ_1}.

A closed form expression for the estimator $\hat{\gamma}_1$ and its variance can be derived; see Moerbeek et al. [23] and Spybrook et al. [24]. Assuming equal-sized clusters of size n_1 and $n_2/2$ clusters per condition, the intervention effect can be estimated as the difference in average log odds between conditions and the variance of $\hat{\gamma}_1$ can be estimated as

$$Var\left(\hat{\gamma}_1\right) = \frac{4\left(\sigma_u^2 + \tau^2/n_1\right)}{n_2} \tag{11.24}$$

where

$$\tau^2 = \frac{1}{2}\left[\frac{1}{\pi_1\left(1-\pi_1\right)} + \frac{1}{\pi_2\left(1-\pi_2\right)}\right] \tag{11.25}$$

is a measure of variability at the individual level.

There are various algorithms for fitting mixed-effects logistic models and obtaining parameter estimates, standard errors, and confidence intervals [25]. The expression for the likelihood of a mixed-effects model is an integral over the random-effects space. For a linear mixed-effects model, this integral can be evaluated exactly. For a generalized linear mixed-effect models, the integral must be approximated. Different approximation methods can give slightly different results.

11.3.2.4 Example

To illustrate analysis for a cluster randomized trial with a dichotomous outcome, we simulate data based on the Breast Cancer Education Program for Samoan Women [26]. This study evaluated the effectiveness of a breast cancer education program tailored to women with Samoan ancestry in the United States. In the trial, 61 Samoan churches were randomized to the intervention or control condition. Women from churches in the intervention arm participated in culturally tailored interactive group discussion sessions with a health educator; the control condition was usual care. The primary outcome was self-reported receipt of a mammogram within eight months.

The trial had a variable number of women per church, ranging from 1 to 42 with a median of 13. For simplicity, we simulated data with 30 churches in each condition and 12 participants per church. The trial had a rather high ICC of 0.19 on the proportions scale (ρ_d). We simulated data with $\rho_d = 0.1$. We assume that the proportions of participants with self-reported mammogram receipt are 0.30 and 0.50 in the control and intervention arms, respectively. The overall population proportion is thus 0.40, and using Equation 11.21, we find that we need to set $\sigma_u = 1.318$. Using Equation 11.22, the ICC on the log odds scale is $\rho_{d(l)} = 0.345$ (note that this value differs substantially from the ρ_d of 0.1). The regression coefficients needed to reflect the success proportions in each arm are $\gamma_0 = -0.4237$ and $\gamma_1 = 0.8473$. R code to simulate data is

```
# set parameter values
n2 <- 60
n1 <- 12
gamma_0 <- -0.4236
gamma_1 <- 0.8473
sigma_u <- 1.318


# simulate outcome data
set.seed(32410)
linprobs1 <- gamma_0 - gamma_1/2+rep(rnorm(n2/2, sd=sigma_u), each=n1)
linprobs2 <- gamma_0+gamma_1/2+rep(rnorm(n2/2, sd=sigma_u), each=n1)
linprobs <- c(linprobs1, linprobs2)
probs <- 1/(1+exp(-linprobs))
y <- sapply(probs, function(x) sample(0:1, 1, prob = c(1 - x, x)))
j <- rep(seq(1:n2), each=n1)
w <- c(rep(-0.5, n1*n2/2), rep(0.5, n1*n2/2))
mamm.data <- data.frame(y, j, w)


# average success proportions in each condition
mean(mamm.data$y[mamm.data$w==0.5])
mean(mamm.data$y[mamm.data$w== - 0.5])
```

The success proportions in the simulated data are 0.36 and 0.49 in the control and intervention arms, respectively. Figure 11.1 shows the distribution of the cluster proportions and the logits of the cluster proportions in the two study arms. The proportions and logits are shifted somewhat lower for the control clusters. The logits of the cluster proportions are specified as normally distributed with different means in each condition; the normality assumption is not fully apparent in the figure, particularly in the control arm, due to the relatively low number of clusters in each condition.

R code to fit the model using glmer in R and abbreviated output are below:

```
# fit model
summary(glmer(y ~ w + (1 | j), data = mamm.data, family = binomial))
...
Random effects:
 Groups Name          Variance Std.Dev.
 j      (Intercept) 1.71       1.31
Number of obs: 720, groups:  j, 60
Fixed effects:
            Estimate Std. Error  z value  Pr(>|z|)
(Intercept)  -0.418       0.192    -2.18     0.029 *
w             0.694       0.384     1.81     0.071 .
```

The test statistics for the fixed effects have approximately a normal distribution rather than a t distribution under the null, so there are no df to consider. The estimates differ from the

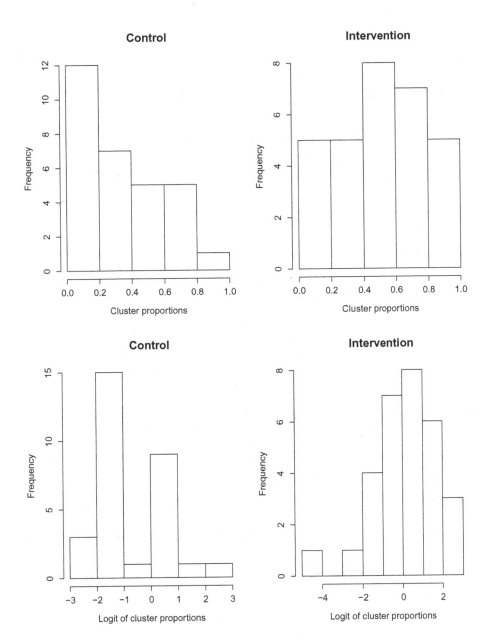

FIGURE 11.1
Distribution of cluster-level proportions and logits of the cluster-level proportions from simulated data, by intervention condition.

true values due to sampling variability. The p-value exceeds the benchmark of 0.05. The odds ratio for the treatment effect is $e^{0.694} = 2.0$. The estimated standard deviation of the random effect is 1.31. Using Equations 11.21 and 11.22, we can calculate the estimated ICCs as 0.342 on the log-odds scale and 0.098 on the proportions scale. These are close to the true values. The discrepancy between these two ICCs serves as a reminder that these two quantities are on different scales and should not be confused.

11.3.3 Other Analysis Methods

We have discussed analysis of data from CRTs using multilevel modeling (linear or generalized linear mixed models). This approach is statistically efficient and easily accommodates regression adjustment for covariates or specification of additional hierarchical data structure. Other methods may be useful for specific studies. For example, a two-sample t test comparing cluster-level summary statistics (means, proportions) is robust to departures from the normality assumption [6]. Other robust options include nonparametric tests on cluster-level statistics and permutation tests. Some of these methods allow for a limited amount of covariate adjustment. For further information, see Hayes and Moulton [6].

Another approach is generalized estimating equations (GEE) [27,28]. GEE assumes a linear or generalized linear model for the expected values of the dependent variable, conditional on the explanatory variables, but does not fully specify a probability model for the data. Rather, the parameters are estimated under a "working model" for the covariance structure; for a cluster randomized trial, an exchangeable correlation structure is typically assumed. Standard errors are obtained using a robust sandwich estimator. For our example, the R code and abbreviated output are

```
summary(geeglm(y ~ w, id=j, data=mamm.data, family=binomial,
    corstr= "exchangeable"))
...
 Coefficients:
            Estimate Std.err Wald Pr(>|W|)
(Intercept)  -0.319    0.148 4.64    0.031 *
w             0.527    0.296 3.17    0.075 .
...
Estimated Correlation Parameters:
       Estimate Std.err
alpha     0.253  0.0476
```

The coefficient estimates in a GEE model are population-average estimates; here the estimated population-average odds ratio is $e^{0.527} = 1.7$. In contrast, mixed-effects logistic models provide odds ratios conditional on cluster. In general, population-averaged odds ratios are closer to the null than are cluster-conditional odds ratios. However, the p-values tend to be similar. Here, the p-values are close. The GEE estimated correlation parameter is the Pearson correlation between observations in the same cluster. For further discussion of population-average versus cluster-specific approaches, see Gardiner et al. [29] and Hu et al. [30].

11.4 Sample Size and Power

When designing a cluster randomized trial or other study, we typically want to ensure that the sample size will be adequate to achieve the study's objectives. The primary objective is usually to detect a clinically meaningful and statistically significant difference between outcomes in the intervention and control conditions.

This section discusses how to calculate statistical power for a cluster randomized trial with a continuous or binary outcome, and how to find the sample size required to achieve a desired level of power. For both types of outcomes, results are first derived for the case of balanced data, with equal numbers of clusters in each condition and clusters of equal size. Subsequent sections consider unequal allocation of clusters, varying cluster sizes, and unequal ICCs in the two arms.

We restrict attention to cluster randomized trials with two-level data structure and a continuous or binary outcome. Discussion of power and sample size for other cluster randomized trial designs and other types of outcomes can be found in several articles [6,8,9,31–34].

11.4.1 Continuous Outcomes

11.4.1.1 Power

We begin by assuming that our data follow the two-level normal Model 11.11, under which the observation Y_{ij}, for individual i in cluster j, follows

$$Y_{ij} = \gamma_0 + \gamma_1 w_j + u_j + \epsilon_{ij}$$

with $u_j \sim N(0, \sigma_u^2)$, $e_{ij} \sim N(0, \sigma_\epsilon^2)$, and u_j and ϵ_{ij} independent. We test for an intervention effect by testing $H_0 : \gamma_1 = 0$ using test statistic in Expression 11.14. When the null hypothesis is true, the test statistic follows a t distribution with $n_2 - 2$ degrees of freedom. When $\gamma_1 \neq 0$, the test statistic follows a noncentral t distribution (see box), which has two parameters, a df and a noncentrality parameter. Here, the df are $n_2 - 2$ and when there are $n_2/2$ clusters in each condition, n_1 individuals in each cluster and equal ICCs in each arm, the noncentrality parameter is

$$\lambda = \frac{\gamma_1}{\sqrt{\left(4\left(\sigma_\epsilon^2 + n_1\sigma_u^2\right)\right)/(n_1 n_2)}} = \frac{\gamma_1}{\sqrt{\left(4\sigma_y^2\left[1 + (n_1 - 1)\rho\right]\right)/(n_1 n_2)}}. \tag{11.26}$$

The numerator is the true difference in means between conditions, and the two versions of the denominator are the square root of the variance of $\hat{\gamma}_1$; see Section 11.3.1.2.

NONCENTRAL T DISTRIBUTION

A random variable of the form

$$\frac{Z + \lambda}{\sqrt{\dfrac{\chi_\nu^2}{\nu}}} \tag{11.27}$$

where Z is a standard normal random variable, χ_ν^2 is a chi-square random variable with ν degrees of freedom and λ is a constant, has a noncentral t distribution with ν degrees of freedom and noncentrality parameter λ, denoted $t_{\nu, \lambda}$. The standard (central) t distribution is the special case of $\lambda = 0$.

It is often convenient to work with standardized effect sizes, which give the difference between the intervention and control condition means in units of the standard deviation of the outcome variable. Here, the standardized effect size is $\delta = \gamma_1/\sigma_y$, where σ_y^2 is the total variance of the outcome. The noncentrality parameter can then be expressed as

$$\lambda = \frac{\delta}{\sqrt{\left(4\left[1+(n_1-1)\rho\right]\right)/(n_1 n_2)}}. \tag{11.28}$$

Benchmarks for standardized effect sizes are given by Cohen [35], who suggested that 0.2, 0.5, and 0.8 represent small, moderate, and large effect sizes, respectively.

The power $1-\beta$ of a hypothesis test is the probability that the value of the test statistic is more extreme than the critical value(s) given that some specified scenario is true. For a two-sided test with type I error rate α, the power for a CRT following the two-level normal Model 11.11 can be calculated as

$$P\left[t_{n_2-2,\lambda} > t_{n_2-2,0}\left(1-\frac{\alpha}{2}\right)\right] + P\left[t_{n_2-2,\lambda} < t_{n_2-2,0}\left(\frac{\alpha}{2}\right)\right] \tag{11.29}$$

where $t_{\nu,0}(a)$ denotes the ath quantile of the standard t distribution νdf. One of these tail probabilities will typically be very small and can be neglected. When the noncentrality parameter λ is expressed as in Equation 11.26, the parameter values required to compute power are n_1, n_2, γ_1, and either σ_ϵ^2 and σ_u^2 or σ_y^2 and ρ. When λ is expressed as in Equation 11.28, the parameter values required are n_1, n_2, δ, and ρ.

An R function to calculate power for a two-level, normal outcome CRT with balanced data and a two-sided test is

```
power.crt.bal <- function(delta, rho, n1, n2, alpha){
  ta <- qt(1-alpha/2, n2-2)
  tb <- qt(alpha/2, n2-2)
  deff <- 1 + (n1-1)*rho
  lambda <- delta/sqrt(4*deff/(n1*n2))
  df <- n2-2
  pow <- pt(ta, df, lambda, lower.tail=FALSE) + pt(tb, df, lambda)
  print("Deff is")
  print(deff)
  print("Power is")
  return(pow)
  }
power.crt.bal(0.4, 0.02, 10, 20, 0.05)
```

Using this function, we can find that the power for a trial with an effect size of $\delta = 0.4$, $\rho = 0.02$, $n_1 = 10$ individuals per cluster, and $n_2 = 20$ total clusters is 0.69 and the design effect is 1.18. The relationship between the distribution of the test statistic under the null and alternative hypotheses for this example is depicted in Figure 11.2.

What factors affect the power of a cluster randomized trial? As $|\lambda|$ increases, the noncentral t distribution moves farther away from zero and power increases. The factors affecting power can thus be gleaned from the expressions for the noncentrality parameter

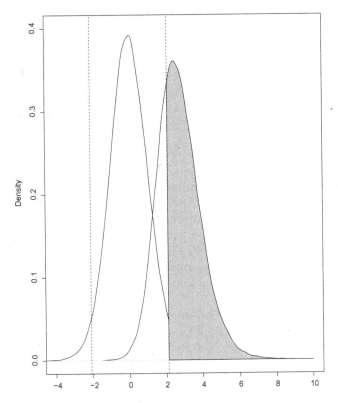

FIGURE 11.2
Illustration of power calculation. Parameter values are $\delta = 0.4$, $\rho = 0.02$, $n_1 = 10$, and $n_2 = 20$. Density on the left represents the distribution of the test statistic under the null, a (central) t distribution with 18 df. Density on the right represents its distribution under the alternative, a noncentral t with 18 df and noncentrality parameter 2.604. Dashed vertical lines indicate the critical values. Shaded area represents power.

in Equations 11.26 and 11.28. Power increases as the treatment effect $|\gamma_1|$ increases and decreases as any of the variance parameters σ_y^2, σ_u^2, or σ_ϵ^2 or ρ increase, all else being equal. What happens to power as we increase cluster size n_1 or number of clusters n_2? To investigate this, we rewrite $Var(\hat{\gamma}_1)$ as

$$Var\left(\hat{\gamma}_1\right) = 4\left(\frac{\sigma_\epsilon^2}{n_1 n_2} + \frac{\sigma_u^2}{n_2}\right). \tag{11.30}$$

As n_2 increases, both components of the variance decrease; as $n_2 \to \infty$, $Var\left(\hat{\gamma}_1\right) \to 0$, $\lambda \to \pm\infty$ and power $\to 1$. However, increasing cluster size n_1 only reduces the first component; it has no effect on the influence of the variance of the cluster means. As $n_1 \to \infty$, $Var(\hat{\gamma}_1) \to \sigma_u^2/n_2$. Thus at some point, increasing the number of individuals per cluster will have a negligible effect on power. In general, power for CRTs is driven more by number of clusters than by cluster size.

> ## POWER: NUMBER OF CLUSTERS VERSUS CLUSTER SIZE
>
> In general, the power of a cluster randomized trial is influenced more strongly by the number of clusters than by the number of individuals per cluster. To increase power, increasing the number of clusters is usually a more effective strategy than increasing cluster size. ·

11.4.1.2 Sample Size: Number of Clusters

Suppose that we wish to determine the number of clusters required to achieve a desired level of power. The size of the clusters is assumed known and constant. Equation 11.29 provides power as a function of total number of clusters n_2; however, we cannot simply invert the equation and solve for n_2 as a function of power because n_2 appears in both the noncentrality parameter of the t distribution and the degrees of freedom. However, the equation can be solved iteratively until the minimum n_2 that provides sufficient power is identified. Note that for equal allocation of clusters to study arms, n_2 must be an even number.

When the number of clusters and therefore the df are sufficiently large ($n_2 \geq 30$ or so), the normal approximation to the t can be used. Using this approach, the minimum total number of clusters required, with $n_2/2$ in each condition, can be calculated as

$$n_2 \geq 4\frac{(z_{1-\alpha/2} + z_{1-\beta})^2}{n_1\delta^2}\left[1 + (n_1 - 1)\rho\right], \tag{11.31}$$

where z_p represents the pth quantile of the standard normal distribution and we assume a two-sided test with type I error rate α. The calculated value of n_2 will need to be rounded to the next highest even integer to achieve an equal number of clusters in each arm. The following R function computes the sample size:

```
sampsize.crt.bal <- function(delta, rho, n1, beta, alpha){
  za <- qnorm(1-alpha/2)
  zb <- qnorm(1-beta)
  n2 <- 4*(za+zb)^2*(1+(n1-1)*rho)/(n1*delta^2)
  print("Total number of clusters required calculated as")
  print(n2)
  print("Required clusters per arm is")
  print(ceiling(n2/2))
  }
```

EXAMPLE 11.1

Suppose we wish to find the minimum number of clusters required to detect a small effect size of 0.2 with 80% power assuming an ICC of 0.05, clusters of size 25, and type I error rate of 0.05. Using the command

```
sampsize.crt.bal(0.2, 0.05, 25, 0.2, 0.05)
```

we find that n_2 is calculated as 69.1 clusters, which we round up to 70 total clusters (35 per condition) to ensure at least 80% power.

Some authors suggest that when the number of clusters and therefore the df for the t distribution are small, one additional cluster per arm should be added to the value calculated by Equation 11.31 to account for using the normal rather than the t distribution; see Hayes and Moulton [6]. In general, a more accurate calculation can be performed using the power Equation 11.29 iteratively.

EXAMPLE 11.2

To detect a larger effect size of $\delta = 0.6$, using Equation 11.31, we calculate $n_2 \geq 7.7$. This number is low enough that we are concerned about using the normal approximation to the t distribution. Using power formula Equation 11.29 to get a more precise result, we calculate that for eight total clusters the power is 67%, whereas for 10 total clusters the power is 80%. Thus we need 10 clusters in total.

If we drop the design effect from sample size formula Equation 11.31 and rearrange terms, we arrive at

$$N = n_1 n_2 \geq 4\frac{(z_{1-\alpha/2} + z_{1-\beta})^2}{\delta^2},\tag{11.32}$$

which provides the minimum required total sample size N for a two-sample t test with equal variances and equal-sized samples $n_1 n_2/2$ in each group, assuming large samples. The difference is the design effect. A common rubric for calculating the sample size required for a cluster randomized trial is to calculate the sample size for a trial involving independent observations and then inflate by the design effect. This approach leads to the formula in Equation 11.31.

VARIANCE INFLATION APPROACH TO SAMPLE SIZE CALCULATION FOR CLUSTER RANDOMIZED TRIALS

A common approach for calculating the sample size for a cluster randomized trial to achieve a desired level of power is to calculate the sample size requirement assuming independent observations and then inflate this number by the design effect:

Total N for cluster randomized trial

$\approx$ Total N for individually randomized trial $\times$ Deff. $\qquad(11.33)$

While this approach can give a good approximation, it is recommended that other factors that may affect power be considered, such as varying cluster sizes, unequal ICCs, and covariates.

EXAMPLE 11.3

Using the variance inflation approach, for $\delta = 0.2$ for an individually randomized trial, we need 785 total subjects to achieve 80% power. For $n_1 = 25$ and $\rho = 0.05$, the design effect is $1 + (n_1 - 1)\rho = 2.2$, so we inflate the total sample size to 1,727. Clusters are of size 25, so this total number of observations corresponds to 69 clusters, which we round up to a total of 70, or 35 in each arm.

11.4.1.3 Sample Size per Cluster

In some situations, we may have a fixed number of clusters available but have a choice as to the number of observations to sample from each cluster. To determine the number of observations to sample from each cluster to achieve a desired level of power, we can solve Equation 11.31 for n_1, which yields

$$n_1 = \frac{4(1-\rho)}{\dfrac{n_2 \delta^2}{\left(z_{1-\alpha/2} + z_{1-\beta}\right)^2} - 4\rho}. \tag{11.34}$$

An R function to calculate sample size per cluster is

```
sampsizeper.crt.bal<-function(delta, rho, n2, beta, alpha){
  za <- qnorm(1-alpha/2)
  zb <- qnorm(1-beta)
  n1 <- 4*(1-rho)/((n2*delta^2/(za+zb)^2) - 4*rho)
  return(n1)
  }
```

> **EXAMPLE 11.4**
>
> Suppose we have 12 hospitals willing to participate in a study and we wish to know how many patients to sample from each hospital to achieve 80% power to detect an effect size of $\delta = 0.5$, assuming $\rho = 0.05$. The R command
>
> ```
> sampsizeper.crt.bal(0.5, 0.05, 12, 0.2, 0.05)
> ```
>
> calculates $n_1 = 20.9$, indicating that we need 21 patients per cluster.
>   It will not always be possible to achieve a desired level of power with a fixed number of clusters, even when using an arbitrarily large cluster size. Indeed, the solution for n_1 will be negative if $n_2 < 4\rho(z_{1-\alpha/2} + z_{1-\beta})^2 / \delta^2$, reflecting the impossibility of always achieving desired power by increasing cluster size. The influence of cluster number versus cluster size on power is discussed in Section 11.4.1.1.

11.4.1.4 Unequal ICCs in Treatment Arms

Thus far we have made the implicit assumption that σ_y^2 and ρ, or equivalently σ_u^2 and σ_e^2, are the same across arms. In some studies, we might expect the variance or correlation parameters to differ across arms. For example, an intervention that encourages interaction among cluster members may result in a higher ICC in the intervention arm, or heterogeneity in the uptake of the intervention may increase the variance of the outcome in the intervention arm.

When we expect different ICCs in the two treatment arms, the variance of the treatment effect estimate is, assuming balanced data,

$$Var(\hat{\gamma}_1) = \frac{2\sigma_y^2}{n_1 n_2} \left\{ \left[1 + (n_1 - 1)\rho_1\right] + \left[1 + (n_1 - 1)\rho_2\right] \right\}, \tag{11.35}$$

which simplifies to Equation 11.12 when $\rho_1 = \rho_2$. This expression can be used in the formula for the noncentrality parameter to calculate power. Using the normal approximation, the total number of clusters required (with $n_2/2$ in each condition) can be calculated as

$$n_2 \geq 2 \frac{(z_{1-\alpha/2} + z_{1-\beta})^2}{n_1 \delta^2} \left\{ \left[1 + (n_1 - 1)\rho_1\right] + \left[1 + (n_1 - 1)\rho_2\right] \right\}. \tag{11.36}$$

11.4.1.5 Unequal Allocation

Thus far, we have considered only trials with equal numbers of clusters in each arm. Unequal allocation can also be considered. Reasons that investigators may choose to use unequal allocation include reducing overall costs when one condition is more expensive to implement than the other, or to make trial participation more attractive by having a higher probability of being assigned to the intervention condition.

Let r denote the proportion of clusters allocated to arm 1. Then the numbers of clusters allocated to arms 1 and 2 are rn_2 and $(1-r)n_2$, respectively, and the variance of the treatment effect estimate becomes

$$Var\left(\hat{\gamma}_1\right) = \frac{\sigma_y^2\left[1+(n_1-1)\rho\right]}{n_1 n_2}\left(\frac{1}{r}+\frac{1}{1-r}\right) = \frac{\sigma_\epsilon^2 + n_1\sigma_u^2}{n_1 n_2}\left(\frac{1}{r}+\frac{1}{1-r}\right). \tag{11.37}$$

The noncentrality parameter can be computed by using the square root of this expression for the denominator in Equation 11.26 or Equation 11.28.

Optimal allocation: For a given effect size, power will be maximized when $Var\left(\hat{\gamma}_1\right)$ is minimized. If we minimize Equation 11.37 with respect to r, we obtain $r = 0.5$, which corresponds to equal allocation. However, the formula assumes that σ_y^2 and ρ, or σ_u^2 and σ_ϵ^2, are the same across arms. If we allow different ICCs and unequal allocation, the variance of the treatment effect estimate is

$$Var\left(\hat{\gamma}_1\right) = \frac{\sigma_y^2}{n_1 n_2}\left[\frac{1+(n_1-1)\rho_1}{r} + \frac{1+(n_1-1)\rho_2}{1-r}\right]. \tag{11.38}$$

The optimal allocation, optimal in the sense of minimizing the variance, can be found to be $\sqrt{d_1}/(\sqrt{d_1}+\sqrt{d_2})$, where the d_k are the design effects, $d_k = 1+(n_1-1)\rho_k, k = 1,2$. For example, with ICCs of 0.02 and 0.06 for control and intervention and a cluster size of 50, the variance (and standard error) are minimized and the power is maximized if we allocate 41% of the clusters to the control condition and 59% to the intervention condition. In general, optimal allocation will involve allocating more clusters to the condition with the higher design effect (and thus higher variance).

Another optimal design strategy is to maximize cost efficiency, defined as the precision (inverse variance) of the treatment effect estimate divided by total study cost. Discussion of such designs can be found in the article by Wu et al. [36]. Additional discussion of cluster randomized designs involving costs can be found in the book by Moerbeek and Teerenstra [33].

UNEQUAL ALLOCATION OF CLUSTERS TO CONDITIONS

When the ICCs are equal in the two treatment arms, maximal power for a cluster randomized trial with a continuous outcome can be achieved by allocating clusters equally to each treatment arm. When the ICCs are different in the two arms, an optimal unequal allocation that maximizes power can be found. The optimal allocation will involve allocating more clusters to the condition with the higher design effect.

11.4.1.6 Covariates

Regression adjustment for covariates is often used in randomized trials to improve precision. In a linear regression model for independent observations, the effect of covariate adjustment is

$$\sigma_\epsilon^2 = \sigma_y^2\left(1 - R_{Y|X}^2\right) \tag{11.39}$$

where σ_y^2 is the total variance of Y, σ_ϵ^2 is the residual variance, and $R_{Y|X}^2$ is the proportion of the variance of Y that is explained by the covariates, represented by X. Because the residual variance is a key component of the standard errors of the regression coefficients, covariates that are strongly associated with the outcome variable can increase precision by reducing the residual variance. Covariates might include demographic characteristics such as age and sex or clinical factors associated with prognosis such as cancer stage. One particularly notable covariate is the outcome variable measured at baseline; for example, a trial might administer a symptoms scale to patients at baseline and at follow-up. In such trials, the data are typically analyzed using an analysis of covariance (ANCOVA), which tests for a mean difference between conditions in the outcome variable controlling for the baseline value of the variable.

Covariate adjustment can also be used in cluster randomized trials, but its effects are more complicated because of the multilevel data structure. In particular, the effects of covariates at the individual level and cluster level are different. We discuss covariates at each level.

Cluster-level covariates: A cluster-level covariate might be the gender or specialty of a healthcare provider in a trial randomizing providers, or the percentage of pupils below the poverty line in a trial randomizing schools. When a cluster-level covariate is added, the model becomes

$$Y_{ij} = \tilde{\gamma}_0 + \tilde{\gamma}_1 w_j + \tilde{\gamma}_2 z_j + \tilde{u}_j + \tilde{\epsilon}_{ij} \tag{11.40}$$

where tildes are used to indicate that the value of the regression coefficients and random terms may be different in the adjusted model. We denote the variance components in the adjusted model as $\tilde{\sigma}_u^2$ and $\tilde{\sigma}_\epsilon^2$. How do $\tilde{\sigma}_u^2$ and $\tilde{\sigma}_\epsilon^2$ compare to σ_u^2 and σ_ϵ^2? Since a cluster-level covariate takes the same value for all members of a cluster, it cannot explain variation among individuals within a cluster. Therefore the within-cluster variance is unchanged and $\tilde{\sigma}_\epsilon^2 = \sigma_\epsilon^2$. For the cluster-level variance, we have

$$\tilde{\sigma}_u^2 = \left(1 - \rho_B^2\right)\sigma_u^2, \tag{11.41}$$

where ρ_B is the between-cluster residual correlation between the outcome and the covariate. Using Equation 11.41 and the relationships $\sigma_u^2 = \rho\sigma_y^2$, $\sigma_\epsilon^2 = (1-\rho)\sigma_y^2$, and $\sigma_y^2 = \sigma_u^2 + \sigma_\epsilon^2$, we can get an expression for the standard error of the treatment effect estimator as

$$SE\left(\hat{\tilde{\gamma}}_1\right) = \sqrt{\frac{4\left(\sigma_\epsilon^2 + n_1\tilde{\sigma}_u^2\right)}{n_1 n_2}} = \sqrt{\frac{4\sigma_y^2\left\{1 + \left[n_1\left(1 - \rho_B^2\right) - 1\right]\rho\right\}}{n_1 n_2}}, \tag{11.42}$$

where the quantity in braces is the design effect when a cluster-level covariate is included, which simplifies to the standard design effect when $\rho_B = 0$. The total number of clusters required to achieve power of $1 - \beta$ is

$$n_2 \geq 4 \frac{(z_{1-\alpha/2} + z_{1-\beta})^2}{n_1 \delta^2} \left\{ 1 + \left[n_1 \left(1 - \rho_B^2 \right) - 1 \right] \rho \right\}. \tag{11.43}$$

Because $1 - \rho_B^2 \leq 1$, we have $1 + \left[n_1 \left(1 - \rho_B^2 \right) - 1 \right] \rho \leq 1 + (n_1 - 1) \rho$; that is, the design effect is reduced. Thus adjusting for a cluster-level covariate can be beneficial in reducing the required sample size. For example, for $\rho_B = 0.6$, $n_1 = 500$, and $\rho = 0.05$, the design effect is reduced from 25.95 to 16.95 by including the covariate, a reduction of 35%. When $n_1 = 5$ or $n_1 = 50$, the reductions are 7.5% and 26%, respectively.

Including a cluster-level covariate comes at the expense of reducing the df in the t statistic for the test of the intervention effect. For each covariate term, one additional df is lost (recall that the df are $n_2 - q - 1$, where q is the number of cluster-level covariates, including the covariate encoding intervention assignment). This effect is negligible if the number of clusters is large, but can be important if the number of clusters is limited.

Individual-level covariates: Now we consider adjusting for an individual-level baseline covariate, denoted c_{ij}. The covariate could be the baseline measurement of the outcome variable, in which case the analysis becomes an ANCOVA, or it could be other covariates associated with the outcome. In this case, the relationships between the adjusted and unadjusted variances are [37]

$$\tilde{\sigma}_\epsilon^2 = \left(1 - \frac{n_1}{n_1 - 1} \rho_W^2 \right) \sigma_\epsilon^2 \quad \text{and} \quad \tilde{\sigma}_u^2 = \left(1 - \rho_B^2 + \frac{1}{n_1 - 1} \rho_W^2 \right) \sigma_u^2 \tag{11.44}$$

where ρ_W and ρ_B are the within-cluster and between-cluster residual correlations between the outcome Y_{ij} and the covariate c_{ij}, respectively [37]. These results show that the addition of the covariate can decrease the within-cluster variance but could increase or decrease the between-cluster variance; an increase occurs when $\rho_B^2 < \left(1 / (n_1 - 1) \rho_W^2 \right)$. This counterintuitive result is discussed by Snijders and Bosker [5,38]. However, when $\sigma_\epsilon^2 > \sigma_u^2$, any increase in σ_u^2 will be outweighed by the decrease in σ_ϵ^2 and overall the variance will be reduced [37]. This condition is expected to hold true for most CRTs.

The total required number of clusters can be found by using, as the design effect,

$$1 + \left[n_1 \left(1 - \rho_B^2 \right) - 1 \right] \rho - \frac{n_1 \rho_W^2 (1 - 2\rho)}{n_1 - 1}. \tag{11.45}$$

When individual-level covariates are entered into the model, no df are lost from the test of the intervention effect.

As an example, suppose the outcome variable is a symptoms score and the planned outcome analysis is an ANCOVA that adjusts for the baseline value of the score. The correlation between the baseline and follow-up scores is expected to be about $\rho_W = 0.7$. If you lack a good estimate for the residual correlation between clusters (ρ_B), a conservative estimate is 0. For $n_1 = 10$ and $\rho = 0.05$, inclusion of the covariate would reduce the design effect from 1.45 to 1.045, corresponding to a reduction of 28% in the required sample size.

Somewhat different formulas for the effect of individual-level covariates have been presented by other authors [39–41].

COVARIATE ADJUSTMENT IN CLUSTER RANDOMIZED TRIALS

Covariate adjustment can be an effective strategy to increase power in a cluster randomized trial. Both cluster-level and individual-level covariates that are correlated with the outcome variable can reduce the residual variance of the outcome and thereby increase the precision of the treatment effect estimator. Adding cluster-level covariates reduces the degrees of freedom for the test of the intervention effect. If the df available for the test of the intervention effect are limited (e.g., <30), cluster-level covariates should be restricted to those that are highly prognostic of the outcome to avoid loss of power.

11.4.1.7 Varying Cluster Sizes

Until now, we have assumed that all clusters have the same number of members. In practice, the sizes of clusters (schools, hospitals, villages, patients with the same healthcare provider) are likely to vary naturally. In addition, clusters that were of equal size at the beginning of a trial may experience non-response or dropout that leads to unequal cluster sizes in the final data set.

It has been shown that, given the same total number of clusters and number of participants, unequal cluster sizes are less efficient for estimating treatment effects than are equal cluster sizes [42]. Thus when cluster sizes vary, efficiency and power are reduced, and to achieve the same power, the sample size must be enlarged.

Let θ and τ denote the mean and standard deviation of the distribution of cluster sizes, respectively. An approximation of the relative efficiency of unequal versus equal cluster sizes is

$$RE \approx 1 - \lambda(1-\lambda)CV^2 \tag{11.46}$$

where $\lambda = \left((\theta)/(\theta + (1-\rho)/\rho)\right)$ and CV is the coefficient of variation of the cluster size distribution, $CV = \tau/\theta$ [43]. The required total number of clusters can be obtained as the sample size assuming equal cluster sizes multiplied by $1/RE$ (note that the design effect is the inverse of the relative efficiency, and $1/RE$ is the design effect here). Other authors [44–46] have derived somewhat different estimates of the relative efficiency, all of which also depend on the ICC and CV of the cluster size distribution.

Investigators typically have a good idea of θ, the expected mean cluster size. To estimate the standard deviation of cluster size, a strategy is to estimate the minimum and maximum cluster size and approximate the standard deviation as one fourth of the range, that is, $\tau \approx (\max - \min)/4$.

As an example, suppose that the mean cluster size is 16 and the ICC is 0.04. Standard deviations of cluster size of 0 (equal sizes), 4, 8, and 16 correspond to CVs of cluster size distribution of 0, 0.25, 0.5, and 1, and estimated relative efficiencies of 1, 0.985, 0.94, and 0.76, respectively. The inverse REs are 1, 1.015, 1.064, and 1.32, meaning the sample sizes need to be inflated by 0%, 1.5%, 6.4%, and 32%.

UNEQUAL CLUSTER SIZES

When cluster sizes vary, efficiency and power are reduced. The loss of efficiency increases as the dispersion of cluster sizes, as measured by the coefficient of variation of the cluster size distribution, increases. The sample size required to achieve the desired level of power can be found by inflating the sample size requirement calculated assuming equal cluster sizes by the inverse of the relative efficiency, which can be approximated using Equation 11.46.

11.4.1.8 Matching and Stratification

In Section 11.2.1, we discussed how matching and stratification can be helpful in promoting balance between arms on prognostic factors. Stratification or matching prior to randomization can also improve power when these designs are used in conjunction with a stratified or matched analysis. In such an analysis, comparisons between conditions are made within strata. If clusters within strata are very similar, these comparisons will be akin to comparing the same experimental units under two different conditions. This reduces the between-cluster variability in the estimation of the intervention effect, reducing the standard error and increasing power.

The main impact of stratification in a CRT is to reduce the between-cluster variance component, σ_u^2 [8]. Because clusters in different conditions are now compared within strata, σ_u^2 is replaced with the variance among clusters within strata, which we denote σ_{um}^2. The variance of the treatment effect estimator becomes

$$Var\left(\hat{\gamma}_1\right) = \frac{4\left(\sigma_\epsilon^2 + n_1\sigma_{um}^2\right)}{n_1 n_2} \tag{11.47}$$

and the total number of clusters required is

$$n_2 \geq \frac{4(z_{1-\alpha/2} + z_{1-\beta})^2}{n_1\gamma_1^2}\left(\sigma_\epsilon^2 + n_1\sigma_{um}^2\right); \tag{11.48}$$

see Crespi [32]. Alternatively, we can define ρ_m as the correlation between cluster-level means within strata or matched pairs, equal to $\sigma_{um}^2/(\sigma_{um}^2 + \sigma_\epsilon^2)$, and the formula becomes

$$n_2 \geq \frac{4\sigma_y^2(z_{1-\alpha/2} + z_{1-\beta})^2}{n_1\gamma_1^2}\left[1 + (n_1 - 1)\rho - n_1\rho_m\rho\right]. \tag{11.49}$$

Thus the design effect is reduced by a factor of $n_1\rho_m\rho$. These formulas apply to studies with matched pairs and with larger strata.

Degrees of freedom will be lost when the stratification factors are used as covariates in a multilevel analysis model, which is needed in order to estimate the treatment effect within strata. Some authors suggest adding two additional clusters per study arm in matched-pair designs and 1–2 additional clusters per arm in designs with larger strata to account for the loss of df [6]. Because of the loss of df, for CRTs with a small number of clusters, pair matching should only be done if the clusters can be matched on factors that are highly correlated with the outcome variable; see Hayes and Moulton [6] for further discussion, including a table showing the break-even values of the matching correlation above which pair matching will provide greater power than an unmatched trial.

EXAMPLE 11.5

Suppose we wish to detect a small effect size of $\delta = 0.2$ with power of 80% and α of 0.05 two-sided, where clusters are of size $n_1 = 25$ and $\rho = 0.05$. We found in Section 11.4 that for an unstratified design, the design effect was 2.2 and the total number of clusters needed was 70. Suppose we consider stratification on cluster type and anticipate a correlation of $\rho_m = 0.3$ within strata. In this case, the design effect is reduced by $25 \times 0.3 \times 0.05 = 0.375$ and becomes 1.825. We will need a total of 1,433 individuals, or 58 clusters. Thus the number of clusters is reduced by 17%.

The matching correlation ρ_M can be difficult to predict. In the face of uncertainty about ρ_M, a conservative approach is to ignore any gain in power due to stratification.

EFFECT OF MATCHING AND STRATIFICATION ON POWER

In a stratified design, comparisons between conditions are made within strata. If the strata are homogeneous, the between-cluster variance relevant to estimating the treatment effect is reduced. Thus matching and stratification can increase power and reduce sample size requirements, in addition to promoting balance on baseline covariates. However, this variance reduction comes with a loss of df for estimating the treatment effect.

11.4.2 Dichotomous Outcomes

In this section, sample size and power formulas for cluster randomized trials with dichotomous outcomes are derived using logic similar to that for CRTs with continuous outcomes. In the section on analysis, we discussed two hierarchical models for binary data, the cluster-level proportions model and the cluster-level log-odds model. For sample size and power, we focus on the cluster-level proportions model, which has parameters that are easier to understand. Since many concepts were previously discussed for continuous outcomes, our discussion of dichotomous outcomes is more brief.

11.4.2.1 Sample Size and Power

Using the cluster-level proportions model, the intervention effect can be estimated as the difference in sample proportions, $\hat{\pi}_1 - \hat{\pi}_2$, and its variance, also given in Equation 11.19, is

$$Var(\hat{\pi}_1 - \hat{\pi}_2) = \left[\frac{\pi_1(1-\pi_1)}{n_1 n_2/2} + \frac{\pi_2(1-\pi_2)}{n_1 n_2/2} \right] \left[1 + (n_1 - 1)\rho_d \right]. \tag{11.50}$$

where ρ_d is as defined in Equation 11.15. For large samples, the test statistic $(\hat{\pi}_1 - \hat{\pi}_2) / \sqrt{Var(\hat{\pi}_1 - \hat{\pi}_2)}$ has a standard normal distribution under the null hypothesis. Using this approach, the total number of clusters required to achieve power of $1 - \beta$ with two-sided α of 0.05 when cluster size is n_1 is

$$n_2 \geq \frac{2(z_{1-\alpha/2} + z_{1-\beta})^2 \left[\pi_1(1-\pi_1) + \pi_2(1-\pi_2) \right] \left[1 + (n_1 - 1)\rho_d \right]}{n_1(\pi_1 - \pi_2)^2}. \tag{11.51}$$

This formula assumes equal allocation, constant cluster sizes and equal ICCs in each arm. These assumptions are relaxed in later sections. This formula can also be derived by calculating the total sample size requirement across both arms, N, for independent observations,

$$N = \frac{2(z_{1-\alpha/2} + z_{1-\beta})^2 \left[\pi_1(1-\pi_1) + \pi_2(1-\pi_2) \right]}{(\pi_1 - \pi_2)^2}, \tag{11.52}$$

and inflating by the design effect. Some authors recommend adding one extra cluster per treatment arm [6]. Power for a given cluster size and number of clusters can be obtained by solving the equation for $z_{1-\beta}$ and applying the standard normal cumulative distribution function. R functions for sample size and power computation are given below.

```
sampsize.crt.bal.bin <- function(p1, p2, rho, n1, beta, alpha){
  za <- qnorm(1-alpha/2)
  zb <- qnorm(1-beta)
  num <- 2*(za+zb)^2*(p1*(1-p1)+p2*(1-p2))*(1+(n1-1)*rho)
  denom <- n1*(p1-p2)^2
  n2 <- num/denom
  print("Total number of clusters required is")
  print(n2)
  print("Required clusters per arm is")
  print(ceiling(n2/2))
}
power.crt.bal.bin <- function(p1, p2, rho, n1, n2, alpha){
  za <- qnorm(1-alpha/2)
  num <- n1*n2*(p1-p2)^2
  denom <- 2*(p1*(1-p1)+p2*(1-p2))*(1+(n1-1)*rho)
  zb <- sqrt(num/denom) - za
  pnorm(zb)
}
```

EXAMPLE 11.6

Suppose that we anticipate proportions of 0.3 and 0.5 in the two arms and clusters each have 25 members. The ICC is estimated to be 0.03. Using the command

```
sampsize.crt.bal.bin(.3, .5, .03, 25, 0.2, 0.05)
```

the trial is estimated to need seven clusters per condition to achieve at least 80% power with two-sided α of 0.05. The actual power can be computed using the command

```
power.crt.bal.bin(.3, .5, .03, 25, 14, 0.05)
```

which indicates that the actual power is 84.5%.

11.4.2.2 *Sample Size per Cluster*

Given a fixed total number of clusters n_2, the required cluster size is

$$n_1 = \frac{(1-\rho_d)\left[\pi_1(1-\pi_1) + \pi_2(1-\pi_2)\right](z_{1-\alpha/2} + z_{1-\beta})^2}{(\pi_1 - \pi_2)^2 n_2 / 2 - \rho_d\left[\pi_1(1-\pi_1) + \pi_2(1-\pi_2)\right]}. \tag{11.53}$$

As mentioned for continuous outcomes, it will not always be possible to achieve desired power with a fixed number of clusters, even if the sample size per cluster is extremely large.

11.4.2.3 Unequal ICCs in Treatment Arms

In some cases we may wish to use separate ICC estimates in the two arms. For example, we may anticipate more dispersion of cluster-level proportions in the intervention arm, or we may want to account for the fact that the ICC depends on the underlying proportion as $Var(\hat{p})/[\pi(1-\pi)]$; see Section 11.14. Separate ICCs by arm can be incorporated into the sample size formula to yield

$$n_2 \geq \frac{2(z_{1-\alpha/2}+z_{1-\beta})^2[\pi_1(1-\pi_1)Deff_1]+\pi_2(1-\pi_2)Deff_2]}{n_1(\pi_1-\pi_2)^2}, \tag{11.54}$$

where $Deff_k = 1+(n_1-1)\rho_{dk}$ is the design effect in arm k.

11.4.2.4 Unequal Allocation

If the number of clusters allocated to arms 1 and 2 are rn_2 and $(1-r)n_2$, respectively, the variance of the risk difference becomes

$$Var(\hat{\pi}_1-\hat{\pi}_2) = \frac{\pi_1(1-\pi_1)d_1}{rn_1n_2} + \frac{\pi_2(1-\pi_2)d_2}{(1-r)n_1n_2} \tag{11.55}$$

where the d_k are the design effects, $d_k = 1+(n_1-1)\rho_k, k = 1,2$; we have allowed unequal ICCs and thus unequal design effects for generality. As discussed for continuous outcomes, the optimal allocation that minimizes the variance of the treatment effect estimator is $\sqrt{d_1}/(\sqrt{d_1}+\sqrt{d_2})$.

11.4.2.5 Covariates

The impact of cluster-level and individual-level covariates on the power of CRTs with continuous outcomes was discussed in Section 11.4.1.6. The impact of adjusting for covariates in trials with binary outcomes is more complex. Due to the nonlinearity of the logistic regression model, it is difficult to derive tractable expressions for the variances in adjusted models [47]. Furthermore, in a logistic regression model, the unadjusted and adjusted treatment effect parameters differ. Unadjusted analyses yield marginal estimates that compare an intervention subject with a randomly selected control subject. Adjusted analyses yield conditional estimates that compare an intervention subject to a control subject with the same covariate values. For continuous outcomes, the adjusted and unadjusted treatment effects are the same, but this is not generally true for binary outcomes [48]. In a logistic model, the adjustment typically increases the estimated treatment effect; that is, estimated odds ratios will be further from 1 (where an odds ratio of 1 indicates no treatment effect). Furthermore, including covariates in a logit model tends to *increase* the variance of the estimated treatment effect in log-odds terms [49].

For these reasons, simple formulas that account for the impact of covariates on power in a CRT with a binary outcome analyzed using a mixed effects logit model are lacking. However, Schochet [47] provides an approach that is based on using a GEE estimator rather than the mixed effects logit model. A key finding is that gains in power due to covariate adjustment are likely to be smaller for binary outcomes than they are for continuous outcomes.

11.4.2.6 *Varying Cluster Sizes*

Varying cluster sizes are less efficient for estimating treatment effects than are equal cluster sizes, as previously discussed. The same approach for accounting for this reduction in efficiency can be used for both continuous and binary outcomes; see Section 11.4.1.7 for more details.

11.5 Additional Resources

Books on the design and analysis of cluster randomized trials include Murray [9], Donner and Klar [8], Eldridge and Kerry [50], Campbell and Walters [31], and Hayes and Moulton [6]. Some of these books, for example, Hayes and Moulton [6], discuss time-to-event outcomes, rates, and counts. Survival outcomes are also discussed by Jahn-Eimermacher et al. [51]. Power analysis for trials with multilevel data, including cluster randomized trials, multicenter trials, and individually randomized group treatment trials, is discussed in Moerbeek and Teerenstra [33]. Sample size calculation for clustered and longitudinal outcomes are discussed in Ahn et al. [44]. Some recent reviews summarize key results for sample size calculations for CRTs [34,52].

Many journals require that reports of trials conform to the guidelines in the Consolidated Standards of Reporting Trials (CONSORT) statement. There is a CONSORT statement extension specifically for cluster randomized trials [14] that provides a checklist of items to include in the trial report, including the ICC, which researchers often neglect to report [53].

11.5.1 Resources for Other Designs

This chapter has covered outcome analysis and sample size and power for some common designs of cluster randomized trials. We discuss several other major designs and provide references.

Individually randomized group treatment trials: In an individually randomized group treatment trial, individuals are randomized to study conditions but receive their intervention with other participants, typically in a group setting, or through a change agent shared with other participants. For example, in a mindful awareness intervention for breast cancer survivors, participants randomized to the intervention were assigned to groups who attended classes together [54]. In these studies, there is little or no group-level ICC at baseline, but a positive ICC is expected among the outcomes of individuals within the same group. Special methods are needed for analysis and sample size estimation for these studies. Literature can be found that discusses these studies much [33,55–58].

Cluster randomized crossover trials: In a simple crossover trial, each subject receives each treatment in random order. Because each subject serves as his or her own control, a crossover design can be quite powerful. In a cluster randomized crossover trial, clusters are randomly allocated to a sequence of interventions. Two designs can be distinguished: crossover at the cluster level, in which each subject is included in only one of the treatment periods, and crossover at the subject level, in which each subject is observed in both periods [59]. Crossover CRTs are discussed by several authors [59–61], and a brief summary of sample size formulas is provided in Crespi [32].

Stepped-wedge trials: A stepped-wedge trial is similar to a crossover trial except that the crossovers are all in one direction, from control to intervention condition, and are staggered over time. Clusters are randomized to cross over to the intervention at time points called steps, and all clusters end the trial in the intervention condition. Stepped-wedge trials are discussed in several articles [62–65].

11.5.2 Resources for Power and Sample Size Calculation

The National Institutes of Health has a website with guidance on research methods related to studies that randomize groups or clusters or that deliver interventions to groups at https://researchmethodsresources.nih.gov. The website includes a sample size calculator.

The free software program Optimal Design Plus Empirical Evidence includes power and sample size calculation for more complex design elements such as three or four levels. The program and documentation are available at http://www.wtgrantfoundation.org.

Moerbeek and Teerenstra [33] describe the SPA-ML (Statistical Power Analysis for Multi-Level designs) program, which is available for free download at http://tinyurl.com/SPAML. Their book describes the use of the program, which currently only handles continuous outcomes.

Campbell and Walters [31] discuss both data analysis and power and sample size for CRTs and provide code in R, Stata, and SPSS. The code is available at their website: http://sheffield.ac.uk/scharr/sections/dts/statistics.

References

1. Wagner EH, Ludman EJ, Aiello Bowles EJ, Penfold R, Reid RJ, Rutter CM, Chubak J, McCorkle R. Nurse navigators in early cancer care: a randomized, controlled trial. *Am J Clin Oncol.* 2014;32(1):12–8.
2. Bastani R, Glenn BA, Maxwell AE et al. Cluster-Randomized Trial to Increase Hepatitis B Testing among Koreans in Los Angeles. *Cancer Epidemiol Biomarkers Prev.* 2015;24(9):1341–9.
3. Sankaranarayanan R, Nene BM, Shastri SS et al. HPV screening for cervical cancer in rural India. *N Engl J Med.* 2009;360(14):1385–94.
4. Hox JJ. *Multilevel Analysis: Techniques and Applications*, Second Edition. New York, NY, USA: Routledge, 2009.
5. Snijders TAB, Bosker RJ. *Multilevel Analysis: An Introduction to Basic and Advanced Multilevel Modeling*, Second Edition. London, England, UK: Sage Publications Ltd., 2012.
6. Hayes RJ, Moulton LH. *Cluster Randomised Trials*, Second Edition. Boca Raton, FL, USA: Taylor & Francis Group, LLC, 2017.
7. Ivers NM, Halperin IJ, Barnsley J, Gromshaw JM, Shah BR, Tu K, Upshur R, Zwarenstein M. Allocation techniques for balance at baseline in cluster randomized trials: A methodological review. *Trials.* 2012;13:120.
8. Donner A, Klar N. *Design and Analysis of Cluster Randomization Trials in Health Research.* New York, NY, USA: Oxford University Press, 2000.
9. Murray DM. *Design and Analysis of Group-Randomized Trials.* New York, NY, USA: Oxford University Press, 1998.

10. Raab GM, Butcher I. Balance in cluster randomized trials. *Stat Med*. 2001;20(3):351–65.
11. Maxwell AE, Danao LL, Cayetano RT, Crespi CM, Bastani R. Implementation of an evidence-based intervention to promote colorectal cancer screening in community organizations: A cluster randomized trial. *Transl Behav Med*. 2016;6:295–305.
12. Taves DR. Minimization: a new method of assigning patients to treatment and control groups. *Clin Pharm Ther*. 1974;15:443–53.
13. Ivers NM, Taljaard M, Dixon S et al. Impact of CONSORT extension for cluster randomised trials on quality of reporting and study methodology: Review of random sample of 300 trials, 2000-8. *BMJ*. 2011;343:343.
14. Campbell MK, Piaggio G, Elbourne DR, Altman DG, Group C. CONSORT 2010 statement: extension to cluster randomised trials. *BMJ*. 2012;345:e5661.
15. Cook JA, Bruckner T, MacLennan GS, Seiler CM. Clustering in surgical trials-database of intracluster correlations. *Trials*. 2012;13:2.
16. Hade EM, Murray DM, Pennell ML et al. Intraclass Correlation Estimates for Cancer Screening Outcomes: Estimates and Applications in the Design of Group-Randomized Cancer Screening Studies. *JNCI Monographs*. 2010;2010:97–103.
17. Murray DM, Blitstein JL. Methods to reduce the impact of intraclass correlation in group-randomized trials. *Eval Rev*. 2003;27(1):79–103.
18. Kish L. *Survey Sampling*. New York, NY, USA: John Wiley, 1965.
19. Manor O, Zucker DM. Small sample inference for the fixed effects in the mixed linear model. *Comput Stat Data Anal*. 2004;46:801–17.
20. Jahn P, Kuss O, Schmidt H, Bauer A, Kitzmantel M, Jordan K, Krasemann S, Landenberger M. Improvement of pain-related self-management for cancer patients through a modular transitional nursing intervention: A cluster-randomized multicenter trial. *Pain*. 2014;155:746–54.
21. Eldridge SM, Ukoumunne OC, Carlin JB. The Intra-Cluster Correlation Coefficient in Cluster Randomized Trials: A Review of Definitions. *Int Stat Rev*. 2009;77:378–94.
22. Turner RM, Omar RZ, Thompson SG. Bayesian methods of analysis for cluster randomized trials with binary outcome data. *Stat Med*. 2001;20:453–72.
23. Moerbeek M, Van Breukelen MPF, and Berger GJP. Optimal experimental designs for multilevel logistic models. *Statistician*. 2001;50(1):1–14.
24. Spybrook J, Bloom H, Congdon R, Hill C, Martinez A, Raudenbush S. *Optimal Design Plus Empirical Evidence: Documentation for the Optimal Design Software*. 2011.
25. Kim Y, Choi Y-K, Emery S. Logistic Regression with Multiple Random Effects: A Simulation Study of Estimation Methods and Statistical Packages. *Am Stat*. 2013;67:171–82.
26. Mishra SI, Bastani R, Crespi CM, Chang LC, Luce PH, Baquet CR. Results of a randomized trial to increase mammogram usage among Samoan women. *Cancer Epidemiol Bisomarkers Prev*. 2007;16(12):2594–604.
27. Diggle PJ, Heagerty PK, Liang KY, Zeger SL. *Analysis of Longitudinal Data*, Second Edition. Oxford: Oxford University Press, 2002.
28. Liang KY, Zeger SL. LongitudinL data analysis using generalized linear models. *Biometrika*. 1986;73:13–22.
29. Gardiner JC, Luo ZH, Roman LA. Fixed effects, random effects and GEE: What are the differences? *Stat Med*. 2009;28:221–39.
30. Hu FB, Goldberg J, Hedeker D, Flay BR, Pentz MA. Comparison of population-averaged and subject-specific approaches for analyzing repeated binary outcomes. *Am J Epidemiol*. 1998;147:694–703.
31. Campbell MJ, Walters SJ. *How to Design, Analyse and Report Cluster Randomised Trials in Medicine and Health-Related Research*. Chichester, UK: John Wiley & Sons Ltd, 2014.
32. Crespi CM. Improved designs for cluster randomized trials. *Annu Rev Public Health*. 2016;37:1–16.
33. Moerbeek M, Teerenstra S. *Power Analysis of Trials with Multilevel Data*. Boca Raton, FL, USA: Taylor & Francis Group, LLC, 2016.
34. Rutterford C, Copes A, Eldridge S. Methods for sample size determination in cluster randomized trials. *Int J Epidemiol*. 2015;44(3):1051–67.

35. Cohen J. *Statistical Power Analysis for the Behavioral Sciences*, Second Edition. Hillsdale, New Jersey: Lawrence Erlbaum Associates, 1988.
36. Wu S, Wong WK, Crespi CM. Maximin optimal designs for cluster randomized trials. *Biometrics*. 2017;73(3):916–26.
37. Moerbeek M. Power and money in cluster randomized trials: When is it worth measuring a covariate? *Stat Med*. 2006;25:2607–17.
38. Snijders TAB, Bosker RJ. Modeled variance in two-level models. *Sociol Methods Res*. 1994;22(3):342–63.
39. Bloom HS, Richburg-Hayes L, Black AR. Using covariates to improve precision for studies that randomized schools to evaluate educational interventions. *Eval Policy Anal*. 2007;29:30–59.
40. Raudenbush SW. Statistical analysis and optimal design for cluster randomized trials. *Psychol Methods*. 1997;2:173–85.
41. Teerenstra S, Eldridge S, Graff M, de Hoop E, Borm GF. A simple sample size formula for analysis of covariance in cluster randomized trials. *Stat Med*. 2012;31:2169–78.
42. Ankenman BE, Aviles AI, Pinheiro JC. Optimal designs for mixed-effects models with two random nested factors. *Stat Sin*. 2003;13:385–401.
43. van Breukelen GJP, Candel MJJM, Berger MPF. Relative efficiency of unequal versus equal cluster sizes in cluster randomized and multicentre trials. *Stat Med*. 2007;26:2589–603.
44. Ahn C, Heo M, Zhang S. *Sample Size Calculations for Clustered and Longitudinal Outcomes in Clinical Research*. Boca Raton, FL: CRC Press/Taylor and Francis Group, 2015.
45. Kerry SM, Bland JM. Unequal cluster sizes for trials in English and Welsh general practices: Implications for sample size calculations. *Stat Med*. 2001;20:377–90.
46. Manatunga AK, Hudgens MG, Chen S. Sample size estimation in cluster randomized studies with varying cluster size. *Biom J*. 2001;43:75–86.
47. Schochet PZ. Statistical power for school-based RCTs with binary outcomes. *J Res Educ Eff*. 2013;6:263–94.
48. Hauck WW, Anderson S, Marcus SM. Should we adjust for covariates in nonlinear regression analyses of randomized trials? *Control Clin Trials*. 1998;19:249–56.
49. Robinson LD, Jewell NP. Some surprising results about covariate adjustment in logistic regression models. *Int Stat Rev*. 1991;58:227–40.
50. Kerry S, Eldridge S. *A Practical Guide to Cluster Randomized Trials in Health Research*. London: Arnold, 2012.
51. Jahn-Eimermacher A, Ingel K, Schneider A. Sample size in cluster-randomized trials with time to event as the primary endpoint. *Stat Med*. 2013;32:739–51.
52. Gao F, Earnest A, Matchar DB, Campbell MJ, Machin D. Sample size calculations for the design of cluster randomized trials: a summary of methodology. *Contemp Clin Trials*. 2015;42:41–50.
53. Crespi CM, Maxwell AE, Wu S. Cluster randomized trials of cancer screening interventions: Are appropriate statistical methods being used? *Contemp Clin Trials*. 2011;32:477–84.
54. Bower JE, Crosswell AD, Stanton AL, Crespi CM, Winston D, Arevalo J, Ma J, Cole SW, Ganz PA. Mindfulness meditation for younger breast cancer survivors: A randomized controlled trial. *Cancer*. 2015;121(8):1231–40.
55. Baldwin SA, Bauer DJ, Stice E, Rohde P. Evaluating models for partially clustered designs. *Psychol Methods*. 2011;16(2):149–65.
56. Pals SP, Murray DM, Alfano CM, Shadish WR, Hannan PJ, Baker WL. Erratum. *Am J Public Health*. 2008;98(12):2120.
57. Pals SP, Murray DM, Alfano CM, Shadish WR, Hannan PJ, Baker WL. Individually randomized group treatment trials: a critical appraisal of frequently used design and analytic approaches. *Am J Public Health*. 2008;98(8):1418–24.
58. Roberts C, Roberts SA. Design and analysis of clinical trials with clustering effects due to treatment. *Clin Trials*. 2005;2:152–62.
59. Rietbergen C, Moerbeek M. The design of cluster randomized crossover trials. *J Educ Behav Stat*. 2011;36(4):472–90.

60. Giraudeau B, Ravaud P, Donner A. Sample size calculation for cluster randomized corss-over trials. *Stat Med*. 2008;27:5578–85.
61. Reich NG, Myers JA, Obeng D, Milstone AM, Perl TM. Empirical power and sample size calculations for cluster-randomized and cluster-randomized crossover studies. *PLoS One*. 2012;7:e35564.
62. Baio G, Copas A, Ambler G, Hargreaves J, Beard E, Omar RZ. Sample size calculation for a stepped wedge trial. *Trials*. 2015;16(1):354.
63. Hemming K, Haines TP, Chilton PJ, Girling AJ, Lilford RJ. The stepped wedge cluster randomised trial: Rationale, design, analysis, and reporting. *BMJ*. 2015;350:h391.
64. Hemming K, Taljaard M. Sample size calculations for stepped wedge and cluster randomised trials: A unified approach. *J Clin Epidemiol*. 2016;69:137–46.
65. Woertman W, de Hoop W, Moerbeek M, Zuidema SU, Gerritsen DL, Teerenstra S. Stepped wedge designs could reduce the required sample size in cluster randomized trials. *J Clin Epidemiol*. 2013;66(7):752–8.

12

Statistical Monitoring of Safety and Efficacy

Jay Herson and Chen Hu

CONTENTS

12.1 Introduction

This chapter describes methods of *statistical monitoring*. The term *monitoring* is used often in clinical trials. It sometimes refers to *site monitoring*, i.e., the process of clinical monitors visiting clinical sites to check on protocol adherence and data quality. The term statistical monitoring will refer to the ongoing process of data review during the trial by the *sponsor* and *data monitoring committee* (DMC). Both groups are seeking safety signals. The sponsor will review data for all treatments pooled while the DMC will review safety data for treatments at least partially unmasked. For efficacy data the DMC may be administering a planned interim analysis of superiority, futility, or non-inferiority. These analyses might result in early termination of the trial.

We will review statistical methods used in the monitoring of safety and efficacy in phase III confirmatory trials for pharmaceutical and radiotherapy interventions in oncology. This chapter will not cover statistical methods for analysis of safety and efficacy that might appear in a final trial report. Although the methods might be the same as or similar to those used at trial completion, the objectives differ.

Section 12.2 will deal with safety monitoring first by the sponsor team and then by the DMC. Section 12.3 will address efficacy monitoring for superiority, futility, and non-inferiority. In Section 12.4 we will look at the newly developed adaptive designs beginning to appear in oncology trials. Section 12.5 will introduce the new methodology of centralized risk-based monitoring. In Section 12.6 we present concluding remarks and some thoughts about the future of monitoring in oncology trials.

12.2 Monitoring of Safety

12.2.1 Introduction

Statistical methods used for final data analysis are described in a *statistical analysis plan* prepared before the trial begins and are published in the final trial report. Statistical methods are used in monitoring safety but are never published. Hence there is no standard of best practice. It is clear that the purpose of statistical analysis in safety monitoring is to find potential signals for further clinical discussion. These signals will guide monitoring during the trial, form the basis for product labeling and a post-market surveillance plan.

Statistical significance ($p < 0.05$) in safety monitoring does not play the analytic role it does for efficacy analysis. Indeed, observation of three cases of renal failure in the gemcitabine arm of a pancreatic cancer protocol versus none in the control group would be of clinical concern even if the comparison did not reach statistical significance. In addition, such a comparison would, most likely, not be pre-planned, raising questions of multiplicity; although, as we will see, this might not be the inferential problem it would be in efficacy analysis. Safety monitoring decisions of safety are made in the absence of knowledge of efficacy and, thus, risk-benefit discussions are premature. Thus, statistical methods and clinical observation are both routes to bringing safety concerns to clinical discussion. Clinical discussion will take place by the sponsor (protocol team at a pharmaceutical company, cooperative oncology group, or contract research organization) and by the DMC.

A comprehensive review of statistical methods in safety monitoring is provided by Herson [1].

12.2.1.1 Planning for Safety Monitoring

The Safety Planning, Evaluation and Reporting Team (SPERT) [2] and the Council for International Organizations of Medical Sciences (CIOMS) Working Group VI [3] have recommended that safety monitoring be formalized and that a program safety analysis plan (PSAP) be formulated early in the clinical program. The PSAP would provide the details of the *treatment-emergent adverse event* (AE) definition, collection, and analysis. One of the most important sections would be what dictionary would be used for adverse event classification. While the pharmaceutical industry generally uses the Medical Dictionary for Regulatory Activities (MedDRA) [4], oncology trials generally use the US National Cancer Institute's Common Terminology Criteria for Adverse Events (CTCAE) [5]. Pharmadhoc [6] provides MedDRA-CTCAE mapping tools.

There has long been a feeling that investigator descriptions and grading of adverse events do not reflect the true impact of the event on the patient, i.e., while the classification might be clinically correct, it does not carry information on the emotional toll or the interference with daily activities that are best described by the patient. The US National Cancer Institute has collaborated with investigators and patients to develop a *patient-reported outcomes* version of the CTCAE known as the PRO-CTCAE [7]. PRO-CTCAE is an online real-time system for patients to report adverse events as they occur. Patients report occurrence, frequency, and severity on a 1–5 scale. Dueck et al. [8] have reported on the validity and reliability of the PRO-CTCAE with positive results.

Adverse events must be graded as to degree of *severity*. The usual severity levels are none, mild, moderate, severe, and life-threatening. CTCAE provides grading definitions for the types of AEs encountered in oncology trials. As a compromise, oncologists sometimes use CTCAE grading and MedDRA AE terms. It is important to distinguish between a severe AE and a *serious adverse event* (SAE). The grading is a clinical term while SAE is a clinical/regulatory term. An SAE, briefly, is an AE that results in death, is life-threatening, requires inpatient hospitalization or prolongation of existing hospitalization, etc. A cancer patient might have severe diarrhea but not be hospitalized, so this would not be an SAE but would be a severe AE. A patient might have a mild transient case of dehydration but, nevertheless, be hospitalized, in which case the mild AE would also be reported as an SAE.

SPERT also recommended that AEs be classified according to tiers. Tier 1 AEs would be those for which specific hypotheses and analysis methods are described before the trial begins, e.g., in a trial for sorafenib, hypotheses for the level of diarrhea, alopecia, and anorexia might be specified. Tier 2 consists of AEs not pre-specified but that have appeared in the trial, e.g., we might see vision problems in patients treated with rituximab for non-Hodgkin's lymphoma. Tier 3 events are infrequent events that do not lend themselves to statistical analysis but should be listed in any report of the trial, e.g., one case of aplastic anemia in a trial for a new chemotherapeutic agent.

In a recent guidance [9], the FDA asked sponsors to make expedited reports of Serious Unexpected Suspected Adverse Reactions (SUSARs). The *SUSAR* is a sponsor's designation of an unexpected severe AE being causally related to the experimental drug. This rule was intended to reduce the number of expedited reports received since "causality" must meet a higher standard than "associated." FDA also defined the adverse event of special interest (AESI). This is a list of AEs that are always of concern regardless of drug or indication. Chuang-Stein and Xia [10] list cardiac, renal, liver, bone marrow toxicity, polymorphic metabolism, etc. as AESIs but suggest adding treatment-specific AEs throughout the clinical program. It will be the task of the sponsor team and the DMC to update the Tier 2 list and the AESI list as the trial progresses.

12.2.1.2 Safety Monitoring: Sponsor View (Masked, Treatment Groups Pooled)

The sponsor staff must remain masked (blind to treatment identity) throughout the trial, but this does not mean that they ignore type and frequency of AE occurrence during the trial. Sponsor staff will prepare tables of reason for discontinuation and type and frequency of AEs in detail in preparation for the periodic DMC meetings. All these analyses will be compiled for both treatment groups combined. The staff must process SUSARs and other regulatory real-time reporting requirements throughout the trial. Additionally, at the outset of the trial, the sponsor has, most likely, created an AESI list and Tier 1 AE list. The sponsor will monitor AEs to detect if levels in the combined treatments are greater than what might be expected from previous trials. To compute a plausible range of values for an AE incidence or severe or higher AE incidence, the normal approximation binomial confidence interval can be calculated, although the Clopper and Pearson [11] confidence interval is becoming common because of its inclusion in popular statistical software.

The incidence rate does not take exposure to treatment into account. A low AE incidence might be only due to patients discontinuing treatment early and, thus, not on treatment long enough to experience an AE. Thus, the AE rate per 100 patient years is preferable. The rate is calculated as AE frequency multiplied by 100 and divided by total patient years on treatment. The reference distribution rate per 100 patient years is Poisson. Herson [12] describes methods for computing one-sample confidence intervals on this rate.

Sponsors might also want to make a Kaplan–Meier life table graph of the accumulation of a first occurrence of a specific AE type over time [13]. This analysis will provide "landmark" estimates of incidence at various time points per treatment group (e.g., the 6-month nausea/vomiting rate). In addition, the graph will allow for the study of temporal patterns in AE incidence, i.e., do the AEs tend to occur early in treatment and then level off or do they occur constantly (exponentially) over time.

Herson [1] provides a comprehensive review of statistical methods of monitoring by the sponsor view, including Bayesian and likelihood applications.

12.2.1.3 Safety Monitoring: Data Monitoring Committee View (Partially or Completely Unmasked)

The DMC members review accumulating safety data periodically. This is done through review of tabular and graphical displays of safety data as well as narrative summaries of AE experiences submitted by investigators. DMCs in the pharmaceutical industry typically consist of two physicians and a biostatistician. The members are typically recruited from academia and must conform to the sponsor's conflict of interest policies at the outset and during the trial. DMCs meet periodically and in closed session after the sponsor team presents the current sponsor view of the safety data. In closed session they can be partially unmasked to treatment, meaning they know the treatments only as A and B or completely unmasked where they know the identities of A and B. DMCs may switch from partially to completely unmasked at any time during the trial, but most DMCs prefer to be completely unmasked from the start of the trial, as the latter being the only way they can fulfill their mission of protecting patient safety. A *Data Analysis Center* (DAC) independent of the sponsor team will prepare the reports for the DMC. The statistician from the DAC is often called the *reporting statistician* and attends the DMC closed sessions but does not vote. Hence the reporting statistician is sometimes called the *non-voting statistician*.

During their meetings, the DMC members will consider the Tier 1 AEs and update the Tier 2 and Tier 3 AEs and the AESIs. These recommendations are made to the sponsor solely on the basis of incidence with both groups pooled. There is no implication that the DMC has observed a differential incidence between treatments.

DMCs review treatment differences in reason for trial discontinuation, deaths, overall AE incidence, and overall incidence of severe or greater AEs. The DMC is also presented a table organized by body system with preferred terms for AEs within body system. This table can also be produced for only moderate or worse AEs. This table generally has many AEs with 0 or 1 observation. This is due to the granularity induced by so many categories. Moreover, many AE types can be spread over several body systems, e.g. patients who develop "congestive heart failure" during the trial might be spread out over terms such as "dyspnea," "orthopnea," and "fatigue," thus evading recognition as congestive heart failure. This problem has been overcome by the development of *standardized MedDRA queries* (SMQs). The SMQs combine terms over several body systems, e.g., MACE—major adverse cardiac event—combines MedDRA terms all-cause mortality, myocardial infarct, and target vessel revascularization. Other SMQs exist for convulsions, depression, and hypersensitivity [10,14]. Schactman and Wittes [15] discuss various strategies for combining MedDRA terms for DMC use.

In reviewing safety data, DMC members will be attempting to separate signal from noise, i.e., to find the true safety signals in the face of random and disease-related events. The role of statistics in this process is to call potential safety issues to the attention of the clinical members. This involves computing metrics for differences in AE incidence between treatments. Statistical significance between treatments is not of paramount importance in safety analysis. When statistically significant, the difference may not be of clinical significance and may be a consequence of multiplicity. The latter term refers to the phenomenon of more testing the more likely it is to find a spurious difference (elevated overall type I error).

Metrics for difference in AE incidence between treatments would include the *odds ratio, relative risk, the Poisson rate ratio,* and Kaplan–Meier life table methods. The odds ratio and relative risk do not take exposure into account and there can be differential exposure between treatment groups due to discontinuations for various reasons. The Poisson rate ratio and Kaplan–Meier life table methods take exposure into account. The Kaplan–Meier graphs are computed as time to first occurrence within patient. They are typically computed only when a potential safety signal is found. The graphs enable the DMC members to review the patterns of onset of the AE of interest.

Let

n_E, n_C = number of patients in experimental and control groups

r_E, r_C = number of patients reporting the event in experimental and control groups

$p_E = r_E/n_E, p_C = r_C/n_E$ incidence in experimental and control groups

The odds ratio OR $= p_E(1 - p_C)/p_C(1 - p_E)$
Relative risk $= p_E/p_C$

The MedCalc online calculator provides efficient computation of the odds ratio and relative risk and their confidence intervals MedCalc [16,17]. These calculations are also available from common statistical software programs. Gibbons and Amatya [18] provide formulas for computation of approximate confidence intervals for the odds ratio and relative risk.

Epidemiologists prefer use of the relative risk, considering it appropriate for cohort studies and for its intuitive appeal. Biostatisticians prefer use of the odds ratio because of its relation to logistic regression analysis, often used to model safety data, and because of its favorable statistical properties as a minimally sufficient statistic for the multinomial likelihood.

The *Poisson rate ratio* is the ratio of incidence per 100 patient years in the experimental group divided by the same for the control group. Herson [12] provides the formula for confidence intervals for the Poisson rate ratio.

In practice the biostatistical member of the DMC will eyeball the odds ratios or the other two measures prior to the DMC meetings and call attention to those ratios that are greater than some threshold such as 3 or 5 regardless of confidence interval inclusion of unity, which would indicate no evidence of statistical significance. Multiple testing of AE differences and the threshold criteria all lead to the criticism of multiplicity. Herson [12] reviews some methods of correcting for multiplicity in safety monitoring, but ignoring multiplicity allows more differences for further consideration than would be under tight statistical type I error control, and this is seen as preferable from a safety standpoint.

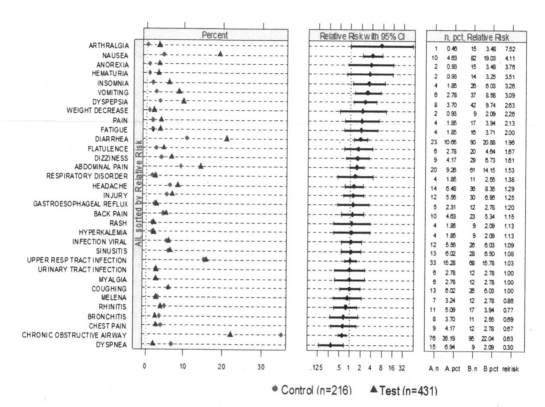

FIGURE 12.1
Adverse event incidence and relative risk in descending order. (Gould A.L. 2013, personal communication.)

Graphical methods are becoming common in safety monitoring. Figure 12.1 was first developed by Amit et al. [19]. It conveniently displays relative risks of the AEs in descending order with their 95% confidence intervals as well as differences in incidence and raw data. Zink et al. [20], Gould [21], and Duke et al. [22] describe additional graphical methods for safety monitoring.

A comprehensive description of the sponsor's and DMC's role in safety monitoring is given by Herson [12].

Of course, the mere finding of a safety signal of concern does not immediately terminate the clinical trial. The DMC will not immediately call the safety concern to the protocol team at the sponsor but rather inform a pharmacovigilance group, a safety unit maintained by the sponsor not related to the protocol team. The issues would be discussed and a plan of action short of immediate termination would be drafted.

12.3 Efficacy and Futility Monitoring

12.3.1 Introduction

Efficacy and futility monitoring in randomized phase III oncology clinical trials plays a critical role in efficiently advancing clinical development while rigorously maintaining the statistical integrity to provide evidence on the benefit-risk ratio. Many, if not all, of these trials use an independent DMC, which is informed by statistical monitoring rules set up prior to trial initiation and makes recommendations regarding continuation as planned, protocol modification, or termination. These statistical monitoring rules repeatedly evaluate the primary hypothesis (based on some pre-planned schedule) over the trial conduct period as efficacy data accumulate. Typically, the interim data and their analysis is not to be considered definitive and suitable for wider dissemination unless and until some pre-specified criteria are met.

12.3.2 Superiority Monitoring

As patients enter into the trial sequentially over time, there is a need to monitor the accumulating efficacy data. From an ethical perspective, it is desirable to provide a mechanism to terminate the trial in the face of mounting evidence of superiority. However, there are a number of pitfalls to conducting multiple analyses, which lead to serious interpretation issues. The key issue is the inflation of type I error rates as a result of multiple comparisons. In addition, due to the random variability, the estimation of true treatment effect can be rather unstable at the early stage of any trial regardless of design, which makes early termination decisions based on an extreme positively biased finding problematic. Furthermore, once early termination occurs, the planned estimation precision and secondary objectives are unlikely to be achieved, which in turn leads to lack of credibility and failure to influence medical practice.

To properly account for these issues, *group sequential design* and analysis has been widely used and arguably becomes the default approach to rigorously monitoring almost all randomized phase III clinical trials. Proschan et al. [23] provide a comprehensive summary of methods. We provide a high-level summary of some frequently used approaches and provide additional references for more in-depth discussions.

In conventional group sequential analyses, trialists prospectively plan for a fixed number, say K, of interim and one final analysis ($k = 1, 2, \ldots, K$). Based on certain criteria, we can also predetermine a boundary b_k such that once the test statistics at each interim analysis exceed the predetermined boundary b_k, we can reject the null hypothesis and consider stopping the trial. The methodology preserves the intended overall type 1 error level and provides the desired statistical power, whether or not the trial proceeds to the planned definitive analysis. The analysis index k is formulated in terms of the so-called information fraction or *information time* t_k, defined as the fraction of "full" information K, the maximum sample size (or number of events in time-to-event analysis). The test statistics may also take on a different form for different types of outcomes (Gaussian, binary, time-to-event data, etc.).

There are numerous group sequential monitoring rules that can be used to monitor trials. Among them, the following are more well-known and probably used more frequently in practice. All methods use adjusted significance levels for early looks at the data, a process known as *alpha spending*.

- Pocock [24] proposed performing each test at a fixed, more stringent test criterion, chosen so that the overall type 1 error is controlled at the planned level, usually 0.05. For example, if there are to be three hypotheses tests, use critical value $Z = 2.289$ corresponding to significance level 0.0221 for each test. For five tests, use $Z = 2.413$ and significance level 0.0155 for each test.

- Haybittle [25] proposed an even simpler solution—test at any time at, say, $\alpha' = 0.001$ and then, if the final analysis is reached, test at the planned 0.05 level because mathematically the interim tests will not affect the overall significance level before the third decimal place. This method is sometimes used for unplanned interim analyses, such as an unmasking needed for a safety decision, but rarely for standard use.

- O'Brien and Fleming [26] proposed a non-constant boundary (function of the information time) that is extreme for early analyses and decreases toward the conventional significance level by the final analysis. This is appealing to trialists because they only would want to terminate a trial early if there is an extreme expression of superiority. The critical value at each successive interim analysis becomes less stringent and converges on a final critical value close to 0.05.

- Lan and DeMets [27] proposed a simplification of the O'Brien–Fleming group sequential. Using methods of Brownian motion, they provide a continuous function of critical values for alpha spending. This allows the selection of critical values at times not precisely specified, as is required by strict application of O'Brien–Fleming. This is the method most used today in phase III oncology trials because DMCs cannot always meet at precisely the information time specified by the statistical analysis plan and the protocol. The DMC biostatistician merely reads the critical value off of the Lan–DeMets curve at the information time of the actual interim analysis.

  Figure 12.2 is a graphical display of critical values used by each of the group sequential methods by information time. With the exception of Pocock, all methods converge to a critical value close to 0.05 at information time 1.0. The Lan–DeMets method is not shown on the graph because it is merely a smoothed version of O'Brien–Fleming.

- Wang and Tsiatis [28], Emerson and Fleming [29], and Pampallona et al. [30] generalized the Pocock and O'Brien–Fleming methods to the power family, which includes the Pocock and O'Brien–Fleming methods as special cases.

There are several problems in the use of interim analysis. First, when a trial terminates early in favor of superiority, the estimate of effect size observed at interim is likely to be inflated

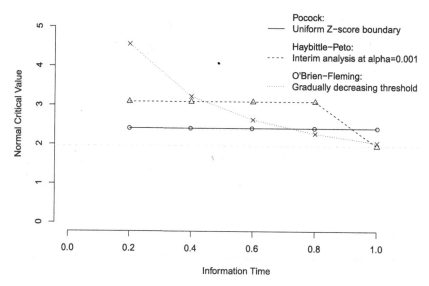

FIGURE 12.2
Group sequential critical values by information time. (Hu C. 2018, personal communication.)

because the condition for termination relied on an extreme value sometimes referred to as a "random high." Thus some form of *shrinkage* should be applied to the estimate, but there is no agreement on methodology and it is rarely done [31,32]. One such example occurred in United Kingdom Medical Research Council's (MRC) AML12 trial [33]. This adequately powered ($n = 1,078$) acute myeloid leukemia trial found one additional course of consolidation therapy (five vs. four courses) did not confer survival benefit ($HR = 1.09$, 95% CI 0.87–1.37, $p = 0.4$). Interestingly, the first two interim analyses presented to the independent DMC, while the accrual was still ongoing and with limited follow-up on those already enrolled, suggested an unexpected, large benefit for the additional course, with hazard ratios of 0.47 (95% CI 0.29–0.77, $p = 0.003$) and 0.55 (95% CI 0.38–0.80, $p = 0.002$), respectively. The main reason that the DMC did not recommend early study closure was that the observed treatment effect was much greater than plausible and long-term impacts were unclear. This decision was vindicated as the final analysis showed no treatment effect. The only explanation found for the early extreme values were not differential patient characteristics but just random fluctuation. Such potential overestimation of the magnitude of treatment effect in oncology trials is also nicely demonstrated in Korn et al. [34], Freidlin and Korn [35], and Zhang et al. [36].

Second, there may not be sufficient safety data collected at the time of interim analysis to allow termination of the trial. Regulators might want the trial to continue at least in the experimental group in order to be sure an adequate safety profile has been attained. Indeed, early termination does not always answer the efficacy question originally posed. In the National Cancer Institute of Canada MA-17 early-stage breast cancer trial evaluating letrozole in postmenopausal women after five years of tamoxifen therapy, patients were randomized to five years of letrozole or placebo [37]. The trial was terminated by the DMC with median follow-up of 2.4 years when a letrozole-to-placebo DFS hazard ratio of 0.43 was observed with $p = 0.00008$. In addition to the problem of extreme value, the trial did not answer the question regarding the efficacy of five-year letrozole therapy—a question of considerable interest to practicing oncologists. In a somewhat similar situation, the DMC for the cardiovascular and cancer prevention trial of estrogen plus progestin of the Women's Health Initiative had observed data for elevated incidence of breast cancer in the hormone group but felt that these early results were not persuasive enough to change practice, so they delayed their termination decision until more data arrived [38].

Improvements in overall survival in cancer trials have caused sponsors to use interim analysis endpoints other than *overall survival* (OS) in order that the interim analysis can take place in a reasonable period of time after trial commencement. Although OS is the primary efficacy endpoint in most pivotal cancer trials, sponsors have been using *disease-free survival* (DFS) or *progression-free survival* (PFS) as the interim analysis endpoint. However, DFS or PFS is not always a predictor of OS. In the SPARC trial for satraplatin in men with castrate-refractory prostate cancer, although a 33% reduction in PFS was achieved by satraplatin, there was no difference in OS observed at the end of the trial [39].

The problem for long follow-up time as a contraindication for selecting OS as an interim analysis endpoint also occurs in newly diagnosed breast cancer. Recently, the FDA issued a guidance indicating that they would be willing to accept *pathologic complete response* (pCR) as an endpoint in neoadjuvant treatment of high-risk, early-stage breast cancer for accelerated approval [40]. However, there is no agreement about the clinical significance of pCR.

In a superiority trial, the final efficacy analysis is often stratified by covariates, and effect size estimate is often adjusted for covariates by proportional hazards or logistic regression methods. In oncology trials it is rare for any of these methods to be applied to an interim analysis. However, it is important that the statistical analysis plan, prepared before the trial begins, specifies precisely how the effect size will be estimated at interim.

12.3.3 Futility Monitoring

Futility monitoring refers to monitoring studies for early evidence of lack of benefit, which includes both harm and absence of tangible benefit. There are oncology trials where a futility guideline is desirable and those where it would be less desirable. Among the desirable reasons are (1) in oncology it is common today for there to be a full pipeline of drugs ready for testing, and sponsors might want to cut their losses with borderline efficacy shown at interim; (2) in add-on trials comparing treatment *A* versus *A + B* where *B* is the experimental treatment and A is an approved drug, there is the potential for a toxicity burden with little efficacy benefit afforded by the experimental treatment; and (3) any trial where safety issues are anticipated.

Futility guidelines would be less desirable in trials where (1) PFS or OS endpoints where delays in efficacy differences are expected, (2) PFS or OS endpoints where enrollment is rapid so that all patients may have already been enrolled at the time of the interim analysis, or (3) trials where there is a learning curve for investigators such as a trial involving immunotherapy where clinics are enrolled in regions unfamiliar with this intervention.

Once a decision is made to include a futility guideline, there are still reasons why a DMC and/or sponsor might want or not want to terminate the trial when a futility guideline is reached. Good reasons to terminate would include (1) a disturbing difference in safety profile between treatment arms, (2) secondary endpoints also show a consistent trend against the experimental treatment, (3) competing therapies are achieving effect sizes greater than what can be achieved by the drug under investigation, e.g., the emergence of AAP therapy in prostate cancer, (4) side effects observed are not severe but are not present in competing therapies, and (5) financial reasons.

Reasons not to terminate would include (1) patient characteristics are beginning to change either because of protocol amendment or the termination of competing trials that admitted a broader range of patients, (2) data might be of questionable quality or suspected delays in reporting events, (3) site monitoring might be insufficient, (4) some positive results in secondary endpoints, (4) information arrives that similar treatments are providing efficacy benefits for this population, (5) analysis of futility is based on PFS but DMC members think there will ultimately be an OS benefit, (6) continue the trial to gather data for planning a new trial or to possibly get approval in other countries, and (7) the ability to gain insight to how efficacy might vary over subgroups of patients. In oncology trials, the results of a futility analysis are almost always non-binding. However, the statistical analysis plan should clearly specify if the results are binding or non-binding.

A complete description of futility analysis in clinical trials is given by Herson et al. [41].

The level of evidence required, and the associated process to stop an ongoing trial, is different between superiority and futility. We generally do not have to provide the same degree of evidence as in superiority monitoring. Often such an asymmetric relationship between superiority and futility monitoring are explicitly acknowledged during trial planning and conduct, and reflected in the futility monitoring tools used.

Before proceeding to describe some futility monitoring methods, it is important to note that futility methods do not spend type I error, rather they spend type II error. Hence no alpha adjustment is needed for their application. This makes futility monitoring easier to implement than efficacy monitoring.

The principal methods of futility interim analysis used in oncology trials today include:

- Conditional power (CP): As information accumulates, trends toward some conclusion may emerge and become sufficiently strong to suggest that some hypothesized effect may be unlikely to be observed. Such projection can be framed in terms of conditional power, which is defined heuristically as Pr(eventually reject H0|current data and some assumption about treatment effect). For decision-making, the assumption used in the conditioning could be either null hypothesis, i.e., no treatment effect, when the observed data trend to the positive results, or the planned alternative hypothesis (or some other favorable alternative) for futility monitoring. The conditional power calculations can aid DMCs in deciding to stop a trial early. This approach is most often undertaken for ad hoc futility monitoring, to assess the evidence that the trial is highly unlikely to show a benefit at conclusion given the current data. Using conditional power, one could set up an *a priori* monitoring boundary, e.g., stop for futility if conditional power falls below a specific value, say 10% or 30%, under the planning treatment effect and possibly some other values near it.

Figure 12.3 presents the relationship between effect-size trajectory after interim analysis and conditional power at the final analysis. The observed effect-size trajectory is shown up to information time 0.50. If the current trend continues, the power in final analysis would be 13%. If the alternative hypothesis of effect size were to prevail going forward, the power for rejection of the null hypothesis would be only 25%. We should not be surprised to see that if the null hypothesis of no treatment effect were to prevail, the power would be only 3%. To recover the full power of 80%, there would have to be substantial improvement in effect size of future patients.

More details on conditional power can be found in Proschan et al. [23].

- Wieand et al. [42] proposed to conduct a futility monitoring when half of the total required events have occurred from both groups, and stop the trial if the relative risk of failure is greater than 1.0, e.g., trend toward harm of new treatment. Under specific assumptions, such rule essentially has no effect on the size of a nominal 0.05-level test, and results in a loss of power of less than 2% for any H1 indicating a

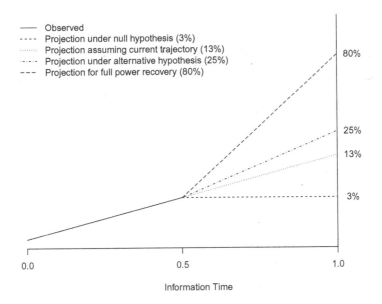

FIGURE 12.3
Conditional power for final analysis under the assumption of various effect size trajectories. (Hu C. 2018, personal communication.)

treatment benefit, as compared to a test at scheduled definitive analysis. Freidlin et al. [43] extended the rule over the entire monitoring period. Rather than considering a single interim analysis at 50% information time, they proposed to find the earliest time point at which the estimated hazard ratio >1.0 (e.g., on logHR scale, 0) would imply that the associated two-sided 95% confidence interval would exclude the alternative used to design the trial. Such earliest (information) time point, denoted as only t_0, can be shown to depend only on the type I and type II errors assumed, but not the design alternative. Furthermore, to accommodate the needs of favoring continuing if the benefit observed is smaller than the planning alternative but nonetheless clinically interesting (say 20% of logHR under alternative, denoted as $\log\theta$), Freidlin et al. [43] suggested a Linear Inefficacy Boundary that connect from $(t_0,0)$ and $(1,\log\theta)$.

NSABP B-14 [44] was one of the seminal trials evaluating the use and optimal duration of tamoxifen in the treatment of early-stage breast cancer. NSABP B-14 demonstrated the benefit of 5-year tamoxifen therapy compared to placebo. An extension trial [45] re-randomized those patients who had completed 5 years of tamoxifen therapy and remained disease free to either an additional 5-year use of tamoxifen or placebo. Based on the interim results at the third of four scheduled interim analyses, the DMC recommended to disseminate the study findings and discontinue prolonged tamoxifen. As elaborated in Dignam et al. [46], at the time of terminating, the conditional power for the protocol-specified alternative hypothesis was 15%, and the estimated HR (placebo/tamoxifen) was 0.59 (95% CI 0.38–0.90) at 77% information time. These interim findings suggested that a benefit for continuing tamoxifen was unlikely to be realized and contributed to the decision of early termination prior to planned final analysis.

Zhang et al. [47] conducted an empirical comparative study of multiple futility rules, including conditional power at 10% and 30% (CP10 and CP30), and Freidlin et al. [43] with a Linear 20% Inefficacy Boundary (20% of logHR under alternative is considered clinically interesting). Based on 52 published superiority phase III trials with survival endpoints across different disease sites, the authors found that none of the futility rules would have stopped positive trials; for negative trials, conditional power rules can be overly aggressive during the second half of a trial as data mature, as well as toward the end of the trial, while the linear inefficacy boundary provided best results overall. For example, in RTOG 9802 [48], a randomized phase III trial evaluating the addition of chemotherapy to radiotherapy for low-grade glioma, the overall survival of the more intensive regimen was significantly better over than radiotherapy alone ($p = 0.001$, $HR = 0.59$), while initially slightly worse (Figure 12.4). Such a hazard-crossing phenomenon is actually not rare when the treatment effect is not anticipated immediately (especially if the experimental treatment is quite aggressive). In the case of RTOG 9802, as shown by Zhang et al. [47], all futility monitoring rules they evaluated would have recommended stopping for futility when evaluated early (information time is less than 50%).

12.3.4 Non-Inferiority Monitoring

The goal of a *non-inferiority* (NI) randomized trial is to demonstrate that the experimental regimen is not inferior to a standard regimen (not inferior, not superior). The NI trial is used for the approval of less invasive, less toxic, and less expensive regimens but is now becoming increasingly used in biosimilar trials. In non-inferiority trials, an experimental therapy is compared to an active control using a pre-specified efficacy margin and

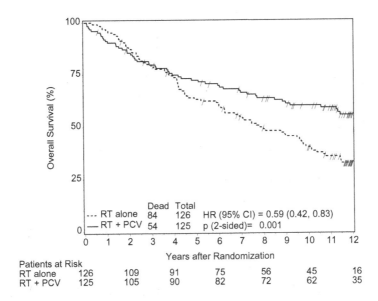

FIGURE 12.4
Kaplan–Meier plot of overall survival in RTOG 9802. (Hu C. 2018, personal communication.)

confidence level. Monitoring NI trials also requires somewhat different considerations. For example, futility monitoring (e.g., early demonstration that the experimental regimen is inferior to standard treatment) may be more popular and feasible than efficacy monitoring (e.g., early demonstration that the experimental regimen is non-inferior) in most local or locally-advanced cancers, where NI trials often involve long follow-up when the primary outcome is time-to-event data.

An example of futility monitoring in a biosimilar trial is the recent trastuzumab trial in metastatic breast cancer [49]. In this trial, as well as the recent superiority trial of adjuvant sunitinib in renal cell carcinoma [50], the supplementary materials online reveal that both trials had planned interim analyses. However there is no mention of the interim analyses in these publications. This is very common in clinical trial reports where the trial did not terminate at the interim analysis. It is important that the compete protocol be included in the text of a clinical trial report including the adjusted critical value at the end of the trial. There is a suspicion that a report of a superiority trial might indicate that statistical significance was achieved at the end of the trial with $p = 0.046$. However due to the adjustment for type I error control, the critical value for statistical significance might have been $p = 0.043$. More importantly, a situation where the sponsor overruled a DMC decision for early termination should be reported.

12.4 Adaptive Designs

Adaptive designs allow for modification of the original trial design based on accumulating data. At planned interim analyses, adaptive designs allow sample size re-estimation, dropping a dose or treatment group, adaptive assignment to treatment, changing objective from superiority to non-inferiority, seamless transition from phase II to phase III, etc.

Dragalin [51] and Gallo and Krams [52] provide a useful overview of the various types of adaptive designs.

The most frequently used adaptive design is for change in effect size of interest. We discuss this method under sample size re-estimation in Section 12.4.1.

12.4.1 Sample Size Re-Estimation

We have seen that at a planned interim analysis, a trial may be terminated for superiority or futility. Another option at interim is *sample size re-estimation*. If, at the interim analysis, we find a conditional power of, say, 20% to find the effect size of interest statistically significant at the end of the trial, we might want to increase the sample size/number of required events to increase the power to reject the null hypothesis.

The most common method of sample size re-estimation today is the *CHW method* described by Cui et al. [53]. This method computes the sample size required to deliver the power to find the effect size observed at interim statistically significant at the end of the trial. It is believed to be more practical than increasing the sample size to find the original effect size of interest statistically significant. A problem arises because the original effect size was agreed to be clinically significant at the start of the trial. Now power is being calculated for a smaller (less favorable) effect size to be statistically significant. As an example, sponsors might have reached agreement with regulators that a hazard ratio for PFS of 0.70 would be clinically significant. At interim, we might observe a hazard ratio for PFS of 0.80 and apply the CHW method. Sponsors include CHW sample size re-estimation in protocols should the interim analysis find that the original effect size is untenable. It is a strategy to salvage the trial with CHW providing power to a more realistic effect size.

As in other interim analysis operations, the sample size re-estimation is administered by the DMC. At this point in the trial, the DMC will usually have a good handle on the safety profile for the experimental treatment. The DMC must now make a risk-benefit decision. If the safety profile was acceptable when the effect size of interest was 0.70, will the members still find it acceptable for an effect size of 0.80? Risk-benefit analysis is difficult during or at the end of a single trial, but it is certainly more difficult when the effect size of interest is a moving target and when it is not clear how regulators and the oncology community would view risk-benefit with a less-favorable effect size.

In terms of statistical theory, the CHW method down weights observations that occur after the interim analysis. Hence there is a departure from the customary "one patient-one vote" data analysis and, more important to many, this down weighting is a direct violation of the likelihood principle. Despite all of its flaws, it is the method of choice for sample size re-estimation today.

A complete description of methods of sample size re-estimation is provided by Chuang-Stein et al. [54].

12.4.2 Adaptive Design Challenges to DMCs

The unorthodox nature of adaptive designs creates some challenges for DMCs. These issues are discussed in detail by Gallo [55], Herson [12,73] and Turnbull [56]. The main challenge is maintaining the masking of sponsor and investigators with adaptations such as changing effect size of interest, changing randomization to favor treatment appearing more effective without sufficient safety data, dropping a treatment or dose arm when the DMC is not comfortable with the safety profile of the remaining arm(s). For the seamless transition from phase II to phase III, the sponsor must allow the DMC "white space" to make a final assessment of the safety profile before the transition can take place.

12.4.3 Master Protocol Designs

In order to create more efficiency in oncology drug development, *umbrella designs* and *basket designs* have been developed. The umbrella design is for a single tumor type, but several experimental drugs, each targeting a different biomarker or genetic mutation, are rotated in and out according to efficacy criteria. This is a cooperative venture among several sponsors. Examples of umbrella designs are ISPY-2, a neoadjuvant breast cancer trial [74], Lung-MAP for advanced squamous-cell lung cancer [75], and STAMPEDE for prostate cancer [57]. The challenge to DMCs in the umbrella design is that the trial has no definitive ending, so a scheme will be necessary for DMC members to rotate in interlocking terms.

The other master protocol design is the basket trial. This design concentrates on the genetic mutation regardless of organ site. As in the umbrella trial, drugs will be rotated through. The goal of the basket design is to develop drugs to treat a genetic mutation rather than a specific organ. Examples of basket trials are the NCI-MATCH [58] and the American Society of Clinical Oncology TAPUR study [59]. The DMC responsibilities are similar to those in the umbrella trials but a specific challenge in the basket trial is there may not be a DMC member with site-specific expertise to advise on mutation-site issues. For example, a renal cell patient may be treated but, at the time, there is no serving DMC member with renal cell expertise.

12.5 Centralized Risk-Based Monitoring

For many decades, data quality and protocol adherence were monitored at each clinical site by *clinical research associates* (CRAs) dispatched by the sponsor or its contractor. As the number of sites expanded globally, this type of surveillance was clearly no longer scalable and, due to lack of focus, not effective. FDA issued a guidance indicating that statistical and graphical methods, centralized at the sponsor, could replace routine site monitoring [60]. Shortly thereafter, sophisticated statistical software began to appear for use in risk-based monitoring. Methodology and software are described by Venet et al. [61], Krikwood et al. [71], and Timmermans et al. [62]. The SAMIT trial, a phase III gastric cancer trial, used centralized statistical monitoring [63]. The specific methodology used in this trial is described by Timmermans et al. [72]. Centralized monitoring does not replace site monitoring completely. Instead it directs fewer CRAs on fewer visits to address issues of concern.

This technology now enables DMC members to review data quality and protocol adherence metrics for each site on the trial. Updated metrics will be supplied throughout the trial. DMC members can recommend improvements to the sponsor after joint review of these metrics.

Centralized statistical monitoring can also detect patterns of data that cannot be attributed to innocent errors but rather to fraud. George and Buyse [64] and George [65] have written about fraud in clinical trials and research misconduct in general. Fanelli [66] performed a meta-analysis of relevant literature and estimated that 14% of scientific studies had some form of data fabrication or falsification. Herson [67] describes procedures for salvaging a clinical trial once fraud is detected. He advocates for a Fraud Recovery Plan to be prepared prior to the start of every clinical trial, which would indicate what actions would be taken should evidence of fraud arise. Having such a plan in place would reduce the likelihood of bias in dealing with fraud.

12.6 Conclusion

We have introduced the art and science of interim monitoring for both safety and efficacy. These methods are always in flux partly due to new methodology and technology and the maturing of the DMC process. Changes are expected as wearable technologies become mainstream. Safety and efficacy data will then be monitored 24/7. In the US, the Twenty-First Century Cures Act has gone in to effect. This act directs FDA to do whatever it takes to speed the drug-approval process. Surely this will affect monitoring methodology.

A recent draft guidance from FDA on safety reporting [68] introduced the notion of a *Safety Assessment Committee* (SAC). The role of this committee is broader than that of a DMC. The SAC will consist of both internal and outside experts and will be charged with integrating safety information from a development program with information from clinical trials of the same or similar drugs, preclinical studies, epidemiologic studies, etc. in order to form a complete safety impression during the development process. It is not clear how the SAC will interact with the DMC. Surely at least the Chair of the DMC should be an ex officio member of the SAC or perhaps the typical three-person DMC should be members of the SAC as well. The role of the SAC is still evolving and it is not clear how sponsors, especially the small emerging biotech companies, will implement the SAC.

Although the role of interim monitoring will evolve, it will remain a vital part of oncology drug development, providing an ethical balance and efficiency in drug development.

References

1. Herson J. Safety monitoring. In *Statistical Methods for Evaluating Safety in Medical Product Development*. AL Gould, ed. Chapter 11. West Sussex: John Wiley & Sons, 2015.
2. Crowe BJ, Xia HA, Berlin JA et al. Recommendations for safety planning, data collection, evaluation and reporting during drug, biologic and vaccine development: A report of the safety planning, evaluation and reporting team. *Clinical Trials*. 2009;6:430–40.
3. CIOMS Working Group. *Management of Safety Information from Clinical Trials: Report of CIOMS Working Group VI*. Geneva: CIOMS, 2005.
4. MedDRA Maintenance Support and Services Organization 2016. www.meddra.org (accessed October 21, 2017).
5. U.S. National Cancer Institute. Cancer Therapy Evaluation Program, *Common Terminology Criteria for Adverse Events v4.0 (CTCAE)*, 2017. https://ctep.cancer.gov/protocoldevelopment/electronic_applications/ctc.htm (accessed October 21, 2017).
6. Pharmadhoc. Coding Solutions for the Pharmaceutical Industry, 2017. www.pharmadhoc.com (accessed October 21, 2017).

7. U.S. National Cancer Institute. Patient Reported Outcomes Version of the Common Technology Criteria for Adverse Events, 2016. https://healthcaredelivery.cancer.gov/pro-ctcae/ (accessed October 21, 2017).

8. Dueck AC, Mendoza TR, Mitchell SA et al. Validity and reliability of the U.S. National Cancer Institute's Patient Reported Outcome Version of the Common Terminology Criteria for Adverse Events. *JAMA Oncology*. 2015;1:1051–9.

9. U.S. Food and Drug Administration. Guidance for Industry and Investigators: Safety Reports Required for INDs and BA/BE Studies, 2012. https://www.fda.gov/downloads/Drugs/Guidances/UCM227351.pdf (accessed October 21, 2017).

10. Chuang-Stein C, Xia HA. The practice of pre-marketing safety assessment in drug development. *J Biopharm Stat*. 2013;23:3–25.

11. Clopper CJ, Pearson ES. The use of confidence or fiducial limits illustrated in the case of the binomial distribution. *Biometrika*. 1934;26:404–13.

12. Herson J. *Data and Safety Monitoring Committees in Clinical Trials*, Second Edition. Boca Raton: Chapman and Hall/CRC, 2017.

13. Kaplan EL, Meier P. Nonparametric estimation from incomplete observations. *J Am Stat Assoc*. 1958;53:457–80.

14. MedDRA. MedDRA Maintenance and Support and Services Organization. *Introductory Guide for Standardized MedDRA Queries, v16.0*, 2013. www.meddra.org (accessed October 21, 2017).

15. Schactman M, Wittes J. Why a DMC safety report differs from a safety section written at the end of the trial. In *Quantitative Evaluation of Safety in Drug Development: Design, Analysis and Reporting*. Q Jiang, HA Xia, eds. Boca Raton: Chapman and Hall/CRC, 2015.

16. MedCalc. Relative risk calculator, 2016a. https://www.medcalc.org/calc/relative_risk.php (accessed October 21, 2017).

17. MedCalc. Odds ratio calculator, 2016b. https://www.medcalc.org/calc/odds_ratio.php (accessed October 21, 2017).

18. Gibbons RD, Amatya AK. *Statistical Methods for Drug Safety*. Boca Raton: Chapman and Hall/CRC, 2016.

19. Amit O, Heisberger RM, Lane PW. Graphical approaches to the analysis of safety data from clinical trials. *Pharm Stat*. 2008;7:20–35.

20. Zink RC, Wolfinger RD, Mann G. Summarizing incidence of adverse events using volcano plots and time intervals. *Clinical Trials*. 2013;10:398–406.

21. Gould AL. Safety graphics. In *Statistical Methods for Evaluating Safety in Medical Product Development*. AL Gould, ed. Chapter 2. West Sussex: John Wiley & Sons, 2015.

22. Duke SP, Jiang Q, Huang J et al. Safety graphics. In *Quantitative Evaluation of Safety in Drug Development: Design, Analysis and Reporting*. Q Jiang, HA Xia, eds. Boca Raton: Chapman and Hall/CRC, 2015.

23. Proschan MA, Lan KKG, Wittes JT. *Statistical Monitoring of Clinical Trials: A Unified Approach*. New York: Springer, 2006.

24. Pocock SJ. Group sequential methods in the design and analysis of clinical trials. *Biometrika*. 1977;64:191–9.

25. Haybittle JL. Repeated assessment of results in clinical trials of cancer treatment. *Br J Radiol*. 1971;44:793–7.

26. O'Brien PC, Fleming TR. A multiple testing procedure for clinical trials. *Biometrics*. 1979;35:549–56.

27. Lan K, DeMets D. Discrete sequential boundaries for clinical trials. *Biometrika*. 1983;70:659–63.

28. Wang SK, Tsiatis AA. Approximately optimal one-parameter boundaries for group sequential trials. *Biometrics*. 1987;43:193–9.

29. Emerson SS, Fleming TR. Symmetric group sequential test designs. *Biometrics*. 1989;45:905–23.

30. Pampallona S, Tsiatis AA, Kim K. Group sequential designs for one-sided hypothesis testing with provision for early stopping in favor of the null hypothesis. *J Stat Plan Inference*. 1994;42:19–35.

31. Pocock SJ, Hughes MD. Practical problems in interim analyses, with particular regard to estimation. *Control Clin Trials.* 1989;10:209–21.

32. Hughes MD, Pocock SJ. Stopping rules and estimation problems in clinical trials. *Stat Med.* 1988;7:1231–41.

33. Wheatley K, Clayton D. Be skeptical about unexpected large apparent treatment effects: The case of an MRC AML 12 randomization. *Control Clin Trials.* 2003;24:66–70.

34. Korn EL, Freidlin B, Mooney M. Stopping or reporting early for positive results in randomized clinical trials: The National Cancer Institute Cooperative Group experience from 1990 to 2005. *J Clin Oncol.* 2009;27:1712–21.

35. Freidlin B, Korn EL. Stopping clinical trials early for benefit: Impact on estimation. *Clinical Trials.* 2009;6:119–25.

36. Zhang JJ, Blumenthal GM, He K. Overestimation of the effect size in group sequential trials. *Clin Cancer Res.* 2012;18:4872–6.

37. Goss PE, Ingle JN, Martino S et al. A randomized trial of letrozole in postmenopausal women after five years of tamoxifen therapy for early-stage breast cancer. *N Engl J Med.* 2003;349:1793–802.

38. Wittes J, Barrett-Connor E, Braunwald E et al. Monitoring the randomized trials of the Women's Health Initiative: The experience of the Data and Safety Monitoring Board. *Clinical Trials.* 2007;4:218–34.

39. Sternberg CN, Petrylak DP, Sartor O et al. Multinational, double-blind, phase III study of prednisone and either satraplatin or placebo in patients with castrate-refractory prostate cancer progressing after prior chemotherapy: The SPARC trial. *J Clin Oncol.* 2009;27:5431–8.

40. U.S. Food and Drug Administration. Guidance for Industry: Pathologic Complete Response in Neoadjuvant Treatment of High-Risk Early-Stage Breast Cancer: Use as an Endpoint to Support Accelerated Approval, 2014. https://www.fda.gov/downloads/drugs/guidances/ucm305501.pdf (accessed October 21, 2017).

41. Herson J, Buyse M, Wittes JT. On stopping a randomized clinical trial for futility. In *Designs for Clinical Trials: Perspectives on Current Issues.* D Harrington, ed. New York: Springer, 2012.

42. Wieand S, Schroeder G, O'Fallon JR. Stopping when the experimental regimen does not appear to help. *Stat Med.* 1994;13:1453–8.

43. Freidlin B, Korn EL, Gray R. A general inefficacy interim monitoring rule for randomized clinical trials. *Clinical Trials.* 2010;7:197–208.

44. Fisher B, Constantino J, Redmond C et al. A randomized clinical trial evaluating tamoxifen in the treatment of patients with node negative breast cancer who have estrogen-receptor-positive tumors. *N Engl J Med.* 1989;320:479–84.

45. Fisher B, Dignam J, Bryant J et al. Five verses more than five years of tamoxifen for lymph note-negative breast cancer: Updated findings from the National Surgical Adjuvant Breast and Bowel Project B-14 randomized trial. *J Natl Cancer Inst.* 2001;93:684–90.

46. Dignam JJ, Bryant J, Wieand HS et al. Early stopping of a clinical trial when there is evidence of no treatment benefit: Protocol B-14 of the National Surgical Adjuvant Breast and Bowel Project. *Control Clin Trials.* 1998;19:575–88.

47. Zhang Q, Freidlin B, Korn EL et al. Comparison of futility monitoring guidelines using completed phase III oncology trials. *Clinical Trials.* 2017;14:48–58.

48. Buckner JC, Shaw EG, Pugh SL et al. Radiation plus procarbazine, CCNU, and vincristine in low-grade glioma. *N Engl J Med.* 2016;374:1344–55.

49. Rugo HS, Barve A, Waller CF et al. Effect of a proposed trastuzumab bisimilar compared with trastuzumab on overall response rate in patients with ERBB2 (HER2)-positive metastatic breast cancer. A randomized clinical trial. *JAMA.* 2017;317:37–47.

50. Ravaud A, Motzer RJ, Pandha HS et al. Adjuvant sunitinib in high-risk rental-cell carcinoma after nephrectomy. *N Engl J Med.* 2016;375:2246–54.

51. Dragalin V. Adaptive designs: Terminology and classification. *Drug Inf J.* 2006;40:425–35.

52. Gallo P, Krams M. PhRMA working group on adaptive designs: Introduction to the full white paper. *Drug Inf J.* 2006;40:421–3.

53. Cui L, Hung MHJ, Wang SJ. Modification of sample size in group-sequential clinical trials. *Biometrics*. 1999;55:853–7.
54. Chuang-Stein C, Anderson K, Gallo P et al. Sample size re-estimation: A review and recommendations. *Drug Inf J*. 2006;40:475–84.
55. Gallo P. Confidentiality and trial integrity issues for adaptive designs. *Drug Inf J*. 2006;40:445–59.
56. Turnbull BW. Adaptive designs from a data safety monitoring board perspective: Some controversies and some case studies. *Clinical Trials*. 2017;14:462–9.
57. Sydes MR, Parmar MKB, James ND et al. Issues in applying multi-arm multi-stage methodology to a clinical trial in prostate cancer: The MRC STAMPEDE trial. *Trials*. 2009;10:39.
58. McNeil C. NCI-MATCH launch highlights new trial design in precision medicine era. *J Natl Cancer Inst*. 2015;107:djv193.
59. Clinicaltrials.gov. TAPUR: Testing the use of Food and Drug Administration approved drugs that target a specific abnormality in a tumor gene in people with advanced stage cancer, 2016. https://clinicaltrials.gov/ct2/show/NCT02693535 (accessed October 21, 2017).
60. U.S. Food and Drug Administration. Guidance for Industry: Oversight of Clinical Investigations. *A Risk-Based Approach to Monitoring*, 2013. https://www.fda.gov/downloads/Drugs/Guidances/UCM269919.pdf (accessed October 21, 2017).
61. Venet D, Doffagne E, Burzykowski T et al. A statistical approach to centralized monitoring of data quality in clinical trials. *Clinical Trials*. 2012;9:705–13.
62. Timmermans C, Venet D, Burzykowski T. Data-driven risk identification in phase III clinical trials using centralized statistical monitoring. *Int J Clin Oncol*. 2016;21:38–45.
63. Tsuburaya A, Yoshida K, Kobayashi M et al. Sequential paclitaxel followed by tegafur and uracil (UFT) or S-1 monotherapy as adjuvant chemotherapy for T4a/b gastric cancer (SAMIT): A phase III factorial randomized controlled trials. *Lancet Oncology*. 2014;15:886–93.
64. George SL, Buyse M. Data fraud in clinical trials. *Clinical Investigation*. 2015;5:161–73.
65. George SL. Research misconduct and data fraud in clinical trials: Prevalence and causal factors. *Int J Clin Oncol*. 2016;21:15–21.
66. Fanelli D. How many scientists fabricate and falsify research? A systematic review and meta-analysis of survey data. *PLOS One*. 2009;4:e5738.
67. Herson J. Strategies for dealing with fraud in clinical trials. *Int J Clin Oncol*. 2016;21:23–7.
68. U.S. Food and Drug Administration. Safety Assessment for IND Safety Reporting—Draft Guidance for Industry, 2015. https://www.fda.gov/downloads/Drugs/GuidanceComplianceRegulatoryInformation/Guidances/UCM477584.pdf (accessed October 21, 2017).
69. Gould AL. 2013, personal communication.
70. Hu C. 2018, personal communication.
71. Kirkwood A, Cox T, Hackshaw A. Application of methods for centralized statistical monitoring in clinical trials. *Clinical Trials*. 2013;10:783–806.
72. Timmermancs C, Doffague E, Venet D et al. Statistical monitoring of data quality and consistenc in the stomach cancer adjuvant multi-institutional grouptrial. *Gastric Cancer* 2016;19:24–30.
73. Herson J. Coordinating data monitoring committees and adaptive clinical trial designs. *Drug Information Journal* 2008;42:297–301.
74. Barker AD, Sigman CC, Keloff GJ et al. ISPY2:j An adaptive breast cancer trial design in the setting of neoadjuvant chemotherapy. *Clinical Pharmacology and Therapeutics* 2009;86:97–100.
75. Herbst RS, Gandara DR, Hirsch FR et al. Lung master protocol (Lung-MAP): A biomarker-driven protocol for accelerating development of therapies for squamous cell lung cancer. *Clinical Cancer Research* 2015;21:1514–24.

Section III

Personalized Medicine

13

Biomarker-Based Phase II and III Clinical Trials in Oncology

Shigeyuki Matsui, Masataka Igeta, and Kiichiro Toyoizumi

CONTENTS

13.1 Introduction

Recent advances in biotechnology, such as genome sequencing, have revolutionized the molecular oncology field and fostered the development of molecularly targeted agents that inhibit specific targeted molecules related to carcinogenesis and tumor growth. More recently, the role of the cancer abbreviations and immunogenic neoantigen generation has increasingly been appreciated, leading to the development of immune checkpoint inhibitors. At the same time, substantial molecular heterogeneity has been identified within histologically defined cancers. All of these new perspectives on cancer biology are expected to be utilized for the development of personalized or precision medicine. Indeed, the methodologies of clinical trials in oncology are in the midst of an evolution that is accelerating the realization of precision medicine [1].

A fundamental component in the evolution of clinical trials is the shift of systemic cancer treatments from traditional cytotoxic agents to a new type of therapy described above. For molecularly targeted treatments, the anticancer effects are likely to be restricted to patient subgroups in which alterations in the treatment target are driving the growth of the cancer. That is, for such treatments, we expect a substantial inter-patient heterogeneity in treatment responsiveness when compared with cytotoxic agents. Therefore, the traditional randomized clinical trial design (mainly for cytotoxic agents), which evaluates an average treatment effect in a broad patient population, would no longer be effective for this new type of treatment.

Another important component in the evolution of clinical trials is closely linked with that discussed above. It involves the use of *predictive markers* to capture inter-patient heterogeneity in treatment responsiveness or to identify a subgroup of patients who are likely to benefit from a new treatment. Since predictive markers will always be paired with a particular treatment, they are often called *companion* markers (for the new treatment). An example is the V600E BRAF point mutation, a companion marker for the BRAF enzyme inhibitor, vemurafenib, in melanoma patients [2]. Another but more complex example is that of graded or continuous markers, for instance the overexpression of the HER2 protein or amplification of the *HER2* gene as a companion marker for the monoclonal antibody, trastuzumab, in metastatic breast cancer patients [3,4]. Similar arguments may also apply to many checkpoint inhibitor-based immunotherapies.

As such, modern clinical trials that aim to achieve precision medicine should involve the co-development of new treatments and predictive markers. However, the additional task of developing predictive markers may inevitably complicate the clinical development of new treatments as a whole [5–11].

Firstly, predictive markers need to be analytically validated. This is to confirm accuracy in measuring the status of binary markers (e.g., the presence or absence of a point mutation), and robustness and reproducibility, especially when measuring levels of ordered or continuous markers (e.g., gene amplification or gene/protein expressions) [5,10,11]. In addition, the latter type of ordered or continuous markers may require determination of a cut-off point to define marker positivity so that the subgroup of patients who will benefit from the treatment can be identified.

Secondly, another requirement for predictive markers is that they need to be clinically validated. This is to assess their ability to predict treatment responsiveness or identify a patient subgroup that is likely to benefit from the treatment [5,10,11]. In the early phases of clinical trials, this assessment is typically based on short-term, surrogate endpoints, for example, pharmacodynamic endpoints in proof-of-concept trials or tumor shrinkage/progression-free survival endpoints in phase II trials. In the confirmatory, phase III setting, a marker-stratified randomized trial (see Section 13.3.2.3) can assess clinical validity based on true endpoints, such as overall survival, by evaluating whether the effect of the new treatment is sufficiently different from that of its control treatment across marker-defined subgroups (i.e., treatment-by-marker interaction).

Thirdly, the clinical utility of associated markers and treatments ultimately needs to be evaluated. This is generally evaluated through a confirmatory phase III trial to ensure that the use of a new treatment combined with a predictive marker results in improved patient outcomes in terms of true endpoints [5,10,11]. When planning such a clinical trial, however, the status of marker development and validation may vary widely, and candidate markers may have differing levels of credibility. This complicates the methodology for the design and analysis of clinical trials that evaluate clinical utility.

As part of the recent evolution of clinical trials based on predictive markers, a new innovative framework based on the *master protocol* has streamlined clinical development of targeted treatments [12–16]. For a particular histologically defined cancer, umbrella/ platform trials may allow for evaluation of marker-treatment hypotheses or for efficient screening of promising marker-treatment combinations. Meanwhile, basket trials aim to define the patient population based on a particular molecular aberration that occurs across histologically different cancers, each of which may represent uncommon or rare subtypes within each histology (see Chapter 14 for more details).

In this chapter, we provide an overview of phase II and III clinical trial designs that utilize predictive markers. In Section 13.2, we introduce various marker-based phase II designs, including the new framework based on the master protocol. In Section 13.3, we discuss various designs to evaluate the clinical utility of a new treatment and a predictive marker, including enrichment, all-comers, and marker-strategy designs. Adaptive designs for marker development and validation are also discussed. Lastly, we present concluding remarks in Section 13.4.

13.2 Phase II Trials

The main objective of phase II clinical trials is to identify promising experimental treatments worth testing in definitive phase III trials. In recent years, the role of this phase as *treatment screening* has become more important as a number of experimental targeted agents have emerged with the advances of molecular oncology. Accordingly, screening efficiency, in terms of time and resources, may be pursued when designing phase II trials, while ensuring ethical conduct with regard to enrolled patients. Generally, short-term, surrogate endpoints, such as pharmacodynamic markers, tumor shrinkage, and progression-free survival, are used in phase II trials.

The use of a predictive marker for patient selection may require many special considerations in clinical trial design. First, every process, from collecting and storing specimens to evaluating the marker, have to be standardized. The performance of the assay in terms of its analytical validity should be guaranteed (e.g., by using commercially available or well-established kits), or at least well evaluated, prior to the trial. The cut-off point for graded or continuous marker values to define marker positivity should also be determined before the trial, although an optimal cut-off point could be explored during or at the end of the phase II trial. Given the specification of the marker with its cut-off point, the marker prevalence, i.e., the proportion of marker-positive (M+) patients in the general patient population, is a critical factor related to the design of the trial; the smaller the prevalence, the more patients must be screened to ensure recruitment of a sufficient number of M+ patients. A short turn-around time for assessing the marker is another important factor; ideally, this should be 2–3 days, or at most about a few weeks, so as not to delay the start of each patient's treatment. See References [5–9,17] for more information about various considerations that are necessary when incorporating predictive markers in clinical trials.

In designing marker-based clinical trials, there are two main strategies, namely the enrichment and all-comers approaches. In an enrichment trial, patients are only eligible if they are M+. This approach is most suitable for situations where there is compelling evidence that marker measurement can effectively identify a subgroup of patients who are responsive to the treatment. Generally, the enrichment approach is characterized as efficient in terms of assessing treatment efficacy in small numbers of eligible patients (given compelling evidence regarding the marker).

On the other hand, when the evidence for the marker is less compelling and it cannot be ruled out that marker-negative (M−) (as well as M+) patients could benefit from the treatment, the all-comers approach is a reasonable choice. With this method, eligibility is not restricted to a particular marker-defined subpopulation; instead, the entire patient population (including the marker-defined subpopulations) is assessed. An important concern, however, is the possibility of limited treatment efficacy in M− patients. This motivates an adaptive design that curtails the initial enrollment of M− patients and restricts the patient enrollment to M+ patients midway through the trial if interim trial data indicate no or clinically meaningless treatment efficacy in the M− patients. This type of adaptation is called *adaptive enrichment*.

The factors that determine whether the enrichment or all-comers approach is used in phase II trials to preliminarily assess treatment efficacy and biomarker development may differ from those used in phase III trials when evaluating the clinical utility of marker-treatment combinations. See References [8,9] for discussions on the selection of the two approaches during the clinical development of targeted treatments in industry-sponsored clinical trials.

In Sections 13.2.1 and 13.2.2 we discuss, for single-arm and randomized trials, respectively, marker-based phase II designs, including adaptive-enrichment designs. We also introduce the new framework based on the master protocol in Section 13.2.3.

13.2.1 Single-Arm Trials

The single-arm design without a control treatment arm is the standard design approach in traditional phase II trials of cytotoxic agents. In this design, all enrolled patients receive the experimental treatment. The observed response rate, typically based on tumor shrinkage, is compared to a pre-specified "reference" level that reflects that of an established or standard treatment. This is typically formulated as a one-sided statistical test of the null hypothesis, that the response rate is equal to or less than the reference level, against the alternative hypothesis, that the response rate is equal to or greater than an expected level of clinical importance worth further treatment development under specified type-I or alpha error, α, and type-II or beta error, β. The single-arm design generally requires relatively small numbers of patients, so it can be seen as an efficient tool for treatment screening, not only of traditional cytotoxic agents, but also of molecularly targeted agents.

The efficiency can be further improved by introducing an interim analysis that allows for an earlier rejection of unpromising treatments. Simon's two-stage design [18] is one of the most popular and basic multi-stage, single-arm designs. Specifically, a first stage enrolls n_1 patients and assesses the number of responders R_1. If $R_1 \leq c_1$, the trial is stopped at the end of this stage with acceptance of the null hypothesis or a declaration of no efficacy of therapeutic interest (futility stopping); here, c_1 is a pre-specified integer representing the futility boundary. Otherwise, a second stage follows in which an additional n_2 patients are enrolled and the number of responders R_2 in these patients is assessed. At the end of the second stage, the total number of responders $R_{1,2}$ ($= R_1 + R_2$) is compared to a pre-specified integer $c_{1,2}$. If $R_{1,2} \geq c_{1,2}$, the treatment is declared efficacious; otherwise, it is not. As explained in Chapter 4, in this two-stage design, the design parameters n_1, n_2, c_1, and $c_{1,2}$ are pre-specified to satisfy an optimality criterion, such as minimization of patients under the null hypothesis, as well as the usual constraints on alpha and beta, such as $\alpha \leq 5\%$–10% and $\beta \leq 10\%$–20% [18]. The calculation of these parameters can be performed using the web-based program at https://brb.nci.nih.gov/.

With a predictive marker, Simon's two-stage design can easily be used in the enrichment approach where only M+ patients are eligible. As a fairly large effect may be expected in these patients, the enrichment single-arm design may further improve the efficiency of the single-arm design.

For the all-comers approach with a predictive marker, we can consider extensions of Simon's two-stage design to assess treatment efficacy both in the overall patient population and in marker-defined subpopulations. With the all-comers approach, we can also assess the clinical validity of the marker by comparing treatment effects across the marker subpopulations.

Generally, statistical analysis methods in the marker-based, all-comers approach can be roughly divided into *top-down* and *bottom-up* methods. The top-down approach begins with an overall analysis that is possibly followed by a subgroup analysis. For example, if the overall treatment effect observed in the first stage is greater than a pre-specified threshold, the overall population remains the subject of analysis in the second stage; otherwise, treatment efficacy in the M+ subgroup is assessed in the second stage.

On the other hand, the bottom-up approach starts with a subgroup-based analysis that is possibly followed by an overall analysis. For example, treatment effects within M+ and M− subgroups are assessed, possibly to determine whether these effects differ between the marker subgroups, i.e., treatment-marker interaction. If the observed effects in both subgroups are greater than some threshold or if there is no evidence of treatment-marker interaction, the overall population is evaluated in the second stage. If the first-stage data indicate treatment-marker interaction, which might represent efficacy in M+ patients, the second stage assesses treatment efficacy in that subgroup only.

Both the top-down and bottom-up approaches can be applied to adaptive-enrichment designs with the option of curtailing the enrollment of M− patients at the end of the first stage (when the treatment effect in these patients is deemed small based on interim data from the first stage). Figure 13.1 shows examples of top-down and bottom-up versions of adaptive enrichment in a two-stage, single-arm design [19–21].

Although the single-arm design (without a control arm) is generally efficient, as noted previously, the main concern in this design is the appropriateness of the specified reference level for the response variable, reflecting historical response data on standard treatments. This is also essentially true for the Bayesian approach (e.g., [22]), where prior distributions of the response rate under null treatment effects must be appropriately specified. Specification of the reference value may be particularly difficult for progression-free survival endpoints (e.g., the progression-free survival rate at 6 months), which may largely depend on patients' prognostic characteristics [23]. For marker-based trials, a general lack of historical data for marker-defined subgroups, not for the overall population, may make the specification more difficult [24] (especially when the predictive marker has some prognostic effect on progression-free survival). In addition, irrespective of the use of a predictive marker, the specification may become difficult when evaluating a combination therapy that includes an established treatment that is deemed efficacious. Therefore, the single-arm design is generally more suitable for evaluating monotherapies [25]. In the next section, we discuss introduction of a control treatment arm, i.e., randomized phase II trials, to address the issue of unreliable historical reference values for the response rate.

13.2.2 Randomized Phase II Trials with a Control Arm

A concurrent, control treatment arm may serve as an appropriate reference in comparison with the experimental treatment arm under a broad range of situations with various patient

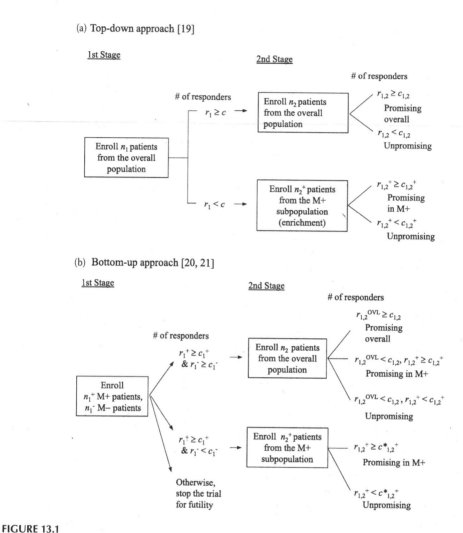

FIGURE 13.1

Two-stage single-arm designs for M+ and M− patients as examples of adaptive enrichment designs using the top-down approach (Panel a) and bottom-up approach (Panel b).

populations, treatments, and endpoints. The randomization would generally allow for unbiased estimation of treatment effects in the study patients, by producing treatment arms that are comparable with respect to any patient baseline characteristic. Randomized phase II trials have been increasingly performed in recent years, especially for evaluating molecularly targeted agents based on progression-free survival endpoints.

However, as the reference value of the response variable is now estimated with some estimation error, rather than pre-specified as a fixed value, a randomized trial may suffer from a lack of efficiency, requiring relatively large numbers of patients for treatment comparison. Some authors propose the alleviation of alpha and beta errors in treatment comparison, such as $\alpha = \beta = 10\%$–20%, by explicitly regarding the purpose of such trials to be treatment screening (rather than confirmation of treatment efficacy) [26]. Such trials, however, may still need large numbers of patients, e.g., at least twice as many patients as single-arm trials. See References [27,28] for further discussion on the comparison between single-arm and randomized designs in oncology phase II trials.

As in single-arm trials, a predictive marker can be incorporated in randomized phase II trials in both the enrichment and all-comers approaches. For the all-comers approach, both top-down and bottom-up analysis plans across the overall patient population and marker-defined subgroups can be applied and are similar to those applied in marker-stratified phase III trials (see Section 13.3.2). Freidlin et al. [29] propose a decision algorithm in a marker-stratified, all-comers phase II trial to recommend the selection of either the enrichment or all-comers approaches in designing the subsequent phase III trial, based on a pseudo-bottom-up method, similar to the marker sequential test (MaST) design (see Section 13.3.2.3).

Methodologies for all-comers, adaptive-enrichment designs have been extensively studied with various criteria for enrichment in the middle of clinical trials with short-term endpoints. Many suppose the use of a single predictive marker [30–32], but some tackle the development of genomic signatures or classifiers based on large numbers of candidate markers [33].

13.2.3 Master Protocol

A new framework of clinical trials has emerged in which multiple parallel clinical trials with different marker-treatment combinations within or across histologically defined cancers are performed under one overarching *master protocol* [12–16]. Prominent features of this framework include the establishment of a common infrastructure, possibly across multiple institutes or centers, for recruiting various cancers, including rare subtypes; the use of a common system to assay genomic alterations or markers related to multiple potential targets for treatment; and prompt assignment of enrolled patients to the appropriate treatment arms or sub-trials with regard to combinations of marker and treatment. In addition, the ability to flexibly add or drop treatment arms or sub-trials (with particular marker-treatment combinations) after initiating the master protocol can be planned as one component of the protocol. Currently, this framework has been applied mainly in the phase II setting, but it is expected to cover phase III trials, probably with the enrichment design (see Section 13.3.1). The master protocol framework aims to enhance the efficiency of screening marker-treatment combinations as well as the efficiency of evaluating their clinical utility (through phase III enrichment trials). Actually, this framework has attracted significant expectations in terms of improving the efficiency of clinical development of targeted treatments as a whole [34].

Based on ongoing clinical trials, we can identify several objectives for utilizing the master protocol, yielding categories of master protocol trials such as umbrella/platform trials and basket trials.

13.2.3.1 Umbrella/Platform Trials

For a particular histologically defined cancer, one objective of the master protocol is to expand the indications of specific marker-treatment combinations whose efficacies have already been demonstrated in particular patient subsets with the same histologic cancer type. This kind of trial has been called an *umbrella* trial [13,16]. For example, in non-small cell lung cancer, the Adjuvant Lung Cancer Enrichment Marker Identification and Sequencing Trial (ALCHEMIST) evaluates two targeted treatments, erlotinib and crizotinib, which were already developed for advanced-stage cancers, in respective sub-trials with early-stage patients; one sub-trial includes patients with EGFR mutations while the other includes those with ALK translocations [35].

Another type of master protocol trial, again, for a particular histologic cancer type, is a phase II trial that uses the master protocol to identify the most promising marker-treatment combinations from among a large pool of *experimental* candidates. This type of trial is often called a *platform* trial [13,16]. Examples include the BATTLE-1 and -2 trials [36–38] and the LungMAP trial [39] in non-small cell lung cancer and the I-SPY 2 trial in early breast cancer [40–41]. In these trials, multiple experimental agents are tested across pre-specified subpopulations or strata based on a set of putative markers.

In order to cope with a number of possible marker-treatment combinations, the BATTLE and I-SPY 2 trials employ outcome-dependent, adaptive randomization of agents, that is, adaptively changing the ratios of the probabilities of assigning patients to respective agent arms within marker strata, so that the chance of assigning a more promising agent increases within marker strata based on interim trial data. Bayesian hierarchical models are used to calculate a posterior distribution of the response rate for an agent (as well as one for a control treatment) for each of the marker strata, which provides the basis for calculating the treatment assignment probabilities used in the adaptive randomization. In the I-PSY 2 trial, the Bayesian model is also used to determine whether to drop a particular agent arm, according to pre-specified efficacy or futility criteria based on a Bayesian predictive probability of the agent being superior to the control treatment.

Although the platform trial has the attractive features discussed above, as well as demonstrated efficiency compared with separate trials for multiple agents [42,43], several issues have been identified [44–47]. Establishing a framework for conducting platform trials involves many challenges and may require the support of multiple pharmaceutical companies, regulatory agencies, and academic communities. Furthermore, special clinical trial infrastructures for handling complex operations (such as those for adaptive randomization) based on molecular assays must be constructed.

Concerning the trial results, the Bayesian posterior distribution and predictive probability may be more difficult to interpret than the standard analysis in randomized trials [44]. Note that simple overall summary statistics between treatment arms employed in traditional randomization trials with equal treatment assignment probabilities are no longer valid because under the adaptive randomization, prognostic factors (related to the markers) may no longer be balanced when comparing treatment arms without adjustment for these factors. Related to this point, as the Bayesian analysis relies on complex hierarchical models and prior specifications, interpretation of such an analysis may be complicated owing to possible model misspecification [44].

Concerning the design aspect, adaptive randomization has been criticized for possible biases and inefficiency [45,46]. Another issue is that the marker strata can be inappropriately specified, including the use of inappropriate component markers, assays, and cut-offs for defining marker positivity [47]. However, introducing a preliminary, exploratory stage of predictive analysis using multiple candidate markers (even genomic data) can address this issue, as was done in the BATTLE-2 trial (even though this approach failed in that trial) [38]. Lastly, by its very nature, the platform trial requires a long period of time, during which the standard treatment would evolve. Possible complexity due to updating or changing the common control arm is a critical issue [39].

Despite the large number of issues discussed above and elsewhere, platform trials are relatively new and have significant potential for revolutionizing phase II treatment screening trials, specifically in terms of dramatically increasing the number of experimental targeted agents analyzed in future clinical trials.

13.2.3.2 Basket Trials

Another objective of the master protocol is similar to that of umbrella trials, namely to expand the indications of specific marker-treatment combinations. However, the expansion is considered across multiple histologically defined cancers, rather than within a particular cancer. *Basket trials* are *histology-agnostic* trials where the trial population is defined based on a particular molecular aberration across histologically different cancers, each of which may represent relatively rare subtypes within each histology. Typically, a basket trial to evaluate a single agent is designed as a phase II single-arm trial with multiple cancer strata with different histologies. Vemurafenib is a BRAF inhibitor whose efficacy was previously demonstrated in melanoma with the BRAF V600E mutation [2]. Hyman et al. [48] performed a basket trial to investigate the efficacy of vemurafenib in other various solid tumors with the same mutation.

Basket designs can also be used to screen new targeted agents. An example is the NCI MATCH trial that involves sub-trials with the single-arm design to assess 20 or more targeted agents in advanced metastatic cancer of many histologic types [49]. Importantly, such trials may aim to develop new targeted agents where the treatment indication is defined based on a particular molecular alteration, rather than on the cancer histology, as is traditionally the case.

Although many basket trials are relatively simple, employing the single-arm design with multiple strata based on cancer histology, the statistical analysis is complicated by possible heterogeneity in treatment responsiveness across cancer strata. For example, in the basket trial by Hyman et al. [48], vemurafenib was shown to be efficacious in some types of cancer with the BRAF mutation, but not in colorectal cancer.

In cases with sufficient numbers of patients within cancer strata, we can analyze individual strata separately. Standard single-arm designs, such as Simon's two-stage analysis, can be applied to each stratum to assess within-stratum treatment efficacy as well as inter-strata heterogeneity in treatment responsiveness. Another approach, which can be more efficient when there are relatively homogeneous effects across strata, is to apply the top-down or bottom-up approaches described in Section 13.2.1 across the pooled population and cancer strata. In this vein, some multi-stage analyses have been proposed [50–53].

Another direction is to use statistical models to allow for information sharing across cancer strata. Several authors proposed Bayesian methods based on hierarchical models with a prior distribution for the underlying stratum-specific response rates [54–56]. Another, but simpler, Bayesian method is to directly specify the prior probability that the underlying stratum-specific response rates are homogeneous, as well as the prior probability that the treatment has a clinically meaningful effect within or across strata [57,58].

Currently, the relative performance of the aforementioned methods has not been well studied, especially in basket trials with small strata representing relatively rare cancer subtypes. Further research in this respect, as well as the development of new statistical methods, is warranted.

13.3 Phase III Trials

The general objective of phase III trials that use a predictive marker is to confirm the clinical utility of the combination of the treatment and marker based on a true endpoint, such as overall survival. Two main approaches to designing such trials have been identified, with the optimal method depending on the specific objectives.

One approach is to confirm the efficacy of the treatment with the aid of the predictive marker. Marker-based, enrichment, or all-comers designs would be the choice in this scenario. As we have seen so far, the status of marker development and validation varied widely in earlier clinical trials, so the developed markers may have differing levels of credibility with regard to launching a phase III trial. The choice between the enrichment and all-comers design would thus depend on the level of credibility of the marker.

Another possible objective of marker-based phase III trials is to confirm the clinical utility of the new marker(s) in treating patients. *Strategy designs* have been proposed and implemented to compare a new treatment strategy based on a new marker or set of markers to that based on the standard practice of care (i.e., not based on the marker). Here, the new markers can be those developed as *prognostic markers*, where the goal is to predict patient outcomes under no treatment or standard treatment, rather than being paired with a particular treatment. Examples of prognostic markers are the gene-expression signatures MammaPrint [59] and Oncotype-Dx [60] for recurrence risk stratification in early-stage breast cancer.

In Sections 13.3.1 through 13.3.3, we discuss the enrichment, all-comers, and strategy designs, respectively. Note that many of the discussions on enrichment and all-comers strategies given in the phase II setting in Section 13.2 also apply to phase III designs, but the control of alpha error will be emphasized more strongly. Unless otherwise noted, in these discussions we suppose the comparison of two treatments, a new treatment and its control (two treatment arms) based on survival-time outcomes and situations where a single binary marker is available at the launch of phase III trials.

13.3.1 Enrichment Designs

When the marker is analytically validated and there are compelling biologic or early trial data that support the so-called Marker-Assumption (I) that the benefit of the new treatment is limited only to M+ patients, it is best to consider an enrichment or targeted design that limits the eligibility for treatment randomization to these patients [10,61–66] (see Figure 13.2a). Recent examples of phase III enrichment trials include those of vemurafenib, for

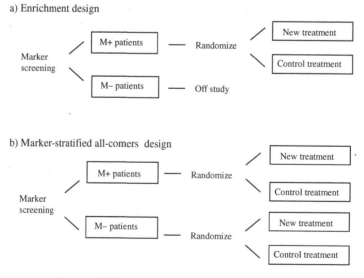

FIGURE 13.2
Enrichment and marker-stratified all-comers designs for M+ and M− patients.

melanoma patients with the V600E BRAF point mutation [2], and of the ALK and ROS1 inhibitor crizotinib, for non-small cell lung cancer patients with a translocation in the ALK gene [67].

Compared with the traditional all-comers design (which does not measure the marker), the enrichment design is shown to be efficient in terms of the number of randomized patients, especially under the conditions that the marker prevalence (the proportion of M+ patients) in the general population is small (e.g., <0.5), the treatment is relatively ineffective in M− patients, and the diagnostic accuracy of the assay (in terms of positive/negative predictive values) is high [66]. If these conditions are satisfied, the number of randomized M+ patients can be fairly small because relatively large treatment effects can be expected in this population. With the emerging phase II master protocol trials, the use of the efficient enrichment design in definitive phase III trials is essential for establishing a new (enrichment-oriented) framework of clinical trials to streamline the clinical development of new targeted treatments (see Section 13.2.3).

The efficiency of the enrichment design, however, does not necessarily imply that fewer patients will undergo marker screening or that the study duration will be shortened. In particular, when the marker prevalence is low, a substantial number of patients might be required for marker screening until the required number of M+ patients is enrolled to ensure treatment randomization [68].

A major drawback to the enrichment design relates to the correctness of the strong Marker-Assumption (I) that only M+ patients benefit from the treatment. Possible factors that can threaten this assumption include imperfections in measuring the molecular target, alternative threshold points that could better define M+ patients (especially for graded or continuous markers), and possible off-target effects of the treatment [65,69]. When information regarding these factors is limited in the design stage of a phase III trial, we cannot rule out the possibility that the remaining M− patients might also benefit from the treatment. The major limitation of the enrichment design is that it does not provide data to evaluate treatment efficacy in M− patients so as to check the validity of Marker-Assumption (I) in the confirmatory phase of clinical development.

To address this issue, some authors have proposed a sequential or tandem enrichment approach that conducts a second trial for M− patients when an initial trial for M+ demonstrates treatment efficacy [9,70]. It is believed that this approach allows for quicker assessment of the treatment in the patient population that is most likely to benefit from it [9]. However, this tandem approach could require a long period of clinical development overall. Also, sequential assessment across marker subpopulations is not necessarily efficient when treatment effects across these groups are relatively homogeneous [68].

13.3.2 Marker-Based, All-Comers Designs

Marker-Assumption (I) may not be sufficiently accurate in clinical trials of many single- or multi-targeted agents, immunotherapies, and other treatments. In the situation where we should consider the possibility that M− patients could also benefit from the treatment, *concurrent* all-comers designs to compare the treatment in both M+ and M− negative patients could be a reasonable choice [10,61–65]. In particular, marker-stratified designs are a standard approach in marker-based, all-comers trials (see Figure 13.2b). Note that the stratification based on the marker will ensure observation of the marker status for all randomized patients and can also incorporate the marker's possible prognostic effects.

When planning a marker-stratified trial, it is natural to adopt Marker-Assumption (II), namely that the treatment is more effective in M+ than in M− patients. This assumption is less restrictive than Marker-Assumption (I) in the enrichment design. However, with Marker-Assumption (II), the statistical analysis plan will become complicated because of assessment across the marker subpopulations, and probably of the overall population. Again, both top-down and bottom-up approaches in Section 13.2 are applicable, but more attention is required for strict control of the study-wise alpha rate in the definitive phase III setting.

13.3.2.1 Null Hypothesis

In considering strict alpha control, we first have to identify appropriate null hypotheses. Associated with Marker-Assumption (II), we denote $H_0^{(+)}$ and $H_0^{(-)}$ representing null hypotheses in the M+ and M− subpopulations, respectively. With Marker-Assumption (II) based on subpopulation-specific treatment effects, we regard subpopulation-specific null hypotheses, $H_0^{(+)}$ and $H_0^{(-)}$, as *genuine* null hypotheses. Under Marker-Assumption (II), we identify the two possible null effect scenarios [71,72]:

- Null-Scenario 1: true $H_0^{(+)}$ and true $H_0^{(-)}$, i.e., $HR^{(+)} = 1$ and $HR^{(-)} = 1$, and
- Null-Scenario 2: false $H_0^{(+)}$ and true $H_0^{(-)}$, i.e., $HR^{(+)} \neq 1$ and $HR^{(-)} = 1$.

Here $HR^{(+)}$ and $HR^{(-)}$ denote hazards ratios of the new treatment relative to the control treatment in the M+ and M− subpopulations, respectively. Null-Scenario 1 is what is called the *global null hypothesis*, while Null-Scenario 2 corresponds to a qualitative interaction (see Section 13.3.2.4).

The null hypothesis of no overall treatment effect, that is, no treatment effect in the overall population, denoted by $H_0^{(o)}$, is also used in many statistical analysis plans (see Figure 13.3). However, $H_0^{(o)}$ can differ from the genuine null hypotheses, $H_0^{(+)}$ and $H_0^{(-)}$. To see this, let $HR^{(o)}$ denote hazards ratios of the new treatment relative to the control treatment in the overall population. In two-sided testing, $HR^{(o)}$: $HR^{(o)} = 1$ holds under Null-Scenario 1: $HR^{(+)} = 1$ and $HR^{(-)} = 1$, but not necessarily under Null-Scenario 2: $HR^{(+)} \neq 1$ and $HR^{(-)} = 1$. Similarly, in one-sided testing, $H_0^{(o)}$: $HR^{(o)} > 1$ does not always hold under Null-Scenario 2: $HR^{(+)} < 1$ and $HR^{(-)} > 1$. Importantly, rejection of $H_0^{(o)}$ may not necessarily represent rejection of Null-Scenario 2. This is linked with the problem that a significant result in an overall test (rejection of $H_0^{(o)}$) can lead to the false assertion of treatment efficacy in M− patients when in fact there is no efficacy in these patients [73–76].

These arguments may raise the question of the role of the overall test in marker-stratified trials. As discussed in the next section, although the overall test does not directly assess the genuine null hypotheses, $H_0^{(+)}$ and $H_0^{(-)}$, it can be positioned as a *working or operational* criterion to expand the treatment indication to the M− subpopulation (in addition to the M+ subpopulation), that is, to the overall population. This is in contrast to traditional clinical trials that are not based on markers, where the overall test serves as the primary analysis to evaluate treatment efficacy in the overall population.

13.3.2.2 Weak or Strong Control

We now consider the control of the study-wise alpha rate based on the genuine null hypotheses, $H_0^{(+)}$ and $H_0^{(-)}$. We can consider two types of control, i.e., weak or strong controls, across these two null hypotheses (see Chapter 10 on the types of control). Weak control is to control the study-wise alpha only under $H_0^{(+)}$ and $H_0^{(-)}$, i.e., Null-Scenario 1 of

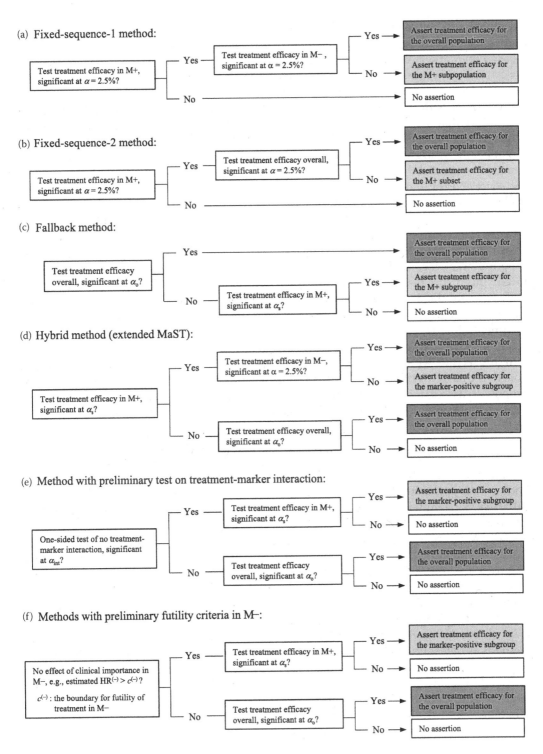

FIGURE 13.3
Various statistical analysis plans across M+ and M− subgroups in the marker-stratified design.

the global null, while strong control is to control the alpha under any combination of $H_0^{(+)}$ and $H_0^{(-)}$, which reduces to control under Null-Scenario 1 and also under Null-Scenario 2 based on Marker-Assumption (II). It should be noted that with Marker-Assumption (II), we can see the weak control as a strict alpha control for asserting treatment efficacy, *at least* for the M+ subpopulation. Also, the further control under Null-Scenario 2, in addition to that under Null-Scenario 1 (i.e., strong control), corresponds to the control of the error probability that the treatment indication is expanded to M− patients (thus, to the overall population) even if the treatment is ineffective in M− patients.

Some authors argue the need for the more stringent, strong control [71,72]. However, it can also be argued that given the strict alpha control under Null-Scenario 1, the need for the strict control under Null-Scenario 2 is unnecessary. Practically, the degree of control under this scenario can vary on a case-by-case basis, taking into account many external factors such as the analytical performance of the marker, marker prevalence, possible adverse effects, prognosis of the disease, availability of other treatments, and treatment costs [73]. For example, given a demonstration of a very large treatment effect in the M+ subpopulation, it could be worthwhile to consider a less stringent control for testing $H_0^{(-)}$ in advanced diseases with no established treatments. This is similar to the approach that separates the demonstration of treatment efficacy and consideration of an indication classifier to identify the marker-based characteristics of the patients in whom the new treatment should be used [74]. Here, the former component is accomplished by a significant result with the strict alpha control under the global null hypothesis.

In this section, we shall permit statistical analysis plans with weak control. However, we note that this may not imply that alpha control is unnecessary under Null-Scenario 2. We propose to assess or monitor the error probability under this scenario, supposing possible treatment effects across the marker subpopulations (see Section 13.3.2.4).

13.3.2.3 Statistical Analysis Plans in the Marker-Stratified Design

Various hierarchical or split-alpha multiple testing methods based on the top-down or bottom-up approaches have been proposed for marker-stratified trials (see Figure 13.3) [10,62–65,73,77]. In what follows, we suppose using one-sided statistical tests to spend the study-wise alpha, $\alpha = 0.025$.

Under Marker-Assumption (II), when there is relatively strong evidence on the marker that a treatment is efficacious in M+ patients, hierarchical, *fixed-sequence-1 and -2 methods* (Figure 13.3a,b) that first test treatment efficacy in M+ patients would be reasonable. If this test in M+ patients is significant at $\alpha = 0.025$, then the treatment effect is also tested using the same significance level α in M− patients in the fixed-sequence-1 method or in the overall population in the fixed-sequence-2 method. Of note, the fixed-sequence-1 method may achieve the strong control (for Null-Scenario 1 and 2). The power of the fixed-sequence methods is determined by the first test in the M+ subgroup, like in the enrichment and sequential or tandem enrichment designs. Accordingly, the fixed-sequence methods are expected to perform well under situations with an excellent marker where there is a fairly large treatment effect in M+ patients [73].

In more common situations where there is no strong evidence on the predictive value of the marker and it is considered that the treatment has the potential to be broadly effective, split-alpha methods that allocate some portion of the total alpha $\alpha = 0.025$ across tests in the overall population and marker subgroups would be reasonable. One simple approach is to apply the Bonferroni method. For example, in the SATURN trial [78] to assess the use of erlotinib as maintenance therapy in patients with non-progressive disease following first-line platinum-doublet chemotherapy, progression-free survival after randomization was tested in all patients at a significance level of $\alpha_o = 0.015$ and at a level of $\alpha_s = 0.01$ in patients whose tumors showed EGFR protein overexpression. The efficiency of the analysis can be improved using less stringent significance levels that incorporate the correlation between the overall and subgroup tests [79,80].

With the splitting of alpha across these two tests, we can position them as constituting a *co-primary* analysis, or alternatively, the subgroup test can be prepared as a fallback test. In the latter *fallback method*, the overall population is tested first and the subgroup test is performed if the first overall test is not significant [10,65] (Figure 13.3c). The operating characteristics of the fallback method would be close to those of the co-primary analysis when a significant overall test is prioritized more highly when interpreting results in the co-primary analysis. Irrespective of the co-primary or fallback analyses, the overall test is likely to be significant if there is a large treatment effect in M+ patients, but no effect in M− patients. Thus, the unconditional use of the overall test has the issue of possible over-assertion of treatment efficacy in M− patients, as noted in Section 13.3.2.1, or the inflation of alpha under Null-Scenario 2. Note that this cannot be resolved by invoking a strong control (e.g., by employing a close testing procedure) for testing $H_0^{(o)}$ and $H_0^{(+)}$. Accordingly, a post-hoc assessment on treatment efficacy in M− patients is therefore warranted to protect future M− patients from overtreatment [73–76].

To address this issue, Freidlin et al. [72] proposed a sequential method called the marker sequential test (MaST) to achieve strict alpha control under Null-Scenario 2, as well as under Null-Scenario 1, like in the fixed-sequence-1 method. Specifically, treatment efficacy in the M+ patients is first tested using a reduced significance level, such as $\alpha_s = 0.022$. If this test is significant, treatment efficacy in the M− patients is tested using a significance level of $\alpha = 0.025$. Otherwise, an overall test is performed using a significance level of $\alpha_o = 0.003 (= \alpha - 0.022)$ (Figure 13.3d). The basic idea underlying alpha control in Null-Scenario 2 is to restrict the implementation of the overall test only when the treatment is found to be ineffective in M+ patients, while the use of the overall test may improve the power to detect homogeneous effects across the marker subgroups. This analysis plan can be seen as a hybrid of the fixed-sequence and split-alpha methods. Like the co-primary or fallback methods, the efficiency of this method can be improved by taking into account the correlation between the overall and subgroup tests. As discussed in Section 13.3.2.2, we have permitted the weak control with no strict alpha control under Null-Scenario 2, so that we can consider a "relaxed" MaST or a hybrid analysis plan with different significance levels, particularly $\alpha_s < 0.022$ and $\alpha_o > 0.003$.

Another split-alpha approach involves using preliminary criteria related to whether there is a treatment-marker interaction [73] or no effect of clinical importance in M− patients [77] to determine whether treatment efficacy is tested in the M+ subgroup or in the overall population (Figure 13.3e and f). The advantage of these methods is that the preliminary criteria used for patient selection can also serve as those for clinical validation of the marker; this will probably provide some evidence for the use of the marker to restrict the treatment indication to M+ patients when treatment efficacy is demonstrated in the M+ subgroup.

The important feature of the aforementioned analysis plans is that they can make two kinds of assertions regarding treatment efficacy, one pertaining to the overall population and the other regarding the M+ subpopulation. In the following section, we provide numerical evaluations on the alpha control under Null-Scenario 2, as well as the power under typical profiles of non-null treatment effects across the marker subpopulations.

13.3.2.4 Numerical Evaluations of Statistical Analysis Plans

We define the three probabilities, $P_{overall}$, P_{M+}, and $P_{success}$, for asserting treatment efficacy in the overall population, for the M+ subgroup only, and for either of them, respectively [73]. For the analysis plans in Section 13.3.2.3, $P_{success} = P_{overall} + P_{M+}$. The total probability, $P_{success}$, can be interpreted as the *probability of success* in treatment development, representing the probability that we can claim treatment efficacy in *at least* the M+ subgroup. There is a trade-off between $P_{overall}$ and P_{M+} for a given value of the total probability $P_{success}$.

The study-wise alpha under Null-Scenario 2 may correspond to $P_{overall}$ under non-null treatment effects in M+ patients and to no effect in M− patients. Smaller $P_{overall}$ but larger P_{M+} may be preferable under such a scenario. On the other hand, under scenarios where the treatment has effects of clinical importance in both M+ and M− subgroups, one may desire larger $P_{overall}$ but smaller P_{M+} so as to protect M− patients from undertreatment. In the meantime, there is a "gray" zone with relatively small effects in M− patients that may or may not be clinically meaningful. See Reference [77] for power criteria incorporating such scenarios.

All numerical evaluations of $P_{overall}$, P_{M+}, and $P_{success}$ were based on asymptotic distributions of log-rank statistics for the overall population or marker-based subgroups. We supposed the use of a stratified log-rank test on treatment efficacy in the overall population and an unstratified log-rank test in marker-defined subgroups. The interaction test was based on the difference between the M+ and M− subgroups in terms of the within-subgroup log-hazard ratios (see Appendix at the end of this chapter).

Concerning the statistical analysis plans, we consider the six methods in Figure 13.3 with strict alpha control, $\alpha = 0.025$, under Null-Scenario 1 of the global null based on the asymptotic distributions of the test statistics. For all alpha-split methods (Figure 13.3c–f), the correlations between the test statistics are incorporated in determining the significance levels of the respective tests. For comparison, we also considered the traditional design that spends all of α for the overall test, where $P_{M+} = 0$ and $P_{overall} = P_{success}$. We set the significance level for the overall test at $\alpha_o = 0.015$ (60% of $\alpha = 0.025$) in the fallback method, the hybrid method, the method with the preliminary test on the treatment-marker interaction, and the method with the preliminary futility criteria in M− patients (Figure 13.3c–f).

The association of $P_{overall}$, P_{M+}, and $P_{success}$ with the total number of events, E, in the clinical trial may provide guidance for the power analysis mentioned above or for sample size determination to identify the alpha and power requirements (Figure 13.4). R-codes for this association analysis in Figure 13.3 are available upon request to the authors. Here, for survival-time outcomes, the marker prevalence pertains to the number of events, not the number of patients, in the clinical trial. We can suppose that the marker proportion in terms of the number of events is equivalent to that in terms of the number of patients in situations where the number of events is slightly fewer than the number of patients under adequate follow-up for advanced diseases or where the event rates are comparable across marker subgroups. In the following illustration, we set the marker prevalence as $p = 0.4$.

We first ascertained alpha control under Null-Scenario 1 of the global null. Table 13.1 summarizes the three probabilities under this scenario. $P_{success}$ corresponds to the study-wise alpha with $P_{success} = \alpha = 0.025$ for all six methods. It is worthwhile to check the component probabilities, $P_{overall}$ and P_{M+}, which can be regarded as alpha *spent* for claiming treatment efficacy for the overall population and that spent for the M+ subgroup, respectively. These probabilities may characterize the operating characteristics under non-null treatment effects to some degree, i.e., a larger $P_{overall}$ (or smaller P_{M+}) under the global null may lead to a larger $P_{overall}$ (or smaller P_{M+}) under non-null treatment effects. The fallback, hybrid, and

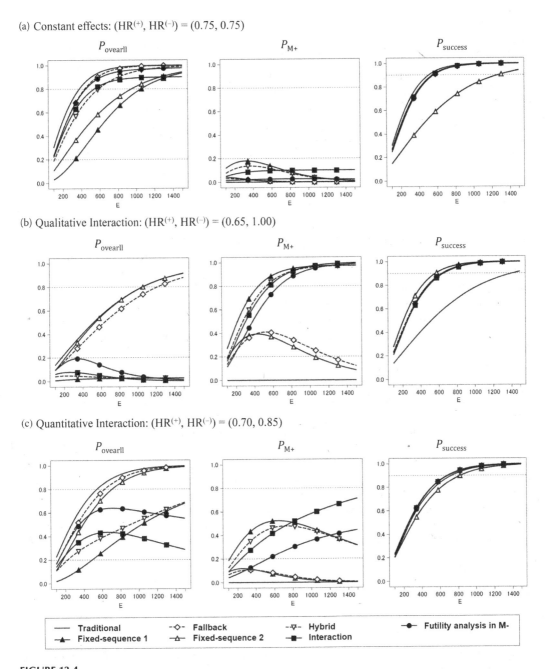

FIGURE 13.4

Curves of the three probabilities for the total number of events E under $(HR^{(+)}, HR^{(-)}) = (0.75, 0.75)$, $(0.65, 1.00)$, and $(0.70, 0.85)$: the significance level for the overall test $\alpha_o = 0.015$ in the fallback and hybrid methods, the method with the preliminary test on the treatment-marker interaction, and the method with the preliminary futility analysis in M− patients. The significance level $\alpha_{int} = 0.1$ was used for the interaction test. The futility boundary $c^{(-)} = \exp(0.0447) = 1.05$ was used for the observed HR for M− patients derived from the condition that a Bayesian posterior probability that the treatment has at least a minimum effect, $HR = 0.8$, under a non-informative prior is less than 0.15 (see Reference [77]). All the methods controlled the study-wise alpha $\alpha = 0.025$ under Null-Scenario-1 of the global null (see Table 13.1).

TABLE 13.1

$P_{overall}$, P_{M+}, $P_{success}$ Under the Global Null (HR$^{(+)}$, HR$^{(-)}$) = (1.0, 1.0) Under a Marker Prevalence of 0.4.

Prob.	Traditional	Fixed-Sequence-1	Fixed-Sequence-2	Fallback	Hybrid	Preliminary Interaction Test	Preliminary Futility Criteria in Marker Negatives
$P_{overall}$	0.025	0.001	0.007	0.015	0.015	0.015	0.010
P_{M+}	0.000	0.024	0.018	0.010	0.010	0.010	0.015
$P_{success}$	0.025	0.025	0.025	0.025	0.025	0.025	0.025

interaction-based methods had the largest $P_{overall}$ of 0.015 under the global null (reflecting the specification of $\alpha_0 = 0.015$).

Concerning the profiles of non-null treatment effects across the marker subgroups, we first consider two extreme scenarios, i.e., constant effects, (HR$^{(+)}$, HR$^{(-)}$) = (0.75, 0.75), and qualitative interaction with no treatment effect in M$-$ patients, (HR$^{(+)}$, HR$^{(-)}$) = (0.65, 1.0). Figure 13.4a and b show curves of $P_{overall}$, P_{M+}, and $P_{success}$ for various values of the total number of events, E, under these scenarios.

For constant effects, many methods provided relatively large $P_{overall}$, but the fixed-sequence methods suffered from low $P_{overall}$ (and $P_{success}$). Meanwhile, all the methods generally maintained fairly small P_{M+}, representing good control of the risk of undertreatment in M$-$ patients.

Under the scenario with qualitative interaction (or Null-Scenario 2), the fixed-sequence-1 method performed best, providing the largest P_{M+} and very small $P_{overall}$, reflecting strong control. In contrast, the fixed-sequence-2 and fallback (and traditional) methods performed worst, with very large $P_{overall}$ (or high risk of overtreatment in M$-$ patients) and small P_{M+}.

The hybrid method and methods based on the preliminary interaction test or futility criteria in M$-$ patients generally performed well in both of the extreme scenarios.

Under the scenario of quantitative interaction with (HR$^{(+)}$, HR$^{(-)}$) = (0.7, 0.85), where the effect size in M$-$ patients has borderline clinical meaning, the characteristics of the respective methods became clearer (Figure 13.4c). The power curves for the fallback and fixed-sequence-2 methods were relatively similar to those in the previous extreme scenarios, while those for the other methods were more sensitive to the change in the underlying treatment effect profiles.

Lastly, we note that many methods, with the exception of the fixed-sequence-2 and traditional methods, could provide fairly large and comparable $P_{success}$ in all the scenarios investigated here.

13.3.2.5 Interim Analysis

In marker-stratified trials, an interim analysis for non-efficacy or futility would be particularly warranted for M$-$ patients with presumably limited treatment efficacy, in order to protect them from unnecessary treatments and follow-ups (see Figure 13.5a). On the other hand, for M+ patients with presumably high treatment efficacy, an interim analysis for efficacy or superiority could be worthwhile in order to quickly deliver superior treatment to these patients.

In addition to its associated ethical benefits, interim analysis can enhance the efficiency of the marker-stratified trial. In particular, the statistical power can be improved by using interim trial data to adaptively narrow the patient population down to a patient

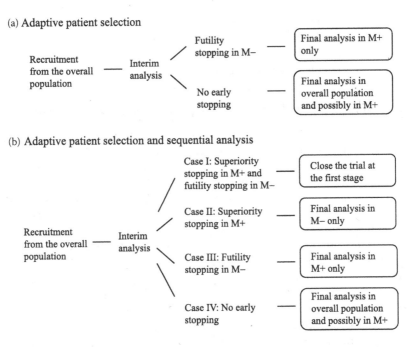

FIGURE 13.5
Two-stage, marker-stratified designs involving adaptive patient selection across M+ and M− patients.

subpopulation that can benefit from the treatment. Note that if the enrollment and follow-up are curtailed in M− patients, they may not be affected in M+ patients, unlike in the adaptive-enrichment design where the criteria for patient enrollment can be changed to recruit more M+ patients.

Special attention is needed for handling time-to-event endpoints typically evaluated in the primary analysis of phase III trials. In this situation, flexible interim adaptations are generally precluded due to the specific difficulty in assuring the independence of an adaptation from any subsidiary information derived from censored patients who have not experienced an event at the time of the interim analysis [81]. Special techniques may be needed to address this difficulty [82,83]. In many cases, the adaptation rule, including the information fraction at the interim analysis, may be pre-specified based on the primary test statistic on the time-to-event endpoint. This is essentially equivalent to the traditional group sequential analysis.

Brannath et al. [84] considered a futility stopping rule for M− patients to determine which, of the overall patients or those who are only M+, are followed and analyzed at the end of the trial. Karuri and Simon [85] proposed a marker-based, two-stage Bayesian design involving interim analysis to stop accrual of M− patients or accrual of all patients. Taking into account the complicated nature of interim monitoring across subgroups under possible marker assumptions, Redman et al. [86] considered subgroup-focused interim monitoring with a futility-monitoring plan to determine early stopping in M− patients based on evaluation within the M+ and M− subgroups as well as the entire study population.

Subgroup-based interim analysis with distinct stopping criteria within the marker subgroups can motivate the reconstruction of the all-comers, marker-stratified trial based on a combination of two concurrent enrichment trials, one for M+ patients and the other for

M− patients. Recently, Matsui and Crowley [68] proposed such a design, shown in Figure 13.5b. It allows for both sequential assessment across the subgroups (Case II), as well as adaptive patient selection (Cases III–IV), while retaining assessment of the entire patient dataset at the final analysis stage (Case IV).

13.3.2.6 *Unstratified Trials with Adaptive Designs for Marker Development and Validation*

Although we have discussed phase III trials that use a completely pre-specified predictive marker, it is quite common for such a marker to be unavailable when initiating the definitive phase III trial. This is the case when a single marker may be available, but no threshold for marker positivity is defined before the phase III trial. Another case is when a single predictive marker is unavailable, but data on several candidate markers or even tens of thousands of genomic markers are measured for pre-treatment tissue specimens before the trial or are scheduled to be measured during the trial. When high-dimensional, genomic marker data are available, there are no *a priori* credentials for predictive markers from among the large number of genomic markers. One approach to these cases is to prospectively design and analyze the randomized trial in such a way that both developing a predictive marker (or signature) and testing treatment efficacy based on the developed marker are conducted in a valid manner.

With this approach, unlike a marker-stratified trial with a pre-specified marker, all-comers trials may not be stratified by any markers. Of note, unstratified randomization does not necessarily diminish the validity of inference regarding treatment efficacy within marker-defined subgroups with moderate to large sizes. Besides, unstratified randomization does not require that the marker value be specified before randomization, so the marker evaluation can be delayed until the time of the primary analysis. However, careful consideration of missing marker data is needed to ensure collection of sufficient numbers of patients with observed marker values and also to prevent selection bias, i.e., dependence of missing measurements on the treatment and other clinical variables.

In the absence of a pre-specified marker at the initiation of the trial, the analysis for marker development and validation must be positioned as a fallback option that spends a small portion of $\alpha = 0.025$, α^*, say 0.005, following no significance in an overall test that spends the main portion of $\alpha - \alpha^*$, say 0.02. The outstanding features of this approach are the application of an optimization or prediction algorithm to develop a marker or signature to identify an appropriate marker subgroup in the development stage and the implementation of a *single* test on treatment efficacy within an identified "M+" subgroup based on the developed marker in the validation stage. The latter feature is in contrast to the traditional exploratory subgroup analysis with multiple tests across subgroups. All the elements in the analysis for marker development and validation must be prospectively defined and specified in the statistical analysis plan.

One application of such an approach, called the *adaptive threshold design* [87], has been considered in determining the appropriate threshold for positivity for a single marker in the fallback stage.

Another application, called the *adaptive signature design* [88], is to develop a predictor or predictive signature using a large number of covariates, possibly high-dimensional genomic markers, in the fallback stage. In the fallback analysis, the full set of patients in the clinical trial is partitioned into a training set and a validation set by the split-sample method. An algorithmic analysis plan is applied to the training set to generate a predictor. The predictor developed using the training data is used to make a prediction for each patient in the validation set. Then, the treatment efficacy is tested in the subset of patients who are predicted to be responsive to the treatment in the validation set. The efficiency of this design, based on the split-sample prediction analysis, can be enhanced by applying cross-validation rather than the split-sample method [89]. A variant of this design that estimates a function of treatment effect based on a continuous, cross-validated predictive score has also been proposed [90].

Another direction is to delay the evaluation of treatment efficacy within a marker-based subgroup until a future time when a predictive marker or signature is developed for the treatment [69,91]. With this approach, one could reserve a small portion of the total alpha for a single test of treatment effect in the subgroup to be determined in the future [69]. This approach is applicable to clinical trials that archive pre-treatment specimens for marker evaluation. At the time of evaluating the new marker in the future, the analysis plan will be prospectively specified to retrospectively utilize and analyze the archived specimens, so it is called the *prospective-retrospective approach*. See Reference [91] for appropriate conditions for conducting this approach.

13.3.3 Strategy Designs

The strategy designs are characterized as a comparison of two strategies in which treatment determination is performed with or without the use of the new marker (Figure 13.6a). One example is a randomized trial for non-small cell lung cancer that compares two strategies: the first exclusively uses a standard treatment (cisplatin + docetaxel); the second is a biomarker-based strategy in which patients diagnosed as resistant to the standard treatment based on the marker are treated with an experimental treatment (gemcitabine + docetaxel), while the rest are treated with the standard treatment [92]. Another example is a randomized trial that compares, on the one hand, a strategy for determining targeted treatments based on tumor chemosensitivity assays or a set of predictive markers and, on the other, with a strategy of using physician's choice of chemotherapy based on standard practice [93,94].

One limitation of these trials is that the marker-based arm can perform better if an experimental treatment guided by a marker is efficacious, regardless of whether the marker is predictive or not. A modification is proposed in which patients in the non-marker-based arm undergo a second randomization to receive one of the same treatments being used in the marker-based arm [62] (Figure 13.6b). By measuring the marker status in all of the patients, this modified design allows clinical validation of the marker as a predictive marker through comparing treatment effects across the marker-based subsets of patients.

Strategy designs may include patients treated with the same treatment in both the marker-based and the non-marker-based arms, resulting in a large overlap in the number of patients receiving the same treatment within the two strategies being compared. Thus, a very large number of patients are required to be randomized to detect a diluted, small overall difference in the endpoint between the two arms.

One modification is to randomize the two strategies to only the patients for whom the two treatments guided by the two strategies differ (Figure 13.6c). This modification requires measurement of the biomarker in all of the patients before randomization. The modified design was employed in a randomized clinical trial called the MINDACT study [59]. In this trial, a marker-based strategy based on the MammaPrint prognostic signature was compared to a strategy based on standard clinical prognostic factors in order to determine whether or not to utilize chemotherapy in women with node-negative, estrogen-receptor-positive breast cancer; discordant cases between the two strategies were subject to randomization [59]. Another possible approach to evaluating the clinical utility of prognostic markers in early-stage cancers is to demonstrate an excellent outcome, e.g., sufficiently low rates

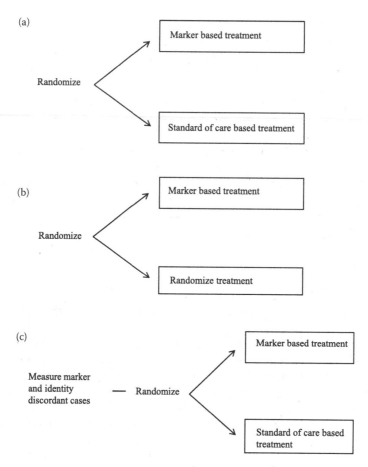

FIGURE 13.6
Strategy designs.

of cancer recurrence that prevent the need for additional chemotherapy, in a particular
marker-defined (low-risk) subpopulation [59,60].

13.4 Conclusion

As we have seen so far, with rapid advances in biotechnology and molecular oncology,
the methodologies of cancer clinical trials have dramatically evolved in recent years
and have included the advent of a new framework based on master protocols. However,
owing to the complex biology of human cancer, we may also need to develop innovative
approaches to multi-directionally characterize the biology and heterogeneity within and
across histologically defined cancers. The incorporation of various molecular data derived
from the large number of markers obtained through these approaches will be of great
importance in the future evolution of clinical trial methodologies and infrastructures. In
clinical trials with sequential or concurrent combinational regimens, these molecular data
will also be crucial for understanding the mechanism of resistance to targeted treatments

and for developing optimal treatment regimens that address the treatment resistance, possibly based on predictive markers.

Appendix: Asymptotic Distributions of the Test Statistics

In a marker-stratified trial with survival outcomes, we assume a proportional hazards model between treatment arms. We employ a log-rank test for treatment comparison and assume an asymptotic distribution for the log-rank statistic, $S \sim N(\theta, 4/E)$, under equal treatment assignment and follow-up [95]. Here, θ is the logarithm of the ratio of the hazard function under the new treatment relative to that under the control treatment (a negative value indicates that the treatment is efficacious), and E is the total number of events observed. For a given number of events, we express a standardized test statistic for testing treatment efficacy in M+ patients as $Z^{(+)} = \hat{\theta}^{(+)}/\sqrt{V^{(+)}}$, where $\hat{\theta}^{(+)}$ is an estimate of $\theta^{(+)}$ and $V^{(+)} = 4/E^{(+)}$, where $E^{(+)}$ is the number of events in the M+ subgroup. We also consider a similar standardized statistic, $Z^{(-)}$, for the M− patients. For testing overall treatment efficacy, we use the standardized test statistic, $Z^{(o)} = \hat{\theta}^{(o)}/\sqrt{V^{(o)}}$, where $\hat{\theta}^{(o)}$ is an estimate of $\theta^{(o)}$ and $V^{(o)} = 4/E^{(o)} = 4/(E^{(+)} + E^{(-)})$. By using an approximation,

$$\hat{\theta}^{(o)} \approx \frac{\{(1/V^{(+)})\hat{\theta}^{(+)} + (1/V^{(-)})\hat{\theta}^{(-)}\}}{(1/V^{(+)} + 1/V^{(-)})} = \frac{(E^{(+)}\hat{\theta}^{(+)} + E^{(-)}\hat{\theta}^{(-)})}{(E^{(+)} + E^{(-)})},$$

we have a stratified statistic for the overall test that incorporates possible prognostic effects of the marker:

$$Z^{(o)} = \frac{\sqrt{V^{(o)}}}{V^{(+)}}\hat{\theta}^{(+)} + \frac{\sqrt{V^{(o)}}}{V^{(-)}}\hat{\theta}^{(-)}$$

For testing a treatment-by-marker interaction in the interaction-based analysis plan, we use the following test statistic:

$$Z^{(int)} = \frac{\hat{\theta}^{(+)} - \hat{\theta}^{(-)}}{\sqrt{V^{(+)} + V^{(-)}}}$$

We assume normality for the aforementioned standardized test statistics with variance 1. The means of $Z^{(+)}$, $Z^{(-)}$, $Z^{(o)}$, and $Z^{(int)}$ are $\theta^{(+)}/\sqrt{V^{(+)}}$, $\theta^{(-)}/\sqrt{V^{(-)}}$, $\sqrt{V^{(o)}}(\theta^{(+)}/V^{(+)} + \theta^{(-)}/V^{(-)})$, and $(\theta^{(+)} - \theta^{(-)})/\sqrt{V^{(+)} + V^{(-)}}$, respectively.

Regarding the covariance (or correlation) between the test statistics, $Z^{(+)}$ and $Z^{(-)}$ can be assumed to be mutually independent. The covariance between $Z^{(+)}$ and $Z^{(o)}$ may reduce to $\sqrt{p}$ [79,80], where it is generally reasonable to use an observed marker prevalence for p. The covariances between the interaction test and overall/subpopulation tests in the interaction-based method can be expressed as

$$\text{cov}(Z^{(int)}, Z^{(o)}) = 0 \quad \text{and} \quad \text{cov}(Z^{(int)}, Z^{(+)}) = \sqrt{\frac{V^{(+)}}{(V^{(+)} + V^{(-)})}} = \sqrt{\frac{E^{(-)}}{(E^{(+)} + E^{(-)})}} = \sqrt{\frac{R}{(1 + R)}},$$

where $R = E^{(-)}/E^{(+)}$ [73]. When $E^{(+)} = pE$ or $R = (1 - p)/p$ can be supposed, we have $\mathrm{cov}(Z^{(\mathrm{int})}, Z^{(+)}) = \sqrt{1-p}$. Generally, we search for the significance levels based on the covariance $\mathrm{cov}(Z^{(\mathrm{int})}, Z^{(+)}) = \sqrt{R/(1+R)}$ using an expected event ratio R under the global null effects, which will depend on the respective (baseline) event rates (possibly with some prognostic effects) and the censoring distributions across marker subpopulations, rather than using the approximations $R = (1 - p)/p$ and $\mathrm{cov}(Z^{(\mathrm{int})}, Z^{(+)}) = \sqrt{1-p}$ [73].

Under the global null hypothesis of no treatment efficacy in the M+ and M− patients (and thus no efficacy in the overall population), we search for the significance levels α_o (for $Z^{(o)}$), α_s (for $Z^{(+)}$), and α_{int} (for $Z^{(\mathrm{int})}$) to control the study-wise type I error rate in the fallback, hybrid, and interaction-based methods.

In the method with preliminary futility criteria in the M− patients [77], the criterion using the boundary point $c^{(-)}$ for the observed HR in the M− patients (see the caption in Figure 13.4) can be expressed as

$$Z^{(-)} = \frac{\hat{\theta}^{(-)}}{\sqrt{V^{(-)}}} > \frac{c^{(-)}}{\sqrt{V^{(-)}}} = c_Z^{(-)}$$

The study-wise alpha rate under the global null hypothesis, $H_{\mathrm{G},0}$, is given by

$$\Pr\left(Z^{(-)} > c_Z^{(-)} \,\&\, Z^{(+)} < c_Z^{(+)} \mid H_{\mathrm{G},0}\right) + \Pr\left(Z^{(-)} \leq c_Z^{(-)} \,\&\, Z^{(o)} \leq c_Z^{(o)} \mid H_{\mathrm{G},0}\right).$$

The former and latter components correspond to $P_{\mathrm{M+}}$ and P_{overall}, respectively. The calculation of $P_{\mathrm{M+}}$ is based on independence between $Z^{(+)}$ and $Z^{(-)}$, while that of P_{overall} is based on a multivariate normal distribution with $\mathrm{cov}(Z^{(o)}, Z^{(-)}) = \sqrt{1-p}$. We can also similarly calculate these probabilities for non-null effects [77].

Acknowledgments

This research was supported by a Grant-in-Aid for Scientific Research (16H06299) and JST-CREST (JPMJCR1412) from the Ministry of Education, Culture, Sports, Science and Technology of Japan.

References

1. Matsui S, Buyse M, Simon R. *Design and Analysis of Clinical Trials for Predictive Medicine*. Boca Raton, FL: Chapman and Hall/CRC Press, 2015.
2. Chapman PB, Hauschild A, Robert C et al. Improved survival with vemurafenib in melanoma with BRAF V600E mutation. *N Engl J Med*. 2011;364:2507–16.
3. Slamon DJ, Leyland-Jones B, Shak S et al. Use of chemotherapy plus a monoclonal antibody against HER2 for metastatic breast cancer that overexpresses HER2. *N Engl J Med*. 2001;344:783–92.
4. Wolff AC, Hammond ME, Hicks DG et al. Recommendations for human epidermal growth factor receptor 2 testing in breast cancer: American Society of Clinical Oncology/College of American Pathologists clinical practice guideline update. *J Clin Oncol*. 2013;31:3997–4013.

5. McShane LM, Cavenagh MM, Lively TG et al. Criteria for the use of omics-based predictors in clinical trials: Explanation and elaboration. *BMC Med.* 2013;11:220.

6. Le Tourneau C, Kamal M, Tsimberidou AM et al. Treatment algorithms based on tumor molecular profiling: The essence of precision medicine trials. *J Natl Cancer Inst.* 2015;108:1–10.

7. Rodon J, Soria JC, Berger R et al. Challenges in initiating and conducting personalized cancer therapy trials: Perspectives from WINTHER, a Worldwide Innovative Network (WIN) Consortium trial. *Ann Oncol.* 2015;26:1791–8.

8. Fridlyand J, Yeh RF, Mackey H, Bengtsson T, Delmar P, Spaniolo G, Lieberman G. An industry statistician's perspective on PHC drug development. *Contemp Clin Trials.* 2013;36:624–35.

9. Fridlyand J, Simon RM, Walrath JC et al. Considerations for the successful co-development of targeted cancer therapies and companion diagnostics. *Nat Rev Drug Discov.* 2013;12:743–55.

10. Simon R. Clinical trial designs for evaluating the medical utility of prognostic and predictive biomarkers in oncology. *Pers Med.* 2010;7:33–47.

11. Teutsch SM, Bradley LA, Palomaki GE, Haddow JE, Piper M, Calonge N, Dotson WD, Douglas MP, Berg AO; EGAPP Working Group. The Evaluation of Genomic Applications in Practice and Prevention (EGAPP) Initiative: Methods of the EGAPP Working Group. *Genet Med.* 2009;11:3–14.

12. Biankin AV, Piantadosi S, Hollingsworth SJ. Patient-centric trials for therapeutic development in precision oncology. *Nature.* 2015;526:361–70.

13. Simon R. Genomic alteration-driven clinical trial designs in oncology. *Ann Intern Med.* 2016;165:270–8.

14. Berry DA. The Brave New World of clinical cancer research: Adaptive biomarker-driven trials integrating clinical practice with clinical research. *Mol Oncol.* 2015;9:951–9.

15. Lacombe D, Burock S, Bogaerts J, Schoeffski P, Golfinopoulos V, Stupp R. The dream and reality of histology agnostic cancer clinical trials. *Mol Oncol.* 2014;8:1057–63.

16. Renfro LA, Sargent DJ. Statistical controversies in clinical research: Basket trials, umbrella trials, and other master protocols: A review and examples. *Ann Oncol.* 2017;28:34–43.

17. Mandrekar SJ, An MW, Sargent DJ. A review of phase II trial designs for initial marker validation. *Contemp Clin Trials.* 2013;36:597–604.

18. Simon R. Optimal two-stage designs for phase II clinical trials. *Control Clin Trials.* 1989;10:1–10.

19. Pusztai L, Anderson K, Hess KR. Pharmacogenomic predictor discovery in phase II clinical trials for breast cancer. *Clin Cancer Res.* 2007;13:6080–6.

20. Jones CL, Holmgren E. An adaptive Simon two-stage design for phase 2 studies of targeted therapies. *Contemp Clin Trials.* 2007;28:654–61.

21. Parashar D, Bowden J, Starr C, Wernisch L, Mander A. An optimal stratified Simon two-stage design. *Pharm Stat.* 2016;15:333–40.

22. Thall PF, Simon R. Practical Bayesian guidelines for phase IIB clinical trials. *Biometrics.* 1994;50:337–49.

23. Korn EL, Liu PY, Lee SJ et al. Meta-analysis of phase II cooperative group trials in metastatic stage IV melanoma to determine progression-free and overall survival benchmarks for future phase II trials. *J Clin Oncol.* 2008;26:527–34.

24. McShane LM, Hunsberger S, Adjei AA. Effective incorporation of biomarkers into phase II trials. *Clin Cancer Res.* 2009;15:1898–905.

25. Seymour L, Ivy SP, Sargent D et al. The design of phase II clinical trials testing cancer therapeutics: Consensus recommendations from the clinical trial design task force of the national cancer institute investigational drug steering committee. *Clin Cancer Res.* 2010;16:1764–9.

26. Rubinstein LV, Korn EL, Freidlin B, Hunsberger S, Ivy SP, Smith MA. Design issues of randomized phase II trials and a proposal for phase II screening trials. *J Clin Oncol.* 2005;23:7199–206.

27. Gan HK, Grothey A, Pond GR, Moore MJ, Siu LL, Sargent D. Randomized phase II trials: Inevitable or inadvisable? *J Clin Oncol.* 2010;28:2641–7.

28. Rubinstein L, Crowley J, Ivy P, Leblanc M, Sargent D. Randomized phase II designs. *Clin Cancer Res.* 2009;15:1883–90.

29. Freidlin B, McShane LM, Polley MY, Korn EL. Randomized phase II trial designs with biomarkers. *J Clin Oncol.* 2012;30:3304–9.

30. Follmann D. Adaptively changing subgroup proportions in clinical trials. *Stat Sin.* 1997;7: 1085–102.

31. Wang SJ, O'Neill RT, Hung HM. Approaches to evaluation of treatment effect in randomized clinical trials with genomic subset. *Pharm Stat.* 2007;6:227–44.

32. Rosenblum M, van der Laan MJ. Optimizing randomized trial designs to distinguish which subpopulations benefit from treatment. *Biometrika.* 2011;98:845–60.

33. Simon N, Simon R. Adaptive enrichment designs for clinical trials. *Biostatistics.* 2013;14:613–25.

34. Bates SE, Berry DA, Balasubramaniam S, Bailey S, LoRusso PM, Rubin EH. Advancing clinical trials to streamline drug development. *Clin Cancer Res.* 2015;21:4527–35.

35. Govindan R, Mandrekar SJ, Gerber DE et al. ALCHEMIST Trials: A golden opportunity to transform outcomes in early-stage non-small cell lung cancer. *Clin Cancer Res.* 2015;21:5439–44.

36. Kim ES, Herbst RS, Wistuba II et al. The BATTLE trial: Personalizing therapy for lung cancer. *Cancer Discov.* 2011;1:44–53.

37. Zhou X, Liu S, Kim ES, Herbst RS, Lee JJ. Bayesian adaptive design for targeted therapy development in lung cancer—a step toward personalized medicine. *Clinical Trials.* 2008;5:181–93.

38. Papadimitrakopoulou V, Lee JJ, Wistuba II et al. The BATTLE-2 study: A biomarker-integrated targeted therapy study in previously treated patients with advanced non-small-cell lung cancer. *J Clin Oncol.* 2016;34:3638–47.

39. Redman MW, Allegra CJ. The master protocol concept. *Semin Oncol.* 2015;42:724–30.

40. Park JW, Liu MC, Yee D et al. Adaptive randomization of neratinib in early breast cancer. *N Engl J Med.* 2016;375:11–22.

41. Rugo HS, Olopade OI, DeMichele A et al. Adaptive randomization of veliparib-carboplatin treatment in breast cancer. *N Engl J Med.* 2016;375:23–34.

42. Saville BR, Berry SM. Efficiencies of platform clinical trials: A vision of the future. *Clinical Trials.* 2016;13:358–66.

43. Yuan Y, Guo B, Munsell M, Lu K, Jazaeri A. MIDAS: A practical Bayesian design for platform trials with molecularly targeted agents. *Stat Med.* 2016;35:3892–906.

44. Harrington D, Parmigiani G. I-SPY 2 – A glimpse of the future of phase 2 drug development? *N Engl J Med.* 2016;375:7–9.

45. Korn EL, Freidlin B. Outcome-adaptive randomization: Is it useful? *J Clin Oncol.* 2011;29:771–6.

46. Thall P, Fox P, Wathen J. Statistical controversies in clinical research: Scientific and ethical problems with adaptive randomization in comparative clinical trials. *Ann Oncol.* 2015;26:1621–8.

47. Berry DA, Herbst RS, Rubin EH. Reports from the 2010 Clinical and Translational Cancer Research Think Tank meeting: Design strategies for personalized therapy trials. *Clin Cancer Res.* 2012;18:638–44.

48. Hyman DM, Puzanov I, Subbiah V et al. Vemurafenib in multiple nonmelanoma cancers with BRAF V600 mutations. *N Engl J Med.* 2015;373:726–36.

49. Abrams J, Conley B, Mooney M et al. *National Cancer Institute's Precision Medicine Initiatives for the new National Clinical Trials Network.* ASCO Educational Book, 2014, pp. 71–6.

50. Leblanc M, Rankin C, Crowley J. Multiple histology phase II trials. *Clin Cancer Res.* 2009;15:4256–62.

51. Roberts JD, Ramakrishnan V. Phase II trials powered to detect tumor subtypes. *Clin Cancer Res.* 2011;17:5538–45.

52. Cunanan KM, Iasonos A, Shen R, Begg CB, Gönen M. An efficient basket trial design. *Stat Med.* 2017;36:1568–79.

53. Liu R, Liu Z, Ghadessi M, Vonk R. Increasing the efficiency of oncology basket trials using a Bayesian approach. *Contemp Clin Trials.* 2017;63:67–72.

54. Thall PF, Wathen JK, Bekele BN, Champlin RE, Baker LH, Benjamin RS. Hierarchical Bayesian approaches to phase II trials in diseases with multiple subtypes. *Stat Med.* 2003;22:763–80.

55. Berry SM, Broglio KR, Groshen S, Berry DA. Bayesian hierarchical modeling of patient subpopulations: Efficient designs of Phase II oncology clinical trials. *Clinical Trials.* 2013;10:720–34.

56. Freidlin B, Korn EL. Borrowing information across subgroups in phase II trials: Is it useful? *Clin Cancer Res.* 2013;19:1326–34.

57. Simon R, Geyer S, Subramanian J, Roychowdhury S. The Bayesian basket design for genomic variant-driven phase II trials. *Semin Oncol.* 2016;43:13–8.

58. Simon R. New designs for basket clinical trials in oncology. *J Biopharm Stat.* 2018;28:245–55.

59. Cardoso F, van't Veer LJ, Bogaerts J et al. 70-gene signature as an aid to treatment decisions in early-stage breast cancer. *N Engl J Med.* 2016;375:717–29.

60. Sparano JA, Gray RJ, Makower DF et al. Prospective validation of a 21-gene expression assay in breast cancer. *N Engl J Med.* 2015;373:2005–14.

61. Hoering A, Leblanc M, Crowley JJ. Randomized phase III clinical trial designs for targeted agents. *Clin Cancer Res.* 2008;14:4358–67.

62. Mandrekar SJ, Sargent DJ. Clinical trial designs for predictive biomarker validation: Theoretical considerations and practical challenges. *J Clin Oncol.* 2009;27:4027–34.

63. Freidlin B, McShane LM, Korn EL. Randomized clinical trials with biomarkers: Design issues. *J Natl Cancer Inst.* 2010;102:152–60.

64. Buyse M, Michiels S, Sargent DJ, Grothey A, Matheson A, de Gramont A. Integrating biomarkers in clinical trials. *Expert Rev Mol Diagn.* 2011;11:171–82.

65. Freidlin B, Korn EL. Biomarker enrichment strategies: Matching trial design to biomarker credentials. *Nat Rev Clin Oncol.* 2014;11:81–90.

66. Simon R, Maitournam A. Evaluating the efficiency of targeted designs for randomized clinical trials. *Clin Cancer Res.* 2004;10:6759–63.

67. Shaw AT, Kim DW, Nakagawa K et al. Crizotinib versus chemotherapy in advanced ALK-positive lung cancer. *N Engl J Med.* 2013;368:2385–94.

68. Matsui S, Crowley J. Biomarker-stratified phase III clinical trials: Enhancement with a subgroup-focused sequential design. *Clin Cancer Res.* 2018;24:994–1001.

69. Simon R, Matsui S, Buyse M. Clinical trials for predictive medicine: New paradigms and challenges. In: *Design and Analysis of Clinical Trials for Predictive Medicine.* S Matsui, M Buyse, R Simon, eds. Boca Raton, FL: Chapman and Hall/CRC Press, 2015:3–10.

70. Liu A, Liu C, Li Q, Yu KF, Yuan VW. A threshold sample-enrichment approach in a clinical trial with heterogeneous subpopulations. *Clinical Trials.* 2010;7:537–45.

71. Rothmann MD, Zhang JJ, Lu L, Fleming TR. Testing in a prespecified subgroup and the intent-to-treat population. *Drug Inf J.* 2012;46:175–9.

72. Freidlin B, Korn EL, Gray R. Marker Sequential Test (MaST) design. *Clinical Trials.* 2014;11:19–27.

73. Matsui S, Choai Y, Nonaka T. Comparison of statistical analysis plans in randomize-all phase III trials with a predictive biomarker. *Clin Cancer Res.* 2014;20:2820–30.

74. Simon RM. *Genomic Clinical Trials and Predictive Medicine.* Cambridge: Cambridge University Press, 2013.

75. Millen BA, Dmitrienko A, Ruberg S, Shen L. A statistical framework for decision making in confirmatory multipopulation tailoring clinical trials. *Drug Inf J.* 2012;46:647–56.

76. Millen BA, Dmitrienko A, Song G. Bayesian assessment of the influence and interaction conditions in multipopulation tailoring clinical trials. *J Biopharm Stat.* 2014;24:94–109.

77. Nonaka T, Igeta M, Matsui S. Statistical testing strategies for assessing treatment efficacy and marker accuracy in phase III trials. *Pharm Stat.* 2019, in Press.

78. Cappuzzo F, Ciuleanu T, Stelmakh L et al. Erlotinib as maintenance treatment in advanced non-small-cell lung cancer: A multicentre, randomised, placebo-controlled phase 3 study. *Lancet Oncol.* 2010;11:521–9.

79. Song Y, Chi GY. A method for testing a prespecified subgroup in clinical trials. *Stat Med.* 2007;26:3535–49.

80. Spiessens B, Debois M. Adjusted significance levels for subgroup analyses in clinical trials. *Contemp Clin Trials.* 2010;31:647–56.

81. Bauer P, Posch M. Letter to the editor: Modification of the sample size and the schedule of interim analyses in survival trials based on data inspections, by H. Schäfer and H.-H. Müller, Statistics in Medicine 2001; 20: 3741–51. *Stat Med.* 2004;23:1333–4.

82. Jenkins M, Stone A, Jennison C. An adaptive seamless phase II/III design for oncology trials with subpopulation selection using correlated survival endpoints. *Pharm Stat.* 2011;10:347–56.

83. Mehta C, Schäfer H, Daniel H, Irle S. Biomarker driven population enrichment for adaptive oncology trials with time to event endpoints. *Stat Med*. 2014;33:4515–31.

84. Brannath W, Zuber E, Branson M, Bretz F, Gallo P, Posch M, Racine-Poon A. Confirmatory adaptive designs with Bayesian decision tools for a targeted therapy in oncology. *Stat Med*. 2009;28:1445–63.

85. Karuri S, Simon R. A two-stage Bayesian design for co-development of new drugs and companion diagnostics. *Stat Med*. 2012;31:901–14.

86. Redman MW, Crowley JJ, Herbst RS, Hirsch FR, Gandara DR. Design of a phase III clinical trial with prospective biomarker validation: SWOG S0819. *Clin Cancer Res*. 2012;18:4004–12.

87. Jiang W, Freidlin B, Simon R. Biomarker-adaptive threshold design: A procedure for evaluating treatment with possible biomarker-defined subset effect. *J Natl Cancer Inst*. 2007;99:1036–43.

88. Freidlin B, Simon R. Adaptive signature design: An adaptive clinical trial design for generating and prospectively testing a gene expression signature for sensitive patients. *Clin Cancer Res*. 2005;11:7872–8.

89. Freidlin B, Jiang W, Simon R. The cross-validated adaptive signature design. *Clin Cancer Res*. 2010;16:691–8.

90. Matsui S, Simon R, Qu P, Shaughnessy JD Jr, Barlogie B, Crowley J. Developing and validating continuous genomic signatures in randomized clinical trials for predictive medicine. *Clin Cancer Res*. 2012;18:6065–73.

91. Simon RM, Paik S, Hayes DF. Use of archived specimens in evaluation of prognostic and predictive biomarkers. *J Natl Cancer Inst*. 2009;101:1446–52.

92. Cobo M, Isla D, Massuti B et al. Customizing cisplatin based on quantitative excision repair cross-complementing 1 mRNA expression: A phase III trial in non-small-cell lung cancer. *J Clin Oncol*. 2007;25:2747–54.

93. Cree IA, Kurbacher CM, Lamont A, Hindley AC, Love S; TCA Ovarian Cancer Trial Group. A prospective randomized controlled trial of tumour chemosensitivity assay directed chemotherapy versus physician's choice in patients with recurrent platinum-resistant ovarian cancer. *Anticancer Drugs*. 2007;18:1093–101.

94. Le Tourneau C, Delord JP, Gonçalves A et al. Molecularly targeted therapy based on tumour molecular profiling versus conventional therapy for advanced cancer (SHIVA): A multicentre, open-label, proof-of-concept, randomised, controlled phase 2 trial. *Lancet Oncol*. 2015;16:1324–34.

95. Tsiatis AA. The asymptotic joint distribution of the efficient score test for the proportional hazards model calculated over time. *Biometrika*. 1981;68:311–5.

14

Genomic Biomarker Clinical Trial Designs

Richard Simon

CONTENTS

14.1 Introduction

The discovery of recurrent somatic alterations in human tumors has had a major impact on drug discovery and development. Cancer drugs are now developed to inhibit the constitutively activated proteins that are the product of mutated oncogenes. New clinical trial designs have been developed in recognition of the importance of genomic alteration instead of the focus on histologic type of the tumor cell. This chapter reviews some of these clinical trial designs.

14.2 Phase II Designs

Many cancer drugs today are developed to inhibit a protein that is the product of a gene that is mutated or amplified in cancer cells. The drug should be effective against tumors for which the mutation is driving the tumor invasion. Patients whose tumors do not contain that mutation are expected to be unresponsive to the drug. The *enrichment design* is a phase III design in which patients are randomized to the test drug or control regimen, but eligibility is restricted to patients whose tumors contain the altered gene.

In some cases, the linkage between the drug and mutation is not so direct. For example, the drug may inhibit a pathway that is thought to be activated by an alteration in a gene that is upstream of the drug target. For example, MEK is a kinase that can be inhibited and is downstream of the RAS proteins which are frequently mutated but not directly "druggable." In these cases, a phase II trial will probably be conducted before the phase III trial.

In some cases, it may be sufficient to conduct a standard Simon Optimal Two-Stage Design [1] for patients containing the mutation. In other cases, it may be best not to exclude patients whose tumors do not contain the mutation. Separate Optimal Two-Stage Designs can be conducted for the "marker-positive" and "marker-negative" cohorts. Freidlin et al. [2] have described a phase II design including both marker-positive and marker-negative cohorts with a decision structure for deciding whether or not the phase III trial should be restricted to marker-positive patients or whether or not a phase III trial should be pursued at all.

14.2.1 Basket Designs

There are an increasing number of molecularly targeted drugs which have been approved for marketing for patients with a specified histologic type of cancer in which the tumor contains a specific genomic alteration. The question then becomes whether that same drug would be effective in patients whose tumors contain the same genomic alteration but in different histologic types of cancer than the one for which the drug was approved. The basket clinical trial was developed for this setting [3]. Generally they are not randomized clinical trials; all patients with the genomic alteration needed for eligibility are assigned the targeted therapy. Basket trials are also often conducted not in the context of off-label use. In that case, there is a biological rationale why the drug should be active against tumors bearing a particular genomic alteration, but it has not been approved for any histologic type of cancer.

In multi-drug basket trials, patients have their tumors tested for a panel of genomic alterations and are then triaged to the drug which targets the most prominent alteration in their tumor. Multi-drug basket trials also provide a mechanism for unified tumor screening and treatment of patients with advanced disease who have subsequently failed standard therapy [4]. Basket trials are, however, very inefficient in that many patients who are screened typically do not contain a genomic alteration that is targeted by any of the available drugs. They are difficult to organize because cooperation from many drug manufacturers are needed.

Newer statistical designs are now available for basket designs [5–7]. These designs are more appropriate and efficient for determining which histologies are sensitive to the test drug compared to the convention of conducting an Optimal Two-Stage Design for the drug pooling and ignoring histologic differences.

14.2.2 Platform Designs

Basket designs pre-specify an "actionable drug" for each genomic alteration. This is usually based on biological knowledge. Platform designs involve multiple drugs and multiple biomarkers without pre-specifying which drugs should work for which biomarker-characterized tumors. For example, suppose that the response probabilities can be characterized as $\leq p_0$ or $\geq p_1$. Platform trials are often conducted in the context of a single histologic type, but suppose that there are K binary biomarkers; x_k denotes the value of the kth biomarker for a patient, and X denotes the vector $(x_1, \dots, x_K)$. We assume that there are I different drugs and $T = i$ means that the drug used for a patient is drug i. To learn from the data accumulated, we might use a logistic model:

$$\text{Logit} \{\Pr[p = p_0 \mid X, T = i]\} = \beta_{0i} + \beta_{1i}x_1 + \cdots + \beta_{Ki}x_K \tag{14.1}$$

where logit{P} means $\log(P)/\log(1 - P)$, and the vertical bar indicates conditional probability. The β_{ki} value is the unknown regression coefficient for the effect of biomarker k when the patient is treated with treatment i.

During the trial, we can compute estimates of the regression coefficients. The treatment predicted to be most effective for a patient with biomarker vector X is the one with the largest value of score

$$S_i = \hat{\beta}_{0i} + \sum_{k=1}^{K} \hat{\beta}_{ki} x_k \tag{14.2}$$

Because maximizing the logit is equivalent to maximizing the probability that $p = p_1$.

Part of the specification of a platform design is specification of how patients are to be treated during the trial. One very simple approach is to ignore biomarker values and conduct I separate two-stage optimal designs. At any point patients would be randomized among the drug-specific optimal designs which are still open for accrual. An optimal design is closed if the number of first-stage responses observed for the drug is insufficient.

Alternatively, we could assign the treatment with the largest value of S_i. Another possibility would be to randomize the treatment assignment among all treatments but weight the randomization so that treatments with the largest scores have the greatest probability of being assigned.

Another important part of the specification of the platform design is specification of how the regression coefficients are to be estimated during and after the trial. If the regression coefficients are considered fixed unknown constants, then the data might as well be analyzed as I separate logistic models; the data for treatment i would tell us nothing about the regression coefficients for treatment j. This may be appropriate in some cases. Suppose that there are two treatments, one a BRAF inhibitor and the other an EGFR inhibitor. If x_1 is an indicator of whether the tumor contains a BRAF mutation, then the effect of that mutation on sensitivity to the BRAF inhibitor is quite different from the effect of that mutation on sensitivity to the EGFR mutation. For that reason, we definitely would not want to assume that $\beta_{11} = \beta_{21}$, where x_1 is taken as the indicator of BRAF mutation.

Bayesian hierarchical modeling [8] does not consider the regression coefficients as fixed independent constants. It also does not assume that the regression coefficients for the effect of two different treatments in a biomarker stratum are equal. It assumes however that those values are related in that they represent independent samples from a common distribution of values. It then places a prior distribution of the parameters of that distribution. With Bayesian hierarchical modeling, point estimates of the regression parameters are not used, and the treatments are scored for a patient with covariate vector X by computing the posterior probability that the response probability $\Pr(p = p_1 | X, T = i, D)$, where D denotes the data available at that point from all of the patients. In that way, the model attempts to "share information" from response data for different treatments in estimating the regression coefficients. The degree of sharing can be overly determined by the initial assumption about the variance of the higher level prior if the number of treatments is small. This can distort the estimation process and the scoring of different treatments.

14.3 Enrichment Design

The enrichment design is a randomized phase III design in which eligible patients are randomized to the test treatment or control but eligibility is restricted to the patients who are "biomarker positive," i.e., have the genomic alteration which makes their tumors more likely to respond to the test drug. This approach was used for the development of trastuzumab and many other drugs approved in the past decade.

Simon and Maitournam [9,10] found that the enrichment design can be much more efficient than the standard design of randomizing all patients and ignoring the biomarker. The efficiency depends on the prevalence of test-positive patients and on the effectiveness of the test treatment in test-negative patients. For trials, they showed that the ratio of number of events needed for the enrichment trial compared to the standard trial is approximately proportional to

$$\{(p_+\delta_+ + p_-\delta_-)/\delta_+\}^2 \tag{14.3}$$

where p_+ is the proportion of patients who are biomarker positive, p_- is the proportion biomarker negative, δ_+ is the log hazard ratio of treatment effect for biomarker-positive patients and δ_- is the log hazard ratio of treatment effect for biomarker-negative patients. In cases where the new treatment is completely ineffective in test-negative patients, the formula above simplifies to approximately $(1/p_+)^2$, which is 16 when only one-quarter of the patients are biomarker positive. If the new treatment is half as effective in biomarker-negative patients as in biomarker-positive patients, then the right-hand side of expression (14.3) equals about 2.56 when 25% of the patients are test positive, indicating that the enrichment design reduces the number of required patients to randomize by a factor of 2.56.

If the enrichment design establishes that the drug is effective in test-positive patients, the drug could be later developed in test-negative patients if the test was considered to have a substantial false-negative error rate. This is preferable to testing new drugs in heterogeneous populations resulting in false-negative results for the overall population.

14.3.1 Umbrella Design

Umbrella designs are multiple enrichment designs of the same histologic type conducted with a common infrastructure for characterization of the patients' tumors [3]. If the patient's tumor has a genomic alteration which makes him/her eligible for one of the enrichment designs, then he/she is triaged to that trial. Once triaged, the patient will be asked to provide informed consent to that particular enrichment trial. Enrichment trials are generally randomized trials of patients with the same histological type of cancer bearing the same genomic alteration. In an umbrella trial, all the patients have the same histological type of cancer but different genomic alterations corresponding to the different component enrichment designs. Each enrichment design is sized independently. The enrichment designs may or may not use the same control treatment regimen but there is no pooling of information because patients with different genomic alterations must be analyzed separately. Two recent examples of umbrella clinical trials are the Lung MAP for patients with advanced squamous cell lung cancer [11] and FOCUS4 for patients with advanced colorectal cancer [12].

14.4 Adaptive Enrichment Designs

When there is uncertainty about the appropriateness of the candidate biomarker, the biomarker can be measured on all patients but not used as an eligibility criterion; i.e., both biomarker-positive and biomarker-negative patients are enrolled in the randomized phase III trial. A number of authors have introduced two-stage designs in which an interim analysis is performed after the first stage of accrual [13,14]. If the overall results are promising, accrual of both biomarker-positive and -negative patients continues. If the overall results are not promising but results for biomarker-positive patients are, the accrual for biomarker-positive patients continues but accrual of biomarker-negative patients ceases. The accrual target for the biomarker-positive patients may be increased by the reduced accrual of marker-negative patients. If neither the overall results nor the results for marker-positive patients are promising, then the trial may be terminated. Here, we will describe the general framework of adaptive enrichment described by Simon and Simon [15], which can be applied to settings other than the case of a single binary biomarker.

The adaptive enrichment design has a pre-specified number of interim analyses, usually one or two. At the interim analysis, the treatment outcomes are investigated with regard to treatment and the candidate biomarkers. As the result of an interim analysis, the eligibility criteria may be restricted based on the candidate biomarkers. One cannot broaden eligibility based on an interim analysis; that is, we assume that the eligibility criteria used for subsequent periods of accrual are "nested."

At the final analysis, a single significance test is performed comparing the test treatment to control. It includes all randomized patients but has the specific form described in the following way. Let Z_j denote the z value for comparing the test treatment to control for patients accrued during period j. If there is a single interim analysis, then $j = 1$ or 2. If there are two interim analyses, then $j = 1, 2,$ or 3. We assume that Z is normalized to have a null distribution that is approximately normal with mean 0 and variance 1. The single statistical significance test conducted is based on the test statistic

$$Z = \frac{\sum_{j=1}^{J} w_j Z_j}{\sum_{j=1}^{J} w_j^2} \tag{14.4}$$

where the weights are pre-specified, non-negative, and add up to 1. Under the null hypothesis, this statistic should be approximately normal with mean 0 and variance 1.

14.4.1 Adaptive Enrichment with Single Binary Covariate

In the case of a single binary covariate, the only possible limitation of eligibility at an interim analysis is to exclude future patients with biomarker negativity. Once this is done, it cannot be undone. There are many possible bases for making this decision. We will describe here the Bayesian monitoring approach of Simon and Simon [16]. The adaptive enrichment design is itself frequentist, involving a single significance test using test statistic (14.4), but the adaption strategy can be Bayesian and the significance test is still valid.

Let p_{--} denote the prior probability that the test treatment is no better than control for either biomarker-negative or biomarker-positive patients. We call this global null H_{00}. Let p_{++} denote the prior probability that the test treatment is more effective than control for both marker-negative and marker-positive patients. Let p_{+-} denote the prior probability that the test treatment is more effective than control for marker-positive patients but not for marker-negative patients, and p_{-+} for the opposite scenario. If we specify these prior probabilities, then at any interim analysis, we can compute the posterior probabilities. If the posterior probability that the test treatment is not effective at all is too large, the trial should be terminated for futility. Otherwise, some part of the trial should be continued. If the posterior probability that the test treatment is not effective in marker-negative patients is large, then such patients should be excluded in the subsequent accrual periods.

At the end of the trial, if the global null hypothesis is rejected based on the test statistic (14.4) is rejected, then we use the prior parameters and the data to compute the posterior probability that the test treatment is effective in test-negative patients and the posterior probability that the test treatment is effective in marker-positive patients. From these posterior probabilities, we can propose a rational labeling indication for the drug.

One point of clarification with regard to endpoints. Survival or disease-free survival may be the primary endpoint of the trial and be used for the significance test (14.4). For interim decision-making about how to potentially restrict eligibility, however, it may be necessary to use a shorter-term endpoint.

14.4.2 Adaptive Enrichment with a Quantitative Biomarker

Another frequently occurring situation is that we have a single quantitative or semi-quantitative biomarker that we think is relevant in selecting use of the test treatment but we don't have a satisfactory "cut point" for the biomarker. We can always select a cut point for use in the phase III trial based on our phase II database, but we may not have enough information to do that reliably.

The biomarker does not enter into the final significance test, so test statistic (14.4) is still used here. For interim decision-making about restricting eligibility using a shorter-term endpoint, a new model is needed when there is a single quantitative biomarker. Suppose we have a binary intermediate endpoint in which the response probability can be either p_0 or p_1. Our model specifies that for the control arm, all patients have response probability p_0. For the treatment arm, patients with biomarker value b less than the true cut point B^* have response probability p_0, whereas if b is greater than B^*, then the response probability is p_1.

We specify a prior distribution for B^* based on having K candidate cut points. Suppose that the biomarker values are normalized to lie on the interval 0–1. Candidate cut point B_0 is 0 corresponding to the setting where the treatment is effective for all patients. The largest candidate biomarker B_K is taken to be 1.01, corresponding to the situation where the test treatment is effective for none of the patients. Between B_1 and B_K, we might define $B_2 = 0.25$, $B_3 = 0.5$, and $B_4 = 0.75$. We might specify that the prior probability that no one benefits, i.e., $\Pr[B^* = 1.01] = 0.5$, that the probability that $\Pr[B^* = B_k] = 0.5/4$ for $k = 1,2,3,4$.

With this model and the prior distribution for B^*, one can easily compute the posterior distribution of B^*. If, at an interim analysis, the posterior probability that the true cut point B^* is less than a pre-specified candidate cut point B_k is too small (i.e. $\Pr[B^* \le B_k | D] < \varepsilon$), then subsequent accrual can be limited to those with biomarker values greater than B_k.

If the frequentist statistical test based on (14.4) rejects the global null hypothesis, then one can compute the posterior distribution of the true cut point B^* and recommend an appropriate drug labeling.

14.5 Conclusion

A variety of new clinical trial designs have been introduced for biomarker driven clinical trials. In this chapter we have described only some of them viewed as most general and easiest to implement. The single most important factor determining the efficiency of a clinical trial is targeting the most appropriate population. Diluting treatment effect by including patients who are unlikely to benefit from the test treatment has a major negative effect on statistical power. This is the reason that the enrichment design has had such an influence in the large number of cancer-drug approvals over the past decade. The new biomarker-driven trial designs can result in efficient and informative trials, but only if the designs are used and the candidate biomarkers are strong. Clinical trialists will need to re-examine some of their existing assumptions about clinical trial design in order to move forward.

References

1. Simon R. Optimal two-stage design for phase II clinical trials. *Control Clin Trials.* 1989;10:1–10.
2. Freidlin B, McShane LM, Polley MYC, Korn EL. Randomized phase II trial designs with biomarkers. *J Clin Oncol.* 2012;30:3304–9.
3. Simon R. Genomic driven clinical trials in oncology. *Ann Intern Med.* 2016;165:270–8.
4. Conley BA, Dorshow JH. Molecular analysis for therapy choice: NCI MATCH. *Semin Oncol.* 2014;41:297–9.
5. Berry SM, Broglio KR, Groshen S, Berry DA. Bayesian hierarchical modeling of patient subpopulations: Efficient designs of phase II oncology clinical trials. *Clinical Trials.* 2013;10:720–34.
6. Simon R. New designs for basket clinical trials in oncology. *J Biopharm Stat.* 2018;28:245–55.
7. Cunanan KM, Iasonos A, Shen R, Begg CB, Gonen M. An efficient basket trial design. *Stat Med.* 2017;36:1568–79.
8. Saville BR, Berry SM. Efficiency of platform clinical trials: A vision of the future. *Clinical Trials.* 2013;13(3).
9. Simon R, Maitournam A. Evaluating the efficiency of targeted designs for randomized clinical trials. *Clin Cancer Res.* 2005;10:6759–63.
10. Simon R, Maitournam A. Evaluating the efficiency of targeted designs for randomized clinical trials: Supplement and correction. *Clin Cancer Res.* 2006;12:3229.
11. Herbst RS, Gandara DR, Hirsch FR et al. Lung Master Protocol – A biomarker-driven protocol for accelerating development of therapy for squamous cell lung cancer. *Clin Cancer Res.* 2015;21:1514–24.
12. Kaplan R. The FOCUS4 design for biomarker stratified trials. *Chinese Clinical Oncology.* 2015;4(3).
13. Wang SJ, O'Neill RT, Hung HMJ. Approaches to evaluation of treatment effect in randomized clinical trials with genomic subset. *Pharm Stat.* 2007;6:227–44.
14. Brannath W, Zuber E, Branson M et al. Confirmatory adaptive designs with Bayesian decision tools for a targeted therapy in oncology. *Stat Med.* 2009;28:1445–63.
15. Simon NR, Simon R. Adaptive enrichment designs. *Biostatistics.* 2013;14:13–25.
16. Simon NR, Simon R. Using Bayesian models in frequentist adaptive enrichment designs. *Biostatistics.* 2017;19:27–42.

15

Trial Designs for Rare Diseases and Small Samples in Oncology

Robert A. Beckman, Cong Chen, Martin Posch, and Sarah Zohar

CONTENTS

15.1 Introduction

In the last decades, common cancers have been stratified into rare molecular segments based on genomic and phenotyping analyses. Next-generation sequencing (NGS) technology has allowed researchers to identify tens of thousands of mutations in patient tumors [1,2]. Indeed, cancers that seemed to define a homogenous population are now known to be heterogeneous, dividing the patient population into many subgroups with their own prognostics, comorbidities, and targeted treatments. For instance, patients suffering from lung adenocarcinomas are stratified in clinical trials according to EGFR mutation, EML4-ALK translocation, K-Ras mutation, or HER2 amplification [3]. As a consequence, statisticians and investigators are facing the methodological problem of planning, conducting, and analyzing clinical trials with small sample sizes.

Conventionally, standard designs for clinical trials are based on (excluding early phases) the large number statistical inference that requires the inclusion of several hundred patients to bring a high level of evidence in the favor of novel treatments. However, these approaches are not suitable to evaluate therapies if the sample size is unavoidably small. In 2013, the European Union launched a specific call dedicated to the development of novel methodology for small sample trials. Three consortiums were selected—Asterix, IDeAl, and InSPiRe—which have proposed many innovative designs for all phases of clinical trials [4]. These consortiums have also collaborated with the Drug Information Association Adaptive Design Scientific Working Group (DIA-ADSWG), an international group of statisticians and clinicians from academia, government, and the pharmaceutical and biotechnology industries devoted to the use of complex adaptive designs, including some members of Asterix, IDeAl, and InSPiRe.

In this chapter, we will only present a subset of the tremendous work that has been achieved by these statisticians. Three topics will be addressed; each will be focusing on a specific phase of drug development. The first section will introduce how to maximize the utilization of available information during early-phase clinical trials, the second section will present innovative approaches to basket trials under small sample constraints with emphasis on the confirmatory setting, and the third section will focus on the application of decision analysis to the optimal design of phase 2/3 seamless adaptive trials in the presence of biomarker subset effects.

15.2 Using External Information and Extrapolation for Planning and Conducting Early-Phase Clinical Trials: Case Study in Pediatric Oncology

Early phase dose-finding studies are designed to obtain reliable information on a drug's safety, tolerability, pharmacokinetics (PK) or/and pharmacodynamics (PD). Usually, it is common to represent these studies as the first translation from laboratory work to clinical settings. However, in most cases information about the drug is already available from previous clinical trials in other populations, schedules, or indications.

Recent regulatory agencies emphasized the need for modeling and for innovative methodology in pediatric clinical trials (ICH E11[R1] guideline [56]). This modeling should be associated with multidisciplinary collaboration to support the choice of doses and design, as well as the need of using the existing knowledge to build better established designs.

These multidisciplinary objectives of exploratory studies are challenging in children for which planning clinical trials is more complex than in adults. Reluctance of children to participate in invasive procedures, difficulty of recruitment, population heterogeneity, stratification by age, and small sample sizes, especially in cases of rare diseases, are major hurdles to overcome. The usefulness of clinical trials in children has largely been debated in the last decades [5–7]. Some authors have argued that the information in adults should be better exploited, both quantitatively and qualitatively, before administering a new drug or dose regimen in children. Moreover, in the small population guideline [8,9], modeling and simulation are emphasized as being more informative than descriptive statistics.

Recently Petit et al. [10], have proposed a unified approach for extrapolation and bridging of adult information in early-phase pediatric studies. In their article, the authors have used different kinds of data sources in order to plan and conduct early-phase clinical trials.

Pharmacokinetic, toxicity, and efficacy data from adults gathered from several published articles were utilized at different stages and under different modeling processes.

The aim of this section is to bring an insightful description of the proposed methods that can seem obscure to readers uninitiated in early-phase biostatistics.

The authors have proposed how to use external information for three axes in the clinical trial planning: (1) the specification of the dose range, (2) the specification of the working model of the dose-finding allocation design, and (3) for dose-finding designs under Bayesian inference, prior distribution calibration methods were proposed.

15.2.1 Specification of the Dose Range

Children are not "small adults" largely because of the immaturity of renal and metabolic systems and because of growth, as it isn't a linear variable. When extrapolating the dose range from adults to children, these features need to be kept in mind. Until now, in oncology, dose-finding pediatric clinical trials are starting at 80% from the adult maximum tolerated dose, or MTD (without any statistical or scientific justification). An alternative is using an allometry adjustment for the calculation of the dose range in children. This is based on the adult dose, $d_{ad,i}$ ($i = 1, \dots, K$); average child weight (according to age range), W_{ch}, the average adult weight, W_{ad}, and a scale parameter describing the rate at which the weight increases. Pharmacology studies have proposed to use the value of 0.75 for the scale parameter.

$$d_i = d_{ad,i} \times \left(\frac{W_{ch}}{W_{ad}} \right)^{0.75} \tag{15.1}$$

We use an illustrative example of a dose-finding study of erlotinib in pediatric brainstem glioma and relapsing/refractory brain tumors as in Petit et al. [10]. In our case study, adult erlotinib dose-finding studies used doses ranging from 100 to 300 mg, and the MTD was found to be 150 mg. In this illustration, 100, 150, 200, 250, and 300 mg were chosen as references for the calculation of pediatric doses under Equation 15.1 for patients aged 2–5 years, with $W_{ad} = 70$ kg and $W_{ch} = 10$ kg; the dose range computed for children (rounded at the closest 5 mg) is as follows: 35, 50, 65, 80, and 100 mg/kg.

15.2.2 Specification of the Working Model

When using a model-based dose-finding approach (for further details, please refer to Chapter 3), and after selecting the dose range for the study, the next step is to parametrize the working model or the initial response guesses associated with each dose.

Assuming equal exposure in adults and children, Petit et al. have proposed to estimate the toxicities associated with each child's dose using adult information. Dose-toxicity studies (phase I and I/II clinical trials) were gathered using a retrospective design of pooled data [11]. This approach proposes to pool several datasets from published dose-finding trials and, through simulation using a down-weighting method, to estimate the overall dose-toxicity relationship. In practice, this method is done in several steps:

1. Select the published phase I trials evaluating the drug under the same setting in terms of schedule or combination.

2. Gather the number of observed dose-limiting toxicities (DLTs) at each dose level, and the number of patients included at each dose level, from all available clinical trials.

3. Calculate the empirical probability of toxicity associated with each dose level by devising the number of observed DLTs at each dose level by the number of patients included at each dose level.

4. Calculate the weights for all of the available doses. These weights are provided by a simulation study based on the model of interest and marginal frequencies provided by the observations. To calculate these weights, we simulate CRM (Continual Reassessment Method) studies of fixed size under the scenario generated by the empirical probability of toxicities. The weights are, thus, the resulting percentage of the total allocation for each dose level.

5. Solve the equation $Wn(a) = 0$ given in Zohar et al. [11], to estimate the dose-toxicity model parameter.

6. On the basis of the selected model using the estimated parameter in step (5), the pooled MTD can be computed.

Seven clinical trials evaluating the erlotinib dose in adults have been gathered (Table 15.1) for the computation of the overall dose-toxicity relationship. The method proposed by Zohar et al., described above, can be computed with the R package "dfped" [12] as follows:
R code:

```
require(dfped)

pardos_2006 <- rbind(c(100,0/3, 3), c(150, 1/3,3), c(200, 0/3, 3), c(250, 3/6, 6))
thepot_2014 <- rbind(c(100, 0/5, 5), c(150,3/25, 25))
calvo_2007 <- rbind(c(150, 1/25, 25))
raizer_2010 <- rbind(c(150,11/99, 99))
vanDenBent_2009 <- rbind( c(200, 6/54, 54))
sheikh_2012 <- rbind(c(150, 0.544, 307))
rocheNTC00531934 <- rbind(c(150, 0.186, 59))


dataTox <- rbind(pardos_2006, thepot_2014, calvo_2007, raizer_2010,
vanDenBent_2009, rocheNTC00531934, sheikh_2012)
dataTox <- data.frame(dataTox)
colnames(dataTox) <- c("doses", "proba", "nbPatients")
nbTox <- dataTox$proba*dataTox$nbPatients
dataTox <- data.frame(dataTox, nbTox)
doses <- c(100,150,200, 250)
nbSimu <- 1000 #higher is the number of simulations, longer it is to compute
metaPhase(dataTox, doses, nbSimu)
```

TABLE 15.1

Number of Toxicity Outcomes and Number of Treated Adults in Erlotinib Clinical Trials

Nb tox (nb Patients) Paper	100 mg	150 mg	200 mg	250 mg
Prados et al.	0 (3)	1 (3)	0 (3)	3 (6)
Raizer et al.	–	11 (99)	–	–
Thepot et al.	0 (5)	3 (25)	–	–
Calvo et al.	–	1 (25)	–	–
Van den Bent et al.	–	–	6 (54)	–
Sheikh and Chambers	–	167 (307)	–	–
Clinical trial ROCHE NTC0053193	–	11 (59)	–	–

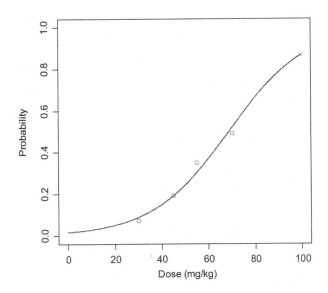

FIGURE 15.1
The estimated dose-toxicity relationship estimated from adult data.

The resultant dose-toxicity relationship associated to the dose levels 100, 150, 200, and 250 mg are 0.07, 0.19, 0.35, and 0.49, respectively. Thus, under the assumption of equal exposure in adults and children, these toxicity probabilities are associated with the doses 35, 50, 65, and 80 mg/kg. However, as given above, the computed doses for the pediatric trial were 35, 50, 65, 80, and 100 mg/kg. So we are missing the toxicity probability associated with the dose 100 mg/kg. In this case, a logistic model was fitted in order to extrapolate the toxicity probability associated with each dose in children. The resultant intercept of logistic model was −4.17328 and the slope 0.06071 (Figure 15.1).

The estimated toxicity probability associated with the dose 100 mg/kg is 0.87.

15.2.3 Calibration of the Prior Distribution

The dose-toxicity model parameter can also be calibrated based on external information. The idea is to use information about the compound or the drug in the calibration of the prior distribution. It can be useful if done carefully, justified, and well documented. It should not be too informative to overpower the clinical trial data, but it shouldn't be completely non-informative so the external data do not bring any add-on to the analyses. Zhang et al. [13] have proposed to adaptively calibrate the prior variance during a trial. The first prior variance $\left(\sigma_{a,\text{NIP}}^2\right)$, $\pi_{\text{NIP}}(a) \sim N\left(\mu_a, \sigma_{a,\text{NIP}}^2\right)$ (where a is the dose-toxicity model parameter, non-informative prior—NIP) is considered as non-informative and the second prior variance $\left(\sigma_{a,\text{LIP}}^2\right)$, $\pi_{\text{LIP}}(a) \sim N\left(\mu_a, \sigma_{a,\text{LIP}}^2\right)$ is informative (least informative prior—LIP) and can be defined from previous data such as described in the Section 15.2.2. In this case, the mean, μ_a, of both distributions can be centered on 0 or based also on previous data. In the package "dfped," the function priorChoice can be used for computing these variances.

15.2.4 Conclusions

The extrapolation and bridging external information such as previous clinical trials in other populations or indication as well as observational data and literature for the design of early-phase clinical trials may improve the results of these studies. When dealing with small sample sizes, any relevant additional information can be useful. However, this information should be carefully added and should be weighted according to its importance and quality.

15.3 A General Design for a Confirmatory Basket Trial

The molecular approach to cancer is transforming cancer biology and therapy, and a corresponding transformation is required in drug-development approaches. Cancer drug development was originally designed with the notion that cancer is a common disease, and traditional large sample sizes for confirmatory clinical trials were feasible. But now, common diseases such as lung cancer are being carved up into small subsets based on their molecular characteristics, particularly at the DNA level. Not only is it increasingly difficult to enroll large sample sizes from these small subsets, but the large number of subsets and large number of experimental therapies have combined to drive a combinatorial explosion of potentially worthy clinical hypotheses. The cost and patient requirements of conventional one indication at a time development are becoming increasingly unsustainable if used exclusively.

There are three broad approaches for developing targeted agents, based on molecular predictive biomarkers as outlined in Chapters 13 and 14.

1. Efficiency-optimized, single-indication co-development of targeted agents and their companion or complementary diagnostics for identifying patients whose tumors are members of a molecular subset [14,15] provides a very clear hypothesis and answer, and is recommended as a foundational study in one indication when feasible. However, this will not always be feasible as discussed above.

2. "Umbrella" trials combine multiple targeted agents and molecular subgroups within a single histology [16,17] or multiple histologies [18], providing operational efficiencies and, under some circumstances, a common control group. Umbrella trials require collaboration among multiple stakeholders, a significant challenge.

3. "Basket" or "bucket" trials feature a single targeted agent and subgroup across multiple histologic indications. The underlying assumption is that molecular classification of cancer is more fundamental than histological classification [19]. From this, it follows that similar subgroups may be studied together, and even pooled, across multiple histologies, as they may share a phenotype of potential clinical benefit from a single targeted agent.

While umbrella trials offer considerable operational efficiencies, basket trials offer the potential for studying multiple indications with a sample size traditionally appropriate for only one indication, a potential savings of several hundred percent in cost and patient enrollment requirements. Furthermore, basket trials can easily be performed by a single sponsor. This section will focus on basket trials. Although this chapter focuses on oncology, basket trials may also be particularly useful in the infectious disease space, where an antimicrobial agent may be studied for infections in different areas of the body by the same organism simultaneously. Furthermore, the ability of basket trials to facilitate dramatic reductions in patient enrollment requirements can be very helpful for rare diseases when a common pathogenesis can be found across rare diseases, i.e., rare autoimmune diseases [20].

Basket trials have previously been used primarily in exploratory settings, and, if used for confirmation, in settings where there is exceptional scientific evidence, observed transformational levels of benefit from the therapy in related applications, and no or minimal other therapeutic options. Under the latter circumstances, there is justification for reducing evidentiary requirements, and indeed in the case of a pioneering basket trial of imatinib in rare tumors with the known targets of imatinib, one indication was approved based on tumor shrinkage in 1 of 4 patients in an unrandomized study [21–24]. While early basket trial designs did not pool indications, but considered each indication separately, more recent designs utilize various degrees of weighted pooling based on techniques such as Bayesian hierarchical modeling [25, 26] or weight hypotheses that all indications are independent, or that all indications are from a common probability distribution, based on the data [27].

In this section, we focus on a proposed basket-trial design that may be suitable in the confirmatory setting [28,29]. By far the greatest costs and patient enrollment requirements are associated with the confirmatory phase of development, and therefore the greatest impact on the problems alluded to above will come from applying the substantial savings potential of basket trials to the confirmatory phase. While applications of non-rigorous designs to exceptional situations are both valuable and justified, therapies with ordinary levels of efficacy are likely to be more common.

A general approach to confirmatory basket studies would be beneficial in several regards. For patients in indications with small populations, for which development by conventional methods is not cost-effective, and thus often does not occur, access to therapies would be more likely. The design below also features the possibility of accelerated approval, providing even earlier access in some instances. Health authorities might have helpful datasets for evaluation of risks and benefits across molecular niche indications. For sponsors, development of niche indications would be facilitated due to lower cost and easier enrollment. Confirmatory basket trials may also be a very valuable component of an organized pipeline of linked adaptive platform trials (i.e., basket and umbrella trials) [30].

15.3.1 Overview of General Confirmatory Basket-Trial Design

In this design, individual histologic subtypes (indications) are grouped together. In order to reap the full efficiency advantage of a basket-trial design, we pool indications in the final analysis so that the total sample size across multiple indications is adequate for formal confirmation by frequentist criteria. Such a design has an inherent risk, in that if indications are included for which the therapy is ineffective, they may dilute the signal from effective indications, resulting in failure of the entire basket. This risk is minimized by "pruning," by which we mean removal of indications that do not show significant evidence of efficacy at an interim analysis, or removal of indications determined to be high risk based on external data, prior to the final pooled analysis. Pruning using data internal to the study creates complications with regard to type I error control, and an overview of this is given in the next subsection. Furthermore, pruning based on internal study data may contribute to a small estimation bias, i.e., a slight tendency to overestimate the benefits associated with approved indications [57].

The recommended study design is randomized, each indication with its own control group, unless there is a common standard of care, in which case a shared control group may be used for the relevant indications. The strong preference for a randomized controlled design follows from the desire that the design be generally applicable in the confirmatory setting. Single arm designs using a concurrent registry control may be considered, preferably only in instances involving ultra-rare tumors with no standard of care, or in which a transformational effect featuring durable responses and a large survival benefit are expected [24]. Concurrent registries control for several factors that affect the validity of historical controls, related to improvements over time in diagnostic sensitivity (resulting in patients being diagnosed and classified with less disease over time) and supportive care (resulting in less favorable outcomes in historical datasets). However, it is a far greater challenge to overcome selection bias, the tendency of clinical research physicians to select patients who will do well, resulting in an overestimate of benefit in single-arm trials. This issue can be partially mitigated by matching algorithms, but physicians have a considerable ability to assess patients in an intuitive manner that is difficult to quantify and thus challenging to adjust for. The use of registry data should be pre-agreed with health authorities.

Each indication cohort would be sized for individual accelerated approval based on a predetermined surrogate endpoint (i.e., response rate, RR, or progression-free survival, PFS) reasonably likely to predict clinical benefit (i.e., overall survival, OS). It is critical that the surrogate endpoint be more sensitive than the final clinical benefit endpoint, i.e., the expected effect size for an effective therapy should be greater for the interim surrogate endpoint than for the final clinical endpoint. Thus, the individual indications may be fully powered for the more sensitive interim endpoint and, when pooled at the end, will have adequate sample size for the less sensitive final endpoint. We note that the endpoints named above should be viewed as examples only, representing a typical case. For example, either of the surrogate endpoints listed here can at times be primary-approval endpoints [31,32]. This approach is an accepted route to accelerated approval in the United States [33]. Agents receiving accelerated approval using this approach have a commitment to demonstrate clinical benefit using the final clinical-benefit endpoint and would receive full approval at that point.

Tumor indications failing to meet the surrogate hurdle for accelerated approval would be "pruned" (removed from the basket). In so far as the surrogate endpoint is expected to correlate to some degree (albeit imperfectly) with the final clinical endpoint, the risk of ineffective indications being included in the final pool is mitigated. For the remaining indications, the patients who contributed data at the interim analysis will be included in the final analysis, which may affect type I error rate, resulting in the need to operate at a lower nominal type I error rate to control the resulting type I error rate of the pooled study, amounting to a statistical penalty for the reuse of internal study information.

Additional indications may be pruned based on external data such as maturing early stage data involving the definitive clinical-benefit endpoint [14], real-world data from off-label use of the therapy (discretionary use by physicians for non-approved indications, permitted in the United States) [20], or data from other agents in the class. Pruning based on external data does not affect type I error and does not incur a statistical penalty. Moreover, use of maturing data from previous early-stage studies allows the definitive endpoint to govern adaptation, removing uncertainties due to imperfect correlation between interim and definitive endpoints.

After pruning, sample size readjustment is required to maintain the power of the final pooled analysis. The sample size adjustment strategy must be pre-specified, pre-agreed with health authorities, and ideally should be managed by an independent data monitoring committee. Chen et al. [29] compared several sample size adjustment strategies and noted that the most aggressive strategy provided the best maintenance of reasonable power with increasing ineffective indications in the study prior to pruning. This strategy consisted of increasing the numbers of patients in the remaining indications to keep the size of the final pooled analysis as originally planned, which corresponds to a larger total study sample size than would have occurred without pruning. With this option, the power is clearly superior (i.e., false-negative rate lower) with pruning than without [29]. For pruning to be effective in mitigating the risk of pooling (e.g., false negatives due to dilution of active indications by inactive ones, as well as false positives in which inactive indications are carried along to a positive pooled result due to active indications), the bar for being included in the final pooled analysis is set relatively high. As such, pruning does not indicate that the indication is not worthy of further investigation, only that the indication is too high risk to be included in the basket.

The remaining indications would be eligible for full approval if the pooled analysis for the definitive clinical benefit endpoint reached statistical significance at the reduced nominal type I error level required by the design. Individual indications would not be required to show statistical significance but rather to demonstrate a sufficient trend for the definitive clinical benefit endpoint to be judged as having a positive benefit-risk balance (pre-agreed for each indication). Those indications not showing a sufficient consistent trend would not be approved, a final "pruning" step. Other data sources could potentially be used to make this judgement in borderline cases.

The above design concept can also be varied to use a smaller amount of data from the definitive clinical endpoint to govern pruning at the interim analysis. In this case, typically the type I error threshold must be set higher at interim due to the smaller amount of data accumulated. The bar for passing pruning and the fraction of available definitive endpoint information available at interim are important design parameters in this instance.

15.3.2 Pruning, Random High Bias, and Type I Error Control

In order for a confirmatory basket trial to meet acceptance from health authorities, it will be necessary for the type I error of the pooled analysis to be rigorously controlled. Thus the type I error is of particular interest in the context of a confirmatory basket trial.

Pruning of indications using data from within the study may inflate the type I error of the pooled analysis. In the setting of the global null hypothesis, where all indications are ineffective, there will still be random variation of the observed interim therapeutic effect around the expected mean of zero. If, in the global null case, we "prune" those indications randomly showing a negative or no therapeutic effect, and retain those randomly showing a positive therapeutic effect, we are in effect "cherry picking" those indications that got off to a good start. As this preliminary data may partly correlate with the final result, there is a greater chance that the remaining indications will randomly generate positive final data, resulting in a type I error. This effect is termed *random high bias*.

A counterbalancing effect, however, is the similarity pruning has with a binding futility analysis, which will have the effect of lowering type I error [29].

It is possible to calculate exactly how much the type I error of the pooled analysis is increased by random high bias, or decreased by binding futility, and to adjust the nominal type I error of the study such that the actual type I error of the final study pooled analysis remains in control after adjustment for these effects [29]. In order to do this, we need to know the rules for pruning, the number of indications pruned, and which indications they were, given that the sample sizes may vary from indication to indication, as well as the percent of the events collected at interim if the interim analysis is based on the same endpoint as the final. We calculate the chance, if all indications truly have no therapeutic effect, of pruning the indications suggested for pruning by random chance and then of observing an apparently positive effect in the pooled analysis of the remainder by random chance. Usually this will be greater than the desired type I error due to inflation by random high bias, and we must pay a penalty by lowering the nominal type I error until the actual false-positive rate of the pooled analysis adjusted for random high bias is as desired. It should be noted that the type I error control applies only to the global null hypothesis (so called "weak control") and not to "strong" control of the family-wise error rate (FWER), i.e., all cases in which at least one of the indications in the basket is negative. The type I error control assumes that the interim stopping rules and other adaptation rules are strictly followed. The rules for these adaptations can be prospectively pre-specified and administered by an independent data monitoring committee. Although a statistical penalty must be paid for pruning based on internal data, the benefit of being able to investigate multiple indications in one study far outweighs the penalty. If indications are pruned due to data external to the study, there is no concern about random high bias affecting the pooled analysis and no need for a statistical penalty. Hence, it may be possible to improve the performance of this design using external data sources such as real-world data [20].

15.3.3 An Application Example

Below we give an application example of the basket trial using PFS as the interim and OS as the final endpoints, respectively. Additional examples involving OS as interim and final endpoints, and a single-arm design involving tumor response rate (RR) as interim and final endpoints are discussed in Beckman et al. [28].

In the example, we assume median PFS on the control therapy is 3 months and median OS is 7 months. We further assume a moderate but imperfect correlation between PFS and OS (0.5).

We study six indications, powering each on an improvement in PFS from 3 to 6 months at interim, corresponding to a hazard ratio of 0.5 and assuming an exponential distribution. We enroll 110 patients in each indication, collecting 88 PFS events shortly after all patients are enrolled. Each indication has 90% power for PFS and a one-sided type I error of 2.5%, standard for confirmatory studies. Indications meeting the PFS endpoint will be included in the final pooled analysis, and may be eligible for accelerated approval.

After pruning, the sample size of the remaining indications is adaptively increased to maintain a sample size of 660 in the final pooled analysis and an OS endpoint is evaluated in the pooled remaining indications. For example, if half the indications are pruned, an additional 330 patients are added. The nominal type I error rate (prior to adjustment for pruning) is set at 0.8%, which is 2.5% after adjustment for pruning. The power is approximately 90% for detecting an increase in OS from 7 to 10 months (exact power depends on the number of positive indications in the basket).

In summary, in this randomized example, a 660–990 patient, randomized confirmatory study has the potential to lead to accelerated and full approval of up to six indications. This is about 1–1.5 times the number of patients in a standard confirmatory oncology trial of a single advanced-disease indication.

15.3.4 Conclusions

We have demonstrated that it is feasible to design a basket trial that may be generally acceptable for confirmatory study of efficacious therapies without requiring exceptional levels of efficacy or scientific evidence. The design we present is based on frequentist principles, but application of Bayesian techniques is an important area for further research, as has been developed for exploratory studies [25–27]. For example, the design might be enhanced by applying a data-driven weighting scheme between the hypothesis that each indication should be evaluated separately compared to the hypothesis that they should all be pooled, the latter as in our design [34].

A confirmatory basket trial design has the potential to dramatically lower the cost of development and the required patient numbers for rare indications, thus making access to life-altering medications possible for patients with these conditions.

15.4 Decision Theoretic Methods for Small Populations and Biomarker Trials

Conventionally, the analysis and design of confirmatory clinical trials to demonstrate a treatment effect of an experimental therapy compared to a control is based on frequentist hypothesis testing. Type I error rate control guarantees that the probability of a false-positive decision in favor of the experimental treatment is bounded under the null hypothesis of no differential treatment effect. Sample size planning based on power considerations ensures a high probability of success if the experimental treatment is actually effective. However, the conventionally applied thresholds (as 80% or 90% for the power and 5% or 1% for the false-positive rate) have little justification besides being a common standard. This becomes particularly evident in difficult experimental situations, where resources are scarce. There, the trade-offs between trial size, costs and complexity, and losses and gains due to incorrect and correct decisions become especially relevant. Chen and Beckman examined portfolios of proof of concept (PoC) trials using a frequentist approach and a utility function that accounted for probability weighted benefits and costs of both the PoC trials and resulting confirmatory trials. Introducing a total budget constraint for the PoC trial portfolio, they showed that the common standard for powering PoC trials is wasteful, even under normal circumstances [35–38]. To account for the actual losses and gains resulting from the correct or incorrect rejection of null hypotheses, as well as the costs of clinical trials, also Bayesian decision theoretic approaches have been proposed (see, e.g., Hee et al. [39] for a review). These approaches rely, as well, on utility functions that quantify costs, losses, and gains.

For example, the utility can be represented by the overall health outcome, assuming that future patients will be treated with the experimental treatment if the trial demonstrates its superiority and with the control treatment otherwise. In addition, to account for trial and treatment costs, the utility can be correspondingly adjusted. To this end, health outcomes and costs have to be mapped to a common scale such that for each trial outcome, a single "utility," given by a real number, can be specified. The utility function cannot be directly used to optimize decision criteria and designs of clinical trials because it typically depends on unknown variables as the trial outcome (which is only known after the trial is completed) and the true effect sizes, for which even after the trial, only estimates are available. However, expected utilities can be computed if a prior distribution on the effect sizes is assumed. Based on these expected utilities, decision criteria and trial designs can be compared and optimized.

In the subsections below we discuss two scenarios. The first scenario is a rare disease, where both the size of the trial and the significance level that is applied for decision-making are determined to optimize an expected utility. The second scenario considers the development of a targeted therapy, where there is prior information that a treatment may work in a subgroup of patients only. Here we consider a scenario, where decision-making is based on classical frequentist hypothesis testing and the trial design is optimized. Furthermore, utilities derived under the perspective of different stakeholders are considered.

15.4.1 Simultaneously Optimizing Trial Designs and Decision Rules: A Case Study in Small Populations

Stallard et al. [40], Pearce et al. [41], and Miller et al. [42] studied the setting of a clinical trial in a rare disease, where the total number of patients to be treated in the lifetime of a drug is small. In this context, there is a trade-off between the number of patients who can be recruited for the trial and the number of remaining patients, who can benefit from the treatment that is recommended based on the trial results. Then, performing a too large trial not only leads to unnecessary high trial costs, but entails that there are only few patients left outside of the trial that can benefit from the new treatment. On the other hand, a too small study may result in a high probability that, due to low power and imprecise treatment-effect estimates, the inferior treatment is selected based on the trial result. Similarly, when specifying how strict the decision criteria for the recommendation of the experimental treatment should be, there is a trade-off between the probability of false positive-decisions under the null hypothesis and the statistical power under the alternative.

To apply the decision theoretic approach, a utility function is specified, where the gain is defined as the health outcome of the patients in the trial plus the health outcome of the remaining patients (assuming they are treated with the treatment selected based on the trial results). This gain is then adjusted by the cost of the trial and the treatments. The health outcome in the trial is assumed to be proportional to the true effect size of the applied treatments and the trial size. The health outcome for the patients outside of the trial is proportional to the effect size of the selected treatment and the number of remaining patients. Based on the resulting utility function and an appropriate prior distribution of the effect sizes, the sample size and the significance level to be applied can be optimized to maximize the expected utility. Pearce et al. [41] found, for example, that for ultra-rare diseases, where the total number of patients is very small, the optimal strategy can be to perform no confirmatory trial at all, but to rely on vague prior information only. As the total population size increases, so does the size of the optimal confirmatory clinical trial. The optimal significance levels, in contrast, decrease as the size of the population increases: the larger the number of future patients, the higher the loss associated with the selection of a non-effective treatment and the lower the false-positive rate of the optimized trial.

15.4.2 Optimizing Trial Designs Only: A Case Study in the Development of Targeted Therapies

A challenge in the application of the decision theoretic approach to clinical trial design is that it requires not only to quantify losses and gains on a single scale, but also the specification of prior distributions on the effect sizes (and safety parameters, if their impact on decision-making is considered as well). As the design and interpretation of the optimized clinical trials may depend sensitively on these prior distributions, their choice may be controversial, especially if different stakeholders and interests are involved in their specification.

If one wants to avoid that the decision rules depend on the choice of a prior distribution, the decision theoretic approach can still be used to optimize trial designs assuming that conventional hypothesis testing procedures using standard thresholds are applied for decision-making. This approach has been investigated in the context of confirmatory trials for the development of targeted therapies, where there is prior evidence that a treatment effect may be stronger (or only present) in a subgroup defined by a biomarker (see Chapters 13 and 14 and also Chen and Beckman [43], Beckman et al. [14], Rosenblum et al. [44,45], Graf et al. [46], Krisam and Kieser [47], and Ondra et al. [48]). Decision-making in the presence of subpopulations is challenging because different types of risks need to be accounted for. If the treatment effect is only present in a subpopulation, it will be diluted in the full population and maybe missed if the subpopulation is not investigated separately. In addition, even if the diluted treatment effect can be demonstrated in the overall population, it is not ethical to treat patients that do not benefit from the treatment and could be identified in advance. On the other hand, selecting a spurious subpopulation increases the risk to erroneously conclude that a treatment is efficacious, or may wrongly lead to restricting an efficacious treatment to a too narrow fraction of a potential benefiting population. To model the trade-off of the potential losses and gains that result from false-positive and false-negative decisions for different populations, as above, utility functions can be applied.

Graf et al. [46], Ondra et al. [48], and Ondra et al. [15] considered a setting where trial sponsors optimize utility functions in the current regulatory setting, assuming that conventional frequentist tests are applied as the basis for marketing authorization. If several hypotheses are tested to demonstrate a treatment effect in either the overall population or in one of the considered subgroups, current guidance [49–51] requires the use of appropriate multiple testing procedures [52–54] that guarantee control of the FWER in the strong sense. This ensures that the probability to erroneously reject one or more true null hypotheses is bounded with the specified significance level, regardless of how many and which null hypotheses are true (see Chapter 10).

We review the model derived in Ondra et al. [15] in which a randomized controlled trial comparing an experimental treatment to a control in the full population and a single subgroup S is considered (see Figure 15.2). The means of a normally distributed endpoint are compared and the overall treatment effect is given by $\delta_F = \lambda \delta_S + (1-\lambda)\delta_{S'}$, where λ

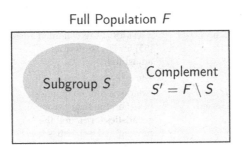

FIGURE 15.2
Clinical trial with a subgroup defined by a biomarker.

denotes the prevalence of subgroup S, and δ_S, $\delta_{S'}$ the effect sizes in the subgroup S and its complement S', respectively. We are interested in the test of the hypotheses $H_F : \delta_F \leq 0$ and $H_S : \delta_S \leq 0$.

In the definition of the utilities, the perspectives of different stakeholders can be accounted for. For example, from a public health perspective, the utility may be given by the overall health outcome adjusted for trial costs, while for a commercial trial sponsor, the utility maybe represented by the net present value. This can be modeled with a utility function where the gains are proportional to the size of the population for which a treatment effect is demonstrated with an appropriate hypothesis test. In addition, for the public health view, the utility depends on the *actual* treatment effect in the respective population, adjusted by treatment costs (which may represent monetary costs as well as side effects). For a commercial sponsor on the other hand, the gain is assumed to be proportional to the *observed* effect estimate (adjusted for some minimal relevant effect size). This corresponds to a setting, where a higher price can be achieved if the pivotal trial shows a large treatment effect. For both the sponsor as well as the public health view, the utility is adjusted for the trial costs, which are not only dependent on the size of the trial, but may also depend on the trial design.

The trial designs considered are summarized in Table 15.2 and include fixed sample trials with fixed subgroup prevalence, where it is assumed that prevalence of the subgroup in the trial corresponds to the prevalence in the total patient population; partial enrichment designs, where the prevalence of the subgroup in the trial may differ from the population prevalence; full enrichment designs, where only patients from the subgroup are recruited; and adaptive partial enrichment designs, where, as in the fixed sample partial enrichment design, in the first stage patients from S and S' are recruited and, after the first stage is completed, an interim analysis is performed. Then, based on the interim results, the second stage sample sizes for S and S' are chosen. This includes the options to stop the trial for futility (second stage sample sizes of zero) or to recruit only patients from the subgroup in the second stage.

To compute the expected utility for a specified trial design, a prior distribution on the effect sizes has to be specified. To illustrate the impact of the prior distribution on the optimal design, two priors are considered: a "strong biomarker prior," which puts a large

TABLE 15.2

Summary of the Considered Trial Designs

Design	Hypotheses	Testing Procedure	Design Parameters
Partial Enrichment with fixed trial prevalence	H_S and H_F	z-test for H_S, stratified z-test for H_F, Bonferroni adjustment	Sample size in the full population
Partial Enrichment	H_S and H_F	z-test for H_S, re-weighted z-test [55] for H_F, Bonferroni adjustment	Sample sizes in S and S'
Full Enrichment	H_S	z-test	Sample size in S
Adaptive Partial Enrichement	H_S and H_F (if selected)	Adaptive combination test based on the stage wise z-statistics for H_S and the re-weighted z-statistics for H_F, Bonferroni adjustment	First stage sample sizes for S and S' and second stage sample size functions for S and S' that depend on interim results

Note: For all trial designs, a modified test of H_F is applied: in addition to the (weighted) z-test, it is assumed that a sufficiently positive trend in the subpopulations S and S' has to be observed to reject H_F. This avoids that efficacy in F is concluded, while only in one subgroup a treatment effect is observed.

weight on the scenario that the treatment is only effective in the subpopulation, and a "weak biomarker prior," which gives more weight to the scenario of a homogeneous treatment effect in the overall population. Expected utilities can now be obtained by integrating the utility functions over the sampling distribution given specific effect sizes in S and S' and then averaging over the prior distribution on the effect sizes. For each of the considered trial designs, the design parameters (see Table 15.2, last column) maximizing the expected utilities can be determined. For the fixed sample designs, the optimal sample sizes can be identified directly via numerical optimization. For the adaptive designs, a dynamic programming approach is applied where, in a first step, for given first-stage sample sizes and each possible interim outcome, optimal second-stage sample sizes are determined that maximize the expected utility conditional on the first-stage outcome. In a second step, the first-stage sample sizes are optimized such that the unconditional expected utility is maximized (under the assumption that the optimal adaptation rule identified in the first step is applied).

For a specific example, Figure 15.3 shows the expected utilities of the optimized designs as function of the prevalence of the subgroup in the overall patient population. If the prior gives only a small weight to the scenario where the treatment is effective in the subgroup only (weak biomarker prior), for both the public health and the sponsor view, an adaptive design is optimal, which can adapt the sample size for the second stage. For the sponsor view, this holds also for the strong biomarker prior. In contrast, for the public health utility function and very low prevalence of the subgroup, the optimal design is to perform no trial at all, because the costs of all trial designs exceeds the expected benefit leading to negative expected utilities. As the prevalence increases, the full enrichment design becomes optimal. The discrepancy between the optimal designs of the sponsor and public health view are due to the differences in the utility functions—the sponsor may benefit from a positive result in the overall population, even if the treatment is only effective in the subgroup. Figure 15.4 illustrates the optimal adaptation rules of the adaptive designs (for the strong biomarker prior). Also, here it can be seen that the optimal designs for the public health view are more conservative, and the region where the trial stops for futility or continues only

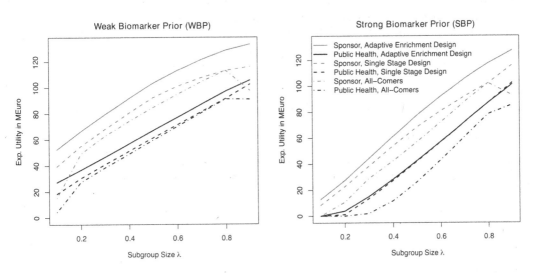

FIGURE 15.3
Utilities of optimized trial designs for different prevalences of the subgroup. (Adapted from Ondra T et al. *Stat Methods Med Res.* 2017;0962280217747312.)

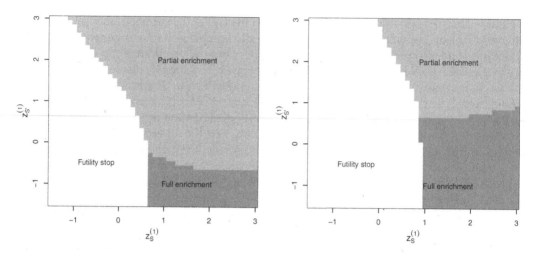

FIGURE 15.4
Optimal adaptation rules for the strong biomarker prior: the sponsor (left) and the public health (right) view. (Adapted from Ondra T et al. *Stat Methods Med Res.* 2017;0962280217747312.)

in the subgroup are larger. Ondra et al. [15] also find that the optimized sample sizes are consistently larger for the optimal designs under the public health view. This may be due to the fact that that the sponsor may benefit from some variability in the effect estimates.

15.4.3 Conclusion

The application of decision theoretic methods for trial design and analysis poses several challenges. The specification of utility functions depends on many, often unknown, parameters. In addition the quantification of health outcomes and their mapping to a common scale with trial and treatment costs can be controversial. Finally, the elicitation of prior distributions can be difficult, especially in an environment with stakeholders with different incentive structures. However, the framework encourages clearer analysis of the impact of decisions for different stakeholders and provides a useful tool to better justify the decision criteria and designs used in clinical trials.

Acknowledgments

The results summarized in Section 15.2 were developed within the FP7 EU project Inspire (Grant Agreement no 602144) and IDEX grant from Université Sorbonne Paris Cité (2013, project 24).

The work in Section 15.3 originated from the Drug Information Association (DIA) workstream on small populations and rare diseases, pathway design subgroup. We are indebted to the following members of the subgroup and DIA workstream: Ohad Amit, Christine Gause, Sebastian Jobjornsson, Lingyun Liu, Robert T O'Neill, Sue Jane Wang, Samuel Yuan, and Yi Zhou.

The results summarised in Section 15.4.1 were developed within the FP7 EU project Inspire (Grant Agreement no 602144) in collaboration with Simon Day, Jason Madan, Frank

Miller, Michael Pearce, Nigel Stallard, and Siew Wan Hee. The results reviewed in Section 15.4.2 are based on a collaboration of the Inspire and FP7 EU Ideal (Grant Agreement no 602552) project in collaboration with Carl-Fredrik Burman, Sebastian Jobjörnsson, Franz König, Thomas Ondra, and Nigel Stallard.

Conflict of Interest Disclosure

RAB is a stockholder in Johnson & Johnson, Inc, and consults for AstraZeneca and EMD Serono. He is the founder and Chief Scientific Officer of Onco-Mind, LLC, founded to foster study of the effect of intratumoral heterogeneity and evolutionary dynamics on optimal personalized treatment strategies for cancer.

CC is a full-time employee of Merck & Co., Inc., and may benefit from expedited drug approvals based on the proposed basket design of clinical trials.

References

1. Cerami E, Gao J, Dogrusoz U et al. The cBio cancer genomics portal: An open platform for exploring multidimensional cancer genomics data. *Cancer Discov.* 2012 May;2(5):401–4.
2. Lawrence MS, Stojanov P, Mermel CH, Robinson JT, Garraway LA, Golub TR, Meyerson M, Gabriel SB, Lander ES, Getz G. Discovery and saturation analysis of cancer genes across 21 tumour types. *Nature.* 2014 Jan 23;505(7484):495–501.
3. Arnedos M, André F, Farace F, Lacroix L, Besse B, Robert C, Soria JC, Eggermont AM. The challenge to bring personalized cancer medicine from clinical trials into routine clinical practice: The case of the Institut Gustave Roussy. *Mol Oncol.* 2012 Apr;6(2):204–10.
4. Hilgers RD, Roes K, Stallard N; Ideal, Asterix and inspire project groups. Directions for new developments on statistical design and analysis of small population group trials. *Orphanet J Rare Dis.* 2016 Jun 14;11(1):78.
5. Brasseur D. Paediatric research and the regulation "better medicines for the children in europe". *Eur J Clin Pharmacol.* 2011;67(Supp.):1–3.
6. Sammons H. Avoiding clinical trials in children. *Arch Dis Child.* 2011;96:291–2.
7. Taylor R, Pizer B, Short S. Promoting collaboration between adult and paediatric clinical trial groups. *Clin Oncol (R Coll Radiol).* 2008;20:714–6.
8. European Medicines Agency (EMA). Guideline on clinical trials in small populations. http://www.ema.europa.eu/docs/en_GB/document_library/n Scientific_guideline/2009/09/WC500003615.pdf, 2006.
9. Hlavin G, Koenig F, Male C, Posch M, Bauer P. Evidence, eminence and extrapolation. *Stat Med.* 2016;35(13):2117–32.
10. Petit C, Samson A, Morita S, Ursino M, Guedj J, Jullien V, Comets E, Zohar S. Unified approach for extrapolation and bridging of adult information in early-phase dose-finding paediatric studies. *Stat Methods Med Res.* 2016 Jan 1;962280216671348.
11. Zohar S, Katsahian S, O'Quigley J. An approach to meta-analysis of dose-finding studies. *Stat Med.* 2011;30:2109–16.
12. dfped, R cran pakage 2018. https://CRAN.R-project.org/package=dfped
13. Zhang J, Braun TM, Taylor JM. Adaptive prior variance calibration in the Bayesian continual reassessment method. *Stat Med.* 2013 Jun 15;32(13):2221–34.

14. Beckman RA, Clark J, Chen C. Integrating predictive biomarkers and classifiers into oncology clinical development programmes. *Nat Rev Drug Discov.* 2011;10(10):735.
15. Ondra T, Jobjörnsson S, Beckman RA, Burman CF, König F, Stallard N, Posch M. Optimized adaptive enrichment designs. *Stat Methods Med Res.* 2017;0962280217747312.
16. Barker AD, Sigman CC, Kelloff GJ, Hylton NM, Berry DA, Esserman LJ. I-SPY 2: An adaptive breast cancer trial design in the setting of neoadjuvant chemotherapy. *Clin Pharmacol Ther.* 2009;86:97–100.
17. Kim ES, Herbst RS, Wistuba II et al. The BATTLE trial: Personalizing therapy for lung cancer. *Cancer Disc.* 2011;1:44–53.
18. Kaiser J. Biomedicine. Rare cancer successes spawn 'exceptional' research efforts. *Science.* 2013;340:263.
19. Ciriello G, Miller ML, Aksoy BA, Senbabaoglu Y, Schultz N, Sander C. Emerging landscapes of oncogenic signatures across human cancers. *Nat Genet.* 2013;45:1127–33.
20. Guinn DA, Madhavan S, and Beckman RA. Harnessing real world data to inform platform trial design. In *Platform Trials*, Antonijevic Z and Beckman RA, eds., CRC Press, Taylor and Francis Group, Boca Raton, Florida, USA, in press, 2018.
21. Meador CB, Micheel CM, Levy MA et al. Beyond histology: Translating tumor genotypes into clinically effective targeted therapies. *Clin Cancer Res.* 2014;20:2264–75.
22. Lacombe D, Burocka S, Bogaertsa J, Schoeffskib P, Golfinopoulosa V, Stuppa R. The dream and reality of histology agnostic cancer clinical trials. *Mol Onc.* 2014;8:1057–63.
23. Sleijfer S, Bogaerts J, Siu LL. Designing transformative clinical trials in the cancer genome era. *J Clin Oncol.* 2013;31:1834–41.
24. Demetri G, Becker R, Woodcock J, Doroshow J, Nisen P, Sommer J. Alternative trial designs based on tumor genetics/pathway characteristics instead of histology. *Issue Brief: Conference on Clinical Cancer Research* 2011; http://www.focr.org/conference-clinical-cancer-research-2011.
25. Berry SM, Broglio KR, Groshen S et al. Bayesian hierarchical modeling of patient subpopulations: Efficient designs of phase II oncology clinical trials. *Clin Trials.* 2013;10:720–34.
26. Cunanan KM, Iasonos A, Shen R et al. An efficient basket trial design. *Stat Med.* 2017;36:1568–79.
27. Simon RM, Geyer S, Subramanian J, Roychowdhury S. The Bayesian basket design for genomic variant-driven Phase II trials. *Seminars in Onc.* 2016;43:13–8.
28. Beckman RA, Antonijevic Z, Kalamegham R, Chen C. Adaptive design for a confirmatory basket trial in multiple tumor types based on a putative predictive biomarker. *Clin Pharmacol Ther.* 2016;100:617–25.
29. Chen C, Li N, Yuan S, Antonijevic Z, Kalamegham R, Beckman RA. Statistical design and considerations of a Phase 3 basket trial for simultaneous investigation of multiple tumor types in one study. *Stat Biopharm Res.* 2016;8:248–57.
30. Trusheim M, Shrier AA, Antonijevic Z et al. PIPELINEs: Creating comparable clinical knowledge efficiently by linking trial platforms. *Clin Pharmacol Ther.* 2016;100:713–29.
31. United States Food and Drug Administration. Guidance for Industry: Clinical trial endpoints for the approval of cancer drugs and biologics, 2007. https://www.fda.gov/downloads/drugs/guidancecomplianceregulatoryinformation/guidances/ucm071590.pdf
32. European Medicines Agency, Guideline on the evaluation of anticancer medicinal products in man, 2013; http://www.ema.europa.eu/docs/en_GB/document_library/Scientific_guideline/2013/01/WC500137128.pdf
33. United States Food and Drug Administration. Guidance for Industry Expedited Programs for Serious Conditions – Drugs and Biologics. http://www.fda.gov/downloads/drugs/guidancecomplianceregulatoryinformation/guidances/ucm358301.pdf
34. Simon RM. Primary site-independent clinical trials in oncology. In *Platform Trials*, Z Antonijevic, RA Beckman, eds. Boca Raton, Florida, USA: CRC Press, Taylor and Francis Group, in press, 2018.
35. Chen C, Beckman RA. Optimal cost-effective designs of Proof of Concept trials and associated Go-No Go decisions. *Proceedings of the American Statistical Association, Biometrics Section.* 2007;394–9.
36. Chen C, Beckman RA. Optimal cost-effective designs of Phase II Proof of Concept trials and associated Go-No Go decisions. *J Biopharm Stat.* 2009a;19:424–36.

37. Chen C, Beckman RA. Optimal cost-effective Go-No Go decisions in late-stage oncology drug development. *Stat Biopharm Res.* 2009b;1:159–69.

38. Chen C, Beckman RA. Maximizing return on socioeconomic investment in phase II proof-of-concept trials. *Clin Cancer Res.* 2014;20:1730–4.

39. Hee SW, Hamborg T, Day S, Madan J, Miller F, Posch M, Zohar S, Stallard N. Decision-theoretic designs for small trials and pilot studies: A review. *Stat Methods Med Res.* 2016;25(3):1022–38.

40. Stallard N, Miller F, Day S, Hee SW, Madan J, Zohar S, Posch M. Determination of the optimal sample size for a clinical trial accounting for the population size. *Biom J.* 2017;59(4):609–25.

41. Pearce M, Hee SW, Madan J, Posch M, Day S, Miller F, Zohar S, Stallard N. Value of information methods to design a clinical trial in a small population to optimise a health economic utility function. *BMC Med Res Methodol.* 2018;18(1):20.

42. Miller F, Zohar S, Stallard N, Madan J, Posch M, Hee SW, Pearce M, Vågerö M, Day S. Approaches to sample size calculation for clinical trials in rare diseases. *Pharm Stat.* 2018;17(3):214–30.

43. Chen C, Beckman RA. Hypothesis testing in a confirmatory Phase III trial with a possible subset effect. *Stat Biopharm Res.* 2009c;1:431–40.

44. Rosenblum M, Liu H, Yen EH. Optimal tests of treatment effects for the overall population and two subpopulations in randomized trials, using sparse linear programming. *J Am Stat Assoc.* 2014b;109(507):1216–28.

45. Rosenblum M, Fang X, Liu H. Optimal, two stage, adaptive enrichment designs for randomized trials using sparse linear programming. Johns Hopkins University. Technical report, of Biostatistics Working Papers. Working Paper 273, 2014a. http://biostats.bepress.com/jhubiostat/paper273 109, 1216–1228.1.

46. Graf AC, Posch M, Koenig F. Adaptive designs for subpopulation analysis optimizing utility functions. *Biom J.* 2015;57(1):76–89.

47. Krisam J, Kieser M. Optimal decision rules for biomarker-based subgroup selection for a targeted therapy in oncology. *Int J Mol Sci.* 2015;16(5):10354–75.

48. Ondra T, Jobjörnsson S, Beckman RA, Burman CF, König F, Stallard N, Posch M. Optimizing trial designs for targeted therapies. *PloS One.* 2016;11(9):e0163726.

49. European Medicines Agency. Points to consider on multiplicity issues in clinical trials. 2002.

50. European Medicines Agency. Guideline on multiplicity issues in clinical trials (draft). 2017.

51. FDA. Multiple endpoints in clinical trials – guidance for industry – draft guidance. 2017.

52. Spiessens B, Debois M. Adjusted significance levels for subgroup analyses in clinical trials. *Contemp Clin Trials.* 2010;31(6):647–56.

53. Placzek M, Friede T. Clinical trials with nested subgroups: Analysis, sample size determination and internal pilot studies. *Stat Methods Med Res.* 2017;0962280217696116.

54. Graf A, Wassmer G, Friede T, Gera R, Posch M. Robustness of testing procedures for confirmatory subpopulation analyses based on a continuous biomarker. 2018.

55. Zhao YD, Dmitrienko A, Tamura R. Design and analysis considerations in clinical trials with a sensitive subpopulation. *Stat Biopharm Res.* 2010;2(1):72–83.

56. ICH11 International Conference on Harmonisation of Technical Requirements for Registration of Pharmaceuticals for Human Use (ICH). Addendum to ich e11: Clinical investigation of medicinal products in the pediatric population - e11(r1), August 2016. http://www.ich.org/fileadmin/Public_Web_Site/ICH_Products/Guidelines/Efficacy/E11/ICH_E11_R1_Step_2_25Aug2016_Final.pdf

57. Li W, Chen C, Li, X, Beckman, RA. Estimation of treatment effect in two-stage confirmatory oncology trials of personalized medicines. *Stat. Med.* 2017; 36: 1843–1861.

16

Statistical Methods for Biomarker and Subgroup Evaluation in Oncology Trials

Ilya Lipkovich, Alex Dmitrienko, and Bohdana Ratitch

CONTENTS

16.1 Introduction

Subgroup-analysis considerations are beginning to attract more attention in late-stage oncology trials due the increasing emphasis on the development of targeted therapies. Targeted therapies are known to benefit only a subset of the overall patient population with certain characteristics. The characteristics are defined in terms of demographic, clinical, and other variables, broadly referred to as *biomarkers*. The role of targeted therapies in oncology development programs has been discussed in multiple publications. A detailed discussion of relevant statistical principles can be found, for example, in Freidlin et al. [1], Freidlin et al. [2], and Sargent and Mandrekar [3]. In addition, relevant regulatory considerations are discussed in the recently released guidelines, including the US Food and Drug Administration's guideline on enrichment strategies in clinical trials [4] and the European Medicines Agency's guideline on subgroup analysis [5].

Large sets of biomarkers are examined in most late-stage trials, which gives rise to a challenging problem of identifying biomarkers that are truly predictive of treatment response, e.g., help predict the treatment effect on the primary efficacy endpoint such as overall survival or progression-free survival. Patient subgroups are defined using individual biomarkers or biomarker signatures (sets of multiple biomarkers). Identification of promising subgroups leads to generation of new hypotheses that can be tested in subsequent confirmatory trials. This class of problems is commonly referred to as *exploratory* or *data-driven* subgroup analysis.

This chapter discusses statistical methods for the investigation of biomarkers and patient subgroups with a beneficial treatment effect in an exploratory subgroup-analysis setting. The chapter emphasizes the importance of using scientifically sound principles in this setting and encourages clinical trial sponsors to employ disciplined approaches to subgroup exploration. It is well known that post-hoc subgroup investigation is very likely to lead to spurious results and it is critical to carefully apply key principles formulated in recent publications, e.g., Lipkovich et al. [6], to enable a reliable quantitative assessment of candidate biomarkers and corresponding patient subgroups. According to these principles, the trial's sponsor needs to prospectively define an analytical evaluation strategy for selecting the most relevant patient subgroups and pre-specify strategies for analyzing treatment effects within these subgroups. For a review of general subgroup-analysis strategies and recent developments in exploratory subgroup analysis, see Dmitrienko et al. [7], Dmitrienko and Millen [8], and Ondra et al. [9], and a comprehensive review is provided by Lipkovich et al. [6]. See also recent review articles focusing on specific types of modeling strategies employed in exploratory subgroup analysis [10–12].

This chapter is organized as follows. Section 16.2 provides an overview of statistical methods aimed at biomarker discovery and subgroup identification. Using a general setting of phase III oncology trials, this section compares traditional approaches that rely on basic univariate models to advanced approaches that utilize modern methods developed in machine learning and related fields. This section also defines a useful taxonomy of principled methods for biomarker and subgroup evaluation and gives examples of popular principled methods. A case study based on a phase III trial in patients with leukemia is introduced in Section 16.3. Section 16.4 introduces the SIDES methodology, a popular approach to biomarker discovery and subgroup identification in late-stage clinical trials. A family of subgroup search algorithms based on the SIDES method is defined and is applied to the case study to illustrate the process of assessing the predictive strength of candidate biomarkers and identify subgroups of patients who experience beneficial treatment in this trial. Conclusions and a discussion of practical considerations in exploratory subgroup analysis are presented in Section 16.5.

16.2 Overview of Methods for Biomarker and Subgroup Evaluation

Methods for biomarker and subgroup evaluation in late-stage clinical trials gained popularity in the last decade with the advances of personalized medicine, also known as precision medicine. These methods emerged from cross-fertilization of several fields, including machine learning, multiple testing, and causal inference, and are replacing various *ad-hoc* approaches to biomarker discovery and subgroup identification that were popular in the past.

A general setting that will be assumed throughout this chapter is that of an oncology phase III clinical trial with a time-to-event endpoint, e.g., overall survival, defined using the following two variables:

- Time to the event of interest or censoring (whichever occurs first) denoted by Y.
- Censoring indicator denoted by δ, where $\delta = 1$ if the event of interest has been observed during the trial and $\delta = 0$ otherwise.

The trial includes N patients who are randomized to the experimental treatment or control. Additionally, suppose that p biomarkers, denoted by $X_1, X_2, \ldots, X_p$, are collected prior to randomization, reflecting various baseline patient characteristics.

An important methodological distinction should be made between *prognostic* and *predictive* biomarkers. A biomarker is said to be prognostic if it helps predict a patent's outcome if the patient is left untreated. A biomarker is considered to be predictive if it *differentially* predicts outcomes for treated versus untreated patients. In other words, a predictive biomarker is predictive of a treatment contrast. In the language of statistical modeling, predictive biomarkers are said to exhibit a treatment-by-biomarker interaction effect, whereas the interaction with treatment is not present for biomarkers that are purely prognostic. For a simple case of a continuous biomarker and a continuous outcome, Figure 16.1 provides a graphical summary of treatment-by-biomarker interactions under four possible scenarios:

- Scenario A: The selected biomarker is prognostic but not predictive.
- Scenario B: The biomarker is both prognostic and predictive.
- Scenario C: The biomarker is predictive but not prognostic.
- Scenario D: The biomarker is neither predictive nor prognostic.

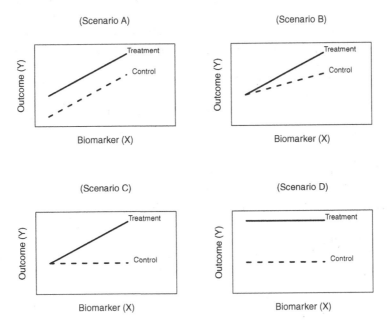

FIGURE 16.1

Predictive and prognostic biomarkers: (Scenario A) X is prognostic but not predictive, (Scenario B) X is both prognostic and predictive, (Scenario C) X is predictive but not prognostic, and (Scenario D) X is neither predictive nor prognostic.

Note that the treatment-by-biomarker interaction is present, i.e., the two lines are not parallel, in Scenarios B and C.

When a biomarker is predictive of treatment response (Scenarios B and C in Figure 16.1), it is fairly common to define a binary classifier based on this biomarker. For example, a cutoff value may be determined, denoted by c, such that a beneficial treatment effect is observed only within the patient subgroup $S = \{X \leq c\}$ and patients in the complementary subgroup may experience no or little treatment benefit. This may be especially useful for settings when a change in treatment difference may be not linear in the biomarker X but rather sharply increases after X reaches some cutoff value c.

16.2.1 Traditional Approaches to Biomarker and Subgroup Evaluation

Historically, the problem of identifying predictive biomarkers and associated subgroup effects was conceptualized within the framework of detecting significant treatment-by-covariate interactions, typically within (generalized) linear models. In the context of oncology trials, a typical exploration of predictive biomarkers would utilize a traditional proportional hazards Cox model including the treatment indicator T, a single candidate biomarker X, and its interaction with treatment $X \times T$. More formally, the hazard function for the time-to-event variable (Y) is modeled as

$$h(y \mid T, X) = h_0(y)\exp(a_1 X + a_2 T + a_3 X \times T),$$

where $h_0(y)$ is the baseline hazard function, which is modeled non-parametrically as a step function, $T = 1$ corresponds to the experimental treatment, and $T = 0$ corresponds to the control.

Typically, a small set of up to 10 candidate biomarkers would be specified in the exploratory analysis section of a trial protocol, and the candidate biomarkers would be evaluated one at a time using a basic univariate analysis approach, such as the described Cox model. Biomarkers whose interaction effects with the treatment are found significant at a pre-specified level, e.g., $\alpha = 0.1$, are retained for further investigation. Often, optimal cut-offs for continuous or ordinal biomarkers are further identified using maximally selected chi-square statistics (see, for example, Miller and Siegmund [13]).

An obvious limitation of the approach outlined in this section is that it can only identify patient subgroups based on a single biomarker. Also, the approach is *ad-hoc* in that it does not control the overall type I error rate or false discovery rate across all investigated biomarkers. The cut-off selection procedures provide type I error control only with respect to the patient subgroups defined by the selected biomarker X. At the same time, the procedure suffers from low power of detecting an interaction effect that may be easily mis-specified when modeled parametrically via a single coefficient a_3.

As an alternative to simplistic univariate approaches, multiple candidate biomarkers (often dichotomized to create binary variables) and their higher order interactions with treatment may be evaluated in a single regression model. Subgroups are identified if the significance of interaction terms reach pre-specified levels. However, this analytic strategy also suffers from low power and arbitrary choice of the significance levels. Most importantly, this approach requires pre-specification of interaction terms and covariate cut-offs to form the individual subgroups. In the presence of a large number of candidate subgroups, regression modeling with stepwise selection of the main and interactions effects using pre-defined entry and removal criteria is often employed to select predictive biomarkers. However, stepwise selection methods are notoriously unstable, and their operating characteristics are not well understood.

Furthermore, methods of penalized regression and their extensions, as well as other methods adopted from statistical/machine learning that may efficiently handle a large number of variables and interaction terms, have been proposed. These complex methods are rarely employed in evaluating biomarkers in clinical trials and may require very careful tuning. As one issue, the treatment-by-biomarker interaction effects, i.e., predictive effects, may be weak relative to the main (prognostic) effects and might not be selected using traditional penalized regression methods such as the lasso that are designed to solve a general problem of prediction rather than subgroup selection (see Imai and Ratkovic [14] and Chapter 18).

16.2.2 Taxonomy of Modern Approaches to Biomarker and Subgroup Evaluation

Methods of biomarker discovery and subgroup identification can be grouped into the following three classes [6]:

- *Global outcome modeling* of the data using both prognostic and predictive effects. Typically, this involves fitting a complex "black-box" model in a high-dimensional space of covariates and covariate-by-treatment interactions. Once the outcome function is estimated, predictive effects can be teased out from the black-box model, e.g., the Virtual Twins method [15], that uses random forests and classification and regression trees.

- *Global treatment effect modeling* focuses on predictive effects only (obviating the need of fitting-in prognostic effects). Methods in this class use global modeling in the sense that individual treatment effects are estimated across the entire biomarker space, which may include vast regions that are relatively flat. This includes various adaptations of tree-based methods to the task of subgroup identification, e.g., the interaction trees (IT) method [16,17], proposed extensions of the GUIDE framework [18], and the model-based recursive-partitioning platform developed by Seibold et al. [19]. An important subclass of global treatment-effect modeling methods is the set of methods for estimating optimal individualized treatment regimes that have recently emerged and gained popularity in clinical trial applications (see, for example, [20,21]). These approaches aim at identifying a direct mapping from biomarker values to an optimal treatment assignment across the entire biomarker space without explicitly modeling the treatment effect.

- *Local modeling of subgroup effects* uses a direct search for treatment-by-covariate interactions and subgroups with desirable characteristics (e.g., with enhanced treatment effect). This approach obviates the need to estimate the response function over the entire covariate space and focuses on identifying specific regions with large differential treatment effects. Some of the methods in this class, e.g., the methods developed by Kehl and Ulm [22] and Chen et al. [23], were inspired by bump hunting methods, also known as patient rule induction methods (PRIM), of Friedman and Fisher [24]. Another strategy for direct subgroup search was implemented via recursive partitioning in the SIDES method introduced in Lipkovich et al. [25] and later extended to the SIDEScreen method [26].

Global outcome modeling methods will be briefly discussed in Section 16.2.3. Sections 16.2.4 and 16.2.5 provide an overview of global treatment effect modeling methods in an oncology setting. Methods that rely on local modeling of subgroup effects such as the SIDES family of subgroup search algorithms will be discussed in detail in Section 16.4.

Additional information and links to available freeware for subgroup identification (including R functions and packages developed by the authors and independent researchers) can be found at

http://biopharmnet.com/subgroup-analysis-software/

The freeware is currently provided for a number of key methods including several implementations of the SIDES [25], the interaction tree [16], Virtiual Twins [15], GUIDE [18], QUINT (qualitative interaction trees [27]), Blasso [28], ROWSi [21], and a model-based recursive partitioning (mob), implemented within R package partykit (A Toolkit for Recursive Partitioning) by Hothorn, Seibold, and Zeileis [42].

16.2.3 Global Outcome Modeling Methods

Biomarker evaluation methods in this class are easy to implement using modern statistical software and they rely on the following general two-step algorithm. At the first step, an appropriate "black-box" model is fitted, for example, using random forest or gradient boosting. As a result, in the context of time-to-event outcomes, for each patient with the covariate vector $x = (x_1, \ldots, x_p)$, one can obtain the ratio of two estimated hazard functions: one predicting the hazard if the patient is treated ($T = 1$) and the other, if the patient is left untreated ($T = 0$):

$$\hat{z}(x) = \hat{h}(y \mid T = 1, X = x)/\hat{h}(y \mid T = 0, X = x).$$

In other words, $\hat{z}(x)$ represents the treatment contrast as a function of covariates.

It is assumed here that the common baseline hazard cancels out and thus $\hat{z}(x)$ is a function of covariates alone and is not time-dependent. At the second step, a new outcome variable is defined for each patient, namely, the treatment effect

$$Z_i = \hat{z}(x_i), \quad i = 1, \ldots, N.$$

The resulting individual treatment contrast values are modeled using any method of predictive modeling to identify important predictors of a favorable treatment effect and, hence, the corresponding patient subgroups. For example, a popular classification and regression trees (CART) [29] method can be used to construct a regression tree where terminal nodes represent subgroups of patients with a homogeneous treatment effect.

16.2.4 Global Treatment Effect Modeling with Interaction Trees

A tree-based model induces a partitioning on the covariate space represented by non-overlapping regions (terminal nodes or "leaves") that define distinct patient subgroups. Hence a tree-based regression model is a natural modeling framework for the subgroup identification problem. A classical tree regression, e.g., the CART method [29], constructs trees with terminal nodes representing subgroups of patients with a homogeneous outcome. One can think of a tree model as a piecewise constant fit to the data, thus predicting a constant outcome within each leaf. For the purposes of subgroup identification, one would need a tree where the predicted treatment effect is constant for patients within each leaf.

A natural extension of a tree-based approach to identifying subsets of a trial's overall population with a homogeneous treatment effect is the interaction trees method [16,17]. The IT method for time-to-event outcomes extends the ideas from CART [29] and the procedure of Leblanc and Crowley [30]. Like many other tree-based procedures, IT uses a three-step algorithm:

- Step 1. Grow an initial unpruned tree (of large size).
- Step 2. Construct a nested sequence of pruned trees (of diminishing size).
- Step 3. Select the best-sized subtree from the sequence.

The splitting (recursive partitioning) process through which the tree nodes are identified is similar to that used in CART in the sense that every covariate is considered as a candidate splitter and the best binary split is evaluated across all potential cut-offs using an appropriate splitting criterion. The difference between IT and CART lies in the splitting criterion. CART relies on the criterion based on the reduction in the residual sums of squares obtained by the model fitted to the parent node with a candidate splitting variable as the single covariate as compared to the intercept-only model. One can use the deviance in place of the residual sum of squares for outcomes involving non-normal likelihood functions.

With the IT method, the criterion is based on the improvement in the model including a split-by-treatment interaction term versus a model without this interaction term. Therefore, in the course of recursive partitioning, each parent node is split based on a covariate having (locally) the largest predictive effect. Therefore, the two resulting child subgroups exhibit a differential treatment effect implying that individual treatment effects within each subgroup are homogeneous compared to the parent node. More formally, two models are entertained for each candidate split $(S = \{0,1\})$ at each parent node, i.e.,

$$h_1(y \mid T, S) = h_{01}(y)\exp(a_{11}S + a_{21}T + a_{31}S \times T)$$

and

$$h_2(y \mid T, S) = h_{02}(y)\exp(a_{12}S + a_{22}T).$$

Here, the predictive effect is captured by the coefficient for the split-by-treatment interaction in the first model, i.e., a_{31}.

The likelihood-ratio test statistic is computed as $G(S) = -2(l_2 - l_1)$, where l_1 and l_2 are the partial log-likelihood values for the models h_1 and h_2, respectively. Using this test statistic, the IT method proceeds as follows:

- The best split S is selected as the split that maximizes $G(S)$, and therefore it is associated with larger estimated interaction effects ($\hat{a}_{31}$).
- The tree is pruned (reduced in size) in order to optimize a trade-off between goodness of fit to the data and model complexity. A sequence of nested pruned trees is formed by applying the "weakest-link" pruning method, as in CART, to the interaction-complexity criterion constructed for a given tree I. This criterion is defined as $G_a(I) = G(I) - a|I|$, where $G(I)$ is the sum of the $G(S)$ statistics over all internal nodes (splits) in the tree I, $|I|$ is the number of splits in the tree and a is a penalty, i.e., the price per split.

- The best-sized subtree from the sequence is selected by further constructing a bias-adjusted estimate $\hat{G}(I)$ that is corrected for over-optimism in $G(I)$ due to the selection of best splits over all possible choices, in the process the tree growing. The amount of correction is determined using the bootstrap. The best tree in the pruning sequence from Step 2 is selected as the one that maximizes $\hat{G}_a(I) = \hat{G}(I) - a|I|$, where a is typically chosen as in the BIC criterion, i.e., $a = \log n$, where n is the sample size.

The output of the IT method is a tree that defines regions in the covariate space (patient subgroups) with a homogeneous treatment effect. A trial's sponsor may be interested in the subgroup with the largest apparent treatment difference or further use resampling methods to obtain "honest" estimates of treatment effects in the selected node. The IT methodology supports various enhancements such as "amalgamation" (merging nodes with similar treatment effects) and evaluating the predictive strength of candidate biomarkers using random forests based on interaction trees.

For more information on the IT method, see Su et al. [16], as well as an example in Lipkovich et al. [6].

16.2.5 Modeling Optimal Treatment Regimes

Recent years saw a burst of research on estimating individualized treatment regimes from both randomized and observational studies that came as a cross-pollination from the literature on machine learning and causal inference ([31,32]).

Generally, a treatment regime is a function that maps a vector of patient data, possibly including baseline patient characteristics and evolving outcomes, into a treatment choice. Here we limit our consideration to treatment regimes $d(X)$ that are defined as functions mapping a p-dimensional vector of baseline patient characteristics or biomarkers denoted by X to a binary treatment choice, i.e., $T = 0$ (control) or $T = 1$ (experimental treatment). In this setting, we are considering an overall (one-time) treatment assignment without dynamic treatment adjustments over time.

In what follows we present a simple method known as the outcome weighted learning (OWL) method that essentially casts the problem of estimating an optimal regime as a prediction problem [20,21]. To introduce the notation, let $Y_i(0)$ and $Y_i(1)$ denote two potential outcomes, e.g., times to an event of interest in a clinical trial, that would have been observed if a randomly selected patient i had been assigned to the control arm ($T = 0$) or treatment arm ($T = 1$), respectively. The value function [31] of an arbitrary treatment regime $d(X)$ is defined as the expected value over potential outcomes for all possible regimes translating a random vector X into a treatment choice, i.e., $V[d(X)] = E[Y(d(X))]$. The value function can be represented in terms of observed rather than potential outcomes using the inverse probability weighting applied to the subset of patients who *actually followed* a given regime $d(X)$, i.e.,

$$V[d(X)] = E\left(\frac{I(T = d(X))}{P(T = d(X))} Y\right).$$

Here, the expectation is evaluated with respect to the random triple (X, T, Y). Note that the probability of following the treatment regime $d(X)$, i.e., $P(T = d(X))$, in the denominator of this expression is applied to the subset of patients who actually followed the regime. Therefore, this probability is simply the probability of taking the observed treatment given the patient-specific covariates, i.e., $P(T = t \mid X)$. In the context of a randomized clinical trial, this probability is equal to the proportion of patients randomized to a given treatment and, in a observational setting, this probability is termed the propensity of treatment and can be estimated using a logistic regression model.

An optimal regime can be obtained as a regime that maximizes the value function or, equivalently, minimizes the complementary expression on a subset of patients who did not follow a given regime, i.e.,

$$\tilde{d}(X) = \text{argmin } E\left(\frac{I(T \neq d(X))}{P(T = t \mid X)} Y\right).$$

The expression on the right-hand side is minimized with respect to an appropriate family of regimes defined as functions of the candidate covariates X, as will be explained later.

A key insight [20] was to recognize that this optimization problem is in fact equivalent to *minimizing* a weighted classification loss with respect to a classifier $d(X)$, i.e., a mapping that assigns (classifies) patients to optimal treatment based on their baseline characteristics. Note that the patient weight used in this problem, $w = Y/P(T = t \mid X)$, is outcome-based as it involves the outcome Y. The objective is to find a classifier $\tilde{d}(X)$ which misclassifies (with respect to the treatment actually received) fewer patients with good observed outcomes compared to patients with poor outcomes. In other words, patients who experienced a beneficial outcome under their actual treatment (assuming that a longer time to the event Y is beneficial) would be assigned a larger weight and the optimal classifier would be more likely to assign such patients to the same treatment as actually received. Conversely, patients with poor outcomes under their actual treatment would receive smaller weights, in which case the optimal classifier may be more likely to assign them to the other treatment, as the misclassification error bears little costs.

It needs to be pointed out that the process of fitting a classifier by directly minimizing a zero-one empirical loss function, such as the one suggested by the expression for optimal treatment regime, is a cumbersome task. To address this problem, various proposals aimed at minimizing an empirical weighted loss using a smooth loss functions have been made in the literature. A natural loss function is an exponential loss used in the familiar logistic regression [21] or a hinge loss used in support vector machines [20]. This loss function gives rise to the OWL method that essentially fits a weighted logistic regression model to the observed treatment indicator, $t_i = \{0,1\}$, with the outcome-based patient-specific weights, $w_i = y_i/\pi$ for patients in the treatment arm and $w_i = y_i/(1-\pi)$ for patients in the control arm. In the context of an oncology trial, y_i is the time-to-event value which may be censored and π is the randomization ratio, which is equal to the probability of being assigned to the experimental treatment arm.

More formally, a treatment regime that optimizes a particular loss function is estimated via a linear function of the candidate biomarkers:

$$\tilde{d}(X) = I\left(\sum_{j=1}^{p} \beta_j X_j > 0\right).$$

Because the underlying problem is often high-dimensional, a penalized regression model can be utilized (see Chapters 17 and 18).

A popular choice is lasso [33]; alternatively, one can use other types of penalized regression, e.g., the elastic net [34] combining the benefits of the lasso with the ridge penalty.

The estimated linear predictor

$$\sum_{j=1}^{p} \hat{\beta}_j x_{ij}$$

can be interpreted as an individual predictive score with larger positive values favoring the experimental treatment and larger negative values favoring the control. A subgroup of patients with an enhanced treatment effect can be defined by applying a clinically meaningful margin c, i.e.,

$$S = \left\{ i : \sum_{j=1}^{p} \hat{\beta}_j x_{ij} \geq c \right\}.$$

Note that c may also reflect the treatment burden or costs.

While the treatment assignment rules as smooth functions of biomarkers were proposed in the original publications on OWL [20,21], proposals for individualized regimes based on tree-structured rules have also been made [35,36].

16.3 Case Study

This section introduces a case study based on a phase III trial in patients with leukemia which affects the body's blood-forming tissues, i.e., the bone marrow and lymphatic system. This clinical trial example will be used throughout the chapter to illustrate biomarker evaluation and subgroup identification techniques.

A phase III trial was conducted to examine the efficacy and safety of an experimental treatment for leukemia versus the standard of care. The trial utilized a two-arm design with 207 patients in the experimental treatment arm and 198 patients in the control arm. The primary endpoint was overall survival. It is important to point out that the treatment effect in the trial's overall population was not statistically significant with a one-sided $p = 0.2096$. The effect was not clinically relevant since the hazard ratio computed from the proportional hazards Cox model was quite close to 1 (a lower hazard ratio would have indicated a beneficial treatment effect in this trial). The estimated hazard ratio was 0.91 and the 95% confidence interval was given by (0.72, 1.15). Figure 16.2 presents the Kaplan–Meier estimates of overall survival in the control and treatment arms.

Since the overall effect in the trial was not clinically meaningful, the trial's sponsor was interested in a thorough evaluation of relevant biomarkers with the ultimate objective of identifying a subset of the overall population with a desirable efficacy profile. Table 16.1 provides information on the seven biomarkers, denoted by $X_1, \ldots, X_7$, that were included in the candidate set. The biomarkers were defined using relevant demographics and clinical variables such as patient's age, gender, ECOG (Eastern Cooperative Oncology Group) performance status and several laboratory parameters. These variables are likely to have prognostic effect on the overall survival, i.e., these baseline characteristics can predict poor outcomes irrespective of the treatment assignment; however, it is unclear whether the selected biomarkers have any predictive

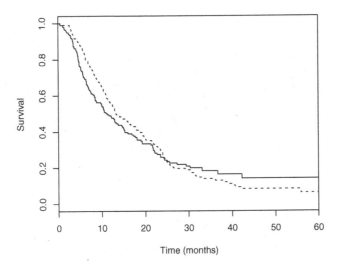

FIGURE 16.2
Kaplan–Meier plots for overall survival in the overall population (solid curve, experimental treatment arm; dotted curve, control arm).

TABLE 16.1

Candidate Biomarkers in the Leukemia Clinical Trial

Biomarker	Description	Type	Range/Nb. of Patients
X_1	Patient's age (years)	Continuous	65–94
X_2	ECOG performance status	Ordinal	1 (302) 2 (103)
X_3	Patient's gender	Nominal	Males (235) Females (170)
X_4	Absolute neutrophil count (10^9/L)	Continuous	1.0–371.5
X_5	Hemoglobin (g/dL)	Continuous	6.8–15.6
X_6	White blood cell count (10^9/L)	Continuous	0.9–48.1
X_7	Platelet count (10^9/L)	Continuous	4–1,632

ability and can be used to reliably identify subgroups of responders. A comprehensive approach to assessing the predictive strength of the candidate biomarkers and selecting subgroups of patients who experienced a beneficial effect in this trial will be described in Section 16.4.

16.4 SIDES Methodology for Subgroup Identification

In this section we will introduce the SIDES methodology for biomarker evaluation and subgroup identification in late-stage clinical trials and illustrate it using the case study from Section 16.3. As indicated in Section 16.2.2, this methodology is an example of a local modeling approach that focuses on direct search for patient subgroups with desirable characteristics such as a beneficial treatment effect. SIDES subgroup search algorithms have been successfully applied to multiple phase III clinical trials, e.g., in a large phase III trial in patients with type 2 diabetes mellitus [37], in a phase III development program for the treatment of nosocomial pneumonia [38], and in a phase III oncology trial [39].

The family of SIDES subgroup search algorithms includes the base SIDES procedure that was proposed by Lipkovich et al. [25], as well as more advanced procedures that were developed in recent publications, including multi-stage procedures such as the Adaptive and Stochastic SIDEScreen procedures [40]. These algorithms employ a number of common components. The following key components will be discussed in detail throughout this section:

- Component 1. Subgroup generation tools.
- Component 2. Subgroup pruning tools, including restrictions on the search space, treatment effect constraints and biomarker screening.
- Component 3. Subgroup interpretation tools, including subgroup proximity assessments, multiplicity adjustments and derivation of "honest" treatment estimates effects.

We will begin with a review of subgroup generation tools (Section 16.4.1) and then examine subgroup pruning tools aimed at improving the efficiency of subgroup search in clinical trials. The most basic approach to subgroup pruning that utilizes restrictions on the magnitude of the treatment effect in child subgroups will be briefly described in Section 16.4.2. More advanced and powerful approaches to constructing subgroup search algorithms with a biomarker screening step will be discussed in Sections 16.4.3 and 16.4.4. Lastly, subgroup interpretation tools will be reviewed in Sections 16.4.5 (subgroup proximity assessment), 16.4.6 (adjustment for selection bias), and 16.4.7 (honest treatment estimates effects).

16.4.1 Subgroup Generation

To define the general approach to efficiently examining subsets of a trial's overall population used in all SIDES subgroup search algorithms, we will first introduce the base SIDES procedure [25], which serves as a "backbone" of the SIDES methodology that generates patient subgroups.

The base SIDES method relies on a recursive partitioning algorithm that generates patient subgroups by exhaustively evaluating a binary splitting criterion for all the cut-offs associated with p candidate biomarkers. It starts with a single parent group taken as the entire training sample and determines M best splits: each of the p candidate biomarkers is contributing with its best split of the current patient population and top M biomarkers are selected.

For continuous and ordinal biomarkers, splits are defined for each cut-off value c as $\{X \leq c\}$ versus $\{X > c\}$ and, for nominal biomarkers such as patient's gender or ethnicity splits are formed by considering all partitions of the biomarker's levels into two mutually exclusive and exhaustive subsets. A split is evaluated using a relevant splitting criterion, which is defined as a function of the treatment effect statistics, denoted by Z_1 and Z_2, that are computed within the two child groups. For convenience, SIDES computes the splitting criterion on a p-value scale, and thus a lower value of the criterion corresponds to a stronger predictive effect. For example, the differential effect splitting criterion commonly used in clinical trial applications is defined as follows:

$$D(Z_1, Z_2) = 2\left(1 - \Phi\left(\frac{|Z_1 - Z_2|}{\sqrt{2}}\right)\right).$$

For a time-to-event outcome, as in our case study, the test statistics used in a splitting criterion are defined as familiar signed log-rank test statistics.

Since optimization occurs over all possible candidate cutoffs, biomarkers with a larger number of cut-off values would have a competitive advantage over biomarkers with a smaller number of cut-off values. As an illustration, patient's age in the leukemia trial has 29 unique values, which results in 28 candidate splits, whereas patient's gender supports a single split. This leads to a well-known problem of covariate selection bias [41,42]. To level the playing field and achieve comparable probabilities of chance selection for biomarkers with different numbers of cut-off values, SIDES uses a multiplicity-adjusted version of the splitting criterion based on a version of the Šidák test, which accounts for the correlations among the D-values across all possible cut-offs (for details, see Lipkovich et al. [25]).

Once the best split per biomarker has been determined, a *promising group* is selected as the one of the child groups with the largest treatment effect. Then the procedure is recursively applied for each promising child. Subgroup growing is controlled by several parameters:

- Search width (M), defined as the number of best splitters retained from the p candidate biomarkers at each level of recursion.
- Search depth (L), defined as the number of times the parent group is recursively split. Subgroups from the last recursion are known as the terminal subgroups. Given the search width and depth parameters, it is easy to show that the maximal number of terminal subgroups is M^L unless sample size or similar restrictions are imposed. If all intermediate patient subgroups are retained, the total number of subgroups is given by $M + M^2 + \cdots + M^L = M(M^L - 1)/(M - 1)$.
- Sample size restriction (n_{min}), defined as the smallest acceptable size of a patient subgroup.

Table 16.2 helps illustrate the subgroup generation process using the case study based on the leukemia trial. This table lists subgroups produced by the base SIDES procedure with the differential effect splitting criterion and the following algorithm parameters:

- Search width: $M = 2$.
- Search depth: $L = 2$.
- Sample size restriction: $n_{min} = 50$.

Using these parameters, the base SIDES selected five promising subgroups. At first, the trial's overall population was split based on two biomarkers that corresponded to the strongest differential effects in the candidate set. These biomarkers were X_2 (ECOG performance status) and X_6 (white blood cell count), and the first-level subgroups based on these biomarkers were denoted by S_2 and S_6, respectively. First-level subgroups that could have been set up based on the other five biomarkers were discarded by virtue of retaining only $M = 2$ best biomarkers at that level. This was done to help reduce the size of

TABLE 16.2

Patient Subgroups Identified by the Base SIDES Procedure Without Subgroup Search Restrictions

Subgroup	Size	Hazard Ratio (95% CI)	One-Sided p-Value Raw	One-Sided p-Value Adjusted
$S_6 = \{X_6 \leq 2.4\}$	306	0.71 (0.55, 0.93)	0.00588	0.511
$S_{62} = \{X_6 \leq 2.4, X_2 = 1\}$	228	0.65 (0.48, 0.88)	0.00278	0.370
$S_{64} = \{X_6 \leq 2.4, X_4 > 3.3\}$	230	0.61 (0.45, 0.83)	0.00060	0.173
$S_2 = \{X_2 = 1\}$	302	0.86 (0.66, 1.13)	0.14105	0.981
$S_{21} = \{X_2 = 1, X_1 \leq 77\}$	164	0.66 (0.46, 0.96)	0.01423	0.687

the search space or, in other words, the size of a potential subgroup pool. The two patient subgroups were further split to construct second-level subgroups, denoted by S_{21}, S_{26}, S_{62}, and S_{64} (note that the subgroup S_{26} that was created when splitting the subgroup S_2 is not explicitly included in the subgroup list because it is identical to S_{62}). The hazard ratios and treatment effect p-values within the resulting patient subgroups are shown in Table 16.2. Clinically relevant hazard ratios were observed in all promising subgroups and the treatment effects were statistically significant at a standard one-sided 0.025 level in most of the subgroups, except for S_2. It is worth pointing out that the adjusted p-values shown in the rightmost column indicate that the subgroup effects were far from being significant after accounting for the multiplicity inherent in subgroup search. More information on the multiplicity adjustment used in the SIDES subgroup search algorithms will be provided in Section 16.4.6.

16.4.2 Basic Constrained Subgroup Search: Treatment-Effect Restrictions

As with any method of recursive partitioning, the base SIDES procedure tends to produce many spurious patient subgroups. The treatment effect may appear significant in these subgroups; however, the statistical significance vanishes once we account for the extensive search with numerous selections among candidate biomarkers and associated cutoff values. To help mitigate this problem, constrained subgroup search methods are commonly applied. One way to reduce the size of the search space is to impose an appropriate restriction on the magnitude of the treatment effect in a child subgroup to justify the split of a parent subgroup. To impose treatment-effect restrictions, consider a vector of *child-to-parent ratios* denoted by $\gamma = (\gamma_1, \ldots, \gamma_L)$. The child-to-parent ratios range from 0 to 1 and, when splitting a parent group at the level $l = 1, \ldots, L$ ($l = 1$ corresponds to the overall population), a split is made only if the smaller p-value of the two child groups formed by the split (denoted by p_C) satisfies the constraint

$$p_C \leq \gamma_l p_P,$$

where p_P is the treatment effect p-value in the parent group.

To illustrate the use of treatment effect restrictions in SIDES-based recursive partitioning algorithms, consider the patient subgroups listed in Table 16.2. This subgroup set was generated without accounting for the magnitude of the treatment effect in child subgroups. A conservative approach with low child-to-parent ratios would most likely result in fewer patient subgroups or even no subgroups at all. For example, if the base SIDES procedure was applied with the common child-to-parent ratio of 0.1, i.e., $\gamma_1 = \gamma_2 = 0.1$, only the subgroup $S_6 = \{X_6 \leq 2.4\}$ would be retained. To see this, recall from Section 16.3 that the one-sided p-value in the trial's overall population was $p_P = 0.2096$. The treatment effect p-value in $S_2 = \{X_2 = 1\}$ is greater than $\gamma_1 p_P = 0.0210$ and thus this subgroup would be discarded, whereas the p-value in $S_6 = \{X_6 \leq 2.4\}$ is less than $\gamma_1 p_P$.

16.4.3 Constraints on the Search Space: Biomarker Screening via SIDEScreen

Section 16.4.2 introduced a basic approach to enabling constrained subgroup search. A more efficient method of reducing the size of a search space is based on three-stage procedures known as SIDEScreen procedures [26]. SIDEScreen procedures do not impose a direct constrain on the subgroup-specific treatment effects using child-to-parent ratios but rather capitalize on model averaging, which effectively reduces the impact of non-informative biomarkers and results in a powerful search algorithm.

SIDEScreen procedures are constructed using appropriate biomarker screening rules based on the concept of *variable importance* (VI). The VI score associated with the biomarker X is denoted by VI(X) and is defined as a measure of its predictive ability computed as the average contribution of that biomarker across all terminal subgroups produced by the base SIDES. The contribution of X is set to zero for all subgroups where it is not used and is a function of the splitting criterion D if the subgroup involved a split on X; see Lipkovich and Dmitrienko [26] for more information on VI scores. The procedure is based on the following three-step algorithm:

- Step 1. Generate patient subgroups using the base SIDES procedure. A VI score is computed for each candidate biomarker in this step.
- Step 2. Screen out noise biomarkers with low VI scores.
- Step 3. Apply the base SIDES procedure to the subset of most important biomarkers chosen in Step 2.

To give an example of a SIDEScreen procedure, we will present the Adaptive SIDEScreen procedure that utilizes a biomarker screening rule with a data-driven threshold. With this rule, the trial's sponsor can control the probability of incorrectly selecting at least one biomarker for the final step of the procedure under the null hypothesis of homogeneous treatment effect across all possible subsets of the trial's population. A biomarker is selected in Step 2 if it appears to be a strong predictor of treatment effect, i.e., if the biomarker's VI score is above the threshold.

$$\text{Select } X_i \text{ if } VI(X_i) > \hat{E}_0 + k\sqrt{\hat{V}_0}, \quad i = 1, \ldots, p,$$

where $\hat{E}_0$ and $\hat{V}_0$ are the mean and variance of the maximal VI score across all candidate biomarkers under the null distribution. These quantities are estimated via a permutation-based approach, e.g., by permuting the treatment labels in the clinical trial dataset. The biomarker screening parameter k is a user-defined parameter that is often set to $k = \Phi^{-1}(1-a)$, where a is the probability of incorrectly selecting at least one noise biomarker for Step 3 in the absence of any predictive biomarkers in the dataset.

Using VI scores, the algorithm capitalizes on the fact that the biomarkers with strong predictive properties are typically included in multiple subgroups whereas weaker or non-informative biomarkers contribute to a much smaller number of subgroups. For example, returning to the patient subgroups presented in Table 16.2, it would be fair to treat the biomarker X_6 (white blood cell count) as a strong predictive biomarker if it was selected as one of the top splitters at the first level and if this biomarker was also involved in the splits at the second and third levels (possibly with larger values of the search width and depth parameters). On the other hand, if a certain biomarker is a weak predictor of treatment response, it is very unlikely to be selected at multiple levels of the subgroup search algorithm. Therefore, by applying a model-averaging approach and averaging biomarker contributions across a large pool of subgroups, it is possible to better detect stronger predictive biomarkers while shrinking the overall impact of noise variables down to zero.

Typically, the model-averaging approach outlined above becomes more efficient with a larger search space. This is accomplished by applying no restrictions on the subgroup search process in the subgroup generation stage of a SIDEScreen procedure. Table 16.3 presents the only patient subgroup based on X_6 selected by the Adaptive SIDEScreen procedure in the leukemia trial. This subgroup was found using the following parameters of the SIDEScreen procedure:

- Step 1. Search width: $M = 5$. Search depth: $L = 3$. Sample size restriction: $n_{min} = 50$. No treatment effect restrictions were imposed.
- Step 2. Biomarker screening parameter: $k = 1$.

The screening step (Step 2) of the Adaptive SIDEScreen procedure is illustrated in Figure 16.3. As shown in this figure, the only biomarker that exceeded the threshold based on the predefined biomarker screening parameter was X_6. The biomarker's VI score was 1.85. The estimated null distribution of the largest VI score was normal with the mean $\hat{E}_0 = 0.93$ and variance $\hat{V}_0 = 0.32$. As a result, the probability of observing the VI score of 1.85 or a larger score under the null distribution was

$$1 - \Phi\left(\frac{1.85 - 0.93}{\sqrt{0.32}}\right) = 0.0537.$$

This quantity is conceptually similar to a p-value in a traditional hypothesis testing framework in the sense that the quantity helps the trial's sponsor to assess whether or not the biomarker's VI score is consistent with the null distribution of VI scores. This probability is quite low, which suggests that the white blood cell count is unlikely to be a noise biomarker in this clinical trial example. It is also instructive to apply the same approach to the biomarker X_2 (ECOG performance status). This biomarker had the second highest VI score of 0.47 and it is easy to verify that the probability of observing this score or a larger score under the null distribution was 0.8118. This high probability suggests that the VI score of X_2 is consistent with the null distribution of VI score and thus it would be unwise to choose the biomarker X_2 for the final stage of the Adaptive SIDEScreen procedure.

TABLE 16.3

Patient Subgroup Identified by the Adaptive SIDEScreen Procedure

Subgroup	Size	Hazard Ratio (95% CI)	One-Sided p-Value Raw	Adjusted
$S_6 = \{X_6 \leq 2.4\}$	306	0.71 (0.55, 0.93)	0.00588	0.037

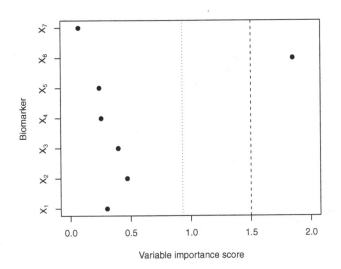

FIGURE 16.3
Variable importance scores for the candidate biomarkers. The dotted vertical line is drawn at the mean of the null distribution of the largest VI score ($\hat{E}_0 = 0.93$) and the dashed vertical line is drawn at the selected threshold ($\hat{E}_0 + k\sqrt{\hat{V}_0} = 1.5$).

16.4.4 Constraints on the Search Space: Biomarker Screening via Stochastic SIDEScreen

It is well known that recursive partitioning procedures are notoriously unstable and a small perturbation in a trial's dataset may lead to a substantial difference in terms of selection of splitting variables and associated cutoffs. The Adaptive SIDEScreen uses variable importance scores computed over a broad set of subgroups in order to stabilize the biomarker selection process and choose biomarkers with strong predictive properties. A further improvement on the Adaptive SIDEScreen method is the Stochastic SIDEScreen introduced in the article by Lipkovich et al. [40].

They key idea behind the Stochastic SIDEScreen procedure is that VI scores are computed over a broader ensemble of subgroups obtained by applying the base SIDES to multiple bootstrap samples from the trial's dataset. Borrowing ideas from *bagging* (bootstrap aggregation), the Stochastic SIDEScreen achieves better performance in terms of

- Suppressing noise biomarkers that now have to compete with additional noise factors induced by bootstrapping.
- Amplifying the effect of strong predictive biomarkers.
- Obtaining measures of uncertainty associated with VI scores.

Unlike the Adaptive SIDEScreen procedure introduced in Section 16.4.3, the Stochastic SIDEScreen procedure uses the smoothed VI scores that are averages across B bootstrap samples for the biomarker X_i, i.e.,

$$\overline{\mathrm{VI}}_B(X_i) = \frac{1}{B}\sum_{b=1}^{B}\mathrm{VI}_b(X_i), \quad i = 1,\ldots,p,$$

rather than the observed VI scores.

To mimic the Adaptive SIDEScreen, the *smoothed* (bagging) versions for the mean and variance of the maximal VI under the null distribution are also obtained. To obtain these scores, K null samples (K could be set tó 1,000) are created by permuting the treatment labels in the original trial's dataset. After that, B_0 bootstrap samples, e.g., 10 samples, are generated from each null set and the bootstrap averages are obtained for the null VI scores as follows:

$$\overline{\mathrm{VI}}_{0,B}^{j}(X_i) = \frac{1}{B_0} \sum_{b=1}^{B_0} \mathrm{VI}_{0,b}(X_i), \quad i=1,\ldots,p, \quad j=1,\ldots,K.$$

Then the maximal null VI scores are computed as

$$\overline{\mathrm{VI}}_{0,B}^{j} = \max_{i=1,\ldots,p} \overline{\mathrm{VI}}_{0,B}^{j}(X_i).$$

The sample mean and variance of $\overline{\mathrm{VI}}_{0,B}^{j}$ across the K null sets are denoted by $\hat{E}_{0,B}$ and $\hat{V}_{0,B}$, respectively. As the final step, we can define the bagging counterpart of the biomarker screening rule used in the Adaptive SIDEScreen procedure. The new biomarker screening rule selects the biomarker X_i, $i=1,\ldots,p$, if

$$\overline{\mathrm{VI}}_B(X_i) > \hat{E}_{0,B} + k\sqrt{\hat{V}_{0,B}},$$

where, as before, k is a predefined parameter of the screening rule.

An illustration of the Stochastic SIDEScreen procedure for biomarker selection is provided in Figure 16.4. The left panel of Figure 16.4 presents a biomarker screening plot similar to

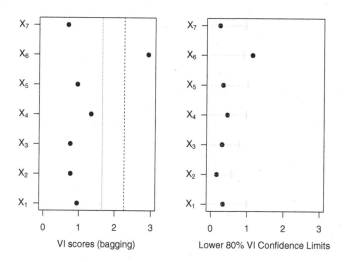

FIGURE 16.4

Biomarker screening rule used in the Stochastic SIDEScreen procedure. The left panel shows the bagging estimators of VI scores. The dotted vertical line indicates the null mean for the maximal VI score ($\hat{E}_{0,B} = 1.653$) and the dashed vertical line is drawn at the selected threshold ($\hat{E}_{0,B} + k\sqrt{\hat{V}_{0,B}} = 2.62$). In the right panel, the circles represent the 10th percentiles of the bootstrap distribution of VI scores and the horizontal dotted lines extend to the null means for each biomarker.

that depicted in Figure 16.3 but, instead of the observed VI scores, Figure 16.4 plots the bagging averages, i.e.,

$$\overline{\text{VI}}_B(X_i), \quad i = 1, \ldots, p.$$

The dotted and dashed lines are drawn in the left panel of Figure 16.4 at the sample mean $\hat{E}_{0,B}$ and at the threshold given by $\hat{E}_{0,B} + \sqrt{\hat{V}_{0,B}}$, respectively. Note that $\hat{E}_{0,B} = 1.65$ and $\hat{V}_{0,B} = 0.37$. The VI score for the top biomarker X_6 in Figure 16.4 is equal to 2.92. A comparison of Figures 16.3 and 16.4 reveals that the smoothed VI scores are somewhat larger than the original VI scores for all biomarkers; however, the gap between the top biomarker X_6 and the other biomarkers is also larger. It is apparent that X_6 stands out further compared to the null distribution of the maximal VI scores.

It is also instructive to compare the biomarker screening rules used in the adaptive and Stochastic SIDEScreen procedures from the following perspective: Using a normal approximation for the maximal VI score, we can compute the probability that a maximal value of the bagging estimator of a VI score from the null set exceeds the observed VI value for the biomarker X_6. This probability is given by

$$1 - \Phi\left(\frac{2.92 - 1.65}{\sqrt{0.37}}\right) = 0.0189.$$

For comparison, the similar probability for the regular (non-bagging) estimate of the VI score presented at the end of Section 16.4.3 was 0.0537. We can see that, in this case, bagging helps a strong predictive biomarker stand out from the set of weaker and potentially non-informative biomarkers. By contrast, for the biomarker X_2, which has the second largest VI score computed from the original data (see Figure 16.3), the probabilities of an incorrect biomarker selection are equal to 0.8118 (with the regular approach) and 0.9275 (with the bagging approach). The latter probability is much higher, and this means that the bagging approach more efficiently suppresses noise biomarkers compared to the regular biomarker screening rule defined in Section 16.4.3.

An alternative approach for utilizing the bootstrap distribution of VI scores when setting up a biomarker screening rule is to compare the lower 10th percentile of the bootstrap distribution for each biomarker with the null mean for that same biomarker. For example, the right panel of Figure 16.4 displays the benchmarks obtained from the null distribution of each individual biomarker. As we can see, setting the lower confidence limit at 10% would support a conclusion to select only one biomarker (X_6) for the last stage of the Stochastic SIDEScreen procedure. Another alternative that may produce tighter confidence limits for bagging estimators can be obtained using the approach developed by Wager et al. [43] and applied in this context by Lipkovich et al. [40].

16.4.5 Interpretation Tools: Subgroup Proximity Measures

We will now discuss the tools that are commonly employed to facilitate the interpretation of patient subgroups identified by the SIDES subgroup search algorithm, including the base SIDES and more advanced SIDEScreen procedures. As many subgroups can be generated by subgroup search methods, especially when the candidate biomarkers are strongly correlated with each other, which is often the case, the sponsor may be interested in selecting a representative subgroup within each "cluster" of highly overlapping subgroups based on criteria of clinical relevance, simplicity, or other subject-matter considerations.

As the very first step, a subgroup proximity assessment is performed to determine whether or not the promising subgroups are close to each other. Proximity assessments of this kind rely on relevant measures of proximity or "distance" between each pair of subgroups. The Jaccard similarity measure is often used in subgroup search exercises. This measure was developed in set theory and, in the context of subgroup search (i.e., measuring the proximity between two subgroups), it is defined as the ratio of the number of patients included in both subgroups to the total number of patients in the two subgroups. More formally, considering two subgroups denoted by S' and S'', the Jaccard similarity measure is defined as

$$J(S', S'') = \frac{|S' \cap S''|}{|S' \cup S''|},$$

where $|S|$ is the number of patients in the subgroup S. This measure ranges between 0 and 1, and a lower value of the Jaccard similarity measure for a pair of subgroups indicates that these subgroups are dissimilar.

Table 16.4 presents the Jaccard similarity matrix for the five patient subgroups identified by the base SIDES procedure in the leukemia trial (these subgroups are defined in Table 16.2). The similarity matrix presented in Table 16.4 supports the following conclusions: First, it follows from the table that the two first-level subgroups in this trial, i.e., S_2 and S_6, are fairly independent of each other with the Jaccard similarity measure of 0.6. Furthermore, the second-level subgroups, i.e., S_{21}, S_{62}, and S_{64}, do not appear to overlap too much, and the corresponding Jaccard similarity measures range between 0.327 and 0.607.

The conclusions presented above obviously rely on fairly subjective criteria, and no definition of what should be considered a low or high value of the the Jaccard similarity measure is pre-specified. A formal analysis of proximity measures can be performed using an appropriate hierarchical clustering method and can help the trial's sponsor uncover important patterns in a Jaccard similarity matrix and clusters of patient subgroups. An example of a subgroup proximity assessment based on hierarchical clustering is provided by Dmitrienko et al. [38].

16.4.6 Interpretation Tools: Adjustment for Selection Bias

An important feature of SIDES procedures is the availability of multiplicity-adjusted p-values to assess the significance of the treatment effect within the promising subgroups.

TABLE 16.4

Jaccard Similarity Matrix for the Subgroups
Identified by the Base SIDES Procedure

Subgroup	S_6	S_{62}	S_{64}	S_2	S_{21}
S_6	1	0.745	0.752	0.600	0.358
S_{62}	0.745	1	0.607	0.755	0.463
S_{64}	0.752	0.607	1	0.482	0.327
S_2	0.600	0.755	0.482	1	0.543
S_{21}	0.358	0.463	0.327	0.543	1

As with any exploratory analysis, multiplicity plays a key role in subgroup identification settings, although it may be obscured by the fact that the set of null hypotheses of no treatment effect arising in subgroup exploration is not pre-specified but "random", i.e., the hypothesis set is data-driven. A multiplicity adjustment serves as an important tool for correcting for the selection bias in subgroup search and multiplicity-adjusted treatment p-values are easily derived using permutation tests, which is explained in detail by Lipkovich et al. [25,39].

There are several approaches to defining the null distribution of no effect across all possible subgroups and, consequently, for generating null sets when performing multiplicity adjustment. It is very common to generate null sets by permuting treatment labels in a trial's dataset. The resulting null sets retain correlations among biomarkers as well as their prognostic effects. However, this type of permutation removes any predictive biomarker effects as well as the overall treatment effect. This scheme is attractive for failed clinical trials where the overall treatment effect may be close to 0. As an alternative approach, one can simultaneously permute the treatment and outcome columns in the dataset. This scheme also retains correlations among biomarkers and eliminates any predictive effects. However, unlike the previously described approach, it also retains the overall treatment effect while removing prognostic effects. This scheme may be useful when trying to identify predictive effects from a clinical trial with a positive overall effect.

A fundamental point we would like to emphasize is that, to obtain correct adjusted p-values, the *same* subgroup search procedure that was applied to the original trial's dataset needs to be applied to each of the null datasets. It is also important to understand that the degree of multiplicity correction depends on the "greediness" of the subgroup search procedure. A procedure that uses a more constrained search space would in general have smaller inherent multiplicity and require less adjustment compared to a procedure that examines every possible subset of the trial's overall population. As an illustration, we can return to the adjusted p-values displayed in the rightmost columns of Tables 16.2 and 16.3. As we can see, the subgroup $S_6 = \{X_6 \leq 2.4\}$ based on the biomarker X_6 was identified using the base SIDES procedure as well as the Adaptive SIDEScreen procedure. It is very important to note that, while the adjusted p-value within this subgroup was highly non-significant ($p = 0.511$) in the context of the base SIDES procedure, the adjusted p-value within the same subgroup suggested that the treatment effect in patients with $X_6 \leq 2.4$ was in fact significant ($p = 0.037$) when the Adaptive SIDEScreen procedure is applied. The major difference between the conclusions based on the two procedures is explained by a substantial reduction of multiplicity burden by the Adaptive SIDEScreen procedure (note that the ratio of the adjusted p-value to unadjusted p-value is $0.037 / 0.00588 = 6.29$ for the Adaptive SIDEScreen versus $0.511 / 0.00588 = 86.9$ for the base SIDES). However, this reduction comes at a cost. The Adaptive SIDEScreen procedure aggressively shrinks the search space by screening out noise biomarkers at the last stage, and since this procedure applies the same data-driven threshold to the original and null datasets, if none of the candidate biomarkers is a strong predictor of treatment response, the analysis would most likely return no promising subgroups. By contract, unconstrained search using the base SIDES procedure almost surely returns a large set of patient subgroups, and thus we ought to pay a higher price in multiplicity adjustments when this procedure is invoked. This example illustrates an important trade-off between the complexity of a subgroup search procedure and the degree of multiplicity adjustment.

It is also worth noting that, when computing the adjusted p-value to assess the treatment effect within a patient subgroup selected by the Adaptive SIDEscreen, we apply the entire biomarker screening process to each null dataset afresh, rather than evaluating only the biomarker X_6 that was identified in the observed data. As a result, the uncertainty associated with biomarker search is fully accounted for when the multiplicity-adjusted p-value is derived.

16.4.7 Interpretation Tools: Honest Treatment Effect Estimates

Another manifestation of multiplicity inherent in subgroup search is presented by possibly exaggerated treatment effects in the promising subgroups if data re-substitution is utilized, i.e., if the treatment effect was computed using the same dataset that was utilized in the subgroup search procedure. It should be emphasized that, even if the identified biomarkers and cutoffs may be correct (owing to carefully tuned parameters of a subgroup search procedure), the re-substitution estimates of treatment effects within these subgroups may still be inflated. Several methods for obtaining "honest" (bias-adjusted) estimates of treatment effects in exploratory subgroup-analysis settings have been proposed in the literature [15,44–47,54].

In this section, we illustrate the process of computing honest estimates of treatment effect by focusing on the leukemia trial example. We will obtain an honest estimate for the subgroup identified by the Adaptive SIDEScreen procedure using the method of cross-validation (CV). The resulting estimates help us quantify the amount of treatment effect in excess of the overall treatment effect that the procedure is capable of recovering from the data at hand. This is the effect that is expected to be replicated in an independent dataset, i.e., a clinical trial to be conducted in the future, if this dataset is sampled from the same population as the current dataset.

In general, cross-validation methods use subsets of a given dataset as "test sets" to mimic the data from a future trial. Briefly, obtaining an "honest" estimate of some performance measure R, e.g., the classification or prediction error, associated with an algorithm A using cross-validation proceeds as follows: First, the analysis dataset is divided into K subsets of equal size, termed "folds," either randomly or using stratified randomization, e.g., by the treatment arm. After this, each fold is used in turn as a test set while the rest of the data are used as the training set. The algorithm A is trained on the training set and then applied to the test set, and the performance measure R is computed from the test data. The process is iterated K times, as explained and an overall measure of performance is computed across the K folds, which essentially reflects the expected performance of the algorithm A with an independent dataset. In the context of computing an "honest" estimate of treatment effect, the treatment effect serves as the performance measure R.

To compute honest treatment effect estimates in the leukemia trial example, we applied the following CV-based approaches:

- Pooled CV method. We first identify "biomarker-positive" subjects from each of the K test sets by applying the "biomarker signature" (i.e., classifying subjects as belonging to a subgroup or not) identified by the Adaptive SIDEScreen procedure from the training dataset with that test set removed. The final biomarker-positive subset is constructed by combining (pooling) across all such sets. A single estimate of the treatment effect is computed from the resulting subset. If the subgroup identification procedure returns no subgroups for any of the test sets, all patients in that set are considered biomarker-positive and contribute to the final subset. In other words, the subgroup is considered to be equal to the overall population.

- Averaged CV method. We first find an estimate of the treatment effect from each of the K test sets based on the appropriate signature that was identified with that set removed from the original dataset, and the final estimate is obtained by averaging the K individual estimates. Again, if no subgroup is returned on any of the test sets, the entire test set is considered to be the "subgroup." This convention results in an automatic shrinkage of the CV estimate to the overall treatment effect if no non-trivial subgroup can be identified.

It is important to understand that the treatment effects uncovered by a CV method cannot be attributed to any specific subgroup signature but rather to a hypothetical "biomarker-positive" subset of patients that a given subgroup search procedure, e.g., the Adaptive SIDEScreen with a pre-specified set of parameters, is capable of detecting.

The two CV methods were applied to 10 replicated random splits of the dataset into K sets, and the resulting treatment effect estimates were averaged. All random splits were stratified by trial arm to preserve the treatment allocation ratio within each CV fold. We expect that, with just a few folds, the CV methods may cause negative bias, i.e., the treatment effect is underestimated. On the other hand, with a larger number of folds, higher variability in the estimated effects will be induced (however, this may be mitigated by using a large number of replications).

The results based on the two CV methods are presented in Figure 16.5. The figure shows the hazard ratio estimate within the patient subgroup identified by the Adaptive SIDEScreen procedure as a function of the number of CV folds (K ranged from 2 to 10). With a large number of folds, the hazard ratio estimate based on the averaged CV method seems to be closer to the estimate of the overall treatment effect, which is equal to 0.91. By contrast, the pooled CV method suggests a stronger positive effect within the chosen subgroup with the hazard ratio estimate approaching 0.84 with 10 folds. As some experiments with simulated data have demonstrated, the pooled CV method may provide more appropriate estimates when at least one strong predictive biomarker is present in the analysis dataset.

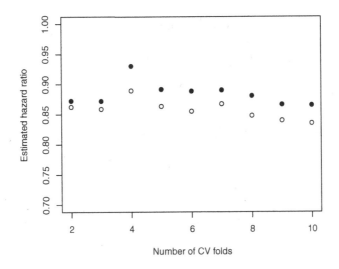

FIGURE 16.5
Cross-validated estimates of the hazard ratio in the patient subgroup identified by the Adaptive SIDEScreen procedure (open circles, pooled CV method; closed circles, averaged CV method).

16.5 Conclusion

This section provides a summary of key considerations presented in this chapter, including a discussion of the recommended data-driven approach to evaluating biomarkers and examining patient subgroups in late-stage clinical trials, arange of applications for principled subgroup-analysis strategies in modern drug development and their comparison, and remaining challenges.

16.5.1 Principles of Data-Driven Subgroup Analysis

A variety of advanced statistical methods have been proposed in the literature over the past 10 years to address important problems arising in the area of biomarker and subgroup evaluation. Many of the articles have stressed the need for a disciplined or principled approach to subgroup discovery [7,48]. Lipkovich et al. [6] contrasted *guideline-driven* and *data-driven* approaches to exploratory or data-driven subgroup analysis. Examples of the former approach can be found in numerous publications [49–51]. Sun et al. [51] provided a battery of checklists that need to be followed in order to improve the level of confidence in exploratory subgroup findings. The common thread of these guidelines is that the results of any subgroup exploration should be "taken with caution." In addition, these guidelines often state that subgroups to be investigated should be pre-specified or data-driven elements of subgroup exploration should be minimized. Needless to say, these recommendations go against the discovery spirit of the science in general and the current emphasis on evidence-driven medicine.

As a more recent example of a guideline-driven approach that encompasses the needs of personalized medicine, the draft guidance from the European Medicines Agency (EMA) on the investigation of subgroups in confirmatory clinical trials [5] should be mentioned. This document attempts to balance two seemingly conflicting concerns. On the one hand, the EMA's draft guideline appears to discourage *post-hoc* subgroup evaluations by trial sponsors but, on the other hand, the guideline emphasizes the value of characterizing the heterogeneity of treatment effects across patient subgroups based on key baseline characteristics. Unfortunately, the draft guideline and other guidelines on subgroup analyses have not succeeded yet to produce a set of operationalizable rules for data-driven biomarker and subgroup evaluation.

By contrast, the data-driven approach to biomarker and subgroup evaluation is based on recent advances in machine leaning, causal inference, and multiple testing. This framework offers a principled way of evaluating accumulated evidence on predictive biomarkers and patient subgroups in clinical trials. Key principles of the data-driven approach were set forth in [6]:

1. Evaluating the type I error rate or false discovery rate for the entire subgroup search strategy (which should be pre-specified).
2. Using complexity control to prevent data overfitting.
3. Controlling (reducing) selection bias when defining candidate subgroups.
4. Accounting for uncertainty of the entire subgroup search strategy.
5. Reproducibility assessment.
6. Obtaining "honest" estimates of treatment effects in the patient subgroups.

For more information on these principles and their application to principled methods of subgroup exploration, the interested reader is referred to the cited publication.

16.5.2 Application of Principled Subgroup Analysis Methods

Trial sponsors often view subgroup analysis as a tool for salvaging clinical trials at later stages of drug development. But the reality is that, in drug development programs that are characterized by a large number of candidate compounds in the absence of an established cure (which is the case in oncology), the sponsor may be most interested in eliminating most of the candidates at early stages rather than proceeding to the phase III trial and failing to meet its efficacy or safety objectives. Principled approaches to subgroup analysis can help facilitate this process as they help establish the futility of a compound by a careful examination of its efficacy and/or safety profile across all meaningful patient subgroups.

While most recent publications have focused on applications of principled subgroup-analysis strategies to efficacy assessments in clinical trials, disciplined subgroup-analysis methods will play an increasingly important role in safety evaluations. It is well known that brute-force subgroup search algorithms would inevitably "detect" a safety signal in any clinical trial database. The general subgroup search framework presented in this chapter can be extended to tackle challenging problems of uncovering patient subgroups with a desirable risk/benefit profile. This framework can be applied to set up a structured approach to investigating safety signals within key patient subgroups and will lead to reducing the probability of false-positive findings.

In addition, as was stated by Lipkovich et al. [6], exploratory subgroup analysis can be viewed from two different perspectives, namely, from the sponsor's perspective and from the perspectives of the society in general. Briefly, most sponsors focus on identifying the best patient for their new treatments, and from the society's perspective, it would be ideal to select the best treatment for a given patient. Different methods for subgroup discovery and analysis have been motivated by these two perspectives. For example, the SIDES methods follow the sponsor's perspective, whereas the methods for choosing individual treatment regimes target the society's perspective.

16.5.3 Relative Advantages of Principled Subgroup Analysis Methods

Different methods that emerged under the general class of "subgroup identification" may have different domains of applicability. To appreciate this, a finer characterization of principled subgroup-analysis methods with respect to different features (than the taxonomy provided in Section 16.2.2) is needed. Here we reproduce some of the key features of subgroup identification methods presented in the Table XV of Lipkovich et al. [6] that would help the reader get oriented in a diverse world of subgroup identification methods and guide in selecting a method or a group of methods most appropriate for the task at hand:

1. Dimensionality of the covariate space, e.g., "low" (less than 10), "medium" (10–50), or "high" (above 50).
2. Results produced by the method, e.g., selected biomarkers or biomarker ranking, individual predictive scores, optimal treatment assignment, subgroup signatures.
3. Evaluation of the type I error rate or false discovery rate (FDR) for the entire subgroup search strategy.
4. Implementation of "complexity control" to prevent data overfitting.
5. Control of selection bias when evaluating candidate patient subgroups.
6. Availability of "honest" estimates of treatment effects in the patient subgroups selected by the method.

One of the most important considerations when selecting the most appropriate subgroup-analysis strategy is the size of the search space (Feature 1). For example, tree-based methods are most applicable in a clinical trial with a few (10–50) biomarkers and a relatively large dataset (e.g., 1,000–2,000 patients). Methods that utilize penalized regression and ensemble learning (that provide excellent complexity control, Feature 4) can handle clinical trial datasets with a large number of candidate biomarkers. As a consequence, these methods can be used in settings where the sample size is rather small, including early-stage trials, where the main focus is on selecting biomarkers, rather than identifying specific patient subgroups, that can be utilized in subsequent phase III trials (Feature 2). Similarly, methods that focus on screening biomarkers using variable importance, such as the Adaptive SIDEScreen and Stochastic SIDEScreen procedures, are suitable for smaller datasets typical for phase II trials.

Furthermore, subgroup-analysis methods that evaluate optimal treatment regimes are useful in large phase III or IV trials that compare several active treatments in a diverse patient population. Methods that allow the sponsor to evaluate the type I error rate or FDR (Feature 3) are especially important for retrospective analyses of failed trials from phase III programs. On the other hand, availability of "honest" estimates of treatment effects (Feature 6) for subpopulations identified for tailoring may be critical for developing enrichment strategies using findings from early phase II trials. Finally, methods that provide selection bias control [18] are important when evaluating candidate subgroups based on biomarkers with drastically different numbers of possible splits. On the other hand, if all candidate biomarkers are binary or trinary (e.g., genotype markers), selection bias is unlikely to be much of a concern.

16.5.4 Challenges and Extensions

Subgroup identification is an area of active research with contributors from several related fields. Perhaps the most challenging task, as we have illustrated in this chapter, is related to integrating machine learning for subgroup discovery with the framework of multiple testing and multiplicity adjustment into a single analytical strategy. Other challenges are presented in the following list, which is necessarily incomplete:

1. Integrating prior information on candidate biomarkers and subgroups in the inference.
2. Incorporating multiple outcomes and repeated measures.
3. Dealing with missing data.
4. Applying subgroup identification to observational (non-randomized) data.

Here, we only provide a brief discussion and refer the interested reader to the literature on the general topic of principled biomarker evaluation and subgroup identification.

Subgroup identification strategies heavily rely on available subject matter or expert information, which includes the plausible mechanisms of actions, relative importance of candidate biomarkers and subgroups, restrictions on the type of predictive effects (e.g., monotonic or non-monotonic), and restrictions on the subgroups space (e.g., restrictions that rely on limiting possible splits to clinically relevant cutoffs). While some of the subject matter considerations can be easily integrated into the subgroup search algorithm by pre-processing the dataset of interest (e.g., the cutoff restrictions can be accommodated by converting the raw biomarker data into percentiles or creating dummy variables), others may require redesigning the underlying method. Bayesian methods for subgroup analysis have an advantage here, as they use a natural way of integrating expert information via specifying relevant prior distributions. As an example, Berger et al. [52] proposed a Bayesian method of recursive partitioning based on the ideas of model averaging that automatically assigns prior probabilities to each candidate subgroup using the subgroup complexity. For a review of Bayesian approaches to subgroup identification, see Henderson et al. [10].

In this chapter, we illustrated subgroup identification methods assuming a *single* outcome variable. In practice, sponsors may often be interested in evaluating patient subgroups using a variety of related (e.g., complementary) outcomes. While most subgroup identification tools are easily extended to tackle multiple outcomes, the issue of multiple testing across the outcome variables arises. This source of multiplicity is unaccounted for within each single analysis, hence more advanced strategies that simultaneously evaluate subgroups based on multiple outcomes and take advantage of the correlations among the endpoints are needed. A special type of multiple outcome data is longitudinal or repeated measures data. Clearly, incorporating standard methods for the analysis of repeated measures such as mixed-effects models within the framework of subgroup identification methods, e.g., recursive partitioning methods, may result in very computationally intensive procedures. Some heuristic approaches are proposed within the regression- and classification-tree modeling framework of GUIDE [45].

Another important issue that arises in subgroup identification, as in many other types of analysis methods in clinical trials, is the issue of incomplete data. Missing values can be encountered in biomarkers as well as outcomes. Some of the methods illustrated in this chapter do not explicitly handle missing outcomes. For time-to-event outcomes, the sponsor can simply censor the outcome at the time the patient was lost to follow-up, which may cause selection bias unless missingness is at random. For continuous and binary outcomes, the last observed value is often used, which essentially assumes missingness completely at random. However, some general methods, such as inverse probability (of censoring) weighting or multiple imputation, can be used in conjunction with many of the proposed subgroup identification methods. This requires additional modeling steps to be completed prior to data analysis and leads to additional challenges. It is not clear how to best integrate the results of subgroup analysis across multiply imputed datasets. One approach may be to combine imputation of missing data with data resampling. For example, one can perform a single imputation from a Bayesian predictive distribution within each bootstrap sample when implementing the recently proposed Stochastic SIDEScreen [40].

Handling missing biomarker values is a more challenging problem for subgroup identification methods given that they thrive on covariate information. Methods based on parametric modeling would have to dispense with the entire patient record as long as a single missing biomarker is present ("case-wise" deletion), unless imputation techniques are used to complete the biomarker profile. Some methods may have "built-in" imputation strategies. For example, the Virtual Twins method imputes missing data using a generic imputation method for random forests. Tree-based methods are less affected by missing covariates in that they do not require case-wise deletion of incomplete data. However, their performance may be severely affected by simply ignoring missing values when evaluating candidate splits. Some tree-based methods, including the methods developed within the GUIDE framework [18], use missingness as a distinct category of a biomarker (that the tree algorithm can split on), which may be a better alternative to ignoring missing observations.

Most methods for subgroup identification were motivated by randomized clinical trials and assume that all biomarkers are independent of the treatment assignment by design. However, analysis of data from non-randomized trials, e.g., by combining data from single-arm trials or using observational data, brings additional challenges since the distribution of biomarkers may differ across the treatment arms. As a result, using standard methods for subgroup identification, e.g., those based on the SIDES recursive partitioning platform, may result in selection bias. One way of dealing with this problem is by using a generic method of inverse probability of treatment weighting based on propensity scores that should be computed using an additional modeling step prior to subgroup search. This approach is similar to using inverse probability of censoring weighting for handling missing data. Of note, estimating optimal treatment regimes via OWL readily accommodates propensity-based weighting as it can be naturally incorporated within the subject weights, see implementation examples in the article by Fu et al. [35]. Also, there are approaches that simultaneously evaluate the biomarkers predictive of treatment effect (for subgroup identification) and biomarkers predictive of treatment assignment (for reducing bias due to lack of randomization); see, for example the facilitating score method developed by Su et al. [53].

Finally, we end this chapter with some cautionary remarks. The general goal of subgroup identification in clinical trials is to generate hypotheses, and thus the findings need to be confirmed in subsequent trials. The approaches presented in this chapter can work with a large number of candidate biomarkers without prior understanding of whether they have predictive power. However, once significant predictors of treatment effect have been discovered, it is important to understand the role of those biomarkers in the experimental treatment's mechanism of action.

References

1. Freidlin B, McShane LM, Korn EL. Randomized clinical trials with biomarkers: Design issues. *J Natl Cancer Inst.* 2010;102:152–60.
2. Freidlin B, McShane LM, Polley MYC, Korn EL. Randomized Phase II: Trial designs with biomarkers. *J Clin Oncol.* 2012;30:3304–9.
3. Sargent DJ, Mandrekar SJ. Statistical issues in the validation of prognostic, predictive, and surrogate biomarkers. *Clin Trials.* 2013;10:647–52.
4. Food and Drug Administration. 2012. Guidance for industry: Enrichment strategies for clinical trials to support approval of human drugs and biological products.
5. European Medicines Agency. 2014. Guideline on the investigation of subgroups in confirmatory clinical trials.
6. Lipkovich I, Dmitrienko A, D'Agostino BR. Tutorial in biostatistics: Data-driven subgroup identification and analysis in clinical trials. *Stat Med.* 2017;36:136–96.
7. Dmitrienko A, Muysers C, Fritsch A, Lipkovich I. General guidance on exploratory and confirmatory subgroup analysis in late-stage clinical trials. *J Biopharm Stat.* 2016;26:71–98.
8. Dmitrienko A, Millen B, Lipkovich I. Multiplicity considerations in subgroup analysis. *Stat Med.* 2017;36:4446–54.
9. Ondra T, Dmitrienko A, Friede T, Gradf A, Miller F, Stallard N, Posh M. Methods for identification and confirmation of targeted subgroups in clinical trials: A systematic review. *J Biopharm Stat.* 2016;26:99–119.
10. Henderson NC, Louis TA, Wang C, Varadhan R. Bayesian analysis of heterogeneous treatment effects for patient-centered outcomes research. *Health Serv Outcomes Res Methodol.* 2016;16:213–33.
11. Janes H, Brown M, Pepe M, Huang Y. An approach to evaluating and comparing biomarkers for patient treatment selection. *Int J Biostat.* 2014;10:99–121.

12. Lamont A, Lyons MD, Jaki T, Stuart E, Feaster DJ, Tharmaratnam K, Oberski D, Ishwaran H, Wilson DK, Horn MLW. Identification of predicted individual treatment effects in randomized clinical trials. *Stat Meth Med Res*. 2018;27:142–57.

13. Miller R, Siegmund D. Maximally selected chi-square statistics. *Biometrics*. 1982;38:1011–6.

14. Imai K, Ratkovic M. Estimating treatment effect heterogeneity in randomized program evaluation. *Annal Appl Statist*. 2013;7:443–70.

15. Foster JC, Taylor JMC, Ruberg SJ. Subgroup identification from randomized clinical trial data. *Stat Med*. 2011;30:2867–80.

16. Su X, Zhou T, Yan X, Fan J, Yang S. Interaction trees with censored survival data. *Int J Biostat*. 2008;4(1): Article 2.

17. Su X, Tsai CL, Wang H, Nickerson DM, Li B. Subgroup analysis via recursive partitioning. *J Mach Learn Res*. 2009;10:141–58.

18. Loh W-Y, He X, Man M. A regression tree approach to identifying subgroups with differential treatment effects. *Stat Med*. 2015;34:1818–33.

19. Seibold H, Zeileis A, Hothorn T. Model-based recursive partitioning for subgroup analyses. *Int J Biostatistics*. 2016;12:45–63.

20. Zhao Y, Zheng D, Rush AJ, Kosorok MR. Estimating individualized treatment rules using outcome weighted learning. *J Am Stat Assoc*. 2012;107:1106–18.

21. Xu Y, Yu M, Zhao YQ, Li Q, Wang S, Shao J. Regularized outcome weighted subgroup identification for differential treatment effects. *Biometrics*. 2015;71:645–53.

22. Kehl V, Ulm K. Responder identification in clinical trials with censored data. *Comput Stat Data Anal*. 2006;50:1338–55.

23. Chen G, Zhong H, Belousov A, Viswanath D. PRIM approach to predictive-signature development for patient stratification. *Stat Med*. 2015;34:317–42.

24. Friedman JH, Fisher NI. Bump hunting in high-dimensional data. *Stat Comput*. 1999;9:123–43.

25. Lipkovich I, Dmitrienko A, Denne J, Enas G. Subgroup identification based on differential effect search (SIDES): A recursive partitioning method for establishing response to treatment in patient subpopulations. *Stat Med*. 2011;30:2601–21.

26. Lipkovich I, Dmitrienko A. Strategies for identifying predictive biomarkers and subgroups with enhanced treatment effect clinical trials using SIDES. *J Biopharm Stat*. 2014;24:130–53.

27. Dusseldorp E, Van Mechelen I. Qualitative interaction trees: A tool to identify qualitative treatment-subgroup interactions. *Stat Med*. 2014;33:219–37.

28. Gu X, Yin G, Lee JJ. Bayesian two-step lasso strategy for biomarker selection in personalized medicine development for time-to-event endpoints. *Contemp Clin Trials*. 2013;36:642–50.

29. Breiman L, Friedman JH, Olshen RA, Stone CJ. *Classification and Regression Trees*. Belmont, CA: Wadsworth, 1984.

30. Leblanc M, Crowley J. Survival trees by goodness of split. *J Am Stat Assoc*. 1993;88:457–67.

31. Qian M, Murphy SA. Performance guarantees for individualized treatment rules. *Ann Stat*. 2011;39:1180–210.

32. Schulte PJ, Tsiatis AA, Laber EB, Davidian M. Q- and A-learning methods for estimating optimal dynamic treatment regimes. *Stat Sci*. 2014;29:640–61.

33. Tibshirani R. Regression shrinkage and selection via the lasso. *J R Stat Soc Ser B*. 1996;58:267–88.

34. Zou H, Hastie T. Regularization and variable selection via the elastic net. *J R Stat Soc Ser B*. 2005;67:301–20.

35. Fu H, Zhou J, Faries DE. Estimating optimal treatment regimes via subgroup identification in randomized control trials and observational studies. *Stat Med*. 2016;35:3285–302.

36. Laber EB, Zhao YQ. Tree-based methods for individualized treatment regimes. *Biometrika*. 2015;102:501–14.

37. Hardin DS, Rohwer RD, Curtis BH, Zagar A, Chen L, Boye KS, Jiang HH, Lipkovich IA. Understanding heterogeneity in response to antidiabetes treatment: A post hoc analysis using SIDES, a subgroup identification algorithm. *J Diab Sci Technol*. 2013;7:420–9.

38. Dmitrienko A, Lipkovich I, Hopkins A, Li YP, Wang W. Biomarker evaluation and subgroup identification in a pneumonia development program using SIDES. In *Applied Statistics in*

Biomedicine and Clinical Trials Design, Chen Z, Liu A, Qu Y, Tang L, Ting N, Tsong Y, eds. New York: Springer, 2015: 427–466.

39. Lipkovich I, Dmitrienko A, Muysers C, Ratitch B. Multiplicity issues in exploratory subgroup analysis. *J Biopharm Stat*. 2018;28:63–81.

40. Lipkovich I, Dmitrienko A, Patra K, Ratitch B, Pulkstenis E. Subgroup identification in clinical trials by Stochastic SIDEScreen methods. *Stat Biopharm Res*. 2017;9:368–78.

41. Loh WY, Shih YS. Split selection methods for classification trees. *Stat Sin*. 1997;7:815–40.

42. Hothorn T, Hornik K, Zeileis A. Unbiased recursive partitioning: A conditional inference framework. *J Comput Graphical Stat*. 2006;15:651–74.

43. Wager S, Hastie T, Efron B. Intervals for random forests: The Jackknife and the Infinitesimal Jackknife. *J Mach Learn Res*. 2014;15:1625–51.

44. Faye LL, Sun L, Dimitromanolakis A, Bulla SB. A flexible genome-wide bootstrap method that accounts for ranking and threshold-selection bias in GWAS interpretation and replication study design. *Stat Med*. 2011;30:1898–912.

45. Loh W-Y, Fu H, Man M, Champion V, Yu M. Identification of subgroups with differential treatment effects for longitudinal and multiresponse variables. *Stat Med*. 2016;35:4837–55.

46. Simon RM, Subramanian J, Li MC, Menezes S. Using cross validation to evaluate the predictive accuracy of survival risk classifiers based on high dimensional data. *Brief Bioinform*. 2011;12:203–14.

47. Huang X, Sun Y, Trow P, Chatterjee S, Chakravatty A, Tian L, Devanarayan V. Patient subgroup identification for clinical drug development. *Stat Med*. 2017;36:1414–28.

48. Ruberg SJ, Shen L. Personalized medicine: Four perspectives for clinical drug development. *Stat Biopharm Res*. 2015;7:214–29.

49. Brookes ST, Whitley E, Peters TJ, Mulheran PA, Egger M, Davey Smith G. Subgroup analyses in randomised controlled trials: Quantifying the risks of false-positives and false-negatives. *Health Technol Assessment*. 2001;5:1–56.

50. Rothwell PM. Subgroup analysis in randomized controlled trials: Importance, indications, and interpretation. *Lancet*. 2005;365:176–86.

51. Sun X, Briel M, Walter SD, Guyatt GH. Is a subgroup effect believable? Updating criteria to evaluate the credibility of subgroup analyses. *Br Med J*. 2010;340:850–4.

52. Berger J, Wang X, Shen L. A Bayesian approach to subgroup identification. *J Biopharm Stat*. 2014;24:110–29.

53. Su X, Kang J, Fan J, Levine RA, Yan X. Facilitating score and causal inference trees for large observational Studies. *J Mach Learn Res*. 2012;13:2955–94.

54. Rosenkranz GK. Exploratory subgroup analysis in clinical trials by model selection. *Biomet J*. 2016;58:1217–28.

17

Developing and Validating Prognostic Models of Clinical Outcomes

Susan Halabi, Lira Pi, and Chen-Yen Lin

CONTENTS

17.1 Introduction

The evaluation of prognostic factors is one of the fundamental tasks in clinical research and will continue to play a critical role in twenty-first century patient management and decision making [1]. Prognostic factors in oncology associate host and tumor variables to clinical outcomes independent of treatment [2]. On the other hand, predictive factors are dependent on the treatment and are described by the interaction between the treatment and the factor in predicting outcomes [2]. Gospodarowicz et al. classify factors as host, tumor-related, or environmental factors [3]. Host-related factors are ones that are related to the patients' characteristics, such as age, performance status, and comorbities. Tumor-related factors are variables that are related to the presence of the tumor and reflect its biology and pathology, such as size of tumor, lymph node involvement and metastasis, and molecular markers (overexpression of HER2). Lastly, environmental factors are ones that are external to the patient, such as access to healthcare or to a cancer-control program [3]. Prognostic factors play a vital role as they influence clinical outcomes, and the American Joint Committee on Cancer has established three criteria that define a prognostic factor [3]. These are: (1) significant, which indicates that the factor does not occur by chance; (2) independent, which means that the factor remains statistically significant in a multivariable analysis; and (3) clinical importance, or the relevance of the factor in managing the patient. The reader is referred to Gospodarowicz et al. for more detail discussions on prognostic factors [3].

There are several reasons why the identification of prognostic factors is important [4]. Understanding the complex relationship between host and tumor-related factors and its impact on clinical outcomes is critical so that (1) clinicians use the prognostic factors in the assessment of stage of disease, (2) investigators gain insights into the disease process, (3) patients and their families are informed about prognosis, (4) enrichment strategies can be developed to evaluate novel therapies and (5) clinical trialists employ these factors in the design and analysis of clinical trials.

This chapter covers topics related to both the design and analysis of prognostic factors, focusing on factors that are relevant at the time of diagnosis or initial treatment. We concentrate on prognostic models, which are created by including several prognostic factors. The chapter first presents examples of how prognostic factors are implemented in the design and analysis of clinical trials and then discusses the design of prognostic factors studies. The focus then shifts to sample size computation and an exploration of the different shrinkage methods of variable selection. Examples from randomized phase III trials are provided. The chapter then considers how to construct risk groups from prognostic models. Next, common problems that are encountered in modeling prognostic factors are highlighted, followed by a succinct review discussion of the advantages of the internal validation approaches and methods for assessing prognostic models. The chapter then offers an overview of pre-screening methods in ultra-high-dimensional space and a genome-wide association study (GWAS) example is provided. The chapter concludes by endorsing the established criteria set by the American Joint Committee on Cancer [5], the Transparent Reporting of a Multivariate Prediction Model for Individual Prognosis or Diagnosis [6] and the Critical Appraisal and Data Extraction for Systematic Reviews of Prediction Modelling [7], in developing and validating prognostic models of clinical outcomes.

17.2 Use of Prognostic Factors in Trials

Prognostic factors are increasingly used in the design, conduct, and analysis of clinical trials. It is common to use a few prognostic factors in the randomization process. As indicated above, prognostic factors can be the standard clinical and pathologic factors, but they may also be genetic and molecular markers from the tumor. The prognostic factors may be combined to create prognostic models that predict clinical outcomes. In certain situations it may be advantageous to use prognostic models in the randomization as opposed to several prognostic factors. For example, in an ongoing phase III trial (A031201), 1,331 men with metastatic castrate-resistant prostate cancer were randomized with equal probability to enzalutamide or enzalutamide plus abiraterone acetate plus prednisone (NCT01949337). The primary endpoint was overall survival. Randomization was stratified by the predicted survival probability determined by a prognostic model of overall survival [8]. The main objective of using a prognostic model in A031201 is to balance the important baseline factors by the treatment assignment.

Risk models have been also employed for screening eligible patients on recent trials. In CALGB 90203, 750 men who are at high risk of recurrence were randomized to either prostatectomy alone or docetaxel plus hormones followed by prostatectomy [9,10]. High-risk men were identified by the Kattan nomogram if their predicted probability of being disease free 5 years after surgery <60% [11]. Another example is the TAILORX trial (NCT00310180), where women with breast cancer had their recurrence score determined by OncotypeDx [12–14]. The OncotypeDx is a 21-gene score that predicts likelihood of recurrence and has been extensively validated [15]. Women with intermediate risk of recurrence (that is an OncotypeDx risk score 11–25) were randomized to endocrine therapy or endocrine therapy plus chemotherapy. The primary objective was testing whether intermediate-risk women randomized to endocrine therapy plus chemotherapy have non-inferior disease-free survival to women treated with endocrine therapy alone.

Prognostic factors have been also used in enrichment trial design with a targeted therapy; these designs are also referred to as targeted designs. In the TOGA trial, over 3,800 gastric cancer patients were screened for HER2 expression, and only patients with HER2-positive cancers were randomized to either trastuzumab plus chemotherapy or chemotherapy alone [16]. The primary question of interest was whether HER2-positive gastric patients treated with trastuzumab plus chemotherapy would have superior overall survival compared to patients treated with chemotherapy alone [16].

Historically, the motivation for the identification of prognostic factors is to estimate the effect of treatment accurately. Therefore, another critical application of prognostic factors is adjusting prognostic factors in order to minimize bias in estimating the treatment effect. It is always recommended that the statistical test for the primary analysis is pre-specified in the statistical analysis plan. The stratified log-rank test is the optimal test for estimating the treatment effect if the randomization is blocked on stratification factors. In the IMpower133 trial, 201 patients with extensive-stage small-cell lung cancer were randomized in 1:1 allocation ratio to atezolizumab plus carboplatin and etoposide or placebo plus carboplatin and etoposide [17]. The stratification factors were sex, ECOG performance-status score, and presence of brain metastases. The primary analysis for the co-primary endpoints overall survival and progression-free survival were based on the stratified log-rank test. In the analysis, the investigator adjusted for sex and ECOG performance status, but not the presence of brain metastases as the latter factor had fewer than 10 events. The stratified hazard ratio for overall survival and progression-free survival were 0.70 (95% CI 0.54–0.91) and 0.77 (95% CI 0.62–0.96), respectively [17].

17.3 Design of Prognostic Studies

Most prognostic factors are identified from retrospective data analysis using small sample sizes, and as a result the data will be unreliable and of poor quality [18,19]. These small studies often lead to conflicting results in predicting outcomes. It is highly recommended that investigators interested in identifying prognostic factors design the study prospectively. In other words, they need to follow the scientific research paradigm by stating a primary hypothesis, defining the primary outcome *a priori*, and justifying the sample size. Importantly, the sample size should be large enough because of the considerable number of potential biases that may occur when conducting such analyses. These issues include missing data of the variables or outcomes, variable-selection methods, multiple comparisons, and assessment of different models [18]. The reader is referred to Simon and Altman, who provide a rigorous review of statistical aspects of prognostic factors studies in oncology [20].

17.3.1 Sample Size Justification

Several articles have tackled the sample size required for identifying prognostic factors of clinical outcomes [4,21]. To facilitate the discussion in estimating the sample size, we utilize CALGB 90206, a phase III trial of interferon alpha with or without bevacizumab in advanced renal cell cancer for illustrative purposes [22]. One of the primary objectives of this correlative study was to determine whether HGF and IL-6 are prognostic biomarkers of overall survival in patients with metastatic renal cell carcinoma. The primary endpoint for the phase III trial was overall survival, defined as the interval between date of randomization and date of death due to any cause. To design this correlative science study, the prevalence of positive HGF and IL-6 are required. Five hundred forty-nine patients with available specimens consented that their plasma be used in this analysis. Table 17.1 presents the hazard ratio that can be detected assuming the following: overall survival follows an exponential distribution, 85% power, a two-sided test of 0.025 (overall type I error rate = 0.05, adjusting for two biomarkers as testing for IL-6 and HGF), prevalence of positive IL-6 and HGF of 0.15–0.50, and events rates of 75%–90%. The power computation

TABLE 17.1

Detectable Hazard Ratio under a Range of Assumptions for the Prevalence of Positive IL-6 and Correlation with Two-Sided Type I Error Rate 0.025% and 80% Power

Prevalence of Prognostic Factor	Correlation with Other Factors	Proportion of Events Observed among 549 Patients with Plasma Available		
		75% Events	80% Events	90% Events
10%	0	1.66	1.63	1.59
	0.2	1.68	1.65	1.60
	0.5	1.79	1.76	1.71
	0.7	2.03	1.99	1.91
25%	0	1.42	1.40	1.38
	0.2	1.43	1.41	1.42
	0.5	1.50	1.48	1.45
	0.7	1.63	1.66	1.57
35%	0	1.37	1.36	1.34
	0.2	1.38	1.37	1.38
	0.5	1.44	1.43	1.40
	0.7	1.56	1.54	1.50
40%	0	1.36	1.35	1.33
	0.2	1.37	1.36	1.34
	0.5	1.43	1.41	1.39
	0.7	1.54	1.52	1.49
50%	0	1.35	1.34	1.32
	0.2	1.36	1.35	1.33
	0.5	1.42	1.40	1.38
	0.7	1.53	1.51	1.47

did not take into account the correlation with other variables. If the correlation between the prognostic variables is factored in, then one would use the sample size formula developed by Schmoor et al. [21]. It is worthwhile noting from Table 17.1 that when the correlation is considered, larger hazard ratios can be detected.

Building prognostic models may be of interest for an investigator. There is an *ad-hoc* rule of thumb where 15 events per variable are required for time-to-event endpoints; and 10 patients per variable are needed for a binary endpoint [4,23]. While the above rule may work initially for exploring the concept of building a model, it is not rigorous for a prospective study. A better approach is to compute the standard error for the concordance index (c-index) [24]. We demonstrate how we conducted simulations to estimate the c-index and the variance of the c-index for the primary objective of developing a prognostic model that will predict overall survival in men with metastatic castrate-resistant prostate cancer who failed first line chemotherapy [25]. The model was to be validated for predictive accuracy using an external dataset (SPARC trial [26]). The goal was that once the final model was chosen, the prediction error will be estimated on the new dataset of patients [27]. Briefly, a proportional hazards relationship was generated from the Weibull regression model $t_i = \exp(\beta'x_i)^* \varepsilon_i$ where β is a vector of the regression coefficients for the covariates, x_i is a vector for the covariates for each subject i. The ε_i were independent and identically distributed (i.i.d.). Weibull random variables with scale parameters $\lambda = \log (2)/12.7$ and $\log (2)/15.1$ (similar to what was reported in the *Lancet* article is that the median survival times were 12.7 and 15.1 months on the mitoxantrone plus prednisone and cabazitaxel plus prednisone arms, respectively). Various shape parameters ranging from 0.25 to 3 (a shape parameter $= 1$ is equivalent to an exponential distribution) were utilized. The censoring distribution is assumed to be random and non-informative and is uniform on the interval [24,30], similar to the number of patients enrolled on the TROPIC trial [25]. The observed survival time is the minimum between failure time and the censoring time. In all of the simulations, the sample size was assumed to be 755, the number of patients in the TROPIC trial [25]. Because of the intensive computing time, we considered a maximum of 10 prognostic covariates. A dataset from a CALGB phase II trial of men who failed first-line chemotherapy was used to study the shape and estimate the distribution the predictor variables. The distribution of the predictors followed either the normal distribution (after transformation) or uniform distribution. β_j ranging from 0.1 to 2 were considered. Using procedures from Gonen and Heller [24], the variance of the c-index denoted as $\tilde{K}(\hat{\beta})$ was estimated by:

$$\widehat{Var}(\tilde{K}(\hat{B})) = \frac{4}{\{n(n-1)\}^2} \sum_i \sum_j \sum_{k \neq j} \{u_{ji} + u_{ij} - \tilde{K}(\hat{\beta})\}\{u_{ki} + u_{ik} - \tilde{K}(\hat{\beta})\},$$

where

$$u_{ji} = \frac{\Phi(-\hat{\beta}'x_{ij}/h)}{1 + \exp(\hat{\beta}'x_{ij})}\} \quad \text{and} \quad \tilde{K}(\hat{\beta})$$

is a smoothed approximation to the concordance probability and is expressed as

$$\tilde{K}(\hat{\beta}) = \frac{2}{n(n-1)} \sum_i \sum_j \left[\frac{\Phi(-\hat{\beta}'x_{ji}/h)}{1 + \exp(\hat{\beta}'x_{ji})} + \frac{\Phi(-\hat{\beta}'x_{ij}/h)}{1 + \exp(\hat{\beta}'x_{ij})} \right],$$

where n is the sample size, Φ is the local distribution function, h is a bandwidth $= 0.5\hat{\sigma}n^{-1/3}$, $\hat{\sigma}$ is the estimated standard deviation of the linear combination $\hat{\beta}'x_i$ computed for each subject, let x_{ij} is the pair-wise difference between $x_i - x_j$ for individuals i and j, respectively. For each of the above scenarios, 10,000 simulated datasets were generated. The results of these simulations showed that the c-index ranged from 0.70 to 0.99 depending on β_j and the shape parameter of the Weibull distribution. Moreover, the c-index can be estimated with simulated variance of 0.000011.

Sample size computation for testing for a predictive biomarker is typically based on a test of interaction between the biomarker and the treatment in a regression model [28,29]. The sample size required for identifying predictive biomarkers are always very large, far exceeding that for prognostic biomarkers. We employ an example from CALGB 90401 to demonstrate how to compute the power for a predictive biomarker of overall survival. The CALGB 90401 trial is a randomized, double-blind, phase III trial in which 1,050 men with metastatic castrate-resistant prostate cancer were randomly assigned to receive bevacizumab, docetaxel, and prednisone or docetaxel, prednisone, and placebo [30]. The primary endpoint is overall survival with a target of 748 deaths. The trial was designed to detect a hazard ratio of 0.79 assuming the log-rank test had power of 85% and a two-sided type I error rate of 0.05. The following assumptions were made to achieve the target 748 deaths: an accrual rate of 29 patients per month over a 35-month enrollment period, with 25-months of follow-up after study closure [30].

The investigators are interested in testing that IL-6 and HGF are predictive biomarkers of overall survival. Of 1,050 patients enrolled on CALGB 90401, 778 (74%) patients have consented and have available plasma samples at baseline. For computing the power for an interaction term, the accrual rate, accrual period, and the follow-up period are needed [28,29]. Or, alternatively, one can estimate the number of events. For CALGB 90401, we assumed that the number of events were 700 deaths (90% of 778), which is similar to what has been observed in the clinical trial [30]. For simplicity in the power computation, IL-6 and HGF levels were assumed to be dichotomized at the median cut-off value derived from a prior analysis, and patients were classified as having either negative (low values) or positive (high) levels. Table 17.2 presents the power for testing the null hypothesis of no treatment-biomarker interaction using a two-sided type I error rate of 0.025 (adjusting for the two predictive biomarkers IL-6 and HGF) and assuming positive biomarker prevalence ranging from 10% to 50%. We assumed that the median survival duration is 21.5 months for patients treated

TABLE 17.2

Power for Testing the Interaction between a Biomarker and Treatment in Predicting Overall Survival

Proportion of Patients with Biomarker	Hazard Ratio in Negative Biomarker Stratum (Δ_1)	Hazard Ratio in Positive Biomarker Stratum (Δ_2)	Hazard Ratio (Δ_2/Δ_1)	Power
0.10	1.0	2.00	2.00	0.63
	1.0	2.15	2.15	0.73
	1.0	2.20	2.20	0.76
	1.0	2.30	2.30	0.80
0.20	1.0	1.75	1.75	0.73
	1.0	2.00	2.00	0.89
	1.0	2.10	2.10	0.93
0.50	1.0	1.50	1.50	0.66
	1.0	1.65	1.62	0.85
	1.0	1.75	1.75	0.91

with docetaxel, similar to what has been observed in the trial [30]. No discrepancy in overall survival distributions by marker level is expected for the docetaxel arm. With 778 patients, there is sufficient power to detect moderation interaction term. For power computations for a predictive biomarker of a time-to-event endpoint, the reader is referred to this useful link: https://stattools.crab.org/Calculators/interactionSurvivalColored.html

17.4 Identification of Prognostic Factors

There are several strategies for the identification of prognostic factors. These approaches include variable-selection, shrinkage methods, and dimension-reduction methods. Standard variable-selection approaches such as the Akaike information criterion and the Bayesian information criterion combined with the logistic regression for binary endpoints [31], proportional hazards regression for time-to-event endpoints [32,33] have been used. In addition, recursive partitioning for both binary and time to event endpoints [34–37] have been utilized extensively in identifying prognostic factors of clinical outcomes [38–42]. It is assumed that the reader is familiar with these standard variable-selection methods. For more detailed information on this topic, the reader is referred to these classical textbooks [4,31,43–45].

17.4.1 Shrinkage Methods

There is a class of shrinkage methods that fit p predictors by shrinking the coefficient estimates toward zero and thus reducing the variance of the coefficient estimates [46]. Consequently, these methods would improve the accuracy of the model [46]. The least absolute shrinkage and selection operator (LASSO) and adaptive LASSO (ALASSO) have been widely employed to build prognostic models of clinical outcomes [47,48]. Ridge regression minimizes this function:

$$\sum_{i=1}^{n}\left(y_i - \beta_0 - \sum_{j=1}^{p}\beta_j x_{ij}\right)^2 + \lambda\sum_{j=1}^{p}\beta_j^2$$

The first term is the residual sum of squares where y_i is the response, x_{ij} is the jth covariate value ($j = 1, 2, \ldots, p$) corresponding to the ith individual, β_j is the regression coefficient jth covariate, and λ is a tuning parameter [46]. Ridge regression has one caveat as it does not shrink all the coefficients exactly to zero and this will impact the prediction accuracy [46]. LASSO is an improvement over ridge regression and it minimizes this function with respect to β:

$$\sum_{i=1}^{n}\left(y_i - \beta_0 - \sum_{j=1}^{p}\beta_j x_{ij}\right)^2 + \lambda\sum_{j=1}^{p}|\beta_j|$$

Similar to ridge regression, the first term is the residual sum of squares and the second term is the L_1 penalty function. Choosing a large tuning parameter causes the coefficients estimates to be equal to zero [46]. Then the LASSO will achieve the sparsity property. LASSO however has some limitations as it selects too many unimportant variables, it selects only one variable from a paired correlation variables, and in $p > n$ situations, performs poorly [38,49–51].

Adaptive LASSO (ALASSO) has been proposed as a modification of the LASSO in order to overcome the limitation of LASSO [52]. The ALASSO enjoys the oracle property [48,52] and minimizes the following penalized function:

$$\sum_{i=1}^{n}\left(y_i - \beta_0 - \sum_{j=1}^{p}\beta_j x_{ij}\right)^2 + \lambda\sum_{j=1}^{p}w_j\,|\,\beta_j\,|$$

ALASSO uses a weighted penalty term in the L_1 penalty, where $w = (w_1, w_2, \dots, w_p)$ is the weight vector. If $\hat{\beta}$ is a $\sqrt{n}$-consistent estimator, e.g., $\hat{\beta}$ (OLS), of $\beta = (\beta_1, \beta_2, \dots, \beta_p)$, then an appropriate choice of the weight w is $1/|\hat{\beta}|$. ALASSO is considered to be an improvement over LASSO as it has consistent variable selection as well as lower prediction error. Parameter tuning is a critical step in determining non-zero coefficients. and ALASSO tends to select fewer non-zero coefficients than the LASSO despite having smaller prediction error.

Elastic net regression utilizes a combination of L_1 penalty and ridge L_2 penalty and is a compromise between LASSO and ridge regression [53]. The elastic net approach minimizes the following penalized function:

$$\sum_{i=1}^{n}\left(y_i - \sum_{j=1}^{p}x_{ij}\beta_j\right)^2 + \lambda\sum_{j=1}^{p}\left((1-\alpha)\,|\,\beta_j\,| + \alpha\,|\,\beta_j\,|^2\right),\quad \alpha \in (0,1).$$

The second term in the function is known as the elastic penalty, and α is a mixing parameter such that when $\alpha = 0$, this leads to LASSO regression while $\alpha = 1$ leads to ridge regression. One of the main advantages of elastic net over LASSO in $p > n$ is elastic net retains more than n variables in the model. In the $p > n$ situation, there exists a group of highly correlated variables and LASSO tends to select only one of the correlated variables in each group by shrinking the coefficients of the other correlated variables to zero [53]. Hastie et al. provide a thorough comparison of these shrinkage techniques [46].

17.4.2 Prostate Cancer Example Predicting Overall Survival

We utilize CALGB 90401, a phase III clinical trial in advanced prostate cancer to illustrate how LASSO and ALASSO were utilized to identify the prognostic factors of the primary endpoint overall survival [8]. Our goal was to select the best prognostic model of overall survival from all sets of candidate models. Because we were in a data-rich situation, we used a rigorous statistical approach to meet this objective as described by Hastie et al. [54] Figure 17.1 presents our overall strategy for the model development and validation. We needed to estimate the performance of all the models and choose the best one. Second, we needed to assess the performance of the final chosen model by estimating the prediction error. It is difficult to give a general rule on how best to split the data [54]. We therefore

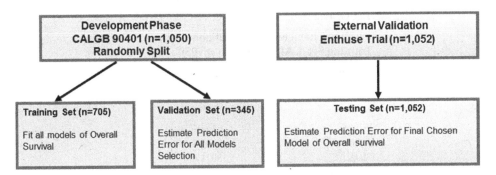

FIGURE 17.1
Overall strategy for model development and validation for overall survival.

randomly divided the 1,050 men in CALGB 90401 in a 2:1 allocation ratio to the *training* ($n = 705$) and *validation* ($n = 345$) sets. Using the *training dataset*, we fit models from all sets of potential models of overall survival. Using the *validation dataset*, the prediction error was estimated for all sets of models for overall survival. The final step was to use the *testing dataset* (Enthuse trial) as an independent dataset for external validation [55]. The Enthuse trial randomized 1,052 men to docetaxel and prednisone with and without zibotentan [55]. Using the *testing dataset*, we were able to estimate the prediction error on new patients independent of the development phase.

Twenty-two common variables were identified in CALGB 90401 and the Enthuse trial. CALGB 90401 collected tumor-related factors, such as alkaline phosphatase, albumin, LDH, and PSA. In addition, host-related factors, such as WBC, platelets, hemoglobin, body mass index, performance status, time since diagnosis, comorbidity, and use of opioid, were also collected. From Table 17.3, it can be noted that some of the prognostic factors were missing. We followed similar methods as White and Royston to impute the missing data [56]. Some of the laboratory variables (such as testosterone, PSA, and alkaline phosphatase) were highly skewed, and we used the logarithm function to transform these variables (Figure 17.2). Adaptive weights were given by the reciprocal of absolute parameter estimates from an unpenalized Cox's model fitted with all 22 baseline prognostic factors (Table 17.4). We considered both LASSO and ALASSO as selection methods. We applied both the Akaike information criterion and the Bayesian information criterion for choosing the optimal model of overall survival. We selected though the regularization parameter to minimize the Bayesian information criterion, as our goal was to have a sparse model. LASSO and ALASSO selected eight and nine variables, respectively. It is interesting that there was no difference in the variables selected between LASSO and ALASSO, although as expected the solution paths are different (Figure 17.3a and b). We determined ALASSO model as the final optimal model since it included the site of metastases for bone. Figure 17.3a presents the solution path for ALASSO, and we observe that LDH greater than upper limit of normal and ECOG performance status were selected early in the L1 path compared to the other variables. This is followed by visceral disease, alkaline phosphatase, albumin, hemoglobin, pain, bone metastases, and then PSA (the BIC stopped at PSA). The final model selected the following prognostic factors: LDH greater than upper limit of normal, ECOG performance status, presence of visceral disease, PSA, alkaline phosphatase, albumin, ECOG performance status, hemoglobin, metastatic site, and analgesic opioid use. We displayed the fitted model in a nomogram (Figure 17.4).

TABLE 17.3

Baseline Characteristics of Patients in the
Training Set (CALGB 90401, $n = 705$)

Age[a] (years)	
Median	69 (62,75)
Race	
White	613 (87%)
Asian	5 (1%)
Black	80 (11%)
Other/missing	7 (1%)
ECOG performance status	
0	403 (57%)
1	276 (39%)
2	26 (4%)
Disease site	
Lymph node only	75 (11%)
Bone/bone + lymph node	512 (73%)
Any visceral	118 (17%)
Measurable disease	
Yes	363 (51%)
No	342 (49%)
Opioid analgesic use	
Yes	213 (30%)
No	341 (48%)
Missing	151 (21%)
LDH >1 ULN	
Yes	265 (38%)
No	437 (62%)
Missing	3 (0%)
LDH U/L	
Median[a]	288.06 (277.89,298.23)
PSA ng/mL	
Median[a]	266.77 (238.77,294.77)
Hemoglobin g/dL	
Median[a]	12.64 (12.62,12.73)
Missing	0
Albumin g/dL	
Median[a]	3.93 (3.91,3.95)
Missing	5
Alkaline phosphatase U/L	
Median[a]	202.32/117.00
Treatment arm	
Docetaxel + Bevacizumab	361 (51%)
Docetaxel + placebo	344 (49%)

[a] Median (inter-quartile range).

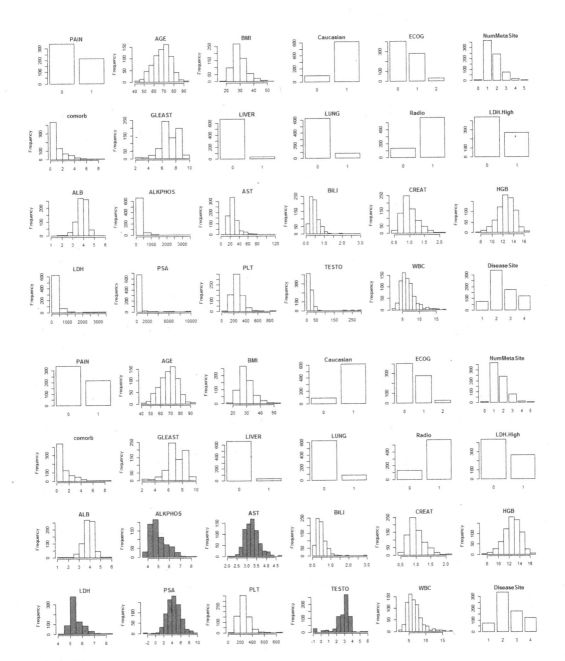

FIGURE 17.2
Distribution of baseline variables in the training set (CALGB 90401).

17.4.3 Non-Parametric Approaches

Non-parametric procedures are worthwhile pursuing when prognostic factors are non-linear.

Random forest is a non-parametric variable-selection method that has been widely used in identifying prognostic factors of clinical outcomes [57,58]. Random forest is built on a user-specified number of decision trees where for each tree, $1-1/e \approx 63\%$

TABLE 17.4

Identified Prognostic Factors by LASSO and ALASSO

Description of Variable (Variable Name)	$\hat{\beta}$ Cox's Model	$\hat{\beta}$ LASSO	$\hat{\beta}$ ALASSO
Site of metastases: Bone (BONE)	0.234	0.000	0.058
Visceral (VISECRAL)	0.400	0.056	0.293
Liver (LIVER)	0.013	0.000	0.000
Lung (LUNG)	0.199	0.000	0.000
Opioid analgesic use (PAIN)	0.136	0.077	0.088
Age in years (AGE)	−0.003	0.000	0.000
Body mass index (BMI)	−0.021	0.000	0.000
Race (Caucasian)	0.034	0.000	0.000
ECOG performance status (ECOG)	0.278	0.190	0.305
Comorbidity (Comorb)	0.070	0.000	0.000
Gleason score (GLEAST)	0.052	0.000	0.000
Prior treatment with radiotherapy (Radio)	0.105	0.000	0.000
LDH ≥ ULN (LDH.High)	0.325	0.203	0.335
Albumin (ALB)	−0.133	−0.080	−0.122
Bilirubin (BILI)	−0.017	0.000	0.000
Hemoglobin (HGB)	−0.094	−0.085	−0.065
Platelets (PLT)	0.000	0.000	0.000
White blood cells (WBC)	0.055	0.000	0.000
Alkaline phosphatase (ALKPHOS)	0.145	0.138	0.145
Aspartate aminotransferase (AST)	0.025	0.000	0.000
Prostate specific antigen (PSA)	0.063	0.026	0.015
Testosterone (TESTO)	−0.088	0.000	0.000
Training C-index		0.662	0.660
Integrated time AUC		0.742	0.740

of observations are randomly used. And for each split in each tree, there are a user-specified number of variables randomly used. Such random selections enable it to have a built-in cross-validation. A decision tree (classification) is built from a root node given by the split that yields the best reduction in an impurity measure among all possible splits, and splits will be conducted at each such node until some convergence criterion is satisfied. For example, the Gini index is used as the impurity function, and for a binary classification and a possible split (condition A), is given by Gini (A) $= 2p(1-p)$, where p is the proportion of observations that satisfies condition A under the mother node. Then random forest makes a prediction of an observation based on the votes from all trees in the forest. It is noted that a split is invariant to a monotonic transformation of the variable. In practice, the number of variables selected at each split does not have much impact on the final forest classifier.

Random forest has been shown to be effective and robust in the dependence among variables. In addition, random forest yields the variable importance score in classification, a measure that reflects each variable's importance in classification, which is a useful tool for dimension reduction [59]. A value of variable importance score of zero denotes no prognostic importance whereas a value of greater than 0 signifies moderate or strong importance. While random forest provides high predictive accuracy based on training samples, it has the main disadvantage of increasing overfitting [60].

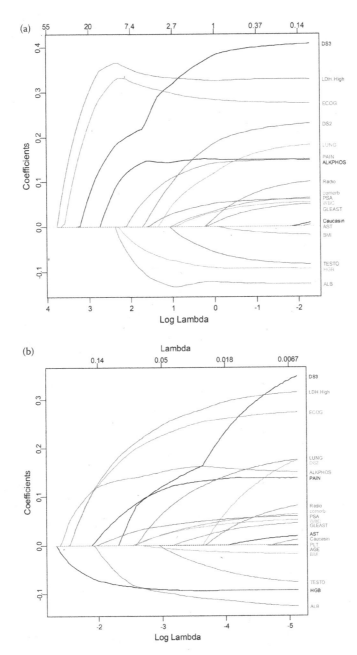

FIGURE 17.3
(a) Solution path for ALASSO and (b) solution path for LASSO.

Other non-parametric methods have been shown to outperform the LASSO-type regression method [61]. Lin and Zhang proposed a model selection and estimation in the smoothing spline (SS-ANOVA) [62]. The component selection and smoothing operator (COSSO) is the non-parametric counterpart of the LASSO, and it reduces to LASSO when COSSO expression is utilized in linear models [62]. One of the most widely used choices for the tuning parameter for the SS is generalized cross-validation suggested by Craven and

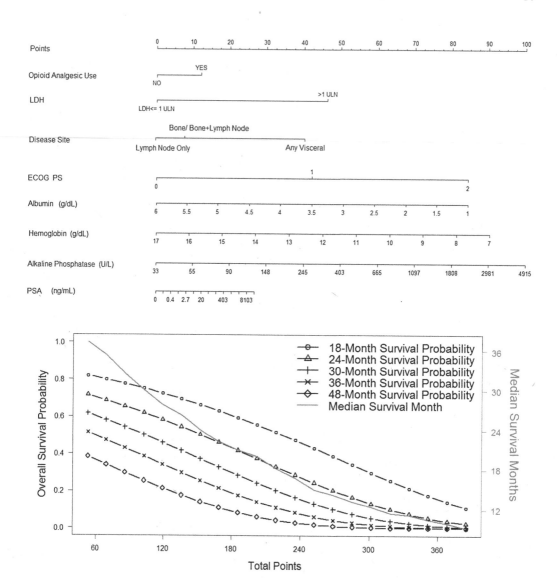

FIGURE 17.4

Pretreatment nomogram predicting probability of survival in men with CRPC. (Printed with permission from Halabi S et al. *J Clin Oncol.* 2014;32(7):671–7.)

Wahba [63]. The COSSO method was extended to non-parametric regression in exponential families [64].

17.4.4 Prostate Cancer Example Predicting PSA Decline

We provide an example to illustrate the application of a random forest. Our motivation was to develop and validate predictive models of ≥30% and ≥50% PSA decline from baseline in men who were enrolled on the TROPIC trial. The TROPIC trial was a randomized, open-label, multicenter, phase III trial of 755 men with mCRPC previously treated with a docetaxel-containing regimen [25]. Participants were randomly assigned to receive either 12 mg/m² mitoxantrone plus 10 mg oral prednisone daily or cabazitaxel 25 mg/m² plus prednisone. PSA decline was defined per the PSA Working Group 2 [65]. We followed a similar logic as presented in Figure 17.1 for building a prognostic model and validating it. We planned to utilize a subset of 488 men who were randomized on the SPRAC trial and who were previously treated with docetaxel for external validation [26]. The SPARC trial was a double-blind, placebo-controlled study comparing the efficacy and safety of satraplatin plus prednisone versus placebo plus prednisone in mCRPC men previously treated with one cytotoxic regimen [26]. We considered 14 variables that were common in the two trials TROPIC and SPARC (external validation set): arm, race, age, body mass index (BMI), time on hormone, years since diagnosis, pain, ECOG performance status (ECOG PS), presence of measurable disease, site of metastases (that is, lymph nodes, bone, visceral), time from docetaxel use to second line chemotherapy <6 months, hemoglobin, PSA, and alkaline phosphatase. We used the random forest not solely as a variable-selection method, but also to impute the missing data as seven out of 14 variables had missing values. To impute the missing values of the variables, the random forest is called with the complete case data. The proximity matrix from the random forest, based on the frequency that pairs of data points are in the same terminal nodes, is used to update the imputation of the missing values. For continuous predictors, the imputed value is the weighted average of the non-missing observations, where the weights are the proximities. For categorical predictors, the imputed value is the category with the largest average proximity. In our analyses, we considered both complete cases and imputed datasets. Of the 14 variables, the random forest procedure selected seven as significant in predicting ≥30% and ≥50% decline in PSA. The significant predictors of PSA decline were: age, body mass index (BMI), years on hormone therapy, years since initial diagnosis, hemoglobin (HGB), alkaline phosphatase (ALK) and PSA (Figure 17.5). The importance of the selected variable is indicated as they are in the far right position in Figure 17.5. Unfortunately the model for ≥30% PSA decline had poor performance in the imputed and complete case (Table 17.5). Despite the low

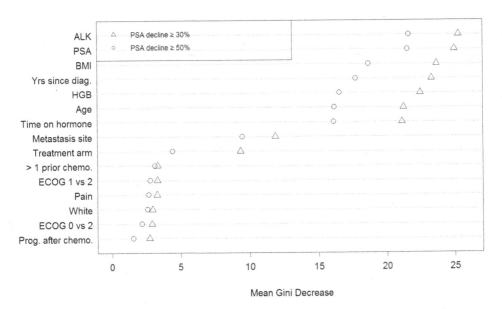

FIGURE 17.5
Gini predictor plot for ≥30% and ≥50% decline in PSA from baseline.

TABLE 17.5

Performance of the PSA Decline Models

Outcome	AUC (95% CI)	Misclassification Error
PSA decline $\geq$ 30% from baseline		
Complete training data	0.57 (0.51,0.63)	38%
Imputed training data	0.56 (0.50,0.61)	38%
Validation set (TROPIC trial)	0.54 (0.46,0.62)	43%
Testing (SPARC trial)	0.53 (0.46, 0.60)	39%
PSA decline $\geq$ 50% from baseline		
Complete training data	0.58 (0.50,0.65)	25%
Imputed training data	0.56 (0.50,0.62)	25%
Validation set (TROPIC trial)	0.51 (0.40,0.58)	29%
Testing (SPARC trial)	0.55 (0.46, 0.63)	21%

misclassification rates for $\geq$50% PSA decline, the area under the curve was low, indicating potential independence between the outcomes and the covariates.

17.5 Common Pitfalls with Modeling

We present common drawbacks in building prognostic models so that statisticians can be wary of using them. These caveats have been extensively discussed by Harrell [4]. It is a common practice that investigators categorize a continuous prognostic factor [4,18]. Regardless of the variable-selection methods used, dichotomizing a continuous variable in the regression model would result in considerable loss of information [66]. It is recommended to apply several categories for the continuous prognostic factor to understand the relationship of it and the hazard of death [66]. Fractional polynomial and cubic splines have been implemented in studying the relationship between continuous prognostic factors and the hazard function [67,68].

Variable selection is a vital step in building a prognostic model [61]. It is astonishing to note that some investigators implement the stepwise methods for selecting prognostic factors. These approaches are not optimal as they will produce overly optimistic regression estimates, low estimated standard error and do not correct for multiplicity that will yield low predictive accuracy [69,18].

Another common caveat is identifying a prognostic factor based on the optimal cut point based on the minimal p-value [70]. This approach, however, does not correct for the multiplicity of comparisons, and the cut point is subjective and arbitrary. There are several algorithms that correct for multiplicity [71], and if the cut point for the prognostic factor is identified, it should be considered as exploratory [72,73]. A prospective study may be needed to confirm the cut point with larger sample size in order to increase the precision of the estimate.

As statisticians, it is critical to check the assumptions of the regression model. If assumptions are not held, then accurate interpretation of the results may be challenging and lead to wrong conclusions. For example, in fitting a prognostic model of time-to-event endpoints, an investigator may miss the important step of checking the proportional hazards assumption. In building a prognostic model of overall survival, a proportional hazards method was initially used to build a prognostic model [74]. However, several factors did not meet the proportional hazards assumption of a constant hazard over time. Subsequently, the authors considered and applied an accelerated failure time model. The reader is referred to an insightful and a thorough review of strategies involved in building models that is provided by Harrell [4].

17.6 Constructing Risk Groups

Combining multiple variables to form a prognostic score is a powerful approach that facilitates the creation of groups of patients with differing risks of progression or death. For this reason, prognostic models will be always implemented in the design, conduct, and analysis of clinical trials. Once the final model is chosen, the next step is to construct risk groups. The estimated survival function at time t is

$$\hat{S}(t) = [\hat{S}_0(t)]^{\exp R}$$

where R is the estimated linear predictor or risk score for the ith individual $\hat{\beta}'x_i = \sum_{j=1}^{p} \hat{\beta}_j' x_{ij}$, $\hat{S}_0(t) = \exp(-\hat{\Lambda}_0(t))$ is the baseline survival function, and $\hat{\Lambda}_0(t)$ is the cumulative baseline hazard function. Risk groups can be formed based on their quartiles or tertiles from the estimated linear predictor. Going back to our overall survival model for prostate cancer, we computed a risk score from the estimated regression coefficients in the training set. We were interested in constructing three risk groups and determined the cut points from the training set based on tertiles (33th and 67th percentiles) and the optimal cut point that adjusted on multiple comparisons [72].

We validated the risk groups using the validation set from CALGB 90401 and the testing set (ENTHUSE 33). Figure 17.6 presents the Kaplan–Meier survival by three risk groups in the validation and testing sets. There were 95, 94, and 92 patients in the high-, intermediate- and low-risk groups, respectively, and their estimated median survival times were 15.1 months (95% CI = 13.7–18.9), 21.6 months (95% CI = 19.9–25.4). and 33.0 months (95% CI = 28.5–37.7, log-rank test p-value <0.0001, left panel Figure 17.6), respectively. The

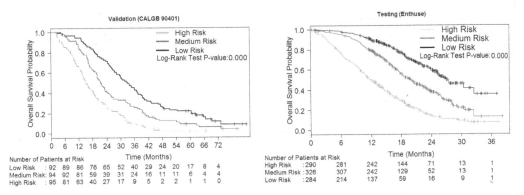

FIGURE 17.6
Kaplan–Meier overall survival curves by the three risk groups in the validation and testing sets. (Modified and printed with permission from Halabi S et al. *J Clin Oncol.* 2014;32(7):671–7.)

estimated hazard ratios for the high- and intermediate-risk groups versus the low-risk group were 2.91 (95% CI = 2.13–4.0) and 1.61 (95% CI = 1.18–2.18), respectively.

In the external validation set, there were 290, 326, and 284 patients in the high-, intermediate-, and low-risk groups. The observed median survival time was 12.1 months (95% CI = 10.9–13.8), 19.9 months (95% CI = 18.1–22.2), and 27.0 months (95% CI = 25.3–NA, p-value <0.0001, right panel Figure 17.6), respectively. The hazard ratios for the high- and intermediate-risk groups versus the low-risk group were 4.27 (95% CI = 3.35–5.43) and 1.92 (95% CI = 1.50–2.46), respectively.

Recursive partitioning is another powerful approach that allows for the construction of risk groups. The main advantages of recursive partition are: it can be non-parametric, uses a linear combination of factors to determine splits, controls for the global type I error rate, can be also employed to both binary and time-to-event endpoints, can handle missing covariate values, and is effective at modeling complex interactions [36]. The major disadvantages of tree models are that they can create complicated structures, which makes the interpretation of the data difficult, they may cause unstable tree structures, and they produce overfitting [2].

17.7 Validation and Assessment of Prognostic Models

One of the most challenging issues in building models is to avoid overfitting. Overfitting refers to a situation when a high predictive accuracy is observed for a prognostic model that has been fit to the training set, but has low accuracy when evaluated in an independent dataset. Thus, the associations between the prognostic factors and the outcomes will be spurious. We illustrate the concept of overfitting assuming a small sample size and large number of prognostic variables [75]. We consider a binary endpoint objective response rate and assume that it is 30%. Let us suppose that this outcome has been observed for 100 patients along with $K = 50$ baseline variables. Our primary objective is to build a prognostic model based on a subset of the variables. We arbitrarily assign the first 30 patients as responders and the remaining 70 as non-responders. It should be noted that by design there is absolutely no relationship between the outcome and any of the variables. Therefore any model that we develop based on these data would be a typical situation of noise-discovery. It is neither practical nor advisable to build the model based on all variables.

One standard approach is to carry out the variable selection after reducing the number of variables and then build a model based on the subset of variables. A common feature-selection approach is to establish the marginal significance of each variable and then build a model based on the top important variables. We therefore select the top 10 variables from the 50 variables based on the rank of the absolute value of the two-sample t-statistic. We next build a model using the logistic regression method. We carry out this simulation analysis once and establish the performance of the resulting model using the area under the curve based on the predicted probabilities from the logistic model (left panel in Figure 17.7). The corresponding AUC is 0.74, which may suggest that we have come up with a good prognostic model. We repeat this simulation 10,000 times and summarize the area under the curve in a boxplot (right panel in Figure 17.7). This confirms our previous observation that we have come up with a good prognostic model. We know, however, that these observations should be rejected since we have trained these models on noise. What went wrong? It is likely that just by chance some of the simulated variables were able to discriminate the responders from the non-responders. Although each model provided a

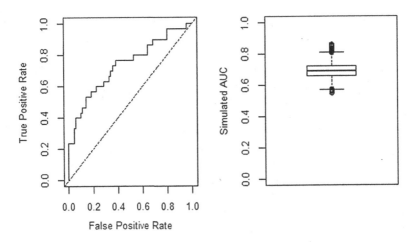

FIGURE 17.7
An illustration of noise-discovery due to overfitting using the logistic regression model. (Printed with permission from Halabi S, Owzar K. *Semin Oncol.* 2010;37(2):e9–18.)

reasonable fit to the data it was trained on, it will unlikely provide a satisfactory fit if we use an external dataset. To reduce the likelihood of overfitting, cross-validation needs to be applied to the variable-selection process and the logistic model needs to be utilized.

The primary goal of a prognostic model is to provide accurate estimates in predicting outcomes in new patients [4,76]. It is therefore important to underscore that validating a prognostic model will always be a vital step after a model has been built [77]. There are two types of validation: external and internal. External validation is the most rigorous approach where the frozen model is applied to an independent dataset from the development data. Ideally, investigators would have an independent dataset available for validation purposes, although this may not be an option. Other types of internal validation such as cross-validation, bootstrapping and bootstrapping using 0.632+ are used to obtain an unbiased estimate of the predictive accuracy of the model [76,78,79]. These validation approaches are considered acceptable [50].

Cross-validation is a generalization of data splitting, where an investigator fits a model based on a random sample before testing it on the sample that was omitted. In 10-fold cross-validation, 90% of the original sample is reserved to building a model and the remaining 10% of the sample is kept to test the model. This process is iterated 10 times, so that all participants have been selected once in testing the model. It is recommended that 200 models need to be fitted and the results averaged over the 200 repetitions in order to have accurate estimates with cross validation. One of the main advantages of cross-validation is that it reduces variability by not relying on a single sample split.

Bootstrapping is a superior method of internal validation without making any assumptions about the distribution of the data [80,81]. The bootstrap does with a computer what the experimenter would do in practice if it were possible, that is, repeat the experiment endless times. Bootstrapping is a process of sampling from the population, and samples are drawn with replacement from the original sample. The samples are of the same size as the original sample. For example, when 300 patients are available for developing a model, bootstrap samples should also have 300 patients, but some patients may not be sampled at all, others patients may be sampled once, others twice, others three times, etc. Similar to cross-validation, the drawing of bootstrap samples needs to be repeated many times in order to obtain stable estimates.

We have reviewed two common re-sampling approaches for assessing the discriminative ability of a prognostic model. Although powerful, these approaches may not be sensitive enough to reveal noise-discovery. If the number of cases compared to the number of controls is relatively small, the cross-validation procedure may produce overly optimistic results. Another layer of re-sampling may be needed to further investigate the result from the validation study. We recommend to randomly permute the outcome and then repeat the validation process [82]. Thus, we break any potential relationship between the outcome and the variables; as a result we expect that any model trained on these random outcomes would not show good predictive ability. In order to carry this analysis, the predictive performance of the noisy data is compared to that of the observed data by repeating the permutation process several times. We should suspect noise if the performance under the noisy data is comparable to that of the observed data.

Assessing the performance of the model is usually conducted by examining calibration and discriminative ability of the model. Calibration signifies the extent of the match between the predicted and observed outcome. Often investigators plot the predicted versus the observed outcome. The model will be well calibrated if the predicted and observed data fall on a 45° line.

Discrimination describes the ability of a prognostic model to distinguish between patients with and without the outcome of interest [4]. Several useful measures can be used to report on the performance of a model [4]. A widely used measure is the c-index, which refers to the proportion of agreement between outcomes and prediction. A value of 0.5 indicates random prediction whereas a value of 1 indicates perfect discrimination. Somers' D rank correlation is another measure and is related to the c-index and is computed as $2*(c-0.5)$. Both c and Somers' D are computed using standard statistical software.

Another widely used measure of predictive accuracy is the time-dependent area under the receiver operating characteristic curve (tAUROC), and this is calculated for each time point. The tAUROCs at different time points can be combined to form an integrated area under the receiver operating characteristic curve for the whole range of the study (iAUROC). Several methods are available in computing the tAUROC [83–85] which would produce different results. The iAUROC is computed as a sum of weighted tAUROCs [83–85], and a value of 0.5 for a model indicates random prediction while an iAUROC > 0.7 may signify a meaningful model predicting a survival endpoint.

A measure of overall performance is R^2_{BS} which evaluates the percentage gain in predictive accuracy compared to the null model, that is a model with no variables, at a single time point [86]. A positive value of R^2_{BS} indicates that a prognostic model predicts well relative to the null model, whereas a negative value signifies that the model predicts poorly.

17.7.1 External Validation of Overall Survival Model

We applied the parameter estimates to the validation and testing sets, and we computed a predicted score for every patient. We assessed the performance of the model by utilizing Uno's integrated measure for the time-dependent area under the receiver operating characteristic curve. We computed the 95% confidence interval for the time-dependent area under the receiver operating characteristic curve based on the bootstrapped method. We modeled risk score as a continuous variable in the proportional hazards model, and it was a significant predictor of overall survival ($p < 0.0001$). The time-dependent area under the curve for risk score as a continuous variable was 0.73 (95% CI = 0.70–0.73) and 0.76 (95% CI = 0.72–0.76) in the validation and testing sets, respectively.

TABLE 17.6

Confusion Matrix from the Random Forest

	Complete Training Data		Imputed Training Data		Validation Data		Testing Data	
	No	Yes	No	Yes	No	Yes	No	Yes
PSA decline $\geq$ 30% from baseline								
Predicted/observed								
No	203	33	235	33	100	65	178	74
Yes	109	32	132	33	21	16	58	28
Sum	312	65	367	66	121	81	236	102
PSA decline $\geq$ 50% from baseline								
Predicted/observed								
No	281	6	320	8	143	56	266	59
Yes	87	3	101	4	3	0	11	2
Sum	368	9	421	12	146	56	277	61

17.7.2 External Validation Predicting PSA Decline

Turning back to our PSA decline models, we observe that the misclassification errors and the area under the receiver operating characteristic curve for $\geq$30% decline and $\geq$50% decline in PSA were high in the validation (TROPIC) and testing (SPARC) sets. This demonstrates the poor performance of the models (Table 17.5). The random forest has a built-in cross-validation procedure and will output a confusion matrix, which is a frequency table based on the predicted versus observed outcomes. When we examined the confusion matrix (Table 17.6), it indicated that the classifiers produced a highly one-sided preference for non-events (no PSA declines).

17.8 High Dimensional Space, Low Sample Size

The emphasis has been thus far on variable selection when the number of predictors is large relative to the sample size. Several pre-screening methods are useful in both the large *p*, small *n* problem and in ultra-high-dimensional space [49,50,87]. There are two main challenges in identifying potential prognostic factors in high-dimensional space: computational intensity and a high false discover rate (FDR), see Chapter 18. Meinshausen and Bühlmann were successful in computing conservative *p*-values for testing hypothesis for high-dimensional space [51].

In ultra-high-dimensional space, the widely used pre-screening procedures are the sure independence screening (SIS), the iterative version of the sure independence screening (ISIS), and principled sure independence screening (PSIS) [49,50,87]. Fan et al. defined ultra-high dimensionality by the exponential growth of the dimensionality in the sample size [50]. The main emphasis of these pre-screening procedures is to select a number of factors that are much lower than the sample. Thus, reduction in dimensionality makes these methods appealing as they perform more accurately in ultra-high-dimensional settings. Applying sequentially pre-screening methods and then variable-selection methods or p-value cut off, such as the FDR, and variable-selection methods are other practical approaches that can result in increased predictive accuracy of clinical accuracy [60,88].

TABLE 17.7

c-Indices Based on the Training and Testing Sets for the GWAS Example

Selection Approach	No. of SNPs Selected	Training Set ($n = 419$)		Testing Set ($n = 204$)
		Original c-Index	Corrected c-Index[a]	c-Index[a] (95% CI)
ISIS-LASSO	2	0.649	0.646	0.664 (0.621–0.707)
ISIS-ALASSO	0	0.649	0.650	0.671 (0.624–0.719)
ISIS-RSF	2	0.650	0.645	0.669 (0.618–0.720)
SIS	2	0.650	0.646	0.669 (0.624–0.714)
SIS-LASSO	0	0.649	0.649	0.671 (0.626–0.717)
SIS-ALASSO	0	0.649	0.650	0.671 (0.620–0.723)
SIS-RSF	2	0.650	0.648	0.669 (0.623–0.716)
PSIS	40	0.749	–	0.572 (0.527–0.617)
PSIS-LASSO	28	0.746	0.727	0.568 (0.524–0.613)
PSIS-ALASSO	24	0.744	0.727	0.573 (0.528–0.617)
PSIS-RSF	35	0.748	–	0.575 (0.529–0.622)
LASSO	16	0.716	–	0.586 (0.540–0.632)
ALASSO	13	0.653	0.634	0.647 (0.601–0.693)

Source: Printed with permission from Pi L, Halabi S. *Diagn Progn Res.* 2018;2.
[a] Based on 200 bootstrapped samples.

17.8.1 Example of SNPs Identification in Ultra-High Dimension

We utilize genome wide association study (GWAS) data from CALGB 90401 to demonstrate the sequential use of pre-screening variables and variable-selection methods in ultra-high-dimensional space. We are interested in identifying germline single-nucleotide polymorphisms (SNPs) that would predict overall survival in patients with metastatic prostate cancer who were genotyped on CALGB 90401. The GWAS data included 498,081 SNPs that were processed from blood samples from 623 Caucasian patients with prostate cancer.

We randomly divided the data into a 2:1 allocation ratio to training ($n = 419$) and testing ($n = 204$) sets, respectively. We applied sequentially pre-screening methods (SIS, ISIS, PSIS) with variable-selection methods (LASSO, ALASSO) to identify the important SNPs of overall survival in the training set. In all the proportional hazards models, we assumed additive models for the SNPs, and we adjusted for risk score based on the predicted survival probability [29]. As expected, we observed a slight indication of overfitting in the training set when the c-index from the original sample was compared to the bootstrapped samples (Table 17.7). The reader is referred to Li et al. for more information [60]. An example of the R code is provided on the companion website https://www.crcpress.com//9781138083776.

17.9 Conclusion

Prognostic studies focus on fundamental questions that are pertinent to patient outcomes. It is strongly recommended that investigators design such studies as prospective ones so that they are identical to the rigor implemented in a clinical trial. And as such, these studies should be thoroughly planned and carefully designed in order to obtain accurate and reliable answers. Similar to any scientific research, prognostic studies should begin by asking a hypothesis, defining *a priori* the primary endpoint, justifying the sample size, and describing appropriate methods for variable selection. In addition, they should include approaches for adjusting for multiplicity, dealing with missing data, testing the robustness of the models applied, and presenting the validation approaches. The emphasis in this chapter has been the development of prognostic models from clinical trial data. These methods could be used for other sources of data such as epidemiologic studies and electronic health records.

It is encouraging to see criteria for evaluating prognostic models published by the Precision Medicine Core of the American Joint Commission on Cancer [5], the Transparent Reporting of a Multivariable Prediction Model for Individual Prognosis or Diagnosis [6] and the Critical Appraisal and Data Extraction for Systematic Reviews of Prediction Modelling Studies [7]. Investigators are recommended to adhere to these guidelines as more rigor will be implemented in building prognostic models and validating them. Thus, the overall quality of the prognostic models will result in a large number of validated models that are vetted. It is anticipated that these models will become part of patient care and patient management as well as in the design and conduct of future trials.

Acknowledgments

This work was funded in part by the United States Army Medical Research Awards W81XWH-15-1-0467 and W81XWH-18-1-0278. This chapter draws and elaborates upon references [2,75].

References

1. Osullivan B, Brierley J, Gospodarowicz M. *Prognosis and Classification of Cancer.* 2015: 23–33.
2. Halabi S, Pi L. Statistical considerations for developing and validating prognostic models of clinical outcomes. In *Oncology Clinical Trials: Successful Design, Conduct, and Analysis.* WKDO Kelly, SP Halabi, eds. New York: Springer Publishing Company. 2018: 313–22.
3. Gospodarowicz MK, O'Sullivan B. *Prognostic Factors in Cancer.* Hoboken: Wiley-Liss, 2006.
4. Harrell FE. *Regression Modeling Strategies with Applications to Linear Models, Logistic and Ordinal Regression, and Survival Analysis Introduction. Regression Modeling Strategies: With Applications to Linear Models, Logistic and Ordinal Regression, and Survival Analysis,* Second Edition, 2015: 1–11.
5. Kattan MW, Hess KR, Amin MB et al. American Joint Committee on Cancer acceptance criteria for inclusion of risk models for individualized prognosis in the practice of precision medicine. *CA Cancer J Clin.* 2016;66(5):370–4.
6. Moons KG, Altman DG, Reitsma JB et al. Transparent Reporting of a multivariable prediction model for Individual Prognosis or Diagnosis (TRIPOD): Explanation and elaboration. *Ann Intern Med.* 2015;162(1):W1–73.
7. Moons KG, de Groot JAH, Bouwmeester W et al. Critical appraisal and data extraction for systematic reviews of prediction modelling studies: The CHARMS checklist. *PLOS Med.* 2014;11(10):e1001744.
8. Halabi S, Lin C-Y, Kelly WK et al. Updated prognostic model for predicting overall survival in first-line chemotherapy for patients with metastatic castration-resistant prostate cancer. *J Clin Oncol.* 2014;32(7):671–7.

9. Beltran H, Wyatt AW, Chedgy EC et al. Impact of therapy on genomics and transcriptomics in high-risk prostate cancer treated with Neoadjuvant Docetaxel and Androgen Deprivation therapy. *Clin Cancer Res*. 2017;23(22):6802–11.

10. Eastham JA, Kelly WK, Grossfeld GD, Small EJ. Cancer and Leukemia Group B (CALGB) 90203: A randomized phase 3 study of radical prostatectomy alone versus estramustine and docetaxel before radical prostatectomy for patients with high-risk localized disease. *Urology*. 2003;62:55–62.

11. Kattan MW, Eastham JA, Stapleton AMF, Wheeler TM, Scardino PT. A preoperative nomogram for disease recurrence following radical prostatectomy for prostate cancer. *J Natl Cancer Inst*. 1998;90(10):766–71.

12. Sparano JA, Gray RJ, Makower DF et al. Adjuvant chemotherapy guided by a 21-gene expression assay in breast cancer. *N Engl J Med*. 2018;379(2):111–21.

13. Paik S, Tang G, Shak S et al. Gene expression and benefit of chemotherapy in women with node-negative, estrogen receptor positive breast cancer. *J Clin Oncol*. 2006;24(23):3726–34.

14. Albain KS, Barlow WE, Shak S et al. Prognostic and predictive value of the 21-gene recurrence score assay in postmenopausal women with node-positive, oestrogen-receptor-positive breast cancer on chemotherapy: A retrospective analysis of a randomised trial. *Lancet Oncol*. 2010;11(1):55–65.

15. Paik S. Multigene assay to predict recurrence of tamoxifen-treated, node-negative breast cancer. *N Engl J Med*. 2004;351(27):2817–26.

16. Bang YJ, Van Cutsem E, Feyereislova A et al. Trastuzumab in combination with chemotherapy versus chemotherapy alone for treatment of HER2-positive advanced gastric or gastro-oesophageal junction cancer (ToGA): A phase 3, open-label, randomised controlled trial. *Lancet*. 2010;376(9742):687–97.

17. Horn L, Mansfield AS, Szczęsna A et al. First-line atezolizumab plus chemotherapy in extensive-stage small-cell lung cancer. *N Engl J Med*. 2018;379(23):2220–9.

18. Altman, D.G., Studies investigating prognostic factors: Conduct and evaluation. In *Prognostic Factors in Cancer*. MK Gospodarowicz, B O'Sullivan, LH Sobin, eds. Hoboken, New Jersey: Wiley-Liss, 2006: 39–54.

19. McShane LM, Sauerbrei W, Taube SE, Gion M, Clark GM. Reporting recommendations for tumor marker prognostic studies (REMARK). *J Natl Cancer Inst*. 2005;97(16):1180–4.

20. Simon R, Altman DG. Statistical aspects of prognostic factor studies in oncology. *Br J Cancer*. 1994;69(6):979–85.

21. Schmoor C, Sauerbrei W, Schumacher M. Sample size considerations for the evaluation of prognostic factors in survival analysis. *Stat Med*. 2000;19(4):441–52.

22. Rini BI, Halabi S, Rosenberg JE et al. Bevacizumab plus interferon alfa compared with interferon alfa monotherapy in patients with metastatic renal cell carcinoma: CALGB 90206. *J Clin Oncol*. 2008;26(33):5422–8.

23. Harrell FE Jr, Lee KL, Califf RM et al. Regression modelling strategies for improved prognostic prediction. *Stat Med*. 1984;3(2):143–52.

24. Gonen M, Heller G. Concordance probability and discriminatory power in proportional hazards regression. *Biometrika*. 2005;92(4):965–70.

25. de Bono JS, Oudard S, Ozguroglu M et al. Prednisone plus cabazitaxel or mitoxantrone for metastatic castration-resistant prostate cancer progressing after docetaxel treatment: A randomised open-label trial. *Lancet*. 2010;376(9747):1147–54.

26. Sternberg CN, Petrylak DP, Sartor O et al. Multinational, double-blind, phase III study of prednisone and either satraplatin or placebo in patients with castrate-refractory prostate cancer progressing after prior chemotherapy: The SPARC trial. *J Clin Oncol*. 2009;27(32):5431–8.

27. Halabi S, Lin C-Y, Small EJ et al. Prognostic model predicting metastatic castration-resistant prostate cancer survival in men treated with second-line chemotherapy. *J Natl Cancer Inst*. 2013;105(22):1729–37.

28. Peterson B, George SL. Sample size requirements and length of study for testing interaction in a 2 x k factorial design when time-to-failure is the outcome [corrected]. *Control Clin Trials*. 1993;14(6):511–22.

29. Royston P, Sauerbrei W. A new approach to modelling interactions between treatment and continuous covariates in clinical trials by using fractional polynomials. *Stat Med*. 2004;23(16):2509–25.
30. Kelly WK, Halabi S, Carducci M et al. Randomized, double-blind, placebo-controlled phase III trial comparing docetaxel and prednisone with or without bevacizumab in men with metastatic castration-resistant prostate cancer: CALGB 90401. *J Clin Oncol*. 2012;30(13):1534–40.
31. Hosmer DW, Lemeshow S, Sturdivant RX. *Applied Logistic Regression*. 3rd Edition. Wiley & Sons, 2013.
32. Hosmer DW, Lemeshow S, May S. *Applied Survival Analysis: Regression Modeling of Time-to-Event Data*. Hoboken, N.J.: Wiley-Interscience, 2008.
33. Cox DR. Regression Models and Life-Tables. *J R Stat Soc Series B Stat Methodol*. 1972;34(2):187.
34. Breiman L. *Classification and Regression Trees*. Boca Raton: Chapman & Hall/CRC, 1984.
35. Leblanc M, Crowley J. Survival trees by goodness of split. *J Am Stat Assoc*. 1993;88(422):457–67.
36. Hothorn T, Hornik K, Zeileis A. Unbiased recursive partitioning: A conditional inference framework. *J Comput Graph Stat*. 2006;15(3):651.
37. Banerjee M, George J, Song EY, Roy A, Hryniuk W. Tree-based model for breast cancer prognostication. *J Clin Oncol*. 2004;22(13):2567–75.
38. Zhou X, Liu K-Y, Wong STC. Cancer classification and prediction using logistic regression with Bayesian gene selection. *J Biomed Inform*. 2004;37(4):249–59.
39. Sparano JA. Prognostic gene expression assays in breast cancer: Are two better than one? *NPJ Breast Cancer*. 2018;4:11.
40. Wishart GC, Azzato EM, Greenberg DC, Rashbass J, Kearins O, Lawrence G, Caldas C, Pharoah PD. PREDICT: a new UK prognostic model that predicts survival following surgery for invasive breast cancer. *Breast Cancer Res*. 2010;12(R1):3–10.
41. Rudloff U, Jacks LM, Goldberg JI et al. Nomogram for predicting the risk of local recurrence after breast-conserving surgery for ductal carcinoma in situ. *J Clin Oncol*. 2010;28(23):3762–9.
42. Halbesma N, Jansen DF, Heymans MW, Stolk RP, de Jong PE, Gansevoort RT. Development and validation of a general population renal risk score. *Clin J Am Soc Nephrol*. 2011;6(7):1731–8.
43. Therneau TM, Grambsch PM. *Modeling Survival Data: Extending the Cox Model*. New York; London: Springer, 2011.
44. Kalbfleisch JD, Prentice RL. *The Statistical Analysis of Failure Time Data*. New York: John Wiley & Sons, 2011.
45. Fleming TR, Harrington DP. *Counting Processes and Survival Analysis*. Hoboken: Wiley, 2005.
46. Hastie T, Tibshirani R, Friedman JH. *The Elements of Statistical Learning Data Mining, Inference, and Prediction.*, New York, NY: Springer-Verlag, 2003.
47. Tibshirani R. The lasso method for variable selection in the Cox model. *Stat Med*. 1997;16(4):385–95.
48. Zhang HH, Lu WB. Adaptive lasso for Cox's proportional hazards model. *Biometrika*. 2007;94(3):691–703.
49. Fan J, Feng Y, Wu Y. *High-Dimensional variable selection for Cox's proportional hazards model*. IMS Collections (Borrowing Strength: Theory Powering Applications) 2010;6:70–86.
50. Fan J, Lv J. Sure independence screening for ultrahigh dimensional feature space. *J R Stat Soc Series B Stat Methodol*. 2008;70(5):849–911.
51. Meinshausen N, Bühlmann P. High-dimensional graphs and variable selection with the Lasso. *Ann Stat*. 2006;34(3):1436–62.
52. Zou H. Theory and methods: The adaptive lasso and its oracle properties. *J Am Stat Assoc*. 2006;101(476):1418.
53. Zou H, Hastie T. Regularization and variable selection via the elastic net. *J R Stat Soc Series B Stat Methodol*. 2005;67(2):301–20.
54. Hastie T, Tibshirani R, Friedman JH. *The Elements of Statistical Learning: Data Mining, Inference, and Prediction*, 2nd Edition. New York: Springer, 2017.
55. Fizazi K, Higano CS, Nelson JB et al. Phase III, randomized, placebo-controlled study of docetaxel in combination with zibotentan in patients with metastatic castration-resistant prostate cancer. *J Clin Oncol*. 2013;31(14):1740–7.

56. White IR, Royston P. Imputing missing covariate values for the Cox model. *Stat Med.* 2009;28(15):1982–98.
57. Breiman L. Random forests. *Mach Learn.* 2001;45(1):5–32.
58. Erho N, Crisan A, Vergara IA et al. Discovery and validation of a prostate cancer genomic classifier that predicts early metastasis following radical prostatectomy. *PLOS One.* 2013;8(6).
59. Carolin S, Boulesteix A-L, Kneib T et al. Conditional variable importance for random forests. *BMC Bioinformatics.* 2008;9:307.
60. Pi L, Halabi S. Combined performance of screening and variable selection methods in ultra-high dimensional data in predicting time-to-event outcomes. *Diagn Progn Res.* 2018;2.
61. Lin CY, Halabi S. On model specification and selection of the Cox proportional hazards model. *Stat Med.* 2013;32(26):4609–23.
62. Lin Y, Zhang HH. Component selection and smoothing in multivariate nonparametric regression. *Ann Stat.* 2006;34(5):2272–97.
63. Craven P, Wahba G. Smoothing noisy data with spline functions estimating the correct degree of smoothing by the method of generalized cross-validation. *Numer Math Numerische Mathematik.* 1978;31(4):377–403.
64. Leng C, Helen Zhang H. Model selection in nonparametric hazard regression. *J Nonparametr Stat.* 2006;18(7–8):417–29.
65. Scher HI, Halabi S, Tannock I et al. Design and end points of clinical trials for patients with progressive prostate cancer and castrate levels of testosterone: recommendations of the Prostate Cancer Clinical Trials Working Group. *J Clin Oncol.* 2008;26(7):1148–59.
66. Royston P, Altman DG, Sauerbrei W. Dichotomizing continuous predictors in multiple regression: a bad idea. *Stat Med.* 2006;25(1):127–41.
67. Royston P, Reitz M, Atzpodien J. An approach to estimating prognosis using fractional polynomials in metastatic renal carcinoma. *Br J Cancer.* 2006;94(12):1785–8.
68. Halabi S, Small EJ, Kantoff PW et al. Prognostic model for predicting survival in men with hormone-refractory metastatic prostate cancer. *J Clin Oncol.* 2003;21(7):1232–7.
69. Altman DG, Royston P. What do we mean by validating a prognostic model? *Stat Med.* 2000;19(4):453–73.
70. Hilsenbeck SG, Clark GM. Practical p-value adjustment for optimally selected cutpoints. *Stat Med.* 1996;15(1):103–12.
71. Lausen B, Schumacher M. Evaluating the effect of optimized cutoff values in the assessment of prognostic factors. *Comput Stat Data Anal.* 1996;21(3):307–26.
72. Hothorn T, Lausen B. On the exact distribution of maximally selected rank statistics. *Comput Stat Data Anal.* 2003;43(2):121–37.
73. Benjamini Y, Hochberg Y. Controlling the false discovery rate: A practical and powerful approach to multiple testing. *J R Stat Soc Series B.* 1995;57(1):289.
74. Smaletz O, Scher HI, Small EJ et al. Nomogram for overall survival of patients with progressive metastatic prostate cancer after castration. *J Clin Oncol.* 2002;20(19):3972–82.
75. Halabi S, Owzar K. The importance of identifying and validating prognostic factors in oncology. *Semin Oncol.* 2010;37(2):e9–18.
76. Steyerberg EW. *Clinical Prediction Models: A Practical Approach to Development, Validation, and Updating.* The Netherlands: Springer, 2010.
77. Altman DG, Royston P. What do we mean by validating a prognostic model? *Stat Med.* 2000;19(4):453–73.
78. Steyerberg EW, Harrell FE. Prediction models need appropriate internal, internal-external, and external validation. *J Clin Epidemiol.* 2016;69:245–7.
79. Efron B, Tibshirani R. Improvements on cross-validation: The.632+ bootstrap method. *J Am Stat Assoc.* 1997;92(438):548–60.
80. Efron B. Model Selection and the Bootstrap. *Math Soc Sci.* 1983;5(2):236–236.
81. Efron B, Gong G. A Leisurely Look at the Bootstrap, the Jackknife, and Cross-Validation. *Am Stat.* 1983;37(1):36–48.

82. Simon RM. *Design and analysis of DNA microarray investigations.* New York; London: Springer, 2011.
83. Heagerty PJ, Zheng Y. Survival model predictive accuracy and ROC curves. *Biometrics.* 2005;61(1):92–105.
84. Uno H, Cai T, Tian L, Wei LJ. Evaluating prediction rules for t-year survivors with censored regression models. *J Am Stat Assoc.* 2007;102(478):527–37.
85. Kamarudin AN, Cox T, Kolamunnage-Dona R. Time-dependent ROC curve analysis in medical research: current methods and applications. *BMC Med Res Methodol.* 2017;17(1):2.
86. Graf E, Schmoor C, Sauerbrei W, Schumacher M. Assessment and comparison of prognostic classification schemes for survival data. *Statist Med.* 1999;18(17–18):2529–45.
87. Zhao SD, Li Y. Principled sure independence screening for Cox models with ultra-high-dimensional covariates. *J Multivar Anal.* 2012;105(1):397–411.
88. Kim S, Halabi S. High dimensional variable selection with error control. *Biomed Res Int.* 2016;2016:8209453.

18

High-Dimensional, Penalized-Regression Models in Time-to-Event Clinical Trials

Federico Rotolo, Nils Ternès, and Stefan Michiels

CONTENTS

18.1 Development and Validation of Multimarker Signatures

Molecular signatures are becoming increasingly important for anticipating the prognosis of individual patients ("prognostic" biomarkers) or for predicting how patients will respond to specific treatments ("predictive" biomarkers, more generally called "treatment–effect modifiers"). A voluminous literature of more than 150,000 articles documenting thousands of claimed biomarkers has been produced in medicine, of which fewer than 100 have been validated for routine clinical practice [1]. Indeed, less than 20 prognostic or predictive biomarkers are recognized with variable levels of evidence in the 2014 European Society of Medical Oncology (ESMO) clinical practice guidelines for lung, breast, colon, and prostate cancer [2].

When developing a valid gene signature, attention must be paid to numerous methodological issues, from bench to bedside [3]. In this chapter, we focus on statistical methods for developing and validating prognostic and predictive multimarker signatures with penalized Cox regression models when the biomarker data have been measured on samples from patients included in cancer clinical trials. Section 18.2 presents the general ideas of penalized regression. Sections 18.3 and 18.4 describe specific methods for prognostic and predictive signatures, respectively. Section 18.5 shows how to implement such methods in R.

18.2 Penalized Regression for Survival Endpoints

18.2.1 Statistical Framework

Since the main endpoint in oncology clinical trials is usually a time-to-event endpoint, we focus on regression models where the response variable is the time to an event of interest. While there are different approaches to the analysis of possibly right-censored data, we will restrict ourselves to the Cox model. This is one of the most popular survival regression models in medical research in general and in oncology, in particular.

In order to set the scene, we define the Cox model for patient $i \in \{1, \ldots, n\}$ in terms of the hazard function

$$h(t, \boldsymbol{X}_i) = h_0(t) \exp\left(\boldsymbol{\beta}^\top \boldsymbol{X}_i\right), \tag{18.1}$$

where $\boldsymbol{X}_i = (X_{i1}, \ldots, X_{ip})^\top$ is a vector of p biomarkers and $h_0(t)$ is the baseline hazard function, i.e., the hazard function when all the covariates equal 0.

Let $t_{(1)} \leq \cdots \leq t_{(n_e)}$ be the ordered times of the n_e observed (uncensored) events. The Cox semi-parametric estimation approach [4] consists of profiling out the baseline hazard function $h_0(\cdot)$ from the log likelihood and obtaining estimates of the regression coefficients $\boldsymbol{\beta}$ by maximizing the partial log likelihood

$$\ell(\boldsymbol{\beta}; \boldsymbol{X}) = \sum_{t_{(i)}=t_{(1)}}^{t_{(n_e)}} \left\{ \boldsymbol{\beta}^\top \boldsymbol{X}_{(i)} - \log \sum_{i^+ \in \mathcal{R}\left(t_{(i)}\right)} \exp\left[\boldsymbol{\beta}^\top \boldsymbol{X}_{i^+}\right] \right\}, \tag{18.2}$$

where $\mathcal{R}(t_{(i)}) = \left\{ i^+ | t_{i^+} \geq t_{(i)} \right\}$ is the risk set at time $t_{(i)}$, i.e., the set of patients who have neither experienced the event of interest nor have been censored at time $t_{(i)}$.

With the evolution of knowledge and technology in genomics, an increasing amount of data can be obtained at a decreasing cost: mutations, gene expression, copy number aberrations, etc. The availability of these high-dimensional datasets poses many statistical challenges, since the number of biomarkers p gets close to, greater than, or even much greater than the sample size n or, more precisely, the total number of events n_e. Possible issues encountered in this context include non-identifiability of the model, high collinearity between biomarkers, and instability of the estimated regression coefficients. Sparse model selection is then desirable, and multiple testing becomes very important when applying classical inference methods.

Several alternative strategies are available to perform such a selection, estimation, and prediction. The most used ones include univariate regressions with type I error control, stepwise model selection, dimension reduction, machine learning techniques, boosting, and penalized regression. This last approach is a very wide family of approaches which allow selecting the most significant features as well as stabilizing the estimates, making inference possible also in the case of $p > n$. In the rest of this chapter, we will primarily focus on methods based on penalized regression, and we will only marginally mention other approaches.

18.2.2 Maximum Penalized Likelihood Inference

Maximum likelihood estimators are known to be asymptotically unbiased under some hypotheses, but they often come with a large variance. Models which are sparser or which shrink the coefficient estimates toward zero introduce bias so as to decrease the variance, aiming at improving the prediction accuracy.

A penalized Cox model is a model where the parameter estimates are obtained by maximizing the partial log likelihood minus a penalty term,

$$\ell_{\text{pen}}(\boldsymbol{\beta},\lambda;\boldsymbol{X}) = \ell(\boldsymbol{\beta};\boldsymbol{X}) - p_\lambda(\boldsymbol{\beta}) \tag{18.3}$$

which is non-decreasing in the coefficients:

$$p_\lambda(\boldsymbol{\beta}) \leq p_\lambda(\boldsymbol{\beta}+\boldsymbol{v}) \, \forall \boldsymbol{v} = (v_1,\ldots,v_p)^\top, |\beta_j + v_j| \geq |v_j|. \tag{18.4}$$

Several penalties have been proposed in the literature (see also Chapter 17), each having different properties. In the rest of this section, Section 18.2.2, we review the main ones.

18.2.2.1 Ridge Penalization

One of the best known penalties in regression models is the ridge penalty, which consists in penalizing the (partial) log likelihood by means of the squared ℓ_2-norm of the regression coefficients:

$$p_\lambda(\boldsymbol{\beta}) = \lambda\boldsymbol{\beta}^\top\boldsymbol{\beta} = \lambda\sum_{j=1}^{p}\beta_j^2. \tag{18.5}$$

The ridge penalty performs a regularization of the coefficient estimates by shrinking them toward zero, without setting any of them exactly to zero. Hence, ridge regression allows estimating high-dimensional regression models but does not provide sparse model selection.

18.2.2.2 Lasso Penalization

The lasso penalty [5–7] is defined as the ℓ_1-norm

$$p_\lambda(\boldsymbol{\beta}) = \lambda \mathbf{1}^\top |\boldsymbol{\beta}| = \lambda \sum_{j=1}^p |\beta_j|, \tag{18.6}$$

with $|\boldsymbol{\beta}| = \left(|\beta_1|, \ldots, |\beta_p|\right)^\top$. As the lasso penalty is not differentiable at the origin, some of the parameters are set exactly to zero if the shrinkage parameter λ is large enough.

> **EXAMPLE 18.1**
>
> Figure 18.1 shows an example of data from a clinical trial evaluating the combination of the adjuvant levamisole and fluorouracil in large-bowel carcinoma [8]. The data are publicly available in the R package `survival` [9].
>
> In the top panel, one can see the values of the estimated regression coefficients when using the ridge penalty, for different values of the shrinkage parameter λ. The bottom panel shows the estimation obtained with the lasso penalization. As the shrinkage parameter increases, some of the coefficients are set exactly to zero.

18.2.2.3 Adaptive Lasso Penalization

In the lasso penalty, the same parameter λ is used for all the regression coefficients. More recently, a modification of the lasso has been proposed to take into account the different importance of the coefficients, by using variable-specific penalties λw_j. This so-called adaptive lasso [10,11] is defined by

$$p_{\lambda,w}(\boldsymbol{\beta}) = \lambda w^\top |\boldsymbol{\beta}| = \lambda \sum_{j=1}^p w_j |\beta_j|, \tag{18.7}$$

where, usually, each weight w_j is defined as the inverse of the absolute value of a preliminary estimate: $1/|\tilde{\beta}_j|$. Although several methods have been employed in the literature to obtain these preliminary estimates, there is no consensus. The simplest method consists in estimating each $\tilde{\beta}_j$ in a variable-specific Cox model [12,13]; this approach based on univariate estimates can be applied both in low- and high-dimensional settings. In a context where the full model can be fitted, which is what we call the low-dimnsional context, the $\tilde{\beta}_j$s can alternatively be set to the estimates obtained in such a full model. Other approaches perform a preliminary estimation of the full model subject to the ridge penalty (18.5) or to the lasso penalty (18.6) [14–16]. In our experience, the best performance in high-dimensional Cox models is achieved using a preliminary simple lasso penalty [17]. Furthermore, this approach has the desirable property of selecting a submodel of the one obtained with the simple lasso.

18.2.2.4 Group Lasso Penalization

Sometimes one needs to select a sparse model subject to some further constraints regarding groups of variables which have a presumed common role, as in the case of genes in the same putative biological pathway. In such a case, it is interesting to either jointly keep in the model or jointly discard a whole group of variables. The group lasso [18] achieves this goal

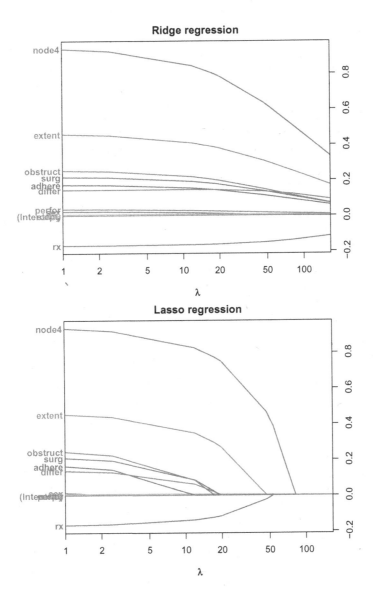

FIGURE 18.1
Coefficient estimations subject to the ridge penalty (top) and lasso penalty (bottom), with different shrinkage parameters λ, for the dataset `survival::colon` in R.

through a penalty which consists in the sum of the ℓ_2-norms of the coefficient vectors of each group. Let us assume that each biomarker j belongs to one group $group(j) = g \in \{1,\dots,G\}$. Then the group lasso penalty is

$$p_\lambda(\boldsymbol{\beta}) = \lambda \sum_{g=1}^{G} \sqrt{\boldsymbol{\beta}_g^\mathsf{T} \boldsymbol{\beta}_g} = \lambda \sum_{g=1}^{G} \sqrt{\sum_{j|group(j)=g} \beta_j^2}, \qquad (18.8)$$

where $\boldsymbol{\beta}_g$ is the vector of all the coefficients in the gth group.

18.2.3 The Choice of the Shrinkage Parameter

As shown in the example of colon cancer (Figure 18.1), the results of a penalized regression can vary dramatically, depending on the value of the shrinkage parameter λ, especially for those penalties, like the lasso, which perform sparse model selection.

In the class of penalties presented above, the higher λ is, the more the partial likelihood is penalized. In methods performing sparse selection, one can define $p(\lambda)$ as the number of biomarkers retained in the model for a given value of λ. Hence, $p(0) = p$, i.e., the model with $\lambda = 0$ corresponds to the unpenalized (full) model. The upper limit of the values of interest is $\lambda_0 = \min\{\lambda \mid p(\lambda) = 0\}$, the first value of λ that drops all the biomarkers from the model. Note that $p(\lambda_0) = 0$ holds trivially by the definition of λ_0.

The choice of the most appropriate value of λ in the interval $[0, \lambda_0]$ can be made according to different criteria. As discussed below, cross-validation is the most commonly employed method in the literature, but several alternatives have been developed in order to overcome some of its drawbacks.

18.2.3.1 Cross-Validation

Cross-validation consists of iteratively developing a model for a random subsample of the data at hand and evaluating its goodness-of-fit or its prediction performance on the left-out subsample [19,20]. For a fixed value of λ, given a random splitting of the data into K folds, the cross-validated likelihood of a penalized Cox model is defined as

$$cvl(\lambda) = \sum_{k=1}^{K} \left\{ \ell_{\text{pen}}\left(\hat{\beta}_{(-k)}, \lambda; X \right) - \ell_{\text{pen}}\left(\hat{\beta}_{(-k)}, \lambda; X_{(-k)} \right), \right\} \tag{18.9}$$

that is, the sum over all the K folds of the contribution of the kth fold to the penalized partial likelihood $\ell_{\text{pen}}(\cdot)$ as predicted by the rest of the data, i.e., with coefficients $\hat{\beta}_{(-k)}$ maximizing $\ell_{\text{pen}}(\cdot)$ in $X_{(-k)}$. As given in detail by Verweij and Van Houwelingen [20], the components of the partial log likelihood (18.2) are interrelated, as each subject i belongs to the risk set of all subjects dying before that subject's event-or-censoring time. Hence, the contribution of fold k is obtained as the difference between the penalized partial log likelihood of the full sample X and that of the training subsample $X_{(-k)}$.

The $cvl(\cdot)$ can be computed for several values of λ and the value $\hat{\lambda}_{cvl} = \text{argmax}_\lambda cvl(\lambda)$ is chosen to maximize $\ell_{\text{pen}}(\cdot)$ over the whole data sample. Several studies [14,17,21,22] have shown that a very high proportion of the variables selected in the model are false positives when using $\hat{\lambda}_{cvl}$, and other improvements have been proposed.

18.2.3.2 One-Standard-Error Rule

Because $\hat{\lambda}_{cvl}$ proves to be an anticonservative choice of λ, choosing a higher value is expected to decrease the proportion of false discoveries. A recent empirical proposal [23,24] consists of considering $D(\lambda) = -cvl(\lambda)$, the deviance of a model with given λ as compared to the saturated model, together with its standard error (SE). The deviance of the values of λ greater than $\hat{\lambda}_{cvl}$ increases with increasing distance from $\hat{\lambda}_{cvl}$ (Figure 18.2). The value of λ based on the so-called one-standard-error rule is the highest one among those with deviance less or equal to $D(\hat{\lambda}_{cvl}) + SE(D(\hat{\lambda}_{cvl}))$.

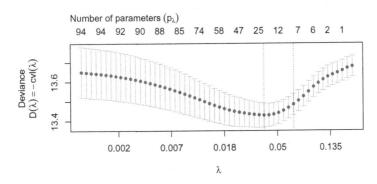

FIGURE 18.2

One-standard-error rule for the choice of the optimal shrinkage parameter λ. The left-hand vertical dotted line shows the $\hat{\lambda}_{cvl}$ minimizing the deviance; the dots show the deviance together with confidence intervals. The right-hand vertical dotted line shows the minimum λ which increases the deviance by no more than one standard error as compared to $\hat{\lambda}_{cvl}$.

18.2.3.3 Percentile-Lasso

It is well known that the optimal value of λ selected via cross-validation can vary considerably, depending on the assignment of the subjects to the folds [25]. The percentile-lasso is an approach based on replicating several times the random assignment of subjects to the K folds [22]. The advantage of such a procedure is that it does not end up with only one optimal value $\hat{\lambda}_{cvl}$, but with its empirical distribution across the replicates. In order to achieve a lower false-positive selection than the simple cross-validation, the percentile-lasso selects a high-rank percentile of the empirical distribution of $\hat{\lambda}_{cvl}$.

18.2.3.4 Stability Selection

Stability selection is a general method to deal with the variable results of any selection method which depends on the variability of random samples. In the case of the cross-validated lasso, stability selection consists of running this latter in random subsamples (without replacement) of the data and recording which biomarkers are retained in the model [26]. This provides an empirical estimation of the selection probability of each biomarker. The final model includes only the biomarkers with selection probabilities higher than a pre-specified threshold and provides finite sample control of the false discovery rate (FDR).

18.3 Signatures Predicting a Survival Outcome

18.3.1 False Positives

Prognostic biomarkers, i.e., biomarkers predicting the outcome, are of primary importance in patient and disease management. These biomarkers can discriminate between patients with a good or bad prognosis in the absence of treatment or in the context of a standard therapy. This means that the probable natural course of the disease can be forecasted by using the biomarker values. In the high-dimensional setting, false positives are becoming one of the major problems in biomedical research [27]. Even though penalized regressions perform

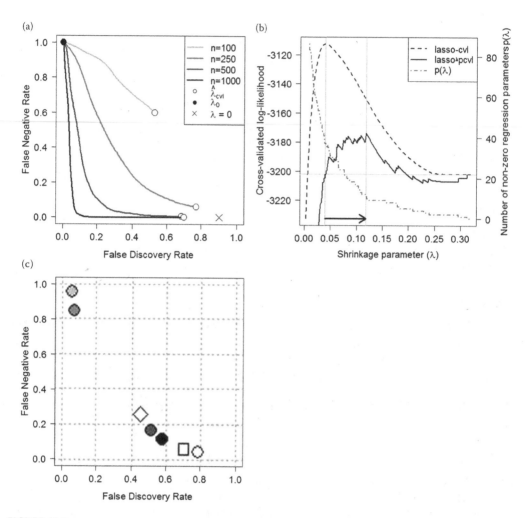

FIGURE 18.3
A motivating illustration of the penalized cross-validated likelihood (*pcvl*) approach. (a) False-negative rate (FNR) and false discovery rate (FDR) according to the values of the shrinkage parameters λ, varying from the one maximizing the cvl ($\hat{\lambda}_{cvl}$, $\bigcirc$) to the smallest value selecting the null model (λ_0 •). Empirical results in simulations (averages across 250 replications) with $p = 100$ biomarkers, among which $q = 10$ are actually prognostic. (b) Cross-validated likelihood (*cvl*), *pcvl*, and number of non-zero regression parameters $p(\lambda)$ according to the shrinkage parameter in a simulated dataset of $n = 500$ patients and $p = 100$ biomarkers, among which $q = 10$ are actually prognostic. (c) FNR vs. FDR for the seven main methods, from simulations with $n = 500$ patients and $q = 20$ biomarkers out of $p = 1,000$ are generated as truly prognostic. Average quantities across 250 replications. $\Diamond$: stability selection, $\square$: adaptive lasso and $\bigcirc$: lasso (white: standard lasso-*cvl*, light gray: lasso-*RIC*, medium gray: lasso-*AIC*, dark gray: lasso-*cvl(1se)*, and black: lasso-*pcvl*). (From Ternès N et al. *Stat Med.* 2017;35(15):2561–73.)

sparse variable selection, several studies have shown that this method can lead to the selection of a high number of false positives. Figure 18.3a illustrates the relationship between the false-negative rate (FNR, i.e., false-negative selection among the truly prognostic biomarkers [28]) and the FDR, (i.e., false-positive selection among the selected biomarkers [29]) for penalized models selected with different values of the shrinkage parameter λ in simulations. These simulations show that (1) the lasso penalty ($\hat{\lambda}_{cvl}$, white dots) well identifies almost all the true prognostic biomarkers (FNR close to zero) when the sample size is sufficiently high but produces a large number of false positives (FDR usually larger than 0.5) and that (2) the FNR-FDR path (in gray) from $\hat{\lambda}_{cvl}$ to λ_0 (black dots) is convex. This means that a moderate increase of the shrinkage parameter λ as compared to $\hat{\lambda}_{cvl}$ can lead to a substantial decrease in the FDR with a comparatively small increase in the FNR, especially for large sample sizes. Therefore, increasing the value of λ favors the selection of truly positive biomarkers, whereas an excessive increase of λ toward λ_0, exceeding the "breaking point" can lead to a large increase of the FNR with negligible additional reduction of the FDR.

18.3.2 Conservative Choices of the Shrinkage Parameter

In the present section, we review methods based on an additive penalization term for a more conservative choice of the shrinkage parameter λ as compared to cross-validation [17]. As previously mentioned in Section 18.2.3, a few methods for the choice of λ have already been proposed in the literature, such as the one-standard-error rule or the percentile-lasso. However, some drawbacks have been noted with these methods. For instance, the one-standard-error rule tends to be extremely conservative (i.e., selection of the null model in most of the cases) in the presence of high censoring. In the low-dimensional setting ($n \gg p$), the estimation of the shrinkage parameter λ is quite stable and therefore no large difference is observed between the percentile-lasso and the standard *cvl*, which both identify a high number of false positives. In a high-dimensional setting, the estimation of the shrinkage parameter λ can be highly variable, often leading to the selection of the null model with the percentile lasso.

For a more conservative choice of the lasso shrinkage parameter λ, several alternative penalties pen(λ) have been proposed [30], based on the maximization of $cvl(\lambda) - pen(\lambda)$. The simplest form of the additional penalty is pen(λ) $= \theta_\lambda f_\lambda(p(\lambda))$ where θ_λ is the penalty multiplier and $f_\lambda(p(\lambda))$ is a function of the number of selected biomarkers $p(\lambda)$. With the trivial choice $f_\lambda(p(\lambda)) = p(\lambda)$, several well-known penalties can be obtained by considering either a fixed penalty multiplier (e.g., $\theta_\lambda = 2$ for the Akaike information criterion, AIC), a penalty multiplier depending on the sample size (e.g., $\theta_\lambda = \log(n)$ for the Bayesian information criterion BIC), or one depending on the number of selected biomarkers for a given λ (e.g., $2\log(p(\lambda))$ for the risk information criterion, RIC). Based on an extensive simulation study [17], we highlighted that some methods, such as the lasso-*BIC*, can be too conservative and greatly increase the FNR, especially when the sample size is low or moderate (namely, $n = 100$ or 500 in our simulations). Other penalized methods such as the lasso-*AIC* or the lasso-*RIC* perform well in general, but they can perform very badly in particular cases: the lasso-*RIC* in the presence of multiple active biomarkers and the lasso-*AIC* in a setting with relatively few number of patients.

As compared to these penalties, we suggested an empirical penalty (Figure 18.3b) which depends on the form of the $cvl(\lambda)$ function: the penalty multiplier θ_λ represents by how much $\hat{\lambda}_{cvl}$ improves the goodness-of-fit and worsens the parsimony of the selected model as compared to λ_0. This multiplier is defined as the ratio between the gain in goodness-of-fit $cvl(\hat{\lambda}_{cvl}) - cvl(\lambda_0)$ and the loss in parsimony $p(\hat{\lambda}_{cvl}) - p(\lambda_0) = p(\hat{\lambda}_{cvl})$, since $p(\lambda_0) = 0$ by definition (see Section 18.2.3). Thus, the non-negative penalty

$$\text{pen}(\lambda) = \frac{cvl(\hat{\lambda}_{cvl}) - cvl(\lambda_0)}{p(\hat{\lambda}_{cvl})} p(\lambda), \forall \lambda \in \left[\hat{\lambda}_{cvl}; \lambda_0\right], \tag{18.10}$$

decreases over $[\hat{\lambda}_{cvl}; \lambda_0]$ as the number of non-null regression parameters $p(\lambda)$ decreases, that is, as λ increases. Our definition of θ_λ allows giving a weight to the variation in parsimony

$p(\lambda)/p(\lambda_{cvl}) \in [0;1]$ that is on the same scale as the magnitude of the *cvl* in the range of $\lambda \in [\hat{\lambda}_{cvl}; \lambda_0]$. Thus, λ is chosen by maximizing the penalized cross-validated log-likelihood $pcvl(\lambda) = cvl(\lambda) - \text{pen}(\lambda)$. Our particular choice of pen(λ) implies that $pcvl(\hat{\lambda}_{cvl}) = pcvl(\lambda_0)$. This forces a trade-off between the goodness-of-fit achieved with smaller values of λ and the parsimony of the selected model achieved with higher values of λ: the two extremes are put on the same level.

In simulations, the lasso-*pcvl* considerably reduces the FDR with a small increase in FNR as compared to the lasso-*cvl* when the sample size is moderate or high ($n = 500$ or $1,000$, Figure 18.3c). When the sample size is small ($n = 100$), in which case the lasso-*cvl* is already conservative (especially for large p), the lasso-*pcvl* does not greatly increase the penalty, leading to results close to those of the lasso-*cvl*.

In Ternès et al. [17], we also considered the prediction accuracy of these penalizations. We evaluated the discrimination of the estimated models through Uno's concordance C-statistic, one of the least biased concordance estimators in the presence of censoring [31]. The average C-statistics of the models estimated through the standard lasso-*cvl* were slightly lower than those of the oracle models (i.e., unpenalized models containing only active biomarkers) especially for low sample size, for which the lasso-*cvl* missed a large number of true positives. No large difference was observed between the lasso-*cvl* and the lasso-*pcvl* in terms of C-statistics.

Some other variants of the lasso penalty, such as the adaptive lasso or the stability selection, have also been tested in the simulation study. Several methods for choosing the biomarker-specific weights of the adaptive lasso have been proposed in the literature, but no consensus has been reached to date. In the simulations, the weights estimated through a preliminary lasso outperformed, in terms of FDR reduction, the use of univariate regression models or ridge regression. These two alternative methods performed extremely poorly, especially in null scenarios. A possible explanation is that the use of a different weight for each biomarker may result in an amplification of the differences between the effects of the biomarkers, whatever the absolute size of such differences. Although this can, in principle, boost the power of the variable selection under alternative scenarios, it could be a drawback in the null scenario, where random differences are artificially amplified, thus inflating the false-positive selection.

The stability selection showed good performances as compared to the standard lasso-*cvl* in most of the alternative scenarios. Nevertheless, it performed alarmingly poor under the null scenarios, with mean FDR across replications reaching even 100% in some of them. Furthermore, the implementation of this approach needs a prespecification of several parameters, which can have a large impact on the results. No guidance exists for the specification of these parameters.

Additional details on the investigated methods and on the extensive simulation study can be found in Ternès et al. [17].

18.3.3 Breast Cancer Application

As an illustration, we show in this section the application of penalized regression to a publicly available dataset from the Gene Expression Omnibus database (www.ncbi.nlm.nih.gov/geo, [32,33]), which contains clinical and gene expression data from 523 patients with breast cancer. The patients were treated by anthracycline-based (19%) or anthracycline plus taxane-based (81%) neoadjuvant chemotherapy. Most of the tumors were HER2-negative (93%), ER-negative (54%), and high grade (59%, grade III). The estimated one-year and five-year distant recurrence-free survival (DRFS) was 91% (95% confidence interval [CI]

88%–93%) and 72% (95% CI 68%–77%), respectively. The objective was to identify prognostic genes for DRFS. Before applying the methods, we first carried out a preprocessing step using the frozen robust multiarray analysis technique [34] to make the samples comparable across patients and using a filtering step based on the interquartile range criterion to reduce the number of candidate genes. We retained for the analysis the expression values of $p = 1,703$ genes (interquartile range ≥ 1).

In this analysis, the standard lasso-*cvl* identified 51 prognostic genes. Some of them are included in several curated breast cancer gene signatures from the Gene Signature DataBase [35]. Among the 51 genes selected by the lasso-*cvl*, nine were also identified by the lasso-*pcvl* and all of them are included in a number of curated breast cancer gene signatures (median: 22, range: 14–53) which is significanlty higher (Wilcoxon test: p-value $= 0.02$) than the 42 genes not identified by the lasso-*pcvl* (median: 16, range: 6–31). Interestingly, the GATA3 gene was identified by the lasso-*pcvl* but not by the lasso-*cvl* and is widely documented in the literature (59 curated signatures [36]).

For a more detailed illustration and discussion of this application, we refer to Ternès et al. [17].

18.4 Signatures Predicting the Treatment Benefit

In the era of stratified medicine, an increasing interest is being given to identify the patients more likely to benefit from the treatment. Hence, it is important to identify in randomized clinical trials (prospectively or retrospectively) the so-called treatment-effect modifiers or predictive biomarkers [37–39]. From a statistical viewpoint, Rothwell [40] put forward that the only reliable approach for assessing the predictiveness of biomarkers is to test their interaction with the treatment. Thus, the general framework for identifying treatment-effect modifiers is a model with the main effects of both the treatment and the biomarkers, and the biomarker-by-treatment interactions. In this section, we propose and describe several methods to (1) select a sparse set of treatment-effect modifiers from among a large number of candidates in a randomized clinical trial and (2) estimate the survival probabilities for individual patients, along with the associated confidence intervals.

18.4.1 Identification of Treatment-Effect Modifiers

In a proportional hazards regression model, the full biomarker-by-treatment interaction model for subject i is

$$h(t, T_i, \boldsymbol{X}_i) = h_0(t)\exp\left(\alpha T_i + \boldsymbol{\beta}^\top \boldsymbol{X}_i + \boldsymbol{\gamma}^\top \boldsymbol{X}_i T_i\right) \tag{18.11}$$

with α, $\boldsymbol{\beta}$, and $\boldsymbol{\gamma}$ the regression coefficients, respectively, for the treatment T_i (+0.5 in the experimental and -0.5 in the control arm), the vector of standardized biomarkers $\boldsymbol{X}_i$, and their interactions $\boldsymbol{X}_i T_i$. The component $\boldsymbol{\gamma}^\top \boldsymbol{X}_i T_i$ estimates the biomarker-dependent treatment effect.

To perform a variable selection and hence identify treatment-effect modifiers, a first penalization could consist of the lasso penalty for both the β_js and the γ_js:

$$p_\lambda(\boldsymbol{\beta}, \boldsymbol{\gamma}) = \lambda\left(\boldsymbol{1}^\top |\boldsymbol{\beta}| + \boldsymbol{1}^\top |\boldsymbol{\gamma}|\right) = \lambda\left(\sum_{j=1}^p |\beta_j| + \sum_{j=1}^p |\gamma_j|\right). \tag{18.12}$$

The main effects and the interactions are equally penalized (same shrinkage parameter λ). We call this approach the full-lasso. Despite the simplicity of this method, the main effects and interactions can have very different sizes. It seems more appropriate to penalize them unequally, using differently weighted penalties for the β_js and the γ_js, an approach similar to the adaptive lasso penalty. In this spirit, we estimate the weights in a preliminary model including the treatment and all the biomarker main effects β_{Rj} and interactions with the treatment γ_{Rj}, with a ridge penalty on the β_{Rj}s and the γ_{Rj}s. Then, the adaptive lasso penalty can be used as follows:

$$p_\lambda\left(\boldsymbol{\beta},\boldsymbol{\gamma}\right)=\lambda\left(\boldsymbol{w}_\beta^\top|\boldsymbol{\beta}|+\boldsymbol{w}_\gamma^\top|\boldsymbol{\gamma}|\right)=\lambda\left(\sum_{j=1}^{p}w_{\beta j}|\beta_j|+\sum_{j=1}^{p}w_{\gamma j}|\gamma_j|\right),\qquad(18.13)$$

with $w_{\beta j}=1/|\tilde{\beta}_{Rj}|$ and $w_{\gamma j}=1/|\tilde{\gamma}_{Rj}|$.

These two approaches performed quite well in almost all the scenarios tested in an extensive simulation study which compared a total of 12 methods [41]. Indeed, in null scenarios, these two methods selected no interactions in most cases, resulting in a low type-I error. However, the full-lasso was highly affected by the presence of prognostic effects. In the alternative scenarios, the full-lasso and adaptive lasso identified most of the treatment-effect modifiers with few false positives. Overall, these two methods performed the best of all the tested approaches in terms of FNR and FDR, especially for a moderate number of biomarkers. However, by increasing the number of candidate biomarkers p, the adaptive lasso performed much worse in the presence of main effects by selecting many fewer biomarkers as treatment modifiers. We also tested an alternative strategy for estimating the weights of the adaptive lasso. This approach consists of estimating a common weight for all β as the average of all the $|\hat{\beta}_{Rj}|$s and similarly a common weight for all γ as the average of all the $|\hat{\gamma}_{Rj}|$s. In practice, this approach was extremely conservative as it selected the null model in most of the cases, even in the alternative scenarios.

Both the full-lasso and adaptive lasso lack the hierarchy constraint: the main effect of a biomarker can be discarded ($\beta_j=0$) irrespective of whether or not the associated interaction is ($\gamma_j=0$). Although this can affect the interpretability of γ_j and the calibration of the model [42], it is of minor importance in the context of selection. Nevertheless, we considered a further approach which forces this constraint. This approach, called ridge+lasso, uses the ridge penalty on the main effects, which are then all kept in the model while controlling for overfitting:

$$p_{\lambda,\lambda_2}\left(\boldsymbol{\beta},\boldsymbol{\gamma}\right)=\lambda_2\boldsymbol{\beta}^\top\boldsymbol{\beta}+\lambda\mathbf{1}^\top|\boldsymbol{\gamma}|=\lambda_2\sum_{j=1}^{p}\beta_j^2+\lambda\sum_{j=1}^{p}|\gamma_j|.\qquad(18.14)$$

Because optimization of both λ and λ_2 is computationally demanding and weakens the generalizability, we favored a precise optimization of λ with a rough selection of λ_2. In our case, we first estimated the β_js in a model without any interaction and then fixed them via an offset in the final model (Equation 18.11). Other methods that do not carry out a selection of the main effects have been investigated and tested, such as dimension reduction approaches (PCA+lasso and PLS+lasso) and the so-called modified covariates approach [43]. In our simulations [41], we found that these methods performed rather similarly in terms of selection accuracy. In null scenarios, their type-I error was moderate to high. In alternative scenarios with no main effects, they identified most of the true treatment-effect modifiers (small FNR), but also many false positives (large FDR). Moreover, keeping all the main effects in the model can be regarded as a major practical drawback, as sparse prognostic signatures can be assessed more easily, more reliably, and at lower cost on different platforms.

In order to enforce the hierarchy constraint while performing selection on both the β_js and the γ_js, one can also consider a group-lasso approach [18], which selects prespecified groups of variables. In this context, p groups (β_j, γ_j) are defined:

$$p_\lambda(\boldsymbol{\beta}, \boldsymbol{\gamma}) = \lambda \sum_{j=1}^{p} \sqrt{(\beta_j, \gamma_j)(\beta_j, \gamma_j)^\top} = \lambda \sum_{j=1}^{p} \sqrt{\beta_j^2 + \gamma_j^2}. \tag{18.15}$$

Based on the results of the simulation study, we concluded that in general, the group-lasso can properly identify most of the true treatment-effect modifiers (high power), but often together with several biomarkers which are just prognostic (high FDR), notably in the case where there are no true treatment-effect modifiers at all. As is obvious by construction of the group-lasso, the presence of main effects increases the FDR in any scenario, due to the fact that groups including the main effect and the interaction of each biomarker are either selected or discarded together. To overcome this drawback, testing strategies could be considered to evaluate interactions in the selected groups [44], or a weighting strategy on the interactions in the spirit of the adaptive lasso. In any case, further arbitrary choices would be required.

For all implemented biomarker-by-treatment interaction models, the predictive score for a patient can be computed by $\hat{\eta}_i = \sum_{j=1}^{p} \hat{\gamma}_j X_{ij}$. Based on these scores, we proposed a gene signature interaction strength criterion similar to [45,46], measuring the concordance between $\hat{\eta}_i$ and the survival time in each treatment arm. We estimated this within-arm concordance via the C-statistic of Uno [32]. Then, we computed the absolute difference of the two C-statistics (ΔC): the larger the difference, the higher the interaction strength.

It is worth noting that computing diagnostic metrics on the same data used to develop the prediction model is known to give overoptimistic results. In the absence of external validation data, cross-validation is a classical solution to get a fairer appraisal of the model performance. In this case, we refer to this procedure as "double cross-validation" in the sense that, for each fold, a first cross-validation is done for the estimation of the penalty parameter λ.

In conclusion, the identification of biomarker-by-treatment interactions is difficult and requires a lot of events, whatever the method employed. Furthermore, it is important to stress that external data for validating the interaction strength is essential, as many of these methods are affected by overfitting. Additional details on alternative methods and their performance in simulations can be found in the article by Ternès et al. [41].

18.4.2 Estimation of the Expected Survival Probabilities

In the present section, we restrict ourselves to the model in Equation 18.11, subject to either the lasso or the adaptive lasso penalty, which are the two approaches with the best selection performance. In this context, the estimation of the individual expected-survival probability of patient i for a given time horizon τ can be obtained by plugging the estimated parameters into the survival function:

$$\hat{S}_i(\tau) = \exp\left[-\hat{H}_0(\tau)\exp\left(\hat{\alpha}T_i + \sum_{j=1}^{p} \hat{\beta}_j X_{ij} + \sum_{j=1}^{p} \hat{\gamma}_j X_{ij} T_i\right)\right] \tag{18.16}$$

with $\hat{H}_0(\tau)$ the cumulative baseline hazard, usually estimated using the non-parametric Breslow estimator [47]. Based on our results in simulations [48], the use of the adaptive lasso provides more precise and accurate estimates of the survival probability than the simple lasso. For both penalizations, the simulations also showed that re-estimating the regression coefficients after selection to obtain unpenalized coefficients reduces the precision of $\hat{S}_i(\tau)$.

In addition to the estimates of the expected survival probabilities for new patients, the construction of their confidence intervals is important to quantify the uncertainty of the estimates. However, constructing valid confidence intervals in penalized regression remains challenging. At least two approaches to estimate the confidence intervals of $\hat{S}_i(\tau)$ at level $1-\theta$ can be thought of: an analytical one based on the normal approximation of the estimator and a non-parametric one based on the bootstrap.

The analytical approach [49] consists of estimating the variance of the cumulative risk $\hat{H}_i(\tau) = \hat{H}(\tau, T_i, X_i)$ based on the Breslow estimator [50] in the Cox model. Thus, the confidence interval of $\hat{S}_i(\tau)$ is approximated by

$$\mathrm{CI}_{1-\theta}\left(\hat{S}_i(\tau)\right) = \left[q_{\theta/2}\left(\hat{S}_i(\tau)\right); q_{1-\theta/2}\left(\hat{S}_i(\tau)\right)\right] \tag{18.17}$$

with $q_\alpha(\hat{S}_i(\tau)) = \exp\left(-\hat{H}_i(\tau) + z_\alpha\sqrt{\hat{\mathrm{V}}(\hat{H}_i(\tau))}\right)$.

The non-parametric approach consists of generating B bootstrap samples of the original data and in estimating the model in Equation 18.11 for each of them. The B estimated models (i.e., $\hat{h}_1(\tau, T, X), \ldots, \hat{h}_B(\tau, T, X)$) are then used to estimate B times the expected survival probability for each patient i at time t, denoted by $\hat{S}_{i,\mathrm{boot}}(\tau) = \left\{\hat{S}_{i,1}(\tau), \ldots, \hat{S}_{i,B}(\tau)\right\}$. Then, the non-parametric confidence interval of the $\hat{S}_i(\tau)$ based on the empirical percentiles $q_\alpha(\cdot)$ of the distribution of $\hat{S}_{i,\mathrm{boot}}(\tau)$ is given by

$$\mathrm{CI}_{1-\theta}\left(\hat{S}_i(\tau)\right) = \left[q_{\theta/2}\left(\hat{S}_{i,\mathrm{boot}}(\tau)\right); q_{1-\theta/2}\left(\hat{S}_{i,\mathrm{boot}}(\tau)\right)\right]. \tag{18.18}$$

In the simulations, the non-parametric bootstrap approach showed empirical coverage probabilities (CPs) of the 95% confidence intervals which were closer to the nominal value as compared to the analytical approach (Figure 18.4). Despite the fact that the former produced slightly overly conservative pointwise confidence intervals (CP: 0.96–0.97), the latter was more biased in the opposite direction (CP: 0.91–0.93 in null scenarios and 0.88 in alternative scenarios).

After estimating expected survivals together with their confidence intervals, one would like to visualize the results graphically. Indeed, a practical way to visualize the predictive effect of a signature can be to plot the expected survival probability of patients according to their treatment-effect modifying score $\hat{\eta}_i$ at a given horizon τ. In order to obtain smoothed average confidence bounds, we considered constrained basis splines (B-splines) $spl(\hat{\eta}_i)$ [51]. Splines are numerical functions that can be used for curve fitting by approximately fitting the data at particular nodes. Obviously, we enforced the constraint that $0 \leq spl(\hat{\eta}_i) \leq 1$. When the model contains also prognostic biomarkers, patients with the same treatment-effect modifying score $\hat{\eta}_i$ may have different survival probabilities $\hat{S}_i(\tau)$ due to different prognostic scores $\sum_{j=1}^{p} \hat{\beta}_j X_{ij}$. As shown in Figure 18.4, despite the fact that the use of splines provides a smoother and clearer graphical representation of the expected probabilities, it introduces a bias in the CPs of the bootstrapped method.

Due to the possible large heterogeneity of $\hat{S}_i(\tau)$ for equal scores $\hat{\eta}_i$, we stratified the treatment-effect modifying plot into four groups according to the prognostic score, using the percentiles proposed by Cox [52]: 16.4%, 50.0%, and 83.6%. Even though categorization

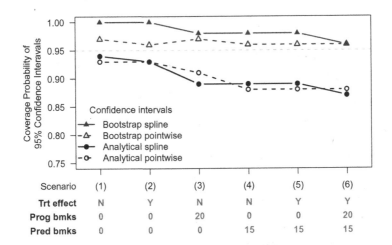

FIGURE 18.4
Coverage probabilities of 95% confidence intervals of the expected five-year survival probability in simulations under three null scenarios (1)–(3) and three alternative scenarios (4)–(6). C Trt: treatment. Prog: prognostic. Pred: predictive. Bmks: biomarkers. (Data drawn from Ternès N et al. *BMC Med Res Methodol.* 2017;17(83):1–12.)

reduces the available information [53], it leads to a more meaningful graphical representation of the prediction model. For a better understanding, a graphical visualization of a biomarker signature in a real application is provided in the Section 18.4.3.

18.4.3 Breast Cancer Application

A retrospective biomarker study was performed on tumor samples from $n = 1,574$ patients in an early breast cancer randomized clinical trial comparing chemotherapy with (arm C+T, $n = 779$) or without (arm C, $n = 795$) adjuvant trastuzumab. A comprehensive description of this dataset is provided by Pogue-Geile et al. [54]. Gene expression data had been collected for $p = 462$ genes and normalized as in the original publication. Clinico-pathological covariates, such as estrogen receptor and nodal status, and tumor size, were also available. The median follow-up time for distant recurrence free survival (DRFS) was 7.1 years and the censoring rate was 73% (i.e., 431 events). The arm-specific five-year DRFS was 84% (95% CI 81%–86%) and 64% (95% CI 61%–68%) for patients in arm C+T and C, respectively. Adding trastuzumab to adjuvant chemotherapy in early breast cancer patients led to significantly better DRFS than adjuvant chemotherapy alone (hazard ratio = 0.46 [95% CI 0.38–0.56]).

The clinico-genomic prediction model, established through the model in Equation 18.11 subject to the adaptive lasso penalty, can be found in Ternès et al. [48]. This model contains 102 prognostic variables (four clinical and 98 genomic variables) and 24 treatment-effect modifiers. Interestingly, some prognostic biomarkers had already been identified in the biomedical literature, e.g., SOX4 [55] or CSNK1D [56]. Some immune genes were also identified in the treatment-effect modifying component, which is consistent with some articles highlighting the involvement of immune pathways in the efficacy of trastuzumab.

Figure 18.5 provides a graphical visualization of the model representing the five-year DRFS according to the treatment-effect modifying score for four prognostic risk groups. The confidence intervals were estimated through the non-parametric bootstrap approach ($B = 200$) and were smoothed via B-splines using two or three nodes chosen by the AIC.

The entire prediction model (126 variables) has a moderate ability to discriminate patients according to their survival probability (C-statistics = 0.67 with double cross-validation).

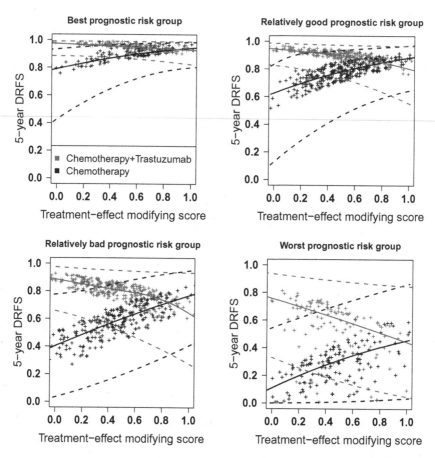

FIGURE 18.5
Five-year distant recurrence-free survival (DRFS) vs. the treatment-effect modifying score of the effect of trastuzumab in early breast cancer. Graphical representation of the 126-covariate prediction model developed using the adaptive lasso penalty. +: point estimates; —: average smoothed splines for point estimates; - - -: average smoothed splines for confidence bounds. (From Ternès N et al. *BMC Med Res Methodol.* 2017;17(83):1–12.)

Regarding the treatment-effect modifying component of the model, it only slightly discriminates between patients in terms of their treatment benefit ($\Delta C = 0.02$ with double cross-validation): the lower the treatment-effect modifying score, the higher the benefit of the trastuzumab. In terms of point estimates, the effect of treatment with trastuzumab on absolute DRFS seems higher for the low treatment-effect modifying scores. However, the width of the confidence intervals is extremely large, and the confidence intervals overlap a great deal between the two arms.

18.5 R Software

The R package `biospear` [57,58] implements the statistical methods presented in this chapter to develop and evaluate prediction models in a high-dimensional Cox regression setting, as well as to estimate the expected survival at a given time point.

In this section, we illustrate the use of the tools which are available in `biospear` by means of a simulated dataset (see also companion website https://www.crcpress. com//9781138083776). The function `simdata()` can be used to generate survival data, together with a set of Gaussian biomarkers.

```
library(biospear)
set.seed(123456)
sdata <- simdata(
n = 500, p = 100, q.main = 5, q.inter = 5,
prob.tt = 0.5, alpha.tt = -0.5,
beta.main = c(-0.5, -0.2),
beta.inter = c(-0.7, -0.4),
b.corr = 0.6, b.corr.by = 10,
m0 = 5, wei.shape = 1,
recr = 4, fu = 2,
timefactor = 1,
active.inter = c("bm003", "bm021", "bm044", "bm049", "bm097"))
```

The dataset `sdata` contains $n = 500$ patients, $p = 100$ biomarkers, of which five (`q.main`) are prognostic and five (`q.inter`) have a true interaction with the treatment. See `?simdata` for details on the other parameters.

The two core functions in the `biospear` package are `BMsel()`, which can be used for developing a prediction model, and `expSurv()`, a tool for estimating the individual expected survival probability based on a prediction model.

The general prediction model in `biospear`, is the proportional hazard model (Equation 18.11). The inclusion of the interaction part $\sum_{j=1}^{p} \gamma_j X_j T$ is under the control of the user via the option `inter` in `BMsel()`, which should be set to `FALSE` to develop a prognostic model or to `TRUE` if a model predicting the treatment effect is sought.

We recommend that the biomarkers X and their interactions with treatment XT be centered and scaled before selection. This can be automatically achieved by setting the options `std.x` and `std.i` to `TRUE`.

18.5.1 Biomarker Selection

The development of a prediction model can be performed with the `BMsel()` function. This function performs variable selection on the main effects β and, if `inter = TRUE`, also on the biomarker-by-treatment interactions γ. The data matrix is entered using the argument `data` and the names or positions of the columns containing the biomarkers, respectively, survival outcome, must be passed by means of the arguments `x`, respectively, `y`.

The `sel` option is used to specify the variable selection methods to be employed. Table 18.1 shows a list of the possible choices, which includes all the methods discussed so far in this chapter. Most of these methods are based on the lasso penalty (Equation 18.6), with the shrinkage parameter chosen by the maximum cvl or its extensions. The elastic-net, ridge, adaptive lasso, and stability selection are available as well. Specific methods for selecting interactions are implemented to control the main effects matrix or to jointly select main effects and interactions (group-lasso) or to fit the modified covariate model [43]. In addition to penalized regression, other methods are available in `BMsel()`, such as univariate selection with FDR control or gradient boosting [59].

TABLE 18.1

List of Selection Methods Implemented in BMsel(), Which can be Entered in the Argument sel

Method	Description
alassoL	Adaptive lasso penalty (lasso weighting)
alassoR	Adaptive lasso penalty (ridge weighting)
alassoU	Adaptive lasso penalty (univariate weighting)
enet	Elastic-net penalty
gboost	Gradient boosting
glasso[a]	Group-lasso penalty
lasso	Lasso penalty (*cvl* criterion)
lasso-1se	Lasso penalty (*cvl* + 1se criterion)
lasso-AIC	Lasso penalty (*cvl* + Akaike information criterion)
lasso-BIC	Lasso penalty (*cvl* + Bayesian information criterion)
lasso-HQIC	Lasso penalty (*cvl* + Hannan–Quinn information criterion)
lasso-pct	Percentile lasso
lasso-pcvl	Lasso penalty (*pcvl* criterion)
lasso-RIC	Lasso penalty (*cvl* + risk information criterion)
modCov[a]	Modified covariates (interactions only)
PCAlasso[a]	Principal component analysis (main effects) + lasso (interactions)
PLSlasso[a]	Partial least squares regression (main effects) + lasso (interactions)
ridge	Ridge penalty
ridgelasso[a]	Ridge (main effects) + lasso (interactions)
stabSel	Stability selection
uniFDR	Univariate selection with false discovery rate control

[a] Only available for the interaction setting.

Using the simulated data from above, we perform the selection using the lasso-*cvl* and lasso-*pcvl* methods:

```
resBM <- BMsel(data = sdata,
method = c('lasso', 'lasso-pcvl'),
inter = TRUE, folds = 5)
```

Note that the options x, y, z, and tt are mandatory unless the data have been simulated using simdata(), in which case the structure of the dataset is standard.

The summary of the results of the selection include the estimated coefficients of (1) the treatment (treat), (2) the selected biomarker main effects (bm047, ..., bm099), and (3) the selected biomarker-by-treatment interactions (bm003:treat, ..., bm079:treat) for each of the two methods. Since the data are simulated and thus the five truly prognostic and the five truly predictive biomarkers are known, the oracle model—i.e., the unpenalized models including only the treatment effect plus these five main effects and five interactions—is also shown:

```
summary(resBM)

           lasso lasso-pcvl   oracle
treat     -4.5e-01   -4.1e-01 -5.1e-01
```

```
bm047         -2.6e-01    -2.2e-01  -3.6e-01
bm063         -4.1e-01    -3.6e-01  -5.8e-01
bm085         -2.4e-01    -2.0e-01  -2.9e-01
bm091         -2.9e-01    -2.3e-01  -3.0e-01
bm094         -2.2e-01    -1.7e-01  -3.1e-01
bm004          2.9e-02     1.8e-02       .
bm024          6.6e-02     3.4e-02       .
bm028          4.1e-02     1.2e-02       .
bm039          2.1e-02     3.2e-04       .
bm041         -2.0e-02    -1.4e-02       .
bm051          6.2e-02     3.9e-02       .
bm058          3.5e-02     5.0e-04       .
bm067         -9.2e-02    -5.8e-02       .
bm075         -6.6e-02    -3.8e-02       .
bm080         -5.1e-02    -1.5e-02       .
bm090          5.4e-02     4.3e-03       .
bm001         -2.2e-02         .         .
bm005          2.6e-02         .         .
bm015          1.9e-02         .         .
bm016          6.7e-03         .         .
bm022         -3.5e-03         .         .
bm033         -3.0e-02         .         .
bm040          3.3e-02         .         .
bm042         -1.6e-02         .         .
bm044         -2.1e-02         .         .
bm055          2.0e-02         .         .
bm060         -2.2e-02         .         .
bm061          8.7e-05         .         .
bm082          6.8e-03         .         .
bm087         -4.0e-02         .         .
bm095          2.0e-02         .         .
bm096          2.3e-02         .         .
bm097          4.8e-03         .         .
bm099         -2.7e-02         .         .
bm003:treat   -3.8e-01    -2.6e-01  -8.0e-01
bm021:treat   -2.7e-01    -8.5e-02  -5.4e-01
bm044:treat   -6.6e-01    -4.8e-01  -8.9e-01
bm097:treat   -3.2e-01    -1.7e-01  -5.4e-01
bm002:treat   -8.2e-02    -4.3e-02       .
bm049:treat   -1.6e-02         .    -3.9e-01
bm011:treat    5.3e-02         .         .
bm016:treat    6.4e-02         .         .
bm030:treat    8.3e-02         .         .
bm051:treat    1.2e-01         .         .
bm064:treat    9.0e-02         .         .
bm079:treat   -5.3e-02         .         .
```

In addition to the treatment indicator and the biomarkers, other clinico-pathological covariates $\mathbf{Z}$ can be forced in the model using the z parameter. For penalized regression methods, this consists of excluding their coefficients from the penalization; for gradient boosting, they are entered via an offset; and for the univariate method, they act as adjustment variables.

18.5.2 Diagnostics

The biospear package provides two functions to assess the performance of a prediction model: predRes() and selRes(). The former, predRes(), computes the prediction accuracy of the developed signature at given time points in terms of: the Uno C-statistic [32], the prediction error (integrated Brier score) [60], and the difference between the arm-specific C-statistics (ΔC-statistic) [41]. These metrics can be computed and plotted (plot not shown) as follows.

```
predAcc <- predRes(res = resBM, traindata = sdata,
int.cv = TRUE, time = 1:5, ncores = 4)
predAcc$'time = 3'

$'Training set'
                  lasso lasso-pcvl oracle
C-index            0.77       0.74   0.75
Prediction Error   0.12       0.13   0.12
Delta C-index      0.38       0.34   0.35

$'Internal validation'
                  lasso lasso-pcvl oracle
C-index            0.70       0.70   0.73
Prediction Error   0.14       0.14   0.13
Delta C-index      0.27       0.27   0.33

plot(predAcc, crit = 'dC')
```

The latter function, selRes(), evaluates the selection accuracy of the developed signature and can be used only with a simulated dataset, for which the truly active biomarkers are known.

```
selRes(resBM)

      lasso lasso-pcvl
FDR   0.583      0.200
FNDR  0.000      0.010
FNR   0.000      0.200
FPR   0.074      0.010
AUC   0.990      0.990
AUPRC 0.895      0.895
```

In addition to the aforementioned FDR and FNR, the output includes several other metrics: the false non-discovery rate (FNDR), the false-positive rate (FPR), the area under the ROC curve (AUC), and the area under the precision-recall curve (AUPRC). These metrics are fully detailed in ?selRes.

Double cross-validation be easily performed in biospear by setting the argument int.cv = TRUE in the function predRes(). Calculations can be sped up by parallel computing, by setting the number of cores (ncores) to a value greater than 1.

18.5.3 Expected Survival Estimation

The expSurv() function implements the methods presented in Section 18.4.2 to estimate the individual patient expected survival probability (Equation 18.16), based on a prediction model computed by the BMsel() function. The argument boot allows choosing between the computation of analytical and bootstrap confidence intervals. Smoothed B-splines (option smooth) and categorization of the prognostic score into risk groups (option pct.group) may be used to obtain a meaningful graphical visualization of a penalized model that includes interactions.

In our example, we compute and plot (plot not shown) the expected survival probabilities at five years as follows.

```
esurv <- expSurv(res = resBM, traindata = sdata,
boot = TRUE, nboot = 100, ncores = 4,
smooth = TRUE, pct.group = 4, time = 5)
plot(esurv, method = 'lasso', pr.group = 3)
```

The confidence intervals are constructed through a 100-replicate bootstrap (boot = TRUE, nboot = 100). We also chose to use B-splines to obtain smoothed confidence intervals (smooth = TRUE) and to split the set of patients into four prognostic groups (pct.group = 4). The code above (figure not shown) plots the estimated survival probabilities according to the treatment arm for the third prognostic group (pr.group = 3) when lasso is used for model selection.

18.6 Conclusion

In this chapter, we have illustrated how to develop prognostic and predictive multimarker signatures in penalized Cox regression models in the context of clinical trials. Although we focused on penalized Cox regression models, many other alternative machine learning techniques are available, such as boosting, random forests, etc.

Different variants of penalization can be used for developing prognostic signatures, depending on the number of false positives one wants to obtain. For predictive signatures, we proposed a unified framework for developing a prediction model with biomarker-by-treatment interactions in a high-dimensional setting, and for internally validating it in the absence of external data; for accurately estimating the expected survival probability of future patients with associated confidence intervals; and for graphically visualizing the developed prediction model. An R package implementing all the methods presented is available.

Cross-validation is a fundamental step in development studies, but external validation on independent trials remains essential in order to reduce the risk of false-positive results.

Declarations

Acknowledgments

The authors acknowledge the National Surgical Adjuvant Breast and Bowel Project (NSABP) investigators of the B-31 trial who submitted data from the original study to dbGaP (dbGaP Study Accession: phs000826.v1.p1) and the NIH data repository. The B-31 trial was supported by: National Cancer Institute, Department of Health and Human Services, Public Health

Service, Grants U10-CA-12027, U10-CA-69651, U10-CA-37377, and U10-CA-69974, and by a grant from the Pennsylvania Department of Health.

Funding

This work was supported by the Foundation Philanthropia Lombard-Odier as a PhD scholarship. The funding sources had no role in the study design, data collection, data analysis, data interpretation, or writing of the manuscript.

References

1. Poste G. Bring on the biomarkers. *Nature.* 2011;469(7329):156.
2. Schneider D, Bianchini G, Horgan D, Michiels S, Witjes W, Hills R, Plun-Favreau J, Brand A, Lawler M, EAP Working Group for Oncology Clinical Research. Establishing the evidence bar for molecular diagnostics in personalised cancer care. *Public Health Genomics.* 2015;18(6):349–58.
3. Michiels S, Ternès N, Rotolo F. Statistical controversies in clinical research: Prognostic gene signatures are not (yet) useful in clinical practice. *Ann Oncol.* 2017;27(12):2160–7.
4. Cox DR. Regression models and life-tables. *J R Stat Soc. Ser B (Methodol).* 1972;34(2):187–220. http://www.jstor.org/stable/2985181.
5. Goeman JJ. L1 penalized estimation in the Cox proportional hazards model. *Biom J.* 2010;52(1):70–84.
6. Tibshirani R. Regression shrinkage and selection via the lasso. *J R Stat Soc. Ser B (Methodol).* 1996;58(1):267–88.
7. Tibshirani R. The lasso method for variable selection in the Cox model. *Stat Med.* 1997;16(4):385–95.
8. Moertel CG, Fleming TR, MacDonald JS et al. Fluorouracil plus levamisole as an effective adjuvant therapy after resection of stage II colon carcinoma: A final report. *Ann Intern Med.* 1991;122:321–6.
9. Therneau TM. A package for survival analysis in S. R package version 2.38, 2015, from https:// CRAN.R-project.org/package=survival.
10. Zhang HH, Lu W. Adaptive lasso for Cox's proportional hazards model. *Biometrika.* 2007;94(3):691–703.
11. Zou H. The adaptive lasso and its oracle properties. *J Am Stat Assoc.* 2006;101(476):1418–29.
12. Huang J, Ma S, Zhang C-H. Adaptive lasso for sparse high-dimensional regression models. *Stat Sin.* 2008;18:1603–18.
13. Sampson JN, Chatterjee N, Carroll RJ, Müller S. Controlling the local false discovery rate in the adaptive lasso. *Biostatistics.* 2013;14(4):653–66.
14. Benner A, Zucknick M, Hielscher T, Ittrich C, Mansmann U. High-dimensional Cox models: The choice of penalty as part of the model building process. *Biom J.* 2010;52(1):50–69.
15. Bühlmann P, Van De Geer S. *Statistics for High-Dimensional Data: Methods, Theory and Applications.* Springer-Verlag, 2011.
16. van de Geer S, Bühlmann P, Zhou S. The adaptive and the thresholded lasso for potentially misspecified models (and a lower bound for the lasso). *Electron J Stat.* 2011;5:688–749.
17. Ternès N, Rotolo F, Michiels S. Empirical extensions of the lasso penalty to reduce the false discovery rate in high-dimensional Cox regression models. *Stat Med.* 2017;35(15):2561–73.
18. Yuan M, Lin Y. Model selection and estimation in regression with grouped variables. *J R Stat Soc: Ser B (Stat Methodol).* 2006;68(1):49–67.
19. Van Houwelingen HC, Le Cessie S. Predictive value of statistical models. *Stat Med.* 1990;9(11):1303–25. ISSN 1097-0258.

20. Verweij PJM, Van Houwelingen HC. Cross-validation in survival analysis. *Stat Med.* 1993;12(24):2305–14.
21. Meinshausen N, Bühlmann P. High-Dimensional graphs and variable selection with the lasso. *Ann Stat.* 2006;34:1436–62.
22. Roberts S, Nowak G. Stabilizing the lasso against cross-validation variability. *Comput Stat Data Anal.* 2014;70:198–211.
23. Friedman J, Hastie T, Tibshirani R. Regularization paths for generalized linear models via coordinate descent. *J Stat Soft.* 2010;33(1):1.
24. Simon N, Friedman J, Hastie T, Tibshirani R. Regularization paths for Cox's proportional hazards model via coordinate descent. *J Stat Softw.* 2011;39(5):1–13.
25. Bøvelstad HM, Nygård S, Størvold HL, Aldrin M, Borgan Ø, Frigessi A, Lingjærde OC. Predicting survival from microarray data—A comparative study. *Bioinformatics.* 2007;23(16):2080–7.
26. Meinshausen N, Bühlmann P. Stability selection. *J R Stat Soc: Ser B (Stat Meth).* 2010;72(4):417–73.
27. MacArthur D. Methods: Face up to false positives. *Nature.* 2012;487(7408):427–8.
28. Pawitan Y, Michiels S, Koscielny S, Gusnanto A, Ploner A. False discovery rate, sensitivity and sample size for microarray studies. *Bioinformatics.* 2005;21(13):3017–24.
29. Genovese C, Wasserman L. Operating characteristics and extensions of the false discovery rate procedure. *J R Stat Soc: Ser B (Stat Methodol).* 2002;64(3):499–517.
30. Müller S, Welsh AH. On model selection curves. *Int Stat Rev.* 2010;78(2):240–56.
31. Uno H, Cai T, Pencina MJ, D'Agostino RB, Wei LJ. On the C-statistics for evaluating overall adequacy of risk prediction procedures with censored survival data. *Stat Med.* 2011;30(10):1105–17.
32. Barrett T, Suzek TO, Troup DB, Wilhite SE, Ngau W-C, Ledoux P, Rudnev D, Lash AE, Fujibuchi W, Edgar R. NCBI GEO: Mining millions of expression profiles–database and tools. *Nucleic Acids Res.* 2005;33(suppl_1):D562–6.
33. Davis S, Meltzer PS. GEOquery: A bridge between the Gene Expression Omnibus (GEO) and BioConductor. *Bioinformatics.* 2007;23(14):1846–7.
34. McCall MN, Bolstad BM, Irizarry RA. Frozen robust multiarray analysis (fRMA). *Biostatistics.* 2010;11(2):242–53.
35. Culhane AC, Schröder MS, Sultana R et al. GeneSigDB: A manually curated database and resource for analysis of gene expression signatures. *Nucleic Acids Res.* 2012;40(D1):D1060–6.
36. Mehra R, Varambally S, Ding L, Shen R, Sabel MS, Ghosh D, Chinnaiyan AM, Kleer CG. Identification of GATA3 as a breast cancer prognostic marker by global gene expression meta-analysis. *Cancer Res.* 2005;65(24):11259–67.
37. Buyse M, Michiels S. Omics-based clinical trial designs. *Curr Opin Oncol.* 2013;25(3):289–95.
38. Michiels S, Koscielny S, Hill C. Interpretation of microarray data in cancer. *Br J Cancer.* 2007;96(8):1155–8.
39. Royston P, Sauerbrei W. Interactions between treatment and continuous covariates: A step toward individualizing therapy. *J Clin Oncol.* 2008;26(9):1397–9.
40. Rothwell PM. Subgroup analysis in randomised controlled trials: Importance, indications, and interpretation. *The Lancet.* 2005;365(9454):176–86.
41. Ternès N, Rotolo F, Heinze G, Michiels S. Identification of biomarker-by-treatment interactions in randomized clinical trials with survival outcomes and high-dimensional spaces. *Biom J.* 2017;59(4):685–701.
42. Bien J, Taylor J, Tibshirani R. A lasso for hierarchical interactions. *Ann Stat.* 2013;41(3):1111.
43. Tian L, Alizadeh AA, Gentles AJ, Tibshirani R. A simple method for estimating interactions between a treatment and a large number of covariates. *J Am Stat Assoc.* 2014;109(508):1517–32.
44. Lockhart R, Taylor J, Tibshirani RJ, Tibshirani R. A significance test for the lasso. *Ann Stat.* 2014;42(2):413.
45. Michiels S, Potthoff RF, George SL. Multiple testing of treatment-effect-modifying biomarkers in a randomized clinical trial with a survival endpoint. *Stat Med.* 2011;30(13):1502–18.
46. Schemper M. Non-parametric analysis of treatment–covariate interaction in the presence of censoring. *Stat Med.* 1988;7(12):1257–66.

47. Breslow NE. Discussion on Professor Cox's paper. *J R Stat Soc. Ser B (Methodol)*. 1972;34:216–7.
48. Ternès N, Rotolo F, Michiels S. Robust estimation of the expected survival probabilities from high-dimensional Cox models with biomarker-by-treatment interactions in randomized clinical trials. *BMC Med Res Methodol*. 2017;17(83):1–12.
49. Therneau TM, Grambsch PM, *Modeling Survival Data: Extending the Cox Model.* Springer-Verlag, 2000.
50. Breslow NE. Covariance analysis of censored survival data. *Biometrics*. 1974;30(1):89–99, from http://www.jstor.org/stable/2529620.
51. Ng P, Maechler M. A fast and efficient implementation of qualitatively constrained quantile smoothing splines. *Stat Modelling*. 2007;7(4):315–28.
52. Cox DR. Note on grouping. *J Am Stat Assoc*. 1957;52(280):543–7, from https://www.jstor.org/stable/2281704.
53. Royston P, Altman DG, Sauerbrei W. Dichotomizing continuous predictors in multiple regression: A bad idea. *Stat Med*. 2006;25(1):127–41.
54. Pogue-Geile KL, Kim C, Jeong J-H et al. Predicting degree of benefit from adjuvant trastuzumab in NSABP trial B-31. *J Natl Cancer Inst*. 2013;105(23):1782–8.
55. Song G-D, Sun Y, Shen H, Li W. SOX4 overexpression is a novel biomarker of malignant status and poor prognosis in breast cancer patients. *Tumor Biol*. 2015;36(6):4167–73.
56. Abba MC, Sun H, Hawkins KA et al. Breast cancer molecular signatures as determined by SAGE: Correlation with lymph node status. *Mol Cancer Res*. 2007;5(9):881–90.
57. Ternès N, Rotolo F, Michiels S. *biospear: Biomarker Selection in Penalized Regression Models*, 2017, from https://CRAN.R-project.org/package=biospear. R package version 1.0.1.
58. Ternès N, Rotolo F, Michiels S. Biospear: An R package for biomarker selection in penalized Cox regression. *Bioinformatics*. 2018;34(1):112–3.
59. Friedman JH. Greedy function approximation: A gradient boosting machine. *Ann S*. 2001;29(5):1189–232.
60. Graf E, Schmoor C, Sauerbrei W, Schumacher M. Assessment and comparison of prognostic classification schemes for survival data. *Stat Med*. 1999;18(17–18):2529–45.

19

Sequential, Multiple Assignment, Randomized Trials

Kelly Speth and Kelley M. Kidwell

CONTENTS

19.1 Motivating Sequential, Multiple Assignment, Randomized Trials by Dynamic Treatment Regimens

Cancers are an insidious set of diseases characterized by uncontrolled cell growth, many of which rapidly adapt to maintain their survival. Cancer treatments differ based on the organ of origin of the cancer, the cancer histology (cell type), tumor stage at diagnosis, whether the cancer is local or has metastasized (spread to other organs), and the patient's personal preferences and characteristics, including age and health status, among others. Depending on the above, standard therapies may include surgery, radiation therapy, chemotherapy, immunotherapy, stem-cell transplantation, and/or targeted therapy. Or, if the cancer is of minimal risk of proliferation, active surveillance may also be used.

Complicating the picture of cancer care is the heterogeneity of cancers and the sequential nature of cancer therapy. Each "line," or stage, of anticancer therapy may include numerous treatment modalities, for example, surgery followed by radiation therapy, and each line may then be followed by a second or third line should the first line be deemed ineffective. Once a treatment line of cancer therapy is successful, many cancers require ongoing therapy to "maintain" the remission. Toxicity to anticancer therapies is also a pervasive issue. Some patients will tolerate therapy reasonably well and complete the full treatment course, whereas others will suffer from unwanted side effects that necessitate a lower, and perhaps less-effective, dose or a change of therapy altogether. A multitude of questions concerning which therapy to give patients at what stage in their treatment, the use of concurrent therapies, timing of therapy (time of initiation, as well as duration), dosage, and order in which the therapy is given, persist. There are an abundance of therapeutic interventions, as well as innumerable response assessments, both formal and informal, which establish *de facto* decision points throughout a patient's cancer care.

Although anticancer therapy is administered as a sequence, conventional clinical trials typically evaluate only one piece in this puzzle. To determine whether outcomes of toxicity and efficacy are comparable among treatments, for example, a novel chemotherapeutic may be evaluated against the standard of care for induction chemotherapy, or two dose fractionations may be compared when used in post-surgical radiotherapy. Whereas the intent of these trials is sound, that is, to develop products that are safe and effective, each trial evaluates a therapeutic option in isolation. As illustrated above, however, cancer care is sequential in nature and treatments may be synergistic or antagonistic when used either in combination or in sequence with other therapies. If the overall objective of cancer therapy is to maximize efficacy over the full course of therapy (i.e., longer survival) with minimal toxicity and burden to the patient, the conventional strategy for evaluating anticancer therapies in isolation may be failing us.

A dynamic treatment regimen (DTR) [1,2] represents a comprehensive course of cancer therapy administered to a patient. DTRs are also known as treatment policies [3], adaptive interventions [4], and adaptive treatment strategies [5]. A DTR is defined as a guideline of a sequence of therapies administered and tailored to a patient. DTRs are "dynamic" in that they are tailored to the patient based on one or more intermediate outcomes such as response to therapy, toxicity, evolving comorbidities, etc. DTRs can be further tailored or personalized based on baseline and time-varying patient and disease characteristics. This is in contrast to a "treatment sequence," which does not change based upon the patient or the patient's disease characteristics. It can be argued that very few, if any, cancer treatments are administered as a treatment sequence and not as a DTR. Patient characteristics and outcomes factor into nearly all patient treatment decisions.

There are several components needed to define a DTR. First, there must be therapeutic options, that is, the ability to select one or more from several viable treatment options or dose levels. For example, therapeutic options for patients with early-stage breast cancer typically include surgery, radiation therapy, chemotherapy, and/or hormonal therapy. Next, there must be critical decision points at which point the regimen begins or can be altered, maintained, or terminated. In the breast cancer example, critical decision points occur with the decision of what to do first for a patient (e.g., give surgery) and the decision of subsequent treatment (e.g., what to give after surgery). Third, DTRs have tailoring variables that are used to personalize treatment. For example, after breast-conserving surgery to remove a lump in the breast for a patient with early-stage disease, lymph nodes (LNs) may be evaluated for the presence of cancerous tissue. If cancer cells are present in one or more LNs, an aggressive systemic chemotherapy may be selected for a young woman in good health, whereas a more mild systemic chemotherapy agent may be selected if the woman were older or otherwise ailing. If there is an absence of cancer cells in the LNs, however, the treatment decision could be to initiate either a more mild chemotherapy or no chemotherapy altogether. In this scenario, "presence of cancerous cells in the lymph nodes," "patient's health status," and "patient's age" are all considered tailoring variables. One specific DTR for the above case is formulated as follows: For women with early-stage breast cancer, first perform surgery. If there are cancerous cells present in the LN, give aggressive systemic chemotherapy. If cancerous cells are absent from the LN, follow the patient. Note that this DTR includes an initial treatment (i.e., surgery) followed by a second treatment (i.e., either chemotherapy or no chemotherapy), where the second treatment depends on a tailoring variable (i.e., the presence or absence of cancerous cells in the LNs). Thus, a DTR is dynamic in its ability to "adapt" based on intermediate clinical outcomes. Other DTRs may include more tailoring variables to further personalize the regimen or different treatment options.

There are several analytic methods through which inference about DTRs can be made. One option is to make inference on the optimality of DTRs using observational data, although inference on observational data is subject to the usual biases of confounding, selection, information, and others. A randomized controlled trial (RCT) can be performed with up-front randomization to one of a number of viable DTRs. A third option is to conduct a sequential, multiple assignment randomized trial (SMART) [6–8], which is a clinical study where patients are randomized to one of a set of treatments at multiple time points and where subsequent randomization depends on intermediate outcomes. Whereas the first two methods mentioned above are not without merit, this chapter focuses on evaluating multiple DTRs within a SMART. SMARTs offer a flexible study design to develop effective DTRs in oncology and distinct advantages over conventional, single-stage trial designs.

19.2 Introduction to Sequential, Multiple Assignment, Randomized Trials

A SMART is a clinical trial design that enables inference to be made on DTRs that are used routinely in clinical practice. Thus, a DTR is a guideline that guides the clinical practice of cancer care and a SMART is a specific trial design used to evaluate DTRs. A SMART can involve any phase of clinical research; the defining characteristic is simply the presence of at least two sequential randomization points where subsequent randomization is based on some intermediate outcome. SMART designs can include any number of stages, although the two-stage SMART is the most common. Each randomization can include two or more treatment options, just as in a conventional RCT. Refer to Figure 19.1 for three of the most common two-stage designs. Note that each design in Figure 19.1 includes a different number of DTRs. There are eight DTRs embedded in design I, which are denoted by the first-stage treatment, second-stage treatment for responders, and second-stage treatment for non-responders: (A,C,E), (A,C,F), (A,D,E), (A,D,F), (B,G,I), (B,G,J), (B,H,I), and (B,H,J). Design II includes four DTRs: (A,C,D), (A,C,E), (B,F,G), and (B,F,H), and design III includes three DTRs: (A,C,D), (A,C,E), and (B,F,G). All of the treatments in Figure 19.1 have unique labels, but some treatments may be repeated across stages or some labels may specify no treatment. Numerous other SMART designs exist with different numbers of stages and treatments at each stage for responders and/or non-responders. The focus of this chapter is the application of SMART design in oncology, although SMARTs are used for the study of other diseases, including in the behavioral and mental-health spheres [9–13].

Sequentially randomized, multi-stage trials in oncology are not new. The number of conventional RCTs in oncology rose starting in the 1960s with the scientific testing of chemotherapeutics to combat the skyrocketing incidence of lung and other cancers, but sequentially randomized trials were not far behind. One of the first sequentially randomized trials in oncology, a SMART "precursor," was initiated in the early 1980s for the treatment of small-cell lung cancer [14]. This and other similar studies in the 1980s and 1990s [14–19] were designed in the same manner of a SMART, i.e., with multiple stages and randomization events. However, because analytic methods for SMARTs were not developed until the early 2000s, these precursor SMARTs did not use efficient methods to compare DTRs, rather they analyzed the data from each stage separately. Since then, a number of SMARTs have been conducted in both hematologic and solid tumors, including acute myeloid leukemia [20],

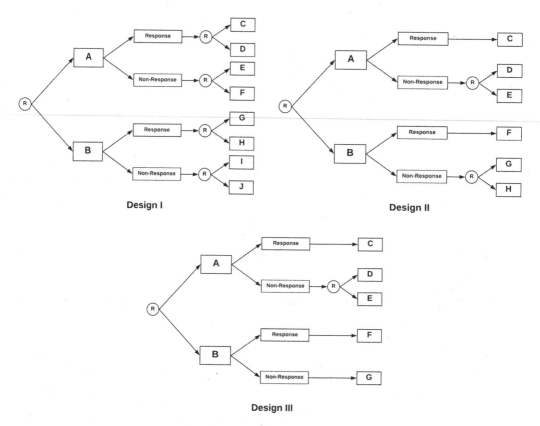

FIGURE 19.1

Three of the most commonly applied Sequential, Multiple Assignment, Randomized Trial (SMART) designs. R denotes randomization and A–J denote treatments which need not be unique. Each SMART design embeds a number of dynamic treatment regimens (DTRs) such that design I includes 8 DTRs, design II includes 4, and design II includes 3.

neuroblastoma [21], diffuse large cell B-cell lymphoma [22], non-Hodgkin's lymphoma [23], multiple myeloma [24], prostate cancer [25], malignant melanoma [26], and follicular and mantle cell lymphomas [27]. At the time of writing this chapter, several SMARTs are being conducted to evaluate DTRs in renal cell carcinoma [28], prostate cancer [29], and smoking cessation [30].

As implied by its name, the key characteristic of a SMART is the sequential randomizations performed on at least some set of study participants, where randomization at intermediate time points is based on the value(s) of intermediate outcomes or other factors, known as tailoring variables. Due to the randomization mechanism incorporated into each treatment stage, questions concerning the differential safety and efficacy among treatments at each stage in the trial can be answered in a similar manner as standard RCTs, but *as part of a treatment regimen*. A benefit of SMART design over the standard, one-stage RCT, however, is the additional ability to make inferences about DTRs as they are used routinely in clinical practice. In this sense, assuming that the overall objective for the spectrum of cancer care is to improve overall survival, a SMART is one trial design that is able to provide potentially more relevant evidence to the current sequential and tailored practice of cancer care.

As with any clinical study, standard study-design practices for a SMART are essential; a study is only as good as the design and the data. SMARTs are often used as developmental study designs to elucidate DTRs that are later tested in a confirmatory RCTs [8]. More and more often, however, SMARTs are being used in a confirmatory context for comparing DTRs. In this case, although it may be easy to conceive of a host of possible DTRs to evaluate, it is important to include only those DTRs that have been vetted within the oncology community such that the optimal DTR is practical for the treatment of cancer patients. These DTRs can be identified through clinical knowledge or experience or pilot studies and should incorporate all expected clinical outcomes.

Reliable early indicators of ultimate endpoints that are practical for clinical use are necessary to define the intermediate outcome in any SMART design. These intermediate endpoints may or may not be (statistical) surrogate outcomes, but should provide an early indication of treatment success or failure and be associated with the overall outcome. In cancer, Response Evaluation Criteria in Solid Tumors (RECIST) response (or Immune-Related Response Criteria) may be appropriate for measurable disease. If there is no measurable disease, changes in reliable tumor biomarker assays, such as serial biopsies or serial circulating protein markers or circulating tumor cells may be considered to define the tailoring variable. Toxicities may also be considered as a standalone intermediate endpoint or in addition to efficacy. The trial protocol should include a plan for patients who experience any common contingencies (e.g., toxicities or other side effects, death, excessive treatment burden, or personal choice). The plans for patients experiencing contingencies should be incorporated into the definition of the embedded treatment regimens [31]. Furthermore, when executing a SMART, standardized and real-time assessments of the chosen intermediate response measure are required.

As with any trial, missing data should be minimized in a SMART. It is important to protocolize how individuals who have missing intermediate outcome information are categorized and analyzed. A pilot SMART may identify the proportion of expected missingness, underlying causes of missingness, and how best to incorporate missing information into the embedded DTRs [32]. There is increased complexity when dealing analytically with missing data for a SMART due to the sequential and longitudinal nature of the design; multiple imputation [33], however, may be used in analysis.

Finally, although a SMART design can accommodate a broad range of research questions pertaining to DTRs and their associated tailoring variables, a disconnect exists between investigating full DTRs used in practice and actual, implementable SMART designs [34]. SMARTs conducted to date have largely included only two stages of randomization, representing two sequential therapy choices and one tailoring variable, whereas clinical practice often includes many treatments over a long period of time tailored to many personal and disease characteristics. A two-stage SMART, however, is a logistically feasible step in the direction of evaluating DTRs that are more similar to clinical practice. Analyzing DTRs from longer-term treatment in the observational setting may provide additional evidence for clinically relevant treatment guidelines.

19.3 SMART Compared to Other Trial Designs

Several trial designs, including factorial, crossover, adaptive, randomized discontinuation or response-guided, and biomarker strategy designs share similarities with SMART designs (Table 19.1). A SMART design may be considered a type of factorial design, i.e., a sequential tailored factorial trial [35]. Factorial designs are used when an investigator

TABLE 19.1

Comparison of Prospective Randomized Clinical Trial Designs

Trial Design	Example Figure[a]	Fixed Trial Characteristics	Treatment May be Personalized on Factors — Baseline	Treatment May be Personalized on Factors — Time-Varying	Sequential Randomization	Interest in Treatment Interactions	Intermediate Response Evaluation and Treatment Course Modified	Considers Sequential Treatment
Standard PRCT		X	X					
Factorial		X	X			X		
Crossover		X	X		X			
Adaptive			X					
Randomized Discontinuation or Response-Guided		X	X				X	X
Biomarker Strategy		X	X		X			
SMART		X	X	X	X	X	X	X

Note: X denotes that the trial design incorporates the trial feature.

[a] Figures are general examples, but each design is not limited by the figure shown here.

is interested in evaluating the effects of two or more treatments used either alone or in combination. In particular, factorial designs can investigate treatment interactions between and among factors. In a standard factorial design, treatment assignment does not depend on intermediate outcomes, in contrast to a SMART. Thus, in a standard factorial design, treatment assignment is simultaneous rather than sequential.

A crossover study is a repeated measures design. In the simplest scenario, patients are exposed to two treatments in succession, allowing for an adequate washout period between treatment administrations to reduce any *carryover effects*. Carryover effects are those observable or unobservable effects of the first agent that remain in the body after the first treatment is administered and that, if unaccounted for, could confound the outcomes observed following administration of the second agent. A pre-specified period of time between the first and second treatment administrations, known as a "washout" period, is generally used to mitigate any carryover effects. The endpoint of interest for crossover studies is evaluated for all patients after each treatment course. In the crossover design, patients are used as their own control. Although crossover designs are infrequently used in oncology clinical trials due to the fact that traditional endpoints are often non-reversible events (e.g., response, time to progression, survival, etc.), there remains some overlap between SMART and crossover design. SMARTs, as described above, include sequential randomizations based on an intermediate outcome whereas, in a designed crossover, each patient is first treated with one agent followed by the other. In crossover designs, however, all participants crossover, so that crossover is not dependent on an intermediate outcome.

Adaptive designs, also termed "flexible designs," allow for modifications to either study or statistical procedures while the clinical trial is ongoing without compromising the integrity of the study. Adaptive designs can be confused with SMART designs given the similar vocabulary used in SMART design (e.g., "adaptive treatment strategies"); however, the designs are distinct. Adaptive designs are defined by the adaptation of the trial operational characteristics during the study based on collected data. Trial adaptations may include changing randomization ratios across treatment arms, removing a therapeutic arm based on poor efficacy, and/or recalculating sample size, among others. Notably, these adaptations apply to future participants based on data from previous participants in the same trial. Adaptive designs may be used during any phase of investigational research, although there is an emphasis on their utility within the exploratory phase. Similar to an adaptive design, a SMART is flexible and can be employed in any study phase. In contrast, a SMART is designed with fixed operational characteristics that remain unmodified throughout the study. Through a series of multiple, sequential randomizations, a SMART allows for patient-specific changes in treatment assignment based on the patient-specific treatment experience and outcomes in the preceding stage. Whereas the two designs are distinct, there have been efforts to integrate adaptive methods within a SMART design [7,9,36–38].

Randomized discontinuation (RD) [39,40] and other response-guided (RG) [41–43] designs have also been used in oncology and share similarities to a SMART. In brief, an RD design is an enrichment design where all participants are given the same drug in a "run-in" period. Following a response assessment, all participants with favorable response data (usually complete response or lack of progression) are randomly assigned to either discontinue or to continue therapy. Conversely all participants who do not exhibit a favorable response, or who suffer unacceptable toxicity, are excluded from the rest of the trial. A SMART modifies this general design by randomly assigning patients to treatments at critical decision points throughout the trial (usually two phases, i.e., not all patients are in the same run-in group or, if there is a run-in period, patients are randomly assigned again in a third phase) and following all patients over time to investigate longitudinal outcomes under different treatment regimens (i.e., as opposed to removing those who do not have stable disease [44]). RG trials are similar to RD designs in that treatment may be sequenced and treatment depends on intermediate response. Similar to a SMART, an RG trial includes pre-specified modification to treatment based on intermediate response evaluations and therefore provides evidence for tailored treatment strategies. Most RG designs differ from SMARTs because the RG design either does not include a random assignment and/or only one random assignment is performed, unlike a SMART which requires sequential random assignments.

Finally, biomarker designs, like the modified marker-based strategy design [45–47], may be similar to SMART designs and even include sequential randomizations. The marker-based strategy designs are distinctive from SMARTs in that patients are initially randomly assigned to receive therapy based either on their marker status or not. Thus, random assignment is not based on an intermediate outcome or changing status like a SMART, but rather on the strategy of using marker status and the fixed value (e.g., positive or negative) of that biomarker.

19.4 Alternative Trial Strategies to a SMART

Two alternative trial strategies, the conduct of multiple one-stage RCTs and an RCT that provides up-front randomization to one of several DTRs, can also develop or assess the efficacy of DTRs. Both of the above-mentioned alternatives to SMART design have merits in their own right but tend to be less efficient than a SMART design for developing and comparing DTRs.

It is tempting to piece together pieces of separate one-stage RCTs to define the best DTRs. For example, one RCT could identify the optimal first-line therapy. For patients who respond to first-line therapy, a second RCT could investigate the best maintenance regimen for these patients. A third RCT could then evaluate the best second-line therapy for first-line non-responders. Thus, an "optimal" DTR could be constructed by combining the results of the three separate, one-stage RCTs to inform a DTR that would include a first-line therapy and a subsequent therapy based on a patient's intermediate response assessment. However, Wolbers and Helterbrand [48] demonstrate through a simulation study that when the overall objective is to determine whether a new agent should be added in combination to the existing standard of care induction regimen, used within the maintenance regimen for responders to the induction therapy, or added to both the induction and maintenance regimens, a SMART more efficiently uses patients, tends to complete the trial in a shorter period of time, and enrolls similar numbers of patients as would be enrolled in three separate, one-stage-at-a-time RCTs. If the research question focuses only on stage-specific outcomes, however, a single-stage RCT is adequate to meet these needs. Thus, the greatest benefit of a SMART comes from the increased validity of the statistical analyses when the overall goal is to evaluate the use of treatments in a multi-sequence regimen and the ability to identify tailoring variables used in assigning treatments or treatment sequences [49].

It is possible that using SMART design can bring products to market faster by identifying more "signal" and less "noise." In other words, SMARTs demonstrate better management of cohort effects that can be seen when combining the results of single, one-stage-at-a-time RCTs. For example, if one were to use the well-accepted intention-to-treat (ITT) method for an overall survival (OS) analysis in a conventional, one-stage RCT, any patient randomized to Treatment A would be analyzed as "Treatment A" regardless of which treatment(s) the patient uses after receiving the investigational treatment in the RCT. It is well known that treatment for cancer patients is both multi-modal and multi-stage and, when one treatment ends, another usually beings. Thus, in effect, an OS analysis of this kind will average the effect over the initial treatments (A and B), as well as all other treatments the patient receives until they die, which may be extensive. If one were to incorporate SMART design from the outset, the effect of Treatment A on OS may be more definitive (either positive or negative) such that a treatment can either advance or be withdrawn from consideration for the indication in question. As stated by Lavori and Dawson [50], a SMART "brings subsequent treatment variation under experimental control, allowing surer inference about the primary treatments and also creating the possibility of inference about optimal choices for the subsequent options."

Additionally, it is anticipated that a SMART is better able to recruit and retain patients throughout the study than single-stage RCTs based on providing additional treatment on study. Due to the fact that a SMART is designed to offer participants treatment options based on their intermediate outcome status, participants may be more likely to follow the designated treatment schema and remain in a SMART (as opposed to dropping out of the study early), because subsequent treatment options have been prospectively addressed in SMART design and are tailored to their individual outcomes [51]. It is also posited the study population better reflects the outcomes and experiences of all patients rather than the subset who remain and complete the study and, hence, the study achieves a higher degree of external validity [51].

With regard to study conduct, a SMART design requires one institutional review board (IRB) in contrast to separate IRB reviews for each protocol. Also, study start-up, management, systems implementation, and monitoring are performed once. This streamlines costs related to study start-up, laboratory and radiology vendors, and management organizations, as opposed to performing a series of one-stage trials.

It is also possible to make inferences about DTRs by performing a single, up-front randomization. Instead of performing sequential randomizations, as is conducted within the SMART design, each DTR is defined up-front and patients are randomized once at the outset of the study to one of several, multi-stage treatment regimens [52]. Whereas this design assesses DTRs, it has one major deficit: its use does not allow one to stratify or balance important variables up to the intermediate time point for second-stage randomization. However, if the intention is to perform a head-to-head comparison of an outcome among several previously-elucidated treatment regimens, then an RCT that up-front randomizes each patient to one of several DTRs may be adequate.

19.5 Research Questions Addressed by SMARTs

SMARTs are extremely flexible both in their design and in the research questions that they can address. Not only can a SMART answer questions similar to those posed in a single-stage RCT, but it can also extend research impact by addressing other questions not accessible through a traditional RCT [53–55]. On account of the multiple, sequential randomizations within a SMART that balance and re-balance the treatment groups with respect to confounding or prognostic variables, SMART designs enable both head-to-head treatment comparisons and effect sizes within each stage *as part of a regimen*, as well as comparisons among the embedded DTRs.

SMARTs can address research questions regarding treatment synergies or antagonisms or delayed treatment effects. For example, consider a case in which a first-line anticancer agent (Agent A) is moderately effective but highly toxic compared with a less effective but more tolerable agent (Agent B). The high rate of toxicity of Agent A, however, precludes many non-responders from receiving second-line therapy. If one were to evaluate the response rates of Agent A compared with Agent B in isolation, Agent A would be selected as the optimal first-line therapy. But if one were instead to evaluate the overall response rate following second-line therapy, one may find that, although Agent B was a modestly less effective first-line therapy than Agent A, starting with Agent B produces a higher overall response. This could be due to the fact that patients who begin first-line therapy with Agent B are better able to tolerate second-line therapy should they experience cancer progression during their first-line therapy [53,56]. Thus, an incorrect treatment conclusion may be reached by restricting investigation to only one line of therapy instead of considering an agent's place within the larger continuum of cancer care that is used in clinical practice. If an investigator's research objective is to optimize the outcome of a treatment sequence, or even if the primary goal is to compare single-stage therapies—but this stage occupies time and space between prior and subsequent therapy decisions—a SMART may be a more appropriate design option than others that cannot investigate DTRs.

Consider design II in Figure 19.1 (questions are similar for designs I and III); the following research questions may be posed from a SMART design:

1. Comparison of first-stage treatments: Is it best to begin DTRs with treatment A or B?
2. Comparison of second-stage treatments: Among non-responders to A, is it best to follow-up with D or E?
3. Comparison of second-stage treatments: Among non-responders to B, is it best to follow-up with G or H?
4. Comparison of all DTRs: Is there a difference between the four embedded DTRs:
   (1) First begin with A, if you respond, receive C; if you do not respond, receive D;
   (2) First begin with A, if you respond, receive C; if you do not respond, receive E;
   (3) First begin with B, if you respond, receive F; if you do not respond, receive G;
   (4) First begin with B, if you respond, receive F; if you do not respond, receive H?
5. Estimation of DTRs and identification of the best DTR: Which of the four DTRs (listed above in number 4) leads to the best outcome?
6. Comparison of two DTRs: Is it best to follow DTR 1 or DTR 3 (two DTRs chosen from the four listed above in number 4)?
7. To develop more deeply tailored DTRs: Are particular baseline characteristics or features up to second-stage randomization associated with best outcomes for those following a particular DTR?

Any of these questions of interest may motivate a SMART design. As with any study, it is important to begin with a relevant research question and select a trial design that meets your scientific needs so that the design is motivated by the science (i.e., the SMART design is not first chosen and retrofitted into the science). Details on powering and analyzing SMARTs with these questions can be found in the following sections.

19.6 Sample Size Calculations

Like a standard RCT, a SMART is powered based on the primary study objective. Depending on the outcome of interest, a maximum sample size can be estimated through standard inputs, including the significance level (i.e., allowable type I error), statistical power, number of therapeutic interventions, randomization probabilities, and practical considerations such as dropout rates. Additional information may also be required to size a SMART, such as an estimate of the response rate (i.e., the proportion of participants who are re-randomized in stage 2) and DTR-specific effect sizes.

Sample size calculations through an online applet are available for pilot studies based on designs I, II, and III from Figure 19.1 to determine SMART feasibility (i.e., that participants will enroll and enough participants will receive each treatment option [32]). The calculation does not depend on the outcome of interest. Rather, for these calculations, a minimum number of subjects desired in one of the second-stage treatment cells, the probability that the minimum number of subjects desired in one of the second-stage treatment cells is greater than that specified, and a guess of the rate of (non-)response to first-stage treatment are required.

For larger, more confirmatory SMARTs, should an investigator be interested primarily in comparing outcomes based on the initial (or subsequent treatment for responders or non-responders) as a part of a treatment regimen, standard sample size calculations for parallel group comparisons can be used. Research questions 1 through 3 from Section 19.5 (Research Questions Addressed by SMARTs) average over subsequent or initial treatment in the SMART context and, thus, reduce to standard parallel group comparisons. It is noted that most recent SMARTs have specified one of these parallel group comparisons (or main effects) of either first- or second-line treatment as a primary study objective [34] and explore DTR comparisons. However, if the primary objective is to compare embedded DTRs or to identify the optimal DTR, SMART-specific sample size calculations are required.

Most of the sample size calculations discussed herein to evaluate DTRs allow for sample size estimation based on a comparison of two embedded DTRs starting with a different initial therapy (see Table 19.2). In practice, investigators often determine the minimum sample size needed for a SMART based on a comparison of the two most clinically divergent DTRs. If the investigator intends to make more than one pairwise comparison of embedded DTRs, an omnibus test may be used or a standard Bonferroni (or other) correction for multiple testing can be used. The Bonferroni-corrected type I error is applied for each pairwise comparison of interest, and the sample size is determined by the maximum of the sample sizes for each of the pairwise comparisons. Alternatively, multiple comparisons may be controlled using another method (e.g., multiple comparisons with the best by Ertefaie et al. [57]), or simulation (as opposed to standard calculations or SMART sample size calculators or applets) may be required to find the appropriate sample size for the SMART.

Sample size calculations to compare DTRs depend on the outcome of interest: continuous (e.g., a biomarker level or test result), binary (e.g., response rate or survival rate at a particular time point), or time to event (e.g., progression-free or overall survival). Formulas and an online applet allow for the comparison of pairs of DTRs for continuous and binary outcomes [55,59] for any SMART design (i.e., you build your own SMART). These calculations allow for different inputs to identify the effect size of interest (e.g., DTR effects or treatment-pathway effects). A guess of the intermediate outcome rate can be provided or the applet will calculate a conservative sample size. Additional, less flexible, formulas exist to compare two or more DTRs or identify the best DTR when the outcome is continuous [55,63–66]

TABLE 19.2

Publicly Available Sample Size Calculators to Power a SMART Study Designed with a DTR-Specific Primary Objective

Outcome/Goal	SMART Design	URL	References
Pilot study/Feasibility	Designs I, II, III	https://moloque1227. shinyapps.io/SMARTsize/	[32,58]
Binary/Compare 2 DTRs	Any [customizable]	https://sites.google.com/a/ umich.edu/kidwell/	[59]
Continuous/Compare 2 DTRs	Any [customizable]	https://sites.google.com/a/ umich.edu/kidwell/	[55]
Continuous/Identify best DTR	Design II [Re-randomize non-responder (or responder) only]	http://methodologymedia.psu. edu/smart/samplesize	[55]
Time to event/Compare 2 DTRs with the same second-stage therapy	Design II [Re-randomize non-responder (or responder) only]	http://www.pitt.edu/~wahed/ Research/Resources/	[60]
Time to event/Compare second-stage therapies	Design II [Re-randomize non-responder (or responder) only]	http://www.pitt.edu/~wahed/ Research/Resources/	[61]
Time to event/ Compare 2 DTRs	Design II [Re-randomize non-responder (or responder) only]	http://methodologymedia.psu. edu/logranktest/samplesize	[62]

for a specific set of SMART designs (generally designs I and II). Cluster-based SMARTs with continuous outcomes can also be sized based on formulas [53,67]. These calculations identify the number of clusters or individuals required assuming an intraclass correlation coefficient and other standard SMART inputs. Formulas and online code or applets are also available for time-to-event outcomes [60–62] assuming design II from Figure 19.1. These calculations require standard inputs, a guess of the estimate of the intermediate outcome rate, and a hazard ratio between the two DTRs of interest. Before selecting a pre-existing sample size calculator, it is recommended that you refer to the specific manuscripts in which the details surrounding the correct usage of the sample size calculations are provided. For other outcomes, questions of interest, or other specific SMART designs, sample size and operating characteristics under a range of parameterizations and clinical scenarios may be explored using computer simulation.

19.7 Methods for Analysis

Given the abundance of data obtained through medical records, ample opportunities exist to study DTRs through longitudinal observational data including marginal structural mean models, G-computation and G-estimation, among other methods [1,13,68–78]. In this section, however, we provide a brief introduction to some of the existing statistical methodology that can be used to estimate and compare DTRs embedded in a SMART design and point readers to the references for more details (including a chapter by Davidian et al. [79] and two textbooks devoted specifically to evaluating DTRs by Chakraborty and Moodie [80] and Kosorok and Moodie [53]). We focus on the practical application of the methodology here as opposed to the underlying theory.

Consider an outcome Y and a set of covariates (O, A), where O includes patient information and A is a binary treatment indicator. Interest is in the expected outcome for each DTR if all subjects were to follow that DTR. When the outcome Y is continuous or binary, one approach to estimate and compare DTRs simultaneously using standard software is through weighted and replication estimation [8,55,59,81]. Data from a SMART are weighted to correct for the bias by design where only some groups of participants are re-randomized (e.g., designs II or III) or there is unequal randomization between treatments (i.e., if randomization occurs at allocation ratios differing from 1:1). For example, in design II, if a simple average of outcomes across those consistent with a DTR was taken, the responders would be over-represented and the non-responders would be under-represented. This over- and under-representation occurs since non-responders are randomized between two treatments and responders all receive one treatment. Weights based on the inverse-probability-of-treatment or the inverse of the randomization probability are applied to participants to correct for the bias by design. For example, for design II, responders receive a weight of 2 (1/0.5) and non-responders receive a weight of 4 (1/(0.5*0.5)). In order for standard software to simultaneously estimate the outcome for each DTR, observations that are consistent with more than one DTR must be replicated. For example, in design II, the responders to treatment A are consistent with the DTRs (A,C,D) and (A,C,E). The data from the responders (of both initial treatment A and B) are replicated so there are two identical observations per responder, but the second-stage treatment assignment as if these participants were non-responders is set to 1 for one observation and 0 for the other observation (assuming dummy variable coding for second-stage treatment). Thus, a newly restructured dataset will have more observations than the number of participants enrolled in the trial. For example, for design II, the dataset used for analysis will include the number of non-responders plus two times the number of responders. Once the data has been restructured, a model depending on the design of the SMART may be estimated using a generalized estimating equation approach with robust variance. For example, for design II, assuming A_1 is the first-stage treatment indicator ($A_1 = 1$ for treatment A and 0 for treatment B), A_2 is the second-stage treatment indicator for non-responders ($A_2 = 1$ for treatment D and G and 0 for treatment E and H), and O denotes any baseline covariates, a saturated model is specified as $E[Y|A_1,A_2,O] = \beta_0 + \beta_1 A_1 + \beta_2 A_2 + \beta_3 A_1 A_2 + \gamma O$. Weighted regression with robust variance estimators can be performed in standard software (e.g., SAS, R, Stata) to estimate the regression coefficients and linear combinations of the coefficients, and Wald tests can be used to estimate and compare DTRs. An example using this type of regression is shown at the end of the chapter.

To further personalize or tailor DTRs, Q-learning or other reinforcement learning methods can be used. Q-learning is a type of reinforcement learning, a topic taken from the machine learning playbook, wherein an optimal sequence of "actions" is selected by maximizing the probability of success at each stage. These methods find baseline and time-varying characteristics that are associated with DTRs so that guidelines can be further personalized to include participant and disease characteristics beyond the intermediate tailoring variable (i.e., like subgroup analyses in standard trials to define the group of subjects for which each DTR is best). Publications discussing Q-learning and its application in the analysis of SMART data with binary or continuous outcomes are numerous [8,9,81–90]. Some of these methods are available through packages including *DTRreg, dyntxregime, iqlearn, qlearn,* and *qlaci* in R (The Comprehensive R Archive Network, CRAN https://cran.r-project.org/).

Estimators for time-to-event outcomes focus on comparing DTRs under various conditions based on mean survival time or the survival distribution [3,60,91–97]. Kaplan–Meier plots can be drawn for each DTR and all, or pairs, of DTRs can be compared. The R package *DTR* estimates and compares DTRs with survival outcomes for SMARTs having a similar design to design II. This package computes estimates and standard errors of the survival function for DTRs at observed event times [3], the weighted risk set estimator of survival function at observed event times [92], compares survival distributions of DTRs using Wald-type tests [3,92] and weighted log-rank tests [93], fits a generalized Cox model and compares survival distributions of DTRs adjusting for covariates [97], and computes and compares estimates for the cumulative hazard ratios between two DTRs [98]. Most of these survival methods also require inverse-probability-of-treatment weighting as mentioned above, but do not replicate the data as in weighted and replicated regression for continuous and binary outcomes. Q-learning has also been applied to censored survival data [99], but has not yet been implemented within a publicly available software package. Finally, various parametric and non-parametric Bayesian model-based estimation approaches exist for time-to-event outcomes [36,37,100,101].

19.8 Illustrations/Data Analysis Examples

We present several examples below to provide illustrations of the topics discussed in this chapter. Each example is hypothetical and chosen to illustrate specific SMART sample size and analytic methods. Thus, the examples may not be clinically relevant and should not be used for clinical decision-making.

EXAMPLE 19.1

Clinical Scenario: Consider patients with Stage IV rectal cancer having a single liver metastasis.

Study Objectives: The primary objective of this pilot study is to evaluate the feasibility of conducting a SMART design in this setting prior to potentially conducting a fully-powered SMART.

SMART Design: Refer to Figure 19.1 design II, with the one caveat in this scenario that patients with no evidence of progressive disease (PD; i.e., responders) are re-randomized at the second treatment stage, whereas patients who have PD (i.e., non-responders) are not.

In the first treatment stage, patients are randomized 1:1 to one of two treatments: Treatment A (intravenous chemotherapy) or Treatment B (chemoradiation to the primary rectal cancer tumor site). Patients are evaluated for PD at Week 8. Patients with no evidence of PD (i.e., Responders) are re-randomized equally to Treatment D/G (partial, sequential surgery to remove the single liver tumor metastasis only) or Treatment E/H (full surgery to remove both the single liver tumor metastasis and the primary rectal tumor). Those patients who have PD at Week 8 receive Treatment C/F (standard of care) and are not re-randomized. In this clinical setting, second-stage treatments are the same across all patients regardless of their first-line treatment and are determined solely by PD status at Week 8.

DTRs: There are four embedded DTRs within this SMART as follows:

1. Treatment A (chemotherapy) followed by Treatment C (standard of care) for patients with PD at Week 8 and Treatment D (partial surgical resection) for patients without PD at Week 8 (A,C,D).
2. Treatment A (chemotherapy) followed by Treatment C (standard of care) for patients with PD at Week 8 and Treatment E (full surgical resection) for patients without PD at Week 8 (A,C,E).

3. Treatment B (chemoradiation) followed by Treatment F (standard of care) for patients with PD at Week 8 and Treatment G (partial surgical resection) for patients without PD at Week 8 (B,F,G).
4. Treatment B (chemoradiation) followed by Treatment F (standard of care) for patients with PD at Week 8 and Treatment H (full surgical resection) for patients without PD at Week 8 (B,F,H).

Tailoring Variables/Intermediate Outcomes: Presence of PD at Week 8 (yes/no) is the tailoring variable used within this SMART design.

Sample size estimate: The web-based sample size calculator designed for pilot studies (available at https://moloque1227.shinyapps.io/SMARTsize/) ensures that a minimum number of patients are observed within each of the six subgroups or treatment pathways. To clarify, a subgroup or pathway is the set of patients who would be treated with each of the following treatments: A-C, A-D, A-E, B-F, B-G, and B-H. A subgroup differs from a DTR in that a DTR is defined as a first-stage treatment followed by two options for the second-stage treatment that depend on whether the patient does or does not have PD at Week 8 (i.e., the tailoring variable).

In the online calculator, Design B is selected with the understanding that, in this example, responders to first-stage therapy (i.e., those patients without PD at Week 8) are randomized to second-stage therapy, whereas patients with PD at Week 8 (non-responders) are not. Based upon a literature review, the probability of a response at the end of the first stage is expected to be 60% for both Treatment A and Treatment B. Although the sample size calculator requests "anticipated rate of non-response at the end of first-stage intervention," the anticipated rate of response at the end of first-stage is provided: 0.60. The minimum number of participants in each subgroup is specified to be 3. The "minimum probability to observe this many participants (3) in each subgroup" is specified as 0.80. Based on the inputs provided, the estimated sample size for the pilot SMART is calculated as 28 subjects.

EXAMPLE 19.2

Clinical Scenario: After conducting the feasibility study described in Example 19.1, a confirmatory trial in patients with stage IV rectal cancer with a single liver metastasis is undertaken.

Study Objectives: The primary objective of this trial is to compare overall survival (OS) between the two most divergent DTRs: (A,C,D) versus (B,F,H) (i.e., chemotherapy followed by standard of care for patients with PD and partial surgical resection for patients without PD versus chemoradiation followed by standard of care for patients with PD and full surgical resection for patients without PD).

SMART Design, DTRs, Tailoring Variables: The SMART design, DTRs, and tailoring variables/intermediate outcomes are the same as described in Example 19.1.

Sample size estimate: To compare two DTRs based on the weighted log-rank test for a time-to-event outcome, the applet available at http://methodologymedia.psu.edu/logranktest/samplesize is used. Note that this applet can be used only with SMART design II and, additionally, that it assumes the re-randomization of non-responders rather than responders. For 1:1 randomization at both the first and second randomization stages, parameters are set as follows: $p_1 = p_{21} = p_{22} = 0.5$. In this example, the strategies compared are specified as "11, 22" to compare DTRs (A,C,D) versus (B,F,H). Type I error is set so that $\alpha = 0.05$ and type II error is set so that $\beta = 0.20$. The hazard ratio that reflects the minimum clinical effect size of interest is set to $\xi = 1.2$. Finally, the probability of observing an event, which in this case is a patient's death, during the study period is set to

$P_{obs} = 0.75$. Based on these assumptions, the required total sample size is 59 patients.

Dataset: The dataset and associated code for this hypothetical SMART are available on the book's companion website https://www.crcpress.com//9781138083776.

Analysis: The *DTR* package [102] in R is used for the comparison of DTRs with a time-to-event outcome.

The original data from this hypothetical SMART include the following variables: Stage 1 and Stage 2 treatment assignments, response status after Stage 1, survival time, and an event indicator. Survival time (S.time) is reported in months. The event indicator (Event) identifies subjects for whom death was observed (1) or censored (0). Response status (Resp1) is coded as 0 for patients who experienced PD and 1 otherwise. Stage 1 treatment (rx1) is entered as 0 for chemotherapy and 1 for chemoradiation and Stage 2 treatment (rx2) is coded as 0 for partial surgical resection, 1 for full surgical resection, or NA if the patient was not re-randomized at the second treatment stage.

First, the original dataset is reformatted to mirror the notation used in the *DTR* package to employ the contrast _ logrank function. There are 38 observations for which NA is recorded for the second-stage treatment (rx2) on account of either censoring or response status following Stage 1. In order for the contrast _ logrank function to process these data correctly, values recorded as NA for the Stage 2 treatment (rx2) must be recoded to 0. Also, note that a vector of 0s is added for column "TR." "TR" is an optional vector indicating the time to response following Stage 1 therapy but was not captured based on our study design.

The test statistics and uncorrected p-values for the omnibus test (i.e., any difference in OS across all DTRs) and the pairwise DTR comparisons obtained from contrast _ logrank are provided in Table 19.3. These data suggest that there is no discernible difference in overall survival across the four DTRs ($p = 0.46$); however, at an uncorrected significance level of 0.05, evidence of differences in overall survival between DTRs (A,C,D) and (A,C,E), (A,C,D) and (B,F,G), and (A,C,D) and (B,F,H) are observed. The survival distributions for the four embedded DTRs: (A,C,D), (A,C,E), (B,F,G), and (B,F,H) are illustrated in Figure 19.2.

EXAMPLE 19.3

Clinical Scenario: Consider women with newly diagnosed, advanced, estrogen-receptor positive (ER+), and HER2 negative (HER2–) breast cancer.

TABLE 19.3

Results from R Package *DTR* and Function Contrast _ Logrank for Comparing Overall Survival in a SMART Like Design II from Figure 19.1 across All DTRs and for All Pairwise Comparisons

	H_0	(Standardized) Test Statistic	Degrees of Freedom	*p*-value
1	A1B1 = A1B2 = A2B1 = A2B2	1.5385	2	0.463
2	A1B1 = A1B2	2.0204	1	0.043
3	A1B1 = A2B1	2.5492	1	0.011
4	A1B1 = A2B2	2.4511	1	0.014
5	A1B2 = A2B1	0.8466	1	0.397
6	A1B2 = A2B2	0.5762	1	0.565
7	A2B1 = A2B2	−0.6824	1	0.495

Note: The notation is given by the package such that in this example, A1B1 denotes DTR (A,C,D), A1B2 denotes (A,C,E), A2B1 denotes (B,F,G), and A2B2 denotes (B,F,H).

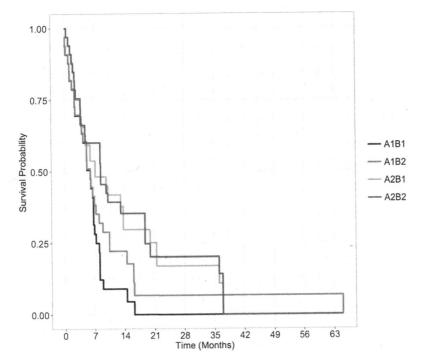

FIGURE 19.2
Kaplan–Meier plot showing the overall survival for all embedded DTRs within the hypothetical SMART considered in Example 19.2. The notation is given by the package such that in this example A1B1 denotes DTR (A,C,D), A1B2 denotes (A,C,E), A2B1 denotes (B,F,G), and A2B2 denotes (B,F,H).

Study Objectives: The primary objective of the SMART is to evaluate and compare the proportion of women alive at 12 months between the two most divergent DTRs: (A,C,D) versus (B,F,G).

SMART Design: Refer to Figure 19.1 design III. Treatment-naive women consenting to the study are randomized to first-line therapy consisting of either Treatment A (letrozole + palbociclib) or Treatment B (letrozole alone), given in 28-day cycles. Ongoing response assessments are performed every 8 weeks for 6 months. In the absence of progressive disease (PD) at 6 months, patients remain on their first-line treatment regimen. For patients randomized to Treatment A (letrozole + palbociclib), patients with evidence of PD at any time point up to 6 months are re-randomized to either Treatment D (fulvestrant alone) or Treatment E (fulvestrant + everolimus). Patients with PD following Treatment B (letrozole alone) would be assigned to Treatment G (fulvestrant + palbociclib).

DTRs: There are three embedded DTRs within this SMART.
1. Treatment A (letrozole + palbociclib) followed by Treatment D (fulvestrant) for patients with PD and continuing first-line treatment for patients maintaining stable disease (A,C,D)
2. Treatment A (letrozole + palbociclib) followed by Treatment E (fulvestrant + everolimus) for patients with PD and continuing first-line treatment for patients maintaining stable disease (A,C,E)
3. Treatment B (letrozole alone) followed by Treatment G (fulvestrant + palbociclib) for patients with PD and continuing first-line treatment for patients maintaining stable disease (B,F,G)

Tailoring Variables/Intermediate Outcomes: Patients are assessed for response to first-line therapy a minimum of every 8 weeks for up to 6 months. At PD, patients begin their second-stage therapy. If there is no PD at 6 months, patients are declared responders and continue with their initial treatment. Only patients initially randomized to Treatment A (letrozole + palbociclib) are re-randomized to second-line therapy following PD based on available and ethical treatment options.

Sample size estimate: The web-based calculator designed for studies with a binary outcome available at https://sites.google.com/a/umich.edu/kidwell/home/tools-for-design-and-analysis is used to calculate sample size comparing the DTRs defined by (A,C,D) and (B,F,G). Based on previous literature [103,104], the proportion of women who are alive at 12 months following DTRs (A,C,D) and (B,F,G) is approximately 0.85 and 0.60, respectively. The assumed probability of response to first-stage therapy is 0.70 for treatment A and 0.40 for treatment B. It is assumed that 1:1 randomizations will be performed at each treatment stage for applicable patients. Further assuming a two-sided test, type I error rate of 0.05, and a desired power of 0.80, the sample size is estimated to be 128 patients. Assuming a 10% dropout rate, a total sample size is chosen of $N = 140$.

Dataset: The dataset and associated code for this hypothetical SMART are available on the book's companion website. This data is illustrated using SAS software (SAS Institute Inc., Cary, NC, USA), although it would work equally well in R or other software.

Analysis: The data have been pre-processed to include a binary outcome indicating whether the patient was alive 12 months following initiation of first-stage treatment ("Alive12mo" = 1), which was determined using survival and censoring times collected during the study. Other variables included in the dataset are patient ID ("id"), first-stage treatment to which the patient was randomized ("rx1"; rx1 = 1 represents Treatment A), a binary indicator of whether the patient experienced a response (i.e., no PD) following first-stage treatment ("Resp' = 1), the second-stage treatment ("rx2") received by the patient in the event of PD following letrozole + palbociclib (rx2 = 1 for fulvestrant), if applicable.

The data must be restructured in order to use weighted and replicated estimation in standard software [4,59]. Although 1:1 randomization is performed for the first-stage treatment, only patients randomized to Treatment A (letrozole + palbociclib) and who have experienced PD are re-randomized to second-stage treatment. Therefore, a weighting step is necessary to account for under- or over-representation of certain groups in the analysis. Patients with PD following Treatment A (letrozole + palbociclib) are randomized twice, so their weight = 1/[(0.5)*(0.5)] = 4. All other patients, including those randomized to Treatment B (letrozole alone) and those patients who have no PD following Treatment A (letrozole + palbociclib), are randomized only once, so their weight = 1/0.5 = 2.

The replication step duplicates observations consistent with two DTRs. Specifically, patients who are randomized to Treatment A (letrozole + palbociclib) and have no PD are consistent with two DTRs: (A,C,D) and (A,C,E). Thus, these observations are duplicated; one observation is assigned to Treatment D ("rx2" = 1) and the second is assigned to Treatment E ("rx2" = 0). Finally, a new interaction variable ("newrx2") is created to represent the interaction effect between the first-stage treatment ("rx1") and the second-stage treatment ("rx2"), which will be present, by design, only in the case that a patient is randomized first to Treatment A. This variable is 0 for those randomized to Treatment B.

The analysis implements the following model:

$$\text{logit}(p) = \beta_0 + \beta_1{}^*\text{rx1} + \beta_2{}^*\text{newrx2}$$

TABLE 19.4

Results from Analysis of Example 19.3 Implementing Weighted and Replicated Regression Model to Estimate the Probability of Overall Survival at 12 Months from the 3 Embedded DTRs within the Hypothetical SMART Design

	Analysis of GEE Parameter Estimates					
	Empirical Standard Error Estimates					
Coefficient	Estimate	Standard Error	95% Confidence Limits		Z	Pr > |Z|
β_0	0.4055	0.2440	−0.0727	0.8836	1.66	0.0965
β_1	1.1239	0.4679	0.2068	2.0410	2.40	0.0163
β_2	0.6190	0.5668	−0.4920	1.7300	1.09	0.2748

Abbreviation: GEE, Generalized Estimating Equation.

TABLE 19.5

The Estimated Probability of Overall Survival at 12 Months from the 3 Embedded DTRs from Example 19.3 Implementing Weighted and Replicated Regression

Label	Proportion Alive at 12 Months	Odds Ratio	p-value
(A,C,D)	0.90	–	–
(A,C,E)	0.82	–	–
(B,F,G)	0.60	–	–
(A,C,D) to (B,F,G)		5.71	0.002

in PROC GENMOD where p represents the probability of survival at 12 months, rx1 is an indicator of the assigned first-stage therapy, and newrx2 represents the second-stage treatment to which non-responders after Stage 1 (i.e., patients with PD) were randomized. Parameter estimates with standard errors and associated 95% confidence intervals are presented in Table 19.4. The intercept (β_0) represents the treatment effect or the log odds of the proportion alive at 12 months for DTR (B,F,G), β_1 represents the difference in the log odds of the proportion alive at 12 months associated with DTR (A,C,E) compared to (B,F,G), and β_2 represents the difference in the log odds of the proportion alive at 12 months between DTRs (A,C,D) and (A,C,E). The probability of survival at 12 months for each of the three DTRs is provided in Table 19.5. In summary, the probability of survival at 12 months for DTR (A,C,D) (letrozole + palbociclib, letrozole + palbociclib, fulvestrant) is 90% whereas the probability of survival for DTR (B,F,G) is estimated to be 60%. There is a statistically significant difference in survival between the two DTRs of interest, $p = 0.002$.

19.9 Conclusion

To advance treatment in oncology and to tackle questions concerning the sequential and personalized use of treatment, SMART designs provide distinct advantages. SMART designs specifically address multiple critical gaps, including the ability to efficiently identify "interactions, duration, sequencing and optimal combinations of therapy for improved individualization of treatment" [105]. SMARTs allow for robust evidence of the sequential, dynamic decision-making that occurs in clinical practice.

SMARTs have the potential to efficiently use participants to provide more information and answer more clinically relevant questions than standard one-stage trials. To help investigators overcome a paradigm shift from standard RCT designs to more novel designs, like a SMART design, investigators should consider a pilot study or, with a statistician, perform simulation studies to better understand the design, methods, and potential benefits. There is much opportunity for the use of SMART designs in the adjuvant, neoadjuvant, and survivorship settings in many cancer types.

References

1. Murphy SA. Optimal dynamic treatment regimes. *J R Stat Soc Series B, Stat Methodol.* 2003;65(2):331–55.
2. Lavori PW, Dawson R. Dynamic treatment regimes: Practical design considerations. *Clin Trials.* 2004;1(1):9–20.
3. Lunceford JK, Davidian M, Tsiatis AA. Estimation of survival distributions of treatment policies in two-stage randomization designs in clinical trials. *Biometrics.* 2002;58 (1):48–57.
4. Nahum-Shani I, Qian M, Almirall D, Pelham WE, Gnagy B, Fabiano GA, Waxmonsky JG, Yu J, Murphy SA. Experimental design and primary data analysis methods for comparing adaptive interventions. *Psychol Methods.* 2012a;17(4):457–77.
5. Dawson R, Lavori PW. Placebo-free designs for evaluating new mental health treatments: The use of adaptive treatment strategies. *Stat Med.* 2004;23(21):3249–62.
6. Lavori PW, Dawson R. A design for testing clinical strategies: Biased adaptive within-subject randomization. *J R Stat Soc. Series A.* 2000;163(1):29–38.
7. Thall PF, Millikan RE, Sung HG. Evaluating multiple treatment courses in clinical trials. *Stat Med.* 2000;19(8):1011–28.
8. Murphy SA. An experimental design for the development of adaptive treatment strategies. *Stat Med.* 2005;24(10):1455–81.
9. Cheung YK, Chakraborty B, Davidson KW. Sequential multiple assignment randomized trial (SMART) with adaptive randomization for quality improvement in depression treatment program. *Biometrics.* 2015;71(2):450–59.
10. Gunlicks-Stoessel M, Mufson L, Westervelt A, Almirall D, Murphy S. A pilot SMART for developing an adaptive treatment strategy for adolescent depression. *J Clin Child Adolesc Psychol.* 2016;5345(4):480–94.
11. Kilbourne AM, Almirall D, Eisenberg D et al. Protocol: Adaptive implementation of effective programs trial (ADEPT): Cluster randomized SMART trial comparing a standard versus enhanced implementation strategy to improve outcomes of a mood disorders program. *Implement Sci.* 2014;9(1):132.
12. August GJ, Piehler TF, Bloomquist ML. Being 'SMART' about adolescent conduct problems prevention: Executing a SMART pilot study in a juvenile diversion agency. *J Clin Child Adolesc Psychol.* 2016;53 45(4):495–509.
13. Shortreed SM, Moodie EEM. Estimating the optimal dynamic antipsychotic treatment regime: Evidence from the sequential multiple-assignment randomized clinical antipsychotic trials of intervention and effectiveness schizophrenia study. *J R Stat Soc. Series C, Appl Stat.* 2012;61(4):577–99.

14. Joss RA, Alberto P, Bleher EA, Ludwig C, Siegenthaler P, Martinelli G, Sauter C, Schatzmann E, Senn HJ. Combined-modality treatment of small-cell lung cancer: Randomized comparison of three induction chemotherapies followed by maintenance chemotherapy with or without radiotherapy to the chest. Swiss group for clinical cancer Research (SAKK). *Ann Oncol.* 1994;5(10):921–28.

15. Stone RM, Berg DT, George SL, Dodge RK, Paciucci PA, Schulman P, Lee EJ, Moore JO, Powell BL, Schiffer CA. Granulocyte-macrophage colony-stimulating factor after initial chemotherapy for elderly patients with primary acute myelogenous leukemia. Cancer and leukemia group B. *N Engl J Med.* 1995;332(25):1671–77.

16. Stone RM, Berg DT, George SL et al. Postremission therapy in older patients with de novo acute myeloid leukemia: A randomized trial comparing mitoxantrone and intermediate-dose cytarabine with standard-dose cytarabine. *Blood.* 2001;98(3):548–53.

17. Estey EH, Thall PF, Pierce S, Cortes J, Beran M, Kantarjian H, Keating MJ, Andreeff M, Freireich E. Randomized phase II study of fludarabine + cytosine arabinoside + idarubicin +/− all-trans retinoic acid +/− granulocyte colony-stimulating factor in poor prognosis newly diagnosed acute myeloid leukemia and myelodysplastic syndrome. *Blood.* 1999;93(8):2478–84.

18. Tummarello D, Mari D, Graziano F, Isidori P, Cetto G, Pasini F, Santo A, Cellerino R. A randomized, controlled phase III study of cyclophosphamide, doxorubicin, and vincristine with etoposide (CAV-E) or teniposide (CAV-T), followed by recombinant interferon-alpha maintenance therapy or observation, in small cell lung carcinoma patients with complete responses. *Cancer.* 1997;80(12):2222–29.

19. Matthay KK, Villablanca JG, Seeger RC et al. Treatment of high-risk neuroblastoma with intensive chemotherapy, radiotherapy, autologous bone marrow transplantation, and 13-Cis-Retinoic acid. Children's cancer group. *N Engl J Med.* 1999;341(16):1165–73.

20. Thomas X, Raffoux E, de Botton S et al. Effect of priming with granulocyte-macrophage colony-stimulating factor in younger adults with newly diagnosed acute myeloid leukemia: A trial by the acute leukemia French association (ALFA) group. *Leukemia.* 2007;21(3):453–61.

21. Matthay KK, Reynolds CP, Seeger RC, Shimada H, Stanton Adkins E, Haas-Kogan D, Gerbing RB, London WB, Villablanca JG. Long-term results for children with high-risk neuroblastoma treated on a randomized trial of myeloablative therapy followed by 13-Cis-Retinoic acid: A children's oncology group study. *J Clini Oncol.* 2009;27(7):1007–13.

22. Habermann TM, Weller EA, Morrison VA et al. Rituximab-CHOP versus CHOP alone or with maintenance rituximab in older patients with diffuse large B-cell lymphoma. *J Clin Oncol.* 2006;24(19):3121–27.

23. van Oers MHJ, Klasa R, Marcus RE et al. Rituximab maintenance improves clinical outcome of relapsed/resistant follicular non-Hodgkin lymphoma in patients both with and without rituximab during induction: Results of a prospective randomized phase 3 intergroup trial. *Blood.* 2006;108(10):3295–301.

24. Mateos M-V, Oriol A, Martínez-López J et al. Bortezomib, melphalan, and prednisone versus bortezomib, thalidomide, and prednisone as induction therapy followed by maintenance treatment with bortezomib and thalidomide versus bortezomib and prednisone in elderly patients with untreated multiple myeloma: A andomized trial. *Lancet Oncol.* 2010;11(10):934–41.

25. Thall PF, Logothetis C, Pagliaro LC, Wen S, Brown MA, Williams D, Millikan RE. Adaptive therapy for androgen-independent prostate cancer: A randomized selection trial of four regimens. *J Natl Cancer Inst.* 2007;99(21):1613–22.

26. Auyeung SF, Long Q, Royster EB, Murthy S, McNutt MD, Lawson D, Miller A, Manatunga A, Musselman DL. Sequential multiple-assignment randomized trial design of neurobehavioral treatment for patients with metastatic malignant melanoma undergoing high-dose interferon-alpha therapy. *Clin Trials.* 2009;6(5):480–90.

27. Forstpointner R, Unterhalt M, Dreyling M et al. Maintenance therapy with rituximab leads to a significant prolongation of response duration after salvage therapy with a combination of rituximab, fludarabine, cyclophosphamide, and mitoxantrone (R-FCM) in patients with recurring and refractory follicular and mantle cell lymphomas: Results of a prospective randomized study of the German Low Grade Lymphoma Study Group (GLSG). *Blood.* 2006;108(13):4003–8.

28. NIH U.S. National Library of Medicine. 2017b. Sequential Two-Agent Assessment in Renal Cell Carcinoma Therapy: The START Trial. Clinicaltrials.gov. August 18, 2017. https://clinicaltrials.gov/ct2/show/NCT01217931?term=sequential+randomization&type=Intr&cond=.

29. NIH National Library of Medicine. 2017. A Dynamic Allocation Modular Sequential Trial of Approved and Promising Therapies in Men with Metastatic Castrate Resistant Prostate Cancer (DynaMO). Clinicaltrials.gov. October 23, 2017. https://clinicaltrials.gov/ct2/show/NCT0270362 3?term=sequential+randomization&type=Intr&cond=Cancer&locn=MD+Anderson&draw= 1&rank=5.

30. NIH U.S. National Library of Medicine. 2017a. SMART for Smoking Cessation in Lung Cancer Screening. Clinicaltrials.gov. June 9, 2017. https://clinicaltrials.gov/ct2/show/NCT02597491.

31. Wang L, Rotnitzky A, Lin X, Millikan RE, Thall PF. Evaluation of viable dynamic treatment regimes in a sequentially randomized trial of advanced prostate cancer. *J Am Stat Assoc.* 2012;107(498):493–508.

32. Almirall D, Compton SN, Gunlicks-Stoessel M, Duan N, Murphy SA. Designing a pilot sequential multiple assignment randomized trial for developing an adaptive treatment strategy. *Stat Med.* 2012;31(17):1887–1902.

33. Shortreed SM, Laber E, Stroup TS, Pineau J, Murphy SA. A multiple imputation strategy for sequential multiple assignment randomized trials. *Stat Med.* 2014;33(24):4202–14.

34. Wallace MP, Moodie EEM, Stephens DA. SMART thinking: A review of recent developments in sequential multiple assignment randomized trials. *Curr Epidemiol Rep.* 2016;3(3):225–32.

35. Murphy SA, Bingham D. Screening experiments for developing dynamic treatment regimes. *J Am Stat Assoc.* 2009;104(458):391–408.

36. Thall PF, Wooten LH, Logothetis CJ, Millikan RE, Tannir NM. Bayesian and frequentist two-stage treatment strategies based on sequential failure times subject to interval censoring. *Stat Med.* 2007;26(26):4687–4702.

37. Wathen JK, Thall PF. Bayesian adaptive model selection for optimizing group sequential clinical trials. *Stat Med.* 2008;27(27):5586–604.

38. Lee JPF. Thall YJ, Müller P. Bayesian dose-finding in two treatment cycles based on the joint utility of efficacy and toxicity. *J Am Stat Assoc.* 2015;110(510):711–22.

39. Stadler WM, Rosner G, Small E, Hollis D, Rini B, Zaentz SD, Mahoney J, Ratain MJ. Successful implementation of the randomized discontinuation trial design: An application to the study of the putative antiangiogenic agent carboxyaminoimidazole in renal cell carcinoma—CALGB 69901. *J Clin Orthod.* 2005;23(16):3726–32.

40. Ratain MJ, Eisen T, Stadler WM et al. Phase II placebo-controlled randomized discontinuation trial of sorafenib in patients with metastatic renal cell carcinoma. *J Clin Oncol.* 2006;24(16):2505–12.

41. Minckwitz G von, Blohmer JU, Costa SD et al. Response-guided neoadjuvant chemotherapy for breast cancer. *J Clin Oncol.* 2013;31(29):3623–30.

42. Smerage JB, Barlow WE, Hortobagyi GN et al. Circulating tumor cells and response to chemotherapy in metastatic breast cancer: SWOG S0500. *J Clin Oncol.* 2014;32(31):3483–89.

43. Naume B, Synnestvedt M, Falk RS et al. Clinical outcome with correlation to disseminated tumor cell (DTC) status after DTC-guided secondary adjuvant treatment with docetaxel in early breast cancer. *J Clin Orthod.* 2014;32(34):3848–57.

44. Almirall D, Compton SN, Rynn MA, Walkup JT, Murphy SA. SMARTer discontinuation trial designs for developing an adaptive treatment strategy. *J Child Adolesc Psychopharmacol.* 2012;22(5):364–74.

45. Sargent DJ, Conley BA, Allegra C, Collette L. Clinical trial designs for predictive marker validation in cancer treatment trials. *J Clin Oncol.* 2005;23(9):2020–27.

46. Freidlin B, McShane LM, Korn EL. Randomized clinical trials with biomarkers: Design issues. *J Natl Cancer Inst.* 2010;102(3):152–60.

47. Freidlin B, Korn EL. Biomarker enrichment strategies: Matching trial design to biomarker credentials. *Nat Rev Clin Oncol.* 2014;11(2):81–90.

48. Wolbers M, Helterbrand JD. Two-stage randomization designs in drug development. *Stat Med.* 2008;27(21):4161–74.

49. Lei H, Nahum-Shani I, Lynch K, Oslin D, Murphy SA. A 'SMART' design for building individualized treatment sequences. *Ann Rev Clin Psychol.* 2012;8:21–48.

50. Lavori PW, Dawson R. Adaptive treatment strategies in chronic disease. *Ann Rev Med.* 2008;59:443–53.

51. Moodie EEM, Karran JC, Shortreed SM. A case study of SMART attributes: A qualitative assessment of generalizability, retention rate, and trial quality. *Trials.* 2016;17(1):242.

52. Ko JH, Wahed AS. Up-front versus sequential randomizations for inference on adaptive treatment strategies. *Stat Med.* 2012;31(9):812–30.

53. Kosorok MR, Moodie EEM. *Adaptive Treatment Strategies in Practice: Planning Trials and Analyzing Data for Personalized Medicine.* SIAM, 2015.

54. Almirall D, Nahum-Shani I, Sherwood NE, Murphy SA. Introduction to SMART designs for the development of adaptive interventions: With application to weight loss research. *Transl Behav Med.* 2014;4(3):260–74.

55. Oetting AI, Levy JA, Weiss RD, Murphy SA. Statistical methodology for a SMART design in the development of adaptive treatment strategies. In *Causality and Psychopathology: Finding the Determinants of Disorders and Their Cures.* Arlington, VA: American Psychiatric Publishing, Inc, 2011.

56. Kidwell KM. SMART designs in cancer research: Past, present, and future. *Clin Trials.* 2014;11(4):445–56.

57. Ertefaie A, Wu T, Lynch KG, Nahum-Shani I. Identifying a set that contains the best dynamic treatment regimes. *Biostatistics* 2016;17(1):135–48.

58. Kim H, Supervisors: Ionides E, Almirall D. n.d. A Sample Size Calculator for SMART Pilot Studies. https://www.siam.org/students/siuro/vol9/S01405.pdf.

59. Kidwell KM, Wahed AS. Weighted log-rank statistic to compare shared-path adaptive treatment strategies. *Biostatistics* 2013;14(2):299–312.

60. Feng W, Wahed AS. Supremum weighted log-rank test and sample size for comparing two-stage adaptive treatment strategies. *Biometrika.* 2008; 95(3):695–707.

61. Feng W, Wahed AS. Sample size for two-stage studies with maintenance therapy. *Stat Med.* 2009;28(15):2028–41.

62. Li Z, Murphy SA. *Sample Size Calculation for Comparing Two-Stage Treatment Strategies with Censored Data.* Ann Arbor, MI: Dept of Statistics. pdfs.semanticscholar.org. 2009. https://pdfs.semanticscholar.org/6ab8/660f8b495e4657ba540b3fb70dc8ca4c4e31.pdf.

63. Murphy SA. A generalization error for Q-Learning. *J Mach Learn Res.* 2005;6(July):1073–97.

64. Ogbagaber SB, Karp J, Wahed AS. Design of sequentially randomized trials for testing adaptive treatment strategies. *Stat Med.* 2016;35(6):840–58.

65. Dawson R, Lavori PW. Efficient design and inference for multistage randomized trials of individualized treatment policies. *Biostatistics.* 2012;13(1):142–52.

66. Dawson R, Lavori PW. Sample size calculations for evaluating treatment policies in multi-stage designs. *Clin Trials.* 2010;7(6):643–52.

67. NeCamp T, Kilbourne A, Almirall D. Comparing cluster-level dynamic treatment regimens using sequential, multiple assignment, randomized trials: Regression estimation and sample size considerations. *Stat Methods Med Res.* 2017;26(4):1572–89.

68. Robins JM, Hernán MA, Brumback B. Marginal structural models and causal inference in epidemiology. *Epidemiology* 2000;11(5):550–60.

69. Murphy SA, van der Laan MJ, Robins JM, CPPRG. Marginal mean models for dynamic regimes. *J Am Stat Assoc.* 2001;96(456):1410–23.

70. Moodie EEM, Richardson TS, Stephens DA. Demystifying optimal dynamic treatment regimes. *Biometrics.* 2007;63(2):447–55.

71. Bembom O, van der Laan MJ. Analyzing sequentially randomized trials based on causal effect models for realistic individualized treatment rules. *Stat Med.* 2008;27(19):3689–716.

72. Robins JM. Optimal structural nested models for optimal sequential decisions. In *Proceedings of the Second Seattle Symposium in Biostatistics*, 189–326. Lecture Notes in Statistics. Springer, New York, NY, 2004.

73. Lavori PW, Dawson R. Improving the efficiency of estimation in randomized trials of adaptive treatment strategies. *Clin Trials.* 2007;4(4):297–308.

74. Krakow EF, Hemmer M, Wang T et al. Tools for the precision medicine era: How to develop highly personalized treatment recommendations from cohort and registry data using Q-Learning. *Am J Epidemiol.* 2017;186(2):160–72.

75. Li Z. Comparison of adaptive treatment strategies based on longitudinal outcomes in sequential multiple assignment randomized trials. *Stat Med.* 2017;36(3):403–15.

76. Lu X, Lynch KG, Oslin DW, Murphy S. Comparing treatment policies with assistance from the structural nested mean model. *Biometrics.* 2016;72(1):10–19.

77. Orellana L, Rotnitzky A, Robins JM. Dynamic regime marginal structural mean models for estimation of optimal dynamic treatment regimes, part I: Main content. *Int J Biostat.* 2010;6(2):Article 8.

78. Zhang B, Tsiatis AA, Laber EB, Davidian M. A robust method for estimating optimal treatment regimes. *Biometrics.* 2012;68(4):1010–18.

79. Davidian M, Tsiatis AB, Laber E. *Dynamic Treatment Regimes. Cancer Clinical Trials: Current and Controversial Issues in Design and Analysis.* CRC Press, 2016, 409.

80. Chakraborty B, Moodie EEM. *Statistical Methods for Dynamic Treatment Regimes: Reinforcement Learning, Causal Inference, and Personalized Medicine.* Statistics for Biology and Health. Springer New York, 2013.

81. Nahum-Shani I, Qian M, Almirall D, Pelham WE, Gnagy B, Fabiano GA, Waxmonsky JG, Yu J, Murphy SA. Q-Learning: A data analysis method for constructing adaptive interventions. *Psychol Methods.* 2012b;17(4):478–94.

82. Chakraborty B, Laber EB, Zhao Y-Q. Inference about the expected performance of a data-driven dynamic treatment regime. *Clin Trials.* 2014;11(4):408–17.

83. Zhao Y, Kosorok MR, Zeng D. Reinforcement learning design for cancer clinical trials. *Stat Med.* 2009;28(26):3294–315.

84. Zhao Y, Zeng D, Socinski MA, Kosorok MR. Reinforcement learning strategies for clinical trials in nonsmall cell lung cancer. *Biometrics.* 2011;67(4):1422–33.

85. Zhao Y, Zeng DRush AJ, Kosorok MR. Estimating individualized treatment rules using outcome weighted learning. *J Am Stat Assoc.* 2012;107(449):1106–18.

86. Song R, Wang W, Zeng D, Kosorok MR. Penalized Q-Learning for dynamic treatment regimens. *Stat Sin.* 2015;25(3):901–20.

87. Lu W, Zhang HH, Zeng D. Variable selection for optimal treatment decision. *Stat Methods Med Res.* 2013;22(5):493–504.

88. Zhang B, Tsiatis AA, Laber EB, Davidian M. Robust estimation of optimal dynamic treatment regimes for sequential treatment decisions. *Biometrika.* 2013;100(3).

89. Schulte PJ, Tsiatis AA, Laber EB, Davidian M. Q- and A-Learning methods for estimating optimal dynamic treatment regimes. *Stat Sci.* 2014;29(4):640–61.

90. Wallace MP, Moodie EEM. Doubly-robust dynamic treatment regimen estimation via weighted least squares. *Biometrics.* 2015;71(3):636–44.

91. Miyahara S, Wahed AS. Weighted Kaplan–Meier estimators for two-stage treatment regimes. *Stat Med.* 2010;29(25):2581–91.

92. Guo X, Tsiatis A. A weighted risk set estimator for survival distributions in two-stage randomization designs with censored survival data. *Int J Biostat.* 2005;1(1).

93. Kidwell KM, Seewald NJ, Tran Q, Kasari C, Almirall D. Design and analysis considerations for comparing dynamic treatment regimens with binary outcomes from sequential multiple assignment randomized trials. *J Appl Stat.* 2017;45:1628–51.

94. Wahed AS, Tsiatis AA. Semiparametric efficient estimation of survival distributions in two-stage randomisation designs in clinical trials with censored data. *Biometrika.* 2006;93(1):163–77.

95. Wahed AS, Tsiatis AA. Optimal estimator for the survival distribution and related quantities for treatment policies in two-stage randomization designs in clinical trials. *Biometrics.* 2004;60(1):124–33.

96. Geng Y, Zhang HH, Lu W. On optimal treatment regimes selection for mean survival time. *Stat Med*. 2015;34(7):1169–84.
97. Tang X, Wahed AS. Comparison of treatment regimes with adjustment for auxiliary variables. *J Appl Stat*. 2011;38(12).:2925–38.
98. Tang X, Wahed AS. Cumulative hazard ratio estimation for treatment regimes in sequentially randomized clinical trials. *Stat Biosci*. 2015;7(1):1–18.
99. Goldberg Y, Kosorok MR. Q-learning with censored data. *Ann Stat*. 2012;40(1):529–60.
100. Saarela O, Arjas E, Stephens DA, Moodie EEM. Predictive Bayesian inference and dynamic treatment regimes. *Biom J*. 2015;57(6):941–58.
101. Xu Y, Müller P, Wahed AS, Thall PF. Bayesian nonparametric estimation for dynamic treatment regimes with sequential transition times. *J Am Stat Assoc*. 2016;111(515):921–35.
102. Tang X, Melguizo M. DTR: An R package for estimation and comparison of survival outcomes of dynamic treatment regimes. *J Stat Softw*. 2015;65(7).:1–28.
103. Finn RS, Crown JP, Lang I et al. The cyclin-dependent kinase 4/6 inhibitor palbociclib in combination with letrozole versus letrozole alone as first-line treatment of oestrogen receptor-positive, HER2-negative, advanced breast cancer (PALOMA-1/TRIO-18): A randomised phase 2 study. *Lancet Oncol*. 2015;16(1):25–35.
104. Cristofanilli M, Turner NC, Bondarenko I et al. Fulvestrant plus palbociclib versus fulvestrant plus placebo for treatment of hormone-receptor-positive, HER2-Negative metastatic breast cancer that progressed on previous endocrine therapy (PALOMA-3): Final analysis of the multicentre, double-blind, phase 3 randomised controlled trial. *Lancet Oncol*. 2016;17(4):425–39.
105. Eccles SA, Aboagye EO, Ali S et al. Critical research gaps and translational priorities for the successful prevention and treatment of breast cancer. *Breast Cancer Res*. 2013;15(5):R92.

Section IV

Advanced Topics

20

Assessing the Value of Surrogate Endpoints

Xavier Paoletti, Federico Rotolo, and Stefan Michiels

CONTENTS

20.1 Introduction

Endpoints to evaluate the treatment benefit in randomized clinical trials should be easy to assess without bias, be reliable, reproducible, sensitive to treatment effects, and reflect a clinical benefit for the patient. The way to measure clinical benefit varies greatly according to the disease. This includes cure measured as a binary variable, modification of some physiological parameters measured as continuous variables, or time-to-event outcomes. Those final endpoints are considered as the gold standard but they can be difficult to assess if they require invasive procedures or extensive exams; they can also be long to collect, in particular for time-to-event outcomes, or be so rare that the statistical power is insufficient to detect meaningful clinical effect sizes with typical sample sizes. Examples include evaluation of cardiovascular events in hypertension disease [1], depression scores, time to death after resectable colon cancer [2], or death in localized prostate cancer [3]. In such situations, it can be an effective strategy to replace the true endpoint by an earlier endpoint (a surrogate) that is easier to measure and that allows the researchers to predict the treatment effect on the true endpoint. For instance, blood pressure might be a surrogate of cardiovascular events; disease-free survival and metastasis-free survival might be surrogates of overall survival in colon and prostate cancer, respectively. Not only can surrogate endpoints potentially address the above-mentioned limits of final endpoints, but they can be also combined to improve the amount of information provided by a clinical trial [4]. Surrogate endpoints, finally, facilitate the design of clinical trials. For instance, in oncology, overall survival is classically considered as the gold standard to evaluate treatments; however, in case of available efficacy results on an intermediate endpoint, such as disease progression, physicians may select the best treatment based on what has been administered or may propose the investigational treatments to patients who were in the control arm leading to biases in the assessment of the final endpoints. Such an example can be found in the HERA breast cancer trial that investigated the benefit of adding trastuzumab to adjuvant chemotherapy in early-breast-cancer patients: more than 50% of the patients allocated to the control arm were proposed to receive trastuzumab once the interim results on disease-free survival (DFS) had been released, leading to a difficult assessment of the effect of treatment on survival in this trial [5]. Consequently, the treatment switch was therefore refused in another trial that investigated the addition of pertuzumab in this setting in order to preserve the possibility to estimate straightforwardly the effect on overall survival (OS) [6]. Had DFS been validated as a surrogate of OS, this may have simplified the design of the pertuzumab trial.

The list of accepted surrogate endpoints is still a matter of debate. The Food and Drug Administration (FDA) announced the publication of such a list for oncology [7]. Several prerequisites to qualify as a surrogate endpoint have been established. Actually, in the literature, two different viewpoints are sometimes confronted: the first one gives the preeminence to the biological or physiological rationale in the assessment of the validity of surrogate endpoints. The surrogate endpoint should be based on a key mechanism that determines the value of the final endpoint: modifying the surrogate will lead to a modification of the final endpoint. Stuart Baker, in a recent article [8], lists five criteria for using surrogate endpoints, among which three of them involve "clinical and biological considerations: similarity of biological mechanisms of treatments between the new trial and previous trials, similarity of secondary treatments following the surrogate endpoint between the new trial and previous trials, and a negligible risk of harmful side effects arising after the observation of the surrogate endpoint in the new trial."

The second viewpoint claims that the deterministic link between the two endpoints may remain obscure as long as the treatment effect on the surrogate enables one to reliably predict the treatment effect on the final endpoints. A strong biological rational is not sufficient to assess the validity of the surrogate, and one needs to quantify the correlation between endpoints as well as between effects on treatment. In this chapter, we will focus on this statistically oriented approach, which is probably the current standard in clinical oncology. The question is to identify the statistical levels of evidence we need in order to say that a surrogate is a valid endpoint.

After introducing some examples (Section 20.2), we review some of the statistical definitions and operational criteria used in different approaches to validate surrogate endpoints based on a single trial (Section 20.3). In Section 20.4, we focus on the statistical methods and the related software in R or SAS that are available when multiple trials are used to validate endpoints, the so-called meta-analytic approach. This gives way to the distinction between the individual and the trial-level surrogacy, a central notion in this field. The case where both the surrogate and the final endpoints are continuous variables serves to present some of the models and concepts. Next, the specific case of both outcomes being time-to-event endpoints is explored. Two approaches are reviewed: the first one, proposed by Burzykowski et al. [9], splits the validation process in two stages while the second uses the Poisson approximation of survival data to provide the statistical measures of surrogacy in a single stage.

Finally, we comment on the robustness of the findings to validate surrogate endpoints.

We use examples in early breast cancer and in stomach cancer that we briefly introduce in the following section.

20.2 Motivating Examples

20.2.1 Top Trial: Pathologic Complete Response in Breast Cancer after Anthracyclines

Neoadjuvant treatment of breast cancer consists of starting systemic treatment before surgery. Pathological complete response (pCR), defined as the absence of residual invasive breast carcinoma after completion of chemotherapy, has been accepted as a surrogate endpoint of disease-free survival to support accelerated approval by the FDA in 2014 [10].

We will illustrate the link between pCR and a long-term survival-based endpoint (distant-metastasis-free survival) using data from 107 patients included in the TOP study, in which patients with estrogen-receptor negative (ER–) tumors were treated with anthracycline (epirubicin) monotherapy [11]. This single trial will be contrasted with the results from Cortazar et al. (2014) [12], who obtained data from 12 international trials and 11,955 patients with early breast cancer to validate pCR as a surrogate endpoint for a long-term endpoint (event-free survival).

20.2.2 The GASTRIC Initiative: Disease-Free Survival, Progression-Free Survival, and Stomach Cancers

Stomach cancer is largely incurable. Resection of the primary tumor and systemic treatments give a median OS around 5 years, and OS is observed to be quite longer in Asia than in Western countries. Advanced/recurrent diseases have a very poor prognosis as most patients eventually die before 2 years. There is then a strong need for better treatments and accelerated clinical trials. Candidate surrogates are DFS in the adjuvant setting and progression-free survival (PFS) in the advanced setting. The GASTRIC (Global Advanced/Adjuvant Stomach Tumor Research International Collaboration) group initiated two individual patient-based meta-analyses of randomized clinical trials of systemic treatment to validate those candidate endpoints [13,14].

20.3 Statistical Requirements to Validate Surrogate Endpoints from Single Trials

20.3.1 What Is Not Sufficient: Patient-Level Correlation

Consider Figure 20.1, where the distant-metastasis-free survival (DMFS) of patients with early breast cancer is plotted according to the pCR response after 4 cycles of neoadjuvant therapy by anthracyclines. DMFS was calculated from the end of the neoadjuvant therapy (date of surgery) in this landmark analysis [15].

Patients who reached pCR had a much better survival than patients who did not. Can we therefore claim that pCR is a surrogate endpoint of DMFS as endorsed by the FDA? Obviously, pCR status is significantly associated with DMFS even though a more appropriate analysis to quantify the quality of this prediction than simple comparison of survival curves is required. At the level of the patient, a strong association between pCR and DMFS is quite probable, and pCR may present an interesting way to predict the patient prognosis. Is this sufficient to use pCR to evaluate treatment benefit? Actually not, as Figure 20.1 does not address the question whether or not modifying the proportion of patients who are pCR negative would translate into a reduction of the hazard of disease relapse. In other words, if one demonstrates that a treatment increases the pCR rate compared to a control (for instance with an odds ratio OR > 1), one cannot infer that the treatment would also

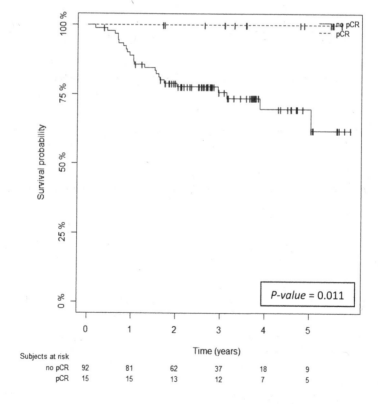

FIGURE 20.1

Distant-metastasis-free survival according to pathological complete response (pCR) at the end of the neo-treatment by anthracyclines of patients with breast cancer. P-value corresponds to the result of a penalized likelihood ratio test for hypotheses within a Cox regression analysis using Firth's penalized likelihood.

reduce the hazard of relapse (hazard ratio HR (DMFS) < 1). This is called the trial-level surrogacy, which is also expressed as the ability to predict the *effect of treatment* on the final endpoint from the effect of treatment on the intermediate endpoint. We shall see in Section 20.4 how this is quantified, in particular, in this specific context.

Note on the immortal time bias. As was reminded more than 30 years ago [16], the analysis of the relationship between response to treatment and time to event can be biased if the response to treatment is measured after initiation of follow-up. This leads to the so-called *immortal time bias* or guarantee-time bias [17]: to evaluate the prognostic value of response (measured for instance at 4 cycles of treatment) on the time-to-event endpoint (such as DMFS), patients developing metastasis during the neoadjuvant treatment, stop treatment and are classified as non-responders, thus inducing a bias in the estimation of the hazard of death. A simple analysis is obtained with the landmark method as presented above. The starting date for the time-to-event endpoint is postponed so that the response is measured before the initiation of the follow-up. Censored observations and events occurring before the landmark time of response evaluation are removed from the analysis.

20.3.2 What Is Controversial: Evidence of No Residual Effect from a Single Trial

In a landmark article, Prentice [18] put forward that a surrogate could be validated if several conditions were satisfied:

1. The surrogate and final endpoints are correlated,
2. The treatment effect on the surrogate is non-null,
3. The treatment effect on the final endpoint is non-null, and
4. The treatment effect on the final endpoint disappears after adjustment on the surrogate endpoint.

In addition to the patient-level correlation (condition 1), Prentice introduced the notion of the surrogate being on the "pathway" between the intervention (the treatment) and the final endpoint.

Operational criteria have then been derived by Freedman, but he also highlighted three limitations, making those theoretical criteria difficult to apply in practice [19]. First, a treatment must have an effect, while in fact, at least half the investigated treatments in randomized clinical trials in oncology conducted by National Cancer Institute–sponsored Cooperative Oncology Groups in the period 1955–2006 did not have significant superior effect compared to the standard treatment [20]. Second, condition 4 implies testing the absence of treatment effect. Accepting the null hypothesis of no treatment effect is always a concern as it is closely related to the power of the experiment. The ability to detect a treatment effect (reject or not the null hypothesis) directly depends on the sample size. Third, one cannot expect to have a perfect surrogate, as defined in condition 4, and the framework of hypothesis testing is probably too restrictive. A measure of the "extent" to which an endpoint meets the Prentice criteria appears more useful in practice.

20.3.3 What Is Controversial: Quantifying the Treatment Effect Explained by the Surrogate Endpoint from a Single Trial

Several authors have proposed to quantify the amount of treatment effect on the final endpoint explained by the surrogate endpoint. Freedman introduced the proportion of the effect of treatment on the final endpoint that is explained by the surrogate. He proposed to compare the lower confidence limit of this proportion to some threshold, e.g., 0.5 or 0.75 to claim an interesting amount of surrogacy. This condition can only be verified if the treatment has a massively significant effect on the final endpoint, a rare situation. Furthermore, in several examples, Molenberghs et al. [21] found this proportion to be larger than 1. As alternative measures, Buyse et al. considered two other quantities: the effect of treatment on the final endpoint relative to that on the surrogate and the association between the two endpoints after adjustment for treatment effect [22].

In the context of time-to-event data, Alonso et al. [23] treated the surrogate endpoint as a time-dependent covariate and used the concept of a likelihood reduction factor to quantify the Prentice criteria. This falls in the framework of measuring information gain and explained randomness. This can be interpreted as a proportion of variation of the final endpoint explained by the surrogate endpoint. However, contrary to the interpretation obtained with continuous endpoints and a linear model, one diffculty is how to adequately deal with censoring, as it would be useful to have a coeffcient that explains the proportion of treatment effect captured by the surrogate and which is not impacted by the censoring mechanisms, especially when the censoring rate is high. Two sets of data with different amounts of censored observation, but having the same distribution of events and leading to the same estimates of treatment effect, may not give the same explained variation. This limit is common to most measures of explained variation applied on censored data. This can be an issue in the context of the evaluation of surrogate endpoints evaluated on several trials with different follow-up. The measure of explained variation may lack some robustness.

Therefore, building on Alonso et al. [23], O'Quigley et al. [24] considered the case of a Cox model to estimate the treatment effect on the true time-to-event endpoint. He proposed using the information gain, $\Gamma(\beta)$, that measures the distance between two models, one indexed by the parameter β associated to the treatment indicator Z, the other indexed by the value 0. The transformation $\rho^2(Z) = 1 - \exp\{-\Gamma(\beta)\}$ can be viewed as the proportion of randomness explained by the regression. The authors provide several estimators that are independent of the censoring mechanism at least when censoring is independent of the event process.

Software: The proportion of effect is straightforward to obtain from any statistical software carrying out regression analysis; the explained variation by O'Quigley et al. can be computed with the SAS macro coded by Abel Tilahun from the University of Hasselt I-BioStat team (https://ibiostat.be/online-resources/surrogate). The measure can be calculated for a single trial or for a collection of trials. In this latter case, a mean value is provided.

20.3.4 Area of Research: Causal Inference

Back to the idea of a deterministic link between the two endpoints, Frangakis and Rubin [25], Conlon et al. [26], and Alonso et al. [27], among others, investigated the amount of treatment effect on the final endpoint that is likely to be "caused" by the treatment effect on the surrogate endpoint. Building on the causal inference framework, several approaches have been proposed. The general idea is to assess the expected causal treatment effects, that is, the average of the individual causal effects over all patients within the trial populations.

To introduce some notation, let us assume that a new treatment ($Z = 2$) is compared to a control ($Z = 1$). Let us assume that in a virtual world, each patient could be treated with both treatments. The two potential surrogate and true outcomes for patient j are noted $S_{2j}, S_{1j}, T_{2j}, T_{1j}$, where the first subscript denotes the treatment. One can then define the treatment effect at the individual level: as $\delta(T_j)$ and $\delta(S_j)$. Of course, in the real world, only the outcomes S and T in the allocated arm are observable and the δs are non-measurable, nor can they be estimated unless untestable assumptions are made. The problem is said to be non-identifiable.

As shown by Alonso et al. [27], if S is a valid surrogate for T, then $\delta(S_j)$ should convey a substantial amount of information about $\delta(T_j)$. The individual causal association (ICA) can be defined as $\rho_\delta = \text{corr}\,(\delta(T_j), \delta(S_j))$. Various publications differ mainly by the type of assumptions postulated to estimate those quantities. It is possible to solve the non-identifiability problem by parameter restriction or the introduction of sensitivity parameters into the relationship between surrogate and final endpoints. For example, Li et al. [4] introduced a monotonicity assumption, that is, $S_{2j} \geq S_{1j}$ for potential binary surrogate endpoints and $T_{2j} \geq T_{1j}$ for potential binary final endpoints (a patient that responds to the control treatment would also have responded to the investigational treatment) and incorporated prior belief as prior distributions on T_{kj} given S_{kj} for each treatment k [4]. Conlon et al. [26] extended the Bayesian framework to multivariate normal models for continuous data and Gaussian copula models for time-to-event data [28]. Finally, Frangakis and Rubin [25], and then Alonso et al. [29], characterized the link between the meta-analytic framework and the causal inference.

The latter authors showed the limitations of the individual causal effect due to the required untestable (and often unrealistic) hypotheses. They proposed an alternative approach relying on the expected causal effects rather than on the individual causal effect. Expected causal effect is simply the usual expected effect of treatment versus the control. In the absence of uncontrolled confounding factors (i.e., in randomized clinical trials), the expected causal effect is $E(T_j|Z_j = 2) - E(T_j|Z_j = 1)$, where the quantities $E(T_j|Z_j = 1)$, $E(T_j|Z_j = 2)$ can be estimated using the observed means of T_j in the control and experimental groups, respectively. Expected causal effect can then be expressed in terms of previously developed measures as the relative effect [22] or explained variations.

Unlike the individual causal effects, expected causal effects are estimable from the data under fairly general conditions. Furthermore, one may argue that in a clinical trial context, expected causal effects are central. In fact, regulatory agencies mostly evaluate expected causal effects for granting commercialization licenses, and surrogate endpoints are primarily used as a tool to speed up this process of approval.

20.4 Statistical Methods and Software to Validate Surrogate Endpoints from Multiple Trials

20.4.1 The Meta-Analytic Approach: Introduction to a Reference Method

To evaluate the association between the treatment effect on the surrogate and on the final endpoint, Buyse et al. [30] used the notion of prediction. Instead of decomposing the treatment effect on the final endpoint between what is due to the surrogate endpoint and what is not, they proposed to quantify the relationship between the two treatment effects. Consider Figure 20.2 where the treatment effect on the candidate surrogate endpoints, here DFS (HR_{DFS}), is shown in terms of the effect on the final endpoint T, OS (HR_{OS}). Each circle corresponds to a trial, the diameter of which is proportional to the sample size. In this example of stomach cancer, there appears to be a strong association between the two measures of treatment effect. Larger effects on the surrogate are associated with larger effects on the final endpoint, which allows the researcher to be confident that a strong

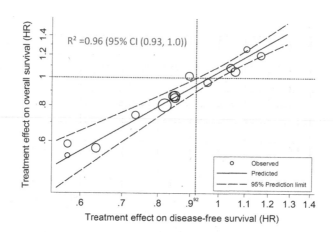

FIGURE 20.2

Relationship between the treatment effect on OS and DFS in adjuvant treatment of stomach cancer. Each circle corresponds to a trial. The size of the circle is proportional to its samples size.

benefit measured in terms of DFS suggests a similarly strong benefit in terms of OS, even before definitive evidence is available for the latter.

This relationship is characterized by a simple linear relationship expressed in the following equation:

$$\ln(HR_{OS}) = 0.047 + 1.239\ln\left(HR_{DFS}\right)$$

This equation enables the researches to predict the treatment effect on the final endpoint from the treatment effect on the surrogate. The linear regression also provides us with a simple measure of trial-level association between the two endpoints: the coefficient of determination, i.e., the part of variation of the treatment effect on T explained by the variation of the treatment effect on S. Finally, this linear regression also provides 95% prediction limits around the treatment effect on T, given the estimated effect on S.

The meta-analytic approach has, for example, served as a basis to validate DFS as a surrogate of OS in colon cancer [31]. However, this approach is not systematically accepted, as illustrated by two recent decisions from the European Medical Agency that did not accept the 5-year complete response rate as a surrogate of OS to approve a new treatment in lymphoma and that did not endorse the results demonstrating that the metastasis-free survival could replace OS in localized prostate cancer [3]. Conversely, coming back to the pCR example in breast cancer (see Section 20.2), the large meta-analyses demonstrated that whilst the individual level association between pCR and OS was strong, the trial-level association between pCR and event-free survival was quite disappointing, as OR_{pCR} had no capacity to predict HR_{DFS} [12], as shown in Figure 20.3.

Interestingly, Vandenberghe et al. [32] used causal inference to investigate the same issue in the EORTC 10994/BIG 1–00 randomized phase III trial in which locally advanced breast cancer patients were randomized to either taxane- or anthracycline-based neoadjuvant chemotherapy. In this analysis, the treatment effect on event-free survival (EFS) was decomposed into an indirect effect via pCR and the remaining direct effect. Only 4.2% of the treatment effect on EFS after 5 years was mediated by the treatment effect on pCR. The authors confirmed the lack of association at the trial level between treatment effects measured by both endpoints.

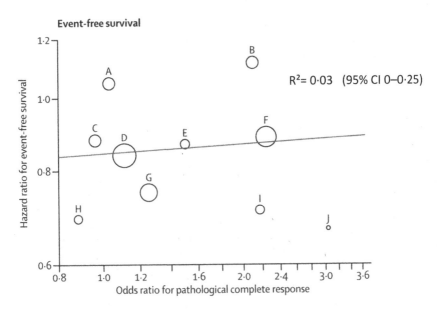

FIGURE 20.3
Relationship between the treatment effect on event-free survival (HR$_{EFS}$) and on pathological complete response (OR$_{pCR}$) in the neoadjuvant treatment setting of early breast cancer. Each circle corresponds to a trial. The size of the circle is proportional to its samples size. (Adapted from Cortazar P et al. *Lancet.* 2014;384(9938):164–72.)

20.4.2 Statistical Implementation

Overall, the statistical sequence underlying the meta-analytic approach is to first estimate the treatment effects on the candidate surrogate and the final endpoint, while accounting for the correlation between the two endpoints and then to characterize the relationship between the two estimated treatment effects; as treatment effects are estimated in a limited number of trials with limited sample sizes, regression of the surrogate treatment effect on the final treatment effect should be adjusted for the measurement errors. This is why this approach is often named a two-stage approach.

This approach is applicable for various types of endpoints. After the landmark article [30] for continuous endpoints, Alonso et al. [33] and Molenberghs et al. [34] have adapted this framework to the cases of endpoints of mixed types, including binary variables and repeated measurements and Burzykovski et al. [9] to the case of survival endpoints. We will start with the simplest case of two continuous outcomes normally distributed, for ease of the presentation, and then describe the case of two time-to-event outcomes. We follow here the notation and the presentation of Buyse et al. [30].

Assume that N trials have been carried out, $i = 1 \ldots N$, to investigate a given treatment (or class of treatments). The sample size of trial i is n_i. As before, T_{ij} and S_{ij} are random variables denoting the true and surrogate endpoints for the jth subject in the ith trial, respectively, and let Z_{ij} be the indicator variable for treatment. One then has two simple linear models relating the treatment effect on the surrogate and on the true endpoint:

$$S_{ij} = \mu_{Si} + \alpha_i Z_{ij} + \varepsilon_{Sij}$$

$$T_{ij} = \mu_{Ti} + \beta_i Z_{ij} + \varepsilon_{Tij}$$

where intercepts and treatment effects are trial-specific. The residual errors ε_{Sij} and ε_{Tij}, assumed to be normally distributed, may be correlated as measured in the same patients/ trials. The correlation matrix of these errors

$$\Sigma = \begin{bmatrix} \sigma_{SS} & \sigma_{ST} \\ \sigma_{ST} & \sigma_{TT} \end{bmatrix}$$

is then used to quantify association at the individual level by

$$R^2_{indiv} = R^2_{\varepsilon_{Tij}|\varepsilon_{Sij}} = \frac{\sigma_{ST}}{\sigma_{SS}\sigma_{TT}}.$$

In the meta-analytic framework, it is assumed that the treatment effects observed in the trials at hand are a random sample from a larger population. In other words, each estimated value μ_{Si} and α_i are random draws from a distribution around the global μ_S and α. More formally, the following model for the second stage is assumed for the intercepts and mean treatment effects:

$$\begin{pmatrix} \mu_{Si} \\ \mu_{Ti} \\ \alpha_i \\ \beta_i \end{pmatrix} = \begin{pmatrix} \mu_S \\ \mu_i \\ \alpha \\ \beta \end{pmatrix} + \begin{pmatrix} m_{Si} \\ m_{Ti} \\ a_i \\ b_i \end{pmatrix},$$

where the second term on the right-hand side is assumed to follow a zero-mean normal distribution with dispersion matrix

$$D = \begin{pmatrix} d_{SS} & d_{ST} & d_{Sa} & d_{Sb} \\ & d_{TT} & d_{Ta} & d_{Tb} \\ & & d_{aa} & d_{ab} \\ & & & d_{bb} \end{pmatrix}.$$

This model can also be regarded as a random-effect model, with fixed effects and random effects giving very similar results in the bivariate normal setting.

The dispersion matrix is then used to quantify the association between the two endpoints at the trial level. The authors proposed using the coefficient of determination $R^2_{b_i|m_{Si},a_i}$ as a measure of surrogacy at the trial level:

$$R^2_{trial} = R^2_{b_i|m_{Si},a_i} = \frac{\begin{pmatrix} d_{Sb} \\ d_{ab} \end{pmatrix}^T \begin{pmatrix} d_{SS} & d_{Sa} \\ d_{Sa} & d_{aa} \end{pmatrix} \begin{pmatrix} d_{Sb} \\ d_{ab} \end{pmatrix}}{d_{bb}}.$$

R^2_{trial} measures the amount of variability of the treatment effect on the true endpoint explained by the treatment effect on the surrogate. In other words, it provides a measure of the quality of the prediction. Good prediction would make it sufficient to measure the surrogate instead of the true endpoint, because residual uncertainty about the treatment effect on the final endpoint is small, given the treatment effect on the surrogate. Although there is no formal consensus on the minimum trial-level R^2 that is needed to validate a surrogate, values closer to 1.00 are desirable. In practice, values larger than 0.80 have been accepted by the FDA to validate a surrogate endpoint in lymphoma trials [35], and in Germany, values above 0.72 are considered acceptable [36].

For all but survival endpoints, a mixed effect formulation of the model has been proposed to estimate all parameters in one stage; furthermore, as introduced earlier, Alonso and Molenberghs [37] showed how to re-express this measure of surrogacy within the Information-Theoretic framework, which unifies the theory with that of other types of endpoints.

Software: The R package Surrogate provides a complete set of tools for the evaluation of surrogate endpoints. Developed by the surrogate research team at the Hasselt University [38], it covers various settings. For instance, the following code is taken from the package documentation (see FixedContContIT).

```
library(Surrogate)
# Example 1, Based on the ARMD data
data(ARMD)
# Assess surrogacy based on a full fixed-effect model
# in the information-theoretic framework:
Sur <- FixedContContIT(Dataset=ARMD, Surr=Diff24, True=Diff52,
Treat=Treat, Trial.ID=Center,
Pat.ID=Id, Model="Full", Number.Bootstraps=50)
# Obtain a summary of the results:
summary(Sur)
# Example 2 (time consuming code), Conduct an analysis based on a
simulated dataset with 2000 patients, 100 trials,
# and Rindiv=Rtrial=.8
# Simulate the data:
Sim.Data.MTS(N.Total=2000, N.Trial=100, R.Trial.Target=.8, R.Indiv.
Target=.8,
Seed=123, Model="Full")
# Assess surrogacy based on a full fixed-effect model
# in the information-theoretic framework:
Sur2 <- FixedContContIT(Dataset=Data.Observed.MTS, Surr=Surr, True=True,
Treat=Treat,
Trial.ID=Trial.ID, Pat.ID=Pat.ID, Model="Full", Number.Bootstraps=50)
# Show a summary of the results:
summary(Sur2)
```

20.4.3 Survival Endpoints

In oncology, endpoints for testing the effect of treatments are commonly time to some event, which can be possibly censored. This censored survival observation entails extra modeling difficulties. As stated above, it is well known that most measures of explained variations strongly depend on the censoring distribution [39]. Furthermore, despite frailty models have been developed to introduce random effects in the survival models, the same joint modeling as the one used for the case of two continuous endpoints requires a model for the correlation structure. In this section, we will review two alternative approaches relying on the copula model for the first one [9] and on the Poisson generalized linear model for the second [40].

20.4.3.1 Copula Approach

The two-stage approach introduced previously serves to circumvent the difficulty of estimating a joint frailty effect model. At the first stage, the correlated treatment effects on each endpoint for each trial are estimated using a copula function, C_θ. A copula is a bivariate distribution function on $(0,1)^2$ that has uniform margins that can be transformed using the well-known inverse-distribution function. A large class of marginal-distribution functions can be used. Furthermore, it allows modeling separately the margins and the correlation structure, providing a versatile tool that can be adapted to numerous situations. Burzykowski and colleagues assumed that the two survival endpoints (S_{ij}, T_{ij}) follow the joint distribution

$$F(s,t) = \Pr(S_{ij} \geq s, T_{ij} \geq t) = C_\theta\{F_{S_{ij}}(s), F_{T_{ij}}(t)\} \text{ with } s, t \geq 0$$

Three copula functions are classically used in survival analysis: the Clayton copula $C_\theta(u,v) = (u^{1-\theta} + v^{1-\theta} - 1)^{1/1-\theta}, \theta > 1$ where positive association, proportional to θ, is assumed. The Hougaard and Plackett copulas are two other functions that can be compared based on some information criteria (typically the Akaike criteria).

In addition to the copula function describing the association between the two time variables, one has also to specify their marginal distributions, i.e., the survival distribution of the two endpoints applicable to each of the N trials. Typically a parametric proportional hazard model is selected. The Weibull distribution has been shown to provide satisfactory fit on various datasets:

$$F_{S_{ij}}(s) = \exp\left\{-\int_0^s \lambda_{Si}(x) \exp(\alpha_i Z_{ij}) dx\right\}$$

$$F_{T_{ij}}(t) = \exp\left\{-\int_0^t \lambda_{Ti}(x) \exp(\beta_i Z_{ij}) dx\right\}$$

where λ_{Si} and λ_{Ti} are the trial-specific marginal baseline hazard functions. Although semi-parametric models can be used, the baseline hazard has then to be estimated, which requires large sample sizes and increases the estimation complexity.

At the second stage, a linear regression between the two treatment effects is built with the model: $\begin{pmatrix} \alpha_i \\ \beta_i \end{pmatrix} = \begin{pmatrix} \alpha \\ \beta \end{pmatrix} + \begin{pmatrix} a_i \\ b_i \end{pmatrix}$ as previously described, where the second term on the right-hand side is assumed to follow a zero-mean normal distribution with dispersion matrix

$$D = \begin{pmatrix} d_{aa} & d_{ab} \\ d_{ab} & d_{bb} \end{pmatrix}.$$

One can see that this regression accounts for the estimation errors at stage 1 through D. Despite providing a better assessment of the uncertainty in prediction of effects on the true endpoint, accounting for estimation error at the first stage also considerably increases the numerical difficulty to obtain proper estimates. In practice, it appears that in many meta-analyses, the estimation of regression model with measurement errors is unstable even in the presence of many and large trials. The estimation and residual errors are difficult to disentangle and one needs to be very cautious in interpreting the results. Numerical computation may lead to convergence issues and non-identifiable models. In particular, this raises the question whether or not the global maximum of the likelihood has been reached.

Once those estimates obtained, and similarly to the case of two normal endpoints, one can derive the determination coefficient R^2_{trial}, which provides a measure of the quality of the prediction of the log-hazard ratio for the true endpoint based on the log-hazard ratio for the surrogate.

The linear relationship at the second stage also gives a simple tool to predict the expected effect on the true endpoint. For a surrogate associated with a high determination coefficient, Burzykowski et al. [41] identified the minimum treatment effect on the surrogate that would enable predicting with a 95% probability a non-null effect on OS, would we have a future very large trial. This surrogate threshold effect (STE) enables design of a trial on a surrogate in order to achieve some desired effect on the true endpoint.

For instance, consider the scatterplot in Figure 20.2 of the treatment on both endpoints. The dotted vertical line corresponds to the minimum effect (i.e., maximum HR) on DFS that should be observed in order to predict with a 95% confidence an HR_{OS} below 1, which is the STE (= 0.92). Of note, it is quite common that the treatment effect on earlier endpoints is stronger than on final endpoints due to the possible rescue treatments between the measure of the surrogate—typically driven by progression/relapse—and that of the true endpoint, due to various risks of events or to competing outcomes. The STE provides an intuitive quantification of the dilution of the treatment effect. In this example, the dilution is mild, which is in line with the very poor prognosis of relapsing patients and the lack of effective rescue treatments. However, one should bear in mind that the STE is estimated with potentially some uncertainty and is applied as if the new trial would have an infinite size.

Evaluation of the correlation at the patient level is complicated by dependency of this measure with the censoring and the event distribution, as introduced earlier. As an alternative, the rank correlation (Spearman's ρ) or the Kendall's τ, which is independent of the marginal distributions, can be estimated.

Software: The R package `surrosurv` [42] allows estimating the three above-mentioned copula functions. For the second stage, regression model with or without adjustment for measurement errors can be fitted. Alternatively, Burzykowski from Hasselt University has made available his set of SAS macros to apply this approach. Compared to `surrosurv`, confidence intervals based on asymptotic approximations are available. Both use maximum likelihood to estimate the parameters.

Correlation structure: The three copulas lead to different correlation structures. In particular, the Clayton copula on the survival distributions is adapted to correlation that would increase over time, contrary to the Hougaard copula; and the Plackett being neutral for this matter. Burzykowski et al. [9] proposed using the Akaike criteria to select the copula providing the best goodness of fit. Renfro et al. [43] also showed that in addition to the choice of copula family, directional misspecification (i.e., to assume a correlation structure that fades away with increasing time, whereas the opposite is true or vice versa) can lead to biased estimates of patient-level and trial-level surrogacy. In other words, the copula can be built on the survival distributions as described above or on the cumulative distribution function of the time-to-event endpoints. The likelihood functions are different, assume very different correlation patterns, and do not lead to the same estimates. For example, the Clayton copula applied to the cumulative distribution function assumes stronger correlation at initial timepoints as compared to survival functions.

20.4.3.2 Simple Weighted Regression

Some authors [31] have proposed replacing the first-step copula model with a set of marginal Cox models. The first stage then consists of estimating the treatment effects using independent Cox semi-parametric models for each trial on each endpoint. The second stage is a weighted regression line where each log-HR is weighted by the size of the trial to reflect the uncertainty around estimates. One of the advantages of this approach is to avoid the strong assumptions of a parametric general meta-analytic model that involves hypotheses on the class of survival models, on the model selection, or on the random effect distributions. The two major limitations of this simplified approach are that (1) the HRs for the true and the surrogate endpoints are estimated independently without accounting for the intra-patient correlations; therefore, individual level correlation cannot be estimated and (2) the estimation errors at the first stage are only partially accounted through the weighting of the trial by its size. As a consequence, the estimation of the confidence interval around the R^2_{trial} has probably insufficient coverage, and surrogacy measures may be biased compared to the joint approach in case the general and parametric meta-analytic model is correctly specified. This approach is probably not to be preferred if the copula model provides high goodness of fit and regression with measurement error model is estimable.

20.4.3.3 Joint Poisson Model

More recently, Rotolo et al. [40] developed an alternative to the copula approach for survival endpoints, using the well-known approximation of the Cox model by Poisson regression.

Let us consider the semi-parametric Cox model, $h(t) = h_0(t)\exp(\beta Z)$. By dividing the time scale into intervals $k = 1, \ldots, K$ corresponding to the intervals between the observed event times, the so-called auxiliary Poisson model for the event indicator $\delta^{(k)}$ that follows a Poisson distribution in the kth interval is $\log(\mu^{(k)}) = \mu_0^{(k)} + \beta Z + \log(y^{(k)})$, where $y^{(k)}$ is the time spent at risk during the period k and $\log(\mu^{(k)})$, the log-hazard for time interval k. Likewise, the parameters $\left\{\mu_0^{(k)}\right\}_{k=1,\ldots,K}$ are the logarithms of the baseline risk rates during each interval k, under the assumption that the risk is constant within the interval. In practice, different time intervals can be employed, leading to an approximation of the estimators of the Cox model parameters. In meta-analyses, each subject j is recruited within a trial. To account for both within-trial and within-subject correlations, the joint survival model for the two endpoints can be expressed as

$$\begin{cases} h_{S_{ij}}(s) = h_{S_i}(s)\exp(u_{ij} + \alpha_i Z_{ij}) \\ h_{T_{ij}}(t) = h_{T_i}(t)\exp(u_{ij} + \beta_i Z_{ij}) \end{cases}$$

where u_{ij} are individual random terms and follow normal distributions of variance σ^2_{indiv}. The parameters of this random-effects model can be estimated equivalently via the auxiliary bivariate mixed Poisson model

$$\begin{cases} \log\left(\mu_{S_{ij}}^{(k)}\right) = \mu_{S_i}^{(k)} + u_{ij} + \alpha_i Z_{ij} + \log y_{S_{ij}}^{(k)} \\ \log\left(\mu_{T_{ij}}^{(k)}\right) = \mu_{T_i}^{(k)} + u_{ij} + \beta_i Z_{ij} + \log y_{T_{ij}}^{(k)} \end{cases}$$

where $y_{S_{ij}}^{(k)}$ and $y_{T_{ij}}^{(k)}$ are the time spent at risk by subject i in trial j with respect to each endpoint during the period k, and where $\left\{\mu_{S_i}^{(k)} = \mu_S^{(k)} + m_{S_i}\right\}$ and $\left\{\mu_{T_i}^{(k)} = \mu_T^{(k)} + m_{T_i}\right\}$ are the baseline log-hazard rates.

In the context of Gaussian linear models, Buyse et al. [30] assumed correlated but distinct individual error terms $u_{S_{ij}}$ and $u_{T_{ij}}$ to derive R^2_{indiv}. In the case of failure time endpoints, such R^2_{indiv} would not express the association between S and T, but between the random effects which modulate their hazard functions, which is a more indirect way of measuring dependence. Therefore, the shared frailty (random) effect is a better measure of association. Kendall's τ can be derived from the variance of the random effect.

The Poisson model naturally provides the correlation ρ_{trial} between the treatment effect on the surrogate endpoint S and the effect on the true endpoint T, which gives the coefficient of determination $R^2_{trial} = \rho^2_{trial}$.

Software: This model formulation does not need any parametric assumption for the marginal baseline hazard functions, contrary to the case for the two-step copula approach. The auxiliary mixed-Poisson model fits naturally into the generalized linear mixed model (GLMM) theory and can accommodate complex structures of random effects. The surrosurv R package provides a nice implementation of this approach, together with customable figures of the regression between both treatment effects as well as the STE. A complete manual has been published by Rotolo et al. [42].

An example of the application of this package on the GASTRIC data, where both the Poisson and the Clayton copula are fitted, is provided below:

```
# Advanced GASTRIC data
# Long computation time!
data('gastadv')
allSurroRes <- surrosurv(gastadv, c('Clayton','Poisson'), verbose =
TRUE)
convergence(allSurroRes)
allSurroRes
predict(allSurroRes)
plot(allSurroRes)
```

Estimation of the parameters may still be tricky in particular meta-analysis datasets. Rotolo et al. [40] proposed several models with decreasing complexity (for instance, assuming common baselines risk across all trials) to facilitate the estimation procedure.

20.4.3.4 GASTRIC Example

The example on the GASTRIC data illustrates the range of values for the estimates of correlation that are obtained with various models and copulas (see Table 20.1). Model Poisson

TABLE 20.1

Surrogacy Results in the Advanced Gastric Setting Extracted

	R^2_{trial}	Kendall's τ
Clayton unadj	0.45 (0.28–0.74)	0.61 (0.59–0.62)
Clayton adj	0.41 (0.15–1.00)	0.61 (0.59–0.62)
Plackett unadj	0.45 (0.28–0.74)	0.62 (0.60–0.63)
Plackett adj	0.40 (0.04–1.00)	0.62 (0.60–0.63)
Hougaard unadj	0.45 (0.28–0.74)	0.32 (0.32–0.33)
Hougaard adj	0.38 (0.01–1.00)	0.32 (0.32–0.33)
Poisson TI	0.63 (0.32–1.00)	0.51 (0.50–0.52)
Poisson TIa	0.83 (0.24–1.00)	0.51 (0.50–0.52)

Source: Rotolo F et al. *Stat Methods Med Res.* 2019;28(1):170–83.

TI incorporates both random trial-treatment interactions (α_i, β_i) and individual random effects u_{ij}, but still has common baselines between trials. It provides both individual-level and trial-level measures of surrogacy τ and R^2_{trial}. Model Poisson TIa extends the model Poisson TI by accounting for trial-specific baseline risks, using shared random effects at the trial level: $\mu_{Si} = \mu_S + m_i$, $\mu_{Ti} = \mu_T + m_i$, with $m_i \sim N(0, \sigma^2 m)$.

Table 20.1 shows the estimates of the trial-level R^2_{trial} and of the individual-level Kendall's τ obtained with the various models described previously. The estimates of trial-level surrogacy from adjusted copula models (0.41, 95% CI 0.15–1.00; 0.40, 95% CI 0.04–1.00; and 0.38, 95% CI 0.01–1.00 for Clayton, Plackett, and Hougaard, respectively) were lower than those from their unadjusted equivalents (0.45, 95% CI 0.28–0.74 for all three). The estimate from model Poisson TI was $R^2 = 0.63$ (95% CI 0.32–1.00). The model Poisson TIa, which had the best convergence metrics of all models, provided the highest estimate of $R^2 = 0.83$ (95% CI 0.24–1.00), which is markedly different from the estimates obtained with the copula.

Surrogacy evaluation in the case of failure times data requires careful consideration of the data, analysis tools and optimization techniques. The definition of convergence is not straightforward. As discussed by Rotolo et al. [40], various criteria can be used leading to acceptation of various values as estimates.

In the advanced setting, we have an example where the correlation at the patient level is relatively strong while the correlation at the trial level is relatively low. For a given patient, PFS is prognostic of the OS, but treatment effect on the PFS in the population does not entail treatment effect on OS.

20.5 Practical Difficulties with the Evaluation of Surrogate Endpoints

20.5.1 Validation of Surrogate and Expected Gain

The primary aim of the evaluation of surrogate is to save time and resources to carry out future randomized clinical trials. How much do we gain is a matter of debate. Consider the GASTRIC project in the adjuvant setting. A very strong association between DFS and OS has been reported at the trial level ($R^2_{\text{trial}} = 0.96$) and DFS was recommended as a primary endpoint for subsequent trials [44]. However, a subsequent analysis investigated the minimum follow-up requested to achieve a high level of association with 5-year OS. Data were artificially censored after 2, 3, and 4 years and the trial-level association was recalculated as shown in Table 20.2.

Despite 4-year DFS achieving a high association with 5-year OS, a somewhat lower association was obtained with 2-year and 3-year DFS, suggesting that DFS still required quite a long follow-up to provide good prediction. This limits the practical interest of using DFS. A counterexample is provided by the assessment of complete response at 30 months as a surrogate of PFS in lymphoma. The FLASH initiative demonstrated high association on multiple trials, which may entail significant benefit in terms of follow-up for future trials [35].

20.5.2 Heterogeneity of Treatment Effects and Surrogacy Evaluation

In the breast cancer example (Figure 20.3), there appears to be a low amount of variation of treatment effects on event-free survival across trials. Actually, if one assumes that

TABLE 20.2

Surrogacy Evaluation from the GASTRIC Project Using Various Timepoints for the Data Analysis of DFS in Order to Predict 5-Year OS, Extracted

Summary Measures	2-year DFS/ 5-year OS	3-year DFS/ 5-year OS	4-year DFS/ 5-year OS	All
Events (DFS/OS)	1135/1489	1379/1489	1511/1489	1763/1705
Rho[a] (95% CI)	0.95 (0.94–0.96)	0.95 (0.95–0.96)	0.96 (0.95–0.96)	0.97 (0.97–0.98)
Unadjusted R^2 (95% CI)[b]	0.78 (0.57–0.98)	0.866 (0.74–1.00)	0.92 (0.84–1.00)	0.96 (0.93–1.00)
STE (HR)	Undefined	Undefined	0.77	0.92

Source: Oba K et al. *J Natl Cancer Inst.* 2013;105(21):1600–7.
Abbreviations: CI, confidence interval; DFS, disease-free survival; HR, hazard ratio; OS, overall survival; STE, surrogate threshold effect calculated on adjusted regression.
[a] Rho represents the Spearman rank correlation coefficient between disease-free survival and overall survival.
[b] R^2 represents the coefficient of determination between treatment effect on disease-free survival and overall survival, not adjusted for measurement error.

one has a meta-analysis model with no treatment by trial interaction at all (no treatment heterogeneity) on the true endpoint, regression and correlation would be non-identifiable, whatever the treatment effects. Even if in theory, random fluctuations will translate in variations in the random effect estimates that serve to derive the surrogacy measurement and should make it possible to obtain R^2 measures; the absence of heterogeneity and the absence of treatment effect impact the accuracy of the measure.

Likewise, due to the limited number of trials, the regression is commonly computed on few points and the weight of outliers may be important. Leave one study out cross-validation as done in Michiels et al. [45] is recommended to assess the robustness of the results. The regression model is rebuilt from scratch on N-1 trials and can serve to predict the effect size on the final endpoint from the effect size on the surrogate endpoint for the left out Nth trial, in a repeated manner. Measures of distance as well as dispersion between observed and predicted HR can be computed [46].

Some authors have suggested splitting the trials in smaller units. For instance, the treatment effect on both endpoints in each country can be estimated as if from different trials [47]. This increases the number of points but decreases the precision of each estimate, possibly dramatically. The choice of the unit is delicate and may lead to different results depending on the observed heterogeneity across the level of splitting. This slicing of the trials is probably not to be recommended.

20.5.3 Extent of Application and New Class of Treatments

A limitation to the validation of surrogate endpoints conditional on treatment effects is the applicability of the results for future trials. Next trials are likely to investigate different classes of agents. The example of the advanced treatment of colon cancer can be noted. PFS was shown to be a reasonably good surrogate endpoint of OS in trials that investigated fluoropyrimidines-based chemotherapies [48], but it turned out to be a poor predictor of OS for bevacizumab, a monoclonal antibody directed against VEGF. Likewise, recent immune checkpoint blockers have patterns of responses different from what has been reported with other treatments; PFS may not be the most appropriate surrogate endpoint for immune checkpoint blockers even in indications where it had been validated. Strictly speaking, surrogate endpoints are applicable only on the same type of agents as those used for validation.

20.5.4 Second-Line Treatments

Likewise, post-progression survival may be impacted by the approval of new lines of treatments. For instance, surrogacy in gastric cancers was assessed on trials of first-line treatments at a time where no subsequent treatments were approved after progression. However, administration of second- and even third-line treatments has become now standard practice. As a consequence, those rescue treatments may either dilute the impact of PFS on OS, or increase the variability in the association due to various degrees of response across patients. Post-progression survival is then seen as an important limitation to the statistical evaluation of surrogacy, at least in tumor types where numerous lines are available, such as in breast or colorectal cancers [49].

20.6 Conclusion

The statistical validation of surrogate endpoints requires evaluating correlation between treatment effects in addition to the association at the patient level, which is particularly challenging in the context of time-to-event data. The hypothesis testing approach has progressively been replaced by an estimation/correlation approach to quantify the amount of surrogacy. The amount of effect on the final endpoint explained by the effect on the surrogate endpoint can be evaluated, or the ability to predict the final endpoint from the surrogate endpoint can be quantified. In this latter case, several trials must be analyzed to characterize the relationship between the two treatment effects. The modeling techniques, as well as their implementations, raise numerous problems that are only partially addressed with the current statistical techniques and software. This is an area of active research. The development of robust packages or software should increase the application of those approaches, while facilitating the evaluation of new proposals.

References

1. Lindholt JS, Sogaard R. Population screening and intervention for vascular disease in Danish men (VIVA): A randomised controlled trial. *Lancet.* 2017;390(10109):2256–65.
2. Allegra C, Blanke C, Buyse M et al. End points in advanced colon cancer clinical trials: A review and proposal. *J Clin Oncol.* 2007;25(24):3572–5.
3. Xie W, Regan MM, Buyse M et al. Metastasis-free survival is a strong surrogate of overall survival in localized prostate cancer. *J Clin Oncol.* 2017;35(27):3097–104.
4. Li Y, Taylor JM, Elliott MR. A Bayesian approach to surrogacy assessment using principal stratification in clinical trials. *Biometrics.* 2011;66(2):523–31.
5. Cameron D, Piccart-Gebhart MJ, Gelber RD et al. 11 years' follow-up of trastuzumab after adjuvant chemotherapy in HER2-positive early breast cancer: Final analysis of the HERceptin Adjuvant (HERA) trial. *Lancet.* 2017;389(10075):1195–205.
6. Swain SM, Baselga J, Kim S-B et al. Pertuzumab, trastuzumab, and docetaxel in HER2-positive metastatic breast cancer. *N Engl J Med.* 2015;372(8):724–34.
7. U.S. Food and Drug Administration. *Surrogate Endpoint Resources for Drug and Biologic Development.* 2018 07/25/2018 [cited 2018 07/26/2018]; Available from: https://www.fda.gov/Drugs/DevelopmentApprovalProcess/DevelopmentResources/ucm613636.htm.

8. Baker SG. Five criteria for using a surrogate endpoint to predict treatment effect based on data from multiple previous trials. *Stat Med*. 2018;37(4):507–18.

9. Burzykowski T, Molenberghs G, Marc B et al. Validation of surrogate end points in multiple randomized clinical trials with failure time end points. *Appl Stat*. 2001;50(4):405–22.

10. U.S. Department of Health and Human Services Center for Drug Evaluation and Research. (CDER), *Pathological Complete Response in Neoadjuvant Treatment of High-Risk Early-Stage Breast Cancer: Use as an Endpoint to Support Accelerated Approval* in *Guidance for Industry*. 2014.

11. Desmedt C, Di Leo A, de Azambuja E et al. Multifactorial approach to predicting resistance to anthracyclines. *J Clin Oncol*. 2011;29(12):1578–86.

12. Cortazar P, Zhang L, Untch M et al. Pathological complete response and long-term clinical benefit in breast cancer: The CTNeoBC pooled analysis. *Lancet*. 2014;384(9938):164–72.

13. Paoletti X, Oba K, Burzykowski T et al. Benefit of adjuvant chemotherapy for resectable gastric cancer: A meta-analysis. *JAMA*. 2010;303(17):1729–37.

14. Gastric Group Oba K, Paoletti X, Bang YJ et al. Role of chemotherapy for advanced/recurrent gastric cancer: an individual-patient-data meta-analysis. *Eur J Cancer*. 2013;49(7):1565–77.

15. Heinze G, Dunkler D. Avoiding infinite estimates of time-dependent effects in small-sample survival studies. *Stat Med*. 2008;27(30):6455–69.

16. Anderson JR, Cain KC, Gelber RD. Analysis of survival by tumor response. *J Clin Oncol*. 1983;1(11):710–9.

17. Giobbie-Hurder A, Gelber RD, Regan MM. Challenges of guarantee-time bias. *J Clin Oncol*. 2013; 31(23): p. 2963–9.

18. Prentice RL. Surrogate endpoints in clinical trials: Definition and operational criteria. *Stat Med*. 1989;8(4):431–40.

19. Freedman LS, Graubard BI, Schatzkin A. Statistical validation of intermediate endpoints for chronic diseases. *Stat Med*. 1992;11(2):167–78.

20. Djulbegovic B, Kumar A, Soares HP et al. Treatment success in cancer: New cancer treatment successes identified in phase 3 randomized controlled trials conducted by the National Cancer Institute-sponsored cooperative oncology groups, 1955 to 2006. *Arch Intern Med*. 2008;168(6):632–42.

21. Molenberghs G, Burzykowski T, Alonso A et al. A perspective on surrogate endpoints in controlled clinical trials. *Stat Methods Med Res*. 2004;13(3):177–206.

22. Buyse M, Molenberghs G. Criteria for the validation of surrogate endpoints in randomized experiments. *Biometrics*. 1998;54(3):1014–29.

23. Alonso A, Molenberghs G, Burzykowski T et al. Prentice's approach and the meta-analytic paradigm: A reflection on the role of statistics in the evaluation of surrogate endpoints. *Biometrics*. 2004;60(3):724–8.

24. O'Quigley J, Flandre P. Quantification of the prentice criteria for surrogate endpoints. *Biometrics*. 2006;62(1):297–300.

25. Frangakis CE, Rubin DB. Principal stratification in causal inference. *Biometrics*. 2002;58(1):21–9.

26. Conlon AS, Taylor JM, Elliott MR. Surrogacy assessment using principal stratification when surrogate and outcome measures are multivariate normal. *Biostatistics*. 2014;15(2):266–83.

27. Alonso A, Van der Elst W, Molenberghs G et al. On the relationship between the causal-inference and meta-analytic paradigms for the validation of surrogate endpoints. *Biometrics*. 2015;71(1):15–24.

28. Tanaka S, Matsuyama Y, Ohashi Y. Validation of surrogate endpoints in cancer clinical trials via principal stratification with an application to a prostate cancer trial. *Stat Med*. 2017;36(19):2963–77.

29. Alonso A, Van der Elst W, Meyvisch P. Assessing a surrogate predictive value: A causal inference approach. *Stat Med*. 2017;36(7):1083–98.

30. Buyse M, Molenberghs G, Burzykowski T et al. The validation of surrogate endpoints in meta-analyses of randomized experiments. *Biostatistics*. 2000;1(1):49–67.

31. Sargent DJ, Wieand HS, Haller DG et al. Disease-free survival versus overall survival as a primary end point for adjuvant colon cancer studies: Individual patient data from 20,898 patients on 18 randomized trials. *J Clin Oncol*. 2005;23(34):8664–70.

32. Vandenberghe S, Duchateau L, Slaets L et al. Surrogate marker analysis in cancer clinical trials through time-to-event mediation techniques. *Stat Methods Med Res*. 2017;27(11):3367–85.
33. Alonso A, Geys H, Molenberghs G et al. Validation of surrogate markers in multiple randomized clinical trials with repeated measurements: Canonical correlation approach. *Biometrics*. 2004;60(4):845–53.
34. Molenberghs G, Geys H. Buyse M. Evaluation of surrogate endpoints in randomized experiments with mixed discrete and continuous outcomes. *Stat Med*. 2001;20(20):3023–38.
35. Shi Q, Flowers CR, Hiddemann W et al. Thirty-month complete response as a surrogate end point in first-line follicular lymphoma therapy: An individual patient-level analysis of multiple randomized trials. *J Clin Oncol*. 2017;35(5):552–560.
36. Validity of surrogate endpoints in oncology Executive summary of rapid report A10-05, Version 1.1. In *Institute for Quality and Efficiency in Health Care: Executive Summaries*. Cologne, Germany, Institute for Quality and Efficiency in Health Care (IQWiG) (c) IQWiG (Institute for Quality and Efficiency in Health Care), 2005.
37. Alonso A, Molenberghs G. Surrogate marker evaluation from an information theory perspective. *Biometrics*. 2007;63(1):180–6.
38. Alonso A, Bigirumurame T, Burzykowski T et al. *Applied Surrogate Endpoint Evaluation Methods with SAS and R*. Chapman & Hall/CRC Biostatistics Series, London: Chapman and Hall, 2017:374.
39. Schemper M, Stare J. Explained variation in survival analysis. *Stat Med*. 1996;15(19):1999–2012.
40. Rotolo F, Paoletti X, Burzykowski T et al. A Poisson approach to the validation of failure time surrogate endpoints in individual patient data meta-analyses. *Stat Methods Med Res*. 2019;28(1):170–83.
41. Burzykowski T, Buyse M. Surrogate threshold effect: An alternative measure for meta-analytic surrogate endpoint validation. *Pharm Stat*. 2006;5(3):173–86.
42. Rotolo F, Paoletti X, Michiels S. Surrosurv: An R package for the evaluation of failure time surrogate endpoints in individual patient data meta-analyses of randomized clinical trials. *Comput Methods Programs Biomed*. 2018;155:189–98.
43. Renfro LA, Shang H, Sargent DJ. Impact of copula directional specification on multi-trial evaluation of surrogate end points. *J Biopharm Stat*. 2015;25(4):857–77.
44. Oba K, Paoletti X, Alberts S et al. Disease-free survival as a surrogate for overall survival in adjuvant trials of gastric cancer: A meta-analysis. *J Natl Cancer Inst*. 2013;105(21):1600–7.
45. Michiels S, Le Maître A, Buyse M et al. Surrogate endpoints for overall survival in locally advanced head and neck cancer: Meta-analyses of individual patient data. *Lancet Oncol*. 2009;10(4):341–50.
46. Bel Hechmi S, Michiels S. Paoletti X et al. Nouvelles mesures de distance pour valider un critère de substitution de type survie. *Revue d'Epidémiologie et de Santé Publique*. 2018;66(S3):S131.
47. Buyse M. Use of meta-analysis for the validation of surrogate endpoints and biomarkers in cancer trials. *Cancer J*. 2009;15(5):421–5.
48. Chibaudel B, Bonnetain F, Shi Q et al. Alternative end points to evaluate a therapeutic strategy in advanced colorectal cancer: Evaluation of progression-free survival, duration of disease control, and time to failure of strategy--an Aide et Recherche en Cancerologie Digestive Group Study. *J Clin Oncol*. 2011;29(31):4199–204.
49. Broglio KR, DA Berry. Detecting an overall survival benefit that is derived from progression-free survival. *J Natl Cancer Inst*. 2009;101(23):1642–9.

21

Competing Risks

Aurélien Latouche, Gang Li, and Qing Yang

CONTENTS

Competing risks are increasingly used in oncology to decipher the effect of a treatment on components of the overall survival. This chapter provides essential tools for summarizing treatment effect with an emphasize in oncology. After presenting the quantities of interest, we detail some new developments with a focus on joint inference. Finally, some case studies are provided with the corresponding R code to reproduce them (available on the book's companion website https://www.crcpress.com//9781138083776).

21.1 Introduction

Competing risks model time to first event and type of first event. The issue of competing risks is prominent in clinical epidemiology and oncology [1,2]. From a motivating breast cancer example originated some of the mostly used two-sample test to compare the probability of relapsing between two treatment arms [3], as well as a regression model [4]. Indeed, the take-home message was that testing the equality of the cause-specific hazards for local relapse was not equivalent to test the equality of the corresponding cumulative incidences of local relapse.

The available material for teaching and applying recent methods for competing risks in oncology reveals a large number of tutorials in solid tumors and hematology. Recent textbooks on competing risks provide a comprehensive overview of relevant quantities to model, regression, two-sample test, and sample size calculation (see Chapter 8 of Klein et al. [5], Young and Chen [6], or Beyersmann et al. [7]). In this chapter we will recall the traditional quantities of interest as well as the years of life lost and its decomposition by cause of death and detail the benefit of employing the so-called joint inference for pairs of quantities of interest.

21.2 Notations and Quantities of Interest

We will consider that two competing events act on the patients, namely *Interest* and *Other*. The failure type, ε, is 1 for *Interest* and 2 for *Other*. If patients can experience more than two events, we will consider *Interest* versus a single endpoint after the competing causes of failure are *aggregated* together. The aggregation of the causes of failure or equivalently the re-coding of $\varepsilon \neq 1$ will not modify the estimation of the cause-specific hazard for the event 1.

The stochastic process is a convenient framework to describe a competing risk model and to formalize what is the cause-specific hazard e.g., a "momentary force," as nicely termed by Beyersmann et al. [7]. This framework will also enable easier understanding of the effect of a covariate on the cause-specific hazard. With two competing events, the competing risks process $(X_t)_{t \geq 0}$ has two absorbing states, thus $X_t \in \{0, 1, 2\}$. For right censored data, we assume that $P(X_0 = 0) = 1$, i.e., every patient will start from state 0 at time origin $t = 0$. Individuals remaining in the state 0 are censored. The observed failure time T is thus $\inf\{t > 0 \mid X(t) \neq 0\}$ and the failure type is X_T.

The cause-specific hazard (CSH) for event k is defined as $\lambda_k(t) = \lim_{h \to 0}(1/h) P\{t \leq T \leq t+h, \varepsilon = k \mid T \geq t\}$. This is the instantaneous rate at which cause k events are occurring $(k = 1, 2)$. Employing the stochastic process notation, the CSH is $\lim_{h \to 0}(1/h)P\{t \leq T \leq t+h, X_T = k \mid T \geq t\}$. We will refer to the cause-specific hazard for event 2 as other cause-specific hazard (OCH) in the case study.

The all-causes hazard (ACH) is the sum over the competing events of the cause-specific hazards, i.e., $\lambda(t) = \lambda_1(t) + \lambda_2(t)$. The all-causes hazard translates into the probability scale since $S(t) = \exp - \int_0^t \lambda(u)du$ as a consequence $\exp - \int_0^t \lambda_k(u)du$ hasn't a probability interpretation. The all-causes hazard completely determines the survival function $S(t) = P(T > t)$ as $S(t) = \exp(-\int_0^t \lambda_1(u) + \lambda_2(u)du$. This relation clearly exemplifies why we have to consider the two competing CSHs for a probability interpretation. Note that neither $\exp(-\int_0^t \lambda_1(u)du)$ nor $\exp(-\int_0^t \lambda_2(u)du)$ have probabilistic interpretations, although they do in terms of overall survival function S.

The cumulative incidence function (CIF) for event k, $F_k(t) = P(T \leq t, \varepsilon = k)$ quantifies the risk of failure from cause k in the time frame $[0, t](k = 1, 2)$. Employing the stochastic process notation, the cumulative incidence expresses as $P(T \leq t, X_T = k)$.

The connection between the cause-specific hazard and the cumulative incidence can be summarized either from the integral or differential form. These connections are useful for understanding how each metric influences the other. Getting familiar with this relationship will help the reader when analyzing competing risks data.

- $S(t) = P(T > t) = 1 - (F_1(t) + F_2(t))$
- $\lambda_k(t) = \dfrac{dF_k(t)}{dt} / S(t)$
- $F_k(t) = \int_0^t S(u)\lambda_k(u)du$

Note that the relation between the cumulative incidence of *Interest* involves the CSH of *Interest* as well as the CSH of *Other*. Thus, for interpreting the cumulative incidence of *Interest*, both CSHs are required.

The subdistribution hazard (SH) is the hazard attached to the cumulative incidence, i.e., the SH of *Interest* is directly related to the cumulative incidence of *Interest*. Indeed, the cumulative incidence function of *Interest* is determined by $F_1(t) = \int_0^t P(T > u)\lambda_1(u)\,du$. The SH, $\tilde{\lambda}_1(t)$, is defined by requiring that

$$F_1(t) = 1 - \exp\left\{-\int_0^t \tilde{\lambda}_1(u)du\right\}$$

as a result, $\tilde{\lambda}_k(t) = (dF_k(t)/dt)/(1 - F_k(t))$.

Previous quantities are quantified in the probability scale either instantaneous for the hazard or absolute for the cumulative incidence. To assess the effect of covariates, a proportionality assumption is often made. In clinical trials the violation of the proportional hazard assumptions, notably for late effect therapy, has motivated the use of other metrics to quantify the benefit of a new treatment. Translating the estimation of a CSH or of a cumulative incidence in the *time* dimension, e.g., a gain or a loss of lifetime, is not straightforward.

For an upper age τ, the τ-restricted mean survival time, $\text{RMST}(0,\tau) = \int_0^\tau S(t)dt$, is the area under the survival curve up to τ. Thus $\text{YL}(0,\tau) = \tau - \int_0^\tau S(t)dt$ is the expected number of life years lost before time τ [8]. $\text{YL}(0,\tau)$ can be decomposed by the k causes of death as follows $\text{YL}(0,\tau) = \sum_k \int_0^\tau F_k(t)dt$. The RMST has gained popularity in oncology clinical trials [9]. An application of the RMST in a competing-risks setting compared to a cause-specific and subdistribution hazards-based analysis can be found in the article by Calkins et al. [10].

21.3 Inference and Joint Inference

We have stressed the interplay of the functionals that drive the competing-risks process. In the context of oncology, a typical pair of functionals would be the cause-specific hazard of relapse and the all-causes hazard. Furthermore, in the context of adverse events, the pair of interest might be the CSH for the event of interest and the OCH. See the Chapter 25 for an application of the competing risks setting for adverse events. With the advent of multiple endpoints and co-primary endpoints, joint inference will be required notably to plan a study with sufficient power [11]. Indeed, multiple primary endpoints become co-primary endpoints when it is necessary to demonstrate an effect on each of the endpoints to conclude that a drug is effective.

21.3.1 Regression Model

Let Z be a binary covariate that could denote a treatment arm or a dichotomized continuous covariate and T be the time to failure. The Cox model assumes that the CSH for a given patient can be factored into a baseline hazard that is common to all patients and a parametric function of the covariates that describes the patients' characteristics.

The CSH for the event of interest is expressed as

$$\lambda_1(t; Z = 1) = \lambda_1(t; Z = 0) \exp \gamma_1. \tag{21.1}$$

Similarly, the CSH for the competing event is

$$\lambda_2(t; Z = 1) = \lambda_2(t; Z = 0) \exp \gamma_2, \tag{21.2}$$

with $\exp(\gamma_\star)$ being the cause-specific hazard ratio (CSHR) for event of type $\star$.

The Fine–Gray model for the event of interest postulates a proportional subdistribution hazard

$$\tilde{\lambda}_1(t; Z = 1) = \tilde{\lambda}_1(t; Z = 0) \exp \beta_1, \tag{21.3}$$

with $\exp \beta_1$ being the subdistribution hazard ratio (SHR).

We stress that there is no reason why β_1 should equal γ_1, although in practice, close estimates can be found [12]. Of note, fitting a Fine–Gray model for the competing risk endpoint *Other* implicitly constrains that the SHR for the competing event *Other* depends on the SHR of *Interest* and the baseline SH. As a result, an increase in the cumulative incidence of *Interest* may be due to either a physiological effect of the exposure or to a decrease in the competing cumulative incidence.

With a multi center clinical trial or a meta-analysis of individual patient data with competing risks, we shall mention two useful extensions.

21.3.1.1 Regression with Clustered Data

The clustering or the dependence within a cluster refers to the potential dependence between the failure times from the cause of interest (1) within a center in a multi center trial or (2) within a trial in a meta-analysis, not to the untestable assumption of the dependence between failure times from different causes. A recent and complete overview of methods can be found in the literature [13,14]. Extension of the Fine–Gray model for clustered data was applied to assess the effect of treatment on the cumulative incidence function of second primary malignancies, accounting for heterogeneity across studies [15].

21.3.1.2 Regression with Missing Cause of Failure

In clinical trials in oncology, there may be a substantial proportion of missing causes of death (for example, in the meta-analysis [16], about 20% of causes of death were missing). When the cause of failure is missing for some of the patients known to have failed, i.e., for some of them, we know that $\varepsilon > 0$ but we do not know whether $\varepsilon = 1$ or $\varepsilon = 2$. Hence, for uncensored patients we observe a missingness indicator M, i.e., $M = 1$ if the cause is missing and $M = 0$ otherwise.

Let $\pi = P(M = 1 \mid T, D, X, U = 0)$ be the conditional probability that the cause of failure is missing among uncensored individuals denoted by $U = 0$, with T, D, and X being the observed failure time, the cause of failure, and a vector of covariates, respectively. Following Moreno-Betancur and Latouche [17], a taxonomy of the mechanism driving missingness is Missing Completely at Random if π is constant, Missing at Random if π does not depend on the cause of failure, and Missing Not At Random otherwise. When the cause of failure is only known to belong to a certain subset of all possible failures, it is referred to as a masked cause of failure.

A common strategy for handling missing cause of failure are either *ad hoc* or model based:

- *Complete case analysis*: The standard terminology with missing data (not missing outcome) is complete case where patients with missing cause of death are discarded. However, for competing risks, this terminology is misleading since complete case is limited to uncensored patients. *Known cause of failure* case only would be more appropriate.

- *Recode*: Recode missing causes according to the observed distribution of the competing events.

- *Extra State*: Adds a new state for missing or unknown cause. This strategy assumes a null missingness probability for events from the cause of interest, which can be seen as a very simplistic type of MNAR mechanism.

- *Model based*: Extensive simulation studies highlighted that the inverse probability weighting (IPW) and multiple imputation (MI) approaches led to approximately unbiased estimates with comparable precision under relaxed assumptions about the missingness mechanism and should therefore be used in the primary analysis [17]. For an in-depth presentation of the implementation of theses strategies, see Moreno-Betancur [18].

21.3.1.3 Regression for Years Lost

For right-censored, time-to-event data, a method based on pseudo-values to perform linear regression for the expected number of life years lost due to any cause and due to cause k was proposed by Andersen [8] and implemented in the R package pseudo [19]. The methodology can be further adapted to consider life expectancy given survival to a certain age t_0 by considering either the conditional survival function $S(t)/S(t_0)$ or the conditional cumulative incidence function $F_k(t)/S(t_0)$ [8].

21.4 Tests

This section recalls the null and alternative hypothesis for CSH and mentions the extension for SH for testing treatment effect in the presence of competing risks. We will assume, for simplicity, that n patients are exposed to two distinct and exclusive failure causes, thus defining a competing risks setting.

The main endpoint of the trial is the occurrence of one of these two failure causes. Patients are randomly assigned to a control group C or an experimental treatment group E. Let τ be some pre-specified fixed time.

The log-rank's test considers the following null hypothesis:

$$H_0 : \lambda_{1E}(t) = \lambda_{1C}(t) \quad \forall 0 < t < \tau.$$

Additionally, when performing cause-specific analysis, we consider that the side condition $\lambda_{2E}(t) = \lambda_{2C}(t)$ is assumed, so that the treatment affects the cause-specific hazard of the event of interest but not that of the competing event [20]. In the presence of early or late effect of the grouping covariate (here, the treatment), a weighted log-rank test can be employed for testing equality of cause-specific hazards.

The Gray test considers the following:

$$H_0 : F_{1E}(t) = F_{1C}(t) \quad \forall 0 < t < \tau.$$

No side condition is postulated since the cumulative incidences sum to 1 [21]. The Gray's test actually compares the cumulative subdistribution hazards in analogy to the log-rank test that compares cumulative cause-specific hazards. The null hypothesis of the Gray's test is equivalent to $H_0 : \tilde{\lambda}_{1E}(t) = \tilde{\lambda}_{1C}(t)$. Finally, motivated by Fleming and Harrington's weighting, a set of new weight functions to compare the cumulative incidence functions between two groups was proposed by Li et al. [22].

As highlighted by Li and Yang [23], given the relationship between the cause-specific hazards, the all-causes hazard, and the cumulative incidence, joint inference for pairs provides equivalent null hypothese; however the alternative hypothesis differs.

21.4.1 Joint Tests

The most common advice when considering the strategy to employ with competing risks is that cause-specific hazard and cumulative incidence reflect different dynamics of the competing-risk process and that depending on the clinical question to address, one approach is more suitable than the other. Joint inference appears as a natural tool to address the advice above, connecting both pairs and facilitating synthesis. Otherwise, the analysis on different scales are somehow disconnected.

This motivates the use of joint tests that consider the following null:

$$H_0 : \lambda_{1E}(t) = \lambda_{1C}(t) \quad \text{and} \quad F_{1E}(t) = F_{1C}(t) \quad \forall 0 < t < \tau.$$

$$H_0 : \lambda_{1E}(t) = \lambda_{1C}(t) \quad \text{and} \quad \lambda_{\bullet E}(t) = \lambda_{\bullet C}(t) \quad \forall 0 < t < \tau,$$

where $\lambda_{\bullet E}(t) = \lambda_{1E}(t) + \lambda_{2E}(t)$ is the all-causes hazard in experimental arm (resp control arm). We further define the all-Causes-hazard ratio (ACHR) as $\text{ACHR} = (\lambda_{\bullet E}(t))/(\lambda_{\bullet C}(t))$.

Two-sample joint tests for $\lambda_1(t)$ and $F_1(t)$ are derived from the asymptotic joint distribution of the weighted log-rank test statistic for $\lambda_1(t)$ and the Gray's test statistic for $F_1(t)$. A chi-square joint test and a maximum joint test can be employed. Similarly, two-sample joint test for cause-specific hazard and all-causes hazard were derived [23].

In the next sections, we illustrate the insights of the joint inference with two case-studies.

21.4.1.1 Example 1: Hodgkin's Disease

The Hodgkin's disease data consists of 865 patients who were diagnosed with Hodgkin's disease and received radio therapy in Princess Margaret Hospital between 1968 and 1986 [24]. Here, we consider *time to second malignancy* after receiving radio therapy, which is an important endpoint for evaluating the side effects of radiotherapy. Death without second malignancy is a competing risk. Among the 865 patients, 93 developed second malignancy, 386 died without second malignancy, and 386 were right censored, i.e., did not experience any of the two events by the end of the study. As an illustration, we investigate whether or not the probability of developing second malignancy were the same among older (≥ 30) and younger (<30) patients.

Figure 21.1a and b depict the cumulative cause-specific hazard and the cumulative incidence functions, respectively, of time to second malignancy for the older (≥ 30) and younger (<30) groups. There appears to be a higher cause-specific hazard for the older patients since the slope of their cumulative cause-specific hazard is noticeably bigger (Figure 21.1a). However, the cumulative incidence functions for the two age groups are barely distinguishable (Figure 21.1b). The two-sample log-rank test for the cause-specific hazard for time to second malignancy yields a p-value = 0.037. The Gray's test for the cumulative incidence for time to second malignancy gives a p-value = 0.770. After applying the Bonferroni adjustment for multiple testing of both CSH and CIF together, none of the two individual tests is statistically significant at level $0.05/2 = 0.025$ at the 5% overall significant level, which is not surprising since the Bonferroni adjustment does not take into account of the joint distribution of the two individual test statistics and is generally conservative.

Next, we apply the chi-square joint test and the maximum joint test [23] to compare between older and younger patients with respect to CSH and CIF jointly. The resulting p-values are

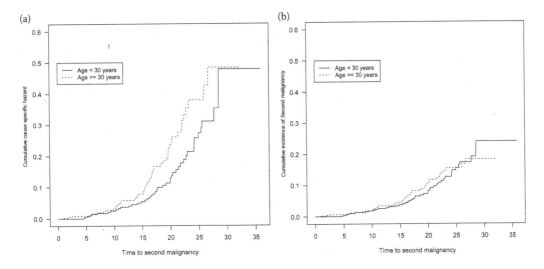

FIGURE 21.1
Power comparisons for three joint tests of a joint regression model with respect to the CSH and CIF pair. (a) Cumulative cause-specific hazard functions for time to second malignancy for older (≥ 30) and younger (<30) patients (log-rank test p-value = 0.037). (b) Cumulative incidence functions for time to second malignancy for older (≥ 30) and younger (<30) patients (Gray's test p-value = 0.770).

TABLE 21.1

Separate and Joint Test Results for the Hodgkin's Disease Example for Three Pairs of Quantities

	Separate Test			Joint Test	
	CSH	CIF	Bonferroni	χ^2	Max
p-value	0.037	0.770	0.074	0.020	0.050
	CSH	ACH	Bonferroni	χ^2	Max
p-value	0.037	$5.2E-8$	$1.0E-7$	$3.4E-7$	$3.0E-8$
	CSH	OCH	Bonferroni	χ^2	Max
p-value	0.037	$4.7E-7$	$9.4E-7$	$3.5E-7$	$8.0E-7$

given in Table 21.1 (line 3), along with the results of the individual tests and the Bonferroni's method. We see that for the (CSH, CIF) pair, the two-sample chi-square joint test yields a p-value 0.02 and the maximum joint test yields a p-value of 0.05. On the contrary, the more conservative Bonferroni's method does not reach the 5% significance level. We also performed joint tests for two other pairs of quantities: (CSH, ACH) and (CSH, OCH) (see lines 2 and 3 of Table 21.1), which show that in addition to an elevated cause-specific hazard of time to second malignancy, the older patients also had a higher risk of dying from other life-threatening diseases without developing second malignancy. This explains why their observed cumulative incidence for time to second malignancy was not significantly different from the younger patients.

In addition to performing hypothesis testing, it is also useful to present confidence region estimates for the hazard ratios of different pairs of quantities of interest using the joint inference methodology developed by Li and Yang [23]. For example, Figure 21.2 depicts the 95% confidence regions for the CSH ratio and the ACH ratio using three different methods (Bonferroni, chi-squared, and maximum). It is clear that the chi-squared joint confidence region reveals that a higher CSH ratio tends to be accompanied by a higher ACH ratio, which cannot be discovered by the Bonferroni or maximum method.

Similarly, Figure 21.3 depicts the 95% confidence region for the CSH ratio and the OCH ratio using the Bonferroni, chi-squared, and maximum methods. Of note, none of the three methods suggest any association between the estimated CSH ratio and the estimated OCH ratio, which is a well known fact, theoretically [23].

On the other hand, the chi-squared confidence region does show that the variance of the OCH ratio becomes smaller as the CSH ratio gets large. Finally, we did not calculate a confidence region for the CSH ratio and CIF hazard ratio because the proportional hazards assumption generally does not hold simultaneously for the CSH and CIF pair [21,23].

21.4.1.2 Example 2: Bone Marrow Transplant (BMT) Study

These data were previously analyzed by Latouche et al. [2] and consist of patients over 50 years of age with acute myeloblastic leukemia. The outcomes of 315 patients treated with a reduced-intensity conditioning regimen (RIC) is compared to those of 407 patients with the standard-of-care myeloablative regimen (MAC). The competing endpoints are relapse ($n = 182$) and treatment-related mortality (TRM, $n = 164$), with 376 patients being censored over two years of follow-up. The main prognostic factor in these analyses is the status of the disease at the time of the transplant. The majority of patients achieve remission before transplantation, but some diseases are resistant to chemotherapy so patients are

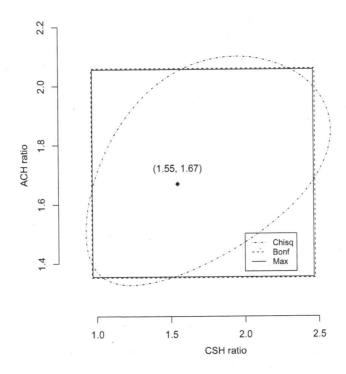

FIGURE 21.2
Confidence region of joint analysis for CSH ratio and ACH ratio (chi-square p-value $= 3.4E - 7$, max p-value $= 0.05$, Bonferroni p-value $= 1.0E - 7$).

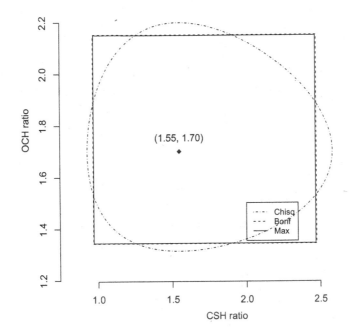

FIGURE 21.3
Confidence region for three joint tests. Confidence region of joint analysis for CSH ratio and OCH ratio (chi-square p-value $= 0.02$, max p-value $= 0.05$, Bonferroni p-value $= 0.074$).

TABLE 21.2

Separate and Joint Test Results for the BMT Example for Three Pairs of Quantities

	Separate Test			Joint Test	
Test	CSH	CIF	Bonferroni	χ^2	Max
p-value	0.00293	$5.75e-05$	0.000115	0.000242	0.00014
Test	CSH	ACH	Bonferroni	χ^2	Max
p-value	0.00293	0.53863	0.005874	$4.411e-06$	0.0071
Test	CSH	OCH	Bonferroni	χ^2	Max
p-value	0.00293	$6.96e-05$	0.000139	$4.403e-06$	$1e-04$

Note: χ^2 and Max are abbreviations for the chi-square joint test and the maximum joint test.

transplanted in the refractory or relapse phase of the disease (referred to as advanced status). The covariates of clinical interest considered in this example are the conditioning regimen (MAC (56%) versus RIC (44%), as well as the disease status at transplantation—other (72%) vs. advanced status (28%).

We first performed three sets of two-sample chi-square joint tests with respect to time to relapse between two different conditioning regimen (MAC vs. RIC) by considering different pairs of quantities: (CSH, CIF), (CSH, ACH), and (CSH, OCH). The p-values are presented in Table 21.2, along with the results of the individual tests and the Bonferroni's method. All the tests are highly significant at 5% significance level.

It is worth noting that the Bonferroni method would always be statistically significant at level $\alpha = 0.05$ if one of the individual tests has a p-value less than 0.025. However, even in this case, joint inference could provide further insight into the effects of variable or condition by examining the joint confidence region of a pair of hazard ratios of interest. For example, Figures 21.4 and 21.5 give 95% confidence regions for the CSH ratio and ACH ratio, and the CSH ratio and OCH ratio, based on their joint Wald test statistics. It is seen that for (CSH, ACH), the areas of confidence region from the chi-square test are much smaller than the Bonferroni and maximum tests. Furthermore, the chi-squared joint confidence region reveals that a higher CSH ratio is associated with a higher ACH ratio, as in the previous example.

21.4.2 Joint Regression Analysis

21.4.2.1 Example 1: Follicular Cell Lymphoma Study (Time to Progression and Progreesion-Free Survival)

The follicular cell lymphoma study [24,25] consists of 541 early-stage (I or II) follicular-type lymphoma patients who were enrolled between 1967 and 1996 and treated with either radiation alone (RT) or with radiation and chemotherapy (CMT). There were 272 events due to disease (relapse or no treatment response), 76 competing-risk events (death without relapse), and 193 censored individuals who didn't experience any of the two events at the end of the follow-up. As in Scheike and Zhang [25], we tested if the CMT group has a longer time to relapse or no treatment response than the RT group. Although one could study different pairs of quantities, we considered joint inference of the cause-specific hazard and the all-causes hazard because they correspond to two commonly used clinical endpoints, namely, time to progression (TTP) and progression-free survival (PFS), in oncology trials.

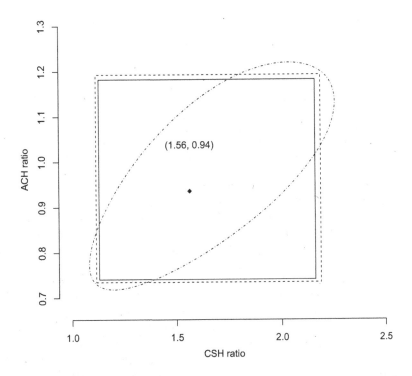

FIGURE 21.4
Confidence region of joint analysis for CSH ratio and ACH ratio (CSH ratio = 1.56, ACH ratio = 0.94, chi-square p-value = $4.411e - 06$, Max p-value = 0.0071, Bonferroni p-value = 0.005874).

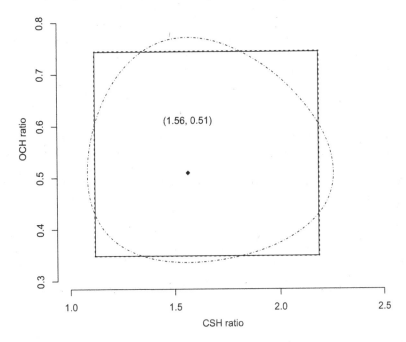

FIGURE 21.5
Confidence region of joint analysis for CSH ratio and ACH ratio (CSH ratio = 1.56, OCH ratio = 0.51, chi-square p-value = $4.403e - 06$, Max p-value = $1e - 04$, Bonferroni p-value = 0.000139).

TABLE 21.3

Separate and Joint Test Results for the Follicular Cell Lymphoma Study

	Separate Test			Joint Test	
Test	CSH	ACH	Bonferroni	χ^2	Max
p-value	0.035	0.037	0.070	0.182	0.047

Note: χ^2 and Max are abbreviations for the chi-square joint test and the maximum joint test.

Here, TTP, defined as time to relapse or no treatment response, is an endpoint for the anti-tumor activity of a treatment, and PFS, defined as time to progression or death before progression, is an endpoint for the overall effects on a patient. In addition to a binary treatment variable (1 for RT, and 0 for CMT), we adjust for patient's baseline age, stage, and hemoglobin level (hgb) by including them as covariates in our models. We conducted the chi-square joint test and the maximum joint test for the treatment variable, and we summarized the results along with the Bonferroni adjustment method and the individual tests in Table 21.3. The maximum joint test (p-value = 0.047) is significant, whereas the chi-square joint test (p-value = 0.182) and the Bonferroni method (p-value = 0.07) are not significant at 5% significance level. The one-sided individual test statistics for CSH and ACH are 1.81 and 1.78, respectively, both exceeding 1.77, the cutoff value of the maximum test. Therefore, at 5% overall significance level, the CMT group has a lower risk of TTP (cause-specific hazard) and a lower risk of PFS (ACH) as compared to the RT group, adjusting for patient's baseline age, stage, and hgb level. Finally, the chi-square joint test has a relatively large p-value because it is actually a two-sided test that is not powered for a one-sided hypothesis, especially when the effect sizes for CSH and ACH are similar.

We also generated confidence regions for the CSH ratio and the ACH ratio in Figure 21.6 using the Wald test statistic from the joint regression models. The confidence regions produced by the chi-square test statistic and maximum test statistic are much tighter around the point estimate than the Bonferroni method.

21.4.2.2 Example 2: BMT Study

In the previous section, we have performed two-sample joint tests with respect to three different pairs of quantities. To determine which conditioning regimen works better after adjusting for other important prognostic factors, we built joint regression models for the CSH and ACH pair by including conditioning regimen as the key group variable and the disease status at transplantation (other vs. advanced status) as the covariate. Again, all three joint test p-values are significant at 5% level (Table 21.4). In Figure 21.7, we present the 95% confidence region for the CSH ratio and the ACH ratio generated using the Wald test statistic from joint regression models for the CSH and ACH pair. Similar to the previous example, the confidence regions produced by the chi-square test statistic and the maximum test statistic are much tighter around the point estimate than the Bonferroni method.

21.4.3 Sample Size

For sample size calculation in the presence of competing risks, there are other contributions [21,26,27] that discuss either the approach relying on the CSHR or the SHR or a combined endpoint in the presence of competing risks.

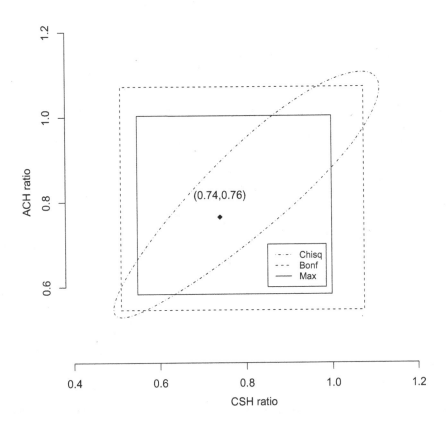

FIGURE 21.6
Confidence region for regression. Confidence region of joint analysis for CSH ratio and ACH ratio with follicular data (CSH ratio = 0.74, ACH hazard ratio= 0.76, chi-square p-value = 0.02, Max p-value = 0.05, Bonferroni p-value = 0.074).

21.4.3.1 Sample Size Calculation for Joint Inference of Cause-Specific Hazard and All-Causes Hazard

The disease-specific survival (DSS), defined as time to death due to a disease of interest, and overall survival (OS), defined as time to death due to all causes, are often used as co-primary endpoints in clinical trials. Sample size determination based on the co-primary endpoints DSS and OS can be facilitated using a competing-risks model under which the treatment effects on DSS and OS are characterized jointly by the cause-specific hazard for the event of interest (CSH_1), the instantaneous risk of failure due to the disease of interest, and the all-causes hazard (ACH) [2,23].

TABLE 21.4

Separate and Joint Test Results for the BMT Study

	Separate Test			Joint Test	
Test	CSH	ACH	Bonferroni	χ^2	Max
p-value	0.012	0.33	0.024	<0.0001	<0.0001

Note: χ^2 and Max are abbreviations for the chi-square joint test and the maximum joint test.

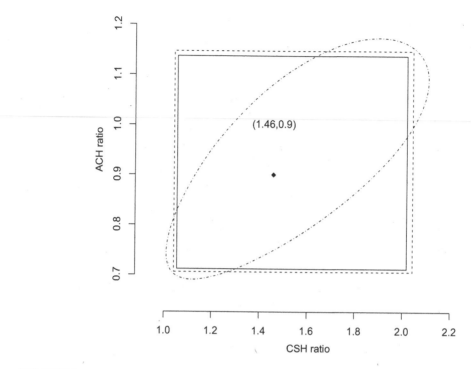

FIGURE 21.7

Confidence region for regression. Confidence region for three joint tests for the CSH and ACH pair with BMT data (CSH ratio = 1.46, ACH ratio = 0.9, chi-square p-value<0.0001, Max p-value<0.0001, Bonferroni p-value =0.024).

A commonly used approach to sample size calculations with respect to a pair of co-primary endpoints is to carry out a power analysis for each endpoint with a Bonferroni adjustment. Power analysis for DSS (CSH_1) has been discussed by Schulgen et al. [20], among others, and power analysis for OS can be done using any method for a single time-to-event outcome. However, a more cost-effective approach is to use the method of Yang et al. [28], who developed an analytical sample size calculation method and an R package powerCompRisk [29] for the co-primary endpoints DSS and OS based on a chi-square joint test and a maximum joint test of CSH_1 and ACH.

Table 21.5 presents the required number of events and required number of patients for the three above-mentioned sample size determination methods from a simulation study [28].

TABLE 21.5

Required Sample Sizes for the Chi-Square Joint Test, the Maximum Joint Test, and the Bonferroni Method, with Nominal Significance Level $\alpha = 0.05$ and Power $1 - \beta = 0.8$

CSH_1 Ratio	ACH Ratio	Number of Cause-1 Failures (D_1)			Number of Patients (N)		
		Chi-Square	Max	Bonferroni	Chi-Square	Max	Bonferroni
1.2	1.2	928	794	916	1266	1082	1248
1.2	1.4	150	248	270	204	338	368
1.2	1.7	42	100	110	56	136	150

Note: The cause-specific hazard for cause 1 in group 1 is $\lambda_{11} = 0.3$, the ratio of cumulative incidence of failure due to cause 1 to all causes is $R = 0.8$, the length of recruitment period is $r = 1$, the length of the follow-up time is $\tau = 10$, and the censoring rate due to loss of follow-up is $R_c = 0.05$.

It is clear from Table 21.5 that in comparison to the Bonferroni adjustment, substantial savings in sample size is achieved by the chi-square joint test if the CSH_1 ratio and the ACH ratio are different and by the maximum joint test if the CSH_1 ratio and the ACH ratio are similar. In practice, it is recommended to perform power analyses for all the three methods and choose the most efficient design, as illustrated later in the case study.

21.4.4 The `powerCompRisk` R Package

The `powerCompRisk` R package can be used to determine

1. The required number of events of interest.
2. The required number of patients based on a two-sample chi-square joint test and a two-sample maximum joint test of CSH_1 and ACH.

Below is a description of the R function:

power.comp.risk (alpha=, beta=, lambda_11=, RR=, HR_1=, HR_all=, attrition=, r=, f=, a1=),

where the input arguments are defined below

alpha: type I error.
beta: type II error.
lambda_11: cause-1 cause-specific hazard in group 1 (control group).
RR: relative risk of cause-1 failure versus all-causes failure in group 1.
HR_1: pre-specified cause-1 cause-specific hazard ratio between groups 1 and 2.
HR_all: pre-specified all-causes hazard ratio between groups 1 and 2.
r: length of patient accrual period.
f: maximum follow-up period.
attrition: attrition rate due to lost to follow-up.
a1: sample allocation proportion for group 1.

The returned values include the *number of events of interest* and the *number of required patients* for both a chi-square test and a maximum joint test. Next, we present an example to illustrate how to obtain the input arguments required by the above R function.

21.4.5 An Example: The 4D Trial

The 4D trial is a randomized, double-blinded, placebo-controlled trial to assess the efficacy of antihyperlipidemic treatment with atorvastatin in reducing occurrence of non-fatal myocardial infarction and cardiovascular mortality [30]. There are three competing risks: non-fatal myocardial infarction, death due to cardiovascular disease, and death due to other causes. In this example, we define the event of interest as the composite event of either non-fatal myocardial infarction or death due to cardiovascular disease, and the competing event as death due to other causes.

Yang et al. [28] demonstrated how to re design this trial based on joint tests of CSH_1 and ACH between the atovastatin and placebo groups. They illustrated how to determine the input arguments for the above R function based on the four-year occurrence of the

TABLE 21.6

Required Number of Cause-1 Failures (D_1) and Number of Patients (N) for the 4-D Trial Based on the Chi-Square Joint Test, the Maximum Joint Test, and the Bonferroni Adjustment, Assuming $R_c = 10\%$ Attrition Rate Due to Lost to Follow-Up, $\tau = 8$ Years of Maximum Follow-Up Time, $r = 2$ years of accrual, $\alpha = 0.05$, and $\beta = 0.20$

Chi-Square Joint Test		Maximum Joint Test		Bonferroni Test	
D_1	N	D_1	N	D_1	N
290	632	252	548	288	624

cause-1 failure and the four-year all-causes incidence. Such information for the control group (group 1) is obtained using information from a previous prospective cohort study from 1985 to 1994 in Germany [31]. Furthermore, assume that the intervention is efficacious if it reduces the four-year occurrence of the cause-1 failure from 52% to 42% or reduces the four-year all-causes incidence from 80% to 70%. The above information can be converted to obtain the following: $\lambda_{11} = 0.26$, RR $= 0.625$, CHR_1 $= 1.44$, HR_all $= 1.33$, as detailed by Yang et al. [28].

Table 21.6 gives the required number of cause-1 events (non-fatal myocardial infarction or death due to cardiovascular disease) and the total number of patients under a given scenario of the attrition rate, maximum follow-up time, and length of accrual period. It is observed that in this example, the design based on the maximum joint test would be the most efficient with the fewest required cause-1 events and total number of patients.

21.5 Conclusion

The R task view Survival provides an up-to-date list of the available packages and is a good source of tutorials https://cran.r-project.org/web/views/Survival.html. Our motivation for this chapter was not to be exhaustive but to highlight some recents contributions that will help to understand and analyze competing-risks data.

References

1. Koller MT, Raatz H, Steyerberg EW, Wolbers M. Competing risks and the clinical community: Irrelevance or ignorance? *Stat Med.* May 2012;31(11–12):1089–97.
2. Latouche A, Allignol A, Beyersmann J, Labopin M, Fine JP. A competing risks analysis should report results on all cause-specific hazards and cumulative incidence functions. *J Clin Epidemiol.* 2013;66(6):648–53.
3. Gray RJ. A class k-sample tests for comparing the cumulative incidence of a competing risk. *Ann Stat.* 1988;116:1141–54.
4. Fine JP, Gray RJ. A proportional hazards model for subdistribution of a competing risk. *J Am Stat Assoc.* 1999;94(446):496–509.

5. Klein JP, van Houwelingen HC, Ibrahim JG, Scheike TH. *Handbook of Survival Analysis*. Chapman and Hall, 2013.

6. Young WR, Chen D. *Clinical Trial Biostatistics and Biopharmaceutical Applications*, CRC Press, 2014.

7. Beyersmann J, Allignol A, Schumacher M. *Competing Risks and Multistate Models with R*, Use R!, First Edition. New York: Springer-Verlag, 2012.

8. Andersen PK. Decomposition of number of life years lost according to causes of death. *Stat Med*. Dec 2013;32(30):5278–85.

9. Trinquart L, Jacot J, Conner SC, Porcher R. Comparison of treatment effects measured by the hazard ratio and by the ratio of restricted mean survival times in oncology randomized controlled trials. *J Clin Oncol*. May 2016;34(15):1813–9.

10. Calkins KL, Canan CE, Moore RD, Lesko CR, Lau B. An application of restricted mean survival time in a competing risks setting: Comparing time to ART initiation by injection drug use. *BMC Med Res Methodol* 18(1):27, 2018.

11. US Food and Drug Administration. Draft Guidance. 2017. Multiple endpoints in clinical trials guidance for industry. Technical report, Food and Drug Administration. Available at: https://www.fda.gov/downloads/drugs/guidancecomplianceregulatoryinformation/guidances/ucm536750.pdf.

12. Beyersmann J, Latouche A, Buchholz A, Schumacher M. Simulating competing risks data in survival analysis. *Stat Med*. 2009;28:956–71.

13. Diao G, Zeng D. Clustered competing risks. In *Handbook of Survival Analysis*, Klein JP, van Houwelingen HC, Ibrahim JG, Scheike TH, eds. Springer, 2013.

14. Ha ID, Christian NJ, Jeong JH, Park J, Lee Y, Analysis of clustered competing risks data using subdistribution hazard models with multivariate frailties. *Stat Methods Med Res*. Dec 2016;25(6):2488–505.

15. Palumbo A, Bringhen S, Kumar SK et al. Second primary malignancies with lenalidomide therapy for newly diagnosed myeloma: A meta-analysis of individual patient data. *Lancet Oncol*. Mar 2014;15(3):333–42.

16. Blanchard P, Lee A, Marguet S et al. Chemotherapy and radiotherapy in nasopharyngeal carcinoma: An update of the MAC-NPC meta-analysis. *Lancet Oncol*. 2015;16(6):645–55.

17. Moreno-Betancur M, Latouche A. Regression modeling of the cumulative incidence function with missing causes of failure using pseudo-values. *Stat Med*. 2013;32(18):3206–23.

18. Moreno-Betancur M. *Regression modeling with missing outcomes: Competing risks and longitudinal data*. PhD thesis, Paris-Sud, 2013. 2013PA11T076.

19. Pohar-Perme M, Gerster M. *Pseudo: Computes Pseudo-Observations for Modeling*, 2017. R package version 1.4.3.

20. Schulgen G, Olschewski M, Krane V, Wanner C, Ruf G, Schumacher M, Sample sizes for clinical trials with time-to-event endpoints and competing risks. *Contemp Clin Trials*. Jun 2005;26(3):386–96.

21. Latouche A, Porcher R. Sample size calculations in the presence of competing risks. *Stat Med*. Dec 2007;26(30):5370–80.

22. Li J, Le-Rademacher J, Zhang MJ. Weighted comparison of two cumulative incidence functions with R-CIFsmry package. *Comput Methods Programs Biomed*. 2014;116(3):205–14.

23. Li G, Yang Q. Joint inference for competing risks survival data. *J Am Stat Assoc*. 2016;111(515):1289–300.

24. Pintilie M, *Competing Risks: a Practical Perspective*. New York: John Wiley & Sons, 2006.

25. Scheike TH, Zhang MJ. Analyzing competing risk data using the R timereg package. *J Stat Softw*. 38(2), 2011.

26. Ohneberg K, Schumacher M. Sample size calculations for clinical trials. In *Handbook of Survival Analysis*. Klein JP, van Houwelingen HC, Ibrahim JG, Scheike TH, eds. Springer, 2013.

27. Rauch G, Beyersmann J. Planning and evaluating clinical trials with composite time-to-first-event endpoints in a competing risk framework. *Stat Med*. Sep 2013;32(21):3595–608.

28. Yang Q, Fung WK, Li G. Sample size determination for jointly testing a cause-specific hazard and the all-cause hazard in the presence of competing risks. *Stat Med*. Apr 2018;37(8): 1389–401.

29. Yang Q, Fung WK, Kawaguchi E, Li G. powercomprisk: Power analysis tool for joint testing hazards with competing risks data. R package version 0.1.0 (https://cran.r-project.org/web/packages/powerCompRisk/index.html), 2017.

30. Wanner C, Krane V, Ruf G, Marz W, Ritz E. Rationale and design of a trial improving outcome of type 2 diabetics on hemodialysis. Die Deutsche Diabetes Dialyse Studie Investigators. *Kidney Int Suppl*. 1999;71:S222–6.

31. Koch M, Kutkuhn B, Trenkwalder E, Bach D, Grabensee B, Dieplinger H, Kronenberg F. Apolipoprotein B, fibrinogen, HDL cholesterol, and apolipoprotein(a) phenotypes predict coronary artery disease in hemodialysis patients. *J Am Soc Nephrol*. 1997;8(12):1889–98.

22

Cure Models in Cancer Clinical Trials

Catherine Legrand and Aurélie Bertrand

CONTENTS

22.1 Introduction

Over the last decades, it has become more and more common in cancer clinical trials to observe patients experiencing long-term relapse-free survival, and cure has become a reality for both patients and clinicians [1]. It is indeed, nowadays, widely accepted that for a number of cancer types, such as early-stage breast cancer [2], colon cancer [3], or childhood acute lymphocytic leukemia [4], among others, treatment can lead to cure for a fraction of the patients.

Therefore, for these types of cancer, the primary goal of cancer therapy can not only be prolongation of survival but should shift toward cure. Maetani and Gamel [1] pointed out that this is especially true for cancers occurring in children, as in this case, a curative treatment will yield many years of healthy life, while a life-prolonging treatment will only offer a limited benefit before relapse takes the child's life. However, it may also be crucial for adult patients to express the benefit of new therapies, not only in terms of delaying death but also in terms of cure, as this can free the patients from cancer-associated sufferings, which could sometimes be more unbearable to patients than death itself [1]. In that new paradigm, the proportion of cured patients (often referred to as the *cure rate*) is an important measure of long-term survival benefit.

A common feature of time-to-event data in clinical trials is *right censoring*, meaning that, at the time of analyzing the trial results, some of the patients have not (yet) experienced the event of interest. Right censoring unfortunately prevents us from distinguishing the cured patients from the patients who will experience the event of interest after the censoring time. Furthermore, a very common assumption is that censored subjects will follow the same survival pattern after withdrawal as non-censored subjects. This leads to the implicit assumption that if the follow-up was long enough, one would observe the event of interest for all patients, which is obviously not the case in clinical trials of curable diseases.

The most commonly used statistical methods for the analysis of clinical trials with a time-to-event type of endpoint are most certainly the (non-parametric) log-rank test [5] and the (semiparametric) Cox proportional hazards (PH) model [6]. Although the latter is widely used, it is well known that it relies on the proportional hazards assumption, that is, while the absolute underlying hazard may vary over time, the hazard ratio (HR) between the two treatment groups remains constant. Although this PH assumption is not necessary for the computation of the log-rank statistic, it is well known that it is required to achieve maximal power of the test and that it is also central to the interpretation of the results [7]. Indeed, if one can assume that the HR between the two treatment groups remains constant, then this single constant can adequately be used to summarize the difference between the two survival curves over time. If the PH assumption is not met, using the semiparametric Cox PH model, and to some extent the log-rank test, may lead to both misleading and uninterpretable conclusions, in particular, if the censoring rate is high [7]. Therefore, while these "classical" survival methods are usually appropriate in clinical settings where we expect few patients to be cured and where the primary goal is to identify treatment allowing prolonging the duration of remission [8], they have been challenged over the last years by the apparition of new treatments having different mechanisms of action on the occurrence of recurrences. Indeed, testing of therapies associated with a *delayed treatment effect* or a *rebound effect* will obviously lead to non-proportional hazards situations. But this is also the case of new therapies having a *curative effect* affecting the proportion of *cured* patients, with or without affecting the timing of occurrence of recurrences for the other patients. In this latter setting, the study sample will consist of a mixture of *"cured"* and *"uncured"* (also often called *"susceptible"*) patients, the latter experiencing disease recurrence after some time following inclusion in the trial. Note that if the primary endpoint is overall survival (OS) or progression-free survival (PFS), as is often the case in oncology trials, it is clear that no patient can be cured from death; one will then be speaking of long-term survivors and it is convenient to think of these long-term survivors as (statistically) cured [9–11]. Such a heterogeneous population of short- and long-term survivors may lead to the PH assumption of the Cox model to be violated. While extensions of the conventional Cox PH model have been proposed to deal with the issue of non-proportionality (e.g., inclusion of time-varying covariate effects [12]), these methods do not adequately allow one to distinguish between the curative and life-prolonging effects of a new treatment [7,11].

Nearly 70 years ago, Boag was the first statistician to draw attention to the presence of cure in the analysis of cancer-related survival data by proposing in a seminal article a parametric cure model allowing estimation of the cure rate as one of its parameters and assuming a log-normal model for the failure time among the uncured patients [13]. A few years later, Berkson and Gage proposed a similar model, but considered an exponential model for the uncured patients [14]. These models were then popularized by Farewell [15,16], who proposed in 1977 to model the cure rate as a dependent variable in a logistic regression. Since then, cure models have been a popular component of the statistical literature. While the advances of statistical research on *cure models* are closely linked to the progress made in the treatment of cancer [10], these models have also been studied in the context of other medical applications, as well as, more broadly, in other fields (such as psychology, sociology, or demography). In all these contexts, interest lies in the impact of one (or more) factor, not only on delaying but also on eradicating the event of interest for a non-negligible part of the population.

Besides medical evidence, a straightforward way to identify whether a particular dataset includes a subset of cured or long-term survivor patients is to inspect the Kaplan–Meier estimate of the survival curve. If a long and stable plateau with heavy censoring is observed in the tail, this can be considered as empirical evidence of a cure fraction [11,17]. Cure models can then be considered as a useful alternative to the standard Cox PH models to explicitly describe the heterogeneity within the patient population and, particularly, whether or not the PH assumption is violated.

Two main families of cure regression models have been proposed: *mixture cure models* and *promotion time cure models*. Mixture cure models follow the lines of the Boag model [13]. They explicitly model the survival function of the population as a mixture of two types of patients: those who are cured and those who are not. Typically, they are composed of two sub-models: a first one for the probability of being (un-)cured, typically modeled via a logistic regression, and a second one which is a survival model for the patients who are not cured, commonly a parametric Weibull or a semi-parametric Cox PH model. As we will briefly review in Section 22.2, many variations of the mixture cure models have been proposed in the statistical literature. A major advantage of these models is the possibility to disentangle the effects of covariates, and in particular of the treatment, on the probability of cure and on the failure time of the uncured patients, resulting in a more accurate picture of the clinical benefit than with a standard Cox analysis. Promotion time cure models, also referred to as *non-mixture cure models*, are based on a totally different approach and have been originally proposed to model the biological evolution of carcinogenic cells [18,19]. They are also sometimes called *bounded cumulative hazard models* and we will show that some specific promotion time cure models can be thought of as a Cox PH model that allows a cure fraction. A number of promotion time cure models have been proposed in the statistical literature; see Section 22.3 for a short overview. As we will discuss later, the interpretation of covariates is different with the promotion time cure models and the mixture cure models. We will also demonstrate in Section 22.4 that depending on the type of data and on the questions to be answered, a promotion time cure model or a mixture cure model may be more appropriate.

While cure models are well known in the field of statistics and have been quite extensively studied in the statistical field for at least the past 20 years, they have not reached the same popularity in a more clinical setting. Cure has now become a reality for both the patients and the clinicians in some types of cancer; however, despite the fact that cure models can therefore be an interesting way to characterize and study patient survival, they are still an underused statistical tool in the context of oncology trials. This may be due to the extreme popularity of the Cox PH model, to the lack of implementation of cure models in standard software, and probably also to a lack of knowledge about these models in this setting. The purpose of this chapter is therefore to introduce the main ideas regarding the use of cure models for survival data analysis in the framework of oncology clinical trials. For a larger-scale and more technical overview of these models, we refer the reader to Peng and Taylor [20] and Amico and Van Keilegom [21] and the references therein.

In this chapter, we focus on right-censored data, unless otherwise specified. We will denote by T the time to the event of interest, with $F(t)$ and $S(t) = P(T > t) = 1 - F(t)$ the associated distribution and survival functions, respectively. Let C be the right-censoring time, with distribution function $G(t)$. We therefore observe $\tilde{T} = \min(T, C)$ and the censoring indicator $\delta = I(T \leq C)$. In Sections 22.2 and 22.3, we give some more details about the two main families of cure models. The motivation on when to use these models, as well as issues related to the choice of the model, are discussed and illustrated with simulation results in Section 22.4. One freely available real database is analyzed in Section 22.5, which allow us to discuss further the interpretation of cure models.

22.2 Mixture Cure Models

22.2.1 Model and Properties

The idea behind mixture cure models is to explicitly take into account the fact that the population is a mixture of two groups of patients: the cured patients, who will never experience the event of interest, and the uncured (susceptible) ones. If we denote by Y the cure indicator, with $Y = 1$ corresponding to a susceptible patient and $Y = 0$ otherwise, we can define the probability of being uncured (or susceptible) as $\pi = P(Y = 1)$. Assuming that, for cured patients, the survival function is $S_c(t) = P(T > t \mid Y = 0) = 1$ for all t (i.e., a degenerate survival function), it is natural to define the mixture cure model by the following unconditional survival function of T:

$$S_{\text{pop}}(t) = P(T > t) = (1 - \pi) + \pi S_u(t), \tag{22.1}$$

where the subscript "pop" indicates that the survival function relates to the whole population and $S_u(t) = P(T > t \mid Y = 1)$ is the survival function of the susceptible patients, which is proper (i.e., $\lim_{t \to \infty} S_u(t) = 0$).

The mixture cure model therefore appears as a combination of two sub-models, one for the probability of cure (often referred to as the *incidence model*) and one for the survival of the uncured patients (the *latency model*). Each of these sub-models can be allowed to depend on (potentially different) covariates, and the mixture cure model hence allows one to disentangle the effect of covariates on the incidence and on the latency. Given a first set of covariates X and a second set of covariates Z, which might be identical to X or partially or completely different from X, the mixture cure model (1) then writes the population survival function as

$$S_{\text{pop}}(t \mid X, Z) = (1 - \pi(X)) + \pi(X) S_u(t \mid Z). \tag{22.2}$$

With such a mixture cure model, the covariates, and in particular the treatment indicator, are therefore allowed to have dissimilar effects on the probability of cure and on the timing of events for the susceptible individuals. This obviously carries additional information about the type of treatment effect compared to a standard Cox PH model. Furthermore, the sets of covariates X and Z may be different, which is in line with the intuition that medical- and patient-related factors associated with short- or long-term effects are not necessarily the same.

Most of the time, the impact of covariates on the incidence is modeled via a logistic regression model, as originally proposed by [15]. The vector of covariates X and the corresponding vector of parameters γ then contain an intercept, and the logistic incidence model for the probability of being uncured can be written

$$\pi\left(X\right) = \frac{\exp\left(X^T \gamma\right)}{1 + \exp\left(X^T \gamma\right)}. \tag{22.3}$$

Few alternatives to the logistic regression have been proposed and without real implementation in practice. Existing ideas consist in considering a more flexible modeling of the incidence, for example, using splines [22] or a single-index structure [23].

On the other hand, various ways to model the latency have been considered and applied in the literature. We may distinguish the parametric and the semi- (or non-) parametric mixture cure models. In the former, the survival times of uncured patients follow a parametric model, while the latter leaves the baseline survival function of the uncured patients unspecified. Fully non-parametric latency models have also been proposed but are not in use in the medical literature (see, for example, Taylor [24] for the case of no covariate and Patilea and Van Keilegom [25] for the case with covariates).

Several parametric models have been considered for the latency (a review can be found in Amico and Van Keilegom [21]), the most popular being probably the Weibull [16] PH model with

$$S_u(t \mid Z) = \exp\left(-\lambda \exp\left(Z^T \beta\right) t^\rho\right), \tag{22.4}$$

with β the vector of parameters corresponding to Z, $\lambda > 0$ the shape parameter, and $\rho > 0$ the scale parameter ($\rho = 1$ in the case of an exponential model, [26]). The hazard function corresponding to this survival function is then

$$\lambda_u(t \mid Z) = \lambda \rho \exp\left(Z^T \beta\right) t^{\rho - 1}. \tag{22.5}$$

The Weibull model is fairly flexible and is often considered to provide a good description of survival times in biomedical applications. However, the assumption of a monotone baseline hazard function may also be problematic. The main advantage of these fully parametric cure models, as we will see in Section 22.2.4, lies in the simplicity of the estimation procedure. Also, it may be useful to rely on parametric models if we are interested in modeling the survival time in the investigated treatment arms, and not only the relative difference between two survival curves [7].

Such parametric models may not be robust to the violation of the distributional assumption for the survival times of the uncured patients. If we still want to rely on a parametric estimation procedure, a possibility is to consider a very flexible parametric form for the baseline hazard function, such as a piecewise constant baseline hazard or the use of splines [27]. However, semi-parametric alternatives, in which the baseline hazard of the latency model is left totally unspecified, have gained popularity in the statistical literature, despite the fact that obtaining semi-parametric estimators and their standard errors can be computationally challenging.

A common alternative to parametric models is indeed to consider a semi-parametric Cox PH model for the latency. This model has been introduced by Kuk and Chen [28] and has then been extensively studied in the literature [17,24,27,29,30]. Such an approach has the advantage of leaving the baseline hazard function $\lambda_{uo}(\cdot)$ of the uncured patients unspecified,

$$\lambda_u(t \,|\, \mathbf{Z}) = \lambda_{uo}(t) \exp(\mathbf{Z}^T \beta),$$

with the corresponding survival function

$$S_u(t \,|\, \mathbf{Z}) = S_{u0}(t)^{\exp(\mathbf{Z}^T \beta)}. \tag{22.6}$$

A very important feature of the (semi-)parametric logistic/PH mixture cure model is that the PH assumption is made at the level of the uncured patients, but not at the level of the population. This appears clearly on Figure 22.1, which displays examples of population survival functions from parametric logistic/Weibull PH mixture cure models including a binary variable (representing, for example, the treatment group). On each plot, the solid line represents the recurrence-free survival (RFS) in the control group, with a cure rate of 40%. The dotted line represents the RFS achieved with an experimental treatment (a) which only has a long-term effect, by increasing the proportion of cure (incidence) but with no impact on delaying recurrences among the uncured patients (latency); (b) which only has a short-term effect, by delaying recurrences among the uncured patients but with no impact on the proportion of cure; and (c) which has both a short- and a long-term effect, affecting both the proportion of cure and the time of recurrences for the uncured patients. As we can see, a curative treatment acting only on the proportion of cure still leads to a PH situation for the population, with the two curves attaining their plateaus by "running parallel to each other" [8]. On the other hand, the other two situations clearly lead to a violation of the PH assumption for the population.

As in classical survival analysis, the accelerated failure time (AFT) and the PH models often confront each other in cure analysis. In the logistic/AFT mixture cure model, the survival times T^u of the uncured patients are modeled as

$$\log(T^u) = \mathbf{Z}^T \beta + \sigma \epsilon, \tag{22.7}$$

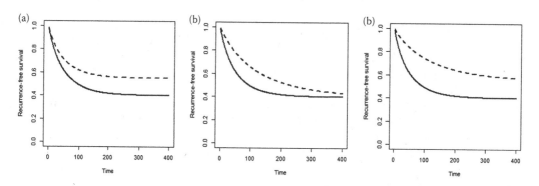

FIGURE 22.1

Survival functions from logistic/Weibull PH models for two treatment groups. The control treatment (solid line) leads to 40% cure, while the experimental treatment (dotted line) has (a) a long-term effect, that is, an impact on the probability of cure (incidence); (b) a short-term effect, that is, an impact on the latency; and (c) both a short- and long-term effect.

with $\sigma > 0$ a scale parameter and ϵ an error term whose density function f_ϵ can be specified in a parametric way or left unspecified, leading to respectively parametric [31–33] or semi-parametric [34,35] logistic/AFT models. These models do not make any PH assumption, neither at the level of the uncured patients (except for a Weibull AFT model) nor at the population level. They are, however, rarely used in the medical literature.

22.2.2 Interpretation

A major advantage of mixture cure models is the fact that each set of parameters can be interpreted separately. As mentioned earlier, the effect of the covariates on both components of the model (incidence and latency) can therefore be clearly disentangled. By quantifying separately the effect of the treatment on the probability of being a long-term survivor and on the event times for those who are not, this model allows one to distinguish a curative from a life-prolonging treatment effect.

Furthermore, the commonly used logistic/(semi-)parametric PH model leads to odds and hazard ratios (OR and HR), the interpretation of which are well known. In the logistic regression incidence sub-model, the parameters γ, representing the impact of the covariates X on the probability to be uncured $\pi(X)$, are interpreted as usual. Parameter values above 0 are associated with covariates which increase the risk to be uncured (and therefore decrease the risk to be cured) when their value increases, with the reverse for parameter values below 0. The quantity $1 - \pi(X)$ can be interpreted as the cure rate for patients with covariate value X.

The parameters β associated with the covariates Z in the latency model represent the impact of the covariates on the time to event for uncured patients and are interpreted according to the model used. With a (semi-)parametric PH model, a positive value is associated to a covariate which increases the hazard of events (and therefore accelerates the events) when its value increases, with the reverse for values below 0. In an AFT model, the covariates act multiplicatively on the time; a positive coefficient indicates a longer time to event when the value of the covariate increases.

22.2.3 Identifiability

A general and informal rule that holds for all cure models requires the follow-up of the study to be sufficiently long: the estimated survival function should exhibit a long plateau containing many censored observations [21]. More formally, the maximum possible event time should be smaller than the maximum possible censoring time.

In a semi-parametric mixture cure model, the latency component $S_u(\cdot)$ (or part of it) is left unspecified and estimated non-parametrically. In this case, some additional information is required for identifiability. In Sy and Taylor [17] and Taylor [24], the survival function (or the baseline survival function) is constrained to reach 0 at the largest observed event time. This *zero-tail constraint* seems natural in contexts in which a cure model is justified, that is, when cured subjects are known to exist and the follow-up is sufficiently long after the largest event time [24].

22.2.4 Model Estimation

The estimation of mixture cure models is classically based on the maximization of the likelihood function. Assume we have independent and identically distributed (i.i.d.) data $(\tilde{T}_i, \delta_i, X_i, Z_i), i = 1, \ldots, n$, where n is the number of patients, with $\tilde{T}_i = \min(T_i, C_i)$ the observed time and $\delta_i = I(T_i \leq C_i)$ the censoring indicator. We have to keep in mind that the cure status is only observed for uncensored observations (who are obviously uncured), not for the censored ones. As in standard survival analysis, the likelihood function is based on the contributions of two types of observations: the uncensored ones ($\delta = 1$), all corresponding to uncured patients (occurring with probability $\pi(X)$), and the censored ones ($\delta = 1$), corresponding either to cured patients (with probability $1 - \pi(X)$) or to uncured patients (with probability $\pi(X)$). The likelihood function is then defined as

$$
\begin{aligned}
L_{\text{MCM}} = \prod_{i=1}^{n} &\left[\pi\left(X_i\right) f_u\left(\tilde{T}_i \mid Z\right) \right]^{\delta_i} \\
&\times \prod_{i=1}^{n} \left[1 - \pi\left(X_i\right) + \pi\left(X_i\right) S_u\left(\tilde{T}_i \mid Z\right) \right]^{1-\delta_i}
\end{aligned}
\tag{22.8}
$$

with $f_u(\tilde{T}_i \mid Z) = -(d/dt)S_u(\tilde{T}_i \mid Z)$.

For fully parametric mixture cure models, the parameters can be estimated by maximizing the likelihood function (Equation 22.8) via numerical techniques, such as the Newton-Raphson algorithm. Asymptotic standard errors can be obtained by inverting the Fisher information matrix of second order derivatives of $\log(L_{\text{MCM}})$. Some adaptations of this likelihood maximization have been proposed for more flexible parametric models; see, for example, Yamaguchi [33] who propose a two-step maximization in the case of a logistic/AFT model with an error distribution from the extended family of generalized gamma.

As mentioned earlier, the semi-parametric logistic/Cox PH mixture cure model does not satisfy the PH assumption at the population level and the partial likelihood approach developed by Cox [6] cannot be applied to estimate this model. Indeed, since the survival function of the uncured patients is conditional on the cure status, one can not eliminate $S_0(t|Y=1)$ in the Cox PH mixture cure model likelihood without losing information about γ [36]. Several estimation procedures have been presented in the literature. Given that the cure status is a latent variable, Sy and Taylor [17] and Peng and Dear [30] have proposed relying on the expectation-maximization (EM) algorithm. An interesting feature of this approach is that the complete-data likelihood, obtained from the explicit contributions of the uncensored observations ($\delta = 1$, $Y = 1$), censored and uncured observations ($\delta = 0$, $Y = 1$), and censored and cured observations ($\delta = 0$, $Y = 0$), can be factorized into two elements. Each element depends only on the parameters of one of the two parts of the model. This obviously simplifies the maximization in the M-step, as it can be performed separately for each set of parameters γ and β. Corbière and Joly [36] advise using non-parametric bootstrap to obtain the standard errors of the estimated parameters. Other proposed estimation approaches include methods based on a marginal likelihood, obtained by integrating out the likelihood function (Equation 22.8) over the distinct event times ($\tilde{Y}_{(j)}, j = 1, \dots, r$) using Monte-Carlo approximation [28]. A penalized likelihood approach, approximating the baseline conditional hazard by a linear combination of cubic normalized B-splines, has been introduced [27].

The estimation procedures proposed until now for the semi-parametric logistic/AFT model are all based on the EM algorithm and follow the same ideas as for the semi-parametric logistic/Cox PH mixture cure model, except for the latency estimation in the M-step, which is then based on extensions of the methods proposed for the classical semi-parametric AFT models.

We refer to Amico and Van Keilegom [21] for a more detailed review of these estimation methods, as well as a discussion on the estimation methods for non-parametric mixture cure models.

22.2.5 Model Implementation

The main R package for the estimation of mixture cure models is the smcure package developed by Cai et al. [37] which allows one to fit a wide range of semi-parametric mixture cure models. Available models for the incidence include the logistic regression model but also other generalized linear models with various link functions, such as the probit. Both semi-parametric PH and AFT models can be fitted for the latency part. The estimation procedure is based on the EM algorithm as developed by Sy and Taylor [17] and Peng and Dear [30] for the PH case and by Zhang and Peng [35] for the AFT case. The variance of the estimated parameters is obtained via bootstrap.

A freely available SAS macro, called PSPMCM, has been developed for the parametric and semi-parametric logistic/PH mixture cure model by Corbière and Joly [36]. The maximization of the likelihood function is performed using the Newton-Raphson procedure (as implemented in PROC NLMIXED) when the latency is modeled via a parametric model (exponential, Weibull, log-normal, or log-logistic) and through an EM algorithm for the semi-parametric case. Standard errors of the estimators are obtained either by inverting the observed Fisher information matrix at the last iteration or via non-parametric bootstrap.

22.3 Promotion Time Cure Models

22.3.1 Model and Properties

In promotion time cure models, the existence of a cure fraction is taken into account by directly choosing an improper form for the survival function of the whole population, instead of separately modeling the survival of cured and uncured patients. This corresponds to considering that cured patients have an infinite survival time. Given a vector of covariates X, these models, introduced by Yakovlev et al. [18] and Yakovlev and Tsodikov [19], have the form

$$S_{\text{pop}}(t \mid X) = \exp\{-\theta(X)F(t)\}, \tag{22.9}$$

where $F(\cdot)$ is the (proper) cumulative distribution function (cdf) of some non-negative random variable such that $F(0) = 0$ and $\theta(X)$ is a known link function with an intercept. The baseline cdf $F(\cdot)$ can either be modeled parametrically, yielding *parametric* promotion time cure models (as is done, among others, by Chen et al. [38]), or left unspecified, which leads to *semi-parametric* promotion time cure models (see, for example, Tsodikov [39]). The cumulative hazard function of this model is $\theta(X) F(t)$, which is bounded: for this reason, promotion time cure models are also sometimes called *bounded cumulative hazard models*.

In the model in Equation 22.9, the cure probability is

$$\lim_{t \to \infty} S_{\mathrm{pop}}(t \mid X) = \exp\{-\theta(X)\}. \tag{22.10}$$

The hazard function corresponding to the survival function (Equation 22.9) is

$$h_{\mathrm{pop}}(t \mid X) = \theta(X)f(t),$$

where $f(t) = dF(t)/dt$ is the baseline density function. Contrary to the mixture cure model considered in the previous section, the promotion time cure model possesses the proportional hazards property:

$$\frac{h_{\mathrm{pop}}(t \mid X_i)}{h_{\mathrm{pop}}(t \mid X_j)} = \frac{\theta(X_i)}{\theta(X_j)}$$

and is therefore sometimes referred to as the *proportional hazards cure model*.

An example of survival functions from a promotion time cure model is displayed and discussed in Section 22.3.3.

22.3.2 Link with Other Models

The semi-parametric promotion time cure model with an exponential link function can actually be seen as a generalization of the Cox PH model [40]. If we assume the link function $\theta(X) = \exp(\beta_0 + X^T\beta)$ with intercept β_0, then Equation 22.9 becomes

$$
\begin{aligned}
S_{\mathrm{pop}}(t \mid X) &= \exp\left\{-\exp\left(\beta_0 + X^T\beta\right)F(t)\right\} \\
&= \exp\left\{-\exp\left(X^T\beta\right)\exp(\beta_0)F(t)\right\} \\
&= \exp\left\{-\exp\left(X^T\beta\right)H(t)\right\},
\end{aligned}
$$

that is, a PH model in which $H(t) = \exp(\beta_0)F(t)$ is a *bounded* cumulative hazard function, taking values in $[0, \exp(\beta_0)]$. In the Cox PH model, the cumulative hazard function is not bounded [6]. However, since, in practice, the estimator of the cumulative hazard function of a Cox PH model is bounded, we have the following links between the estimates obtained from a promotion time cure model with exponential link, $\hat{\beta}_{0,\mathrm{PT}}$, $\hat{\beta}_{\mathrm{PT}}$, and $\hat{F}_{\mathrm{PT}}(t)$, and the estimates from a Cox PH model estimated by maximizing the profile likelihood, $\hat{\beta}_{\mathrm{PH}}$ and $\hat{H}_{\mathrm{PH}}(t)$:

$$\hat{\beta}_{\mathrm{PT}} = \hat{\beta}_{\mathrm{PH}}$$
$$\exp\left(\hat{\beta}_{0,\mathrm{PT}}\right) = \hat{H}_{\mathrm{PH}}\left(T_{(n)}\right)$$
$$\exp\left(\hat{\beta}_{0,\mathrm{PT}}\right)\hat{F}_{\mathrm{PT}}(t) = \hat{H}_{\mathrm{PH}}(t),$$

where $T_{(n)}$ is the largest event time.

This means that, when the exponential link function is used, the estimates of the regression coefficients of a semi-parametric promotion time cure model can, in practice, be obtained from fitting a standard Cox PH model. This explains that, as long as the PH assumption is met, the Cox model also provides reliable results even in the presence of a non-negligible cure fraction. In that situation, however, the parameters should be interpreted in the context of a promotion time cure model, that is, taking into account that cure is a possibility.

Although the model in Equation 22.9 is not equivalent to a mixture cure model, it is possible to re-express the population survival function with a mixture expression as

$$\{1 - p(X)\} + p(X)S_u(t \mid X) \tag{22.11}$$

where $p(X)$ is the probability of being susceptible, and $S_u(\cdot \mid X)$ is the conditional survival function of the uncured patients. We have already seen that in the promotion time model, the cure probability is given by

$$1 - p(X) = \exp\{-\theta(X)\}, \tag{22.12}$$

and one can show that the conditional survival of the susceptible subjects is [38]

$$S_u(t \mid X) = \frac{\exp\{-\theta(X)F(t)\} - \exp\{-\theta(X)\}}{1 - \exp\{-\theta(X)\}}. \tag{22.13}$$

However, we clearly see from Equations 22.12 and 22.13 that in this formulation of the promotion time cure model as a mixture cure model, both the incidence and the latency depend on the same set of covariates X, whereas this is not necessarily the case in a mixture cure model. Moreover, these covariates appear at more than one place in the conditional survival function (Equation 22.13), which is never the case in the classical versions of the mixture cure model.

22.3.3 Interpretation

As mentioned previously, in the promotion time cure model, the covariates X affect both the probability of being cured and the survival of uncured patients. This is best understood by considering the seminal biological interpretation of this model, which was developed with the idea of modeling cancer relapse [38]. Assume that, after an initial treatment, individual i has N_i carcinogenic cells that are left active, that is, cells which could metastasize. N_i is then an unobservable latent variable assumed to follow a *Poisson*($\theta(X_i)$) distribution. The cured individuals are those for whom $N_i = 0$. The kth carcinogenic cell ($k = 1, \ldots, N_i$) takes a time W_{ik} (called the *promotion time*) to produce a detectable tumor mass. Conditionally on N_i, the variables $W_{ik}, k = 1, \ldots, N_i$ are mutually independent, independent of N_i, and have a common cdf $F(t)$. The time until the relapse of the cancer, which is the observed event time, is then $T_i = \min(W_{i1}, \ldots, W_{iN_i})$, which has survival function:

$$S(t \mid X) = P(N = 0 \mid X) + \sum_{j=1}^{\infty} P(T_1 > t, \ldots, T_j > t \mid N = j, X)P(N = j \mid X)$$

$$= \exp\{-\theta(X)\} + \sum_{j=1}^{\infty} \{1 - F(t)\}^j \frac{\theta(X)^j \exp\{-\theta(X)\}}{j!}$$

$$= \exp\{-\theta(X)\} + \exp\left[\{1 - F(t)\}\theta(X) - \theta(X)\right] - \exp\{-\theta(X)\}$$

$$= \exp\{-\theta(X)F(t)\},$$

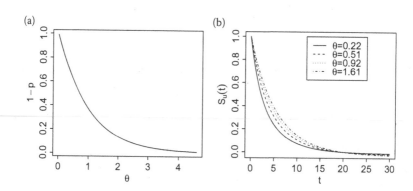

FIGURE 22.2

Representation of the effect of the value of θ on the cure probability and on the conditional survival function. (a) Cure probability as a function of the value of θ. (b) Conditional survival function for different values of θ, corresponding to cure probabilities of 0.20 (when $\theta = 1.61$), 0.40 (when $\theta = 0.92$), 0.60 (when $\theta = 0.51$), and 0.80 (when $\theta = 0.22$).

which corresponds to Equation 22.9. The covariates X have an impact on the number of cells which can metastasize; as a consequence, these covariates directly influence the cure probability, but also the conditional survival of the uncured patients.

The relation between $\theta(X)$ and the cure probability is illustrated in Figure 22.2a, while Figure 22.2b displays the relation between $\theta(X)$ and the conditional survival of the uncured patients. The resulting effect on the population survival function is represented in Figure 22.3.

The parameters of a promotion time cure model can hence be interpreted based on the biological interpretation of the model. A covariate whose increase yields an increase in $\theta(X)$ actually increases the mean number of cells which can metastasize; larger values of this covariate are hence associated with a lower cure probability and a lower survival (at all times) for the uncured patients. Conversely, if an increase in a covariate lowers the value of $\theta(X)$, then larger values of this covariate are associated with a higher cure probability and a better survival (at all times) for the uncured patients.

In the (common) particular case where $\theta(X)$ is assumed to be $\exp(\beta_0 + X^T\beta)$, then a one-unit increase in a continuous covariate X_1 is associated with a multiplication of the mean number of cells that can metastasize by a factor $\exp(\beta_1)$. For a binary covariate X_2

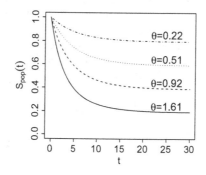

FIGURE 22.3

Survival function for different values of θ, corresponding to cure probabilities of 0.20 (when $\theta = 1.61$), 0.40 (when $\theta = 0.92$), 0.60 (when $\theta = 0.51$), and 0.80 (when $\theta = 0.22$).

(1 = treatment versus 0 = control), the mean of N is $\exp(\beta_2)$ times larger in the treatment group than in the control group.

The biological interpretation thus allows one to easily understand the different components appearing in Equation 22.9 and the covariate effect. However, promotion time cure models can also be used in contexts where such a biological interpretation does not hold. For example, Bremhorst et al. [41] explain that, in a fertility progression study where the transition to second or third birth is analyzed, arguments in favor of the conception of another child can be seen as the latent cells. The time needed for these arguments to be convincing is then the observed time.

Although Equation 22.13 makes it clear that the model parameters are not easily interpreted in terms of the covariate effect on the conditional survival of the susceptible patients, such an interpretation can always (although not directly) be recovered. One example is given in Figure 22.4, in the case of a model with a binary covariate. Figure 22.4a contains the representation of the value of the ratio of the cure probabilities, $1 - p(1)$ and $1 - p(0)$, as a function of the coefficient of the covariate. As can be expected from our previous discussion, a negative coefficient for X corresponds to a higher cure rate in the treatment group than in the control group, while the situation is reversed when the coefficient is positive. The curve is decreasing: the larger the coefficient, the smaller the cure probability in the treatment group (compared to the control group). The interpretation of the treatment effect on the conditional survival function of the susceptible patients can be found in Figure 22.4b. Here again, a negative coefficient for X implies a more favorable situation for the treatment group, compared to the control group: the survival function for the uncured patients of the treatment group is higher, at all times, than the curve of the uncured patients of the control group.

22.3.4 Identifiability

The general principles regarding identifiability in cure models discussed in Section 22.2.3 still hold. In the semi-parametric promotion time cure model, the cdf $F(\cdot)$ is left unspecified and estimated non-parametrically. As was the case in the mixture cure model, identifiability requires additional information. The conditions required

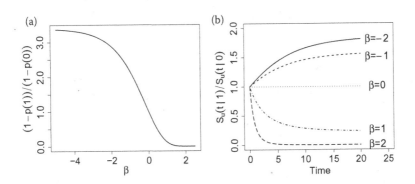

FIGURE 22.4
Representation of the effect of a binary covariate in the promotion time cure model $S_{pop}(t|X) = \exp(-\exp(0.2 + \beta X)F(t))$ with $F(\cdot)$ the cdf of a truncated exponential distribution of mean 6, truncated at 20. (a) Effect of the binary covariate on the ratio of cure probabilities, as a function of β. (b) Effect of the binary covariate on the ratio of survival functions, for different values of β.

for the semi-parametric version of the model in Equation 22.9 with $\theta(X) = \eta(\beta_0 + X^T\beta)$ to be identifiable in the regression parameters and in F can be found in Portier et al. [42], who focus on right-censored data. In addition to classical conditions about the covariates and the function $\eta(\cdot)$, Portier et al. [42] explain that we need a threshold τ, called the *cure threshold*, such that all censored individuals with a censoring time larger than this threshold are treated as known to be cured, that is, $T_i = C_i = \tilde{T}_i = \infty$. However, the estimation methods for this model mentioned in Section 22.3.5 force the estimated baseline cdf to be 1 beyond the largest observed failure time. This amounts to considering that no event can occur after this time: the cure threshold is then determined to be this largest event time. This restriction is the equivalent of the zero-tail constraint imposed in the mixture cure model, explained in Section 22.2.3.

22.3.5 Model Estimation

Promotion time cure models have been studied mainly (but not only) in the Bayesian literature, because the posterior distribution of the regression parameters is often proper when using non-informative improper priors, an attractive property in a Bayesian setting [43]. Most of the Bayesian methods proposed rely on Markov Chain Monte Carlo (MCMC) methods to approximate the posterior distribution of the parameter of interest; see, for example, Chen et al. [38] in a parametric case and Ibrahim et al. [44] and Tsodikov et al. [45] in a semi-parametric setting. A Bayesian approach including a smoothing parameter to control the degree of parametricity in the right tail of the baseline survival distribution $F(t)$ can be found in the article by Ibrahim et al. [46].

Several frequentist estimation procedures have been proposed for the semi-parametric version of the model from Equation 22.10 applied to right-censored data, among which the maximization of a profile likelihood [39,42] and the maximization of the full likelihood through a profiling approach [47] or a backfitting approach [48].

The full log-likelihood of the model in Equation 22.9 is

$$\ell_{PTM} = \sum_{i=1}^{n} \delta_i I\left\{-F\left(\tilde{T}_i\right)\theta\left(X_i\right) + \log F\{\tilde{T}_i\} + \log \theta\left(X_i\right)\right\}$$
$$+ \left(1 - \delta_i\right)I(\tilde{T}_i < \infty)\left\{-F\left(\tilde{T}_i\right)\theta\left(X_i\right)\right\} - I\left(\tilde{T}_i = \infty\right)\theta\left(X_i\right),$$

where $F\{\tilde{T}_i\}$ is the jump size of the step function F at $\tilde{T}_i$.

Note that, in the specific case where $\theta(X) = \exp(\beta_0 + X^T\beta)$ is used, the parameters can be estimated by maximizing the profile likelihood, as in a classical Cox PH model (see Section 22.3.2).

22.3.6 Model Implementation

Two R packages (available on the CRAN website) contain estimation procedures for the semi-parametric promotion time cure model: miCoPTCM, for right-censored data (implementing the method of Ma and Yin [48]), and intercure, for interval-censored data (implementing the profile likelihood approach of Liu and Shen [49]). The coxph function of package survival, implementing the maximization of the profile likelihood, can be used to estimate the version of the promotion time cure model which is equivalent to a classical PH model.

In Stata, the `cureregr` command fits parametric cure models, the `stpm2` [50] command enables the estimation of flexible parametric cure models, and the `strsnmix` [9] command fits parametric non-mixture cure models.

22.3.7 Extensions

A common and important extension of the model from Equation 22.9 consists of making the baseline cdf depend on covariates:

$$S_{\text{pop}}(t \mid X, Z) = \exp\{-\theta(X) F(t \mid Z)\}, \tag{22.14}$$

as considered, for example, by Tsodikov [51]. Contrary to the covariates appearing in the classical model (Equation 22.9), these new covariates Z affect the time needed by a metastatic cell to produce a detectable tumor mass, but not the mean number of such cells. As a result, these covariates influence the short-term survival, but not the cure rate. Including two sets of covariates hence allows one to distinguish between the effect on the cure rate (long-term effect) and the effect on the survival (short-term effect). However, these models do not possess the PH property, and some restrictions on the covariates may be required for identifiability (see Bremhorst and Lambert [52] for an identifiability result in one such model).

Other possible extensions of the promotion time cure model (Equation 22.9) directly motivated through the biological interpretation of this model have also been proposed. See, for example, Zeng et al. [47], who relax the assumption of mutual independence between the promotion times Z_{ik} via the introduction of a subject-specific frailty term; Cooner et al. [53] and Gu et al. [54], who consider other types of distribution for the number N of cells that can metastasize and a number $r > 1$ of cells to be promoted to have an event; and Tournoud and Ecochard [55] who, in addition, suggest to include covariates affecting the mean number of cells, N, as well as the distribution of the promotion time of each cell.

Finally, it is interesting to note that general classes of cure models have been developed, which encompass both mixture and promotion time cure models as special cases. For example, Yin and Ibrahim [43] propose a new class of cure models through a Box-Cox transformation of the survival function of the population, with as special cases the mixture cure models and the promotion time cure models with covariates in the baseline cdf.

22.4 When to Use a Cure Model

Two main questions usually come up when speaking about cure models. First, when should we use a cure model to analyze our data or, put slightly differently, what are the consequences of not taking the cure fraction into account? Second, if we actually are in a situation where a cure model is appropriate, should we rather use a mixture cure model or a promotion time model? To discuss these questions, we present here a short simulation study investigating the consequences of misspecifying the model to be used for estimation, such as assuming a classical Cox PH model when there is actually a cure fraction in the sample, or assuming a mixture cure model when the data actually follow a promotion time cure model.

22.4.1 Presentation of the Simulations

In this simulation study, we consider six different settings based on the way the data were generated. For each setting, 500 datasets, each containing 500 patients, were simulated. These patients are first randomly allocated to one of the treatment arms according to a binary covariate $X \sim Bern(0.5)$ ($X = 1$ for treatment versus $X = 0$ for control). Their time to event is then generated according either to a parametric PH model (no cure), a logistic/parametric PH mixture cure model or a promotion time cure model as described below. We consider right censoring with censoring times following a truncated Weibull distribution whose parameters were chosen to achieve the desired level of censoring.

Settings 1 and 2: The times to event are generated from a parametric PH model with an exponential baseline hazard (shape parameter set to 6). The regression parameter for the treatment indicator was set to -1, corresponding to an HR of 0.37 in favor of the treatment group. For these data, there are therefore no cured patients. We consider both a setting with a (relatively) high censoring rate (Setting 1) and a setting with a (relatively) low censoring rate (Setting 2).

Setting 3: The times to event are generated from a parametric promotion time cure model with exponential link and exponential baseline cdf (shape parameter set to 6, distribution truncated at 20). The regression parameters are set to 0.2 for the intercept and -0.5 for the treatment effect.

Settings 4, 5, and 6: The times to event are generated from a logistic/parametric PH mixture cure model with a (truncated) exponential baseline cumulative hazard function in the latency (shape parameter of 4 for the three settings and truncation limit of, respectively, 50, 20, and 50). These three settings differ by the inclusion of the covariate effect either in both the incidence and latency parts (Setting 4: parameter values of 1 and -1 for the intercept and the treatment effect in the incidence and parameter value of -1 for the treatment effect in latency), in the incidence part only (Setting 5: parameter values of 1 and -1 for the intercept and the treatment effect in the incidence and parameter value of 0 for the treatment effect in latency), or in the latency part only (Setting 6: parameter values of 0.5 and 0 for the intercept and the treatment effect in the incidence and parameter value of -1 for the treatment effect in latency). These settings therefore correspond to a treatment being both curative and life-prolonging (Setting 4), only curative (Setting 5) with no effect on the time of events among the uncured, or only life-prolonging (Setting 6) while ultimately not impacting the long-term outcome of the patients.

The average censoring and cure rates per treatment arm and overall are presented in Table 22.1; Setting 6 is the only one leading to the same proportion of cure in both treatment arms. Figure 22.5 represents, for each setting, the theoretical survival curves by treatment group at the level of the population (obtained without considering censoring), as well as the Kaplan–Meier (KM) estimated survival curves for one random dataset of each setting (accounting for censoring). As expected, the PH assumption is met in Settings 1, 2, and 3, but one could also consider that Setting 5 meets this assumption since both curves are first parallel and then reach their plateaus at the same time. The presence of a

TABLE 22.1

Simulation Settings Characteristics

	Average Censoring Rate			Average Cure Rate		
	Overall	$X = 0$	$X = 1$	Overall	$X = 0$	$X = 1$
Setting 1	0.53	0.39	0.67			
Setting 2	0.27	0.15	0.39			
Setting 3	0.57	0.50	0.64	0.38	0.29	0.48
Setting 4	0.54	0.40	0.69	0.38	0.27	0.50
Setting 5	0.56	0.48	0.64	0.38	0.27	0.50
Setting 6	0.54	0.48	0.60	0.38	0.38	0.38

plateau (and therefore of a cure fraction) is clear from the estimated survival curves from Settings 3 to 6. Setting 2 obviously shows no plateau in the estimated survival curves; however, Setting 1 may be confusing. It is therefore important to keep in mind that it is the combination of a sufficiently long follow-up, a long plateau, and a sufficient number of censored observations in the plateau that can be considered as an indication of the presence of a cure fraction.

For each setting, all datasets were analyzed using a semi-parametric Cox PH model (CM, fitted with the coxph function of the R package survival), a semi-parametric promotion time cure model (PTM, fitted with the PTCMestimBF function of the R package miCoPTCM), and a semi-parametric logistic/Cox PH mixture cure model (MCM, fitted with the smcure function of the R package smcure). All models included the treatment covariate, which was included in both parts of the MCM.

22.4.2 Simulations Results

The results regarding the estimation of the treatment effect in the various settings are presented in Table 22.2; standard errors of the estimated coefficients of the mixture cure model have been estimated using bootstrap with 500 replications. While the estimated coefficients obtained from different models can not be compared together (except for the CM and the PTM) as they do not represent the same quantity, one can compare the estimated cure fractions. They are presented in Table 22.3 and are estimated by $\exp\{-\exp(X^T\hat{\beta})\}$ in the PTM and $1 - \hat{\pi}(X)$ in the MCM. The *overall* cure rate is estimated, for each dataset, by the average of the estimated individual cure probabilities. Another possibility to compare the fit of the PTM and the MCM in each setting is to consider the estimated conditional survival function of the uncured subjects given by Equation 22.13 for the PTM and by Equation 22.6 for the MCM.

The consequences of a model mis-specification can vary largely, depending on the true model underlying the data and on the focus of the estimation: cure probability, conditional survival function, treatment effect size, and significance. Due to the link between the Cox PH model and the semi-parametric promotion time cure model with exponential link function (discussed in Section 22.3.2), the results obtained when using these models are clearly very similar. The only difference is that, since the promotion time cure model assumes a cure fraction in the data, the treatment coefficient is interpreted in terms of both its short- and long-term effects.

When there is actually no cure fraction in the data (Settings 1 and 2), the treatment effect is well recovered by the PTM and quite well by the MCM when the censoring is not

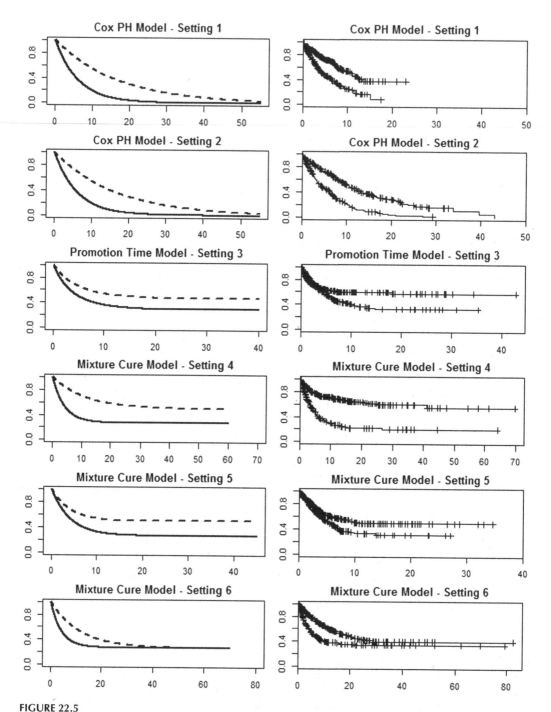

FIGURE 22.5
Survival functions in the simulation settings. Left panel: Theoretical survival functions for each simulation setting. Right panel: Estimated survival functions for a random dataset for each setting. The solid line represents the control group ($X = 0$) while the dotted line represents the treatment group ($X = 1$).

TABLE 22.2

Simulation Results: Estimation of the Coefficients

		Cox Model	Promotion Time Model		Mixture Cure Model		
		β_1	β_0	β_1	γ_0	γ_1	β_1
Set. 1	**True value**	**−1.000**					
	Average	−1.009	1.287	−1.009	3.846	−2.699	−0.692
	Emp. S.E.	0.142	0.314	0.142	1.709	1.725	0.225
	Est. S.E.	0.140	0.190	0.139	0.454	0.552	0.198
	Prop. RH_0	1.000	1.000	1.000	1.000	0.971	0.779
Set. 2	**True value**	**−1.000**					
	Average	−1.004	1.999	−1.004	13.504	−9.822	−0.926
	Emp. S.E.	0.113	0.218	0.113	6.936	7.118	0.124
	Est. S.E.	0.112	0.166	0.111	3.427	3.442	0.125
	Prop. RH_0	1.000	1.000	1.000	0.752	0.721	1.000
Set. 3	**True value**		**0.200**	**−0.500**			
	Average	−0.501	0.192	−0.501	0.890	−0.779	−0.154
	Emp. S.E.	0.136	0.118	0.136	0.236	0.299	0.203
	Est. S.E.	0.139	0.110	0.139	0.199	0.302	0.193
	Prop. RH_0	0.960	0.408	0.960	0.998	0.745	0.133
Set. 4	**True value**				**1.000**	**−1.000**	**−1.000**
	Average	−1.048	0.475	−1.048	1.019	−0.983	−1.000
	Emp. S.E.	0.131	0.131	0.131	0.184	0.295	0.228
	Est. S.E.	0.142	0.122	0.140	0.157	0.292	0.207
	Prop. RH_0	1.000	0.990	1.000	1.000	0.947	0.057
Set. 5	**True value**				**1.000**	**−1.000**	**0.000**
	Average	−0.540	0.225	−0.540	1.055	−1.035	0.010
	Emp. S.E.	0.132	0.106	0.132	0.246.	0.303	0.192
	Est. S.E.	0.138	0.106	0.138	0.213	0.301	0.188
	Prop. RH_0	0.988	0.564	0.988	1.000	0.947	0.057
Set. 6	**True value**				**0.500**	**0.000**	**−1.000**
	Average	−0.504	0.221	−0.504	0.525	−0.001	−0.999
	Emp. S.E.	0.132	0.122	0.132	0.165	0.304	0.194
	Est. S.E.	0.134	0.115	0.133	0.127	0.290	0.190
	Prop. RH_0	0.954	0.500	0.954	0.963	0.083	1.000

Note: Emp. S.E.: empirical standard error; Est. S.E.: estimated standard error; Prop. RH_0: proportion of cases in which the hypothesis H_0: $\beta = 0$ was rejected. Empirical standard errors were computed by taking the standard deviation of the estimated values over all replications.

too high. The estimated coefficients in the incidence part of the MCM are largely biased and accompanied by a very large standard error, showing, as expected, an instability in the estimation of this part of the model. The ability of the models to acknowledge the absence of cure (by estimating a very low cure rate and by appropriately estimating the conditional survival of the uncured patients, data not shown) is highly dependent on the amount of censoring, as can be seen by comparing the results obtained in Settings 1 and 2. This phenomenon is to be understood in the light of the zero-tail constraint, which is

TABLE 22.3

Simulation Results: Estimation of the Cure Rate

		Promotion Time Model			Mixture Cure Model		
		Overall	Control	Treatment	Overall	Control	Treatment
Set. 1	True value	0	0	0	0	0	0
	Average	0.155	0.038	0.273	0.150	0.043	0.258
	Emp. S.E.	0.140	0.030	0.103	0.132	0.038	0.102
Set. 2	True value	0	0	0	0	0	0
	Average	0.038	0.001	0.074	0.029	0.003	0.055
	Emp. S.E.	0.046	0.002	0.040	0.041	0.006	0.044
Set. 3	True value	0.380	0.295	0.477	0.380	0.295	0.477
	Average	0.389	0.298	0.480	0.383	0.293	0.473
	Emp. S.E.	0.100	0.042	0.041	0.101	0.048	0.048
Set. 4	True value	0.380	0.269	0.500	0.380	0,269	0.500
	Average	0.385	0.202	0.568	0.379	0.267	0.491
	Emp. S.E.	0.187	0.041	0.042	0.123	0.035	0.062
Set. 5	True value	0.380	0.269	0.500	0.380	0.269	0.500
	Average	0.384	0.286	0.481	0.378	0.261	0.495
	Emp. S.E.	0.105	0.038	0.040	0.125	0.046	0.045
Set. 6	True value	0.380	0.378	0.378	0.380	0.378	0.378
	Average	0.379	0.288	0.470	0.373	0.373	0.374
	Emp. S.E.	0.101	0.043	0.043	0.051	0.038	0.061

Note: Emp. S.E.: empirical standard error.

used for identifiability purposes and which treats all censored observations after the last event time as belonging to the cure group; this leads to a positive bias in the estimation of the cure probability and a negative bias in the estimation of the survival function of the uncured patients.

When there is actually a fraction of cure patients, we have to distinguish situations were the PH assumption may be considered to hold. This is the case when data are generated from a PTM (Setting 3) or from an MCM with a treatment affecting only the incidence (Setting 5, in which the conditional survival functions for both groups level off at the same time point). In that case, it is interesting to note that, although we can not formally compare their coefficients, the PTM and the MCM seem to recover the treatment effect. However, the PTM does not allow us to disentangle the short- and the long-term effects. When the data are generated from a PTM and fitted with an MCM, the true joint treatment effect on both the cure probability and the conditional survival function of the PTM is split into both parts of the model (the average of both estimated coefficients, $\hat{\beta}_1$ and $\hat{\gamma}_1$, is incidentally close to the true, unique coefficient). As a result, the significant effect was sometimes recovered for the incidence part, but rarely for the latency. The estimation of the cure rate in each arm as well as of the conditional survival curve for the uncured is nearly unbiased with both the PTM and the MCM. Figure 22.6 displays the results obtained when fitting an MCM on PH data generated by a PTM (Setting 3) and vice versa (Setting 5).

The situation is, however, different when data have been generated from an MCM and one can not assume PH anymore, as is the case in Settings 4 and 6. Although, in our simulation

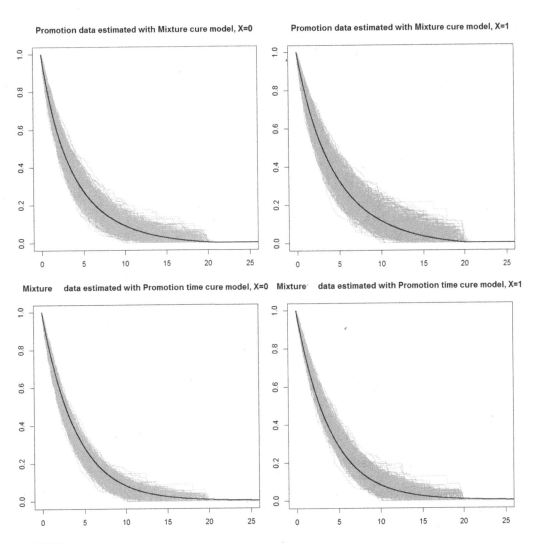

FIGURE 22.6
Estimated conditional survival functions for the PTM data estimated with an MCM (top panel) and for the MCM data estimated with a PTM (bottom panel), for the control arm (left panel) and the treatment arm (right panel).

setting, the PTM seems to recover some part of the treatment effect, the estimated cure rate is biased downward in the control arm and upward in the treatment arm. As shown in the top of Figure 22.7, this leads to an overestimation of the conditional survival in the control group and an underestimation in the treatment group. In the control arm, the PTM assigns too few patients to the cure group and hence estimates a too high survival for the uncured; the opposite holds in the treatment group (too many patients to the cure group and a too low survival for the uncured). These biases are due to the model mis-specification; only one coefficient is estimated, which tries to summarize both effects (on the incidence and the latency). As a result, none of the two effects is correctly estimated. As expected, there are no such problems when estimating these data with the appropriate model; see bottom panel of Figure 22.7.

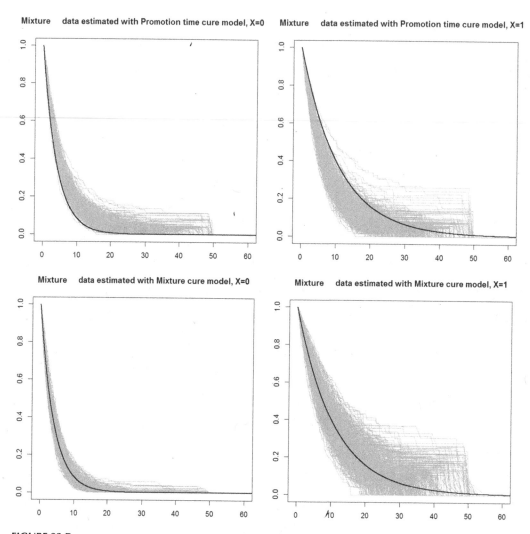

FIGURE 22.7

Estimated conditional survival functions for the MCM data (Setting 4) estimated with a PTM (top panel) and with a MCM (bottom panel), for the control arm (left panel) and the treatment arm (right panel).

22.5 Melanoma Clinical Trial

The ECOG phase III clinical trial e1684 was set up to evaluate the effect on relapse-free survival (RFS) of high-dose interferon alpha-2b (IFN) versus placebo (PBO) as postoperative adjuvant therapy. A total of $n = 285$ patients were randomized to either IFN ($n = 145$) or PBO ($n = 140$). Two additional covariates are included in the freely available database: gender (39.8% female) and age (centered to the mean). The main analysis of this trial, as it appears in the original publication of the trial results, does not take cure into account [63]. However, there is clear evidence, both from a medical point of view and by inspecting the estimated survival curves in each treatment group (see Figure 22.8, left panel), of a presence of a cure fraction. These data have already been extensively used to illustrate publications on cure models [36,46,61].

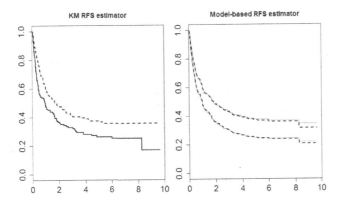

FIGURE 22.8
Melanoma data. Left panel: Kaplan–Meier estimated population RFS curves (the solid line represents the IFM arm and the dotted line the PBO arm). Right panel: Model-based estimated population RFS for male patients of average age as obtained from the PTM model (gray solid lines) and from the MCM (black dotted lines).

The dataset is freely available in the R package smcure and can be loaded in R using the following command (see also the book's companion website for the R code):

```
library(smcure)
data(e1684)
```

The estimated KM survival curve for RFS shows a plateau starting at around six years, with however an event still occurring at around eight years in the control group. Both estimated survival curves run in parallel and reach their plateau at about the same time; in line with Section 22.4, we will consider that the results from both the PTM and the MCM can be trusted.

A classical semi-parametric Cox PH model (CM) can be fitted on these data with the R package survival, as follows:

```
library(survival)
cox <- coxph(Surv(FAILTIME,FAILCENS==1)~TRT+AGE+SEX,e1684)
summary(cox)
```

A semi-parametric promotion time model (PTM) with exponential link function can be fitted with the R package miCoPTCM:

```
library(miCoPTCM)
vc <- matrix(nrow=4,ncol=4,0)
ptcm <- PTCMestimBF(formula=Surv(e1684$FAILTIME,e1684$FAILCENS)
                ~TRT+SEX+AGE,data=e1684,varCov=vc,
                init=runif(4))
summary(ptcm)
```

Finally, a semi-parametric logistic/Cox PH mixture cure model (MCM) can be estimated with the R package smcure:

```
mcm <- smcure(Surv(FAILTIME,FAILCENS)~TRT+SEX+AGE,
            cureform=~TRT+SEX+AGE,data=e1684,model="ph,"
            nboot=500)
printsmcure(mcm)
```

Results (obtained with the three models) for the treatment effect, adjusted for gender and sex, are displayed in Table 22.4.

As expected, the results obtained with CM and PTM are similar, with the interpretation of the β coefficients being linked to both short- and long-term effects. To interpret these results, we have to keep in mind that IFN is coded 1 while PBO is coded 0. The treatment effect, adjusted for gender and age, is -0.365 for the PTM model, indicating an advantage for patients treated with IFN ($p < 0.05$). The MCM model allows us to disentangle the effect of IFN on the occurrence of and on the timing of the event. The OR for the probability to be uncured, adjusted for gender and age, is $\exp(-0.588) = 0.556$, corresponding to a lower risk of being uncured in the IFN group, and thus a higher risk to being cured in this group. Regarding the latency part of the model, the HR for the uncured patients is $\exp(-0.154) = 0.857$ in favor of the IFN group too. It appears from these results that IFN indeed has a beneficial effect on the RFS of the patients but that this effect is mainly a long-term one, acting on the probability to be cured. The fact that this OR is not significant might be explained by the fact that twice the number of parameters have to be estimated in the MCM compared to the PTM. For a male ($x_{male} = 0$) with average age ($x_{age} = 0$), the estimated cure fraction can be recovered from the PTM as

$$\text{PBO arm} : \exp\left(-\exp(0.412)\right) = 0.221$$
$$\text{IFN arm} : \exp\left(-\exp(0.412 - 0.365)\right) = 0.351$$

and about the same values can be retrieved from the MCM:

$$\text{PBO arm} : 1 - \frac{\exp(1.365)}{1 + \exp(1.365)} = 0.203$$
$$\text{IFN arm} : 1 - \frac{\exp(1.365 - 0.588)}{1 + \exp(1.365 - 0.588)} = 0.315$$

TABLE 22.4

Melanoma Data: Results from Semi-parametric Cox PH Model (CM), Semi-parametric Logistic/Cox PH Model (MCM), and Semi-parametric Promotion Time Cure Model (PTM) for Treatment (0: Control and 1: Treatment) Adjusted for Age (Centered to the Mean) and Gender (0: Male and 1: Female)

		CM	MCM Incidence	MCM Latency	PTM
Intercept	Estimate		1.365		0.412
	S.E.		0.322		0.139
	P-value		0.000		0.003
Treatment	Estimate	-0.360	-0.588	-0.154	-0.365
	S.E.	0.144	0.349	0.169	0.154
	P-value	0.012	0.092	0.363	0.017
Age	Estimate	0.005	0.020	-0.008	0.005
	S.E.	0.005	0.016	0.006	0.005
	P-value	0.357	0.205	0.212	0.358
Gender	Estimate	-0.018	-0.087	0.099	-0.018
	S.E.	0.147	0.328	0.172	0.159
	P-value	0.903	0.791	0.564	0.909

Note: S.E.: standard error.

The estimated population RFS curves by treatment arm from the PTM and the MCM for male patients with average age are displayed on the right panel of Figure 22.8. Curves obtained from both models are very similar and we find back at the tails of these curves the results given above about cure rate estimation.

22.6 Conclusion

Cure models are still rarely used in cancer clinical trials, despite the fact that there are nowadays several cancer types for which we can expect a fraction of the population to be cured. One argument against the use of the cure models is that as long as the PH assumption is met, the Cox PH model provides reliable estimates of the treatment effect [7]. Indeed, our simulations show that if the cure fraction is ignored and a Cox PH model is fitted, the treatment coefficient is perfectly recovered in size and significance in cases where we have PH. This is indeed true and actually not surprising, given the mathematical link between the Cox PH model and the semi-parametric promotion time cure model [40]. In the specific context of our simulations, when the PH assumption does not hold, the Cox PH model also allows one to recover a significant treatment effect, whose estimated value appears to be close to the average of both true coefficients. However, using the promotion time cure model instead of the Cox PH model has the advantage of making it clear that the coefficient associated with the treatment should be interpreted both in terms of short- and long-term effects.

One may argue that, unless the disease is always fatal, the proportion of patients being cured should be considered as an important component of the survival benefit, rather than just considering the HR or median time to failure [1]. If the PH assumption is met, both the Cox PH model and the promotion time cure model can actually be used to obtain an estimate of the cure fraction, which is clearly a useful piece of information in the evaluation of curative treatment. However, one has to be careful that this proportion may be overestimated by the promotion time cure model if there is actually no cure fraction and if the censoring is high. So, like many other authors (see, for example, Sy and Taylor [17]), we recommend not using such models when there is no evidence of cure.

Furthermore, the Cox PH and the promotion time models should not be used whenever the PH assumption is not met. If the reason of this non-proportionality is the presence of a cure fraction, which can be assessed from the presence of a long plateau including a sufficient number of censored observations, then the mixture cure model is a flexible alternative to be considered. Both parametric and semi-parametric versions are easily accessible in R and SAS, and both Cox PH and AFT models can be considered in the latency part. The (semi-)parametric logistic/Cox PH model provides parameter estimates which have an easy interpretation and can be translated into ORs or RRs for the probability of cure and in HRs for the event-times among the uncured, providing a clear view on the short- and long-term effects of the treatment.

The statement that "as long as one can assume that not all patients will experience the event of interest, a cure model should be preferred" needs to be nuanced. First, assuming that there is a cure fraction is not enough; we must have evidence of it, through sufficient follow-up, in order for a cure model to perform well. Second, even when there is such evidence from the data, if we still have PH and if we are not particularly interested in splitting the curative effect from an event-delaying effect, nor to emphasize the presence of a cure fraction, then the classical Cox PH model can indeed still be used. However, whenever there exists a fraction of cured or long-term survivors, additional information (or, even, more correct information, in the case of non-PH) can be gained from using an appropriate cure model analysis compared with a standard Cox analysis. Unfortunately, there are, up until now, no clear criteria on what is evidence of a cure fraction and no widely available statistical way to test whether there is "a sufficiently long plateau containing enough censored observations." Some attempts to develop statistical tests on the presence of cure have been proposed [56–60], but they have not been implemented in available software. As a consequence, one has mainly to rely on a visual inspection of the tail of the KM-estimated survival curves.

A key ingredient in clinical trials is the design phase and, in particular, the sample size calculation. All standard procedures for sample size calculation with time-to-event endpoints actually rely on the PH assumption and can therefore not be used if we expect the presence of a cure fraction that could put this assumption in jeopardy. A sample size formula for the logistic/PH mixture cure model has been proposed by Wang et al. [61] and later implemented in the R package NPHMC [62]. This formula can be used to compute the required sample size for testing differences in the short- and/or long-term outcome of the patients and can account for various accrual patterns. Furthermore, the NPHMC package allows one to choose for the latency part of the model between a parametric PH model (exponential or Weibull) or a Cox semi-parametric PH model. Numerical examples and simulation results are presented and show that ignoring the cure rate can lead to either underpowered or overpowered studies [61].

References

1. Maetani S, Gamel W. Parametric cure model versus proportional hazards model in survival analysis of breast cancer and other malignancies. *Adv Breast Cancer Res*. 2013;2:119–15.
2. Rutqvist LE, Wallgren A, Nilsson B. Is breast cancer a curable disease? A study of 14,731 women with breast cancer from the Cancer Registry of Norway. *Cancer*. 1984;53:1793–800.
3. Sargent D, Sobrero A, Grothey MJ et al. Evidence for cure by adjuvant therapy in colon cancer: Observations based on individual patient data from 20,898 patients on 18 randomized trials. *J Clin Oncol*. 2009;27:812–77.
4. Bleyer WA. Acute lymphoblastic leukaemia in children: Advances and prospectus. *Cancer*. 1990;65:689–95.
5. Mantel N. Evaluation of survival data and two new rank order statistics arising in its consideration. *Cancer Chemother Rep*. 1966;50:163–70.
6. Cox DR. Regression models and life-tables. *J R Stat Soc Ser B*. 1972;34:187–220.
7. Paoletti X, Asselain B. Survival analysis in clinical trials: Old tools or new techniques. *Surg Oncol*. 2010;19:55–8.
8. Sposto R. Cure model analysis in cancer: An application to data from the Children's Cancer Group. *Stat Med*. 2002;21:293–312.
9. Lambert PC. Modeling of the cure fraction in survival studies. *Stata J*. 2007;7:1–25.
10. Othus M, Barlogie B, LeBlanc ML, Crowley JJ. Cure models as a useful statistical tool for analyzing survival. *Stat Clin Cancer Res*. 2012;18:3731–6.
11. Yilmaz Y, Lawless JF, Andrulis IL, Bull SB. Insight from mixture cure modeling of molecular markers for prognosis in breast cancer. *J Clin Oncol*. 2013;31(16):2047–54.
12. Thomas L, Reyes EM. Tutorial: Survival estimation for Cox regression models with time-varying coefficients using SAS and R. *J Stat Softw*. 2014;61:1–23.

13. Boag JW. Maximum likelihood estimates of the proportion of patients cured by cancer therapy. *J R Stat Soc Ser B.* 1949;11:15–44.

14. Berkson J, Gage RP. Survival curve for cancer patients following treatment. *J Am Stat Assoc.* 1952;47:501–15.

15. Farewell VT. A model for a binary variable with time-censored observations. *Biometrika.* 1977;64:43–6.

16. Farewell VT. The use of mixture models for the analysis of survival data with long-term survivors. *Biometrics.* 1982;38:1041–6.

17. Sy JP, Taylor JMG. Estimation in a Cox promotional hazards cure model. *Biometrics.* 2000;56:227–36.

18. Yakovlev AY, Asselai B, Bardou V-J, Fourquet A, Hoang T, Rochefedière A, Tsodikov AD. A simple stochastic model of tumor recurrence and its application to data on premenopausal breast cancer. In *Biométrie et Analyse de Données Spatio-Temporelles*, Asselain B, Boniface M, Duby C, Lopez C, Masson J-P, Tranchefort J, eds. Société française de biométrie, ENSA Rennes, 1993: 66–82.

19. Yakovlev AY, Tsodikov AD. *Stochastic Models of Tumor Latency and Their Biostatistical Applications.* Singapore: World Scientific, 1996.

20. Peng Y, Taylor JMG. Cure models. In *Handbook of Survival Analysis,* Klein J, van Houwelingen H, Ibrahim JG, Scheike TH, eds. Boca Raton, FL, USA: Chapman and Hall, 2014: 113–134.

21. Amico M, Van Keilegom I. Cure models in survival analysis. *Annu Rev Stat Appl.* 2018;5(1).

22. Wang J, Ghosh SK. Shape restricted nonparametric regression with Bernstein polynomials. *Computat Stat Data Anal.* 2012;56:2729–41.

23. Amico M, Van Keilegom I, Legrand C. The single-index/Cox mixture cure model. *Biometrics.* 2018 Nov 14; Epub ahead of print.

24. Taylor JMG. Semi-parametric estimation in failure time mixture models. *Biometrics.* 1995;51:899–907.

25. Patilea V, Van Keilegom I. A general approach for cure models in survival analysis. *Submitted.* 2018. Accessible on arXiv via the link: https://arxiv.org/abs/1701.03769

26. Ghitany ME, Maller RA, Zhou S. Exponential mixture models with long term survivors and covariates. *J Multivar Anal.* 1994;49:218–41.

27. Corbière F, Commenges D, Taylor JMG, Joly P. A penalized likelihood approach for mixture cure models. *Stat Med.* 2009;28:510–24.

28. Kuk AYC, Chen C-H. A mixture model combining logistic regression with proportional hazards regression. *Biometrika.* 1992;79:531–41.

29. Lu W. Maximum likelihood estimation in the proportional hazards cure model. *Ann Inst Stat Math.* 2008;60:545–74.

30. Peng Y, Dear KBG. A nonparametric mixture model for cure rate estimation. *Biometrics.* 2000;56:237–43.

31. Peng Y, Dear KBG, Denham JW. A generalized f mixture model for cure rate estimation. *Stat Med.* 1998;17:813–30.

32. Scolas S, El Ghouch A, Legrand C, Oulhaj A. Variable selection in a flexible parametric mixture cure model with interval-censored data. *Stat Med.* 2016;35:1210–25.

33. Yamaguchi K. Accelerated failure-time regression models with a regression model of surviving fraction: An application to the analysis of permanent employment in Japan. *J Am Stat Assoc.* 1992;87(418):284–2.

34. Li C-S, Taylor JMG. A semi-parametric accelerated failure time cure model. *Stat Med.* 2002;21:3235–47.

35. Zhang J, Peng Y. A new estimation method for the semiparametric accelerated failure time mixture cure model. *Stat Med.* 2007;26:3157–71.

36. Corbière F, Joly P. A SAS macro for parametric and semiparametric mixture cure models. *Comput Methods Programs Biomed.* 2007;85(2):173–80.

37. Cai C, Zou Y, Peng Y, Zhang J. Smcure: An R-package for estimating semiparametric mixture cure models. *Comput Methods Programs Biomed.* 2012;108(3):1255–60.

38. Chen M-H, Ibrahim JG, Sinha D. A new Bayesian model for survival data with a surviving fraction. *J Am Stat Assoc*. 1999;94:909–19.
39. Tsodikov A. A proportional hazards model taking account of long-term survivors. *Biometrics*. 1998;54:1508–16.
40. Portier F, El Ghouch A, Van Keilegom I. On proportional hazards cure models. *Bernoulli*. 2017;23(4B):3437–68.
41. Bremhorst V, Kreyenfeld M, Lambert P. Fertility progression in Germany: An analysis using flexible nonparametric cure survival models. *Demogr Res*. 2016;35(18):505–34.
42. Portier F, El Ghouch A, Van Keilegom I. Efficiency and bootstrap in the promotion time cure model. *Bernoulli*. 2017;23:3437–68.
43. Yin G, Ibrahim JG. Cure rate models: A unified approach. *Can J Stat*. 2005;33:559–70.
44. Ibrahim JG, Chen M-H, Sinha D, *Bayesian Survival Analysis*. Boca Raton: Springer-Verlag, 2001.
45. Tsodikov A, Ibrahim JG, Yakovlev AY. Estimating cure rates from survival data: An alternative to two-component mixture models. *J Am Stat Assoc*. 2003;98:1063–78.
46. Ibrahim JG, Chen M-H, Sinha D. Bayesian semiparametric models for survival data with a cure fraction. *Biometrics*. 2001;57:383–8.
47. Zeng D, Yin G, Ibrahim JG. Semiparametric transformation models for survival data with a cure fraction. *J Am Stat Assoc*. 2006;101:670–84.
48. Ma Y, Yin G. Cure rate model with mismeasured covariates under transformation. *J Am Stat Assoc*. 2008;103:743–56.
49. Liu H, Shen Y. A semiparametric regression cure model for interval-censored data. *J Am Stat Assoc*. 2009;104:1168–78.
50. Andersson TM-L, Lambert PC. Fitting and modeling cure in population-based cancer studies within the framework of flexible parametric survival models. *Stata J*. 2012;12:623–38.
51. Tsodikov A. Semi-parametric models of long- and short-term survival: An application to the analysis of breast cancer survival in Utah by age and stage. *Stat Med*. 2002;21:895–920.
52. Bremhorst V, Lambert P. Flexible estimation in cure survival models using Bayesian P-splines. *Computational Statistics and Data Analysis*. 2016;93:270–84.
53. Cooner F, Banerjee S, Carlin B, Sinhaet D. Flexible cure rate modeling under latent activation scheme. *J Am Stat Assoc*. 2007;102:560–72.
54. Gu Y, Sinha D, Banerjee S. Analysis of cure rate survival data under proportional odds model. *Lifetime Data Anal*. 2011;17:123–34.
55. Tournoud M, Ecochard R. Promotion time models with time-changing exposure and heterogeneity: Application to infectious diseases. *Biom J*. 2008;50:395–407.
56. Broet P, De Rycke Y, Tubert-Bitter P, Lellouch J, Asselain B, Moreau T. A semiparametric approach for the two-sample comparison of survival times with long-term survivors. *Biometrics*. 2001;57:844–52.
57. Hsu WW, Todem D, Kim KM. A sup-score test for the cure fraction in mixture cure models for long-terms survivors. *Biometrics*. 2016;76:1348–57.
58. Laska EM, Meisner MJ. Nonparametric estimation and testing in a cure model. *Biometrics*. 1992;48:1223–34.
59. Maller RA, Zhou S. *Survival Analysis With Long Term Survivors*. New York: Wiley, 1996.
60. Zhao Y, Lee AH, Yau KKW, Burke V, McLachlan GJ. A score test for assessing the cured proportion in long-term survivor mixture cure model. *Stat Med*. 2009;28:3454–66.
61. Wang S, Zhang J, Lu W. Sample size calculations for the proportional hazards cure model. *Stat Med*. 2012;31(29):3959–71.
62. Cai C, Wang S, Lu W, Zhang J. Nphmc: An R-package for estimating sample size of proportional hazards mixture cure models. *Comput Methods Programs Biomed*. 2014;113(1):290–300.
63. Kirkwood JM, Strawderman MH, Ernstoff MS, Smith TJ, Borden, EC, Blum RH. Interefron alfa-2b adjuvant therapy of high-risk resected cutaneous melanoma: The Eastern Cooperative Oncology Group Trial EST 1684. *J Clin Oncol*. 1996;14:7–17.

23

Interval Censoring

Yuan Wu and Susan Halabi

CONTENTS

23.1 Introduction

In clinical trials, participants will be followed-up for a pre-specified period of time until the primary endpoint has matured. If the assessments are conducted continuously, participants may experience the event of interest during the follow-up period. Other participant may not have an event and are administratively censored at the end of the follow-up period. The time-to-event endpoint, measured by the assessment will be right censoring, including observed time to event and right-censored time due to loss of follow-up or end of the study. Another frequently occurring right censored endpoint is overall Survival, for which the follow-up does not need to be continuous to obtain the time of death for participants. On the other hand, if the assessment is not performed continuously, then for some endpoints, such as progression-free survival, the data will usually become interval censoring. By interval censoring, we mean that the event is only known to have occurred within a time interval. A special case for interval censoring is group censoring, which means, before the end of the study, the observation times for all patients are fixed. Then the censored intervals, including the event times, will never have overlaps. Both right censoring and interval censoring are incomplete data instead of missing data, as partial information for the event time is available.

Right censoring is a well-accepted concept in clinical studies and standard statistical packages include modules for the analysis of right-censored data. However, the analyses for interval censoring are rarely developed or available in the standard statistical packages. A common but problematic approach to analyzing interval censoring is to convert the interval-censored data to right-censored data and use existing methods for right censoring.

Here, we name this approach as "naively converting." For the naively converting approach, if an event is censored between a finite interval, then the mid point, the two end points of this interval, or a random point on this interval will be used as the event time to convert the original data set to a right-censored one. We must admit that in many situations, using the naively converting method will not cause estimation problems. Nevertheless, this cannot be the reason for not pursuing the correct, although more complicated, statistical method for analyzing the data since researchers often have no way to check the applicability for the naively converting approach.

The chapter intends to promote the use of the appropriate statistical analysis methods for interval-censored data. The chapter discusses the parallels to the three most popular estimation or testing methods for right-censored survival analysis, that is, the Kaplan–Meier estimation, the log-rank test, and the Cox's proportional hazards model. We use a publicly available dataset to illustrate how to use these three methods specifically designed for interval-censored data.

23.1.1 The Estimation Issue for the Naively Converting Approach

We can demonstrate that the naively converting approach is not optimal, methodologically speaking, either by simulation or mathematical derivation. We choose to briefly explain why the naively converting approach is not the optimal methodology using the latter (mathematical derivation). We focus on non-parametric maximum likelihood estimation (MLE) for the survival function with interval censoring as an example. Let $\{(U_i, V_i] : i = 1, \dots, n\}$ be the set of censored intervals, each containing event time T_i. Note that in this chapter, it is assumed that n observations are independent, and $(U_i, V_i]$ and T_i are independent for all $i \in \{1, \dots, n\}$ except in Section 23.2.3. Then the likelihood function can be written as

$$L = \prod_{i=1}^{n} [F(V_i) - F(U_i)] \tag{23.1}$$

for cumulative distribution function $F(\cdot)$. We now construct a set of disjoint intervals whose left and right end points lie in the sets $\{U_i : i = 1, \dots, n\}$ and $\{V_i : i = 1, \dots, n\}$, respectively, and which contains no other members of $\{U_i\}$ or $\{V_i\}$. We write this set of disjoint intervals as $\{(W_j, S_j] : j = 1, \dots, m\}$. Figure 23.1 clearly shows how to get these disjoint intervals from all censored intervals. In Figure 23.1, every rectangle denoted by borders U_i and V_i represents a censored interval $(U_i, V_i]$ for $i \in \{1, \dots, 6\}$, and each filled rectangle denoted by borders W_j and S_j represents a disjoint interval $(W_j, S_j]$ for $j \in \{1, \dots, 4\}$.

Suppose we have the non-parametric MLE based on the likelihood function (Equation 23.1). It is easy to observe that the MLE must assign positive probability mass for each censored interval, otherwise the likelihood function will be zero. The MLE assigns some probability mass for a censored interval, but outside all disjoint intervals, the likelihood function can always be increased by moving the mass to one of the disjoint intervals. Hence, the MLE can only assign positive probability mass on the disjoint intervals. Since the Kaplan–Meier estimation assigns positive mass at each observed event time, then the Kaplan–Meier estimation obtained using the naively converting approach cannot be the non-parametric MLE based on the original interval-censored data, unless the naively converting approach happens to pick every event time on one of the disjoint intervals.

On the other hand, the Kaplan–Meier estimation was shown to be the non-parametric MLE by [1]. This implies that if all censored intervals are completely overlapped or mutually

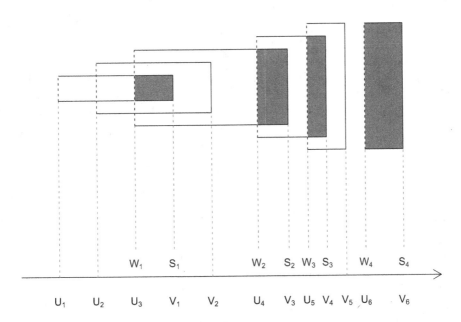

FIGURE 23.1
Censored intervals and disjoint intervals for random interval censoring.

exclusive plus the naively converting approach picks an identical event time on the same censored interval, the Kaplan–Meier estimation through the naively converting approach will become the non-parametric MLE for the original interval-censored data. This is because the set of censored intervals is totally the same as the set of disjoint intervals. For example, Figure 23.2 presents the case for group censoring when each censored interval becomes one of the disjoint intervals. However, in practice, it is really uncommon to see this kind of data in clinical studies, since exact group-censored data are rare and usually some random noise is seen for scheduled observation times.

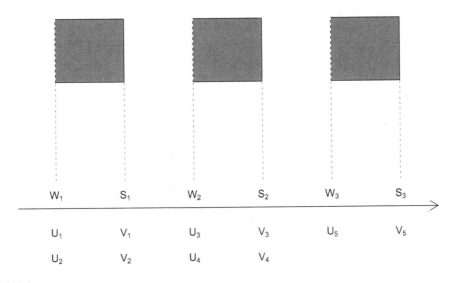

FIGURE 23.2
Censored intervals and disjoint intervals for group censoring.

The estimation issue for naively converting interval censoring to right censoring is not well recognized in clinical studies. We recognize that sometimes it is not possible for statisticians to use the correct statistical methods to analyze interval-censored data. This is because the data collection is designed for right-censored and not interval-censored data. Hence, statisticians are encouraged to work with their collaborators to make sure the data collection during the conduct of the studies is designed so that interval-censored methods can be applied.

23.2 Non-Parametric and Semi-Parametric Approaches for Analyzing Interval-Censored Data

As reviewed by Sun [2], several authors have proposed different approaches for interval censoring. These include parametric, multiple-imputation, non-parametric, and semi-parametric approaches. The parametric approach is easy to understand and implement, but the parametric form is questionable. Usually there is no means to verify the chosen parametric form and it is likely to be wrong. The multiple-imputation approach is similar to the "naively converting" approach in the sense that it will convert the data to right censoring so that existing right-censored methods can be used. Compared to the naively converting approach, the multiple-imputation approach is more sophisticated and based on the well-accepted idea for handling missing data. However, very few asymptotic results have been developed for this approach. Non-parametric and semi-parametric approaches are the most developed techniques for interval-censored survival data, which is a likelihood estimation-based method. There are many articles on the computation and asymptotic properties for this type of estimation [2], and we will discuss the non-parametric and semi-parametric approaches.

This section discusses the general interval censoring, which means the intervals can have 0 as left end points (left censoring) and have ∞ as right end points (right censoring). The general interval censoring is also called case 2 interval censoring. A special case for the general interval censoring is current status data, which has only left-censored or right-censored event times. Current status data occur more often in animal studies than in human studies. It is worthwhile to note that all methods designed for general interval censoring can be also applied to current status data.

23.2.1 Non-Parametric Maximum Likelihood Estimation

In this section, we discuss the non-parametric MLE for the distribution of the survival function for interval-censored data. As we indicated in Section 23.1.1, the Kaplan–Meier estimation is actually a non-parametric MLE for right-censored data. Therefore, the non-parametric MLE for interval censoring can be seen as the extended Kaplan–Meier estimation for interval censoring.

As we explained in Section 23.1.1, the non-parametric MLE can only have probability mass in the resulting set of the disjoint intervals. The likelihood function (Equation 23.1) can be rewritten as

$$L = \prod_{i=1}^{n} \sum_{j=1}^{m} \alpha_{i,j} p_j, \tag{23.2}$$

where $\alpha_{i,j} = I[(W_j, S_j] \subset (U_i, V_i]]$ and $p_j = Pr(W_j < T_i \le S_j)$ for event-time variable T_i. Obviously, we have $\sum_{j=1}^m p_j = 1$. Note that when right censoring appears, there exists at least one $i \in \{1, \ldots, n\}$ such that $V_i = \infty$. Then it is possible that the right end point of last disjoint interval $S_m = \infty$.

The non-parametric MLE based on the likelihood function (Equation 23.2) can be obtained by the expectation-maximization (EM) algorithm [3]. To use the EM algorithm, a full likelihood function based on the unobserved complete data is necessary, and importantly, the complete data should be consistent with the observed data. Following on the discussion in Section 23.1.1, let all T_i's for $i \in \{1, \ldots, n\}$ only have probability mass at members of $\{S_j\}_{j=1}^m$ with $Pr(T_i = S_j) = p_j$ for $i \in \{1, \ldots, n\}$ and $j \in \{1, \ldots, m\}$. Then the full likelihood function for unobserved event time $\{T_i = t_i\}_{i=1}^n$ is

$$L_f = \prod_{i=1}^n Pr(T_i = t_i), \tag{23.3}$$

where for every $i \in \{1, \ldots, n\}$, there exist $j \in \{1, \ldots, m\}$ such that $T_i = S_j$. In this way, the observed data for Equation 23.2 and the complete data for Equation 23.3 will be consistent. On the other hand, if the complete data are chosen such that some events occur outside the set of disjoint intervals, it contradicts the conclusion that events can only occur inside the disjoint intervals based on the censored intervals, which means they are not consistent.

Let $d_j = \sum_{i=1}^n I[T_i = S_j]$ for $j \in \{1, \ldots, m\}$. The full likelihood function can be rewritten as $L_f = \prod_{j=1}^m p_j^{d_j}$ and the corresponding full log likelihood function is $l_f(\mathbf{p}) = \sum_{j=1}^m d_j \log p_j$ for $\mathbf{p} = (p_1, \ldots, p_m)'$. To get the non-parametric MLE, the EM algorithm is given as follows.

1. Let $d = 0$ and choose the initial value as $\mathbf{p}^{(d)} = (1/m, \ldots, 1/m)$.

2. *E-step:*

$$Q(\mathbf{p}, \mathbf{p}^{(d)}) = E\left(l_f(\mathbf{p}) \mid \mathcal{D}, \mathbf{p}^{(d)}\right)$$

$$= \sum_{j=1}^m \left\{ \log p_j E\left[\sum_{i=1}^n I[T_i = S_j] \mid \mathcal{D}, \mathbf{p}^{(d)} \right] \right\}$$

$$= \sum_{j=1}^m \left\{ \log p_j \sum_{i=1}^n Pr\left(T_i = S_j \mid U_i, V_i, \mathbf{p}^{(d)}\right) \right\}$$

$$= \sum_{j=1}^m \left\{ \log p_j \sum_{i=1}^n \frac{\alpha_{i,j} p_j^{(d)}}{\sum_{l=1}^m \alpha_{i,l} p_l^{(d)}} \right\},$$

where $\mathcal{D}$ denotes the observed data.

3. *M-step:* Maximizing $Q(\mathbf{p}, \mathbf{p}^{(d)}) + \lambda(\sum_{j=1}^m p_j - 1)$ yield to the update of $\mathbf{p}$:

$$p_j^{(d+1)} = \frac{1}{n} \sum_{i=1}^n \frac{\alpha_{i,j} p_j^{(d)}}{\sum_{l=1}^m \alpha_{i,l} p_l^{(d)}} \quad \text{for } j \in \{1, \ldots, m\}. \tag{23.4}$$

Let $d = d + 1$ and go back to E-step until convergence.

In addition to the EM algorithm, several authors have proposed other methods to find the non-parametric MLE. These are the iterative convex minorant method and its modification [4,5]. Compared to its alternatives, the EM algorithm may not be efficient in computing time but it is very stable. To the authors' best knowledge, the asymptotic properties for the non-parametric MLE are only partially developed and the limiting distribution is not available yet. A conjecture is given that the bootstrap confidence interval is applicable [6].

23.2.2 Comparing Survival Functions

Sun [7] proposed a test for comparing the survival functions among treatment groups with interval-censored data. The test statistic has the log-rank form for the well known log-rank test designed for right-censored data. Fay [8] established that the test proposed by Sun [7] is actually the score test.

Suppose there are k treatments to compare. If the ith subject receives the lth of m treatments, let x_i be a $k \times 1$ vector of zeros except for the lth row, which is one. As discussed in Section 23.2.1, assume events only occur at members of $\{S_j\}_{j=1}^m$, and let $\mathbf{p} = (p_1, \dots, p_m)'$ with $p_l = \Pr(T = S_l)$ for $l \in \{1, \dots, m\}$. The survival function at time t given x_i is denoted as

$$Pr(T > t \mid x_i) = g(t \mid x_i'\beta, \gamma(\mathbf{p})),$$

where β is a $k \times 1$ vector of treatment parameters, $\gamma(\mathbf{p}) = \{\gamma_1(\mathbf{p}), \dots, \gamma_{m-1}(\mathbf{p})\}'$ is the $(m-1) \times 1$ vector of the nuisance parameter. To simplify derivation, we introduce S_0 with $S_0 < S_1$. It is clear that $g(S_0 \mid x_i'\beta, \mathbf{p}) = 1$ and $g(S_m \mid x_i'\beta, \gamma(\mathbf{p})) = 0$ for any $x_i'\beta$. Let $g(S_j) = g(S_j \mid x_i'\beta, \beta = 0, \gamma(\mathbf{p}))$ for $j \in \{0, \dots, m\}$. It is clear that $1 > g(S_1) > \cdots > g(S_{m-1}) > 0$. Based on $g(t \mid x_i'\beta, \mathbf{p})$, the likelihood function with interval-censored data $\{(U_i, V_i)\}_{i=1}^n$ is

$$L = \prod_{i=1}^{n} \left\{ g(U_i \mid x_i'\beta, \gamma(\mathbf{p})) - g(V_i \mid x_i'\beta, \gamma(\mathbf{p})) \right\}. \tag{23.5}$$

To derive the score, take the derivative of $\log(L)$ with respect to β, and evaluate the derivative at $\beta = 0$ and at the MLE for $\gamma(\mathbf{p})$ given $\beta = 0$. When $\beta = 0$, the likelihood function (Equation 23.5) becomes the non-parametric likelihood function (Equation 23.1), and the EM algorithm described in Section 23.2.1 can be used to get the MLE for $\gamma(\mathbf{p})$. It is clear that if $\hat{\mathbf{p}}$ is the non-parametric MLE obtained from the EM algorithm, then the MLE for $\gamma(\mathbf{p})$ given $\beta = 0$ is $\gamma(\hat{\mathbf{p}})$. Then the score test statistic for β is

$$U = \left\{ \frac{\partial \log(L)}{\partial \beta} \right\}_{\beta=0, \gamma(\mathbf{p})=\gamma(\hat{\mathbf{p}})}.$$

It can be shown that the lth row of score statistic U is given as

$$U_l = \sum_{j=1}^{m-1} \omega_j \left\{ d_{j,l} - \frac{n_{j,l} d_j}{n_j} \right\}, \tag{23.6}$$

where

$$\omega_j = \frac{\hat{g}(S_j)\hat{g}'(S_{j-1}) - \hat{g}(S_{j-1})\hat{g}'(S_j)}{\hat{g}(S_j)\{\hat{g}(S_{j-1}) - \hat{g}(S_j)\}}$$

with $\hat{g}(S_j) = g(S_j)|_{\mathbf{p}=\hat{\mathbf{p}}}$ and $\hat{g}'(S_j) = \left\{ \dfrac{\partial g(S_j \mid \eta = x_i'\beta, \gamma(\mathbf{p}))}{\partial \eta} \right\}_{\eta=0, \gamma(\mathbf{p})=\gamma(\hat{\mathbf{p}})}$,

$$d_j = n\big(\hat{g}(S_{j-1}) - \hat{g}(S_j)\big), \quad n_j = n\hat{g}(S_{j-1}),$$

$$d_{j,l} = \sum_{i=1}^{n} x_{i,l} \frac{\alpha_{i,j}\big(\hat{g}(S_{j-1}) - \hat{g}(S_j)\big)}{\sum_{u=1}^{m} \alpha_{i,u}\big(\hat{g}(S_{u-1}) - \hat{g}(S_u)\big)}$$

and

$$n_{j,l} = \sum_{i=1}^{n} x_{i,l} \frac{\sum_{v=j}^{m} \alpha_{i,v}\big(\hat{g}(S_{v-1}) - \hat{g}(S_v)\big)}{\sum_{u=1}^{m} \alpha_{i,u}\big(\hat{g}(S_{u-1}) - \hat{g}(S_u)\big)}$$

with $\alpha_{i,j} = I[(W_j, S_j] \subset (U_i, V_i]]$ as in Section 23.2.1.

It is clear that d_j is the expected number of events at S_j; n_j is the expected number at risk prior to S_j. By the formula in Equation 23.4 for the EM algorithm given in Section 23.2.1, we have

$$d_j = n\big(\hat{g}(S_{j-1}) - \hat{g}(S_j)\big) = \sum_{i=1}^{n} \frac{\alpha_{i,j}\big(\hat{g}(S_{j-1}) - \hat{g}(S_j)\big)}{\sum_{u=1}^{m} \alpha_{i,u}\big(\hat{g}(S_{u-1}) - \hat{g}(S_u)\big)}.$$

Hence, $d_{j,l}$ is the expected number of events at S_j for group l. By $\hat{g}(S_m) = 0$,

$$n_j = n\hat{g}(S_{j-1}) = \sum_{i=1}^{n} \frac{\sum_{v=j}^{m} \alpha_{i,v}\big(\hat{g}(S_{v-1}) - \hat{g}(S_v)\big)}{\sum_{u=1}^{m} \alpha_{i,u}\big(\hat{g}(S_{u-1}) - \hat{g}(S_u)\big)}.$$

Then $n_{j,l}$ is the expected number at risk prior to S_j for group l. Hence, the score U has the form of a weighted log-rank statistic, in which $(n_{j,l}d_j)/n_j$ is the expectation of $d_{j,l}$ under null hypothesis $\beta = 0$.

To use the score from Equation 23.6, $g(t \mid x_i'\beta, \gamma(\mathbf{p}))$ needs to be specified to evaluate ω_j. For $j \in \{1, \ldots, m-1\}$, assume that

$$\frac{g(S_{j-1} \mid x_i'\beta, \gamma(\mathbf{p})) - g(S_j \mid x_i'\beta, \gamma(\mathbf{p}))}{g(S_j \mid x_i'\beta, \gamma(\mathbf{p}))} = \exp(x_i'\beta) \frac{g(S_{j-1}) - g(S_j)}{g(S_j)}. \tag{23.7}$$

Then for $j \in \{1, \ldots, m-1\}$,

$$g(S_j \mid x_i'\beta, \gamma(\mathbf{p})) = \prod_{u=1}^{j} \big[1 + \exp\{\gamma_u(\mathbf{p}) + x_i'\beta\}\big]^{-1}. \tag{23.8}$$

with $\gamma_j(\mathbf{p}) = \log\{(g(S_{j-1}) - g(S_j))/g(S_j)\}$. Using Equation 23.8, it can be shown that for $j \in \{1, \ldots, m-1\}$,

$$\omega_j = 1.$$

Hence the score from Equation 23.6 proposed by Sun [7] for interval censoring has the log-rank form. Note that the assumption in Equation 23.7 implies that the odds of the discrete hazard for a given x_i is proportional to that for a different x_j value.

To make sure the score test is valid, one has to assume that m does not change with n for using the MLE theory. Let V be the observed information matrix as

$$V = -\left[\frac{\partial^2 \log(L)}{\partial\beta\partial\beta'} - \left[\frac{\partial^2 \log(L)}{\partial\beta\partial\gamma'}\right]\left[\frac{\partial^2 \log(L)}{\partial\gamma\partial\gamma'}\right]^{-1}\left[\frac{\partial^2 \log(L)}{\partial\beta\partial\gamma'}\right]'\right]_{\beta=0, \gamma=\gamma(\hat{p})}.$$

Then $UV^- U'$ has an asymptotically chi-square distribution with $k-1$ degrees of freedom, where V^- is the generalized inverse of V.

Compared to the above score test for interval censoring, the well known log-rank test for right-censored data is purely non-parametric, that is, normality for the log-rank test statistic under the null hypothesis was established without the proportional assumption (Equation 23.7) and m is not required to be fixed [9].

The score statistic U has different weighted log-rank forms as discussed by Fay [8]. Specifically, if the proportional discrete hazards assumption holds, that is

$$\frac{g(S_{j-1} \mid x_i'\beta, \gamma(\mathbf{p})) - g(S_j \mid x_i'\beta, \gamma(\mathbf{p}))}{g(S_{j-1} \mid x_i'\beta, \gamma(\mathbf{p}))} = \exp(x_i'\beta)\frac{g(S_{j-1}) - g(S_j)}{g(S_{j-1})}, \qquad (23.9)$$

then

$$\omega_j = \frac{\hat{g}(S_{j-1})\{\log \hat{g}(S_{j-1}) - \log \hat{g}(S_j)\}}{\hat{g}(S_{j-1}) - \hat{g}(S_j)};$$

if the proportional odds assumption holds, that is,

$$\frac{g(S_j \mid x_i'\beta, \gamma(\mathbf{p}))}{1 - g(S_j \mid x_i'\beta, \gamma(\mathbf{p}))} = \exp(x_i'\beta)\frac{g(S_j)}{g(S_j)}, \qquad (23.10)$$

then

$$\omega_j = \hat{g}(S_{j-1}).$$

23.2.3 Proportional Hazards Model with Interval Censoring

Let x_i be $k \times 1$ covariate vector for subject i and β be the corresponding $k \times 1$ parameter vector. Denote $S(\cdot \mid x_i)$ as the survival function of event time T_i given covariate x_i for all $i \in \{1, \ldots, n\}$. Under the proportional hazards (PH) assumption, $S(t \mid x_i) = \exp\{-\Lambda_0(t)\exp(x_i'\beta)\}$ with baseline cumulative hazard function $\Lambda_0(\cdot)$. If it is assumed that $(U_i, V_i]$ and T_i are dependent given covariate vector x_i for $i \in \{1, \dots, n\}$, then for interval-censored data $\{(U_i, V_i], x_i\}_{i=1}^n$, the likelihood function is

$$L = \prod_{i=1}^n \{S(U_i \mid x_i) - S(V_i \mid x_i)\}.$$

To ease the computing work, the spline-based sieve estimation approach is introduced here, in which $\Lambda(\cdot)$ is assumed to be smooth and only restricted in a spline function class, a subset of the non-parametric function class, for its estimation. The smoothness assumption of the baseline hazard is a reasonable one in practice. The spline function form for $\Lambda(\cdot)$ adopted by Wang et al. [10] is $\Lambda(\cdot) = \sum_{q=1}^{q_l} \gamma_q I_q(\cdot)$, where $\{I_q(\cdot)\}_{q=1}^{q_l}$ is the set of monotone I-spline basis functions and q_l equals the number of interior knots for these basis functions plus degree of the spline functions. The reader is referred to Ramsay [11] for additional details. Now the goal is to find the sieve spline-based MLE for θ, where $\theta = (\beta', \gamma')'$ and $\gamma = (\gamma_1, \dots, \gamma_{q_l})'$.

Wang et al. [10] proposed using EM algorithm to obtain the sieve MLE based on the PH model with interval censoring. To simplify the derivations in Wang et al. [10], we rewrite the observed data $\{(U_i, V_i], x_i\}_{i=1}^n$ as $\mathcal{D} = \{U_i, V_i, \delta_{i,1}, \delta_{i,2}, \delta_{i,3}, x_i\}_{i=1}^n$, where $\delta_{i,1} = I[T_i \le U_i]$, $\delta_{i,2} = I[U_i < T_i \le V_i]$ and $\delta_{i,3} = I[T_i > V_i]$. Then the likelihood function can be rewritten as

$$L = \prod_{i=1}^n \{1 - S(U_i \mid x_i)\}^{\delta_{i,1}} \{S(U_i \mid x_i) - S(V_i \mid x_i)\}^{\delta_{i,2}} S(V_i \mid x_i)^{\delta_{i,3}}. \tag{23.11}$$

The EM algorithm is based on a two-stage data augmentation to get a full likelihood function with latent Poisson random variables. For each subject i with $i \in \{1, \dots, n\}$, given x_i, let $N_i(t)$ be a latent non-homogeneous Poisson process whose first jump is the event time T_i. Then $\Pr(N_i(t) = 0 \mid x_i) = \Pr(T_i > t \mid x_i) = \exp\{-\Lambda_0(t) \exp(x_i'\beta)\}$, which implies the given x_i Poisson random variable $N_i(t)$ has mean $\Lambda_0(t) \exp(x_i'\beta)$.

Let $Z_i = N_i(U_i)$ and $W_i = N_i(V_i) - N_i(U_i)$. Thus, conditional on x_i, Z_i and W_i are Poisson random variables with mean $\Lambda_0(U_i) \exp(x_i'\beta)$ and $\{\Lambda_0(V_i) - \Lambda_0(U_i)\} \exp(x_i'\beta)$, and Z_i and W_i are independent. Like for observed data, it is assumed that $N_i(t)$ is independent of U_i and V_i. Then the likelihood function based on Z_i, W_i, U_i, V_i conditional on x_i can be reduced to

$$L_1(\theta) = \prod_{i=1}^n \mathcal{P}_{Z_i \mid x_i}(z_i) \mathcal{P}_{W_i \mid x_i}(w_i), \tag{23.12}$$

where $\mathcal{P}_{A \mid x}(\cdot)$ is the probability mass function for variable A given x.

For the EM algorithm, as we have discussed in Section 23.2.1, it is critical that the observed data for Equation 23.11 and the latent data for Equation 23.12 are consistent. This can be explained by the following equalities.

$$\Pr(T_i \le U_i \mid x_i) = \Pr(Z_i > 0 \mid x_i),$$

$$\Pr(U_i < T_i \le V_i \mid x_i) = \Pr(Z_i = 0, W_i > 0 \mid x_i) = \Pr(Z_i = 0 \mid x_i)\Pr(W_i > 0 \mid x_i),$$

$$\Pr(T_i > V_i \mid x_i) = \Pr(Z_i = 0, W_i = 0 \mid x_i) = \Pr(Z_i = 0 \mid x_i)\Pr(W_i = 0 \mid x_i).$$

Hence, $\mathcal{D}$ and $\{Z_i, W_i, x_i, U_i, V_i\}_{i=1}^n$ are consistent.

A second stage of data augmentation is also considered in order to handle the spline coefficients γ easily. Specifically, for each i, $N_i(t)$ is decomposed as a sum of q_l mutually independent poisson process as $N_i(t) = \sum_{q=1}^{q_l} N_{i,q}(t)$, where we assume that $\{N_{i,q}(t)\}_{q=1}^{q_l}$ has corresponding mean set $\{\gamma_q I_q(t) \exp(x_i'\beta)\}_{l=1}^{q_l}$. Then $Z_i = \sum_{q=1}^{q_l} Z_{i,q}$ and $W_i = \sum_{q=1}^{q_l} W_{i,q}$, plus conditional on x_i, $\{Z_{i,q}\}_{q=1}^{q_l}$ and $\{W_{i,q}\}_{q=1}^{q_l}$ are two sets of q_l mutually independent Poisson random variables with mean sets $\{\gamma_q I_q(U_i) \exp(x_i'\beta)\}_{q=1}^{q_l}$ and $\{\gamma_q \{[I_q(V_i) - I_q(U_i)] \exp(x_i'\beta)\}_{q=1}^{q_l}$, respectively. It is reasonable to assume that $\{N_{i,q}(t)\}_{q=1}^{q_l}$ is independent of U_i and V_i. Then the full likelihood function for the EM algorithm is based on the distribution of $Z_{i,q}$, $W_{i,q}$, U_i, V_i conditional on x_i, and it is given as

$$L_f(\theta) = \prod_{i=1}^{n} \prod_{q=1}^{q_l} P_{Z_{i,q}|x_i}(z_{i,q}) P_{W_{i,q}|x_i}(w_{i,q}). \tag{23.13}$$

The relationship between the distributions of Z_i and $Z_{i,q}$ for $q \in \{1, \ldots, q_l\}$ is

$$\Pr(Z_i = n_0 \mid x_i) = \Pr\left(\sum_{q=1}^{q_l} Z_{i,q} = n_0 \mid x_i \right)$$

$$= \sum_{\{n_{0,q}\}_{q=1}^{q_l} : \sum_{q=1}^{q_l} n_{0,q} = n_0} \Pr(Z_{i,1} = n_{0,1}, \ldots, Z_{i,q_l} = n_{0,q_l} \mid x_i)$$

$$= \sum_{\{n_{0,q}\}_{q=1}^{q_l} : \sum_{q=1}^{q_l} n_{0,q} = n_0} \prod_{q=1}^{q_l} \Pr(Z_{i,q} = n_{0,q} \mid x_i).$$

Similarly, it can be shown that

$$\Pr(W_i = m_0 \mid x_i) = \sum_{\{m_{0,q}\}_{q=1}^{q_l} : \sum_{q=1}^{q_l} m_{0,q} = m_0} \prod_{q=1}^{q_l} \Pr(W_{i,q} = m_{0,q} \mid x_i).$$

So $\{Z_i, W_i, x_i\}_{i=1}^{n}$ and $\left\{(Z_{i,q})_{q=1}^{q_l}, (W_{i,q})_{q=1}^{q_l}, x_i, \right\}_{i=1}^{n}$ are consistent and the observed data $\mathcal{D}$ and the complete data $\left\{(Z_{i,q})_{q=1}^{q_l}, (W_{i,q})_{q=1}^{q_l}, x_i, U_i, V_i \right\}_{i=1}^{n}$ are consistent. Hence, it is valid to use the full likelihood function (Equation 23.13) for the EM algorithm. For the expectation step, let $Q(\theta, \theta^{(d)}) = E\left[\log\{L_f(\theta)\} \mid \mathcal{D}, \theta^{(d)} \right]$; it is easy to obtain the expectation by removing the summation terms unrelated to θ

$$Q(\theta, \theta^{(d)}) = \sum_{i=1}^{n} \sum_{q=1}^{q_l} \left[\left\{ E(Z_{i,q} \mid \mathcal{D}, \theta^{(d)}) + E(W_{i,q} \mid \mathcal{D}, \theta^{(d)}) \right\} \{ \log(\gamma_q) + x_i'\beta \} \right.$$
$$\left. - \gamma_q I_q(V_i) \exp(x_i'\beta) \right]. \tag{23.14}$$

Since for fixed x_i, $\{Z_{i,q}\}_{q=1}^{q_l}$ is a set of q_l mutually independent Poisson random variables with their corresponding mean set $\{\gamma_q I_q(U_i) \exp(x_i'\beta)\}_{q=1}^{q_l}$. Then as mentioned by Wang et al. [10], it can be shown that $Z_{i,q}$ given Z_i, $\mathcal{D}$ and $\theta = \theta^{(d)}$ follows the binomial distribution $(Z_i, [(\gamma_q^{(d)} I_q(U_i)) / (\sum_{l=1}^{q_l} \gamma_q^{(d)} I_l(U_i))])$.

Then

$$E\left(Z_{i,q} \mid \mathcal{D}, \theta^{(d)}\right) = E\left\{E\left(Z_{i,q} \mid Z_i, \mathcal{D}, \theta^{(d)}\right) \mid \mathcal{D}, \theta^{(d)}\right\}$$

$$= \frac{\gamma_q^{(d)} I_q(U_i)}{\sum_{l=1}^{q_I} \gamma_q^{(d)} I_l(U_i)} E\left(Z_i \mid \mathcal{D}, \theta^{(d)}\right).$$

Similarly,

$$E\left(W_{i,q} \mid \mathcal{D}, \theta^{(d)}\right) = \frac{\gamma_q^{(d)} \{I_q(V_i) - I_q(U_i)\}}{\sum_{l=1}^{q_I} \gamma_q^{(d)} \{I_q(V_i) - I_q(U_i)\}} E\left(W_i \mid \mathcal{D}, \theta^{(d)}\right).$$

And

$$E\left(Z_i \mid \mathcal{D}, \theta^{(d)}\right) = E\left(Z_i \mid \delta_{i,1}, x_i, \theta^{(d)}\right)$$

$$= \frac{\sum_{l=1}^{q_I} \gamma_q^{(d)} I_q(U_i) \exp\left(x_i' \beta^{(d)}\right) \delta_{i,1}}{1 - \exp\left\{-\sum_{l=1}^{q_I} \gamma_q^{(d)} I_q(U_i) \exp\left(x_i' \beta^{(d)}\right)\right\}}$$

$$E\left(W_i \mid \mathcal{D}, \theta^{(d)}\right) = E\left(W_i \mid \delta_{i,1}, \delta_{i,2}, x_i, \theta^{(d)}\right)$$

$$= \sum_{q=1}^{q_I} \gamma_q^{(d)} \{I_q(V_i) - I_q(U_i)\} \exp\left(x_i' \beta^{(d)}\right) \delta_{i,1}$$

$$+ \frac{\sum_{q=1}^{q_I} \gamma_q^{(d)} \{I_q(V_i) - I_q(U_i)\} \exp\left(x_i' \beta^{(d)}\right) \delta_{i,2}}{1 - \exp\left\{-\sum_{q=1}^{q_I} \gamma_q^{(d)} \{I_q(V_i) - I_q(U_i)\} \exp\left(x_i' \beta^{(d)}\right)\right\}}.$$

This completes the details for the expectation step for the EM algorithm.

The maximization step to get updated $\theta^{(d+1)}$ is done by solving for $(\partial Q(\theta, \theta^{(d)}))/\partial \beta = 0$ and solving for $(\partial Q(\theta, \theta^{(d)}))/\partial \gamma_q = 0$ for $q \in \{1, \ldots, q_I\}$. Then the EM algorithm is given as follows.

1. Let $d = 0$ and choose the initial value $\theta^{(d)} = (\beta^{(d)\prime}, \gamma^{(d)\prime})'$
2. Update β by solving for the following equation to get $\beta^{(d+1)}$.

$$\sum_{i=1}^{n} \left\{E\left(Z_i \mid \mathcal{D}, \theta^{(d)}\right) + E\left(W_i \mid \mathcal{D}, \theta^{(d)}\right)\right\} x_i$$

$$= \sum_{i=1}^{n} \sum_{q=1}^{q_I} \gamma_q^{(d),*}(\beta) I_q(V_i) \exp(x_i'\beta) x_i,$$

where

$$\gamma_q^{(d),*}(\beta) = \frac{\sum_{1=1}^{n} \left\{E\left(Z_{i,q} \mid \mathcal{D}, \theta^{(d)}\right) + E\left(W_{i,q} \mid \mathcal{D}, \theta^{(d)}\right)\right\}}{\sum_{i=1}^{n} I_q(V_i) \exp(x_i'\beta)}.$$

3. Update γ by letting $\gamma_q^{(d+1)} = \gamma_q^{(d),*}\left(\beta^{(d+1)}\right)$ for $q \in \{1, \dots, q_l\}$.

4. If $\theta^{(d+1)}$ is close enough to $\theta^{(d)}$, stop and let $\theta^{(d+1)}$ be the sieve MLE for θ. Otherwise, let $d = d + 1$ and go to step 2.

Assuming that the number and the positions of the spline knots do not change with n, the variance-covariance matrix for $\hat{\beta}$ can be obtained by the inverse of the observed information matrix based on the likelihood function (Equation 23.11) evaluated at $\hat{\beta}$.

Another sieve spline approach that is similar to that of Wang et al. [10] was proposed by Zhang et al. [12] for the proportional hazards model for interval-censored data. The main difference is that Zhang et al. [12] used the monotone spline form to replace $\log\{\Lambda_0(\cdot)\}$ instead of $\Lambda_0(\cdot)$. In addition, Zhang et al. [12] proposed to use the generalized gradient projection method [13] to obtain the sieve MLE. Furthermore, the variance estimation for the sieve MLE allows both the number and the positions of the spline knots to change along with n [12].

23.3 Analyzing Interval-Censored Data by R

This section briefly introduces two R packages corresponding to the three estimation or testing methods with interval censoring as we discussed in Section 23.2. Both packages are available for installation from CRAN (The Comprehensive R Archive Network). Specifically, the first package is called *interval* and targets non-parametric MLE by the EM algorithm as discussed in Section 23.2.1 and weighted log-rank tests as discussed in Section 23.2.2. The second package is called *ICsurv* and targets sieve semi-parametric MLE for Cox's PH model based by the EM algorithm as discussed in Section 23.2.3. A data set is provided by the package *ICsurv* and will be used to illustrate both packages. For convenience, how to use the package *ICsurv* is first illustrated.

23.3.1 Package *ICsurv*

The package *ICsurv* provides a dataset called *Hemophilia*, which was collected as a part of a multicenter, prospective study aiming at assessing the HIV-1 infection rate among patients with hemophilia. The times when these patients contracted HIV-1 are not known exactly but fall within an interval. There are 544 patients in the collected dataset. These 544 patients were classified as high, medium, low, or no dose groups based on their average annual doses of blood products. The dataset has eight columns, in which *d1*, *d2*, and *d3* are indicators specifically for left, interval, and right censoring; *L* and *R* are left- and right-censored times; *Low*, *Medium*, and *High* are indicators specifically for low, medium, and high dose groups. Goedert et al. [14] and Kroner et al. [15] provide more details about this dataset.

The main function for the package *ICsurv* is *PH.ICsurv.EM* for fitting the Cox PH model. To use this main function, the following inputs are needed:

- *n.int* for number of interior knots
- *order* for degree of I-spline functions
- *g0* and *b0* for initial values for regression parameters and spline parameters, respectively

- *t.seq* for a sequence of increasing time points to evaluate the baseline hazard function
- *tol* for the convergence criterion of the EM algorithm

The following is the R code with results for using Cox PH model to analyze the data set *Hemophilia* for checking the dose effect of blood products on the HIV-1 infection time.

```
library(ICsurv)

data(Hemophilia)
d1<-Hemophilia[,1]
d2<-Hemophilia[,2]
d3<-Hemophilia[,3]
Li<-Hemophilia[,4]
Ri<-Hemophilia[,5]
Xp<-as.matrix(Hemophilia[,c(6,7,8)])

fit <- PH.ICsurv.EM(d1, d2, d3, Li, Ri, Xp, n.int=7, order=3,
     g0=rep(1,10), b0=rep(0,3), t.seq=seq(0,57,1), tol=0.001)

fit$b
[1] 1.838337 3.007786 3.410810

fit$var.b
               [,1]           [,2]            [,3]
[1,]  0.0055322416  -0.008218135   0.0006773972
[2,] -0.0082181347   0.010267643  -0.0042720283
[3,]  0.0006773972  -0.004272028   0.0426510994
```

We also provide the R code and the corresponding results for the data set *Hemophilia* when it is naively converted to right-censored data. This is done by treating the right end points of all finite censoring intervals as true event times, while keeping all right-censored times. Then R package *Survival* is used to fit the naively converted data set for the blood products dose effect on the HIV-1 infection time.

```
library(survival)

data.set<-Hemophilia

Time<-rep(0,544)
Time[data.set$d3==0]<-data.set[data.set$d3==0,]$R
Time[data.set$d3==1]<-data.set[data.set$d3==1,]$L

Status<-rep(0,544)
Status[data.set$d3==0]<-1
Status[data.set$d3==1]<-0

group<-rep("No",544)
group[data.set$Low==1]<-"Low"
group[data.set$Medium==1]<-"Medium"
group[data.set$High==1]<-"High"

data.set$Time<-Time
data.set$Status<-Status
data.set$group<-group
```

TABLE 23.1

Coefficients Estimation (Coef) and Their Standard Errors
(SE) Based on Semi-parametric Method (Semi) and Naively
Converting Method (Naive)

	Semi			Naive		
	coef	SE	p-value	coef	SE	p-value
Low	1.838	0.074	$< 2e - 16$	1.808	0.220	$2.2e - 16$
Medium	3.008	0.101	$< 2e - 16$	2.978	0.217	$< 2e - 16$
High	3.411	0.207	$< 2e - 16$	3.335	0.227	$< 2e - 16$

```
fit1 <- coxph(Surv(Time, Status) ~ Low + Medium + High,
data=data.set, method="breslow")

fit1$coef
    Low    Medium      High
1.808274 2.977836 3.334816

fit1$var
            [,1]          [,2]          [,3]
[1,] 0.04825846 0.03511239 0.03527402
[2,] 0.03511239 0.04699770 0.03668583
[3,] 0.03527402 0.03668583 0.05140601
```

Table 23.1 compares the results from the Cox's PH model by using the package *ICsurv* for the original data set *Hemophilia* with the results using the package *Survival* for the naively converted right-censored data. Although the estimated coefficients and the p-values from both methods are similar, we observe higher estimated standard error for the naive method. This would have an impact on the confidence interval estimated for each coefficient.

23.3.2 Package *interval*

We use the package *interval* to illustrate how to analyze the data set *Hemophilia* here. The package *interval* has two main functions: *icfit* for non-parametric MLE and *ictest* for score test.

Both functions need a data frame as their main input. Specifically, for both functions, the data frame has to have two columns for left and right end points of censored intervals; for the score tests function *ictest*, the data frame is also required to have one column of categorical treatment names.

In addition, the function *ictest* also requires the input *scores* with value *logrank1*, *logrank1*, or *wmw* corresponding to the assumptions in Equations 23.7, 23.9, or 23.10, respectively. For both functions, the input *maxit* can be set for the maximum number of iterations for the EM algorithm and input *intervals* stands for case 2 interval censoring.

The following R code and the corresponding results demonstrate how to analyze the dataset *Homophilia*. The function *icfit* is used to find the non-parametric MLEs for four dose groups of blood products, which are presented by four survival curves in the left panel of Figure 23.3. The function *ictest* is used to test whether or not the survival functions among the four groups are the same.

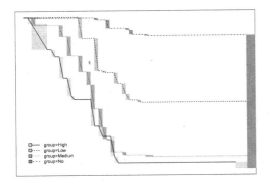 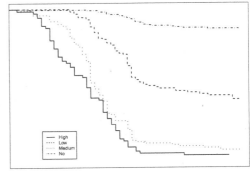

FIGURE 23.3
Estimated survival curves for four dose groups of blood products by two approaches. Left panel is based on the semi-parametric method; Right panel is based on the naively converting method.

```
library(interval)

data.set<-Hemophilia
data.set[data.set$d3==1,]$R<-Inf
group<-rep("No",544)
group[data.set$Low==1]<-"Low"
group[data.set$Medium==1]<-"Medium"
group[data.set$High==1]<-"High"
data.set$group<-group

data.fit<-icfit(Surv(L, R, type = "interval2")~group,
data = data.set, control=icfitControl(maxit=10^6))

plot(data.fit,YLEG=0.2)

ictest(Surv(L, R, type = "interval2") ~ group,

data = data.set, control=icfitControl(maxit=10^6),
socres="logrank1")

# score test results
Chi Square = 287.4916, p-value < 2.2e-16
```

What follows is the R code with results given for using the *Survival* package to fit the naively converted data set as described in Section 23.3.1. Specifically, the non-parametric MLEs are obtained for four dose groups based on this converted right-censored data, which are plotted by the corresponding Kaplan–Meier curves in the right panel of Figure 23.3. In addition, the log-rank test is used to compare the differences among the four survival curves based on the converted right-censored data.

```
library(survival)

surv<-survfit(Surv(Time,Status) ~ group, data=data.set)

plot(surv, lty=c(1,2,3,4))
legend(8,0.2,c("High","Low","Medium", "No"), lty = 1:4)
```

```
survdiff(Surv(Time, Status) ~ group,data=data.set)

\# Log-rank test results
Chisq= 434 on 3 degrees of freedom, p= 0
```

Figure 23.3 implies that by the naively converting approach, the survival rates for all four groups are overestimated. This makes sense since unobserved event times are naively assumed to occur at the right end points for corresponding finite censored intervals.

23.4 Conclusion

This chapter discusses three of the most widely used survival analysis estimation or testing methods for interval-censored data. It also discusses how to adopt these methods by using a statistical software R with a publicly available dataset. The reader is referred to Sun [2] for a more comprehensive review of interval-censored survival data analysis.

In recent years, advanced univariate, interval-censored, time-to-event regression models have been developed. These include time to event with competing risks by Li [16], time to event with a cured portion by others [17–20]. In addition, methods for bivariate (multivariate) interval-censored data have also been developed. These are non-parametric MLE [21,22] and regression [23–25].

References

1. Kaplan EL, Meier P. Nonparametric estimation from incomplete observations. *J Am Stat Assoc.* 1958;53:457–81.
2. Sun J. *The Statistical Analysis of Interval-Censored Failure Time Data.* New York: Springer-Verlag, 2006.
3. Turnbull BW. The empirical distribution function with arbitrarily grouped, censored and truncateddata. *J R Stat Soc Ser B.* 1976;38:290–5.
4. Groeneboom P, Wellner JA. *Information Bounds and Nonparametric Maximum Likelihood Estimation. DMV Seminar, Band 19.* New York: Birkhäuser, 1992.
5. Jongbloed G. The iterative convex minorant algorithm for nonparametric estimation. *J Comput Graph Stat.* 1998;7:310–21.
6. Huang J, Wellner JA. Interval censored survival data: A review of recent progress. *Proceedings of the First Seattle Symposium in Biostatistics: Survival Analysis,* Lin D, Fleming T, eds. New York: Springer, 1997.
7. Sun J. Self-consistency estimation of distributions based on truncated and doubly censored data with applications to aids cohort studies. *Stat Med.* 1996;15:1387–95.
8. Fay MP. Comparing several score tests for interval censored data. *Stat Med.* 1999;18:273–85.
9. Fleming TR, Harrington DP. *Counting Process and Survival Analysis.* New York: John Wiley, 1991.
10. Wang L, McMahan CS, Hudgens MG, Qureshi ZP. A flexible, computationally efficient method for fitting the proportional hazards model to interval-censored data. *Biometrics.* 2016;72:222–31.
11. Ramsay JO. Monotone regression splines in action. *Stat Sci.* 1988;3:425–41.
12. Zhang Y, Hua L, Huang J. A spline-based semiparametric maximum likelihood estimation for the cox model with interval-censored data. *Scand J Stat.* 2010;37:338–54.

13. Jamshidian M. On algorithms for restricted maximum likelihood estimation. *Comput Stat Data Anal.* 2004;45:137–57.

14. Goedert J, Kessler C, Adedort L et al. A prospective study of human immunodeficiency virus type 1 infection and the development of aids in subjects with hemophilia. *N Engl J Med.* 1989;321:1141–8.

15. Kroner B, Rosenberg P, Adedort L, Alvord W, Goedert J. Hiv–1 infection incidence among persons with hemophilia in the united states and western europe, 1978–1990. Multicenter hemophilia cohort study. *J Acquir Immune Defic Syndr.* 1994;7:279–86.

16. Li C. The fine-gray model under interval censored competing risks data. *J Multivar Anal.* 2016;143:327–44.

17. Hu T, Xiang L. Efficient estimation for semiparametric cure models with interval-censored data. *J Multivar Anal.* 2013;121:139–51.

18. Liu H, Shen Y. A semiparametric regression cure model for inverval-censored data. *J Am Stat Assoc.* 2009;104:1168–78.

19. Ma S. Mixed case interval censored data with a cured subgroup. *Stat Sin.* 2010;20:1165–81.

20. Hu T, Xiang L. Partially linear transformation cure models for interval-censored data. *Comput Stat Data Anal.* 2016;93:257–69.

21. Maathuis MH. Reduction algorithm for the npmle for the distribution function of bivariate interval-censored data. *J Comput Graph Stat.* 2005;14:352–62.

22. Wu Y, Zhang Y. Partially monotone tensor spline estimation of the joint distribution function with bivariate current status data. *Ann Stat.* 2012;40:1609–36.

23. Hu T, Zhou Q, Sun J. Regression analysis of bivariate current status data under the proportional hazards model. *Can J Stat.* 2017;45:410–24.

24. Zeng D, Gao F, Lin DY. Maximum likelihood estimation for semiparametric regression models with multivariate interval-censored data. *Biometrika.* 2017;104:505–25.

25. Zhou Q, Hu T, Sun J. A sieve semiparametric maximum likelihood approach for regression analysis of bivariate interval-censored failure time data. *J Am Stat Assoc.* 2017;112:664–72.

24

Methods for Analysis of Trials with Changes from Randomized Treatment

Nicholas R. Latimer and Ian R. White

CONTENTS

24.1 Introduction

It is common for participants in randomized controlled trials to change treatments during trial follow-up. Trial participants could change onto another randomized treatment (i.e., control group participants could switch onto the experimental treatment, or vice versa) or could change onto an entirely different treatment. Treatment changes could occur for a variety of reasons—toxicity or tolerability issues could force a change or disease progression may mean that a participant needs to move on to the next available treatment. Treatment changes can be problematic because they complicate the interpretation of a standard intention-to-treat (ITT) comparison of randomized groups—the distinction between the randomized groups may be somewhat blurred. Therefore, it may be desirable to attempt to adjust for treatment changes in order to estimate outcomes that would have been observed in the absence of such changes.

The issue of treatment changes and making adjustments for them is a challenging field. Whether or not it is useful or desirable to adjust for treatment changes depends on what we are trying to estimate in an analysis—the estimand—which depends upon the question that we are trying to answer—the decision problem. Decision problems are not always straightforward to define, and may differ for different audiences and jurisdictions. Different treatment changes may occur over time and may vary between countries—an issue for multinational trials. Trial participants who change treatments are likely to have different prognoses to those who do not, because treatment decisions are unlikely to be random. These factors are all important when considering how to address the treatment changes problem.

Throughout this chapter, we will consider multiple types of treatment changes, but, for simplicity, we focus mainly on one type: that of treatment switching, whereby participants randomized to the control group of a trial switch onto the experimental treatment during the trial. This type of switching is common in oncology trials [1–3]. For this reason, we also focus on time-to-event outcomes.

The concepts of estimands and decision problems are described in Section 24.2. The problems created by treatment changes are illustrated in Section 24.3. A simulated dataset is introduced in Section 24.4. Simple and more complex methods for adjusting for treatment changes are described in Sections 24.5 and 24.6, respectively, and illustrated in the simulated dataset using the statistical software Stata. Approaches for identifying appropriate adjustment methods on a case-by-case basis are discussed in Section 24.7. In Section 24.8 we consider extensions and further issues, including the applicability of adjustment methods for non-time-to-event outcomes.

24.2 Defining the Question

Before starting an analysis, it is important to define the question being addressed—the decision problem—and the quantity being estimated—the estimand.

If the decision problem requires a comparison of the benefits of the two treatment policies implemented in the trial, then an intention-to-treat (ITT) analysis is ideal. The estimand here is sometimes called "effectiveness." Even here, treatment changes cannot be entirely ignored: a description of the treatment changes that occurred in the trial is needed to describe the treatment policies that were implemented.

At the opposite extreme, a decision problem might require an assessment of what would have happened in the trial had there been no treatment changes. The estimand here is sometimes called "efficacy." This is not usually the aim in a phase III trial, since some treatment changes—for example, stopping a treatment due to adverse events—are an unavoidable aspect of the treatment. Instead, the decision problem often relates to what would have happened in the trial if specific treatment changes had not occurred.

A common question arises in a trial of an experimental treatment against standard care when the standard care arm participants are allowed to switch to receive the experimental treatment during the trial. Here, a funder wants to compare funding the experimental treatment with not funding the experimental treatment, but the ITT analysis compares immediate experimental treatment with deferred experimental treatment. Thus, the funder has a decision problem relating to what would have happened in the trial if no participants on the standard care arm received experimental treatment, but other treatment changes occurred as observed.

In general, we need to record and define treatment changes clearly and specify which treatment changes we wish to adjust for and which we do not want to adjust for to address our decision problem. The methods described in this chapter all implicitly or explicitly refer to potential outcomes, also called counterfactual outcomes [4], which are outcomes that would have occurred if specified treatment changes had not occurred. Our focus is on alternatives to the ITT analysis for circumstances where it is deemed necessary to adjust for treatment switching.

24.3 The Impact of Treatment Switching

Figure 24.1 illustrates the treatment-switching problem, when participants randomized to the control group of a trial are permitted to switch onto the experimental treatment after disease progression, and when an estimate of the overall survival (OS) advantage of the new treatment is required. The horizontal axis represents OS time, which is made up of progression-free survival (PFS) and post-progression survival (PPS). The upper row (Row 1) illustrates the survival experience of the control group in the absence of treatment switching. The middle row (Row 2) illustrates the survival experience of the experimental group, where the experimental treatment extends PFS and PPS and the difference in OS between Row 1 and Row 2 is denoted as the "true" OS difference. The bottom row (Row 3) illustrates what might be observed if a proportion of participants randomized to

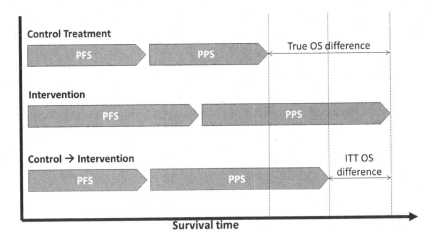

FIGURE 24.1
Illustrating treatment switching. PFS, progression-free survival; PPS, post-progression survival; OS, overall survival; ITT, intention-to-treat. (Source: Reproduced from Latimer NR et al. *Med Decis Making* 2014 with permission of SAGE publications.)

the control group switches onto the experimental treatment after disease progression has been observed. PFS is not affected but, assuming control group participants benefit from switching onto the experimental treatment, PPS is extended. The difference between Row 3 and Row 2 is denoted the "ITT" OS difference. If our decision problem dictates that we estimate the treatment effect that would have been observed in the absence of treatment switching, then we wish to estimate the difference between Row 2 and Row 1, but in the presence of switching, we observe Row 2 and Row 3. The adjustment methods discussed in this chapter attempt to estimate what Row 1 would have looked like, given the observed Row 2 and Row 3.

Treatment switching has most often been acknowledged as an issue in health technology assessment (HTA), where there is a clear need to compare a state of the world in which the new intervention exists to a state of the world where the new intervention does not exist, in order that efficient resource allocation decisions can be made. In fact, switching has been an issue in greater than 50% of cancer technology appraisals [5]. Making adjustments to account for switching can have a large impact. In National Institute for Health and Care Excellence (NICE) Technology Appraisal 321, which considered dabrafenib for melanoma, 57% of participants randomized to the control group of the pivotal trial switched onto dabrafenib at some point during the trial. A standard ITT analysis produced an overall survival hazard ratio (HR) of 0.76 and an incremental cost-effectiveness ratio of over £95,000 per quality-adjusted life year gained, which is much higher than the ratio acceptable to NICE, even taking into account the end-of-life nature of the disease [6]. In addition to the ITT analysis, the manufacturer of dabrafenib conducted an analysis to adjust for the switching observed, which resulted in an overall survival HR of 0.55 and an incremental cost-effectiveness ratio of under £50,000 per quality-adjusted life year gained. Dabrafenib was recommended by NICE.

However, the use of adjustment methods is not straightforward. In NICE Technology Appraisal 215 of pazopanib for renal cell carcinoma, 51% of participants randomized to the control group of the pivotal trial switched onto pazopanib during the trial [7]. The ITT analysis produced an overall survival HR of 1.26 compared to interferon, and in economic terms pazopanib was dominated—it was both more expensive and less effective [7]. The manufacturer of pazopanib presented several analyses that attempted to adjust for the treatment switching, and HRs varied between 0.39 and 0.80 depending on the adjustment method used. Corresponding incremental cost-effectiveness ratios varied between approximately £22,000 per quality-adjusted life year gained and £72,000 per quality-adjusted life year gained [7]. NICE typically concludes that treatments are cost-effective if incremental cost-effectiveness ratios are below £30,000 per quality-adjusted life year gained, or below £50,000 for some cancer treatments [8]. Therefore, not only can adjusting or not for treatment switching potentially change reimbursement decisions for new drugs, but also the choice of method used to adjust for switching can potentially change reimbursement decisions—using one method, a new intervention may appear cost-effective, whereas using another method, the same intervention may appear cost-ineffective.

24.4 Illustrative Dataset

The "NovCEA" data represent a simulated randomized controlled trial studying the effectiveness of a novel cancer treatment. The sample size was 300, with 108 randomized to the placebo group and 192 randomized to the experimental group. Information on the following prognostic characteristics were collected at baseline:

- Prognosis group (either "good" or "bad")
- A biomarker, carcinoembryonic antigen (CEA); this is a protein in the blood which has an increasing level as the cancer becomes further progressed
- Health-related quality of life score (HRQOL), which has a value between 0 and 1, where 1 represents perfect health and 0 represents death

A summary of the baseline variables, event variables, and time-updated variables is provided in Table 24.1.

TABLE 24.1

NovCEA Data, Baseline Variables

Variable	Label	Values	Obs	Mean (SD) or Number (%)
Baseline Variables				
id	Participant ID	1–300	300	–
progb	Bad prognosis indicator	0/1	300	152 (50.7%)
CEAb	Carcinoembryonic antigen	Quantitative	300	20.1 (1.1)
HRQLb	Health-related QOL	Quantitative (0–1)	300	0.44 (0.11)
trtrand	Randomized group	0 for placebo, 1 for experimental	300	192 (64.0%)
Event Variables				
progtime	Progression time	Days	300	172.7 (150.7)
progind	Indicator of disease progression	0/1	300	268 (89.3%)
CEAprog	CEA at progression	Quantitative	265	28.0 (5.2)
xotime	Time of switch	Days	49	132.9 (53.1)
xoind	Indicator of switch	0/1	300	49 (16.3%)
disconexp	Time of treatment discontinuation	Days	244	223.8 (159.3)
deathtime	Survival time	Days	300	274.6 (158.2)
deathind	Death indicator	0/1	300	257 (85.7%)
admin	End study time	Days (always 546)	300	546 (0.0)
Time Updated Variables				
time	Visit time	Days (steps of 21)	4,054	173.6 (135.9)
CEAtdc	CEA	Quantitative (time updated)	4,054	27.4 (6.1)
HRQLtdc	Health-related QOL	Quantitative (0–1) (time updated)	4,054	0.46 (0.11)
progtdc	Whether progression occurred recently	1 (occurred in the previous interval)/2 (occurred in the two intervals before that)/0 (otherwise)	4,054	0.06 (0.24)
xotdo	Whether switch occurs in this interval	0/1 (missing for intervals after switch occurs)	3,666	0.01 (0.11)
deathtdo	Whether death occurs in this interval	0/1	4,054	0.25 (0.56)

Abbreviations: Obs, observations; SD, standard deviation.

Participants in the placebo control group were permitted to switch onto the experimental treatment after disease progression had been observed, as an ethical requirement to offer experimental treatment to all. Treatment switches occurred at clinic visits, and usually at or shortly after progression. Participants receiving the experimental treatment could discontinue it, with discontinuations occurring at the point that disease progression (suggesting treatment failure) was observed.

Figure 24.2 presents a lifeline plot, presenting the durations spent on and off treatment for each participant, as well as disease-progression times.

The key question of interest for this chapter relates to what the overall survival treatment effect associated with the experimental treatment would have been if the control arm did not have access to the experimental treatment. Box 24.1 demonstrates the application of a standard ITT analysis to the dataset, unadjusted for treatment switching. Throughout the chapter, after each adjustment method is introduced, its application to the NovCEA dataset will be demonstrated, and Kaplan-Meier curves associated with each analysis are shown in Figure 24.3. We use Stata software, version 14.2 [9]. Two versions of the dataset are used: NovCEAwide has one record per participant and excludes the time-updated variables, while NovCEAlong has one record per clinic visit per participant. They are available together with STATA code on the companion website https://www.crcpress.com//9781138083776.

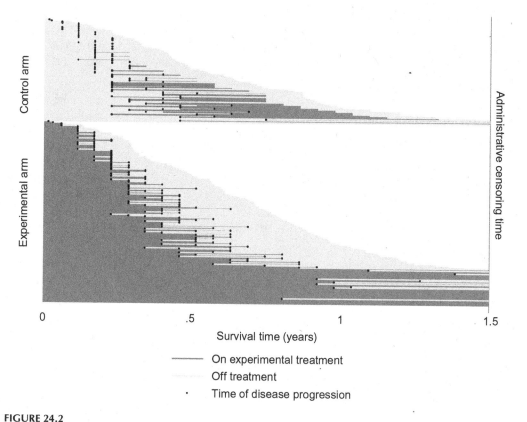

FIGURE 24.2
NovCEA data, Lifeline plot. Each horizontal line represents periods on and off treatment during one trial participant's lifetime.

```
                  BOX 24.1   INTENTION-TO-TREAT ANALYSIS
**************************************************************
*** ITT analysis
**************************************************************

use NovCEAwide, clear

*** Declare the data to be survival data
stset deathtime, failure(deathind) id(id) scale(365.25)

*** Plot the ITT Kaplan Meier curve for overall survival.
sts graph, by(trtrand) risktable title(ITT analysis for OS)

*** Find the ITT hazard ratio for OS ***
stcox trtrand
```

24.5 Simple Methods for Adjusting for Treatment Changes: Per-Protocol Analysis

Per-protocol analysis involves censoring participants at the point where specified treatment changes occur. The specified treatment changes will be those for which we want to adjust, and must be carefully defined. In the context being discussed here, the specified treatment change occurs when participants randomized to the control group begin to receive the experimental treatment: hence, control group participants are censored at the point at which they start experimental treatment. Analysis of the censored data implicitly assumes that the censoring—and therefore the switching—is non-informative; that is, it is not related to any prognostic characteristics, and therefore those who switch at time *t* are comparable at time *t* to those who could have switched at time *t* but did not. This represents a severe limitation of the per-protocol analysis, because switching is unlikely to be non-informative: a clinician faced with a control group trial participant is likely to consider prognostic characteristics (such as disease progression, performance status, quality of life, biomarkers) when deciding whether or not to advise that the participant switches onto the experimental treatment.

Box 24.2 demonstrates the application of a simple per-protocol analysis on the NovCEA dataset.

```
                  BOX 24.2   PER-PROTOCOL ANALYSIS
**************************************************************
*** PP analysis
**************************************************************

use NovCEAwide, clear

*** Censor switchers
replace deathind = 0 if xoind==1
replace deathtime = xotime if xoind==1
```

```
*** Declare the data to be survival data
stset deathtime, failure(deathind) id(id) scale(365.25)

*** Plot the ITT Kaplan Meier curve for overall survival.
sts graph, by(trtrand) risktable title(PP analysis for OS)

*** Find the PP hazard ratio for OS ***
stcox trtrand
```

24.6 Complex Methods for Adjusting for Treatment Changes

For each method, we give "concept", "method," and "extensions". Non-statisticians may want to focus on the "concept" parts.

24.6.1 Inverse Probability of Censoring Weights

24.6.1.1 Concept

The inverse probability of censoring weights (IPCW) method represents an elaboration of the simple per-protocol analysis and is a well-known technique for addressing informative censoring [10,11]. When a specified treatment change occurs, participants are censored as in the PP analysis, but the remaining observations are weighted in an attempt to remove the selection bias introduced through the censoring. The IPCW method involves four main steps. First, as for per-protocol analysis, we decide what event requires censoring. Second, we model the probability of being censored over time using a "switching model." Third, for each individual at each time, we compute the inverse probability of remaining uncensored. Finally, we use these inverse probabilities as weights in a weighted analysis using our "outcome model."

When using the IPCW method, we must assume that, at each time, there are no systematic differences between those who are censored and those who remain uncensored, conditional on variables in our switching model; that is, there must be no unmeasured confounding (NUC). Here, a confounder is a variable which predicts switching *and* predicts the counterfactual time to event (that is, the time to event that would have been observed in the absence of switching). Potential confounders could be clinical progression, performance status, or quality of life. Importantly, variables in the switching model can be time-dependent. For instance, if switching occurs immediately after disease progression in many control-group participants, then those who have switched at a given time (loosely, "switchers") differ systematically from non-switchers who are still being followed up because they have experienced more progressions and therefore have a worse average counterfactual time to event: switching is a time-dependent confounder and PP analysis is subject to selection bias. Among those who have just progressed, however, we might be willing to assume that there are no systematic differences between switchers and non-switchers, and thus a switching model with disease progression as the only covariate might be sufficient; or we might be willing to assume this conditional on further time-dependent confounders, such as performance status and quality of life, which would thus be included as further covariates in the switching model. In these cases, an IPCW analysis would allow us to avoid selection bias.

The validity of IPCW analysis rests on the NUC assumption. Typically, we use many baseline confounders and time-dependent confounders in an attempt to satisfy the NUC assumption, but we can never be sure that we have no unmeasured confounders. Ideally, if it is likely that switching will occur in a forthcoming trial, consideration should be given to the likely confounders at the trial planning stage, in order to ensure that these are measured adequately during the trial [12]. A common issue with oncology trials is that potentially important confounders (such as performance status and quality of life) are measured only until the point of disease progression. In circumstances where switching occurs after disease progression (and potentially an appreciable period of time after disease progression), this is likely to be insufficient—data collection must continue beyond disease progression in order for the NUC assumption to be satisfied. Clinical expert opinion and information from previous studies may provide information on what variables are likely to be confounders. In relatively small datasets with small numbers of events, the inclusion of a large number of confounders may result in problems with switching model convergence. One potential analytical approach involves exploring the impact of adding more and more potential confounders into the model; if there is little impact on the estimated treatment effect this may suggest that the initial model is adequate.

24.6.1.2 Method

IPCW is commonly applied working in discrete time, dividing follow-up into small intervals and using pooled logistic regression. When small intervals are used and where event probabilities are low, pooled logistic regression is almost identical to a Cox regression model [13]. First, we construct our switching model in the control arm. In practice, we are unlikely to have enough data to fit one model at each interval, and instead we fit a model across intervals, such as

$$\text{logit } P(A_{ik} = 1 | Z_i = 0, V_i, A_{i,k-1} = 0, \bar{L}_{ik}, T_i \geq k) = \alpha' V_i + \gamma' L^*_{ik} + \delta' f(k) \tag{24.1}$$

where for individual i, A_{ik} indicates treatment in interval k (regarded as constant through the interval), Z_i indicates the randomized group (0 for control, 1 for experimental), V_i is an array of an individual's baseline covariates, $\bar{L}_{ik}$ represents the history of an individual's time-dependent covariates up to the beginning of interval k, T_i is the interval in which the event (switch) occurs, L^*_{ik} is a suitable concise summary of $\bar{L}_{ik}$, and $f(k)$ represents a spline function of time k. Specifying this model is not straightforward: in particular, we need to select covariates and suitable functional forms in order to define L^*_{ik} from the potentially rich history $\bar{L}_{ik}$.

To construct the inverse probability weights, we use the fitted switching model to predict the probability of switching for individual i in interval k:

$$\hat{\pi}_{ik} = \hat{P}(A_{ik} = 1 | Z_i = 0, V_i, A_{i,k-1} = 0, \bar{L}_{ik}, T_i \geq k) \tag{24.2}$$

where the hat indicates that these probabilities are estimated. Then we construct the weight for individual i in interval t by multiplying together all their probabilities of remaining uncensored (i.e., unswitched) up to interval t, with the weight representing the inverse probability of remaining unswitched up to interval t:

$$\hat{W}_{it} = \prod_{k=0}^{t} \frac{1}{1 - \hat{\pi}_{ik}} \tag{24.3}$$

These represent "unstabilized weights." Often stabilized weights are used instead. These represent an extra step that usually increases the efficiency of an IPCW analysis. The problem is that unstabilized weights can be very variable, and this decreases statistical efficiency [14]. Stabilized weights are designed to be less variable by changing the numerator in Equation 24.3 from 1 to a value which is as similar as possible to the denominator without reintroducing confounding [15]:

$$\hat{W}_{it}^{stab} = \prod_{k=0}^{t} \frac{1 - \hat{\pi}_{0ik}}{1 - \hat{\pi}_{ik}} \qquad (24.4)$$

The simplest choice for the numerator is given by

$$\hat{\pi}_{0ik} = \hat{P}(A_{ik} = 1 \mid Z_i = 0, A_{i,k-1} = 0, T_i \geq k). \qquad (24.5)$$

For example, if V_i and $\bar{L}_{ik}$ are not in fact confounders, then $\hat{\pi}_{0ik} = \hat{\pi}_{ik}$, so the stabilized weights all equal 1, while the unstabilized weights are not all 1 and lose efficiency [16]. If baseline covariates V_i are to be adjusted for in the outcome model, then a further improvement is to include those covariates in the right-hand side of Equation 24.5, thus making the numerator and denominator even more similar [15]. Time-dependent covariates $\bar{L}_{ik}$ cannot be adjusted for in the outcome model so must be used only in the denominator of Equation 24.4.

Finally, we fit our outcome model using the weights. If we use stabilized weights with baseline covariates V_i included in the right-hand side of Equation 24.5, then we must adjust for these covariates in the analysis, because these weights do not correct for confounding by V_i (due to being in the numerator and the denominator) [15]. If unstabilized weights are used, then V_i may or may not be included in the outcome model. The outcome model could be a pooled logistic regression model fitted to both arms of the trial. In a weighted analysis, ordinary standard errors are not valid. Standard errors must instead be computed in a robust way, using the sandwich variance. A weighted analysis is less efficient (has larger standard errors) than an unweighted analysis—the reason to use IPCW is to remove bias, not to gain efficiency.

24.6.1.3 Refinements

Aside from making the no unmeasured confounding assumption, the main problem with IPCW arises when very large weights are estimated. With very high switching proportions, small numbers of non-switching participants may be allocated very high weights, leading to large standard errors. In the most extreme case, if all participants switched—or all participants who became at risk of switching switched (in our example, those control group participants who experienced disease progression)—then IPCW would fail because there are no non-switchers to weight up. This problem is known as non-positivity. If IPCW does not fail, but large weights are estimated, then variable selection and the functional form of the switching model should be reconsidered. Further discussion of the variable selection issue is available in the literature [14,17]. The range of weights estimated should always be considered when reviewing the output of an IPCW analysis.

Instead of working in discrete time, we could work in continuous time. Here, we would model the time to switch using a Cox model with time-dependent covariates and then apply the estimated time-dependent weights in a survival analysis, although survival models with time-dependent weights are difficult to fit in some statistical software.

In this chapter, we focus on the use of methods to adjust for switching from the control group onto the experimental treatment during a randomized controlled trial. However, we acknowledge that other types of switching could occur. The IPCW method can be used to adjust for a variety of types of switching. All relevant switchers could be censored, in either arm of the trial, with remaining participants being upweighted. Different switching models could be fitted to each treatment arm, or even for each different type of switching within each treatment arm, though this may lead to additional problems with small sample sizes. In Box 24.3, we demonstrate the application of the IPCW method to adjust for the treatment switching in the NovCEA dataset.

BOX 24.3   INVERSE PROBABILITY OF CENSORING WEIGHTS ANALYSIS

```
*******************************************************************
*** IPCW step 1: Censor observations and re-format the data (details not given)
*******************************************************************


use NovCEAlong, clear


*******************************************************************
*** IPCW step 2: Model the probability of switching over time
*******************************************************************


*** Install program for restricted cubic splines (only needs to be done once)
ssc install rcsgen

*** Create splines for a time-dependent intercept
*** using 5 knots based upon the event time distribution
rcsgen time, df(4) if2(xotdo==1) gen(timexosp)

*** Use logistic regression to predict switching given baseline covariates
logistic xotdo progb CEAb HRQLb timexosp* if trtrand==0

*** Estimate the probability of switching for each participant-observation
predict pxo1 if e(sample)

*** Use logistic regression to predict switching given baseline and
*** time-updated covariates. Model excludes participants with progtdc==0
*** since they are certain not to switch
logistic xotdo progb CEAb HRQLb progtdc CEAtdc HRQLtdc timexosp* ///
      if trtrand==0 & progtdc>0

*** Estimate the probability of switching for each participant-observation
predict pxo2 if e(sample)

*** Get it right for participants with progtdc==0
replace pxo2 = 0 if trtrand==0 & progtdc==0


*******************************************************************
*** IPCW step 3: For each individual at each time,
*** compute the inverse probability of remaining uncensored
*******************************************************************
```

```
*** Estimate the probabilities of remaining unswitched given baseline
*** covariates
*** Variable firstobs identifies the first observation for each participant
sort id time
gen punswitchbase = 1-pxo1 if firstobs
replace punswitchbase = punswitchbase[_n-1] * (1-pxo1) if !firstobs

*** Estimate the probabilities of remaining unswitched given baseline
***    and time-updated covariates
gen punswitchupdate = 1-pxo2 if firstobs
replace punswitchupdate = punswitchupdate[_n-1] * (1-pxo2) if !firstobs

*** Compute stabilised and unstabilised weights
gen sweight = punswitchbase / punswitchupdate if trtrand==0
gen uweight = 1 / punswitchupdate if trtrand==0

*** summarise the weights, and inspect the data
***    to be confident that you have computed the weights correctly.
summ sweight if xotdo==0
summ uweight if xotdo==0

*** set the weights to 1 in the treatment arm.
replace sweight = 1 if trtrand==1
replace uweight = 1 if trtrand==1

**************************************************************************
*** IPCW step 4: Use these weights in a weighted analysis of the outcome model
**************************************************************************

*** create spline basis for time as variables timesp*
rcsgen time, df(4) gen(timesp)

*** fit the weighted outcome model, first with sweight then uweight
logistic deathtdo trtrand progb CEAb HRQLb timesp* [pw=sweight] if
xotdo==0, ///
        cluster(id)
logistic deathtdo trtrand progb CEAb HRQLb timesp* [pw=uweight] if
xotdo==0, ///
        cluster(id)

*** produce KM for unstabilised analysis
gen tin  = time
gen tout = time+21
stset tout deathtdo if xotdo==0 [iw=uweight], time0(tin) scale(365.25)
sts, by(trtrand) risktable(,format(%4.0f)) title(IPCW analysis for OS)
```

24.6.2 Two-Stage Estimation

24.6.2.1 Concept

The two-stage estimation (TSE) adjustment method was suggested recently in response to the type of treatment switching often seen in cancer trials, whereby switching usually happens soon after disease progression [18]. The time of disease progression is considered to be a time point at which participants are at a similar stage of disease and is referred to as a "secondary baseline." Post-progression survival in switchers and non-switchers is compared to estimate the effect of switching, which is expressed in the form of a time ratio, a factor by which treatment multiplies survival. This time ratio is then used to derive counterfactual survival times for switchers—that is, survival times that would have been observed in the absence of switching. Finally, the adjusted treatment effect is estimated by comparing the observed survival times in the experimental group to the adjusted survival times in the control group.

When using TSE, we must assume that at the point of disease progression (or any other time point used as the secondary baseline for the analysis), there are no systematic differences between those who switch treatments and those who do not, conditional on variables included in the model for the switching effect—that is, there must be NUC at the secondary baseline. In the context of the TSE method, a confounder is a variable which predicts both switching and the counterfactual time to event. We may use confounders measured at the original baseline or at the time of the secondary baseline. As with IPCW, we can never be sure that we have satisfied the NUC assumption, so similar strategies— such as clinical expert opinion and information from previous studies—may be used to help specify a suitable model. Data collection issues are again important, although for TSE, data are only required for confounders up to the point of the secondary baseline; if this is disease progression, this is likely to be satisfied by most cancer trials since data collection is usually continued until this time point.

When switching can occur at some point after the secondary baseline, we must also assume that there is NUC between the secondary baseline and the time of switch. This highlights a key limitation of the TSE method in comparison to IPCW: TSE uses standard regression techniques and therefore cannot handle confounding by time-dependent covariates. If switching happens quickly after the secondary baseline (as is often the case in cancer trials, where switching happens soon after disease progression), the potential for time-dependent confounding may be small; however, in order to assess this, it is important to investigate the distribution of times to switch from the secondary baseline time point.

24.6.2.2 Method

The TSE method is formalized using a "counterfactual" framework. We will first introduce the method in the special case where control group participants who switch treatments do so immediately upon disease progression, before considering a more general case where switching can occur some time after disease progression.

First, a dataset is formed from participants in the control group who experienced disease progression, including as observed outcome the time from secondary baseline to event, denoted T_i^{pps} for individual i. A counterfactual survival model is then used to relate this observed outcome to a potential outcome that would have been observed without treatment, denoted $T_i^{pps}(0)$, through a treatment effect ψ. In this case, $T_i^{pps}(0)$ represents the time from progression to death that would have been observed in the absence of treatment (i.e., if there had been no switching).

We assume that treatment multiplies lifetime by a ratio $\exp(-\psi)$. If ψ is less than 0, then $\exp(-\psi)$ is greater than 1, so treatment extends life. For untreated individuals (non-switchers), $T_i^{pps} = T_i^{pps}(0)$; for treated individuals (switchers), $T_i^{pps} = \exp(-\psi) \times T_i^{pps}(0)$. Equivalently, we can define untreated survival times for switchers as

$$T_i^{pps}(0) = \exp(\psi) \times T_i^{pps} \tag{24.6}$$

We estimate $\exp(\psi)$ using an accelerated failure time (AFT) model fitted to post-progression survival data comparing control group switchers to control group non-switchers. This may be any AFT model, such as a Weibull, log-logistic, log-normal, or generalized-gamma model, adjusted for prognostic covariates measured at or before the time of disease progression. Once $\exp(\psi)$ has been estimated, it can be substituted into Equation 24.6 in order to estimate $T_i^{pps}(0)$ for switchers, and then counterfactual overall survival times for switchers are

$$T_i(0) = T_i^{ttp} + T_i^{pps}(0) \tag{24.7}$$

where T_i^{ttp} is the observed time to progression.

If, instead of occurring immediately upon disease progression, switching is lagged after the secondary baseline, the AFT model used to estimate $\exp(\psi)$ must include a time-dependent switch indicator, and $\exp(\psi)$ is then used to adjust the time from switch to event, rather than the time from the secondary baseline to the event. The AFT model is still fitted to time from the secondary baseline (not to time from switch) in order to retain a common disease-related secondary baseline where the NUC assumption holds, but the parameters included in the counterfactual survival model change. T_i^{ttp} is replaced by time from randomization to time of switch, denoted T_i^{off}, and T_i^{pps} is replaced by time from switch to event, denoted T_i^{on}. As highlighted above, we must assume that there is NUC between the secondary baseline and the time of switch. Counterfactual overall survival times for switchers are then

$$T_i(0) = T_i^{off} + \exp(\psi) \times T_i^{on}. \tag{24.8}$$

Once counterfactual survival times have been obtained for switchers, the resulting dataset can be analyzed using any approach desired, for instance using a Kaplan–Meier plot, a Cox regression model, or a parametric survival model. This is the second stage of the TSE method. Unfortunately, the standard error and confidence intervals obtained will not be appropriate, because they do not account for the adjustments that have been made to the dataset. To obtain valid standard errors and confidence intervals, we must bootstrap the entire two-stage process.

24.6.2.3 Refinements

Censoring by loss to follow-up—whereby not every participant experiences the event of interest—is problematic for the second stage of the TSE method. We use the model in Equation 24.8 to adjust survival times for switchers whether they experienced the event or were censored. Hence, using Equation 24.8, we estimate counterfactual censoring times for switchers, but not in non-switchers. Because we expect that switching will be related to prognostic characteristics, it is likely that counterfactual censoring times will be related to prognostic covariates and therefore may constitute informative censoring. To avoid this problem, it is generally recommended that re-censoring is undertaken, to break the dependence between censoring time and treatment received [19]. This is achieved by re-censoring the counterfactual survival time associated with a given value of ψ for all participants in the group in which switching occurs at the minimum of the administrative censoring time C_i and $C_i \exp \psi$, representing the earliest possible censoring time over all possible treatment trajectories, $D_i^*(\psi)$. $T_i(0)$ for the given value of ψ is then replaced by $D_i^*(\psi)$ if it is greater than $D_i^*(\psi)$ [19–21]. Whilst re-censoring avoids problems with informative censoring, it involves a loss of information, as the survival data are artificially censored at a time point earlier than the follow-up times observed in the trial. This may be problematic if the objective is to estimate a long-term treatment effect if the treatment effect is not constant over time, or to estimate long-term survival if trends in the hazard change over time. Hence, conducting analyses with and without re-censoring represents a sensible approach [22].

The TSE method is relatively simple in terms of computation because standard regression models are used to estimate $\exp(\psi)$. However, model convergence problems can arise in small sample sizes, particularly when switching proportions are very high and/or when there are a large number of covariates included in the model. As for the IPCW, covariate selection is important, and in the context of the TSE, we only require covariates that are prognostic for survival. Different distributions may also be considered for the AFT model. We wish to obtain an accurate estimate of the treatment effect associated with switching observed in the trial period and therefore we may use model fit statistics such as Akaike's Information Criterion (AIC) or the Bayesian Information Criterion (BIC) to determine preferred model distributions. Non-parametric AFT models could also be considered.

The TSE method could be used to adjust for a variety of switching types. If switching occurred both in the experimental group and in the control group, different AFT models could be used to estimate the effect of switching in the two groups, and these different effects could be used to adjust survival times in the respective groups. If switching was to multiple different treatments, theoretically, several AFT models could be used in stage 1 of the method to estimate the treatment effect associated with each different treatment that participants switched to. In practice this may not be possible, as numbers that switched to each different treatment may be small. Instead, adjustments could be made for categories of treatments, or an average treatment effect associated with switching (irrespective of the treatment switched to) could be estimated.

In Box 24.4 we demonstrate the application of the TSE method to adjust for the treatment switching in the NovCEA dataset. Note that we do not show the data manipulation process, where we generate a dataset from which $\exp(\psi)$ can be estimated, and we do not include code for the bootstrap process.

BOX 24.4 TWO-STAGE ESTIMATION ANALYSIS

```
********************************************************************
*** Two-stage estimation
********************************************************************


********************************************************************
*** TSE step 1: Estimate the effect of switching, and estimate counterfactuals
********************************************************************


*** Use the NovCEA_2stage dataset, which is manipulated for the TSE method
*** Experimental group participants and non-progressors are excluded
*** Time prior to progression is excluded
*** All times now represent time since progression
*** New covariates:
***     tong is a 0/1 indicator of switching
***     HRQLprog is HRQL at time of progression
use NovCEA_2stage, clear
```

```
*** Fit accelerated life model
streg tong progb progtime CEAb CEAprog HRQLb HRQLprog, dist(weibull) time

*** record the treatment effect as an acceleration factor
scalar Weib_tg = exp(-_b[tong])

*** Go back to the original dataset
use NovCEAwide, clear

*** generate variable to hold adjusted survival times
gen timecf = deathtime

*** use counterfactual model to estimate adjusted survival times in switchers
replace timecf = xotime + (deathtime - xotime) * scalar(Weib_tg) ///
    if trtrand==0 & xoind==1

*****************************************************************************
*** TSE step 2: analyse the counterfactual dataset
*****************************************************************************

*** Incorporate recensoring, using 'admin' as the 'endstudy' time for each
*** participant
gen trecens = timecf
replace trecens = admin * scalar(Weib_tg) if (trtrand==0)
replace deathind = 0 if (timecf>trecens & trtrand==0)
replace timecf = trecens if (timecf>trecens & trtrand==0)

*** Declare the adjusted OS times to be our survival data
stset timecf, failure(deathind) id(id) scale(365.25)

*** Draw Kaplan-Meier graph
sts graph, by(trtrand) risktable title(TSE with recensoring for OS)

*** Estimate the hazard ratio adjusted for switching
stcox trtrand
```

24.6.3 Rank Preserving Structural Failure Time Model

24.6.3.1 Concept

The rank preserving structural failure time model (RPSFTM) works in a similar way to the TSE method in that a treatment effect associated with the experimental treatment is estimated and then used to derive what survival times would have been had the experimental treatment not been received. However, unlike the TSE method, the RPSFTM does not differentiate between the treatment effect received in the experimental group and the treatment effect received by switchers—ψ is assumed to be the same irrespective of when treatment is received. As a result, the NUC assumption is avoided.

The model presented in Equation 24.8 relates an observed survival time, T_i, to a potential outcome $T_i(0)$ through a treatment effect, ψ. This model is a RPSFTM [23]. Therefore, the TSE method uses a RPSFTM in its second stage. In contrast, the RPSFTM uses the model throughout, to estimate ψ and to simultaneously estimate counterfactual survival times. The estimation procedure used by the RPSFTM adjustment method is based upon two key assumptions. Firstly, it is assumed that randomization has worked perfectly; that is, that if no participants in either treatment arm had received any treatment, average survival times in the randomized groups would be equal. Secondly, it is assumed that there is a common treatment effect associated with the experimental treatment; hence, ψ is the same no matter when treatment is received. Given these assumptions, an estimation procedure called g-estimation is used, with the estimated value $\hat{\psi}$ being that for which counterfactual survival times $T_i(0)$ are independent of randomized group [23]. Once $\hat{\psi}$ has been identified, the adjusted treatment effect can be estimated by comparing observed survival times in the experimental group to the adjusted survival times in the control group.

The common treatment effect assumption represents the main limitation of the RPSFTM method. Whilst the assumption of perfect randomization is also important, it is generally regarded as reasonable—and in fact underpins most analyses undertaken on randomized trials. In contrast, the common treatment effect assumption is generally considered to be problematic, particularly in the context of the switching typically observed in cancer trials, where switching usually happens after disease progression. This may indicate that switchers have a lower capacity to benefit from the new treatment, because their disease is further advanced. In theory, the RPSFTM can be extended to estimate different treatment effects in switchers compared to the experimental group, but in practice this approach has been unsuccessful, with meaningful point estimates for two treatment effects difficult to determine when relying solely on the randomization assumption [19]. Instead, we encourage sensitivity analysis around the common treatment effect assumption, whereby it is assumed that the treatment effect in switchers is $x\%$ lower or higher in switchers than in the experimental group. This allows the sensitivity of the estimated adjusted treatment effect to this assumption to be assessed and may allow the analysis to be used by decision-makers where otherwise it would be rejected. Whilst the common treatment effect assumption is clearly a limitation of the RPSFTM method, together with the randomization assumption and g-estimation, it means that the RPSFTM does not require the NUC assumption, which represents an important advantage compared to the IPCW and TSE methods.

24.6.3.2 Method

The RPSFTM assumes model from Equation 24.8 and the perfect randomization and common treatment effect assumptions, and uses g-estimation to estimate ψ. Hence, for each trial participant the observed event or censoring time, T_i is split into time spent off treatment, T_i^{off}, and time spent on treatment, T_i^{on}. The counterfactual survival model from Equation 24.8 relates observed survival times T_i to counterfactual untreated survival times $T_i(0)$ by assuming that treatment multiplies lifetime by a ratio $\exp(-\psi)$. The g-estimation procedure involves taking a range of possible values of ψ and using them in the model in Equation 24.8 to calculate $T_i(0)$ for every individual in the trial (whether they were originally randomized to the control group or the experimental group, and whether or not they switched treatments). A g-test is then used to assess whether or not $T(0)$ is balanced across randomized groups. The test used for the g-test can be chosen by the analyst, but conventionally is the same test as used in the ITT analysis; typically this is the log-rank test, but other tests such as the Cox model (particularly if baseline covariates are being adjusted for in the analysis) could be used. The test statistic chosen should indicate the direction as well as the magnitude of the difference between groups: for example, the signed log-rank statistic should be used rather than its square. The test statistic is graphed against ψ and the best estimate of ψ is where $Z(\psi)$ crosses 0, since this gives the best balance in $T(0)$ between treatment arms. The 95% confidence interval for ψ is represented by values of ψ where the g-test does not reject $Z(\psi) = \pm 1.96$. In practice, g-estimation can be implemented using a grid search over a specified range of possible values for ψ, or using interval bisection.

Once ψ has been estimated, the observed survival times in the experimental group can be compared to the adjusted survival times in the control group to estimate treatment effects adjusted for treatment switching. However, it is important to note that the p-value and confidence intervals produced by these comparisons are inappropriate, because they do not account for the fact that data have been adjusted for switchers. Instead, it is recommended that the p-value from an equivalent ITT analysis (e.g., a Cox model, with or without covariates) should be used and confidence intervals calculated accordingly. The logic for this is that the null hypothesis is the same for the ITT analysis and the RPSFTM analysis. As an alternative to retaining the ITT p-value, the entire adjustment procedure could be bootstrapped, as for the TSE method, but this gives very similar results [19].

24.6.3.3 Refinements

Censoring is problematic for the RPSFTM, as it is for the TSE method. Re-censoring should be considered but may be problematic when the objective is to estimate long-term treatment effects. Conducting analyses with and without re-censoring represents a sensible approach.

A simple refinement of the RPSFTM is to include prognostic baseline covariates in the model to improve the precision of estimates.

In practice, the g-estimation procedure may not always work as desired. It is possible that a value for ψ that provides a g-test statistic of $Z = 0$, or values for the confidence limits, may not be found. Clearly it is an important problem if $Z(\psi) = 0$ cannot be solved, whereas with confidence limits it may be that the range of possible values for needs to be broadened, or that one or other of the limits has to be reported as $-\infty$.

An alternative approach is to use Branson and Whitehead's iterative parameter estimation (IPE) algorithm in place of g-estimation [24]. This uses the same counterfactual survival model framework as the RPSFTM, but uses an iterative estimation procedure to estimate ψ by assuming that the counterfactual survival times follow a parametric distribution.

The RPSFTM method is best suited to adjusting for switches between randomized treatments. Adjusting for other types of switching is more problematic, because in practice the RPSFTM is only able to estimate one treatment effect. The RPSFTM requires that trial participants are either "on treatment" or "off treatment." The period spent on treatment is allocated the same treatment effect, ψ, and therefore it is logical to expect that the time each participant spends "on treatment" involves the same treatment for each participant. If switching was to a range of different treatments, the RPSFTM is unlikely to be appropriate, unless it can be credibly assumed that all the treatments involved have similar efficacy.

Further, the RPSFTM may be problematic if the control treatment in a trial is an active therapy. The time spent on the experimental treatment is classified as time "on treatment," and time spent on the control treatment is classified as time spent "off treatment." If, later in their survival time, participants do indeed spend time completely off any active therapy, then to accurately model the treatment pathway, the RPSFTM would need to incorporate terms for time spent on treatment A, T_i^{onA}, time spent on treatment B, T_i^{onB}, and time spent off treatment, T_i^{off}. Unfortunately, as previously noted, the RPSFTM is not able to estimate more than one treatment effect. A solution to this is to redefine T_i^{on} and T_i^{off} in order to estimate the survival advantage/disadvantage associated with ever starting the experimental treatment—that is, the effectiveness of the treatment pathway initiated with the experimental treatment compared to the treatment pathway initiated with the control treatment. Then, T_i^{on} equals total time after initiating the experimental treatment,

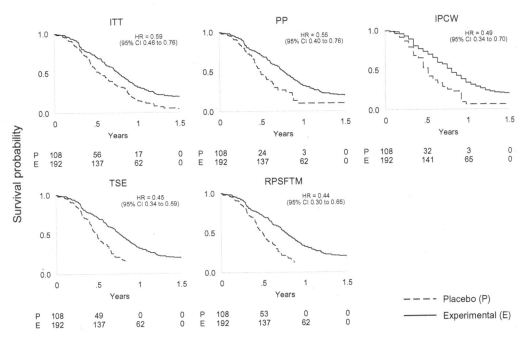

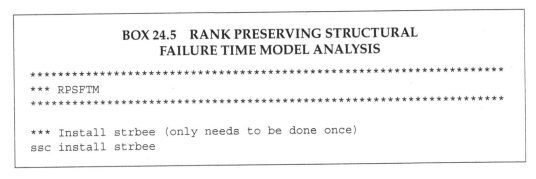

FIGURE 24.3

Overall survival Kaplan–Meier curves. Intention-to-treat (ITT) analysis; per-protocol (PP) analysis; inverse probability of censoring weights (IPCW) analysis; two-stage estimation (TSE) analysis; rank preserving structural failure time model (RPSFTM) analysis.

ignoring treatment discontinuation times. This alternative approach, sometimes referred to as the "ever-treated" or "treatment group" approach, bears similarities to standard ITT analyses of trials, where treatment effects are estimated by comparing randomized groups without accounting for treatment discontinuation. In addition, from an HTA perspective, where the decision problem typically involves an assessment of the effectiveness and cost-effectiveness of inserting a new treatment into the treatment pathway, the "treatment group" approach is appropriate, provided that the treatment pathways observed in the trial (aside from the switching that the RPSFTM is being used to adjust for) represent realistic treatment pathways.

In Box 24.5 we demonstrate the application of the RPSFTM "on treatment" and "treatment group" methods to adjust for the treatment switching in the NovCEA dataset. We use the user-written Stata program strbee [25]. The overall survival Kaplan-Meier curves for each analysis presented in this chapter are illustrated in Figure 24.3.

BOX 24.5   RANK PRESERVING STRUCTURAL FAILURE TIME MODEL ANALYSIS

```
**************************************************************************
*** RPSFTM
**************************************************************************


*** Install strbee (only needs to be done once)
ssc install strbee
```

```
*** Load wide data
use NovCEAwide, clear
stset deathtime deathind, id(id)

*** For participants who do not switch, recode xotime to equal
*** deathtime
replace xotime = deathtime if xoind == 0

*** Fit the "ever-treated" or "treatment group" RPSFTM, with
*** re-censoring
strbee trtrand, xo0(xotime xoind) endstudy(admin)

*** Check 1: graph Z-statistic against psi: graph should move from
*** >1.96 to <-1.96
strbee, zgraph(title(Z graph from RPSFTM))

*** Check 2: plot the observed and counterfactual Kaplan Meier
*** graphs
*** If RPSFTM has worked well, the "trtrand=0 untreated" curve
***      should be close to the "trtrand=1 untreated" curve
strbee, kmgraph(showall title(KM graph from RPSFTM))

*** Estimate the adjusted hazard ratio
strbee, hr
```

24.7 Identifying Appropriate Methods

We can never be certain which adjustment method will produce the least bias in a real-world case. However, through an assessment of the characteristics of the case in question, we may be able to rule out specific methods, or we may be able to determine which methods are potentially appropriate. The following four steps may be of use when attempting to identify an appropriate adjustment method on a case-by-case basis. Table 24.2 presents an overview of the methods, allowing their similarities and differences—and their key assumptions and limitations—to be distinguished.

24.7.1 Assessment of Methodological Assumptions

It is not possible, statistically, to test the key assumptions made by the adjustment methods. However, attempts should be made at assessing whether or not they are plausible. For instance, for the IPCW and TSE methods, comprehensive data collection on prognostic factors is critical. If it is clear that important prognostic characteristics were not measured at all, or were only measured for a portion of the trial period, the NUC assumption may not be credible. Similarly, if there are substantial amounts of missing data amongst the characteristics that were supposed to be measured, problems may arise, although imputation techniques could be used to address this issue.

For the RPSFTM, and for the re-censoring that may be included within the RPSFTM and TSE methods, it is important to consider the nature of the treatment effect. Depending upon the biological mechanism of action of the treatment, it may or may not be appropriate to

TABLE 24.2

Comparison of Methods

	IPCW	TSE	RPSFTM
Identifying assumption	NUC	NUC	Common treatment effect
Requires time-updated confounders	Yes	Yes	No
Pitfalls	Large weights Model convergence	Treatment changes after progression Model convergence Re-censoring	Active control-arm therapy Unsuccessful g-estimation Re-censoring
Details to be specified[a]	Discrete or continuous time Covariates and their functional forms in switching model Stabilized or unstabilized weights	Choice of AFT Covariates and their functional forms in AFT Whether to re-censor	Choice of g-test Whether to re-censor "On treatment" or "ever-treated"
How to check	Range of weights	Existence of secondary baseline Time from secondary baseline to switch	Graphs of counterfactual untreated outcomes
Extension to other treatment changes	Easy: censor at all treatment changes	Fairly easy: estimate effect of each treatment	Rarely possible

Abbreviations: IPCW, inverse probability of censoring weights; TSE, two-stage estimation; RPSFTM, rank preserving structural failure time model; NUC, no unmeasured confounding; AFT, accelerated failure time model.
[a] In addition, the nature of the switching event must be specified for all methods.

assume a common treatment effect. Clinical opinion on this issue is likely to be of high importance. This could also inform sensitivity analyses undertaken to assess the sensitivity of results to violations of the common treatment effect assumption.

24.7.2 Assessment of "Model Diagnostics"

Table 24.2 outlines key pitfalls associated with the adjustment methods. It is important to carefully monitor the output of analyses. For instance, the range of weights produced by an IPCW analysis is an important indicator of the likely performance of the method; the coefficient of variation of the weight has been shown to be particularly important [26]. The output of the g-estimation procedure is also important to consider, in order to assess whether or not multiple solutions for ψ have been found, and whether or not counterfactual untreated outcomes are similar between randomized groups. The exact technique used to apply each of the methods must be reported clearly. Sensitivity in the results to methodological choices should be considered: for instance, whether or not re-censoring was incorporated in RPSFTM and TSE analyses, which g-test was used, whether the RPSFTM was applied on an "on treatment" or "ever-treated" basis, which AFT model was used within the TSE method, and which covariates were included in the switching and outcome models for the IPCW and in stage 1 of the TSE method.

24.7.3 Results of Simulation Studies

It is also possible to learn about the likely performance of adjustment methods in specific scenarios through simulation studies. In a real-world setting, we do not know which adjustment method has performed best because we do not know what would have happened had treatment switching not occurred. However, in a simulation study, the "truth" is known. Here, survival data are simulated and switching is applied to a proportion of simulated control-group participants according to prognostic characteristics. Adjustment methods are then applied and assessed relative to the recorded truth. Scenarios can then be varied according to key factors such as switching proportion, disease severity, treatment effect size (and commonality), and sample size, and patterns in results across different scenarios can be evaluated.

Several such simulation studies have been published [18,22,26–28]. Results suggest that the RPSFTM method produces very low levels of bias in a wide range of scenarios when there is a common treatment effect. Whilst the RPSFTM is sensitive to violations of this assumption, low levels of bias may still be produced if the treatment effect size is relatively low; a proportional violation of the common treatment effect assumption is much more important when the estimated acceleration factor (i.e., $\exp(-\psi)$) is 2.5 than when it is only 1.5. The TSE method has performed well across a wide range of simulated scenarios, though this result should be treated with caution because the reported scenarios have all involved switching that occurred soon after disease progression, fitting well the requirements of the TSE method. In simulated scenarios, the IPCW method has generally produced marginally higher bias than the RPSFTM method (when a common treatment effect has been simulated) and the TSE method. As expected, the IPCW becomes prone to high levels of error when the switching proportion is high, leaving very low numbers of non-switchers in the control group. It is difficult to conclude what number of non-switchers must remain in order for the IPCW method to produce reliable results, but one study suggests a number of 20 [27].

One simulation study has focused on the inclusion or exclusion of re-censoring within the RPSFTM and TSE adjustment methods [22]. It reported that RPSFTM and TSE analyses which incorporated re-censoring usually produced negative bias in the control group restricted mean survival (therefore overestimating the treatment effect), assuming a treatment effect that either fell or remained constant over time and hazards that fell over time. In contrast, RPSFTM and TSE analyses that did not incorporate re-censoring consistently produced positive bias in the control group restricted mean survival (therefore underestimating the treatment effect) which was often smaller in magnitude than the bias associated with the re-censored analyses. The authors recommended that adjustment analyses using these methods should be conducted with and without re-censoring, as this may provide decision-makers with useful information on where the true treatment effect is likely to lie [22].

24.7.4 Validation of Results

In this chapter, we have discussed methods to adjust for switching using data collected within the trial of interest. An alternative could be to make use of external trial or registry data consisting of a similar participant population who received the same treatment as in the control arm of the trial of interest, but in which treatment switching did not occur. Or, such data could be used to validate the estimates produced by the adjustment methods. Making comparisons between estimates provided by the adjustment methods and external data sources has proven a powerful tool in analyses presented to NICE in the UK, where it appears adjustment analyses are more likely to be accepted if this validation task is attempted [5].

24.8 Extensions

HTA agencies such as NICE have adopted adjustment methods as alternatives to the ITT analysis to adjust estimates for time-to-event outcomes in the presence of treatment switching. Other agencies, such as the Pharmaceutical Benefits Advisory Committee (PBAC) in Australia and the Scottish Medicines Consortium (SMC) appear to have followed suit, whilst the German Institute for Quality and Efficiency in Health Care (IQWiG) seems less open to these methods. Regulatory agencies such as the United States Food and Drug Administration (FDA) and the European Medicines Agency (EMA) have been involved in international workshops on this topic [12,29], and the new draft addendum to ICH E9 appears to suggest that switching adjustment analyses may be presented more regularly to regulators in the future, as assessment of "hypothetical" estimands (including treatment effects in the absence of treatment switches) is legitimized.

To the best of our knowledge, all adjustment analyses presented to regulatory or HTA decision-makers have focused upon making adjustments for time-to-event outcomes, rather than other outcomes. If treatment switching is expected to affect time-to-event outcomes, it seems reasonable to assume that it could also affect any other outcomes measured after the point of switch, such as response rates, quality of life, or resource use. Often these outcomes may not be measured after the point of switch, but sometimes they may be. In addition, treatment switching may also be an issue in trials where time-to-event outcomes are not of primary interest. Switching could also occur in trials of treatments that are not expected to alter survival, but which may improve quality of life. Inverse probability weighting methods could be used to analyze such outcomes, or structural mean models in place of the RPSFTM. Instrumental variables methods have also been suggested for analyzing continuous outcomes [30], but at present there has been relatively little focus on these methods. This represents an area for further research.

References

1. Latimer NR, Abrams KR, Lambert PC et al. Adjusting survival time estimates to account for treatment switching in randomised controlled trials – an economic evaluation context: Methods, limitations and recommendations. *Med Decis Making.* 2014. doi:10.1177/0272989x13520192.
2. Jonsson L, Sandin R, Ekman M et al. Analyzing overall survival in randomized controlled trials with crossover and implications for economic evaluation. *Value Health.* 2014;17(6):707–713.
3. Ishak KJ, Proskorovsky I, Korytowsky B, Sandin R, Faivre S, Valle J. Methods for adjusting for bias due to crossover in oncology trials. *Pharmacoeconomics.* 2014;32(6):533–46.
4. Rubin DB. Bayesian inference for causal effects: The role of randomization. *Annals of Statistics.* 1978; 6: 34–58.
5. Latimer NR. Treatment switching in oncology trials and the acceptability of adjustment methods. *Expert Rev Pharmacoecon Outcomes Res.* 2015;1–4:20.
6. GlaxoSmithKline UK. Single technology appraisal (STA) Melanoma (unresectable/metastatic BRAFV600 mutation-positive) – dabrafenib. Submission to the National Institute for Health and Care Excellence, 30 April 2014. Available from https://www.nice.org.uk/guidance/ta321/documents/melanoma-braf-v600-unresectable-metastatic-dabrafenib-id605-evaluation-report2 (accessed 13 Feb 2018).

7. GlaxoSmithKline UK. Pazopanib (Votrient®) for the first-line treatment of patients with advanced renal cell carcinoma (RCC): Addendum to GSK's submission to NICE, 20 July 2010. Available from https://www.nice.org.uk/guidance/ta215/documents/renal-cell-carcinoma-first-line-metastatic-pazopanib-manufacturer-submission-addendum3 (accessed 13 Feb 2018).

8. National Institute for Health and Care Excellence. *Guide to the Methods of Technology Appraisal.* London: NICE, 2013 www. nice.org.uk/process/pmg9 (accessed 2 June 2017).

9. StataCorp. 2015. *Stata Statistical Software: Release 14.* College Station, TX: StataCorp LP.

10. Robins JM, Finkelstein DM. Correcting for noncompliance and dependent censoring in an AIDS clinical trial with inverse probability of censoring weighted (IPCW) log-rank tests. *Biometrics.* 2000; 56(3):779–788.

11. Hernán MA, Brumback B, Robins JM. Marginal structural models to estimate the joint causal effect of nonrandomized treatments. *J Am Statist Assoc.* 2001; 96(454):440–448.

12. Latimer NR, Henshall C, Siebert U, Bell H. Treatment Switching: Statistical and decision making challenges and approaches. *Int J Technol Assess Health Care.* 2016; 32(3):160–166.

13. D'Agostino R. B., Lee M.-L., Belanger A. J. Relation of pooled logistic regression to time dependent Cox regression analysis: The Framingham Heart Study. *Stat Med.* 1990;9:1501–1515.

14. Seaman SR, White IR. Review of inverse probability weighting for dealing with missing data. *Stat Methods Med Res.* 2013;22: 278–95.

15. Cole SR, Hernán MA. Constructing inverse probability weights for marginal structural models. *Am J Epidemiol.* 2008;168(6):656–64.

16. Robins JM, Hernán MA, Brumback B. Marginal structural models and causal inference in epidemiology. *Epidemiology.* 2000;11:550–60.

17. Dodd S, White IR, Williamson P. A framework for the design, conduct and interpretation of randomised controlled trials in the presence of treatment changes. *Trials.* 2017;18:498.

18. Latimer NR, Abrams KR, Lambert PC, Crowther MJ, Wailoo AJ, Morden JP, Akehurst RL, Campbell MJ. Adjusting for treatment switching in randomised controlled trials – A simulation study and a simplified two-stage method. *Stat Methods Med Res.* 2017;26(2): 724–51, doi:10.1177/0962280214557578.

19. White IR, Babiker AG, Walker S, Darbyshire JH. Randomization-based methods for correcting for treatment changes: Examples from the Concorde trial. *Stat Med.* 1999; 18(19):2617–2634.

20. Robins JM. The analysis of randomized and non-randomized AIDS treatment trials using a new approach to causal inference in longitudinal studies. In *Health Service Research Methodology: A Focus on AIDS,* Sechrest L., Freeman H., Mulley A, eds. Washington, D.C.: U.S. Public Health Service, National Center for Health Services Research, 1989, 113–159.

21. Robins JM. Analytic methods for estimating HIV treatment and cofactor effects. In *Methodological Issues of AIDS Mental Health Research.* Ostrow D.G., Kessler R. eds. New York: Plenum Publishing, 1993: 213–290.

22. Latimer NR, White IR, Abrams KR, Siebert U. Causal inference for long-term survival in randomised trials with treatment switching: Should re-censoring be applied when estimating counterfactual survival times? *Stat Methods Med Res.* 25 June 2018 , doi:10.1177/0962280218780856

23. Robins JM, Tsiatis AA. Correcting for Noncompliance in Randomized Trials Using Rank Preserving Structural Failure Time Models. *Commun Stat Theory Methods.* 1991; 20(8):2609–2631.

24. Branson M, Whitehead J. Estimating a treatment effect in survival studies in which patients switch treatment. *Stat Med.* 2002; 21(17):2449–63.

25. White IR, Walker S, Babiker AG. strbee: Randomization-based efficacy estimator. *The Stata Journal* 2002; 2(2):140–150.

26. Latimer NR, Abrams KR, Siebert U. Two-stage estimation to adjust for treatment switching in randomised trials: A simulation study investigating the use of inverse probability weighting instead of re-censoring. Health Economics & Decision Science Discussion Paper Series, The University of Sheffield, No.19.01, 2019. Available from https://www.sheffield.ac.uk/scharr/sections/heds/discussion-papers/19_01-1.831150 (accessed 15 Feb 2019).

27. Latimer NR, Abrams KR, Lambert PC, Morden JP, Crowther MJ. Assessing methods for dealing with treatment switching in clinical trials: A follow-up simulation study. *Stat Methods Med Res.* 2018;27(3):765–84, doi: 10.1177/0962280216642264.
28. Morden JP, Lambert PC, Latimer N, Abrams KR, Wailoo AJ. Assessing statistical methods for dealing with treatment switching in randomised controlled trials: A simulation study. *BMC Med Res Methodol.* 2011;11(4).
29. Henshall C, Latimer NR, Sansom L, Ward RL. Treatment switching in cancer trials: Issues and proposals. *Int J Technol Assess Health Care.* 2016; 32(3):167–174.
30. Dunn G, Maracy M, Tomenson B. Estimating treatment effects from randomized clinical trials with noncompliance and loss to follow-up: The role of instrumental variable methods. *Stat Methods Med Res.* 2005; 14(4):369–95.

25

The Analysis of Adverse Events in Randomized Clinical Trials

Jan Beyersmann and Claudia Schmoor

CONTENTS

In clinical studies, the primary outcome often is a time to event [1]. In oncology, arguably the most common primary outcomes are progression-free survival (PFS) and overall survival (OS). Every patient experiences these outcomes, although possibly after study closure, leading to censored data. Censoring is a major reason why these outcomes are analysed using so-called survival analysis. The proportion of patients who have not yet experienced PFS is approximated using the Kaplan–Meier estimator within treatment groups; the same applies to OS. The log-rank test is the common test to compare treatment groups, and treatment effects are typically quantified in terms of a hazard ratio. If adjustment of the hazard ratio with respect to prognostic or predictive factors is required, the Cox proportional hazards model is often the regression technique of choice.

Use of these advanced statistical techniques is well established when studying efficacy, but the statistical analysis of adverse events when studying safety typically is much more simplistic. This is both strange and inadequate, because the general data structure, characterized by timing of events, censoring, and varying follow-up times, is the same for both efficacy and safety considerations.

In many safety analyses, the two major workhorses to quantify the "risk" for a patient of experiencing at least one adverse event of a specific type within treatment groups are the incidence proportion and the incidence rate [2–8]. The incidence proportion is the number of patients with an observed adverse event divided by size of the group. Although common, the incidence proportion neither accounts for censoring nor for varying follow-up times. The incidence rate has the same numerator as the incidence proportion, but the denominator is the cumulative person-time at risk. The incidence rate is sometimes also called exposure-adjusted incidence rate, if person-time at risk is only counted while exposed. The incidence rate does account for both censoring and for varying follow-up times. The ratio of the incidence rates between treatment groups is an estimator of a hazard ratio under the simplifying assumption that the hazards themselves, and not just their ratio, are constant over time. It is because of this simplifying assumption that use of the incidence rate is often criticized. Perhaps more serious, however, is the restriction that the Kaplan–Meier-type approximation of the proportion of patients still alive and without an adverse events fails. The reason is that Kaplan–Meier estimation is only appropriate for probabilities of all-encompassing composite endpoints such as PFS, which combines progression and death, or OS, which includes death, which every patient will experience at some point in time (although potentially after study closure). In contrast, adverse events (of a specific type) may be precluded by, for example, death. In statistical terminology, death without prior adverse event is called a competing risk or a competing event.

Our starting point for our considerations below will be that we assume some given safety patient set, typically containing patients who have received at least one dose of a drug. Definition of a safety patient set will imply some implicit choices. Time zero will often be time of first intake, possibly after time of randomization, follow-up of adverse events may be stopped after progression of disease in oncological trials, time under subsequent therapies may (or may not) be included, follow-up may be focused on adverse events of special interest, and adverse events may be distinguished between treatment emergent and not treatment emergent. For what follows, we will accept these choices and consider methods of quantifying adverse event risk based on such a safety patient set. We will however comment on censoring by disease progression (or some other disease event initiating subsequent therapy) in Section 25.5.

The remainder of this chapter is organized as follows: Section 25.1 explains the connection between the two commonly used measures of incidence of adverse events, the incidence proportion and the incidence rate. The connection is easily seen for uncensored data with complete follow-up. Although rare in practice, we start with this uncensored case in Section 25.1.1 for ease of presentation. It also allows us to highlight the challenges imposed by censoring, which we consider in Section 25.1.2 and which are not accounted for by using the incidence proportion. The methods discussed in Sections 25.1.1 and 25.1.2 will be illustrated in a data example in Section 25.1.3.

In the data example, we have refrained from considering a too specific example. A major reason is that, in our experience, discussions then get quickly lost in the specifics of the safety analysis at hand; see our comments on our starting point above. For instance, it is not uncommon that regulators and representatives from the pharmaceutical industry discuss whether to further follow up adverse events after disease progression. This is a relevant question, but not a topic of this chapter. We have therefore chosen to simulate data inspired by the data example in Allignol et al. [2] for illustrating the different methodological approaches. R code for simulation is offered in the online supplement in the companion website of the book.

In Section 25.1, we will find that only the incidence rate accounts for censoring and varying follow-up times, but that it also requires a restrictive parametric assumption and a complementary competing-risks analysis. Non-parametric analyses, accounting for competing risks, are therefore discussed in Section 25.2. Section 25.3 explains how the common log-rank test and the Cox model may be used for comparison of treatment groups in the present context. Adverse events may also be recurrent and Section 25.4 discusses to which extent the present methodology will also be relevant for recurrent adverse events. A summary is given in Section 25.5. Here, we discuss the common debate whether occurrence of adverse events is independently or informatively censored by, e.g., disease progression, and connections to the notion of estimands [9].

25.1 The Two Workhorses Incidence Rate and Incidence Proportion and Their Connection

For the time being, we consider a single treatment group with n patients. The safety outcome of interest is time to first occurrence of a specific adverse event. The adverse event may be recurrent, but we start with explaining the difficulties when analyzing time to *first* occurrence. We also account for the possibility that the adverse event does not occur, because some other disease event precludes occurrence of the adverse event. Above, we have considered death without prior adverse event as a competing event. In oncology, it is also common that follow-up of adverse events is concentrated on the duration of progression-free survival. After progression, typically a second-line therapy is initiated and systematic follow-up of adverse events may be stopped. In such a situation, progression is another competing event, while the adverse event under consideration is "adverse event before progression." In the following, we will denote for patient i, $i =, 1,...n,$

> T_i time to adverse event or competing event, whatever occurs first,
>
> ε_i type of event, 1 for adverse, 2 for competing,
>
> C_i time to censoring,

and also

> \# AE number of patients with an observed adverse event,
>
> i.e., $\displaystyle\sum_{i=1}^{n} 1(T_i \leq C_i) \cdot 1(\varepsilon_i = 1),$

where we have written $1()$ for the indicator function. Here, we have also combined all competing events into one composite, $\varepsilon_i = 2$. This is sufficient for studying the risk of an adverse event. However, extensions to more competing events are straightforward, e.g., if one wishes to distinguish between death without prior progression diagnosis and progression. Also note that the censoring time C_i is *not* the time of an observed disease event of the individual patient precluding the subsequent observation of an adverse event, as these events have to be considered as competing. C_i is the time to end of follow-up without former observation of adverse or competing event.

The connection between incidence proportion and incidence rate is not too difficult, but it appears to be not widely known. For clarity of presentation, we start with the complete data case *without* censoring in Section 25.1.1 before we turn to the practically more common situation of censored data in Section 25.1.2.

25.1.1 The Case of Uncensored Data

The incidence proportion, also known as crude rate, simply is

$$\frac{\# AE}{n}. \tag{25.1}$$

In the absence of censoring, which we can formally express as $C_i = \infty$ for all patients i, (Equation 25.1) is the proper estimator of the "absolute risk of an anytime adverse event"

$$P(\varepsilon = 1), \tag{25.2}$$

where we have suppressed the index i in the notation to denote the theoretical modeling quantities. Quantity (Equation 25.2) denotes the *absolute* risk, because it is a probability with values in [0, 1], and it is the risk for an adverse event *anytime*, because Equation 25.2 is the right-hand limit of the cumulative event probability (also known as cumulative incidence function in competing risks) of an adverse event

$$P(T \leq t, \varepsilon = 1). \tag{25.3}$$

The connection between Equations 25.2 and 25.3 is

$$P(T \leq \infty, \varepsilon = 1) = P(\varepsilon = 1). \tag{25.4}$$

In anticipation of Section 25.1.2, we note that estimation of $P(\varepsilon = 1)$ will, in general, not be feasible with censored data because of limited follow-up.

The incidence rate (also known as incidence density) is

$$\frac{\# AE}{\sum_{i=1}^{n} \min(T_i, C_i)} = \frac{\# AE}{\sum_{i=1}^{n} T_i}, \tag{25.5}$$

where the denominator is the cumulative patient-time at risk. Recall, for the time being, that we assume complete follow-up, such that T_i is always smaller than C_i. The incidence rate does not quantify an absolute risk, because it is not bounded by unity— one may easily "tune" the value to something very small or something very large (possibbly exceeding 1) depending on the units used to measure time. Readers should note that there is nothing wrong here. One must, of course, pay attention to the time scale used.

The connection between the two measures of incidence is easy to see and it was already known to Florence Nightingale and William Farr in their nineteenth century cooperative work on adverse events in hospitals [10], but some later researchers appear to have lost sight of it. Recalling that the number of patients with an observed adverse event here is

$$\# AE = \sum_{i=1}^{n} 1(\varepsilon_i = 1),$$

using again and for the time being that we assume complete follow-up and no censoring, it is easy to see that there are also patients with observed competing events,

$$\sum_{i=1}^{n} 1(\varepsilon_i = 2).$$

For later use, and in the absence of censoring, also note that

$$\sum_{i=1}^{n} 1(\varepsilon_i = 1) + \sum_{i=1}^{n} 1(\varepsilon_i = 2) = n.$$

We may now also define the competing incidence rate

$$\frac{\sum_{i=1}^{n} 1(\varepsilon_i = 2)}{\sum_{n=1}^{n} T_i},$$

and find that the incidence proportion is nothing but the relative magnitude of the incidence rate of an adverse event with respect to the sum of all incidence rates. Consider

$$\frac{\# AE \big/ \sum_{n=1}^{n} T_i}{\# AE \big/ \sum_{n=1}^{n} T_i + \sum_{i=1}^{n} 1(\varepsilon_i = 2) \big/ \sum_{n=1}^{n} T_i}.$$

Patient-time at risk cancels, and the last display equals the incidence proportion

$$\frac{\# AE}{\# AE + \sum_{i=1}^{n} 1(\varepsilon_i = 2)} = \frac{\# AE}{n}.$$

This calculation is competing-risks theory in a nutshell and it brings up the question of what incidence rates actually estimate. We have already noted that incidence rates are not estimators of probabilities. Rather, survival methodology [11] finds that the incidence rates estimate the event-specific hazards

$$\alpha_j(t) = \lim_{\Delta t \searrow 0} P(T < t + \Delta t, \varepsilon = j | T \geq t) / \Delta t, \quad j = 1, 2, \tag{25.6}$$

under the assumption that these hazards are constant over time,

$$\alpha_j(t) = \alpha_j \quad \text{for all times } t.$$

We will therefore use the notation

$$\hat{\alpha}_1 \quad \text{incidence rate of adverse events,}$$
$$\hat{\alpha}_2 \quad \text{incidence rate of competing events.}$$

It is now easy to see [11] that under the assumption of constant event-specific hazards, the "survival function" of time T until any first event, adverse or competing, is

$$P(T > t) = 1 - P(T \leq t) = \exp\left(-t \cdot (\alpha_1 + \alpha_2)\right), \tag{25.7}$$

and the cumulative event probabilities (or cumulative incidence functions) are

$$P(T \leq t, \varepsilon = j) = \int_0^t P(T \geq u)\alpha_j \, du = P(T \leq t) \cdot \frac{\alpha_j}{\alpha_1 + \alpha_2}. \tag{25.8}$$

Estimators of these quantities are obtained by replacing the theoretical quantities α_j with the incidence rates $\hat{\alpha}_j$. On the right-hand side of Equation 25.8, one may use both $P(T \geq u)$ and $P(T > u)$. We prefer the former, because it will yield the usual nonparametric estimator later; see Section 25.2.

When the event-specific hazards are time-varying, the formulas for survival function and for the cumulative event probabilities will be more complex, but their form will be similar. Hence, there is something to be learned from Equations 25.7 and 25.8, even if the hazards are not constant. We note:

1. Probabilities depend on all event-specific hazards involved. Hence, computing the incidence rate $\hat{\alpha}_1$ for adverse events without also computing the competing incidence rate $\hat{\alpha}_2$ is a somewhat incomplete analysis.
2. In the absence of a competing event, the cumulative adverse event probability is given by

$$1 - \exp\left(-t \cdot \alpha_1\right), \tag{25.9}$$

but using Equation 25.9 in the presence of a competing event would overestimate the cumulative adverse event probability, because the latter is bounded by $P(\varepsilon = 1)$, which is not the case for Equation 25.9.

3. In the absence of censoring, the incidence proportion estimates the right-hand limit of the cumulative adverse event probability. With the usual limited follow-up and with the usual censoring, this will, in general, not be feasible. Alternatively, it would have to rely on arguably questionable parametric extrapolation.

25.1.2 The Case of Censored Data

We now consider the—much more realistic—situation where patients are subject to censoring times $C_i < \infty$. The incidence proportion still is

$$\frac{\# AE}{n} = \frac{\sum_{i=1}^n 1(T_i \leq C_i) \cdot 1(\varepsilon_i = 1)}{n},$$

which is now less than $\sum_{i=1}^n 1(\varepsilon_i = 1)/n$ as a consequence of censoring. This implies that the commonly used incidence proportion *underestimates* the "absolute risk of an anytime adverse event." Also note that we cannot use $\sum_{i=1}^{n} 1(\varepsilon_i = 1)/n$ with censored data anymore, because $1(\varepsilon_i = 1)$ will be unknown for a censored patient who has $C_i < T_i$.

In contrast, the incidence rates

$$\hat{\alpha}_j = \frac{\sum_{i=1}^{n} 1(T_i \leq C_i) \cdot 1(\varepsilon_i = j)}{\sum_{i=1}^{n} \min(T_i, C_i)} \tag{25.10}$$

are still valid estimators of the event-specific hazards under a constant hazards assumption. Plugging these estimators into formulas in Equations 25.7 and 25.8 will still yield valid estimators of $P(T > t)$ and of $P(T \leq t, \varepsilon = j)$, but one will typically restrict evaluating these functions to times t covered by the length of follow-up in the sample. This is also the reason why one should be cautious, to say the least, about using $\hat{\alpha}_1/(\hat{\alpha}_1 + \hat{\alpha}_2)$ for estimating $P(\varepsilon = 1)$. Such an approach would rely on (often heavy) parametric extrapolation using a parametric model that is typically considered to be rather restrictive.

25.1.3 Data Example

Similar to the "anonymized" data example in Allignol et al. [2], we have chosen not to present a too specific data example. The reason is that, in our experience, discussions then soon digress, addressing questions such as whether the adverse event under consideration is treatment related or not. Such questions are important, but we are here interested in the basic data structure and statistical techniques accounting for it. Another common question is whether the constant hazard assumption can be made. This also is an important question— and in Section 25.2.2 we present a simple technique for checking this assumption—but again our experience is that discussions tend to focus on this question of goodness-of-fit and one loses sight of the basic data structure of competing risks.

We have therefore chosen to simulate "adverse event" data following a constant event-specific hazards model, with and without censoring. For the time being, this allows us to shelve the question whether incidence rates are appropriate estimators and focus on the following:

- Demonstrate that cumulative adverse event probabilities will be overestimated if the competing-risks structure is ignored.
- Demonstrate that incidence proportions are not appropriate in the presence of censoring.

R Code [12] for simulation, data analyses, and plotting is available online.

To begin, we have chosen to simulate from event-specific hazards

$$\alpha_1 = 0.2 \quad \text{and} \quad \alpha_2 = 0.1.$$

Interpreting type 2 events as "death without prior adverse event," this choice implies that eventually—with "infinite" follow-up—two-thirds of all patients first experience an adverse event and one-third of all patients die without prior adverse event. Simulation follows the approach in Beyersmann et al. [13], simulating event times T from Equation 25.7 and deciding on event type 1 in a binomial experiment with probability $0.2/(0.2 + 0.1)$.

Simulating a dataset with 100 individuals *without* censoring, calculating incidence rates $\hat{\alpha}_1$ and $\hat{\alpha}_2$, and substituting the true event-specific hazards by the incidence rates in Equation 25.8 produces the black curve in Figure 25.1. (In our simulated example, $\hat{\alpha}_1 = 0.19$ and $\hat{\alpha}_2 = 0.11$, see the R code in the online supplement.) The true underlying cumulative adverse event probability (Equation 25.8) is indicated by black dots in Figure 25.1. We find that the estimated curve is reasonably close to the true probability. Also shown is the curve (dark grey) which was estimated based on Equation 25.9, and which would be the proper estimator *in the absence of competing risks*. The curve is seen to approach 100%, because it ignores the competing risks structure as explained earlier. This is a clear case of overestimation, because the true cumulative adverse event probability is bounded from above by two-thirds.

Figure 25.2 recalculates these estimates in the presence of censoring, where we have simulated uniform censoring times on the time interval [0, 20], resulting in 15% censored observations. Also included are the uncensored estimates (dashed lines). In this example, censoring does not substantially disturb the point estimates. In fact, assuming constant event-specific hazards, incidence rates are consistent estimators of the cumulative adverse event probabilities (Equation 25.8), both in the uncensored and in the censored case, but the variance is increased with censoring present (results not shown).

However, censoring does disturb validity of the incidence proportion (Equation 25.1). In the absence of censoring, this is identical to $\hat{\alpha}_1/(\hat{\alpha}_1 + \hat{\alpha}_2)$, in our simulated example 0.63. However, with censoring, they differ. In our example, the incidence proportion (Equation 25.1) then was 0.55, while relating the adverse event incidence rate to the all-events incidence rate gave 0.65. Recall that the true value is two-thirds, illustrating that the incidence

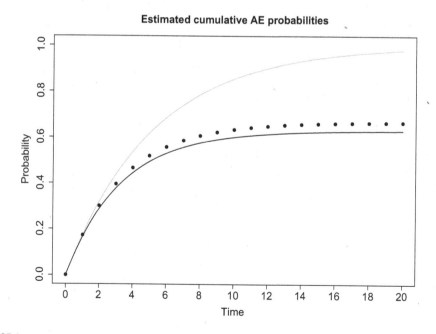

FIGURE 25.1

Estimating cumulative adverse event probabilities based on incidence rates in the absence of censoring. The upper curve (dark grey) overestimates, because it ignores competing events. The lower curve (black) is the estimated probability based on Equation 25.8. The black dots indicate the true probabilities (Equation 25.8).

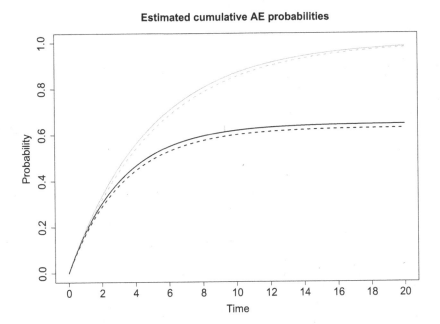

FIGURE 25.2
Estimating cumulative adverse event probabilities based on incidence rates in the presence of censoring. Solid lines: The upper curve (dark grey) overestimates, the lower curve (black) is the empirical counterpart of Equation 25.8. The dashed curves are those from Figure 25.1 without censoring.

proportion underestimates in the presence of censoring. Also recall that the black curves in Figures 25.1 and 25.2 approach $\hat{\alpha}_1/(\hat{\alpha}_1 + \hat{\alpha}_2)$ by construction as time increases.

25.2 Beyond Incidence Rates and the Constant Hazards Assumption: Nelson-Aalen, Aalen-Johansen, Kaplan–Meier

One aspect of the strange gap between efficacy analyses in studies with a time-to-event outcome and common analyses of adverse events for evaluating safety is that the former typically relies on non- and semi-parametric methods (Kaplan–Meier, log-rank, Cox) accounting for censoring, while the latter either does not account for censoring (incidence proportion) or makes a very restrictive parametric assumption (incidence rates). Of course, non- and semi-parametric methods of survival analysis also apply to studying adverse events. One complication is that they must account for competing risks, but this is, of course, also true for parametric methods as seen earlier. Section 25.2.1 explains the common non-parametric methods to quantify the risk of an adverse event, and a practical data analysis is in Section 25.2.2.

25.2.1 Non-Parametric Estimation

The most obvious use of non-parametric methods in the present context is to use the Kaplan–Meier estimator for estimation of the survival function $P(T > t)$ from Equation 25.7. As in Section 25.1, T is neither time to death nor time to adverse event, but it is the time to a composite outcome. That is, T is the time to adverse event ($\varepsilon = 1$) or competing event ($\varepsilon = 2$), whatever occurs first. In the simple setting of constant event-specific hazards, we found that

$$P(\varepsilon = 1) = \frac{\alpha_1}{\alpha_1 + \alpha_2} \quad \text{and} \quad P(\varepsilon = 2) = \frac{\alpha_2}{\alpha_1 + \alpha_2},$$

which also means that every patient experiences the composite event at some point in time (although possibly after closure of the study) because

$$P(\varepsilon = 1) + P(\varepsilon = 2) = 1.$$

This is important because $1 -$ the Kaplan–Meier estimator then estimates the probability of experiencing the composite event, which tends to one with infinite follow-up. As we will see later, this is the reason why the Kaplan–Meier method must not be used when there are additional competing events not being included in the composite event. In such a situation, the estimated probability of experiencing the composite event must not tend to one with ever longer follow-up because there also is a positive probability to experience the additional competing events.

We will need some extra notation:

$\Delta N(t)$ number of observed composite events at time t

$Y(t)$ number of patients at risk just prior time t

$\Delta N_1(t)$ number of observed adverse events at time t

$\Delta N_2(t)$ number of observed competing events at time t

Note that

$$\Delta N_1(t) + \Delta N_2(t) = \Delta N(t).$$

The usual Kaplan–Meier estimator for estimating $P(T > t)$ then is

$$\hat{P}(T > t) = \prod_{u \leq t} \left(1 - \frac{\Delta N(u)}{Y(u)}\right), \tag{25.11}$$

where the product is taken over all unique event times u in $(0, t]$.

In Equation 25.3, we have introduced the cumulative event probability $P(T \leq t, \varepsilon = 1)$. Analogously, $P(T \leq t, \varepsilon = 2)$ is the cumulative event probability of the competing event. Obviously, the following "balance equation" holds true, which we will use in the following to derive an estimator of the cumulative event probabilities:

$$P(T \leq t, \varepsilon = 1) + P(T \leq t, \varepsilon = 2) + P(T > t) = 1. \tag{25.12}$$

Checking the increments, i.e., the step sizes of the Kaplan–Meier estimator,

$$\prod_{s < u} \left(1 - \frac{\Delta N(s)}{Y(s)}\right) - \prod_{s \leq u} \left(1 - \frac{\Delta N(s)}{Y(s)}\right),$$

we find

$$1 - \hat{P}(T > t) = \sum_{u \leq t} \hat{P}(T \geq u) \frac{\Delta N(u)}{Y(u)}, \tag{25.13}$$

where now the sum in the above display is taken over all unique event times u in $(0, t]$ and $\hat{P}(T \geq u)$ is the value of the Kaplan–Meier estimator just prior to the jump at time u. Note that the right-hand side of Equation 25.13 has the interpretation of summing up over all empirical probabilities to have a composite event at time u, $u \in (0, t]$. Recalling balance Equation 25.12, we now decompose the right-hand side of Equation 25.13,

$$\sum_{u \leq t} \hat{P}(T \geq u) \frac{\Delta N(u)}{Y(u)} = \sum_{u \leq t} \hat{P}(T \geq u) \frac{\Delta N_1(u)}{Y(u)} + \sum_{u \leq t} \hat{P}(T \geq u) \frac{\Delta N_2(u)}{Y(u)}$$

$$= \hat{P}(T \leq t, \varepsilon = 1) + \hat{P}(T \leq t, \varepsilon = 2), \tag{25.14}$$

where the terms on the right-hand side of the previous display are the important Aalen-Johansen estimators [14]. Practical implementation of the Aalen-Johansen estimators is described in Beyersmann et al. [15], together with variance estimation [16] and construction of confidence intervals.

Note that a still common mistake [1] is to use a Kaplan–Meier-type method to approximate, e.g., the cumulative adverse event probability:

$$\text{Don't use}: \quad 1 - \prod_{u \leq t} \left(1 - \frac{\Delta N_1(u)}{Y(u)} \right)$$

$$= \sum_{u \leq t} \left(\prod_{v < u} \left(1 - \frac{\Delta N_1(v)}{Y(v)} \right) \right) \frac{\Delta N_1(u)}{Y(u)}, \tag{25.15}$$

which is larger than the proper Aalen-Johansen estimator, because

$$\prod_{v < u} \left(1 - \frac{\Delta N_1(v)}{Y(v)} \right) \geq \prod_{v < u} \left(1 - \frac{\Delta N(v)}{Y(v)} \right).$$

This Kaplan–Meier-type method to approximate the cumulative adverse event probability must therefore not be used. As outlined above, 1 – the Kaplan–Meier estimator will tend to one with infinite follow-up. Doing likewise for the competing cumulative event probability will result in violations of the natural balance in Equation 25.12. The reason is that Kaplan–Meier estimation is only suitable for probabilities of all-encompassing composite endpoints such as PFS or OS, which every patient will experience at some point in time (although potentially after study closure). This does not hold for adverse events which may be precluded by competing events as, e.g., death.

In all the formulas of the present Section so far, the terms $\Delta N(t)/Y(t)$, $\Delta N_1(t)/Y(t)$, and $\Delta N_2(t)/Y(t)$ have been key ingredients. In fact,

$$\hat{A}_j(t) = \sum_{u \leq t} \Delta N_j(u)/Y(u), \quad j = 1, 2, \tag{25.16}$$

is the fundamental Nelson-Aalen estimator of the cumulative event-specific hazard

$$A_j(t) = \int_{(0,t)} \alpha_j(u)du, \quad j = 1, 2,$$ (25.17)

where the event-specific hazards are as in Equation 25.6 and may now vary freely in time. Again, practical implementation is described in Beyersmann et al. [15], together with variance estimation and construction of confidence intervals. Because the Nelson-Aalen estimators are nonparametric, we now estimate cumulative hazards rather than the hazards $\alpha_j(t)$ themselves. This is analogous to calculating empirical cumulative distribution functions rather than estimating densities. One important use of the Nelson-Aalen estimators is to judge goodness-of-fit. For checking the constant hazards assumption underlying the incidence rates of Section 25.1, one may, e.g., plot $\hat{A}_j(t)$ against $\hat{\alpha}_j \cdot t$.

The Nelson-Aalen estimators appear to be simple sums (they are not really "simple," because the times u in Equation 25.16 are random), but they are more than worth a second look. One general trick to be learned from the Nelson-Aalen estimators is how to *code hazard analyses* such as the Cox regression model of Section 25.3. Calculating

$$\hat{A}_1(t) = \sum_{u \leq t} \Delta N_1(u)/Y(u)$$

only counts observed type 1 events as events in the numerator, while observed type 2 events and censorings are handled alike. The latter are kept in the denominator until occurrence of a type 2 event or occurrence of censoring. From a coding perspective, this entails that both observed type 2 events and censorings may be coded as "censored" *for calculating $\hat{A}_1(t)$.* However, roles reverse calculating

$$\hat{A}_2(t) = \sum_{u \leq t} \Delta N_2(u)/Y(u),$$

which now only counts observed type 2 events as events in the numerator, while observed type 1 events and censorings are handled alike. It is important to note that only censorings are coded as "censored" simultaneously in both calculations. We reiterate that, e.g., turning $\hat{A}_1(t)$ into the Kaplan–Meier-type estimator (Equation 25.15) is not a meaningful procedure, although both calculations code observed type 2 events as censorings. The reason is that probabilities also depend on $A_2(t)$, estimated by the Nelson-Aalen estimator $\hat{A}_2(t)$ of the previous display.

Another way of checking that the Aalen-Johansen estimators should be used for estimating the cumulative event probabilities is to consider the simpler case of no censoring as in Section 25.1.1. There, we had found that the incidence proportion (Equation 25.1) is the proper estimator of $P(\varepsilon = 1)$, provided that each patient is followed-up until occurrence of a type 1 or of a type 2 event. In fact, and still in the absence of censoring, the Aalen-Johansen estimator for adverse events, evaluated at time t, equals

$$\frac{\text{Number of adverse events in } [0, t]}{n},$$

which equals the incidence proportion when evaluated at the largest observed event time.

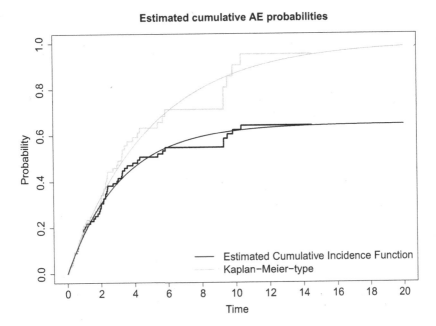

FIGURE 25.3

Non-parametric estimation of cumulative adverse event probabilities in the presence of censoring. Solid lines: The lower step function (black) is the Aalen-Johansen estimator, the upper step function (dark grey) is the Kaplan–Meier-type estimator from Equation 25.15. Also included are the parametric counterparts (smooth curves) from Figure 25.2.

25.2.2 Data Example

We revisit the censored data from Section 25.1.3. Figure 25.3 shows the Aalen-Johansen estimator of the cumulative adverse event probability together with the parametric estimator based on constant incidence rates from Figure 25.2 (black lines). The estimates are reasonably close, but step sizes of the non-parametric estimator become larger for later times, reflecting that for later times there are less data. Also demonstrated is the overestimation produced by using one minus the Kaplan–Meier-type procedure from Equation 25.15 (gray lines), again non-parametrically verifying what we have seen earlier from the parametric estimator in Figure 25.2.

Figure 25.4 illustrates a simple goodness-of-fit procedure, plotting the Nelson-Aalen estimator $\hat{A}_1(t)$ (Equation 25.16) against the cumulative incidence rate $\hat{\alpha}_1 \cdot t$. If the constant hazard model holds, the line in the plot should approximately follow the diagonal. This is, in fact, what we see in the plot, but the plot also indicates the increased variation for later time points in the form of a more erratic behavior.

25.3 Comparison of Treatment Groups

So far, we have discussed quantifying the risk of an adverse event within one treatment group. Typically, researchers will also need to compare treatment groups with respect to an adverse event outcome. It is not uncommon to use risk difference, relative risk,

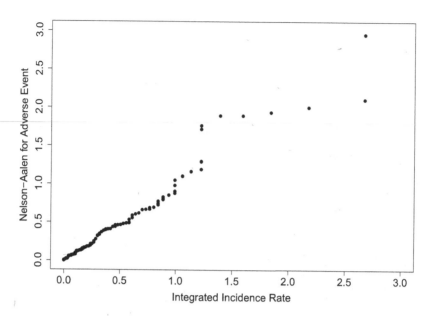

FIGURE 25.4

Plot of $\hat{A}_1(t)$ against $\hat{\alpha}_1 \cdot t$ for observed adverse event times.

or odds ratio based on incidence proportions [17]. However, such comparisons will be compromised in the presence of censoring. We have seen that the incidence proportion, in general, underestimates the "absolute risk of an anytime adverse event," and, e.g., a relative risk will divide something too small by something too small, the outcome of which is uncertain. Considering the ratio of incidence rates would be a natural alternative, and, in fact, this ratio would be an estimator of the event-specific hazard ratio assuming constant event-specific hazards. Again, this assumption is often criticized as being too restrictive. It may be checked using the Nelson-Aalen estimators of Section 25.2.1, and if in doubt, one natural option would be the semi-parametric Cox model and the log-rank test, which we outline below.

25.3.1 Cox Model, Log-Rank Test

Recall that we have two event-specific hazards. Hence, we consider two event-specific, semi-parametric Cox proportional hazards models,

$$\alpha_j(t\,|\,Z) = \alpha_{j;0}(t)\exp(\beta_j' Z), \quad j = 1, 2, \tag{25.18}$$

where $\alpha_{j;0}(t)$ is an unspecified event-specific baseline hazard, β_j is the vector of event-specific regression coefficients and Z a vector of baseline covariates including treatment group. If the only relevant covariate is treatment group, $Z \in \{0, 1\}$, then $\exp(\beta_1)$ is the hazard ratio for the outcome "adverse event," and $\exp(\beta_2)$ is the hazard ratio for the competing outcome. In this situation, $Z \in \{0, 1\}$, one also finds that the score test is the common log-rank test to compare the type j event-specific hazard between groups.

Recalling the coding trick discussed toward the end of Section 25.2.1, any Cox regression software can be used to compare the event-specific hazards for adverse events between groups, $j = 1$, technically censoring the time to adverse events by observed competing events. This analysis should be analogously repeated for the competing outcome, $j = 2$, now technically censoring the time to the competing composite by observed adverse events. We will illustrate this in Section 25.3.2 and refer to Beyersmann et al. [15] for a practical in-depth discussion.

Although straightforward from a software perspective, the *interpretation* of two such Cox analyses may not be without subtleties in terms of probability statements. This is easily seen under a constant hazard assumption. Suppressing treatment group membership in the notation, recall that

$$\frac{\alpha_1}{\alpha_1 + \alpha_2}$$

then is the "absolute risk of an anytime adverse event" $P(\varepsilon = 1)$. Now, assume that the experimental treatment has no effect on the hazard α_1 of an adverse event and a reducing effect on the competing hazard α_2. If the competing event simply is, say, "death without prior adverse event," then this constellation of effects certainly is desirable. However, looking at the "absolute risk of an anytime adverse event," which also is the plateau or upper limit of the cumulative adverse event probability, this will be *increased*, because the numerator remains unchanged, while the denominator is decreased. There is nothing wrong with this finding: If patients remain event-free for a prolonged time, because the "all-events hazard" $\alpha_1 + \alpha_2$ is decreased, and if patients are exposed to an unchanged hazard (or "intensity" or "momentary force") of an adverse event, one will eventually see more adverse events in this group. Grambauer et al. [18] give an in-depth discussion of such phenomena in an analysis of the occurrence of bloodstream infections in patients having received stem-cell transplantation.

Such subtleties have inspired regression modeling approaches to "directly" compare the cumulative event probabilities over time [19,20], but a discussion goes beyond the scope of the present chapter, and we refer readers to Chapter 21 on competing risks; see also References [15,21,22]. Another possibility would be to consider the Aalen-Johansen estimators of the cumulative event probabilities within groups and to quantify the uncertainty of the estimators or of their difference using time-simultaneous confidence bands rather than point-wise confidence intervals. However, this approach will typically require resampling procedures [23,24].

25.3.2 Data Example

We have chosen to illustrate the use of Cox regression in a somewhat idealized situation depicted in Figure 25.5. The scenario is idealized in that a novel treatment is supposed to reduce the hazard of the competing event "death without prior adverse event" by a hazard ratio of 0.6 (a pronounced effect), while leaving the hazard of an adverse event unchanged. The aim of considering such an idealized situation, where we have also chosen to simulate a larger sample size, is to demonstrate that Cox regression works in the presence of competing risks and can recover the underlying event-specific hazard ratios.

To begin, consider the multistate representation [15] of the competing risks situation at hand in Figure 25.5. Patients are considered to be in an initial state 0 upon time origin. Occurrence of an adverse event is modeled by a transition into state 1; competing events are transitions into state 2. The event-specific hazards are interpreted as forces of transition, moving along the arrows in Figure 25.5. Following Beyersmann et al. [25], the width of the arrows is chosen proportional to the value of the event-specific hazards. Taking the

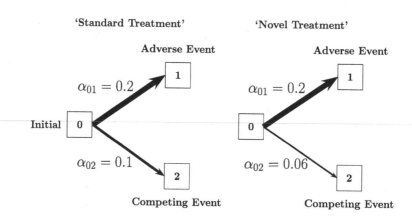

FIGURE 25.5

Multi-state graphic of the event-specific hazards for "standard treatment" (left) and for "novel treatment" (right).

widths of the arrows out of state 0 together illustrates the magnitude of the all-events hazard $\alpha_1 + \alpha_2$. Comparing the right part with the left part of Figure 25.5 illustrates that this all-events hazard is reduced by the novel treatment, prolonging the waiting time T until any first event. Figure 25.5 also illustrates that the reduction is via the competing hazard, leaving the adverse event hazard unchanged. Comparing the relative magnitude of the event-specific hazards within the left part and within the right part also visually conveys the impression that, *eventually*, there will be more adverse events in the "novel treatment" group because there will be fewer competing events.

We have now simulated 500 individuals per treatment group. This would be a larger number of patients for a trial in oncology, but the rationale was to demonstrate that Cox regression "works," and we wanted to ensure that there are enough observed events for this exercise. Censoring was simulated as before and followed the same uniform distribution in both groups. Using standard Cox software (coxph in R) and coding both observed competing events and censored event times as "censored," we found an event-specific hazard ratio of "novel treatment" versus "control treatment" of $_{0.93}1.09_{1.28}$ where the indices give the 95% confidence interval, following a suggestion in Louis and Zeger [26]. The competing analysis, now coding both observed adverse events and censored event times as "censored," found an estimated event-specific hazard ratio of $_{0.46}0.60_{0.78}$. Finally, Figure 25.6 displays the Aalen-Johansen estimators of the cumulative adverse event probabilities in both treatment groups together with the true probability curves. The plateaus of the curves are as discussed above.

The estimated hazard ratios and their confidence intervals reasonably capture the true underlying quantities, at least in this rather large sample, and one even finds a significant "effect" of the "novel treatment" on the competing event "death without prior adverse event." However, this simple example also illustrates some open issues: To begin, the estimated hazard ratios do not reflect the magnitude of the baseline event-specific hazards depicted in Figure 25.5. The relevance of, say, a protective hazard ratio of 0.6 also depends on the magnitude of the baseline event-specific hazard and its magnitude in comparison to the competing event-specific hazard. Secondly, the situation at hand is idealized in the sense that there is a protective effect on the hazard of "death without prior adverse event" while not changing the adverse event hazard. The analysis, however, does not consider "death after adverse event." The discussion in Section 25.5 returns to this issue in the context of the recent debate on estimands. Finally, the "ideal analysis" of our idealized

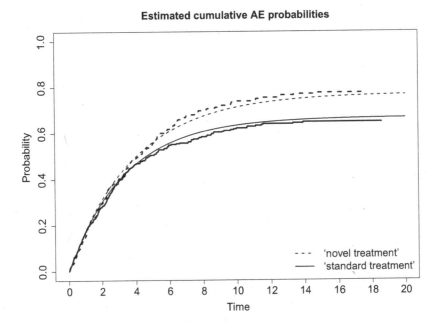

FIGURE 25.6
Aalen-Johansen estimators (step functions) of the cumulative adverse event probabilities within the treatment groups. Also included are the true probability curves (smooth curves).

situation might be investigating superiority of the novel treatment and with regard to the hazard of "death without prior adverse event" while establishing non-inferiority with regard to the adverse event hazard. Arguably, such an analysis would require even larger and presumably unrealistic sample sizes.

25.4 Recurrent Adverse Events

So far, we have considered studying the occurrence of a *first* adverse event of a specific type. However, patients can experience an interesting adverse event more than once, and sometimes the repeated occurrence may be important to analyze in detail. In this section we will briefly discuss how the presented methods can be translated to the analysis of these so-called recurrent adverse events. We have seen in the previous sections that the analysis of adverse events requires survival methods and that survival methods are based on hazards. If we are willing to assume that the hazard is constant in time, the incidence rate may account for recurrent adverse events, considering

$$\frac{\text{Number of observed adverse events}}{\text{Cumulative patient-time at risk}},$$

where the numerator now also counts recurrent occurrence of the adverse event, and a patient may continue to contribute to the person-time at risk in the denominator after occurrence of an adverse event, if the adverse event is recurrent. The underlying assumption of a constant hazard is now an even stronger assumption, because it would entail that the hazard does not depend on whether or not a previous adverse event has occurred in the past. One approach to relax this assumption would be a Cox model similar to Equation 25.18, say,

$$\alpha_1(t \mid Z(t)) = \alpha_{1;0}(t) \exp(\beta_1' Z(t)),$$

where $\alpha_{1;0}(t)$ now is an unspecified baseline hazard of a recurrent adverse event and the time-dependent covariate or covariate vector $Z(t)$ would model the impact of past adverse events. For instance, $Z(t)$ could simply be the number of previous adverse events, such that each additional adverse event would change the hazard by a factor $\exp(\beta_1)$. The assumption that the occurrence of each previous adverse event changes the hazard of a further adverse event by a constant factor may be considered as not realistic. A model relaxing this assumption could include a time-dependent covariate vector $Z(t)$ where initially all entries are set to zero. The first entry then changes to one with the first adverse event, the second entry changes to one with the second adverse event and so forth. The first adverse event would change the hazard by a factor $\exp(\beta_{11})$, the second adverse event would change the hazard by a factor $\exp(\beta_{12})$, et cetera. In practice, the number of entries of $Z(t)$ will need to be determined beforehand and chosen by practical considerations, limiting the dimension of $Z(t)$ to, e.g., five, such that a sixth or seventh adverse event would not further change the hazard in the model. Another option would be to take $Z(t)$ as a scalar with initial value 0 and value 1 only if an adverse event has occurred during a predefined time span prior to time t. Such Cox models for recurrent events are often called Andersen-Gill models, named after the authors of the first complete mathematical treatment of the Cox model from a counting process perspective [27], and extensions to stratifying the baseline hazard by number of adverse event occurrence are also available [28,29].

Clearly, the modeling options for recurrent adverse events are numerous and go beyond the scope of the present chapter. The notion of hazards will, however, remain crucial, and with it will the presence of competing risks remain crucial. Readers are referred to the textbook literature [30].

25.5 Conclusion

The key message of this chapter is that adverse event data are subject to varying follow-up times and censoring and thus require survival analysis methods. However, standard Kaplan–Meier-type procedures are not appropriate, because an adverse event may be precluded by death or some other competing event. A proper analysis may be achieved using incidence rates, provided that the event-specific hazards are constant over time and that not only the incidence rate for adverse events is calculated, but also its counterpart for the competing outcome. If the constant hazards assumption is in doubt, non- and semi-parametric methods are available, and, e.g., the formula in Equation 25.14 describes how the nonparametric analysis of the event-specific hazards can be translated onto the probability scale.

In closing, we wish to comment on two additional aspects that, in our experience, are increasingly discussed in the context of analyzing adverse events. One aspect is whether censoring by an observed disease event as, e.g., progression in oncology is "informative" or not. The second aspect is a connection to the current debate on estimands [9,31].

Starting with the notion of censoring, we reiterate that censoring is omnipresent in time-to-event studies and requires use of specialized techniques—survival analysis or, more generally speaking, event history analysis [32]—irrespective of whether the analysis is one of efficacy or one of safety. Now, in cancer trials it is common that systematic follow-up of adverse events is stopped when patients progress and/or start a subsequent therapy. Censoring the adverse event process by the time of progression then is usually deemed "informative," because one will regularly assume that the hazard of an adverse event is changed by the progression event. In other words, those "uncensored" (alive and neither adverse event nor progression yet) and those "censored" by progression (alive and progressed, no adverse event before progression) are considered to have different hazards for an adverse event.

On the other hand, in Section 25.2.1, we have seen that we may treat observed progression events prior to an adverse event as censoring events, when the aim is to analyze the adverse event hazard in a *time-to-first-event* analysis. In other words, censoring by progression events yields a valid hazard analysis of occurrence of an adverse event *before* progression, but if the hazard of an adverse event *after* progression changes, then the latter hazard can not be the target of estimation. At the end of the day, this is just common sense: if a progression diagnosis stops recording of adverse events, inference for adverse events after progression is not possible.

Assuming that we are content with analyzing the hazard of an adverse event before progression, censoring by diagnosed progression is "independent" in the sense that it allows for a valid analysis of the hazard of an adverse event, as we have seen above. However, we have also seen that such an analysis is incomplete in the sense that it additionally requires an analysis of the hazard of the competing event, including progression and also death, when the aim is inference for the cumulative adverse event probability. In this sense, censoring by diagnosed progression is "informative." We refer to Andersen et al. [33] for an in-depth discussion and to Beyersmann et al. [15] for a practical account.

The discussion on whether the time to adverse event should be censored by disease progression has a connection to the current debate on estimands. Based on the consideration that a proper statistical analysis should start with a translation of the trial objective into a precise definition of the treatment effect that is to be estimated, called the estimand, the new draft addendum R1 to the ICH E9 Guideline on statistical principles in clinical trials entitled "Estimands and Sensitivity Analyses in Clinical Trials" was developed [31].

One aspect of the debate centers around the question of how to handle post-randomization events in randomized clinical trials such as treatment discontinuation, e.g., because of progression. The framework was mainly discussed in the context of efficacy analyses until now. Five classes of estimands are currently considered, which, in the context of the analysis of adverse events, can be described as follows (see also Unkel et al. [34]).

- The "treatment policy estimand" is interested in the comparison of treatment groups with respect to adverse event occurrence until death or end of follow-up, irrespective of any intercurrent events. So, only death has to be treated as a competing event, and the collection of adverse event data after progression or treatment discontinuation is required.

- The "while on treatment estimand" includes the adverse events until discontinuation of treatment and requires the collection of adverse event data up to this event. Now, treatment discontinuation and death have to be treated as competing events in the analysis.

- The "composite estimand" combines the occurrence of the adverse event with the intercurrent events progression or treatment discontinuation and death without prior adverse event. The endpoint is then time to the first of these events, and no competing events are present. In the notation of Section 25.1, this estimand only considers time T, and the information contained in ε is not used.
- The "hypothetical estimand" targets an effect that would occur in the hypothetical scenario, e.g., if no patient experienced the intercurrent event progression. The practical relevance of this scenario is a matter of debate.
- The "principal stratum estimand" defines subsets of patients in whom disease progression or treatment discontinuation occurs either under one of the treatments or under both. Since in general these groups cannot be identified in advance, causal inference methods with additional and usually untestable assumptions are required.

Different stakeholders may be interested in different aspects of treatment comparison and, therefore, in different estimands. However, whatever the estimand, this chapter has demonstrated that statistical techniques from survival analysis are required to investigate the risk of an adverse event in time-to-event studies. Except for special cases such as the composite estimand, the statistical analysis must account for competing events.

References

1. Schumacher M, Ohneberg K, Beyersmann J. Competing risk bias was common in a prominent medical journal. *J Clin Epidemiol*. 2016;80:135–6.
2. Allignol A, Beyersmann J, Schmoor C. Statistical issues in the analysis of adverse events in time-to-event data. *Pharm Stat*. 2016;15:297–305.
3. Chuang-Stein C. Safety analysis in controlled clinical trials. *Drug Inf J*. 1998;32(1_suppl):1363S–72S.
4. Chuang-Stein C, Le V, Chen W. Recent advancements in the analysis and presentation of safety data. *Drug Inf J*. 2001;35(2):377–97.
5. Gait JE, Smith S, Brown SL. Evaluation of safety data from controlled clinical trials: The clinical principles explained. *Drug Inf J*. 2000;34(1):273–87.
6. Ioannidis JA, Evans SW, Gtzsche PC et al. Better reporting of harms in randomized trials: An extension of the consort statement. *Ann Intern Med*. 2004;141(10):781–8.
7. Lineberry N, Berlin JA, Mansi B et al. Recommendations to improve adverse event reporting in clinical trial publications: A joint pharmaceutical industry/journal editor perspective. *BMJ*. 2016;355.
8. Siddiqui O. Statistical methods to analyze adverse events data of randomized clinical trials. *J Biopharm Stat*. 2009;19(5):889–899.
9. Akacha M, Bretz F, Ruberg S. Estimands in clinical trials—broadening the perspective. *Stat Med*. 2017;36(1):5–19.
10. Beyersmann J, Schrade C. Florence Nightingale, William Farr and competing risks. *J R Stat Soc Ser A Stat Soc*. 2017;180(1):285–93.
11. Kalbfleisch J, Prentice R. *The Statistical Analysis of Failure Time Data*, Second Edition. Hoboken: Wiley, 2002.
12. R Core Team. *R: A Language and Environment for Statistical Computing*. Vienna, Austria: R Foundation for Statistical Computing, 2016.

13. Beyersmann J, Latouche A, Buchholz A, Schumacher M. Simulating competing risks data in survival analysis. *Stat Med.* 2009;28:956–71.
14. Aalen O, Johansen S. An empirical transition matrix for non-homogeneous Markov chains based on censored observations. *Scand J Stat.* 1978;5:141–50.
15. Beyersmann J, Allignol A, Schumacher M, *Competing Risks and Multistate Models with R.* New York: Springer, 2012.
16. Allignol A, Schumacher M, Beyersmann J. A note on variance estimation of the Aalen-Johansen estimator of the cumulative incidence function in competing risks, with a view towards left-truncated data. *Biom J.* 2010;52:126–37.
17. Amit O, Heiberger RM, Lane PW. Graphical approaches to the analysis of safety data from clinical trials. *Pharm Stat.* 2008;7(1):20–35.
18. Grambauer N, Schumacher M, Dettenkofer M, Beyersmann J. Incidence densities in a competing events analysis. *Am J Epidemiol.* 2010;172(9):1077–84.
19. Eriksson F, Li J, Scheike T, Zhang M-J. The proportional odds cumulative incidence model for competing risks. *Biometrics.* 2015;71:687–95.
20. Fine J, Gray R. A proportional hazards model for the subdistribution of a competing risk. *J Am Stat Assoc.* 1999;94(446):496–509.
21. Beyersmann J, Scheike T. Chapter competing risks regression models. In *Handbook of Survival Analysis*, Klein J et al. eds. Boca Raton, FL: Chapman & Hall/ CRC, 2014.
22. Schmoor C, Schumacher M, Finke J, Beyersmann J. Competing risks and multistate models. *Clin Cancer Res.* 2013;12:12–21.
23. Bluhmki T, Schmoor C, Dobler D, Pauly M, Finke J, Schumacher M, Beyersmann J. A wild bootstrap approach for the Aalen-Johansen estimator. *Biometrics.* 2018;74(3):977–85.
24. Lin D. Non-parametric inference for cumulative incidence functions in competing risks studies. *Stat Med.* 1997;16:901–10.
25. Beyersmann J, Gastmeier P, Schumacher M. Incidence in ICU populations: how to measure and report it? *Intensive Care Med.* 2014;40:871–6.
26. Louis TA, Zeger SL. Effective communication of standard errors and confidence intervals. *Biostatistics.* 2008;10(1):1–2.
27. Andersen P, Gill R. Cox's regression model for counting processes: A large sample study. *Ann Stat.* 1982;10:1100–20.
28. Andersen P, Borgan Ø. Counting process models for life history data: A review. *Scand J Stat.* 1985;12:97–140.
29. Prentice RL, Williams BJ, Peterson AV. On the regression analysis of multivariate failure time data. *Biometrika.* 1981;68(2):373–9.
30. Cook RJ, Lawless J. *The Statistical Analysis of Recurrent Events.* Springer Science & Business Media. 2007.
31. International Conference on Harmonisation. *Draft Guideline ICH E9 (R1) addendum on estimands and sensitivity analysis in clinical trials to the guideline on statistical principles for clinical trials*, 2017.
32. Aalen O, Borgan Ø, Gjessing H. *Survival and Event History Analysis.* New York: Springer, 2008.
33. Andersen P, Borgan Ø, Gill R, Keiding N. *Statistical Models Based on Counting Processes.* New York: Springer, 1993.
34. Unkel S, Amiri M, Benda N et al. On estimands and the analysis of adverse events in the presence of varying follow-up times within the benefit assessment of therapies. *Pharmaceutical Statistics.* 2019;18:166–83. https://doi.org/10.1002/pst.1915

26

Analysis of Quality of Life Outcomes in Oncology Trials

Stephen Walters

CONTENTS

26.1 Introduction

Some studies using QOL outcomes have repeated assessments over time and are longitudinal in nature. In an Randomized Controlled Trial (RCT) and other longitudinal studies, there may be a baseline QOL assessment and several follow-up assessments over time. This chapter will describe how the QOL data from such studies can be summarized, tabulated, and graphically displayed. These repeated QOL measurements, on the same individual subject, are likely to be related or correlated. This means that the usual statistical methods for analyzing such data which assume independent outcomes may not be appropriate. This chapter will show how repeated QOL measures for each subject can be reduced to a single summary measure for statistical analysis and how standard statistical methods of analysis can then be used. Finally, the chapter will describe a more complex modeling approach, based on an extension of the linear regression model which allows for the fact that successive QOL assessments by a particular patient are likely to be correlated.

26.2 Three Broad Approaches to Analyzing Repeated QOL Assessments

With one QOL observation on each subject or experimental unit, we are confined to modeling the population average QOL, called the marginal mean response; there is no other choice. However, with repeated QOL measurements, there are several different approaches that can be adopted. Three broad approaches are [1]:

1. Time by time analysis
2. Response feature analysis—the use of summary measures
3. Modeling of longitudinal data

An important initial step, prior to analyzing the repeated QOL assessments, is to tabulate the data and/or graphically display it [2]. This will give us an idea of how the QOL outcomes change over time.

26.3 Example Dataset: The AIM-HIGH Trial

To illustrate some of the methods, we will use data from the AIM-HIGH RCT [3,4], which is available on the companion website, https://www.crcpress.com//9781138083776 together with STATA code. The primary objective of the AIM-HIGH trial was to determine the effects of low-dose, extended-duration treatment with interferon alfa-2a as an adjuvant therapy compared to no further treatment (observation) on overall survival (OS) and recurrence-free survival (RFS) in patients with radically resected stage IIB and stage III cutaneous malignant melanoma at high risk of recurrence. The trial design was a pragmatic, unblinded, individually randomized controlled trial with a minimum of 24 months of follow-up.

Six hundred and seventy-four patients with completely resected high-risk melanoma were enrolled onto the study; 338 patients were allocated to receive interferon alfa-2a (IFN group) and 336 were allocated to observation alone (OBS group; no further treatment). Patients within the study were randomized to either interferon alpha-2a at 3 mega units three times per week for 2 years or until recurrence, or observation alone (no further treatment). QOL data in the form of the European Organisation for Research and Treatment of Cancer (EORTC) QLQ-C30 were originally intended to be collected at baseline and 3, 6, 12, 24, 36, 48, and 60 months for a subgroup of patients. However, QOL data were actually collected at a variety of time points, post-randomization from 3 days to 77 months. Over the first two years of follow-up, the time points were classified as baseline, 6 months (QOL assessment between baseline and 6 months post-randomisation), 12 months (QOL assessment between

6 and 12 months), 18 months (QOL assessment between 12 and 18 months), and 24 months (QOL assessment between 18 and 24 months).

The EORTC QLQ-C30 is a 30-item, cancer-specific instrument designed to assess the health-related quality of life (QOL) of cancer patients participating in international clinical trials [5]. The QLQ-C30 version 1.0 used in the AIM-HIGH trial incorporated five functional scales: physical (PF), role (RF), cognitive (CF), emotional (EF), and social (SF); three symptom scales: fatigue (FA), pain (PA), and nausea and vomiting (NV); a global health status/QOL scale (QL); and six single items assessing additional symptoms commonly reported by cancer patients: dyspnea (DY), loss of appetite (AP), insomnia (SL), constipation (CO), diarrhea (DI), and a single item on the perceived financial impact of the disease (FI). All of the scales and single-item measures range in score from 0 to 100. A high scale score represents a higher response level. Thus a high score for a functional scale represents a high/healthy level of functioning; a high score for the global health status/QOL represents a high QOL; but a high score for a symptom scale/item represents a high level of symptomatology/problems [6]. In this chapter, we shall use the global health/QOL scale of the EORTC QLQ-C30.

Figure 26.1 shows that 444 patients (out of 674), or 66%, had a valid baseline QOL assessment; 230/338 (68%) in the IFN group and 214/336 (64%) in the OBS group. There was no evidence of a difference in survival between the interferon and observation group

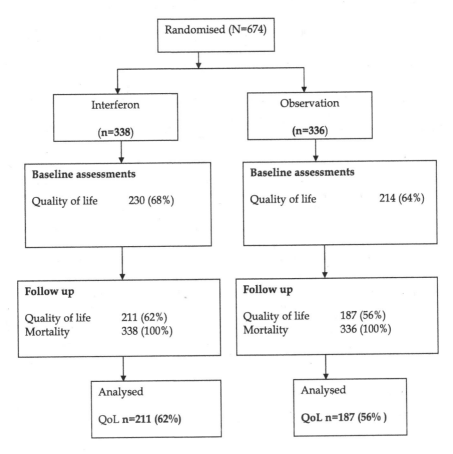

FIGURE 26.1
Flow of patients through the AIM High Randomized Controlled Trial. (Adapted from Dixon S et al. *Br J Cancer.* 2006;94(4):492–8.)

of patients who had quality of life assessments, and the median survival was over 3.8 years in both groups combined.

26.4 Summarizing Repeated QOL Assessments

An important initial step, prior to analyzing the repeated QOL assessments, is to tabulate the data and/or graphically display the data. This will give us an idea of how the QOL outcomes change over time. In the AIM-HIGH trial, the patients' QOL was assessed at several time points post-randomization; using the EORTC QLQ-C30 patient reported outcome measure [4] and summarized as the baseline measurement (0) and 6, 12, 18, and 24 months post-randomization. Table 26.1 shows one way of presenting such data. In Table 26.1, differences between the interferon and observation groups in the mean EORTC QLQ-C30 global health status/QOL scale scores are not tested at each time point. The results of the hypothesis test and confidence intervals are only presented for the two summary measures, in the last two rows of the table. More on how these summary measures are calculated is described later. The sample size at each of the follow-up time points varies, and therefore it is important to report the sample size for each row of the data.

Table 26.2 is in the same format as Table 26.1, but this time we report the results for only those patients who completed all five QOL assessments. This makes it easier to see how the mean QOL scores vary over time. We can plot such data as a line graph (Figure 26.2), with a separate line for each group. Figure 26.2 clearly shows how the QOL outcome varies over the 2-year period. The observation group appears to have slightly better QOL scores at all five time points compared to the interferon group. Because the same numbers of subjects are included at each time point, it is legitimate to join the mean QOL scores together with a solid line. However, if the data in Table 26.1 were presented in a similar figure, it would be misleading to join the observed means at each time point by solid lines, since we are not measuring the same people at each time point. Figure 26.3 shows how the mean EORTC QLQ-C30 global health status/QOL scale scores vary over time using all available QOL data. The number of valid QOL observations per group at each assessment time point are included just below the horizontal axis. This makes clear how the number of valid QOL observations at the various follow-up assessment points declines over time and on how many subjects the calculation of each mean summary measure is based on. Again, the mean profiles and pattern are similar to the complete case analysis with the observation group having slightly better QOL scores at all five time points compared to the interferon group.

The mean EORTC QLQ-C30 global health status/QOL scores could be compared between the interferon and observation groups using a two-independent-samples t test. (Full details of the assumptions and how to carry out a two-independent-samples t test are described in more detail by Campbell et al. [7] and Walters [2].)

26.5 Time by Time Analysis

A series of two-independent-samples t tests (or the non-parametric equivalent) could be used to test for differences in QOL between the two groups at each time point. For example,

TABLE 26.1

Mean EORTC QLQ-C30 Global Health Status/QOL Scale Scores Over Time by Treatment Group with All Valid Patients at Each Time Point

EORTC QLQ-C30 Global Health Status/QOL Outcome[a] Time (Months)	Group Randomized to						Difference in Means[b]	95% CI		P-value[c]
	Interferon			Observation				Lower	Upper	
	N	Mean	SD	N	Mean	SD				
Baseline	230	71.7	20.7	214	73.5	19.2				
6 months	206	67.5	19.9	185	75.1	18.0				
12 months	90	67.7	19.8	92	75.2	18.7				
18 months	74	68.8	19.9	67	72.5	21.8				
24 months	66	69.7	20.4	60	76.7	17.9				
Mean follow-up score	210	67.5	19.1	187	74.6	16.7	−7.1	−10.7	−3.6	<0.001
AUC	39	135.4	33.5	37	149.0	31.6	−13.6	−28.5	1.3	0.072

Source: Data from Dixon S et al. *Br J Cancer.* 2006;94(4):492–8.

Abbreviations: CI, confidence interval; AUC, area under the curve.

[a] The EORTC QLQ-C30 global health status/QOL scale is scored on a 0 (poor) to 100 (high QOL) scale.

[b] A negative difference in means indicates the observation group has the better QOL.

[c] P-value from two-independent-samples *t* test.

TABLE 26.2

Mean EORTC QLQ-C30 Global Health Status/QOL Scale Scores Over Time by Treatment Group with Patients Who Completed all Five QOL Assessments

| EORTC QLQ-C30 Global Health Status/QOL Outcome[a] | Group Randomized to | | | | | | Difference in Means[b] | 95% CI | | P-value[c] |
| | Interferon | | | Observation | | | | | | |
Time (Months)	N	Mean	SD	N	Mean	SD		Lower	Upper	
Baseline	39	66.7	22.2	37	71.6	20.8				
6 months	39	67.7	21.7	37	75.0	17.8				
12 months	39	68.7	19.6	37	75.9	17.6				
18 months	39	66.5	18.5	37	73.6	22.1				
24 months	39	69.3	21.1	37	75.5	19.9				
Mean follow-up score	39	68.6	16.6	37	74.9	15.9	−6.2	−13.6	1.2	0.100
AUC	39	135.4	33.5	37	149.0	31.6	−13.6	−28.5	1.3	0.072

Source: Data from Dixon S et al. *Br J Cancer* 2006;94(4):492–8.
Abbreviations: CI, confidence interval; AUC, area under the curve.
[a] The EORTC QLQ-C30 Global health status/QOL scale is scored on a 0 (poor) to 100 (high QOL) scale.
[b] A negative difference in means indicates the observation group has the better QOL.
[c] P-value from two-independent-samples *t* test.

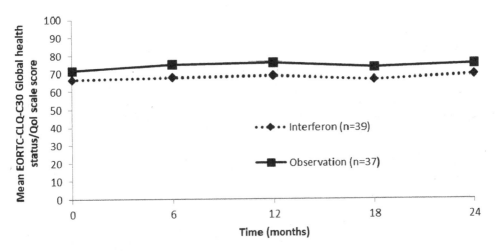

FIGURE 26.2
Profile of Mean EORTC QLQ-C30 Global health status/QoL scale scores over time by treatment group complete case analysis. (Data from Dixon S et al. *Br J Cancer*. 2006;94(4):492–8.)

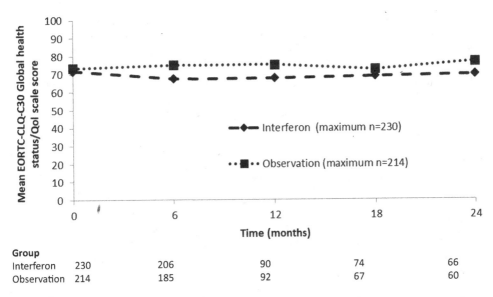

Group					
Interferon	230	206	90	74	66
Observation	214	185	92	67	60

FIGURE 26.3
Profile of mean EORTC QLQ-C30 global health status/QoL scale scores over time by treatment group all available data. (Data from Dixon S et al. *Br J Cancer*. 2006;94(4):492–8.)

**BOX 26.1   RECOMMENDATIONS WHEN PRESENTING
DATA AND RESULTS IN TABLES [8]**

The amount of information should be maximized for the minimum amount of ink.
   Numerical precision should be consistent throughout an article or presentation, as far as possible.
   Avoid spurious accuracy. Numbers should be rounded to two effective digits.

Quantitative data should be summarized using either the mean and standard deviation (for symmetrically distributed data) or the median and interquartile range or range (for skewed data). The number of observations on which these summary measures are based should be included.

Categorical data should be summarized as frequencies and percentages. As with quantitative data, the number of observations should be included.

Each table should have a title explaining what is being displayed, and columns and rows should be clearly labeled.

Gridlines in tables should be kept to a minimum.

Where variables have no natural ordering, rows and columns should be ordered by size.

BOX 26.2   GUIDELINES FOR CONSTRUCTING GRAPHS [8]

Each graph should have a title explaining what is being displayed.

Axes should be clearly labeled.

Gridlines should be kept to a minimum (and drawn in a faded shade).

With a small sample size (<20), plot the individual QOL scores over time.

For larger sample sizes (>20), summarize the data with the mean or median QOL score and plot these over time.

It is preferable to join the summary points by a dotted line if there are a different number of subjects at each time point.

The number of observations (at each time point) should be included.

---

in the AIM-High trial, we could compare mean EORTC QLQ-C30 global health status/QOL scale scores between the interferon and observation groups at 6, 12, 18, and 24 months using a series of four independent sample t tests. The procedure is straightforward but has a number of serious flaws and weaknesses [9]. The QOL measurements in a subject from one time point to the next are not independent, so interpretation of the results is difficult. The large number of hypothesis tests carried out implies that we are likely to obtain significant results purely by chance. We lose information about the within-subject changes in QOL over time. Consequently, it will not be described any further here.

26.6 Response Feature Analysis: The Use of Summary Measures

Here, the repeated QOL measures for each participant are transformed into a single number considered to capture some important aspect of the participant's response [10].

A simple and often effective strategy [11] is to

1. Reduce the repeated QOL values into one or two summaries.
2. Analyze each summary as a function of (p) covariates or explanatory variables, $x_1, x_2, \ldots, x_p$.

TABLE 26.3

Response Features Suggested in Matthews et al.

Type of Data	Property to be Compared between Groups	Summary Measure
Peaked	Overall value of response	Mean or area under the curve
Peaked	Value of most extreme response	Maximum (minimum)
Peaked	Delay in response	Time to maximum or minimum
Growth	Rate of change of response	Linear regression coefficient
Growth	Final level of response	Final value or (relative) difference between first and last
Growth	Delay in response	Time to reach a particular value

Source: Matthews JNS et al. *BMJ*. 1990;300:230–5.

Examples of summary measures include the area under the curve (AUC) or the overall mean of post-randomization measures. Other possible summary measures are listed in Matthews et al. [10] and are shown in Table 26.3. Having identified a suitable summary measure, a simple *t* test (or ANOVA) can be applied to assess between-group differences. If the data for each patient can effectively be summarized by a pre-treatment mean and a post-treatment mean, then the analysis of covariance (ANCOVA) is the preferred method of choice [12]. It is superior to both the analysis of post-treatment means or analysis of mean changes. Diggle et al. [11] suggested that provided the data are complete, then the method of derived variables or summary measures can give a simple and easily interpretable analysis with a strong focus on particular aspects of the mean response.

26.7 The Area Under the Curve (AUC)

The AUC is a useful way of summarizing the information from a series of measurements on one individual [10]. The AUC can also be used to summarize repeated QOL scores over time into a single measure of health for each patient.

26.8 Calculating the AUC

The area (see Figure 26.4) can be split into a series of shapes called trapeziums. The areas of the separate individual trapeziums are calculated and then summed for each patient. The mean AUC in each group can then be calculated.

Let Y_{ij} represent the QOL response variable observed at time t_{ij} for observation $j = 1, \ldots, n_i$ on subject $i = 1, \ldots, m$. The set of repeated QOL outcomes for subject i can be collected into a single row or row vector matrix of length n_i, i.e., $Y_i = (Y_{i1}, Y_{i2}, \ldots, Y_{in_i})$. The AUC is for the ith subject is calculated by

$$AUC_i = \frac{1}{2} \sum_{j=1}^{n_i} (t_{j+1} - t_j)(Y_j + Y_{j+1}). \tag{26.1}$$

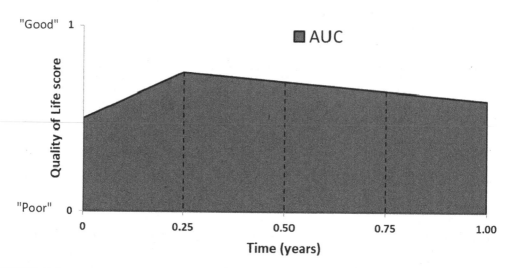

FIGURE 26.4
Summary measure of QOL: the AUC.

The units of AUC are the product of the units used for Y_{ij} and t_{ij} and may not be easy to understand since QOL outcomes have no natural units. So it may be useful to divide the AUC by the total time to get a weighted average level, of QOL, over the time period. We can calculate the AUC even when there are missing data, except when the first and final observations are missing.

In the AIM-HIGH trial, each patient's QOL was assessed five times—at baseline (0) and at 6, 12, 18, and 24 months—using the EORTC QLQ-C30 [4]. The AUCs calculated from the AIM-HIGH trial were based on 24 months or 2 years of follow-up. If the time t_{ij} for each QOL assessment is represented as a fraction of a year, then the AUCs represent the weighted average level of QOL over the 2-year period. An AUC of 200 corresponds to "good health" over the year; conversely, an AUC of 0 corresponds to "poor health" over the period. If we divide by the total time (of 2 years), then we get back to the 0 to 100 scale of the original EORTC QLQ-C30 measurement which may make interpretation of the results easier.

Consider a patient in the AIM-HIGH trial with EORTC QLQ-C30 global health status/QOL scale scores of 33.3, 41.7, 50.0, 50.0, and 41.7 at 0, 0.5, 1, 1.5, and 2 years. The AUC for this patient is calculated as

$$\text{AUC} = 0.5 \times \{[0.5 \times (33.3 + 41.7)] + [0.5 \times (41.7 + 50.0)] + [0.5 \times (50.0 + 50.0)] + [0.5 \times (50.0 + 41.7)]\} = 98.6.$$

The AUC is a useful way of summarizing the information from a series of measurements on one individual. Parametric confidence intervals for the mean difference in AUC between groups can also be calculated as, again, the AUCs are more likely to be a fairly good fit to the Normal distribution. Multiple linear-regression methods can be used to adjust AUCs for other covariates (e.g., age, sex, center).

Table 26.1 shows that the mean AUC was 135.4 (SD 33.5) and 149.0 (SD 31.6) in the interferon and observation groups, respectively, a difference in means of −13.6 (95% CI −28.5–1.3; $p < 0.072$), with a negative difference in means indicating the observation group has the better QOL.

26.9 Other Summary Measures

The figures and tables from the AIM-HIGH study suggest that the mean EORTC QLQ-C30 global health status/QOL scale scores at 6, 12, 18, and 24 months follow-up are fairly similar (the lines in the graph appear to be almost horizontal at these time points). Therefore, another sensible summary measure would be the mean post-randomization follow-up EORTC QLQ-C30 global health status/QOL scale score. For this summary measure, patients need only to have one valid follow-up EORTC QLQ-C30 global health status/QOL scale score.

A simple analysis would be to use the two-independent-sample t test to compare mean follow-up EORTC QLQ-C30 global health status/QOL scale score scores between the interferon and observation groups.

The correlation between baseline and mean follow-up pain scores is 0.62, so a more powerful statistical analysis is an ANCOVA or multiple regression [12]. This involves a multiple-regression analysis with the average follow-up QOL (the mean of the 6-, 12-, 18-, and 24-month assessments) as the dependent variable, Y_i, and the baseline QOL, x_{Base_i}, and treatment group, x_{Group_i}, (coded interferon = 0, observation = 1) as covariates.

The linear-regression model for the ith subject is

$$\bar{Y}_i = \beta_0 + \beta_{Base} x_{Base_i} + \beta_{Group} x_{Group_i} + \varepsilon_i \tag{26.2}$$

where ε_i is a random error term—which is Normally distributed with expected value or mean of zero and a variance σ^2 of 1, i.e., $\varepsilon_i \sim N(0, \sigma^2)$—and β_0 is a constant.

Table 26.4 shows the results on the fitting the ANCOVA model to the data. The adjusted difference is smaller than the unadjusted difference, and the confidence intervals for the adjusted analysis are narrower; but both the simple and adjusted analysis suggest that the observation group has the better average QOL over the follow-up period.

26.10 Longitudinal Models

In longitudinal studies, multiple assessments of QOL on the same subject at different time points are used, and the within-subject responses are then correlated. This correlation must be accounted for by analysis methods appropriate to the data. Several models have been proposed for the analysis of such data. They are usually classified into marginal or random-effects models. Random-effects models are also called generalized linear mixed models or multilevel models or conditional models. The choice of one or the other depends on the objectives of the study.

Diggle et al., Fayers and Machin, and Walters [2,11,13] emphasize the importance of graphical presentation of longitudinal data prior to modeling. Figures 26.2 and 26.3 show the mean levels of QOL in patients with malignant melanoma, before and during treatment, for the global health status/QOL dimension of the EORTC QLQ-C30 by randomized group. The curves do not overlap and there is some evidence to suggest that for later QOL measurements, the curves are parallel and that the mean difference between treatments is now fairly constant.

TABLE 26.4

Unadjusted and Adjusted Differences in Mean Follow-up EORTC QLQ-C30 Global Health Status/QOL Scale Scores between the Interferon and Observation Groups

Global Health Status/QOL Outcome[b] Time (Months)	Group Randomized to						Unadjusted				Adjusted[a]			
	Interferon		Observation					95% CI				95% CI		
	N	Mean	SD	N	Mean	SD	Difference in Means	Lower	Upper	P-value[c]	Difference in Means	Lower	Upper	P-value
Mean follow-up score	210	67.5	19.1	187	74.6	16.7	7.1	3.6	10.7	<0.001	5.9	3.1	8.7	<0.001

Source: Data from Dixon S et al. *Br J Cancer.* 2006;94(4):492–8.

Note: A positive difference in means indicates the observation group has the better QOL.

Abbreviation: CI, confidence interval.

[a] Difference adjusted for baseline EORTC QLQ-C30 global health status/QOL scale score.

[b] The EORTC QLQ-C30 global health status/QOL scale is scored on a 0 (poor) to 100 (high QOL) scale.

[c] P-value from two-independent-samples *t* test.

a) No time effect and no group effect (flat horizontal lines that coincide)

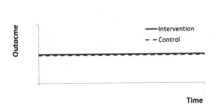

b) Time effect but no group effect (one coincident line with a non-zero gradient)

c) Group effect but no time effect (two flat parallel horizontal lines)

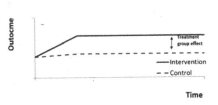

d) Group effect and time effect (two parallel lines with same gradient)

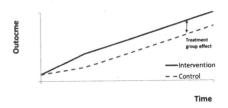

e) Group*time interaction effect (two lines with different gradients)

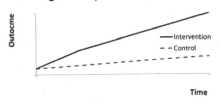

FIGURE 26.5
Five potential profiles and models for analyzing longitudinal outcome data.

Figure 26.5a–e shows some simple example profiles of the possible treatment effects on the QOL outcome over time. These five graphs lead to the specification of five possible statistical models for the QOL outcome [2]:

a. QOL outcome = constant
b. QOL outcome = baseline + time
c. QOL outcome = baseline + group
d. QOL outcome = baseline + time + group
e. QOL outcome = baseline + time + group + group*time interaction

Ideally these models should be investigated in reverse order, i.e., Model e) first. If there is a significant group*time interaction term in Model e), then Figure 26.5e shows an example of the possible pattern of the outcome over time by group. The group-by-time interaction term in the model represents the additional effect of the intervention (compared to the control) on the QOL outcome over time. If there is no significant group*time interaction then we can a fit a simpler Model d) to the QOL outcome data to see if there is both a significant group and time effect in this model. Figure 26.5d shows an example of the possible pattern for the QOL outcome over time by group. If only the group or time effect was statistically significant, but not both, we would then go on to fit Model b) or c). Depending on the results of Model d), we would either go on to fit Model b) if there was no group effect but a significant time effect (see Figure 26.5b) or Model c) (see Figure 26.5c) if there was no significant time effect but a group effect. In the event of no significant group or time effect, then model a) (Figure 26.5a) is most appropriate for the outcome data.

In longitudinal studies, multiple assessments of QOL on the same subject at different time points are used and the within-subject responses are then correlated. This correlation must be accounted for by analysis methods appropriate to the data. Several models have been proposed for the analysis of such data. They are usually classified into *marginal* or *random-effects* models. Random-effects models are also called *generalized linear mixed models* or *multilevel models* or *conditional* models. The choice of one or the other depends on the objectives of the study.

Figures 26.2 and 26.3 suggest that Model c)/Figure 26.5c or Model a)/Figure 26.5a may be appropriate for the EORTC QLQ-C30 global health status/QOL outcome in the AIM-HIGH Trial. The next few sections of this chapter will describe two ways in which we can fit such longitudinal models to QOL data. Marginal and random-effects models can be fitted in most statistical software packages such as R (which is freely available and can be downloaded from https://cran.rproject.org/bin/windows/base/), SAS, SPSS, and STATA. This chapter will use STATA to illustrate the results of fitting marginal and random-effects model to the AIM-HIGH Trial data.

26.11 Autocorrelation

If y_{i1} and y_{i2} represent the values of two successive QOL assessments by the same (*i*th) patient and m represents the total number of patients completing both assessments in the sample, then Equation 26.3 measures the strength of association or autocorrelation between successive longitudinal measurements of QOL on the same patient:

$$r_{T(1,2)} = \frac{\sum_{i=1}^{m}(y_{i1} - \bar{Y_1})(y_{i2} - \bar{Y_2})}{\sqrt{\sum_{i=1}^{m}(y_{i1} - \bar{Y_1})\sum_{i=1}^{m}(y_{i2} - \bar{Y_2})}} \quad (26.3)$$

where $\bar{Y_1}$ and $\bar{Y_2}$ are the sample mean QOL scores at times t_1 and t_2, respectively.

There are several underlying patterns of the autocorrelation matrix, R, that are often used in the modeling of QOL data:

1. Independent (sometimes termed random)
2. Unstructured
3. Exchangeable, uniform, or compound symmetric.
4. Autoregressive structure (sometimes called multiplicative or time series).

1) Independent

$R_{t,s} = 1$ if $t = s$
0 otherwise

Time	Time			
Time	0	1	2	3
0	1			
1	0	1		
2	0	0	1	
3	0	0	0	1

2) Unstructured

Unstructured imposes only the constraint that the diagonal elements of the working correlation matrix be 1.
$R_{t,s} = 1$ if $t = s$
r_{ts} otherwise, $r_{ts} = r_{st}$

Time	Time			
Time	0	1	2	3
0	1			
1	ρ_{10}	1		
2	ρ_{20}	ρ_{21}	1	
3	ρ_{30}	ρ_{31}	ρ_{32}	1

3) Exchangeable, uniform or compound symmetric

- $R_{t,s} = 1$ if $t = s$
- ρ otherwise

Time	Time			
Time	0	1	2	3
0	1			
1	ρ	1		
2	ρ	ρ	1	
3	ρ	ρ	ρ	1

4) Autoregressive structure (sometime called multiplicative or time series)

$R_{t,s} = 1$ if $t = s$
$\rho^{|t-s|}$ otherwise

Time	Time			
Time	0	1	2	3
0	1			
1	$\rho_{|1-0|} = \rho_1$	1		
2	$\rho_{|2-0|} = \rho_2$	$\rho_{|2-1|} = \rho_1$	1	
3	$\rho_{|3-0|} = \rho_3$	$\rho_{|3-1|} = \rho_2$	$\rho_{|3-2|} = \rho_1$	1

FIGURE 26.6
Patterns of autocorrelation for the correlation matrix Rt,s between observations at times t and s.

These are illustrated in more detail in Figure 26.6. The autocorrelation pattern affects the way in which the computer packages estimate the regression coefficients in the corresponding statistical model, and so it should be chosen with care.

The error structure is independent (sometimes termed random) if the off-diagonal terms of the autocorrelation matrix R are zero. If all the correlations are approximately equal or uniform, then the matrix of correlation coefficients is termed exchangeable, or compound symmetric. This means that we can reorder (exchange) the successive observations in any way we choose in our data file without affecting the pattern in the correlation matrix. Unstructured correlation imposes only the constraint that the diagonal elements of the working correlation matrix R be 1. Frequently, as the time between successive observations increases, the autocorrelation between the observations decreases. Thus, we would expect a higher autocorrelation between QOL assessments made only two days apart than between two QOL assessments made one month apart. A correlation matrix of this form is said to have an autoregressive structure (sometimes called multiplicative or time series).

When deciding what correlation matrix to use, in the longitudinal model, why not always use unstructured correlation? It might appear that using this option would be the most sensible one to choose for all longitudinal datasets. This is not the case since it necessitates the estimation of many nuisance parameters. This can sometimes cause problems in the estimation of the parameters of interest, particularly when the sample size is small and the number of time points is large [14].

The correlations in Table 26.5 clearly show the off-diagonal terms are non-zero and that the assumption of an independent autocorrelation matrix for the marginal model is unrealistic. The correlations between the four post-baseline QOL assessments at 6, 12, 18, and 24 months are of similar magnitude and range between 0.50 and 0.72. This suggests the assumption of an exchangeable correlation structure for the repeated QOL assessment for this data in not unrealistic.

TABLE 26.5

Autocorrelation Matrix for the EORTC QLQ-C30 Global Health Status/QOL Outcome in the AIM-HIGH Trial Assessed at Five Time Points

Time (months)	0	6	12	18	24
0	1				
6	0.57	1			
12	0.58	0.63	1		
18	0.44	0.50	0.61	1	
24	0.61	0.55	0.72	0.60	1

Note: The correlations in Table 26.5 clearly show the off-diagonal terms are non-zero and that the assumption of an independent autocorrelation matrix for the marginal model is unrealistic. The correlations between the 4 post-baseline QOL assessments at 6, 12, 18, and 24 months are of similar magnitude and range between 0.50 and 0.72. This suggests the assumption of an exchangeable correlation structure for the repeated QOL assessment for this data in not unrealistic.

26.12 Analyzing Longitudinal Quality of Life Outcome Data with a Marginal Model

Let Y_{ij} be the QOL outcome for the ith subject for observation j measured at time t_{ij}, for observation $j = 1$ to n_i on subject $i = 1$ to m. Then a simple model for the data, assuming independent outcomes, is

$$Y_{ij} = \beta_0 + \beta_1 t_{ij} + \varepsilon_{ij} \tag{26.4}$$

where t_{ij} is the time variable, ε_{ij} is a random error term with $\varepsilon_{ij} \sim N\left(0, \sigma_e^2\right)$ and $\text{Corr}(\varepsilon_{ij}, \varepsilon_{ik}) = 0$, β_0 is the mean outcome at baseline, and β_1 is the time effect.

The basic marginal model takes the same form as the simple (independence) model:

$$Y_{ij} = \beta_0 + \beta_1 t_{ij} + \varepsilon_{ij} \tag{26.5}$$

but the residuals, ε_{ij}, are correlated, i.e., $\text{Corr}(\varepsilon_{ij}, \varepsilon_{ik}) = \rho(x_{ij}, x_{ik}; \Re)$.

The correlation matrix, $\Re$, is usually estimated by an exchangeable correlation matrix, R, that assumes the outcomes for a subject at observation j are equally correlated with the outcomes at observation k. But we can also assume an unstructured matrix. This common correlation, ρ, is the intracluster correlation coefficient (ICC).

In the marginal modeling approach, we only need to specify the first two moments of the responses for each person (i.e., the mean and variance). With continuous normally distributed data, the first two moments fully determine the likelihood, but this is not the case for other generalized linear models.

Since the parameters specifying the structure of the correlation matrix are rarely of great practical interest (they are what is known as nuisance parameters), simple structures (e.g., exchangeable or first order autoregressive) are used for the within-subject correlations, giving rise to the so-called working correlation matrix. Liang and Zeger [15] show that the estimates of the parameters of most interest, i.e., those that determine the mean profiles over time, are still valid even when the correlation structure is incorrectly specified. The marginal generalized linear modeling approach uses generalized estimating equations (GEEs) to estimate the regression coefficients [15].

In statistics, a generalized estimating equation (GEE) is used to estimate the parameters of a marginal generalized linear model with a possible unknown correlation between outcomes. Parameter estimates from the GEE are consistent even when the correlation/covariance structure is miss-specified. The focus of the GEE is on estimating the average response over the population ("population-averaged" effects) rather than the regression parameters that would enable prediction of the effect of changing one or more covariates on a given individual. GEEs are usually used in conjunction with Huber–White standard error estimates, also known as "robust standard error" or "sandwich variance" estimates. GEEs belong to a class of semiparametric regression techniques because they rely on specification of only the first two moments. They are a popular alternative to the likelihood-based generalized linear mixed model, which is more sensitive to variance structure specification. Using GEE, any required covariance structure and link function may be assumed and the parameters estimated without specifying the joint distribution of the repeated observations. Estimation is via a multivariate analogue of a quasi-likelihood approach [16].

26.13 Treatment × Time Interactions and Baseline Measurements

Figures 26.2 and 26.3 and the non-overlapping lines in the graphs imply there is unlikely to be a "Treatment × Time" interaction. However, it is still important to test for any such interaction in any regression model. Fortunately, with the marginal model approach, this is relatively easy to do and simply involves the addition of an extra regression coefficient to the model. If treatment is coded as a 0/1 variable (i.e., 0 = interferon and 1 = observation) and assessment time as a continuous variable, then the additional interaction term is simply the product of these two variables (which will be 0 for all the interferon group patients and equal to the QOL assessment time in the observation group patients).

For RCTs, with a baseline measurement of the outcome variable, at time 0, since it is not an "outcome," it seems sensible to fit this variable as a covariate in the model and treat it like other baseline covariates such as age, gender, and treatment group.

26.14 Example of Using a Marginal Model to Analyze Longitudinal QOL Outcome Data: The AIM-HIGH Trial

The marginal model for the EORTC QLQ-C30 global health status/QOL outcome is

$$\hat{Q}L_{ij} = \hat{\beta}_0 + \hat{\beta}_1 QL\_baseline_i + \hat{\beta}_2 Time_{ij} + \hat{\beta}_3 Group_i + \hat{\beta}_4 Group_i * Time_{ij} + \hat{\varepsilon}_{ij} \qquad (26.6)$$

where QL_{ij} is the QOL outcome for the ith subject for observation j measured at time t_{ij}, for observation $j = 1$ to n_i on subject $i = 1$ to m. The estimates of the regression coefficients for the marginal model given by Equation 26.6 are given in the output in Table 26.6. The interaction term is not statistically significant. Thus there was no reliable evidence of a "Treatment × Time" interaction. Therefore, we can now use a simpler model without the interaction term to test for a group and time effect on QOL. This is shown in Table 26.7.

TABLE 26.6

Estimated Regression Coefficients from a Marginal Regression Model, in STATA 15 Using the xtgee Procedure with Coefficients Estimated by GEE with Robust Standard Errors to Show the Effect of Group on Outcome, EORTC QLQ-C30 Global Health Status/QOL Outcome, from the AIM-HIGH RCT; $n = 397$.

```
. xtgee QL QL_baseline Time  Group Interaction, family(gaussian) link(identity)
corr(exchangeable) vce(robust)

GEE population-averaged model             Number of obs     =         840
Group variable:              Patient_id   Number of groups  =         397
Link:                          identity   Obs per group:
Family:                        Gaussian                          min =         1
Correlation:               exchangeable                          avg =       2.1
                                                                 max =         4
                                          Wald chi2(4)      =      223.98
Scale parameter:              262.2825    Prob > chi2       =      0.0000

                        (Std. Err. adjusted for clustering on Patient_id)
----------------------------------------------------------------------------
             |               Robust
          QL |     Coef.   Std. Err.      z    P>|z|     [95% Conf. Interval]
-------------+--------------------------------------------------------------
 QL_baseline |      0.53       0.04    14.09   0.000       0.46       0.61
        Time |      1.34       1.28     1.05   0.293      -1.16       3.85
       Group |      7.31       2.05     3.57   0.000       3.30      11.32
 Interaction |     -2.16       1.79    -1.21   0.227      -5.67       1.35
       _cons |     28.54       3.15     9.05   0.000      22.36      34.72
----------------------------------------------------------------------------
```

> The interaction term is not statistically significant. Thus there was no reliable evidence of a 'Treatment x Time' interaction. Therefore, we can now use a simpler model without the interaction term to test for a group and time effect on QoL.

26.15 Checking the Assumptions

Table 26.8 shows the estimated within-subject correlation matrices for the EORTC QLQ-C30 global health status/QOL outcome if we assume a compound symmetric or exchangeable correlation structure for the repeated QOL assessments. The upper diagonal gives the observed matrix before the model fitting. The fitted autocorrelation was 0.39 which is lower than the observed correlations. The observed deviation between the fitted model and observed autocorrelations is not too great, suggesting that the assumption of compound symmetry is not unreasonable.

26.16 Random-Effects Models

The random-effects model assumes that the correlation arises among repeated responses because the regression coefficients vary across individuals. Random-effects models are particularly useful when inferences are to be made about individuals, rather than the population average. Thus a random-effects approach will allow us to estimate the QOL status of an individual patient. The regression coefficients, β, represent the effect of the explanatory variables on an individual patient's QOL. This is in contrast to the marginal

TABLE 26.7

Estimated Regression Coefficients from a Marginal Regression Model, in STATA 15 Using the xtgee Procedure with Coefficients Estimated by GEE with Robust Standard Errors to Show the Effect of Group on Outcome, EORTC QLQ-C30 Global Health Status/QOL Outcome, from the AIM-HIGH RCT; $n = 397$

```
xtgee QL QL_baseline Time  Group, family(gaussian) link(identity)
corr(exchangeable) vce(robust)

GEE population-averaged model              Number of obs     =       840
Group variable:              Patient_id    Number of groups  =       397
Link:                          identity    Obs per group:
Family:                        Gaussian                       min =         1
Correlation:                exchangeable                       avg =       2.1
                                                               max =         4
                                           Wald chi2(3)      =    222.62
Scale parameter:              262.9127     Prob > chi2       =    0.0000

                       (Std. Err. adjusted for clustering on Patient_id)
-----------------------------------------------------------------------------
             |               Robust
         QL  |     Coef.   Std. Err.      z     P>|z|    [95% Conf. Interval]
-------------+---------------------------------------------------------------
 QL_baseline |      0.53       0.04    14.07   0.000       0.46        0.61
        Time |      0.32       0.90     0.35   0.723      -1.44        2.07
       Group |      5.34       1.41     3.80   0.000       2.59        8.10
       _cons |     29.57       2.99     9.89   0.000      23.71       35.43
-----------------------------------------------------------------------------

. estat wcorrelation, compact

Error structure: exchangeable
Estimated within-Patient_id correlation: .39422207
```

> The estimated exchangeable correlation between the outcomes

Source: Dixon S et al. *Br J Cancer.* 2006;94(4):492–8.

TABLE 26.8

Observed and Estimated with-Patient Autocorrelation Matrices (Exchangeable Model) from the Cancer Patients in the AIM-HIGH Trial. The Upper Diagonal Gives the Observed Matrix before Model Fitting Whilst the Lower Gives the Exchangeable Form after Model Fitting[a]

Time (months)	6	12	18	24
6	1.00	*0.63*	*0.50*	*0.55*
12	0.39	0.39	*0.61*	*0.72*
18	0.39	0.39	1.00	*0.59*
24	0.39	0.39	0.39	1.00

[a] The model contains time, baseline QOL, and group as covariates.

model coefficients, which describe the effect of the explanatory variables on the population average. It is based on the assumption that the subjects in the study are chosen at random from some wider patient population.

Let y_{ij} be the QOL outcome for the ith subject for observation j measured at time t_{ij}, for observation $j = 1$ to n_i on subject $i = 1$ to m then a random-effects model is

$$Y_{ij} = \underbrace{\beta_0 + \beta_1 t_{ij}}_{Fixed} + \underbrace{w_i + \varepsilon_{ij}}_{Random} \ldots \tag{26.7}$$

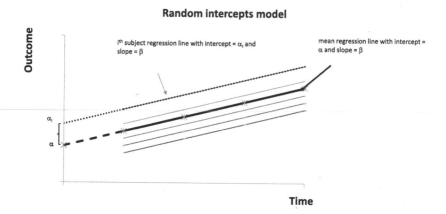

FIGURE 26.7
Graphical illustration of a simple random intercepts model.

where t_{ij} is the time variable and ε_{ij} is a random error term with $\varepsilon_{ij} \sim N\left(0, \sigma_e^2\right)$, β_0 is the mean outcome at time 0 and β_1 is the time effect, and ω_i is the random effect of subject i across all time points with $\omega_i \sim N(0, \sigma_\omega^2)$. Variation in ω_i induces variation in the mean outcome across all subjects. It assumes the treatment effect is homogenous across the subjects and is sometimes known as the *random-intercept* model. The fixed portion of the model, in Equation 26.7, states that we want one overall regression line representing the population average QOL over time. The random effect serves to shift this regression line up or down according to each individual subject. Figure 26.7 illustrates this graphically.

The random-intercept model for the EORTC QLQ-C30 global health status/QOL outcome is

$$\hat{Q}L_{ij} = \hat{\beta}_0 + \hat{\beta}_1 QL\_baseline_i + \hat{\beta}_2 Time_{ij} + \hat{\beta}_3 Group_i + \hat{\omega}_i + \hat{\varepsilon}_{ij} \tag{26.8}$$

where QL_{ij} is the QOL outcome for the ith subject for observation j measured at time t_{ij}, for observation $j = 1$ to n_i on subject $i = 1$ to m; ε_{ij} is a random error term with $\varepsilon_{ij} \sim N(0, \sigma_e^2)$; and ω_i is the random effect of subject i across all time points with $\omega_i \sim N\left(0, \sigma_\omega^2\right)$. The estimates of the regression coefficients for the random-effects model given by Equation 26.8 are given in the output in Table 26.9. The estimates of the two variance components for the two random effects, i.e., σ_ω^2 labeled var(_cons) and σ_e^2 labeled var(Residual), are 107.83 and 155.88, respectively. The estimated ICC from the random-effects model is $107.83/(107.83 + 155.88) = 0.41$, which is very similar to the ICC estimate of 0.39 from the marginal model.

26.17 Random Slopes

If we suspect that individuals will have a different pattern or change in quality of life over time, then to allow for this possibility of a different slope over time for each subject, we need

TABLE 26.9

Estimated Regression Coefficients from Random-Intercept Mixed-Effects Model, in STATA 15 using the `Mixed` Procedure with Coefficients Estimated by ML (Maximum Likelihood) to Show the Effect of Group on Outcome, EORTC QLQ-C30 Global Health Status/QOL Outcome, from the AIM High RCT; $n = 397$

```
mixed QL QL_baseline Time Group || Patient_id:
```

| Mixed-effects ML regression | | | | Number of obs | = | 840 |
| Group variable: Patient_id | | | | Number of groups | = | 397 |

Obs per group:

		min =	1
		avg =	2.1
		max =	4

| | | Wald chi2(3) | = | 259.42 |
| Log likelihood = -3481.2579 | | Prob > chi2 | = | 0.0000 |

QL	Coef.	Std. Err.	z	P>|z|	[95% Conf. Interval]	
QL_baseline	0.53	0.03	15.33	0.000	0.47	0.60
Time	0.29	0.85	0.35	0.730	-1.38	1.97
Group	5.36	1.41	3.80	0.000	2.60	8.12
_cons	29.53	2.77	10.65	0.000	24.10	34.97

Random-effects Parameters	Estimate	Std. Err.	[95% Conf. Interval]	
Patient_id: Identity				
var(_cons)	107.83	14.52	82.82	140.39
var(Residual)	155.88	10.26	137.01	177.35

a random-slopes model. The random-slope effect allows us to have separate, non-parallel, regression lines for each individual subject, with an overall mean regression line for all subjects. Figure 26.8 illustrates this graphically.

The equation for the random-slopes model can be written as

$$Y_{ij} = \underbrace{\beta_0 + \beta_1 t_{ij}}_{Fixed} + \underbrace{\omega_{0i} + \omega_{1i} t_{ij} + \varepsilon_{ij}}_{Random} \ldots \tag{26.9}$$

where t_{ij} is the time variable, ε_{ij} is a random error term with $\varepsilon_{ij} \sim N(0, \sigma_e^2)$, β_0 is the mean (baseline) outcome and β_1 is the time effect, ω_{0i} is the random (intercept) effect of subject i across all time points with $\omega_{0i} \sim N(0, \sigma_{\omega 0}^2)$, and ω_{1i} is the random (slope) effect of subject i over time with $\omega_{1i} \sim N(0, \sigma_{\omega 1}^2)$.

The random-slopes model for the EORTC QLQ-C30 global health status/QOL outcome is

$$\hat{QL}_{ij} = \hat{\beta}_0 + \hat{\beta}_1 QL\_baseline_i + \hat{\beta}_2 Time_{ij} + \hat{\beta}_3 Group_i + \hat{\omega}_{0i} + \hat{\omega}_{1i} time_{ij} + \hat{\varepsilon}_{ij} \tag{26.10}$$

where QL_{ij} is the QOL outcome for the ith subject for observation j measured at time t_{ij}, for observation $j = 1$ to n_i on subject $i = 1$ to m. The estimates of the regression coefficients for the random-effects model given by Equation 26.10 are given in the output in Table 26.10.

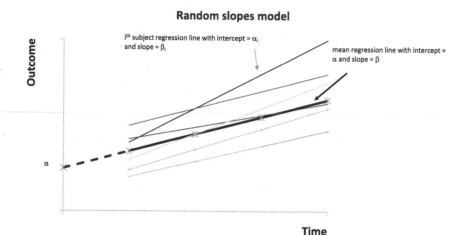

FIGURE 26.8
Graphical illustration of a simple random intercept and slopes model.

The estimates of the variance components for the three random effects, i.e., $\sigma_{\omega1}^2$ labeled var(Time); $\sigma_{\omega0}^2$ labeled var(_cons), and σ_e^2 labeled var(Residual), are 16.99, 94.89, and 149.6, respectively.

Since the random-intercept model is nested within the random-slopes models we can use a likelihood ratio test (LRT) to see whether or not the random slope is significant. The likelihood ratio test statistic is the deviance and is sometimes also referred to as the likelihood ratio statistic or −2 log likelihood for the full model minus the deviance for full model excluding the relevant explanatory variable. This test statistic follows a Chi-squared distribution with 1 degree of freedom (df). The log likelihood for the "full" model with the random-slope term is $L_0 = -3479.61$ (from Table 26.10), and the log likelihood for the simpler random-intercept model is $L_1 = -3481.26$ (from Table 26.9). The difference $(L_1 - L_0)$ is multiplied by 2 and the answer is the LRT statistic value, which is 3.30. This can be interpreted as a Chi-squared statistic with 1 degree of freedom. The LRT value is 3.30, which is compared to a chi-squared distribution on 1 degree of freedom resulting in a p-value of 0.06. The p-value of 0.06 favors the simpler random-intercept model that allows only for a patient-specific shift over the more complex random-slopes model that allows for a random patient-specific regression line. The estimated treatment effect and its associated confidence interval for the simpler random-intercept model is 5.4 (95% CI 2.6–8.1) which is very similar to the estimated treatment effect and its associated confidence interval for the random-slope model of 5.5 (95% CI 2.8–8.3).

Both the marginal model (Table 26.7) and simple random-intercept effect model (Table 26.9) have produced identical estimates of the treatment effect and associated confidence interval of 5.3 (95% CI 2.6–8.1), when rounded to one decimal place. This suggests that the observation group has the better QOL over time compared with the interferon group.

TABLE 26.10

Estimated regression coefficients from random intercept and slopes effects model, in STATA 15 using the mixed procedure with coefficients estimated by ML (Maximum Likelihood) to show the effect of group on outcome, EORTC QLC-C30 Global health status/QoL outcome, from the AIM High RCT (Dixon et al. 2006) $n = 397$.

```
mixed QL QL_baseline Time Group || Patient_id: Time
```

Mixed-effects ML regression			Number of obs	=	840
Group variable: Patient_id			Number of groups	=	397
			Obs per group:		
			min =		1
			avg =		2.1
			max =		4
			Wald chi2(3)	=	261.40
Log likelihood = −3479.6094			Prob > chi2	=	0.0000

QL |	Coef.	Std. Err.	z	P>|z|	[95% Conf. Interval]	
QL_baseline |	0.54	0.03	15.37	0.000	0.47	0.60
Time |	0.15	0.91	0.17	0.869	−1.63	1.93
Group |	5.54	1.41	3.94	0.000	2.79	8.30
_cons |	29.39	2.77	10.61	0.000	23.96	34.82

Random-effects Parameters |	Estimate	Std. Err.	[95% Conf. Interval]	
Patient_id: Independent |				
var(Time) |	16.99	10.36	5.14	56.15
var(_cons) |	94.89	16.23	67.87	132.67
var(Residual) |	149.66	10.64	130.21	172.03

26.18 Missing Data

Many studies which use QOL measures are longitudinal and are therefore likely to have missing data. For QOL measures this can be individual questions, termed missing items, or missing questionnaires, as might occur if a patient did not attend an assessment. *Item non-response* arises when there are single missing item(s) from an otherwise complete questionnaire. *Unit non-response* is when the whole QOL questionnaire is missing when one was anticipated from the patient. There are two important potential consequences of missing data [17]. The first is the decrease in precision (wider confidence intervals) and power caused by the reduction in data. The second, and more serious, is the potential for bias in the estimation of both between (e.g., treatment effect) and within group effects (e.g., change over time). In these circumstances, the results may be biased and not reflect the true state of affairs. If the proportion of missing data is small, then little bias will result. If the proportion of data missing is not small, then a crucial question arises: Are the characteristics of patients with missing data different from those for whom complete data are available? If the answer is yes, then the study results will be biased and will not reflect the truth [2].

Table 26.1 shows that of the 444 patients with valid baseline QOL outcome data, only 29% (166/444) had valid QOL outcome data at 24 months of follow-up. That is, 71% of the possible outcome data is potentially missing.

26.19 Types and Patterns of Missing Data

There are essentially three types of missing data.

1. Missing Completely at Random (MCAR)

   When the probability of response at time t is independent of both the previously observed values and the unobserved values at time t. For example, in the AIM-HIGH trial, this would mean that missing values at, say, 6 months post-randomization follow-up are not associated with, say, the baseline value of quality life and other factors such as age and gender as well as the patient's current unobserved EORTC QLQ-C30 global health status/QOL.

2. Missing at Random (MAR)

   When the probability of response at time t depends on the previously observed values but not the unobserved values at time t. For example, in the AIM-HIGH trial, this would mean that missing values at, say, 6 months post-randomization follow-up would depend on the baseline value of quality of life and other factors such as age and gender but not the current unobserved EORTC QLQ-C30 global health status/QOL.

3. Not Missing at Random (NMAR)

   When the probability of response at time t depends on the unobserved values at time t. For example, in the AIM-HIGH trial, this would mean that missing values at, say, 6 months post-randomization follow-up would depend on the current unobserved EORTC QLQ-C30 global health status/QOL, perhaps because the patient's cancer had progressed and their current health status had worsened considerably since their baseline value.

If a patient drops out of the study or dies, their data from a certain point onward will be unobserved. This pattern of missingness is termed *monotone*. *Intermittent missingness* is defined to be when an outcome is unobserved at one assessment but is observed at a following assessment. Intermittent missing data are more typical in studies of individuals with chronic conditions. In contrast, monotone missing data occur in studies of populations with significant morbidity or mortality and in trials where monotonicity occurs as the result of the design (Table 26.11).

26.20 Describing the Extent and Patterns of Missing Data

After the study has been conducted and it is time to report results, the first step is to describe how many participants were in the study at each time point. A table, such as Table 26.1, or a CONSORT flow diagram (Figure 26.1) is a good way to achieve this.

TABLE 26.11

Pattern of Missing QOL Data from the AIM-HIGH Trial: Mean EORTC QLQ-C30 Scores at Each Time Point by Group

EORTC QLQ-C30 Group	Global Health Status/QOL				
	Time (Months)				
	0	6	12	18	24
Interferon group: baseline only data ($n = 19$)	74				
Observation group: baseline only data ($n = 27$)	71				
Interferon group: complete data to 6 months only ($n = 95$)	73	65			
Observation group: complete data to 6 months only ($n = 75$)	73	74			
Interferon group: complete data to 12 months only ($n = 21$)	74	72	68		
Observation group: complete data to 12 months only ($n = 32$)	73	74	72		
Interferon group: complete data to 18 months only ($n = 16$)	66	66	64	62	
Observation group: complete data to 18 months only ($n = 10$)	75	77	73	74	
Interferon group: complete data to 24 months ($n = 39$)	67	68	69	66	69
Observation group: complete data to 24 months ($n = 37$)	72	75	76	74	75

To explore mechanisms of missingness, a good place to start is a graph of the QOL outcome versus time, stratified by dropout time, as shown in Figure 26.9 for the AIM-HIGH trial data. If the trajectories over time are substantially different, the data are not MCAR. For example, are patients who have lower baseline values more likely to drop out, or are steeper rates of increase or decrease over time associated with dropout? In Figure 26.9 there is little separation between the missingness patterns, providing some evidence that the data are likely to be missing completely at random.

We can explore missing data mechanisms by comparing those who dropout versus those who do not via *t* tests (e.g., comparing mean age of those who drop out vs. those who do

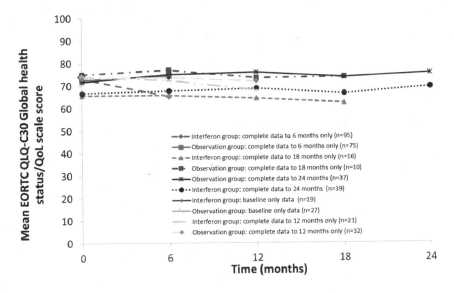

FIGURE 26.9
Profile of mean QOL scores over time stratified by drop out time.

not), cross tabulations, and logistic regression (to look at multiple predictors of drop out). Survival analysis can be used to look at predictors of time to dropout.

Figure 26.1 shows that 444 patients (out of 674), or 66%, had a valid baseline QOL assessment; 230/338 (68%) in the IFN group and 214/336 (64%) in the OBS group ($p = 0.233$). Comparison of the $N = 398$ patients with a valid baseline QOL assessment and at least one valid follow-up QOL assessment and $N = 276$ patients with no baseline or follow-up QOL assessments suggested that the two groups have similar age ($p = 0.151$), gender ($p = 0.349$), histology ($p = 0.078$), and lengths of follow-up ($p = 0.528$) (Table 26.1). There was no interaction between treatment group and follow-up QOL assessment status with regard to overall survival ($p = 0.251$) and no evidence of a difference in overall survival between the no follow-up QOL data and valid follow-up data groups (log-rank $p = 0.84$). Median survival was 4.05 years for patients with no valid follow-up QOL data versus 3.81 years for patients with valid baseline and follow-up QOL data.

The question of how to deal with patients who die, which is likely in studies of cancer patients, is a "vexing issue," which has not been resolved [17]. Some researchers impute a score of 0 (or whatever is the minimum possible score on the scale). While this is reasonable for some scales where 0 is explicitly anchored to death (e.g., utilities, functional well-being), it does not make sense for others such as symptom and physical scales, where a 0 could mean that the deceased patient is experiencing severe nausea, vomiting, and pain.

Analyses which assume the outcome data are MCAR are complete case analysis, repeated univariate (time-by-time) analysis, marginal models with coefficients estimated by "standard" GEE. and summary measures such as AUC. On the other hand, analyses which assume the outcome data are MAR are random-effects models (maximum likelihood methods); multiple imputation (MI); and marginal models with coefficients estimated by extensions to GEE with inverse probability weights (IPW), MI, or both (doubly robust estimation, DR-GEE).

Different assumptions are required for the two models regarding missing data. The marginal model using the GEE requires a missing data process completely at random (MCAR). Under this assumption, missingness does not depend on individual characteristics (observed or not). In contrast, random-effects models only need the less stringent assumption of missing at random (MAR). In this process, the probability of missingness depends only on observed variables (previous covariates or outcomes).

26.21 What If We Think the Data Is MNAR?

Unfortunately we cannot test whether or not the data are MNAR. We have not observed the missing QOL score, so it is not possible to formally test the hypothesis that the "missingness" does not depend on the QOL at the time the assessment is missing. The data we need to test the hypothesis are missing! More complex models are required.

In trials of cancer patients at the end of life or being treated with palliative care, the statistical analysis can be problematic due to high levels of missing data, attrition, and response shift (a change over time in an individual's basis on which they perceive, judge, and value their own QOL) as disease progresses. For palliative care and end-of-life care studies, it is likely that data are MNAR since the likelihood of a patient of not responding to a quality of life (QOL) questionnaire at a particular time point is likely to depend on the unobserved values for their QOL at that time point [18]. Analyses which assume the outcome data are MNAR are mixture models (MM) and pattern mixture models are a special case of MM, shared random-effects parameter models, joint multivariate models, and selection models. These complex models are beyond the scope of this chapter, but the interested reader is referred to Chapter 27, to the article by Bell and Fairclough [17] or the textbooks by Fairclough [19,20] and Diggle et al. [11] for more details.

26.22 Random-Effects versus Marginal Models

In practice, both random-effects and marginal models provide valid methods for the analysis of longitudinal QOL data. The two approaches lead to different interpretations of between-subject effects (particularly for binary outcomes). In the marginal model, the treatment group coefficients from the model represent the average difference between the intervention or control treatments. In the random-effects model the treatment group coefficients from a model represent the difference in effect of offering either the intervention or control treatment on an individual subject.

But for continuous outcomes, using a linear-regression model, the coefficients from a random-effect model can have a marginal interpretation! Ideally one would choose the model which best answers the scientific research question being asked in the study. In RCTs we are clearly interested in the average difference in the treatment effect between the intervention and control groups. For this, a marginal model appears to be appropriate as the treatment effect of a marginal model represents the average difference between the treatment and control groups across the whole population without being specific to the individuals used in the trial. However, in an RCT we may also be interested in the effect of the intervention or control treatment on an individual subject. In these circumstances, the random-effects model would give the effect of either the intervention or control treatment on an individual subject. There is a continuing debate on this subject!

26.23 Conclusion

This chapter has described how QOL data from longitudinal studies can be summarized, tabulated, and graphically displayed. This chapter has shown how repeated QOL measures for each subject can be reduced to a single summary measure for statistical analysis and how standard statistical methods of analysis can then be used. Finally, the chapter has described two extensions of the linear-regression model, marginal and random-effects models, which allows for the fact that successive QOL assessments by a particular patient are likely to be correlated.

References

1. Everitt BS. *A Handbook of Statistical Analyses using S-Plus*, Second Edition. Boca Raton, Florida: Chapman & Hall/CRC, 2002.

2. Walters SJ. *Quality of Life Outcomes in Clinical Trials and Health Care Evaluation: A Practical Guide to Analysis and Interpretation.* Chichester: Wiley, 2009.

3. Hancock BW, Wheatley K, Harris S et al. Adjuvant interferon in high-risk melanoma: the AIM-HIGH study – United Kingdom Coordinating Committee on Cancer Research randomized Study of Adjuvant low-dose extended-duration interferon alfa-2a in high-risk resected malignant melanoma. *J Clin Oncol.* 2004;22:53–61.

4. Dixon S, Walters SJ, Turner L, Hancock BW. Quality of life and cost-effectiveness of interferon-alpha in malignant melanoma: results from randomised trial. *Br J Cancer.* 2006;94(4):492–8.

5. Aaronson NK, Ahmedzai S, Bergman B et al. The European Organisation for Research and Treatment of Cancer QLQ-C30: A quality-of-life instrument for use in international trials in clinical oncology. *J Natl Cancer Inst.* 1993;85:365–76.

6. Fayers, P, Aaronson, N, Bjordal, K, Sullivan, M on behalf of the EORTC Quality of life Study Group. 1995. *EORTC QLQ-C30 Scoring Manual.* Belgium: EORTC Study Group on Quality of Life.

7. Campbell MJ, Machin D, Walters SJ. *Medical Statistics: A Text Book for the Health Sciences,* Fourth Edition. Chichester: Wiley, 2007.

8. Freeman JV, Walters SJ, Campbell MJ. *How to Display Data.* Oxford: BMJ Books, Blackwell, 2008.

9. Everitt BS. *Statistics for Psychologists.* Mahwah, New Jersey: Lawrence Erlbaum Associates, 2001.

10. Matthews JNS, Altman DG, Campbell MJ, Royston P. Analysis of serial measurements in Medical Research. *BMJ.* 1990;300:230–5.

11. Diggle PJ, Heagerty P, Liang K-Y et al. *Analysis of Longitudinal Data,* Second Edition. Oxford: Oxford University Press, 2002.

12. Frison L, Pocock SJ. Repeated measures in clinical trials: analysis using mean summary statistics and its implications for design. *Stat Med.* 1992;11:1685–704.

13. Fayers PM, Machin D. *Quality of Life: the Assessment, Analysis & Interpretation of Patient-Reported Outcomes,* Second Edition. Chichester: Wiley, 2007.

14. Rabe-Hesketh S, Everitt BS. *A Handbook of Statistical Analyses using Stata,* Fourth Edition. Chapman & Hall/CRC: Boca Raton, 2007.

15. Liang KY, Zeger SL. Longitudinal data analysis using generalized linear models. *Biometrica.* 1986;73:13–22.

16. Wedderburn RWM. Quasi-likelihood functions, generalised linear models and the Gaussian method. *Biometrika.* 1974;61:439–47.

17. Bell ML, Fairclough DL. Practical and statistical issues in missing data for longitudinal patient reported outcomes. *Stat Methods Med Res.* 2014;23(5):440–59.

18. Preston NJ, Fayers P, Walters SJ et al. Recommendations for managing missing data, attrition and response shift in palliative and end-of-life care research: Part of the MORECare research method guidance on statistical issues. *Palliat Med.* 2013;27(10):899–907.

19. Fairclough DL. *Design and Analysis of Quality of Life Studies in Clinical Trials.* New York: Chapman & Hall, 2002.

20. Fairclough DL. *Design and Analysis of Quality of Life Studies in Clinical Trials,* Second Edition. New York: Chapman & Hall, 2010.

27

Missing Data

Stephanie Pugh, James J. Dignam, and Juned Siddique

CONTENTS

27.1 Introduction

In many clinical trials, responses or measurements are collected at numerous points over time. A common, albeit unfortunate, issue in analysis of this longitudinal data is when one or more of the sequences of measurements may be incomplete. Missing data are frequently encountered in, for example, measures of quality of life (QOL) data (see Chapter 26), such as the patient-completed Expanded Prostate Cancer Index Composite (EPIC) QOL questionnaire, and laboratory values, such as prostate specific antigen (PSA), collected at intervals during and/or after treatment. The main problems that arise with missing data are that (1) there is loss of sample size and estimation efficiency if patient data is omitted and (2) the distribution of the observed data may not be the same as the distribution of the missing data, so that omission produces biased estimates. How missing data is handled in the analysis depends on what assumptions are made regarding the reasons for missingness, as will be discussed in this chapter.

Reasons for missing data are numerous. A patient missing an office visit or refusing to complete a required form is a common source of missing data, including patient-reported outcome (PRO) data. However, missing data need not be the fault of the patient, as the research team may fail to perform a test or scan, or there may be an insufficient tissue sample or error when conducting a laboratory test. Regardless of the various reasons as to why information is missing, the problem of missing data must be accounted for in the design and analysis of a clinical trial, as it can impact the generalizability of the trial results and introduce bias in estimates of the treatment effect. Details of study design and analysis plans to minimize missing data and its impact on results are presented elsewhere, but in general, every effort should be taken to prevent missing data, and monitoring the amount of (and reasons for) missing data should take place throughout the trial. During the analysis, the amount and type of missing data needs to be assessed and then appropriate analysis techniques used, and the general strategy and methods should be specified *a priori* in the study protocol.

To facilitate this discussion of missing data, a phase III clinical trial is used as an illustration. The trial, conducted by NRG Oncology, is a phase III non-inferiority trial comparing two radiotherapy (RT) fractionation (incremental dose delivery over time) schedules in patients with low-risk prostate cancer [1]. The trial randomized 1,115 men with low-risk prostate cancer to conventional RT (73.8 Gy in 41 fractions over 8.2 weeks) or to hypofractionated RT (70 Gy in 28 fractions over 5.6 weeks). The study found that hypofractionated RT was not inferior to conventional RT in terms of the primary clinical outcome of disease-free survival, although an increase in late gastrointestinal/genitourinary toxicity was observed in men treated with hypofractionated RT.

The ultimate choice to adopt the hypofractionated regimen depends on toxicity and additional patient-reported measures of QOL. Thus, the trial collected PROs related to adverse symptomatic consequences of radiation; these were measured prior to the start of RT and at 6, 12, 24, and 60 months from the end of RT. Although several PROs were collected on this trial, only one will be described here. The EPIC is a 50-item measure designed to evaluate patient function and bother after prostate cancer treatment using a Likert-like scale with responses transformed to a scale of 0–100 [2]. Higher scores indicate better QOL. The instrument includes four separately validated domains: urinary, bowel, sexual, and hormonal. The EPIC bowel score means by treatment arm across time are shown in Figure 27.1. Higher EPIC bowel scores indicate better health-related QOL. Results of the

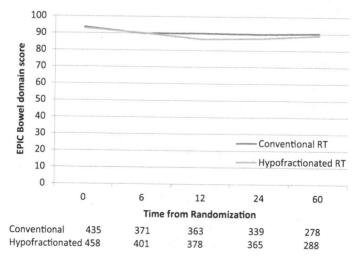

FIGURE 27.1

EPIC Bowel domain scores measured at pre-treatment, 6 months, 12 months, 24 months, and 60 months from the end of radiation.

PRO analysis found no clinically significant differences between hypofractionated and conventional RT in terms of anxiety/depression and bowel, bladder, sexual, and general quality of life [3]. Missing data were encountered as expected, and in the remainder of this chapter, approaches to analysis will be illustrated using NRG Oncology's RTOG 0415 trial.

This chapter encompasses missing data and its effects as it applies to the design and conduct of the trial, its analysis, and interpretation of the results. Missing baseline covariate data will also be briefly covered. The above-mentioned prostate cancer trial will be used as an example throughout to illustrate the various scenarios and analysis methods that will be described. Due to the many analysis methods available, this chapter will narrow its focus to those applicable to continuous data as depicted in the prostate cancer trial.

27.2 Trial Design

Since missing data is such a common problem in clinical trial research that involves repeated longitudinal assessments, steps should be taken to reduce its occurrence and give careful thought to analysis options during the design. The Panel on Handling Missing Data in Clinical Trials describes ideas to implement when designing clinical trials, which was subsequently summarized by Little et al. [4]. Their first recommendation is that the protocol should explicitly state the objective(s) of the trial, the corresponding primary endpoint(s), how and when the outcome(s) will be collected, the measures of the treatment effects, and the potential impact and the handling of missing data. Additionally, shortening the length of time from enrollment to definitive assessment of the primary endpoint can reduce the proportion of patients who are lost to follow-up through attrition over time.

The intent-to-treat principle requires inclusion of randomized patients that did not receive the protocol-specified treatment or stopped treatment early. However, patients who discontinue protocol treatment may be more difficult to maintain in the cohort as they may be more prone to more general non-compliance including dropout (no observations occur after a specific time point). Little et al. recommend some design features or alternate designs that take this into account [4], including utilizing a run-in period prior to randomization in which all patients are assigned to the active treatment, and only those who tolerate and adhere to the treatment are then randomized. These types of designs may not always be feasible, especially in the context of radiation therapy for cancer or treatment consisting of a single dose of an agent, and tend to work best when long-term efficacy is of importance.

Outcome measures that are likely to lead to substantial missing data should be avoided. Repeated longitudinal assessments, while critical in many research questions, are particularly prone to missingness. One possible solution may be to convert longitudinal data to a time-to-event construct, so that censored data methods can be used. This involves defining the event of interest using a threshold for some state, such as health deterioration, calculated from the change from baseline to each follow-up assessment time. An event would occur when a patient first experiences deterioration. The competing risk of death can also be incorporated into the sample size calculation. This approach can be used in populations in which deterioration could lead to dropout. It does require a tool with an appropriate definition of deterioration, and the hypothesis must also be framed in this context. The time of definitive assessment should be chosen to reduce missing data, and the assessments must be frequent enough so as to reasonably represent time to event (i.e., not too infrequent). Specifically, the median survival of the population should be considered to reduce missing data due to death. Another, more simplistic way to account for missing data is to inflate the sample size by the inverse of one minus the projected dropout rate. Although this may guarantee the necessary statistical power at the time of analysis, it can often lead to bias depending on the reason for missingness. Even with sufficient statistical power, detecting the treatment effect is unlikely if the potential bias from the missing data is of similar size to that of the hypothesized treatment effect [5].

Little et al. go on to discuss ways to limit missing data once the trial is actively accruing [4]. Monitoring rates of missing assessments and, whenever possible, the reasons for missingness will allow for interventions to reduce missing data in real time. An example would be in the case of quality of life data. If targets are set to 0% missing pretreatment assessment, 10% at the end of treatment, and 20% at 1 year in follow-up, monitoring the proportion of missing data will allow the study team to determine if there are deviations from these targets that would undermine any meaningful analysis. Monitoring patient compliance by treatment arm is also important to ensure that non-compliance is similar between arms, which would introduce further bias into the analysis. Collecting the reason why the assessment was not completed by the patient (such as patient refusal, patient too ill, unable to contact the patient, institutional error, etc.) can assist the study team in ways to reduce missing data, as well as provide critical information as to the missing data mechanism (discussed in Section 27.3). If the institution is committing errors in carrying out the assessment schedule or having difficulty contacting the patient, the study team can work with the staff at the institution to prevent future errors or on the necessary information to be collected for the staff to contact patients, such as up-to-date phone numbers, emails, and addresses. Limiting the burden and inconvenience of data collection on both the institutional staff collecting the data as well as the patient is a key step to reducing missing data.

27.3 Missing Data Mechanisms

27.3.1 Types of Missing Data Mechanisms

The relationship between the probability that an observation is missing and the value itself is known as the missing data mechanism. Little and Rubin [6] classified missing data into three types: missing completely at random (MCAR), missing at random (MAR), and missing not at random (MNAR). MCAR occurs when the missingness of specific data records is unrelated to the data values, and therefore the subset of patients with observed data is representative of the entire sample. For instance, a study may only follow up on a randomly selected subsample of patients in order to save money, which is not related to any missing or observed data. This type of missing data is the least problematic in terms of analysis, as standard methods of analysis will produce unbiased results, but also the least likely to occur in a clinical trial setting. MAR occurs when missingness depends on information in the observed data and not the missing data (i.e., there are factors that can be identified that relate to missingness, but not to the missing data values). An example would be patients who live farther away from the hospital are more likely to miss visits than those who live closer. The subset of patients with complete data would not be representative of the entire sample as it would consist of more patients who live closer to the hospital. However, within strata defined by distance from the hospital, the missing data are MCAR. The analysis method would need to address this bias, for example, by conditioning on distance via a statistical model. Finally, MNAR occurs when missingness depends on the missing data and possibly the observed data. This can occur if a patient is too sick from the treatment to attend a visit. Appropriate analysis techniques for data that are MCAR or MAR would not be appropriate for MNAR as the observed data are not representative of the entire sample, even after adjusting for other known information. Missing data that are considered MNAR are often called non-ignorable. Ignorable data can be analyzed using the observed data only while non-ignorable data require one to address the form of the missing data mechanism which requires making additional assumptions [7]. The focus of the remainder of this chapter will be on definitions of the missing data mechanism and the analysis techniques for each type of missing data mechanism.

27.3.2 Detecting Missing Data Mechanisms

Before beginning the analysis, assessment of missing data and the detection of the missing data mechanism should take place. Patterns of missing data include dropout and intermittent missing data (an observation at a given assessment time is missing but the next is observed) [8]. Typically a combination of both is seen in clinical trials. An example of missing data patterns in the example prostate study is shown in Table 27.1.

Since solely MCAR data are rare in clinical trials, the first step in detection of the missing data mechanism should be to identify covariates that predict missing observations [8]. Kendall's τ_b statistic can be used to determine the correlation between a dichotomous variable, the missing data indicator, and other variables that can be categorical, ordinal, or continuous. Logistic-regression analysis can be used to model the probability of a missing assessment and assess the effect of various covariates, such as pretreatment characteristics. If the missingness is monotone and there are a large number of assessments, a Cox proportional hazards model can be used to model the time to dropout. Significant covariates should be included in the analysis of the data. Comparing pretreatment characteristics between patients with missing assessments and those with completed assessments can aid in identification of differences between these groups of patients. In the prostate cancer study, patients who completed the EPIC versus those who did not complete the EPIC differed in terms of planned RT modality (as this was a stratification factor identified prior to randomization) at 6, 12, and 60 months but not the pretreatment or 24-month EPIC assessment. Patients who planned to receive intensity-modulated RT (IMRT) were more likely to complete the EPIC assessment than those who were not (80.2% vs. 74.7% prior to treatment, $p = 0.23$; 82.5% vs. 72.0% at 6 months, $p = 0.0006$; 81.9% vs. 75.2% at 12 months, $p = 0.018$; 80.9% vs. 77.7% at 24 months, $p = 0.25$; 77.0% vs. 82.9%, $p = 0.024$).

Determining whether the data are MNAR is more complex as the missing assessment has not been observed. There are no options for direct tests of MNAR. Surrogate measures for missingness, such as caregiver information, disease progression, or toxicity, can be used to test whether there is an association with missingness, which would allude to the presence of MNAR data. This requires an assumption that the relationship between the observed responses and the surrogate measures are the same as the missing responses and the surrogate measures.

TABLE 27.1

Patterns of Completion for EPIC

Drop-Out Pattern[a]	Time Point Prior to RT	6 Months	12 Months	24 Months	60 Months	Hypo-fractionated RT (n = 478)	Conventional RT (n = 484)	Total (n = 962)
N/A	X	X	X	X	X	170	211	381
24 months	X	X	X	X		79	72	151
12 months	X	X	X			31	30	61
Baseline	X					25	25	50
N/A	X	X		X	X	21	22	43
N/A	X	X	X		X	24	17	41
6 months	X	X				11	22	33
N/A	X		X	X	X	17	15	32
24 months	X	X		X		14	8	22
N/A						15	5	20
24 months	X		X	X		10	10	20
N/A		X	X	X	X	8	5	13
12 months	X		X			7	6	13
N/A	X	X			X	9	4	13
N/A	X		X		X	6	5	11
N/A	X			X	X	3	6	9
24 months	X			X		3	5	8
N/A		X	X	X		6	1	7
N/A	X				X	6	1	7
N/A		X		X	X	2	4	6
24 months				X		1	2	3
N/A				X	X	3	0	3
24 months			X	X		2	1	3
24 months		X	X	X		0	3	3
N/A					X	2	0	2
12 months		X	X			0	2	2
N/A		X	X		X	2	0	2
12 months			X			0	1	1
N/A		X			X	1	0	1
6 months		X				0	1	1

Note that data can still be missing within each pattern. N/A indicates that the patient did not drop out or never completed an assessment.

[a] Drop-out pattern refers to the time point in which the patient dropped out of the study (i.e., did not complete any assessments after this time point).

27.4 Analysis of Missing Data

27.4.1 Missing Completely at Random

Generally, any analysis that uses only complete data records, rather than all available data, can be used under MCAR [8]. For example, analysis from pretreatment assessment to each of the follow-up time points (6, 12, 24, and 60 months from the end of RT) in the prostate cancer trial would require that both the baseline and follow-up time point data were observed. This was assessed using change scores, calculated as (follow-up score – baseline score). Since higher EPIC bowel scores indicate better health-related QOL, a negative change score indicates a decline in health-related QOL. Patients in the hypofractionated RT arm experienced a larger decline from baseline in mean EPIC bowel domain change score at 12 months as compared to the conventional RT arm, as indicated by comparison via a two-sample t test (-7.5 vs. -3.7, $p = 0.0007$). This analysis uses only patients who had both baseline and 12-month scores available, which was only 65.9% of the trial primary analysis cohort (328 out of 484 patients on the hypofractionated RT arm and 306 out of 478 on the conventional RT arm). While potentially a valid analysis (if MCAR is in effect), at a minimum, there is a substantial loss in information in omitting this large number of patients, as well as a potential for selection bias. Furthermore, if the data are actually MAR or MNAR, which is typically the case in clinical trial research, these results will be biased and not reflective of the population. In any case, the reduced sample size will result in a loss of statistical power to detect the treatment effect and can cause the benefit of randomization to be lost (missing data may be MCAR, but still lead in imbalances in other factors between treatment arms) [9].

In addition to the simple analysis above, any other appropriate method such as analysis of variance (ANOVA), univariate linear models with a single time point used as an outcome variable (i.e., modeling the effect of treatment arm while adjusting for baseline score on the 12-month EPIC domain score), and binomial tests of proportions (i.e., comparing the percentage of patients who decline by a specific time) can be carried out under the MCAR assumption. Likewise, multivariate ANOVA for repeated measures assumes data are MCAR as only patients with complete data across all repeated measures are included [9]. The generalized estimating equation (GEE), which can be used for categorical, count, and continuous single and repeated measures requires complete data and thus can be used when MCAR is assumed.

27.4.2 Missing at Random

27.4.2.1 Mixed Effects Models

There are many analysis methods that can be used in the presence of data that are MAR. A generalized linear model (GLM), a mixed-effects regression model creates a framework to explain the missing data by inclusion of covariates in the model or by the patient's observed responses [9]. These models make use of all available data from each patient and typically employ maximum likelihood estimation. These are denoted as mixed models because they contain both random and fixed effects: a random subject effect is included to account for the repeated nature of the observations along with fixed effects, which include covariates such as treatment arm and any other patient/disease characteristics. The simplest case of the linear mixed-effects model is as follows:

$$y_{ij} = \beta_0 + \beta_1 t_{ij} + \beta_2 x_k + \beta_2(x_k * t_{ij}) + \varepsilon_{ij}$$

where y is the measurement of interest, $i = 1, 2, \ldots, N$ represents the patient, $j = 1, 2, \ldots, n_j$ time points, $k = 1, 2$ treatment arms. Time is represented by t, treatment arm is represented by x, and the regression coefficients are represented by β. The term $x_k * t_{ij}$ is an interaction term between time and treatment arm to detect if the effect of treatment varies over time. The error term, ε_{ij}, is typically assumed to be normally and independently distributed with a zero mean and common variance. Mixed-effects models are deemed subject-specific since the estimates of the regression coefficients are conditional on the random subject or patient effect. This can be contrasted with fixed effects of GEE models that are deemed population-averaged or marginal since these regression estimates are averaged over the population of patients.

As described by Brown and Prescott [10] and Gibbons [9], there are many advantages to using mixed-effects models. The repeated nature of longitudinal data, specifically repeated measurements taken on the same patient over time, creates correlation between these measurements as well as heterogeneity between patients. Through the introduction of a random effect, a covariance pattern can be fitted that will lead to improved estimates of fixed effects and their standard errors, especially in the presence of missing data. The most basic type of mixed-effect model has a continuous outcome variable assuming a normally distributed residual error structure. Many longitudinal endpoints are not continuous and thus these models have been extended to consider discrete data, such as categorical or ordinal outcome measures [11]. Other distributional assumptions and correlated residual errors of measurement are also possible. Extensions to allow for autocorrelated residuals lead to higher accuracy in estimates of uncertainty [12]. Mixed-effects models can be thought of as having two levels (group level and patient level), because, for example, repeated EPIC assessments are nested within patients. Higher levels of nesting are also possible, such as repeated EPIC assessments that are nested within patients who naturally are nested within hospitals, the effect of which may be of interest in some cases [13].

Examples of various extensions of the linear mixed-effects model are models that include a random intercept term. Using the same notation as previously presented, below is the formulation for the random intercept model, which can be extended to also include a random slope:

$$y_{ij} = \beta_0 + \beta_1 t_{ij} + \beta_2 x_k + \beta_2(x_k * t_{ij}) + v_{0i} + \varepsilon_{ij},$$

where v_{0i} represents the effect of patient i on his repeated measurements. This random intercept allows each patient to deviate from the mean response at baseline and assumes that the errors of measurement are independent conditional on this random patient-specific effect. This model does assume that the slope for each patient is the same. Additionally, assuming a random slope is an extension of this model.

Taking the prostate cancer study as an example: men completed the EPIC questionnaire periodically for 5 years. The outcome variable in a random-intercept, mixed-effects regression model is the bowel domain score at 6, 12, 24, and 60 months. In addition to treatment arm, a time-indicator variable, and baseline EPIC bowel score, other pretreatment characteristics can be considered as fixed effects in order to explain the missing data as seen in Table 27.2A. As one can imagine, the number of fixed effects can become very large. Various selection methods exist to reduce the number of covariates, such as backward selection. In backward selection, all possible covariates are included in the model and those with a p-value greater than a specified value (such as 0.10) are removed from the model until the only covariates remaining are those with a p-value less than or equal to the specified number. Although patients were also enrolled by institutions, a higher level model (in this case three levels to account for the repeated EPIC assessments for each patient within an institution) was not considered for various reasons such as institutional effects weren't relevant to the particular analysis that was conducted on this trial and due to the large number of institutions that enrolled patients (>100) and the fact that many of these institutions only enrolled a single patient. If a limited institution study is conducted, with

TABLE 27.2

Reduced EPIC Bowel Model

Analysis Method	Effect	Estimate	Standard Error	*p*-Value
A. Mixed effects model	Intercept	30.42	5.02	<0.001
	Time	−0.0002	0.01	0.98
	Treatment arm (Conventional RT)	−2.00	0.78	0.010
	Baseline EPIC Bowel Score	0.66	0.04	<0.001
	Zubrod (0)	2.46	1.48	0.097
	RT Modality (3D-CRT)	−0.09	0.98	0.92
	Ethnicity (Hispanic or Latino)	0.36	2.53	0.89
	Age (≤65)	−0.29	0.80	0.71
	Race (Other)	−3.13	1.07	0.004
B. Multiple imputation with mixed effects model	Intercept	31.41	4.60	<0.001
	Time	0.012	0.015	0.40
	Treatment arm (Conventional RT)	−1.64	0.72	0.023
	Baseline EPIC Bowel Score	0.63	0.042	<0.001
	Zubrod (0)	2.67	1.45	0.066
	RT Modality (3D-CRT)	−0.018	0.94	0.98
	Ethnicity (Hispanic or Latino)	0.55	2.39	0.82
	Age (≤65)	−0.44	0.76	0.56
	Race (Other)	−2.37	0.97	0.015
C. Pattern mixture model using multiple imputation with mixed effects model	Intercept	27.90	4.50	<0.001
	Time	0.0082	0.011	0.46
	Treatment arm (Conventional RT)	−2.07	0.69	0.0028
	Baseline EPIC Bowel Score	0.67	0.040	<0.001
	Zubrod (0)	2.45	1.32	0.064
	RT Modality (3D-CRT)	−0.44	0.85	0.61
	Ethnicity (Hispanic or Latino)	0.83	2.11	0.69
	Age (≤65)	−0.19	0.68	0.78
	Race (Other)	−2.00	0.87	0.0022

Note: Reference level is in parentheses.

only say 20 institutions participating, inclusion of another random effect can be included in the model.

Note that if a patient is missing the 12-month assessment, but the 6, 24, and 60 month assessments were observed, this patient can still be included in a mixed-effects model. This solves the issue of reduced statistical power from removing patients without complete data, as well as making the sample generalizable, since all patients are included, making these models very attractive for analysis in the presence of ignorable missing data. However, these models tend to be more computationally complex as compared to analysis methods assuming MCAR, such as a linear fixed-effects model, and depend on correctly specifying the model [9].

In this example, planned RT modality, age, race, and ethnicity were found to differ between men who completed the EPIC and men who did not and thus were included in the reduced model. Significant covariates in the model were baseline EPIC bowel score (patients with higher baseline scores have higher scores in follow-up), and treatment arm (patients on the hypofractionated RT arm have a lower score than those on the conventional RT arm).

Time was not significant, which means that EPIC bowel scores did not change significantly across time after completing treatment. A between-arm test at 12 months, adjusting for baseline EPIC bowel score, was conducted with the hypofractionated arm having an EPIC bowel score 2.00 lower than the conventional arm (estimate $= -2.00$, standard error $= 0.79$, $p = 0.011$).

27.4.2.2 Imputation

Imputation, or replacing missing responses with reasonable values, is another option under MAR that allows use of analysis methods that assume MCAR. These methods preserve the original full dataset. Single imputation methods, whereby a method is used to generate a single substitute value for the missing observation, are not recommended due to unrealistic assumptions or lack of accounting for variability [4]. For example, last observation carried forward, which as the name implies uses the prior observation in place of the missing value, assumes that the patient's outcome does not change after dropping out. Single imputation methods such as mean substitution or using a single predicted value from a regression model do not reflect the uncertainty in such a prediction [14]. Alternatively, multiple imputation is a Monte Carlo technique that replaces missing values by multiple potential values. The imputed values are chosen such that they represent information about the missing value that is encompassed in the observed data [15]. Rubin [16] describes the phases of multiple imputation. First, missing values are replaced M times via the selected imputation method, which creates M complete datasets. These datasets are analyzed separately using standard approaches that assume MCAR, with appropriate adjustment to the variance estimates. The results from the M analyses are combined to make a single inference. Less than 10 imputations are generally needed. Specifically, the efficiency of an estimate based on M imputations is approximately

$$\left(1 + \frac{\gamma}{M}\right)^{-1}$$

where γ is the proportion of missing data for the quantity being estimated [16]. Molenberghs and Kenward [15] show that 10 imputations provide at least 92% relative efficiency for up to 90% of missing information.

Selecting the imputation method is very critical to conducting an analysis with multiple imputation. Methods exist utilizing either regression models and sampling techniques, and regardless of the method chosen, several general principles are as follows. Overall, the method should produce unbiased estimates, account for the variation of the randomness of the chosen values as well as the loss of information due to due the missing values, and leave the covariance structures of the repeated measurements unaltered [8]. An explicit regression model can be used to predict the missing values and can incorporate other information (e.g., covariates and response variables) from the subjects. These models assume that the missing data is an ignorable conditional on the observed data and other covariates included in the regression model. In order to avoid biasing the treatment comparison toward the null hypothesis, imputation should be conducted separately by treatment group or with treatment as a covariate in the model [8]. All variables that will be included in the analysis model should be included in the imputation model, such as stratification factors and potential confounding variables (a variable that influences both the outcome variable, EPIC score in our example, and independent variable of interest, treatment assignment in our example). Including variables that explain a considerable amount of variance of the variable that is being imputed can help reduce the uncertainty of the imputations [17]. These variables can consist of reasons why the data is missing (i.e., patient refused, institutional error, etc.) and any variables that differ significantly between patients who completed the assessment as compared to those who did not.

To continue the previous example, a regression model can be used to impute M datasets containing a non-missing 12-month observation. In addition to using the patient's observations from 6, 24, and 60 months, other variables such as factors that were found to differ between men who completed the EPIC and men who did not (specifically, planned RT modality, age, race, and ethnicity) and local progression status at 6, 12, 24, and 60 months can be used in the imputation model to aid in the imputation. Progression can upset a patient and cause him to not continue with portions of the study in order to move onto other types of treatment. Results from the mixed-effects model using imputed data are shown in Table 27.2B. A test of the between-arm difference at 12 months, adjusting for baseline EPIC bowel score, resulted in a significant difference in favor of the conventional RT arm (estimate $= -1.49$, standard error $= 0.74$, $p = 0.046$).

Limitations of this approach are that when used with a small sample size or when only weak predictors of missingness are available, the variance of the imputation model can increase. Adding more covariates can help decrease the variance, but if these covariates do not greatly improve the efficiency of the model, as measured by R^2, this approach is unlikely to improve the imputation model [8, 17].

Closest predictor [8, 16, 17] and predictive mean matching [8, 16, 18] are variations on the explicit regression model approach. They do not allow imputation of impossible or out of range values and tend to be more robust under departures from the normality assumption. These procedures are available in commonly used statistical software programs such as SAS.

27.4.3 Missing Not at Random

Under MNAR, several modeling approaches are available, however these have some limitations due to strong assumptions that cannot be formally tested [8, 15, 19]. These assumptions may need to be defended from a clinical, rather than statistical, perspective. Estimates from these models tend not to be robust to model mis-specification, which introduces the need for sensitivity analyses, which will also be discussed.

27.4.3.1 Pattern Mixture Models

Pattern mixture models work by grouping patients based on their distribution of responses so that the analysis of all patients is based on a mixture of each group's distribution. The joint distribution of Y, the outcome of interest, and D, the indicator of dropout pattern, given X covariates, are modeled as

$$f(Y, D|X) = f(Y|D, X) f(D|X).$$

These models assume missing data are ignorable within each group. The various dropout patterns for EPIC bowel score is stated in Table 27.1. Due to some of the patterns having a small number of patients and the assumption of ignorable missing data within the groups, some were combined by the timing of the last assessment. If data are still MNAR within a group, it is assumed that the proportion is so small that it will have a minimal effect on the estimates in that group [8]. Therefore, the last observed time point is denoted in the first column of Table 27.1. A pattern mixture model using multiple imputation was performed on the prostate cancer data. Here, a new model, simulated from the posterior predictive distribution of a conditional regression model that was fitted to each pattern, was used to impute missing values [20]. The imputed datasets were then analyzed using a mixed effects model with results combined and shown in Table 27.2C. A test of the between-arm differences was conducted at the 12-month time point adjusting for baseline EPIC bowel score resulted in a significant difference between treatment arms in favor of the conventional arm (estimate $= -1.97$, standard error $= 0.70$, $p = 0.0053$).

27.4.3.2 Shared Parameter Models

As described by Henderson et al. [21], joint models (models that combine a longitudinal marker and time-to-event random variable) with a shared parameter can be used to

1. Make inferences about longitudinal data while adjusting for non-ignorable dropout,
2. Determine the distribution of time to an event conditional on intermediate longitudinal measurements, or
3. Make inferences concerning the development of repeated measurements in conjunction with event time processes.

Tsiatis et al. developed a joint model with shared parameters in the context of characterizing the relationship between a repeated measure and a time-to-event measure, specifically CD4 count in HIV patients with their time to progression to AIDS or death in hopes of using CD4 count as a surrogate endpoint for time to AIDS onset or death [22, 23]. In the prostate cancer example, the focus is on inferences on longitudinal data in the presence of non-ignorable dropout. A joint model can be built with a shared parameter between EPIC bowel score across time and time to dropout in order to assess the effect of radiation dose. This model would allow for appropriate inference if the time to dropout is considered MNAR.

The standard joint model contains a longitudinal submodel and survival (i.e., time-to-event endpoint) submodel and can incorporate covariates of interest into both submodels. Rizopoulos [24] and Ibrahim et al. [25] provide an excellent overview of the joint model including its formulation and applications. In the typical case, the longitudinal component is a linear mixed-effects model, as previously described for missing data that is MAR. In a simple formulation of the joint model, the linear model for outcome Y_{ij} is made up of a random error that is usually assumed to be normally distributed as previously described, ε_{ij}, and a trajectory function, X_{ij}, that is a linear or quadratic function of time and treatment:

$$Y_{ij} = X_{ij} + \varepsilon_{ij}$$

where $i = 1, 2, \ldots, N$ patients and $j = 1, 2, \ldots, n_j$ time points. The trajectory function,

$$X_{ij} = \eta_{0i} + \eta_{1i} * t_{ij} + \gamma Z_i$$

contains η_{0i} and η_{1i} which represent a random intercept and slope, both assuming a multivariate normal distribution and the treatment indicator Z_i with corresponding regression coefficient γ. The survival component is typically a parametric model such as the Weibull which includes the exponential as a special case. The formulation of the hazard function at time t is as follows:

$$h(t) = h_0(t)\exp(\beta X_{ij} + \alpha Z_i)$$

where β measures the association between the longitudinal outcome Y_{ij} and the time to event while α is the coefficient for the treatment effect Z_i. As seen in the above formulation, the two models are connected through the trajectory function, X_{ij}. Ibrahim et al. [25] depict the underlying causal diagram for joint models very nicely (Figure 27.2). The overall treatment effect is the effect of the longitudinal data on the time-to-event data multiplied by the effect of treatment on the longitudinal data added to the effect of treatment on the time-to-event data. If there is no association between the longitudinal data and the time-to-event data, then a joint model is not needed. In this situation, the ideal joint model would produce the same results as separate longitudinal and survival models [21]. If the association between the longitudinal data and the time-to-event data is negative, then the hazard is decreasing which implies that increases in the longitudinal data produce increases in the time to the event of interest [25].

27.4.3.3 Selection Models

Similar to a joint model, the selection model can also be broken down into a joint distribution, but one for the outcome and another for the missing data mechanism. The longitudinal outcome and missing data mechanism given specified covariates is written as $f(Y, M|X) = f(Y|X) f(M|Y, X)$. The model for the outcome does not depend on the missing data mechanism, while the model for the missing data mechanism depends on both observed and missing outcome data [8]. In fact, the joint model can be written as a random-effects selection model in which the missing data mechanism is modeled via the time to dropout. One disadvantage is that assumptions must be made about the missing data process which are untestable [26].

There are various types of selection models. Diggle and Kenward proposed a selection model for continuous outcomes [27] that combines a multivariate normal linear model for the longitudinal response data with a logistic regression model for the dropout process that includes dependencies on previous responses, as well as the current response. This model can be used if the missing data mechanism is assumed to be MNAR, however by setting one or two parameters in the logistic model to 0, it becomes reflective of MAR (only dependent on current data) and MCAR (no dependencies) data, respectively. Intermittent

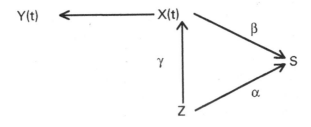

FIGURE 27.2
Causal diagram for joint modeling framework. $Y(t)$, observed longitudinal data; $X(t)$, trajectory function; S, survival; Z, treatment; α, treatment effect on survival; γ, treatment effect on longitudinal process; β, effect of longitudinal process on survival. (Copyright © 2010 by American Society of Clinical Oncology.)

missing data that is non-monotone (e.g., not missing data due to dropout) is considered to be ignorable, as is the case with pattern mixture models. Troxel et al. proposed a selection model when non-monotone missingness is present for continuous longitudinal data [28].

27.4.3.4 Multiple-Model Multiple Imputation

Siddique et al. [29] developed a multiple imputation framework that avoids attempting to correctly model the missing data mechanism by incorporating many ignorability assumptions into a single inference. This analysis is conducted by choosing a model from a distribution of models, each incorporating ignorable or non-ignorable mechanisms. Models included in this distribution depend on choices regarding the variation of the observed and unobserved (missing) values. Multiple imputations are generated using the selected model. This process is repeated, generating multiple-model multiple imputations. Inferences are combined using nested imputation combining rules that integrate between-model uncertainty into the standard errors of the parameter estimates [30]. Pre-specification of the models is recommended as to avoid picking the model that produces the desired results.

Although this approach is complex and, thus, can be difficult to explain to the investigator, imputation allows for complete-data analysis. Another advantage of this approach are the various extensions that can be used. Some studies may experience limited missing data early in the study with more at the longer follow-up times. Models assuming ignorable missing data could be used for the earlier time points and models assuming non-ignorable missing data for the later time points. Different models can be chosen for different groups of patients such as by treatment arm or patients who drop out versus those who do not. Collecting the reason for missing data can provide crucial information to make decisions regarding the model selection.

27.4.4 Sensitivity Analyses

Given the various limitations of models assuming MNAR data, one may wonder how to interpret the results of these analyses. Sensitivity analyses involve running the same analysis, either under different assumptions, using different models, etc., to see how the results vary. If the results lead to similar conclusions, specifically for the estimation of the treatment effect, then one can feel confident in the results. A disadvantage is that it can produce an array of answers rather than a single conclusion.

Using the prostate study as an example, analyses were run using a t test, which assumes MCAR, a mixed-effects model and imputation, both of which assume MAR, and a pattern mixture model, which assumes MNAR due to dropout. The 12-month between-arms differences in EPIC bowel score results in the same conclusion (statistically significant between-arms difference in favor of the conventional RT arm) in all four analysis methods (Table 27.3). The t test, however, has the highest between-arm difference in mean score (−3.73) with the multiple imputation with mixed-effects model having the smallest (−1.49). It should be noted that the t test is comparing the change from baseline to 12 months while the models are testing the difference at 12 months while adjusting for various covariates such as baseline EPIC bowel score.

In terms of the modeling, the treatment effect is statistically significant and the time effect is not for all three models (Table 27.2). The direction of the time estimate does change between the mixed-effects model (Table 27.2A) and the model using imputation and the pattern mixture model (Table 27.2B, C), however the estimate is very small and highly non-significant in all models. Since the results for both the 12-month between-arm difference in

TABLE 27.3

Between Arm Differences at 12 Months in EPIC Bowel Score

Method	Difference between Arms	Standard Error	*p*-Value
t-test	−3.73	13.14[a]	0.0007
Mixed effects model	−2.00	0.79	0.011
Multiple imputation with mixed effects model	−1.49	0.74	0.046
Pattern mixture model using multiple imputation with mixed effects model	−1.97	0.70	0.0053

[a] Standard deviation.

EPIC bowel change from baseline to 12 months and the treatment effect over time appear to be the same for all four analyses considered, one can feel confident in the results without specification of the missing data mechanism. One reason for this may be due to the large sample size. Only 85 patients in each arm are needed to detect a moderate effect size with 90% statistical power, as specified in the trial protocol, while this study had 328 and 306 patients on the hypofractionated and conventional RT arms, respectively. This is why Bruner et al. describe the importance of clinically meaningful differences [3]. When tested against a clinically meaningful difference of 4 points (calculated as the moderate effect size for this study) in the EPIC bowel score between arms ($H_0 = \mu_C - \mu_H \leq 4$), the test was not significant for the t test ($p = 0.51$) and the mixed model ($p = 0.99$).

27.4.5 Missing Baseline Covariates

There is another type of missing data that can arise in clinical trials that will be briefly described here: missing baseline covariate data. There are a variety of approaches for how to handle missing baseline covariate data. One way is to avoid non-response from the outset via how variables are specified. In the prostate cancer trial, patients are asked to complete a form that asks for their ethnicity and race among other demographic variables. One response option is "Unknown" for both of these questions. In these analyses, <10% of patients answered these two questions as "Unknown" and were thus categorized with Hispanics and other race since the patient specifically selected "Unknown." The option to exclude these patients, or likewise patients with a completely missing response, from the analysis is always available, however this will reduce the sample size, decreasing statistical power and possibly increasing bias in the estimates. Imputation methods as described earlier can be performed, providing a ready solution to the commonly encountered problem of missing baseline covariates.

Groenwold et al. [31] describe a method for incorporating a missing-indicator dummy variable ($0 = $ not missing, $1 = $ missing) to indicate if the value is missing or not. All missing values are then set to the same value. This method allows for inclusion of these subjects into the analysis; however a statistical model, rather than a univariate test, must be performed. This method appears to be unbiased in randomized trials, regardless of the missing data mechanism, but biased in non-randomized trials [31, 32].

Multiple imputation with chained equations (MICE) is a frequently used method for missing covariate data and works well for a variety of baseline covariate data types. However, in the context of longitudinal studies, as previously described for missing outcome data, it does require certain specifications be made correctly in order to obtain unbiased results. Erler et al. [33] describe this method, its advantages and disadvantages, along with a Bayesian method, which was their recommended approach. Generally, imputation is a valid approach in most situations, but sensitivity analyses can be conducted using different approaches to assess how each affects the estimates.

27.5 Conclusion

Clinical trials collecting repeated longitudinal assessments are prone to missing data. Methods can be implemented in the design of the trial to reduce this missing data as much as possible. During the conduct of the trial, both study and site staff must work diligently in order to obtain all data, or as much as they can. There are a variety of analysis techniques that can be performed on data that contain missing values, with most depending on the missing data mechanism and whether the mechanism is considered ignorable or non-ignorable. Sensitivity analyses aid in interpreting the results when assumptions are made regarding the missing data mechanism.

References

1. Lee WR, Dignam JJ, Amin MB et al. Randomized phase III noninferiority study comparing two radiotherapy fractionation schedules in patients with low-risk prostate cancer. *J Clin Oncol.* 2016;34(20):2325–32.
2. Wei JT, Dunn RL, Sandler HM et al. Comprehensive comparison of health-related quality of life after contemporary therapies for localized prostate cancer. *J Clin Oncol* 2002;20:557–66.
3. Bruner DW, Pugh SL, Lee WR et al. NRG Oncology/RTOG 0415, Phase III non-inferiority study of 2 fractionation schedules in low-risk prostate cancer: Practice change tipping point of quality of life results? *J Clin Oncol.* 2016; 34(2):1.
4. Little RJ, D'Agostino R, Cohen ML et al. The prevention and treatment of missing data in clinical trials. *N Engl J Med.* 2012 October 4;367(14):1355–60.
5. Panel on Handling Missing Data in Clinical Trials Committee on National Statistics Division of Behavioral and Social Sciences and Education. *The Prevention and Treatment of Missing Data in Clinical Trials.* Washington, DC: The National Academies Press, 2010.
6. Little RJA, Rubin DB. *Statistical Analysis with Missing Data*, 2nd Edition. New York, NY: John Wiley & Sons, 2002.
7. Ibrahim JG. Missing data methods in longitudinal studies: a review. *Test (Madr).* May 1, 2009;18(1):1–43.
8. Fairclough D. *Design and Analysis of Quality of Life Studies in Clinical Trials*, 2nd Edition. Boca Raton, FL: Chapman & Hall/CRC, 2010.
9. Gibbons RD, Hedeker D, DuToit S. Advances in analysis of longitudinal data. *Annu Rev Clin Psychol.* 2010 April 27;6:79–107.
10. Brown H, Prescott R. *Applied Mixed Models in Medicine*, 2nd ed. West Sussex: John Wiley & Sons, Ltd, 2006. ISBN: 9780470023563
11. Goldstein H. Nonlinear multilevel models, with an application to discrete response data. *Biometrika.* 1991;78:45–51.
12. Chi EM, Reinsel GC. Models for longitudinal data with random effects and AR(1) errors. *J Am Stat Soc.* 1989;84:452–59.
13. Raudenbush SW, Bryk AS. *Hierarchical Linear Models. 2.* Thousand Oaks, CA: Sage, 2002.
14. Schafer JL. Multiple imputation: A primer. *Stat Meth in Med Res* 1999;8:3–15.

15. Molenberghs G, Kenward MG. *Missing Data in Clinical Studies.* West Sussex, England: John Wiley, 2007.

16. Rubin DB. *Multiple imputation for Nonresponse in Surveys.* New York: John Wiley, 1987.

17. Van Buuren S, Boshuizen HC, Knook DL. Multiple imputation of missing blood pressure covariates in survival analysis. *Stat Med.* 1999;18:681–94.

18. Heitjan DF. Ignorability in general incomplete-data models. *Biometrika.* 1994;81(4):701–8.

19. Kenward MG. Selection models for repeated measurements with non-random dropout: An illustration of sensitivity. *Statist Med.* 1998;17:2723–32.

20. Ratitch B and O'Kelly M. Implementation of Pattern-Mixture Models Using Standard SAS/STAT Procedures," in *Proceedings of PharmaSUG 2011 (Pharmaceutical Industry SAS Users Group),* SP04, Nashville, 2011.

21. Henderson R, Diggle P, Dobson A. Joint modeling of longitudinal measurements and event time data. *Biostatistics.* 2000;1(4):465–80.

22. Tsiatis AA, DeGruttola V, Wulfson MS. Modeling the relationship of survival to longitudinal data measured with error: applications to survival and CD4 counts in patients with AIDS. *J Amer Statist Assoc.* 1995;90:27–37.

23. Tsiatis AA, Davidian M. Joint modeling of longitudinal and time-to-event data: An overview. *Statistica Sinica.* 2004;14:800–34.

24. Rizopoulos D. *Joint Models for Longitudinal and Time-To-Event Data with Applications in R.* Boca Raton, FL: CRC Press, 2012.

25. Ibrahim JG, Chu H, Chen LM. Basic concepts and methods for joint models of longitudinal and survival data. *JCO.* 2010;28(16):2796–801.

26. Satty A. An analysis of selection models for non-ignorable dropout: An application to multi-centre trial data. *J Biom Biostat.* 2015;6:246.

27. Diggle PJ, Kenward MG. Informative dropout in longitudinal data analysis (with discussion). *Appl Stat.* 1994;43:49–93.

28. Troxel, AB, Harrington DP, Lipsitz SR. Analysis of longitudinal data with non-ignorable non-monotone missing values. journal of the Royal Statistical Society. *Series C (Applied Statistics).* 1998;47(3):425–38.

29. Siddique J, Harel O, Cresp CM. Addressing missing data mechanism uncertainty using multiple-model multiple imputation: Application to a longitudinal clinical trial. *Ann Appl Stat.* 2012;6(4):1814–37.

30. Shen, ZJ. *Nested Multiple Imputation.* PhD thesis, Department of Statistics. Cambridge, MA: Harvard University, 2000.

31. Groenwold RHH, White IR, Donders ART, Carpenter JR, Altman DG, Moons KGM. Missing covariate data in clinical research: When and when not to use the missing-indicator method for analysis. *CMAJ.* 2012;184(11):1265–9.

32. White IR, Thompson SG. Adjusting for partially missing baseline measurements in randomized trials. *Stat Med* 2005;24:993–1007.

33. Erler NS, Rizopoulos D, Rosmalen J, Jaddoe VWV, Franco OH, Lesaffre EMEH. Dealing with missing covariates in epidemiologic studies: A comparison between multiple imputation and a full Bayesian approach. *Statist Med.* 2016;35:2955–74.

Index

Printed in the United States
by Baker & Taylor Publisher Services